Lecture Notes in Computer Science 13020

More information about this subseries at http://www.springer.com/series/7412

Huimin Ma · Liang Wang · Changshui Zhang ·
Fei Wu · Tieniu Tan · Yaonan Wang ·
Jianhuang Lai · Yao Zhao (Eds.)

Pattern Recognition and Computer Vision

4th Chinese Conference, PRCV 2021
Beijing, China, October 29 – November 1, 2021
Proceedings, Part II

Editors
Huimin Ma
University of Science and Technology Beijing
Beijing, China

Changshui Zhang
Tsinghua University
Beijing, China

Tieniu Tan
Chinese Academy of Sciences
Beijing, China

Jianhuang Lai
Sun Yat-Sen University
Guangzhou, Guangdong, China

Liang Wang
Chinese Academy of Sciences
Beijing, China

Fei Wu
Zhejiang University
Hangzhou, China

Yaonan Wang
Hunan University
Changsha, China

Yao Zhao
Beijing Jiaotong University
Beijing, China

ISSN 0302-9743 ISSN 1611-3349 (electronic)
Lecture Notes in Computer Science
ISBN 978-3-030-88006-4 ISBN 978-3-030-88007-1 (eBook)
https://doi.org/10.1007/978-3-030-88007-1

LNCS Sublibrary: SL6 – Image Processing, Computer Vision, Pattern Recognition, and Graphics

This Springer imprint is published by the registered company Springer Nature Switzerland AG
The registered company address is: Gewerbestrasse 11, 6330 Cham, Switzerland

Preface

Welcome to the proceedings of the 4th Chinese Conference on Pattern Recognition and Computer Vision (PRCV 2021) held in Beijing, China!

PRCV was established to further boost the impact of the Chinese community in pattern recognition and computer vision, which are two core areas of artificial intelligence, and further improve the quality of academic communication. Accordingly, PRCV is co-sponsored by four major academic societies of China: the China Society of Image and Graphics (CSIG), the Chinese Association for Artificial Intelligence (CAAI), the China Computer Federation (CCF), and the Chinese Association of Automation (CAA).

PRCV aims at providing an interactive communication platform for researchers from academia and from industry. It promotes not only academic exchange but also communication between academia and industry. In order to keep track of the frontier of academic trends and share the latest research achievements, innovative ideas, and scientific methods, international and local leading experts and professors are invited to deliver keynote speeches, introducing the latest advances in theories and methods in the fields of pattern recognition and computer vision.

PRCV 2021 was hosted by University of Science and Technology Beijing, Beijing Jiaotong University, and the Beijing University of Posts and Telecommunications. We received 513 full submissions. Each submission was reviewed by at least three reviewers selected from the Program Committee and other qualified researchers. Based on the reviewers' reports, 201 papers were finally accepted for presentation at the conference, including 30 oral and 171 posters. The acceptance rate was 39.2%. PRCV took place during October 29 to November 1, 2021, and the proceedings are published in this volume in Springer's Lecture Notes in Computer Science (LNCS) series.

We are grateful to the keynote speakers, Larry Davis from the University of Maryland, USA, Yoichi Sato from the University of Tokyo, Japan, Michael Black from the Max Planck Institute for Intelligent Systems, Germany, Songchun Zhu from Peking University and Tsinghua University, China, and Bo Xu from the Institute of Automation, Chinese Academy of Sciences, China.

We give sincere thanks to the authors of all submitted papers, the Program Committee members and the reviewers, and the Organizing Committee. Without their contributions, this conference would not have been possible. Special thanks also go to all of the sponsors and the organizers of the special forums; their support helped to make the conference a success. We are also grateful to Springer for publishing the proceedings.

October 2021

Tieniu Tan
Yaonan Wang
Jianhuang Lai
Yao Zhao
Huimin Ma
Liang Wang
Changshui Zhang
Fei Wu

Organization

Steering Committee Chair

Tieniu Tan	Institute of Automation, Chinese Academy of Sciences, China

Steering Committee

Xilin Chen	Institute of Computing Technology, Chinese Academy of Sciences, China
Chenglin Liu	Institute of Automation, Chinese Academy of Sciences, China
Yong Rui	Lenovo, China
Hongbing Zha	Peking University, China
Nanning Zheng	Xi'an Jiaotong University, China
Jie Zhou	Tsinghua University, China

Steering Committee Secretariat

Liang Wang	Institute of Automation, Chinese Academy of Sciences, China

General Chairs

Tieniu Tan	Institute of Automation, Chinese Academy of Sciences, China
Yaonan Wang	Hunan University, China
Jianhuang Lai	Sun Yat-sen University, China
Yao Zhao	Beijing Jiaotong University, China

Program Chairs

Huimin Ma	University of Science and Technology Beijing, China
Liang Wang	Institute of Automation, Chinese Academy of Sciences, China
Changshui Zhang	Tsinghua University, China
Fei Wu	Zhejiang University, China

Organizing Committee Chairs

Xucheng Yin	University of Science and Technology Beijing, China
Zhanyu Ma	Beijing University of Posts and Telecommunications, China
Zhenfeng Zhu	Beijing Jiaotong University, China
Ruiping Wang	Institute of Computing Technology, Chinese Academy of Sciences, China

Sponsorship Chairs

Nenghai Yu	University of Science and Technology of China, China
Xiang Bai	Huazhong University of Science and Technology, China
Yue Liu	Beijing Institute of Technology, China
Jinfeng Yang	Shenzhen Polytechnic, China

Publicity Chairs

Xiangwei Kong	Zhejiang University, China
Tao Mei	JD.com, China
Jiaying Liu	Peking University, China
Dan Zeng	Shanghai University, China

International Liaison Chairs

Jingyi Yu	ShanghaiTech University, China
Xuelong Li	Northwestern Polytechnical University, China
Bangzhi Ruan	Hong Kong Baptist University, China

Tutorial Chairs

Weishi Zheng	Sun Yat-sen University, China
Mingming Cheng	Nankai University, China
Shikui Wei	Beijing Jiaotong University, China

Symposium Chairs

Hua Huang	Beijing Normal University, China
Yuxin Peng	Peking University, China
Nannan Wang	Xidian University, China

Doctoral Forum Chairs

Xi Peng	Sichuan University, China
Hang Su	Tsinghua University, China
Huihui Bai	Beijing Jiaotong University, China

Competition Chairs

Nong Sang	Huazhong University of Science and Technology, China
Wangmeng Zuo	Harbin Institute of Technology, China
Xiaohua Xie	Sun Yat-sen University, China

Special Issue Chairs

Jiwen Lu	Tsinghua University, China
Shiming Xiang	Institute of Automation, Chinese Academy of Sciences, China
Jianxin Wu	Nanjing University, China

Publication Chairs

Zhouchen Lin	Peking University, China
Chunyu Lin	Beijing Jiaotong University, China
Huawei Tian	People's Public Security University of China, China

Registration Chairs

Junjun Yin	University of Science and Technology Beijing, China
Yue Ming	Beijing University of Posts and Telecommunications, China
Jimin Xiao	Xi'an Jiaotong-Liverpool University, China

Demo Chairs

Xiaokang Yang	Shanghai Jiaotong University, China
Xiaobin Zhu	University of Science and Technology Beijing, China
Chunjie Zhang	Beijing Jiaotong University, China

Website Chairs

Chao Zhu	University of Science and Technology Beijing, China
Zhaofeng He	Beijing University of Posts and Telecommunications, China
Runmin Cong	Beijing Jiaotong University, China

Finance Chairs

Weiping Wang	University of Science and Technology Beijing, China
Lifang Wu	Beijing University of Technology, China
Meiqin Liu	Beijing Jiaotong University, China

Program Committee

Jing Dong	Chinese Academy of Sciences, China
Ran He	Institute of Automation, Chinese Academy of Sciences, China
Xi Li	Zhejiang University, China
Si Liu	Beihang University, China
Xi Peng	Sichuan University, China
Yu Qiao	Chinese Academy of Sciences, China
Jian Sun	Xi'an Jiaotong University, China
Rongrong Ji	Xiamen University, China
Xiang Bai	Huazhong University of Science and Technology, China
Jian Cheng	Institute of Automation, Chinese Academy of Sciences, China
Mingming Cheng	Nankai University, China
Junyu Dong	Ocean University of China, China
Weisheng Dong	Xidian University, China
Yuming Fang	Jiangxi University of Finance and Economics, China
Jianjiang Feng	Tsinghua University, China
Shenghua Gao	ShanghaiTech University, China
Maoguo Gong	Xidian University, China
Yahong Han	Tianjin University, China
Huiguang He	Institute of Automation, Chinese Academy of Sciences, China
Shuqiang Jiang	Institute of Computing Technology, China Academy of Science, China
Lianwen Jin	South China University of Technology, China
Xiaoyuan Jing	Wuhan University, China
Haojie Li	Dalian University of Technology, China
Jianguo Li	Ant Group, China
Peihua Li	Dalian University of Technology, China
Liang Lin	Sun Yat-sen University, China
Zhouchen Lin	Peking University, China
Jiwen Lu	Tsinghua University, China
Siwei Ma	Peking University, China
Deyu Meng	Xi'an Jiaotong University, China
Qiguang Miao	Xidian University, China
Liqiang Nie	Shandong University, China
Wanli Ouyang	The University of Sydney, Australia
Jinshan Pan	Nanjing University of Science and Technology, China
Nong Sang	Huazhong University of Science and Technology, China
Shiguang Shan	Institute of Computing Technology, Chinese Academy of Sciences, China
Hongbin Shen	Shanghai Jiao Tong University, China
Linlin Shen	Shenzhen University, China
Mingli Song	Zhejiang University, China
Hanli Wang	Tongji University, China
Hanzi Wang	Xiamen University, China
Jingdong Wang	Microsoft, China

Nannan Wang	Xidian University, China
Jianxin Wu	Nanjing University, China
Jinjian Wu	Xidian University, China
Yihong Wu	Institute of Automation, Chinese Academy of Sciences, China
Guisong Xia	Wuhan University, China
Yong Xia	Northwestern Polytechnical University, China
Shiming Xiang	Chinese Academy of Sciences, China
Xiaohua Xie	Sun Yat-sen University, China
Jufeng Yang	Nankai University, China
Wankou Yang	Southeast University, China
Yang Yang	University of Electronic Science and Technology of China, China
Yilong Yin	Shandong University, China
Xiaotong Yuan	Nanjing University of Information Science and Technology, China
Zhengjun Zha	University of Science and Technology of China, China
Daoqiang Zhang	Nanjing University of Aeronautics and Astronautics, China
Zhaoxiang Zhang	Institute of Automation, Chinese Academy of Sciences, China
Weishi Zheng	Sun Yat-sen University, China
Wangmeng Zuo	Harbin Institute of Technology, China

Reviewers

Bai Xiang
Bai Xiao
Cai Shen
Cai Yinghao
Chen Zailiang
Chen Weixiang
Chen Jinyu
Chen Yifan
Cheng Gong
Chu Jun
Cui Chaoran
Cui Hengfei
Cui Zhe
Deng Hongxia
Deng Cheng
Ding Zihan
Dong Qiulei
Dong Yu
Dong Xue
Duan Lijuan
Fan Bin
Fan Yongxian
Fan Bohao
Fang Yuchun
Feng Jiachang
Feng Jiawei
Fu Bin
Fu Ying
Gao Hongxia
Gao Shang-Hua
Gao Changxin
Gao Guangwei
Gao Yi
Ge Shiming
Ge Yongxin
Geng Xin
Gong Chen
Gong Xun
Gu Guanghua
Gu Yu-Chao
Guo Chunle
Guo Jianwei
Guo Zhenhua
Han Qi
Han Linghao
He Hong
He Mingjie
He Zhaofeng
He Hongliang
Hong Jincheng
Hu Shishuai
Hu Jie
Hu Yang
Hu Fuyuan
Hu Ruyun
Hu Yangwen
Huang Lei
Huang Sheng
Huang Dong
Huang Huaibo
Huang Jiangtao
Huang Xiaoming
Ji Fanfan
Ji Jiayi
Ji Zhong
Jia Chuanmin
Jia Wei
Jia Xibin
Jiang Bo
Jiang Peng-Tao
Kan Meina
Kang Wenxiong

Lei Na
Lei Zhen
Leng Lu
Li Chenglong
Li Chunlei
Li Hongjun
Li Shuyan
Li Xia
Li Zhiyong
Li Guanbin
Li Peng
Li Ruirui
Li Zechao
Li Zhen
Li Ce
Li Changzhou
Li Jia
Li Jian
Li Shiying
Li Wanhua
Li Yongjie
Li Yunfan
Liang Jian
Liang Yanjie
Liao Zehui
Lin Zihang
Lin Chunyu
Lin Guangfeng
Liu Heng
Liu Li
Liu Wu
Liu Yiguang
Liu Zhiang
Liu Chongyu
Liu Li
Liu Qingshan
Liu Yun
Liu Cheng-Lin
Liu Min
Liu Risheng
Liu Tiange
Liu Weifeng
Liu Xiaolong
Liu Yang
Liu Zhi
Liu Zhou
Lu Shaoping
Lu Haopeng
Luo Bin
Luo Gen
Ma Chao
Ma Wenchao
Ma Cheng
Ma Wei
Mei Jie
Miao Yongwei
Nie Liqiang
Nie Xiushan
Niu Xuesong
Niu Yuzhen
Ouyang Jianquan
Pan Chunyan
Pan Zhiyu
Pan Jinshan
Peng Yixing
Peng Jun
Qian Wenhua
Qin Binjie
Qu Yanyun
Rao Yongming
Ren Wenqi
Rui Song
Shen Chao
Shen Haifeng
Shen Shuhan
Shen Tiancheng
Sheng Lijun
Shi Caijuan
Shi Wu
Shi Zhiping
Shi Hailin
Shi Lukui
Song Chunfeng
Su Hang
Sun Xiaoshuai
Sun Jinqiu
Sun Zhanli
Sun Jun
Sun Xian
Sun Zhenan
Tan Chaolei
Tan Xiaoyang
Tang Jin
Tu Zhengzheng
Wang Fudong
Wang Hao
Wang Limin
Wang Qinfen
Wang Xingce
Wang Xinnian
Wang Zitian
Wang Hongxing
Wang Jiapeng
Wang Luting
Wang Shanshan
Wang Shengke
Wang Yude
Wang Zilei
Wang Dong
Wang Hanzi
Wang Jinjia
Wang Long
Wang Qiufeng
Wang Shuqiang
Wang Xingzheng
Wei Xiu-Shen
Wei Wei
Wen Jie
Wu Yadong
Wu Hong
Wu Shixiang
Wu Xia
Wu Yongxian
Wu Yuwei
Wu Xinxiao
Wu Yihong
Xia Daoxun
Xiang Shiming
Xiao Jinsheng
Xiao Liang
Xiao Jun
Xie Xingyu
Xu Gang
Xu Shugong
Xu Xun

Xu Zhenghua
Xu Lixiang
Xu Xin-Shun
Xu Mingye
Xu Yong
Xue Nan
Yan Bo
Yan Dongming
Yan Junchi
Yang Dong
Yang Guan
Yang Peipei
Yang Wenming
Yang Yibo
Yang Lu
Yang Jinfu
Yang Wen
Yao Tao
Ye Mao
Yin Ming
Yin Fei
You Gexin
Yu Ye
Yu Qian
Yu Zhe
Zeng Lingan
Zeng Hui
Zhai Yongjie
Zhang Aiwu
Zhang Chi
Zhang Jie
Zhang Shu
Zhang Wenqiang
Zhang Yunfeng
Zhang Zhao
Zhang Hui
Zhang Lei
Zhang Xuyao
Zhang Yongfei
Zhang Dingwen
Zhang Honggang
Zhang Lin
Zhang Mingjin
Zhang Shanshan
Zhang Xiao-Yu
Zhang Yanming
Zhang Yuefeng
Zhao Cairong
Zhao Yang
Zhao Yuqian
Zhen Peng
Zheng Wenming
Zheng Feng
Zhong Dexing
Zhong Guoqiang
Zhou Xiaolong
Zhou Xue
Zhou Quan
Zhou Xiaowei
Zhu Chaoyang
Zhu Xiangping
Zou Yuexian
Zuo Wangmeng

Contents – Part II

Computer Vision, Theories and Applications

Dynamic Fusion Network for Light Field Depth Estimation

Yukun Zhang[1], Yongri Piao[1], Xinxin Ji[1], and Miao Zhang[2](✉)

[1] School of Information and Communication Engineering,
Dalian University of Technology, No. 2 Linggong Road, Dalian 116024, China
{zhangyukun,jxx0709}@mail.dlut.edu.cn, yrpiao@dlut.edu.cn
[2] International School of Information Science and Engineering,
Dalian University of Technology, No. 2 Linggong Road, Dalian 116024, China
miaozhang@dlut.edu.cn

Abstract. Focus-based methods have shown promising results for the task of depth estimation in recent years. However, most existing focus-based depth estimation approaches depend on maximal sharpness of the focal stack. These methods ignore the spatial relationship between the focal slices. The problem of information loss caused by the out-of-focus areas in the focal stack poses challenges for this task. In this paper, we propose a dynamically multi-modal learning strategy which incorporates RGB data and the focal stack in our framework. Our goal is to deeply excavate the spatial correlation in the focal stack by designing the pyramid ConvGRU and dynamically fuse multi-modal information between RGB data and the focal stack in a adaptive way by designing the multi-modal dynamic fusion module. The success of our method is demonstrated by achieving the state-of-the-art performance on two light field datasets.

Keywords: Light field depth estimation · Focal slices · Dynamic fusion.

1 Introduction

Depth estimation is a crucial step for understanding geometric relations within a scene. Accurate and reliable depth information plays an important role in computer vision, including object tracking [3], autonomous driving and pose estimation [10]. Depth estimation can be broadly classified into active and passive acquisition. In contrast to active techniques that involve sending a controlled energy beam and detecting the reflected energy, passive techniques are image-based methods and are more in accord with the human visual perception of the depth, i.e., humans use a great variety of vision-based passive depth cues, such as texture, edges, size perspective, binocular disparity, motion parallax, occlusion effects and variations

This work was supported by the Science and Technology Innovation Foundation of Dalian (#2019J12GX034), the National Natural Science Foundation of China (#61976035), and the Fundamental Research Funds for the Central Universities (#DUT20JC42).

H. Ma et al. (Eds.): PRCV 2021, LNCS 13020, pp. 3–15, 2021.
https://doi.org/10.1007/978-3-030-88007-1_1

in shading. Monocular depth estimation, as low-cost, convenient and efficient passive techniques, has attracted lots of interest lately. However, depth estimation from a single image of a generic scene is an ill-posed problem, due to the inherent ambiguity of mapping an intensity or color measurement into a depth value. On the other hand, inspired by the analogy to human depth perception, multi-view depth estimation has achieved great success, including binocular and multi-view stereo. The similarities and correspondences produced between each pixel in the images produce far superior results. However, these approaches are sensitive to imaging systems and require careful alignment and calibration in the setup.

The light field enables a unique capability of post-capture refocusing. A stack of focal slices are generated as they are taken at different depths, which contain abundant spatial parallax information. Furthermore, focusness information caters to humans visual fixation that allows our eyes to maximize the focus we can give to the object in a scene. In spite of this promising characteristics for focusness information, there are a few studies documenting their efficacy for depth estimation. Early works mainly aimed at determining the depth of a pixel by measuring its sharpness or focus at different images of the focal stack [9]. Later on, several approaches based on deep learning have been proposed [4]. These methods use convolutional neural networks (CNNs) to extract effective focusness information for facilitating depth estimation instead of hand-crafted features.

While these methods demonstrate that focusness information is useful for depth estimation, there is still large room for further improvement in terms of two key aspects: (1) How we deeply excavate the spatial correlation between focal slices for obtaining useful focusness information is critical for depth estimation. Since the different focal slices are focused at different depths, the spatial correlation between the focal slices is closely related to depth variation of the objects in a scene. Most of previous focus-based depth estimation networks used standard 2D CNNs to learns filters that extend across the entire focal stack. However, spatial correlation between the focal slices is likely to be ignored. As a result, the focusness information has not been well-captured. (2) How do we effectively fuse focusness features and RGB information to reduce information loss in the depth map? While focusness information in the focal slices provides implicit depth cues, leading to better depth estimation, out-of-focus areas with unknown sharpness could be prone to information loss error, leading to inaccurate depth estimation. Considering that the RGB image contains high quality sharpness and can be used to compensate for missing data in the out-of-focus area of the focal slices, we believe combining multi-modal information is beneficial to improve the accuracy of depth maps. Albeit most of recent methods performed data fusion by employing some manually set such as, sum fusion, weighted fusion and concatenate fusion. Those methods are unable to take full advantages of multi-modal information between RGB images and focal slices.

Our core insight is that we can leverage RGB data and the focal stack to learn an estimation model of depth by deeply excavating the spatial correlation in the focal stack and fusing multi-modal cues between RGB image and focal slices. Concretely, our contributions are mainly three-fold:

- We proposed a pyramid ConvGRU for correlating the focus and depth. Based on the observe that different focal slices possess focus area of multiple scales and focused at different depth, the pyramid ConvGRU is designed to excavate the spatial correlation between different focal slices and sequently pass multi-scale focusness information along the depth direction.
- We propose a multi-modal dynamic fusion module (MDFM) in which multi-modal features are fused in an adaptive manner. This fusion strategy allows the convolution kernel parameters to dynamically change with the input focusness features in the process of convolution with RGB features, thereby avoiding information loss in the depth map.
- We demonstrate the effectiveness of the proposed model on two light field datasets. The results show that our approach achieves superior performance over the state-of-the-art approaches.

2 Related Work

Recently, deep learning, in particular Convolutional Neural Networks (CNNs), has successfully broken the bottleneck of traditional methods in a wide range of fields. For the light field depth estimation, [11] introduced a fully-convolutional neural network for highly accurate depth estimation. [4] suggested the first end-to-end learning method to compute depth maps from the focal stack. [12] estimated scene depths from a single image, using the information provided by a camera's aperture as supervision. They introduced two differentiable aperture rendering functions that use the input image and predicted depths to simulate the depth-of-field effects caused by real camera apertures. Then they trained a monocular depth estimation network end-to-end to conduct depth estimation from RGB images. [8] proposed a lightweight network to estimate depth in light fields through a hierarchical multi-scale structure, which is trained end-to-end. [14] proposed a view selection module that generates an attention map indicating the importance of each view and its potential for contributing to accurate depth estimation. By exploring the symmetric property of light field views, they enforced symmetry in the attention map and further improve accuracy. These methods demonstrate that focusness cues can greatly contribute to depth estimation, achieving superior results. However, there is still large room for further improvement in terms of spatial correlation excavation in the focal stack and multi-modal fusion between the RGB image and the focal stack.

3 Method

3.1 The Overall Architecture

In this paper, our goal is to excavate the spatial correlation among focal slices and dynamically fuse RGB information and focusness features for focus-based depth estimation. The overall framework is described in Fig. 1. The dynamic fusion network is mainly composed of a two-stream (RGB stream and focal stream)

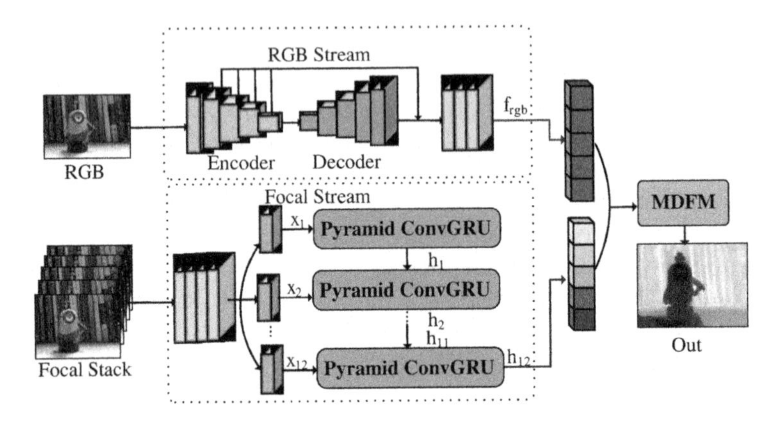

Fig. 1. The whole pipeline.

networks and a multi-modal dynamic fusion module (MDFM). We feed RGB images into the RGB stream and take the focal stacks into the focal stream to extract RGB features and focusness features with ample spatial geometric information of light field, respectively. Then, the RGB features and focusness features are injected into the multi-modal dynamic fusion module to adequately refine and integrate clues from different modalities for predicting depth maps. The whole process is described as follows. **First**, the RGB image I_0 is fed into the RGB stream to extract the effective global structure information in the RGB map. The RGB stream consists of an encoder, a decoder and a refinement module. The encoder is based on the SeNet model that contains 5 convolutional blocks. The decoder employs four up-sample modules to gradually up-scale the final feature from the encoder while decreasing the number of channels. The refinement module concatenates the feature from encoder and decoder in color channels with three 3×3 convolutional layers. The process can be written as

$$f_{rgb} = N_{rgb}(I_0), \tag{1}$$

where f_{rgb} denotes the extracted RGB features and N_{rgb} is the RGB stream network.

Second, all focal slices $\{I_1, I_2, ...I_n\}$ are fed into another focal stream to generate focusness information from focal slices, n is the number of focal slices. We first apply four 5×5 convolutional layers to encode raw focusness features x_i. This procedure can be defined as:

$$x_i = N_{focal}(I_i), \tag{2}$$

where i represents the ith focus slice and N_{focal} is the encoder layers of focal stream network. Then, the raw focusness feature x_i is sent to the proposed pyramid ConvGRU N_{pgru} to excavate the spatial correlation between different focal slices. The pyramid ConvGRU can be formulated as:

$$\{h_1, h_2, ...h_n\} = N_{pgru}(I_1, I_2, ...I_n), \tag{3}$$

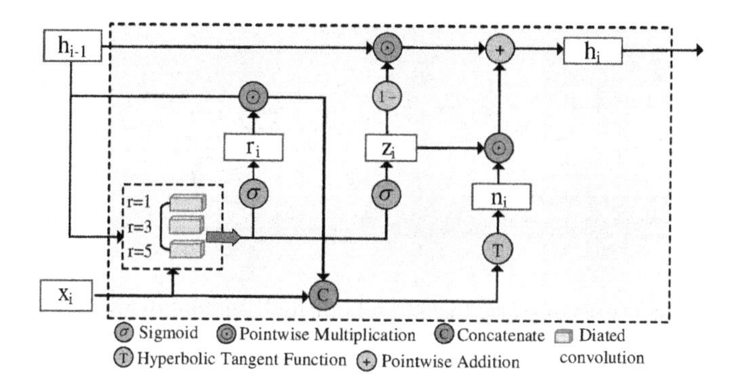

Fig. 2. The architectures of our pyramid ConvGRU.

where $\{h_1, h_2, ...h_n\}$ represents the output features of the pyramid ConvGRU.

Last, we propose a multi-modal dynamic fusion module (MDFM) to dynamically fuse multi-modal information between RGB features and the focusness features in an adaptive way for providing good depth map.

3.2 Pyramid ConvGRU

Considering the stack of focal slices in a scene possess focus-area of multiple scales and focused at different depth, we aim to correlate the depth information with multi-scale focusness features. To do this, we propose a pyramid ConvGRU to excavate the spatial correlation between different focal slices. In this way, multi-scale focusness information in different slices can be transferred along the depth direction. We draw ideas from recent works in [2]. Concretely, they use a typical ConvGRU to capture short and long term temporal dependencies. In our work, we consider the focusness features as a feature map sequence. By modifying the typical ConvGRU structure, our pyramid ConvGRU can not only capture different scale contextual information, but also learn the spatial correlation between adjacent focal slices in a serialized manner.

The detailed structure of our pyramid ConvGRU is shown in Fig. 2. When the encoded raw focusness feature x_i passes through the ith pyramid ConvGRU, the current input feature x_i will be fused with the last output feature of previous focal slice h_{i-1} in the pyramid ConvGRU. The formulation of the proposed pyramid ConvGRU is:

$$r_i = \sigma \begin{bmatrix} W_{r1} * [x_i, h_{i-1}], \\ W_{r3} * [x_i, h_{i-1}], \\ W_{r5} * [x_i, h_{i-1}] \end{bmatrix}, \tag{4}$$

$$z_i = \sigma \begin{bmatrix} W_{z1} * [x_i, h_{i-1}], \\ W_{z3} * [x_i, h_{i-1}], \\ W_{z5} * [x_i, h_{i-1}] \end{bmatrix}, \tag{5}$$

$$n_i = \tanh(x_i * W_{xn} + r_i \odot h_{i-1} * W_{hn} + b_n), \tag{6}$$

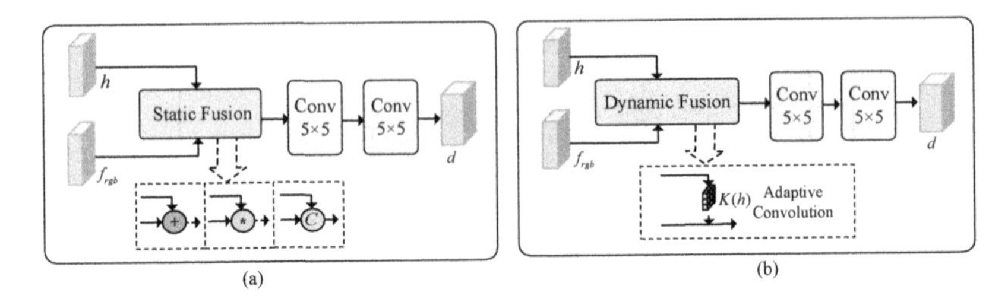

Fig. 3. The architectures of static fusion and our multi-modal dynamic fusion module. (a) Three different static fusion methods: sum fusion, weighted fusion, concatenate fusion. (b) Our multi-modal dynamic fusion module.

$$h_i = (1 - z_i) \odot h_{i-1} + z_i \odot n_i, \tag{7}$$

where all W_* and b_* are model parameters to be learned. σ is sigmoid function. $\odot$ and $*$ are element-wise multiplication and convolution, respectively. For each current input x_i and h_{i-1}, we use the update gate z_i to control how much information from the previous slice feature h_{i-1} and candidate states n_i need to be retained in the output feature h_i. The reset gate r_i controls whether the calculation of the candidate state n_i depends on the previous slice output feature h_{i-1}. In order to pass multi-scale focusness features corresponding to focus areas of different scales, we use an atrous spatial pyramid pooling (ASPP) module [1] for each of the gates instead of traditional convolution, which can encode multi-scale foucsness information by applying dilated convolution at multiple parallel filters with different rates and field-of-views. The dilation rate r is 1, 3 and 5, respectively. Then, the fused new feature h_i are obtained through weighted combination h_{i-1}, the output of the update gate z_i and the candidate state n_i.

3.3 Multi-modal Dynamic Fusion Module (MDFM)

As RGB images and focal slices imply different depth information, we consider that fusing multi-modal information is important for depth estimation task. However, existing fusion schemes usually employ some manually set, including sum fusion, weighted fusion, concatenate fusion. These static fusion schemes not suitable for our task because the RGB features and focusness features are not equivalent quantities, is prone to information loss. As shown in the Fig. 3(a), these static fusion methods processes the entire image. When the network parameters are fixed, the convolution kernel does not change with the input features, ignoring the relationship between multi-modal features. Therefore, a more proper and effective strategy should be considered. To this end, we introduce a multi-modal dynamic fusion module (MDFM) to dynamically fuse the RGB features and focusness features in an adaptive manner. In this module, the filter varies with focusness features is used to convolve with RGB information, thereby avoiding information loss.

Our proposed MDFM is shown in Fig. 3(b). Specifically, the process of our module consists of two steps: 1) we choose an adaptive convolution kernel [13] instead of the spatially invariant convolution. In our work, the standard spatial convolution W is adapted at each pixel using focusness feature f via the adaptive kernel. Therefore, when the focusness features change, the parameters of the convolution kernel also change. The process can be defined as:

$$K(h_i, h_j) = \exp\left(-\frac{1}{2}(h_i - h_j)^T (h_i - h_j)\right) W[p_i - p_j], \tag{8}$$

where $[pi - pj]$ is index of the spatial dimensions of an array with 2D spatial offsets, i and j represent the pixel coordinates, W is a standard spatial convolution.

2) we apply the generated adaptive convolution kernel to RGB feature f_{rgb}, making the whole network dynamically fuse multi-modal information to get an accurate depth map. For the output from the pyramid ConvGRU h, we have:

$$d_i = \sum_{j \in \Omega(i)} K(h_i, h_j) f_{rgb\_j} + b, \tag{9}$$

where f_{rgb_j} represents the output from the RGB stream, b is bias term, d represents the depth prediction map. Before we perform convolution operation, the adaptive convolution kernel has a pre-defined form and depends on the content of focusness features. It is hoped that the prediction map depends on both the RGB features and the reliable focusness information. Finally, we refine the depth map by adding two consecutive 5×5 convolutional layers. Compared to traditional methods of fusing multi-modal information which parameters do not change with input features, our model dynamically selects convolution kernel parameters based on the content of the focusness features, avoiding information loss.

4 Experiments

4.1 Experiments Setup

Datasets. To demonstrate the effectiveness of the proposed approach, we conduct experiments on DUT-LFDD dataset [16] and LFSD dataset [7]. Note that although there are some synthetic light field datasets provided by [12] which consist of multi-view images and corresponding depth maps. The models trained on such synthetic data have trouble generalizing to real-world data.

To prevent the overfitting problem, we augment the training set by the following operations: (1) Flipping: we only consider horizontal flipping of images at a probability of 0.5. (2) Rotating: the RGB image, focal stack and the depth image are rotated by a random degree $r \in [-5,5]$. (3) Color Jitter: brightness, contrast, and saturation values of the sample are randomly scaled by $c \in [0.6,1.4]$.

Evaluation Metrics. We adopt six metrics for comprehensive evaluation, including Root Mean Squared error (RMSE), Absolute Relative Error(Abs Rel), Squared Relative error (Sq Rel), Accuracy with threshold ($\delta = [1.25, 1.25^2, 1.25^3]$). They are universally-agreed and standard for evaluating a depth estimation model and well explained in many literatures.

Table 1. Quantitative results of the ablation analysis for our Pyramid ConvGRU.

Methods	Error metric↓			Accuracy metric↑		
	RMSE	Abs Rel	Sq Rel	$\delta < 1.25$	$\delta < 1.25^2$	$\delta < 1.25^3$
2DCNN	.3667	.1879	.1015	.7102	.9435	.9930
ConvGRU	.3648	.1821	.0976	.7159	.9444	.9931
Pyramid ConvGRU	**.3457**	**.1707**	**.0864**	**.7342**	**.9516**	**.9952**

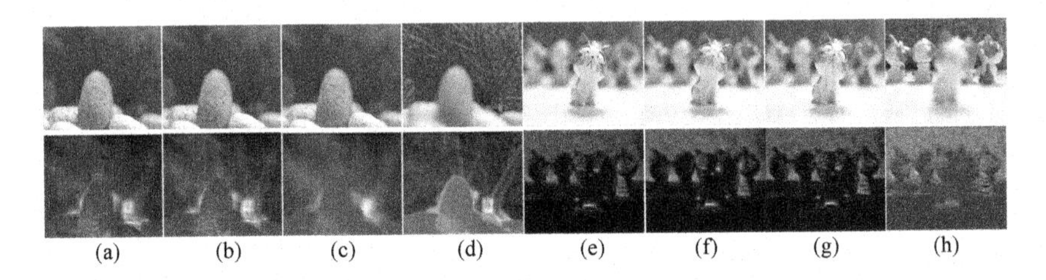

Fig. 4. (a)-(h) represent the output features of pyramid ConvGRU at different stages in different scenes. For each scene, we randomly selected the output features of 4 pyramid ConvGRU for display.

Implementation Details. We implement our network based on Pytorch framework with one Nvidia GTX 1080Ti GPU. We train the RGB stream and focal stream using Adam optimizer with an initial learning rate of 0.0001, and reduce it to 10% for every 5 epochs. We set $\beta1 = 0.9$, $\beta2 = 0.999$, and use weight decay of 0.0001. The encoder module in the RGB stream is initialized by a model pretrained with the ImageNet dataset. The other layers in the network are randomly initialized. The batchsize is 1 and maximum epoch is set 80. The input focal stack and RGB image are uniformly resized to 256×256. During training, we set the number of focal slices n to 12.

4.2 Ablation Studies

Effect of the Pyramid ConvGRU. The pyramid ConvGRU is proposed to excavate the spatial correlation between different focal slices. To verify the effectiveness of the pyramid ConvGRU, we first replace the pyramid ConvGRU with 7 convolution layers (noted as 2DCNN). Table 1 shows that the pyramid ConvGRU improves the RMSE performances by nearly 2% than 2DCNN. We believe that this improvement is due to the cyclic structure retaining the spatial relationship between different focusness features. Then we compare the performance of our pyramid ConvGRU and ConvGRU. We replace pyramid ConvGRU with the ConvGRU. And we can observe from Table 1 that the improvements in performance are achieved by using our pyramid ConvGRU. These improvements are reasonable since our model can adapt to different sizes of focus areas compared to ConvGRU. Furthermore, we randomly select the output features of pyramid ConvGRU at different stages in different scenarios for visualization. We can see from Fig. 4 that the feature maps of different focal slices contain different details and are passed by our pyramid ConvGRU.

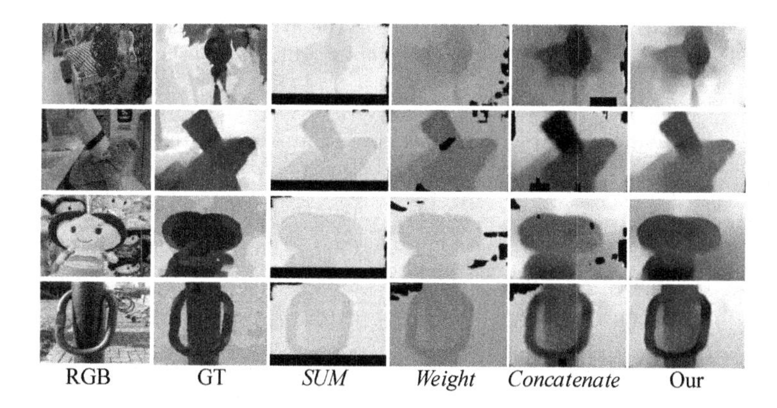

Fig. 5. Visual results of the effect of MDFM. SUM represents sum fusion. $Weight$ represents weighted fusion. $Concatenate$ represents concatenate fusion.

Table 2. Quantitative results of the ablation analysis for our MDFM. SUM represents sum fusion. $Weight$ represents weighted fusion. $Concatenate$ represents concatenate fusion.

Methods	Error metric			Accuracy metric		
	RMSE	Abs Rel	Sq Rel	$\delta < 1.25$	$\delta < 1.25^2$	$\delta < 1.25^3$
SUM	.6600	.3777	.3211	.4186	.6838	.9184
$Weight$	.4877	.2603	.1652	.5190	.8584	.9749
$Concatenate$	.4268	.1917	.1065	.6250	.9122	.9790
Our	**.3457**	**.1707**	**.0864**	**.7342**	**.9516**	**.9952**

Effect of MDFM. The MDFM is proposed for dynamically fusing multi-modal information between RGB features and focusness features. To demonstrate the effectiveness of the MDFM, we compare the MDFM with a variety of conventional fusion methods, including sum fusion(noted as SUM), weighted fusion(noted as $Weight$) and concatenate fusion(noted as $Concatenate$). For better comparison, we replace the fusion block in the framework. The results of the comparison are shown in Fig. 5. It can be seen that the quality of depth map achieves accumulative improvements by a large margin with the MDFM. Especially in regions of depth discontinuities, the MDFM is able to recover the depth more accurately and preserves more structure information compared to other static fusion methods. Numerically, as shown in Table 2, the concatenate fusion has the best effect in the static fusion method, our proposed MDFM reduces the RMSE performances by nearly 8% than concatenate fusion.

4.3 Comparison with State-of-the-arts

We compare our method with 6 other state-of-the-arts light field depth estimation methods, containing both deep-learning-based methods (DDFF [4], EPINet

Table 3. Quantitative comparison with state-of-the-art methods. From top to bottom: DUT-LFDD dataset, LFSD dataset. $*$ respects traditional methods. The best result results are shown in **boldface**.

Type	Methods	Error metric			Accuracy metric		
		RMSE	Abs Rel	Sq Rel	$\delta < 1.25$	$\delta < 1.25^2$	$\delta < 1.25^3$
DUT-LFDD	VDFF*	.7192	.3887	.3808	.4040	.6593	.8505
	PADMM*	.4730	.2253	.1509	.5891	.8560	.9577
	DDFF	.5255	.2666	.1834	.4944	.8202	.9667
	LF_OCC*	.6233	.3109	.2510	.4524	.7464	.9127
	LF*	.6897	.3835	.3790	.4913	.7549	.8783
	EPINET	.4974	.2324	.1434	.5010	.8375	.9837
	LFattNet	.5254	.2344	.1481	.4703	.8166	.9792
	MANet	.4607	.1922	.1044	.5709	.9274	.9947
	Our	**.3457**	**.1707**	**.0864**	**.7342**	**.9516**	**.9952**
LFSD	VDFF*	.5747	.3320	.2660	.4730	.7823	.9359
	PADMM*	.4238	.2153	.1336	.6536	.8880	.9770
	DDFF	.4255	.2128	.1204	.6185	.8916	.9860
	Our	**.3612**	**.1796**	**.0901**	**.6973**	**.9339**	**.9874**

[11], MANet [8], LFattNet [14]) and traditional methods (VDFF* [9], PADMM* [5], LF* [6], LF_OCC* [15]). To verify the generalization of our network, we conduct experiments on two light field datasets.

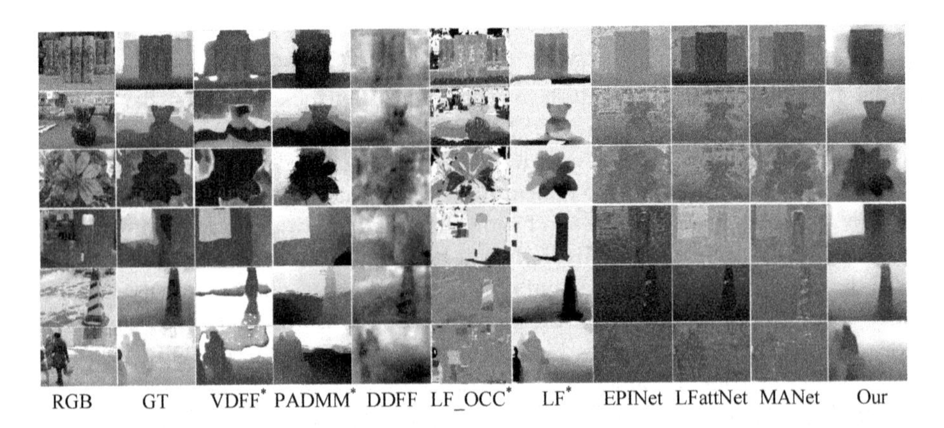

Fig. 6. Visual comparisons of our method over other approaches. Our method produces more accurate depth maps than others, and our results are more consistent with the ground truths (denoted as GT).

Quantitative Evaluation. As shown in Table 3, our method is able to clearly outperform the other state-of-the-art methods on DUT-LFDD dataset in terms of six evaluation metrics. Not only that, we also apply the model parameters trained on DUT-LFDD dataset directly to the LFSD dataset for testing, and our method achieves significant advantages, such as Top-1 accuracies (Sq Rel) and Top-1 accuracies (RMSE). Note that the LFSD dataset only provides the focal slices, RGB images and depth maps, therefore, we only compare our method with 3 focus-based methods on LFSD dataset.

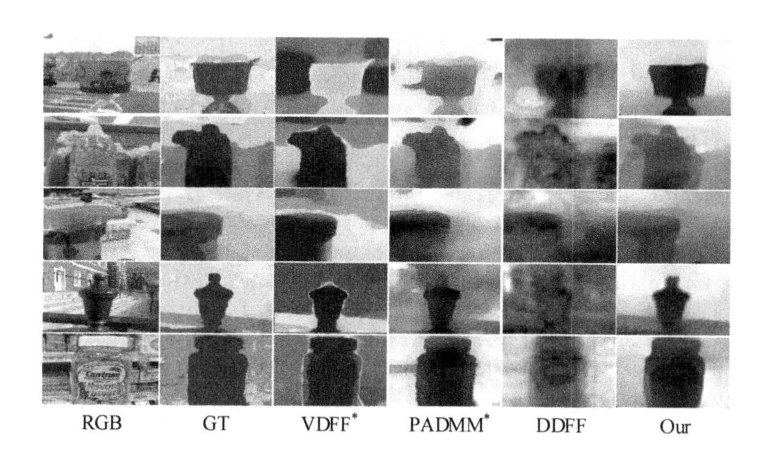

Fig. 7. Comparison with deep learning-based methods on LFSD dataset: RGB images, the corresponding ground truth, our estimated depth maps and other state-of-the-arts light field depth estimation results.

Qualitative Evaluation. Figure 6 provides some challenging samples of results comparing our method with other state-of-the-art methods. It can be seen that our method can achieve accurate prediction. As shown in 1th, 3th and 4th rows of Fig. 6, when the background is complex or the foreground is similar to the background, it is necessary to distinguish objects from the background. Our method has obvious depth levels because the focusness information is effectively transmitted along the pyramid ConvGRU. As shown in 2th row of Fig. 6, when the surface of the object is smooth, the object will be submerged in the background. It is more difficult to capture the edge of the object, our method combine RGB features and focusness features to better preserve the edge information of the object, relying on the multi-modal dynamic fusion module. We further illustrate the visual results of our method on LFSD dataset in Fig. 7. We can find that compared with other methods, our network retains more detailed information and reduces a lot of noise.

5 Conclusion

In this paper, we propose a multi-modal learning strategy which incorporates RGB data and the focal stack for focus-based depth estimation. Our pyramid

ConvGRU excavates the spatial correlation between different focal slices and sequently pass multi-scale focusness information along the depth direction. The MDFM dynamically fuses the RGB features and focusness features in an adaptive manner. In this module, the filter varies with focusness features is used to convolve with RGB information. Our experiments show that the proposed method achieves superior performance, especially in challenging scenes.

References

1. Chen, L.C., Papandreou, G., Kokkinos, I., Murphy, K., Yuille, A.L.: Deeplab: semantic image segmentation with deep convolutional nets, atrous convolution, and fully connected CRFS. IEEE Trans. Pattern Anal. Mach. Intell. **40**(4), 834–848 (2017)
2. Eom, C., Park, H., Ham, B.: Temporally consistent depth prediction with flow-guided memory units. IEEE Trans. Intell. Transp. Syst. **21**, 4626–4636 (2019)
3. Ghasemi, A., Vetterli, M.: Scale-invariant representation of light field images for object recognition and tracking. In: Computational Imaging XII, vol. 9020, p. 902015. International Society for Optics and Photonics (2014)
4. Hazirbas, C., Soyer, S.G., Staab, M.C., Leal-Taixé, L., Cremers, D.: Deep depth from focus. In: Jawahar, C.V., Li, H., Mori, G., Schindler, K. (eds.) ACCV 2018. LNCS, vol. 11363, pp. 525–541. Springer, Cham (2019). https://doi.org/10.1007/978-3-030-20893-6_33
5. Javidnia, H., Corcoran, P.: Application of preconditioned alternating direction method of multipliers in depth from focal stack. J. Electron. Imaging **27**(2), 023019 (2018)
6. Jeon, H.G., et al.: Accurate depth map estimation from a lenslet light field camera. In: Proceedings of the IEEE Conference on Computer Vision and Pattern Recognition, pp. 1547–1555 (2015)
7. Li, N., Ye, J., Ji, Y., Ling, H., Yu, J.: Saliency detection on light field. In: CVPR, pp. 2806–2813 (2014)
8. Li, Y., Zhang, L., Wang, Q., Lafruit, G.: Manet: multi-scale aggregated network for light field depth estimation. In: ICASSP 2020–2020 IEEE International Conference on Acoustics, Speech and Signal Processing (ICASSP), pp. 1998–2002. IEEE (2020)
9. Moeller, M., Benning, M., Schönlieb, C., Cremers, D.: Variational depth from focus reconstruction. IEEE Trans. Image Process. **24**(12), 5369–5378 (2015)
10. Park, K., Patten, T., Prankl, J., Vincze, M.: Multi-task template matching for object detection, segmentation and pose estimation using depth images. In: 2019 International Conference on Robotics and Automation (ICRA), pp. 7207–7213. IEEE (2019)
11. Shin, C., Jeon, H.G., Yoon, Y., So Kweon, I., Joo Kim, S.: Epinet: A fully-convolutional neural network using epipolar geometry for depth from light field images. In: Proceedings of the IEEE Conference on Computer Vision and Pattern Recognition, pp. 4748–4757 (2018)
12. Srinivasan, P.P., Garg, R., Wadhwa, N., Ng, R., Barron, J.T.: Aperture supervision for monocular depth estimation. In: Proceedings of the IEEE Conference on Computer Vision and Pattern Recognition, pp. 6393–6401 (2018)
13. Su, H., Jampani, V., Sun, D., Gallo, O., Learnedmiller, E., Kautz, J.: Pixel-adaptive convolutional neural networks. Vision and Pattern Recognition. arXiv: Computer (2019)

14. Tsai, Y.J., Liu, Y.L., Ming, O., Chuang, Y.Y.: Attention-based view selection networks for light-field disparity estimation. In: Proceedings of the AAAI Conference on Artificial Intelligence, vol. 34, no. 7, pp. 12095–12103 (2020)
15. Wang, T.C., Efros, A.A., Ramamoorthi, R.: Occlusion-aware depth estimation using light-field cameras. In: Proceedings of the IEEE International Conference on Computer Vision, pp. 3487–3495 (2015)
16. Zhang, M., Li, J., Wei, J., Piao, Y., Lu, H.: Memory-oriented decoder for light field salient object detection. Adv. Neural Inf. Process. Syst. **32**, 898–908 (2019)

Metric Calibration of Aerial On-Board Multiple Non-overlapping Cameras Based on Visual and Inertial Measurement Data

Xiaoqiang Zhang[1(✉)], Liangtao Zhong[1], Chao Liang[1], Hongyu Chu[1], Yanhua Shao[1], and Lingyan Ran[2]

[1] School of Information Engineering, Southwest University of Science and Technology, Mianyang, Sichuan, China
{xqzhang,chuhongyu,}@swust.edu.cn, zhongliangtao@mails.swust.edu.cn, syh@cqu.edu.cn

[2] School of Computer Science, Northwestern Polytechnical University, Xi'an, Shaanxi, China
lran@nwpu.edu.cn

Abstract. Recently, the on-board cameras of the unmanned aerial vehicles are widely used for remote sensing and active visual surveillance. Compared to a conventional single aerial on-board camera, the multi-camera system with limited or non-overlapping field of views (FoVs) could make full use the FoVs and would therefore capture more visual information simultaneously, benefiting various aerial vision applications. However, the lack of common FoVs makes it difficult to adopt conventional calibration approaches. In this paper, a metric calibration method for aerial on-board multiple non-overlapping cameras is proposed. Firstly, based on the visual consistency of a static scene, pixel correspondence among different frames obtained from the moving non-overlapping cameras are established and are utilized to estimate the relative poses via structure from motion. The extrinsic parameters of non-overlapping cameras is then computed up to an unknown scale. Secondly, by aligning the linear acceleration differentiated from visual estimated poses and that obtained from inertial measurements, the metric scale factor is estimated. Neither checkerboard nor calibration pattern is needed for the proposed method. Experiments of real aerial and industrial on-board non-overlapping cameras calibrations are conducted. The average rotational error is less than 0.2°, the average translational error is less than 0.015 m, which shows the accuracy of the proposed approach.

Keywords: Non-overlapping field of view · Metric calibration · Camera-imu system · Unmanned aerial vehicle

Supported in part by the National Natural Science Foundation of China under Grant 61902322, in part by the Doctoral Fund of Southwest University of Science and Technology under Grant 19zx7123, in part by the Open Fund of CAUC Key Laboratory of Civil Aviation Aircraft Airworthiness Certification Technology under Grant SH2020112706.

H. Ma et al. (Eds.): PRCV 2021, LNCS 13020, pp. 16–28, 2021.
https://doi.org/10.1007/978-3-030-88007-1_2

1 Introduction

With the continuous development of unmanned aerial vehicle (UAV) and aerial photography technology, various aerial vision applications have been developed. The mobility and the ability of active surveillance make the UAV a good platform for remote sensing, object detection and tracking. In applications like visual surveillance, a larger FoV of the on-board camera would enhance the performance of the algorithm. Monocular large FoV cameras such as the fish-eye camera usually have severe distortions, which may change the appearance of the object and hence affect the performance in object detection or tracking. Recently, multi-camera system with limited or non-overlapping FoVs are becoming popular in the virtual reality community [5]. Compared to a monocular system, multiple non-overlapping cameras would have more FoVs. Besides, fewer cameras would be used in such a system compared to the conventional multi-view system with common FoVs. The reduced weight and power consumption of the multiple non-overlapping cameras system make it suitable for on-board equipment for the UAV. Figure 1(a) illustrates a self-build aerial on-board non-overlapping camera system with 4 cameras. With a carefully designed 3D-printed bracket, the entire system can be easily mounted on a UAV (red dashed bounding box in Fig. 1(a)). The total weight of the system is around 160 grams. Figure 1(b) gives the images captured from the system. It can be seen in Fig. 1 that there is no common FoV between adjacent cameras in the system.

Fig. 1. The aerial on-board non-overlapping camera system and its captured images. **(a)** A self-build non-overlapping camera system on a UAV. **(b)** Images of a parking lot captured from the non-overlapping camera system (Color figure online)

In many vision applications, accurate calibration of the visual system is important to the performance of algorithms. Conventional multi-view system can either be extrinsically calibrated from Zhang's method [23] or from self calibration method [20] using calibration object in the common FoV. However, for a system with multiple non-overlapping cameras, such approaches are difficult to adopt due to the non-overlapping FoVs. In order to calibrate the relative poses of the non-overlapping cameras, different approaches are proposed to establish the pixel correspondences among images with limited or no common FoVs. Some of these approaches would rely extra cooperative calibration objects like a mirror [10,19], multiple checkerboards [9,12,18,22], or designed planar patterns [7,8,11,21]. Other methods are usually based on moving camera pose estimation techniques like structure from motion [17]. However, moving camera based approaches [14,15,24] usually suffers from the lack of metric information, i.e., the physical distance in the real world. The scale ambiguity motivates us to seek a metric calibration approach for multiple non-overlapping cameras.

In this paper, we intend to solve the metric calibration problem of the UAV on-board multiple non-overlapping cameras, that is, the estimation of the relative rotations and translations in metric scale between each cameras and one selected reference camera in the system. A novel visual and inertial data based multiple non-overlapping cameras metric calibration is proposed. By jointly moving the entire system and capturing image sequence of a static scene, the correspondences between 2D pixel locations of feature points and 3D scene points are established. The camera pose of each image can be estimated via solving the perspective-n-point problem. To solve the scale ambiguity problem in the estimated poses, we make use the UAV on-broad inertial measurement unit (IMU) to estimate the metric scale factor. Real experimental results of two different multiple non-overlapping cameras systems quantitatively demonstrates the accuracy of the proposed calibration method, which are comparable with the multiple checkerboards based approaches. The metric scale estimation accuracy is further quantitatively verified in object 3D reconstruction experiments.

2 Related Works

In the literature, the calibration for multiple camera system with limited or no common FoVs can be roughly grouped into two categories, namely cooperative calibration objects based methods and moving cameras based methods.

Early work in [9] utilizes a 3D object with known geometry to estimate the relative poses among multiple cameras. Compared to the carefully designed and fine 3D objects, more commonly used objects in the multiple non-overlapping cameras calibration procedures are the planar checkerboards [9,12,18,22] or designed patterns [7,8,11,21]. Liu et al. [12] propose a method to calibrate the non-overlapping cameras using a compound object consisting of two unknown geometry of checkerboards. Yin et al. [22] introduce an extrinsic calibration method of non-overlapping cameras by solving linear independent equations. By moving the multi-camera system at least twice, enough constraints can be established. The rotations and translations are optimized separately. For pattern based approaches, Li et al. [11] propose a feature descriptor-based calibration pattern, which contains more features of varying scales than checkerboards. Pattens can be recognized and localized even if the pattern is partially observed. Xing et al. [21] design a patten with different types of identity tags or textures in the blank regions of a checkerboard, which also allows calibrations with partially observed patterns. To overcome the problem of non-overlapping FoVs, mirrors are also used for calibration [10,19], which allows cameras to observe the calibration object via reflection. However, sophisticated light path designs are usually needed for these mirror based systems, which restricts the flexibility of the calibration.

Multi-camera calibration based on moving camera method are inspired by Gaspi and Irani [6], which assumes that the lack of common FoV can be compensated by the movements of cameras. In the literature, structure from motion (SfM) has been used to calibrate multiple camera systems [14,15,24]. Zhu et al. [24] extends the single-pair hand-eye calibration used in robotics to multi-camera systems. By utilizing the planar structures in the scene, a plane-SfM is proposed for multiple non-overlapping cameras. High-precision measuring device or specially designed calibration objects are not needed in these approaches. However, because the decomposition of epipolar matrix can only estimate the relative translations between two cameras up to an unknown scale, SfM based approaches usually suffers a scale ambiguity.

3 Metric Calibration Based on Visual and Inertial Measurement Data

3.1 Notation and Problem Formulation

To formally describe the proposed approach, we first introduce coordinate systems utilized in our work. Suppose that n cameras $C_1, C_2, \ldots, C_n$ and an IMU are rigidly connected in the non-overlapping camera system. During the calibration procedure, the system is jointly moving. $C_{i,t}$ denotes the local camera coordinate system of camera C_i at time t, I_t is the inertial coordinate system at time t. We use W to denote the world coordinate system of a static 3D scene.

A notation system of superscripts and subscripts is used for denoting vectors, the relative rotation and translation between different coordinate systems. Vectors in coordinate system A is denoted by the superscripts, and a time varying vector is denoted by a subscripts t. For example, $\mathbf{P}^A$ is the coordinates of a 3D point $\mathbf{P}$ in coordinate system A, $\mathbf{v}_t^A$ is a time varying vector in A. Suppose $\mathbf{P}^A, \mathbf{P}^B$ are the coordinates of a 3D point $\mathbf{P}$ in coordinate system A and B, respectively, we would have $\mathbf{P}^B = R_A^B \mathbf{P}^A + \mathbf{t}_A^B$. $R_A^B, \mathbf{t}_A^B$ denote the relative rotation and translation, respectively. If the relative poses are time varying, a subscripts t would be used. To avoid abuses in notation of t, all translations are in **bold**.

Without loss of generality, one camera of the multiple non-overlapping camera system can be selected as the reference camera C_{ref}. The objective of the calibration approach is to estimate $\{R_{C_{ref}}^{C_i}, \mathbf{t}_{C_{ref}}^{C_i} | i = 1, 2, \ldots, n\}$. For a metric calibration, the metric scaled translation $\mathbf{t}_{C_{ref}}^{C_i}$, which denotes the physical distances in metres, should be estimated. The metric scale factor is denoted as s. An overview of the proposed method is illustrated in Fig. 2.

3.2 Relative Pose Estimation via Structure from Motion

For the non-overlapping camera systems, the lack of common FoV make it difficult to find pixel correspondences between two views, resulting in difficulty of

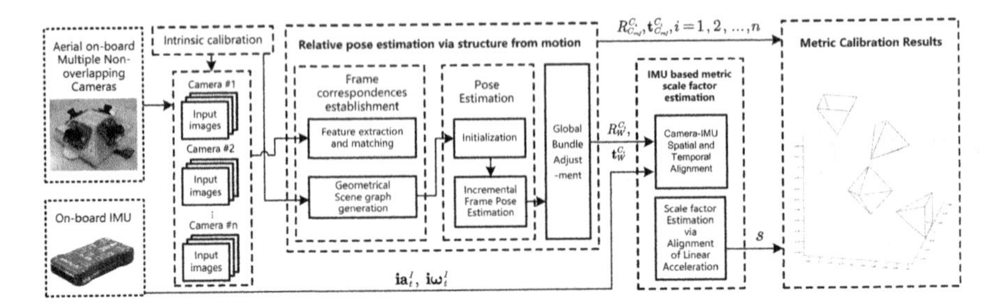

Fig. 2. Calibration framework of the proposed approach. The relative poses of the n cameras are estimated via SfM and global bundle adjustment. The scale ambiguity is then solved by aligning visual and inertial linear accelerations.

direct relative pose estimation. The relative rotation and translation estimation in this section is based on structure from motion, which estimates the poses of moving cameras and sparse scene points 3D locations from unordered images.

It can be assumed that the intrinsic parameters of each non-overlapping cameras are pre-calibrated and the cameras are synchronized. By jointly moving the entire system and observing a static scene, a series of images from different cameras at different times are captured, which form the input for SfM. First, image features like SIFT are extracted and matched on these images to find correspondences between images. By geometrically verifying images pairs via planar homography or epipolar geometry, a scene graph can be constructed. Images are the nodes and verified image pairs are the edges. By selecting two nodes, a series of 3D scene points can be initialized by stereo reconstruction. Other images are then incrementally added and registered to the reconstructed scene from the 2D-3D correspondences. New scene points can be added by solving the triangulation problem. Finally, a global bundle adjustment is preformed by minimizing the re-projection error of all M feature pixels among all images:

$$E_{repj} = \sum_{m=1}^{M} ||\pi(K_i, R_W^{C_{i,k}}, \mathbf{t}_W^{C_{i,k}}, \mathbf{X}_j^W) - \mathbf{x}_m||^2, \tag{1}$$

where $\pi(K_i, R_W^{C_{i,k}}, \mathbf{t}_W^{C_{i,k}}, \mathbf{X}_j^W)$ is the function to project 3D points $\mathbf{X}_j^W$ to camera C_i at time k. x_m is the actual pixel location that corresponding to the projection. After the optimization, the camera poses of the input images can be estimated.

Based on SfM, the poses of multiple cameras (n in total) in the system at different time (t in total) can be estimated with respect to the world coordinate system W, which are denoted by $\{R_W^{C_{i,k}}, \mathbf{t}_W^{C_{i,k}} | i = 1, 2, \ldots, n, k = 1, 2, \ldots, t\}$. Thus, the relative pose between camera C_i and C_{ref} at time k, $R_{C_{ref,k}}^{C_{i,k}}, \mathbf{t}_{C_{ref,k}}^{C_{i,k}}$, can be calculated by

$$R_{C_{ref,k}}^{C_{i,k}} = R_W^{C_{ref,k}} \left(R_W^{C_{i,k}}\right)^{-1}, \tag{2}$$

$$\mathbf{t}_{C_{ref,k}}^{C_{i,k}} = \mathbf{t}_W^{C_{ref,k}} - R_{C_{ref,k}}^{C_{i,k}} \mathbf{t}_W^{C_{i,k}}. \tag{3}$$

Because that cameras in the non-overlapping system are rigidly connected, it can be assumed that $\{R_{C_{ref,k}}^{C_{i,k}}, \mathbf{t}_{C_{ref,k}}^{C_{i,k}} | k = 1, 2, \ldots, t\}$ at different time are constant. Thus, the final estimation of $R_{C_{ref}}^{C_i}, \mathbf{t}_{C_{ref}}^{C_i}$ could be obtained by averaging t estimations at different times. Similar procedure is performed to obtain $\{R_{C_{ref}}^{C_i}, \mathbf{t}_{C_{ref}}^{C_i} | i = 1, 2, \ldots, n\}$. It should be noted that there would be a scale ambiguity in the estimated relative translations, which will be solved in the next subsection.

3.3 Inertial Measurement Data Based Metric Scale Factor Estimation

Based on SfM, the relative translations $\{\mathbf{t}_{C_{ref}}^{C_i} | i = 1, 2, \ldots, n\}$ are determined up to an unknown scale. To complete the metric calibration, the metric scale factor, which corresponds to the physical distances of $\{\mathbf{t}_{C_{ref}}^{C_i} | i = 1, 2, \ldots, n\}$, should be estimated. Although in some recent approaches, physical size of the checkerboard squares [22] can be used for computing the metric scale factor, we plan to solve the scale ambiguity in a pattern-free way. Our scale estimation procedure is based on Mustaniemi et al.'s approach [13], which optimizes the scale factor to align the acceleration obtained from differentiated visual poses and that from the IMU.

The linear acceleration of the system can be obtained in two ways. One can subtract the gravity vector from the raw accelerometer readings to obtain the linear acceleration in I. It is denoted by $\mathbf{ia}_t^I$. In the meanwhile, since the entire system is rigid, it can be assumed that the linear acceleration of the system is the same with that of one selected camera, say camera C_i. Hence, by differentiating the time varying positions of C_i in W, i.e., $\{\mathbf{t}_{C_{i,k}}^W, | k = 1, 2, \ldots, t\}$ twice, the linear acceleration in W can be computed. Which is denoted by $\mathbf{va}_t^W$. Considering the fact that noises in camera position are amplified, Rauch-Tung-Striebel smoother [16] is used when performing the double differentiation [13]. Note that there would usually be a scale ambiguity between $\mathbf{ia}_t^I$ and $\mathbf{va}_t^W$. By aligning the two linear accelerations, the metric scale factor can be estimated. Assuming N linear accelerations are measured in total, the objective function for scale factor estimation can be defined as

$$\arg\min_{s} \sum_{t=1}^{N} ||sR_{W,t}^{C_i}\mathbf{va}_t^W - R_I^{C_i}\mathbf{ia}_t^I||^2, \tag{4}$$

where $R_{W,t}^{C_i}$ is the time varying rotation between W and C_i, $R_I^{C_i}$ the relative rotation between I and C_i.

The two accelerations in Eq. (4) are usually not temporally aligned. To estimated $R_I^{C_i}$ and the time offset t_d, an objective function can be defined as

$$\arg\min_{R_I^{C_i}, \mathbf{b}_\omega^{C_i}, t_d} \sum_{t=1}^{N} ||\mathbf{v}\omega_t^{C_i} - R_I^{C_i}\mathbf{i}\omega_{t-t_d}^I + \mathbf{b}_\omega^{C_i}||^2, \tag{5}$$

where $\mathbf{v}\omega$ and $\mathbf{i}\omega$ are the angular velocity obtained from visual rotation differentiation and gyroscope, respectively. Since the outputs of a real gyroscope or a real accelerometer are usually biased, the gyroscope bias $\mathbf{b}_{\omega}^{C_i}$ in C_i is also estimated. Based on the time offset, Eq. (4) can be written as

$$\arg\min_{s,\mathbf{b}_a^{C_i}} \sum_{t=1}^{N} ||s\mathbf{R}_{W,t}^{C_i}\mathbf{va}_t^W - \mathbf{R}_I^{C_i}\mathbf{ia}_{t-t_d}^I + \mathbf{b}_a^{C_i}||^2, \tag{6}$$

where $\mathbf{b}_a^{C_i}$ denotes the accelerometer bias in C_i. In practice, the optimization in Eq. (5) is minimized using alternating optimization. Equation (6) is optimized in frequency domain. A more formal description of the implementation is given in [13]. After the estimation of metric scale factor s, we amplify the relative translations of $\{\mathbf{t}_{C_{ref}}^{C_i}|i = 1, 2, \dots, n\}$ by s, and completes the metric calibration.

4 Experimental Results

In this section, we describe the details of the experiments to verify the accuracy of the proposed approach. First, we give the details of the two multiple non-overlapping cameras systems, together with the IMU system. Quantitative evaluations of the metric calibration results of the two systems are then given. Finally, the metric scale estimation is quantitatively evaluated via object 3D reconstruction.

4.1 Equipment

In order to verify the feasibility and accuracy of the proposed approach, an aerial non-overlapping camera system is designed and constructed. We also build a non-overlapping camera system based on industrial cameras for further verification of the calibration accuracy.

The aerial non-overlapping cameras system includes four embedded cameras, an image synchronization board, a Raspberry Pi, and a 3D-printed camera bracket. The entire system is shown in Fig. 3(a). The orientation of the four cameras are carefully designed so that there are no common FoVs between adjacent cameras. For the embedded cameras, we use OV9281 with a resolution of 1280 × 800 pixels. An Arducam four-lens camera capture board [1] is used for synchronization. For the IMU system, we directly adopt the MPU6000 chip integrated in the Pixhawk4 UAV flight controller (Fig. 3(b)) [4]. The acquisition frame rate of non-overlapping cameras in this system is designed to be 20 frame per second (FPS), while the acquisition frequency of the IMU 100 Hz. It should be noted that we do not perform the synchronization between the cameras and the IMU. The captured images and the inertial data are saved using a Raspberry Pi. The total weight of the system is around 160 g, and the power supply of the system are directly from the UAV batteries. It is convenient to adopt the system to another UAV platform.

(a) (b) (c)

Fig. 3. Equipment used in the experiment. **(a)** The aerial on-board non-overlapping camera system. **(b)** Pixhawk 4 UAV flight controller, the IMU chip provides the inertial data for both of the two system. **(c)** Industrial cameras based non-overlapping camera system.

For the industrial camera based system, we choose to use the Daheng MER-131-75GM cameras [2]. The system is shown in Fig. 3(c), in which 3 cameras are pointing at different angles for a non-overlapping FoV. The resolution of the captured image is 1280 × 1024 pixels and the frame rate is 20. A STM32ZET6 MCU is used for hardware based synchronization. We also use the IMU in Pixhawk UAV flight controller in this system, with an acquisition frequency 100 Hz. The captured images and inertial data are uploaded to a PC station for further calibration.

4.2 Metric Calibration of the Aerial On-Board Non-overlapping Camera System

In this experiment, our goal is to quantitatively evaluate the accuracy of the proposed metric calibration method. We apply the proposed approach to the aerial on-board non-overlapping camera system mentioned in Sec. 4.1. Since there are 4 cameras in total in the system, we choose one of them as the reference camera. The relative rotations and translations of all the other 3 cameras are estimated w.r.t the reference camera. We name the three camera as camera #1 to #3.

To quantitatively evaluate the accuracy of the calibration, we also perform a multiple camera based approach to obtain the ground truth of the relative rotation and metric translations. Figure 4(a) gives the details when obtaining the ground truth. By using a global camera C_{global}, which shares common FoV with both camera C_1 and C_{ref}, two checkerboards can be used to obtain the relative poses between C_1 and C_{global}, and that between C_{ref} and C_{global}. Then the relative pose between C_1 and C_{ref} can be obtained via coordinate transforms and we perform a bundle adjustment to optimize the relative poses. By utilizing multiple checkerboards, the ground truth of the relative poses between all cameras and the reference camera can be obtained.

Table 1 shows the quantitative evaluation of the metric calibration results for aerial on-board non-overlapping camera system. For each row, the roll angle, yaw angle and pitch angle of the relative orientations between camera $\{C_i | i = 1, 2, 3\}$

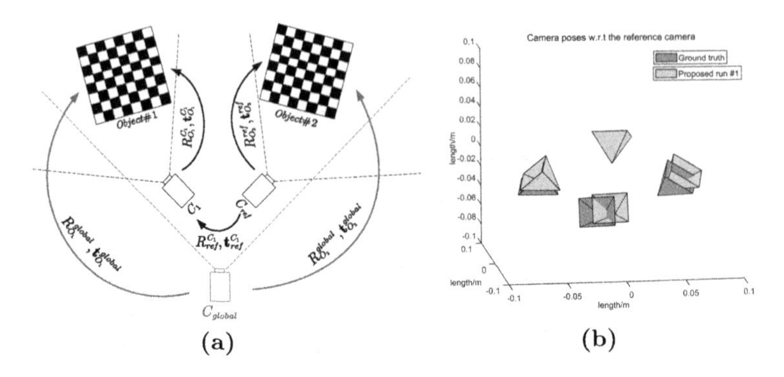

Fig. 4. Calibration results for aerial on-board non-overlapping camera system. **(a)** procedure for obtaining the ground truth **(b)** metric calibration results in run #1. (Color figure online)

and the reference camera C_{ref} is given in degree. The relative translations is given along X, Y and Z axes in metres. We run 3 different calibrations for the same system. The calibration results, together with the ground truth values are given. It can be seen that the average rotational error is less that $\mathbf{0.2}°$ and the average translational error is less than **0.015** m. Figure 4(b) illustrates the comparison between ground truth relative poses (in red color) and the estimated ones in run #1 (in blue color). Please note that in both Table 1 and Fig. 4(b), the errors in scale of the estimated translations and the ground truth are small, which shows the accuracy of the metric scale estimation.

Table 1. Estimated relative orientations and positions of different cameras w.r.t the reference camera in the **aerial on-board** non-overlapping camera system.

Approaches	camera	roll (°)	yaw (°)	pitch (°)	$\mathbf{t_x}$ (m)	$\mathbf{t_y}$ (m)	$\mathbf{t_z}$ (m)
Ground truth values	#1	44.6052	44.9146	−1.4001	0.0283	−0.0425	−0.0364
	#2	90.1890	−1.4332	88.6350	0.0532	−0.0027	−0.0672
	#3	54.7661	29.1199	124.2423	0.0217	0.0411	−0.0389
Proposed run #1	#1	44.5014	44.9908	−1.3208	0.0342	−0.0402	−0.0332
	#2	90.2865	−1.4905	88.8299	0.0535	0.0085	−0.0640
	#3	54.8420	29.0332	124.6838	0.0203	0.0462	−0.0325
Proposed run #2	#1	44.6957	45.0860	−1.3637	0.0298	0.0337	−0.0415
	#2	90.4184	−1.416	88.7724	0.0566	0.0087	−0.0643
	#3	55.1482	28.8704	124.6518	0.0197	0.0445	−0.0316
Proposed run #3	#1	44.5544	44.8269	−1.2525	0.0231	−0.0483	−0.0306
	#2	90.1434	−1.3628	88.7618	0.0531	−0.0025	−0.0654
	#3	54.9328	29.1326	124.2975	0.0181	0.0353	−0.032
Average error		0.1380	0.0921	0.1810	0.0026	0.0135	0.0047

4.3 Metric Calibration of an Industrial Non-overlapping Camera System

To further evaluate the performances of the proposed method on different camera systems, we apply the propose approach to the industrial cameras based non-overlapping camera system mentioned in Sect. 4.1. Multiple checkerboards based approach that is similar with that in Sect. 4.2 is conducted for obtaining the ground truth. In this experiment, we run the calibration for 2 different times. Table 2 gives the quantitative evaluations of the metric calibration results. It can be seen that the average rotational error is less than 0.05°, and the average translational error is less than 0.005 metres. From the comparison between Table 1 and Table 2, it can be seen that cameras with a higher resolution (1280 × 1024 pixels in Table 2 v.s. 1280 × 800 pixels in Table 1) would have a calibration result with less errors. Besides, the proposed method can be applied to different non-overlapping camera systems with a reasonable accuracy.

Table 2. Estimated relative orientations and positions of different cameras w.r.t the reference camera in the **industrial** non-overlapping camera system.

Approaches	camera	roll (°)	yaw (°)	pitch (°)	$\mathbf{t_x}$ (m)	$\mathbf{t_y}$ (m)	$\mathbf{t_z}$ (m)
Ground truth values	#1	2.1209	14.3206	0.8938	0.0147	−0.0581	−0.0148
	#2	3.0223	35.1218	2.1217	0.0252	−0.1075	−0.0252
Proposed run #1	#1	2.0896	14.2859	0.8889	0.0174	−0.0557	−0.0102
	#2	2.9402	35.1025	2.1134	0.0237	−0.1047	−0.0237
Proposed run #2	#1	2.1076	14.2665	0.8773	0.0198	−0.0617	−0.0198
	#2	2.9507	35.0922	2.0832	0.0317	−0.1161	−0.0215
Average error		0.0496	0.0344	0.0170	0.0040	0.0043	0.0037

4.4 Experiments of Applications for Object Metric 3D Reconstruction

Because the metric relative translations are estimated, one can obtain a metric reconstruction of scene objects purely based on images from the systems. No external metric sensors, such as the depth sensor or the inertial sensor, is needed in the metric reconstruction.

In this experiment, we give a quantitatively evaluation of the metric scale in this application. We place 3 different boxes in a static scene (Fig. 5(a)). By moving the on-board non-overlapping cameras and observing several static scene objects, a dense 3D reconstruction can be obtained via SfM [17] (Fig. 5(b)). The length of a object in the scene can be obtained using Meshlab [3] (Fig. 5(c)). Based on the metric scale estimated from the proposed method, the 3D reconstruction can be scaled to the physical scale. We compare the estimated object length with the ground truth length obtained from a ruler (Fig. 5(c)). Table 3 gives the quantitative evaluation of the estimated object lengths or widths in metres. It can be seen that the errors are all less than 2%, which shows the accuracy of the estimated metric scale factor in the proposed approach.

Table 3. Quantitative evaluation of the metric 3D reconstruction on 3 different scene objects. The scale factor is 0.1713 in this experiment.

Object		Ruler (m)	Reconstructed	Estimated (m)	Error
#1	Length	0.7050	4.1589	0.7124	1.06%
	Width	0.5190	3.0546	0.5233	0.81%
#2	Length	0.3110	1.8062	0.3094	0.35%
	Width	0.1950	1.1508	0.1971	1.34%
#3	Length	0.6720	3.9714	0.6803	1.20%
	Width	0.1650	0.9496	0.1627	1.51%

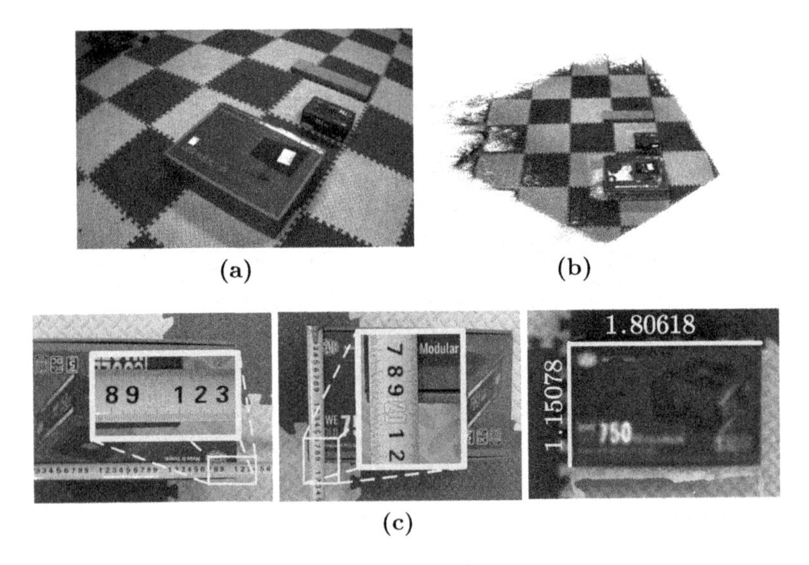

Fig. 5. Results of applications for object metric 3D reconstruction. **(a)** A static scene with 3 boxes. **(b)** Reconstructed point cloud. **(c)** Length and width of object #2 obtained from both rulers and the point clouds.

5 Conclusions

In this work, a novel metric calibration method for aerial on-board multiple non-overlapping cameras is proposed. The 3D geometry and consistence of a static scene is utilized to establish the pixel correspondences between non-overlapping cameras. To overcome the scale problem in SfM based pose estimation, the inertial measurement data is used for estimate the metric scale. Real experiment of both aerial on-board and industrial camera based non-overlapping camera systems shows the accuracy of our approach. No calibration object is used during the calibration procedure. In the future work, we would like to consider the application of aerial non-overlapping cameras for remote sensing, metric scene reconstruction, and active object detections.

References

1. ArduCam. https://www.arducam.com
2. Daheng Imaging cameras. https://www.daheng-imaging.com/
3. Meshlab. https://www.meshlab.net/
4. Pixhawk flight controller. https://pixhawk.org/products/
5. Anderson, R., et al.: Jump: virtual reality video. ACM Trans. Graph. (TOG) **35**(6), 1–13 (2016)
6. Caspi, Y., Irani, M.: Aligning non-overlapping sequences. Int. J. Comput. Vis. (IJCV) **48**(1), 39–51 (2002)
7. Dong, S., Shao, X., Kang, X., Yang, F., He, X.: Extrinsic calibration of a non-overlapping camera network based on close-range photogrammetry. Appl. Opt. **55**(23), 6363–6370 (2016)
8. Gong, Z., Liu, Z., Zhang, G.: Flexible global calibration of multiple cameras with nonoverlapping fields of view using circular targets. Appl. Opt. **56**(11), 3122–3131 (2017)
9. Kassebaum, J., Bulusu, N., Feng, W.C.: 3-D target-based distributed smart camera network localization. IEEE Trans. Image Process. (TIP) **19**(10), 2530–2539 (2010)
10. Kumar, R.K., Ilie, A., Frahm, J.M., Pollefeys, M.: Simple calibration of non-overlapping cameras with a mirror. In: Proceedings of the IEEE Conference on Computer Vision and Pattern Recognition (CVPR), pp. 1–7. IEEE (2008)
11. Li, B., Heng, L., Koser, K., Pollefeys, M.: A multiple-camera system calibration toolbox using a feature descriptor-based calibration pattern. In: Proceedings of the IEEE/RSJ International Conference on Intelligent Robots and Systems (IROS), pp. 1301–1307. IEEE (2013)
12. Liu, Z., Zhang, G., Wei, Z., Sun, J.: A global calibration method for multiple vision sensors based on multiple targets. Meas. Sci. Technol. **22**(12), 125102 (2011)
13. Mustaniemi, J., Kannala, J., Särkkä, S., Matas, J., Heikkilä, J.: Inertial-based scale estimation for structure from motion on mobile devices. In: Proceedings of the IEEE/RSJ International Conference on Intelligent Robots and Systems (IROS), pp. 4394–4401. IEEE (2017)
14. Pagel, F.: Extrinsic self-calibration of multiple cameras with non-overlapping views in vehicles. In: Video Surveillance and Transportation Imaging Applications 2014, vol. 9026, p. 902606. International Society for Optics and Photonics (2014)
15. Pagel, F., Willersinn, D.: Motion-based online calibration for non-overlapping camera views. In: 13th International IEEE Conference on Intelligent Transportation Systems, pp. 843–848. IEEE (2010)
16. Rauch, H.E., Tung, F., Striebel, C.T.: Maximum likelihood estimates of linear dynamic systems. AIAA J. **3**(8), 1445–1450 (1965)
17. Schonberger, J.L., Frahm, J.M.: Structure-from-motion revisited. In: Proceedings of the IEEE Conference on Computer Vision and Pattern Recognition (CVPR), pp. 4104–4113 (2016)
18. Strauß, T., Ziegler, J., Beck, J.: Calibrating multiple cameras with non-overlapping views using coded checkerboard targets. In: 17th International IEEE Conference on Intelligent Transportation Systems, pp. 2623–2628. IEEE (2014)
19. Sturm, P., Bonfort, T.: How to compute the pose of an object without a direct view? In: Narayanan, P.J., Nayar, S.K., Shum, H.Y. (eds.) ACCV 2006. LNCS, vol. 3852, pp. 21–31. Springer, Heidelberg (2006). https://doi.org/10.1007/11612704_3
20. Svoboda, T., Martinec, D., Pajdla, T.: A convenient multicamera self-calibration for virtual environments. Presence Teleoperators Virtual Environ. **14**(4), 407–422 (2005)

21. Xing, Z., Yu, J., Ma, Y.: A new calibration technique for multi-camera systems of limited overlapping field-of-views. In: Proceedings of the IEEE/RSJ International Conference on Intelligent Robots and Systems (IROS), pp. 5892–5899. IEEE (2017)
22. Yin, L., Wang, X., Ni, Y., Zhou, K., Zhang, J.: Extrinsic parameters calibration method of cameras with non-overlapping fields of view in airborne remote sensing. Remote Sens. **10**(8), 1298 (2018)
23. Zhang, Z.: A flexible new technique for camera calibration. IEEE Trans. Pattern Anal. Mach. Intell. (TPAMI) **22**(11), 1330–1334 (2000)
24. Zhu, C., Zhou, Z., Xing, Z., Dong, Y., Ma, Y., Yu, J.: Robust plane-based calibration of multiple non-overlapping cameras. In: 2016 Fourth International Conference on 3D Vision (3DV), pp. 658–666. IEEE (2016)

SEINet: Semantic-Edge Interaction Network for Image Manipulation Localization

Ye Zhu[1,2], Na Qi[1], Yingchun Guo[1], and Bin Li[2,3](✉)

[1] The School of Artificial Intelligence, Hebei University of Technology, Tianjin 300401, China
zhuye@hebut.edu.cn, 201922802018@stu.hebut.edu.cn, gyc@scse.hebut.edu.cn

[2] Guangdong Key Laboratory of Intelligent Information Processing and Shenzhen Key Laboratory of Media Security, Shenzhen University, Shenzhen 518060, China
libin@szu.edu.cn

[3] Shenzhen Institute of Artificial Intelligence and Robotics for Society, Shenzhen 518129, China

Abstract. Previous manipulation detection methods usually utilize semantic to detect suspected regions but the edge is abridged, which poses a greater challenge in manipulation location. We propose a novel dual framework named Semantic-Edge Interaction Network (SEINet) for locating the manipulated regions, including splicing, copy-move and removal. The dual streams and Cross Interaction (CI) pattern aim to extract semantic and edge features under the supervision of semantic and edge Ground-Truth, respectively. In addition, we propose a Bidirectional Fusion Module (BFM) to incorporate the dual stream feature maps with the decoder of U-net. Extensive experiments, which are evaluated on Synthetic, CASIA and NIST16 datasets, prove that the proposed SEINet can locate the manipulated regions more accurate than state-of-the-art methods, and is more robustness to noise, blur, and JPEG recompression attacks.

Keywords: Image manipulation location · Dual-streams network · Cross interaction · Bidirectional fusion

1 Introduction

With the development of image editing technology and user-friendly software, the image manipulation is easier for many tampered operations, such as splicing, copy-move and removal, etc. Traditional methods for image manipulation localization generally use specific attribute features to detect specific types of tampering, including Color Filter Arrays (CFA) [1], Lateral Chromatic Aberration (LCA) [2], Photo Response Non-Uniformity (PRNU) [3], Sensor Linear Pattern (SLP) [4].

While traditional methods focus on manual statistical features, the deep convolutional neural networks based methods learn the tampered features autonomously, which can be classified into patch-based and pixel-based. The patch-based methods divided the image into small patches to locate the genuine and tampered patches through inconsistencies.

The original version of this chapter was revised: acknowledgement of a research grant added. The correction to this chapter is available at https://doi.org/10.1007/978-3-030-88007-1_55

H. Ma et al. (Eds.): PRCV 2021, LNCS 13020, pp. 29–41, 2021.
https://doi.org/10.1007/978-3-030-88007-1_3

The camera model features [5, 6] and resampling features LSTM-En [7] were extracted to locate manipulation region in patch level. GSCNet [8] combined U-Net and LSTM networks to build relationships of image patches, but the computational complexity is slightly high.

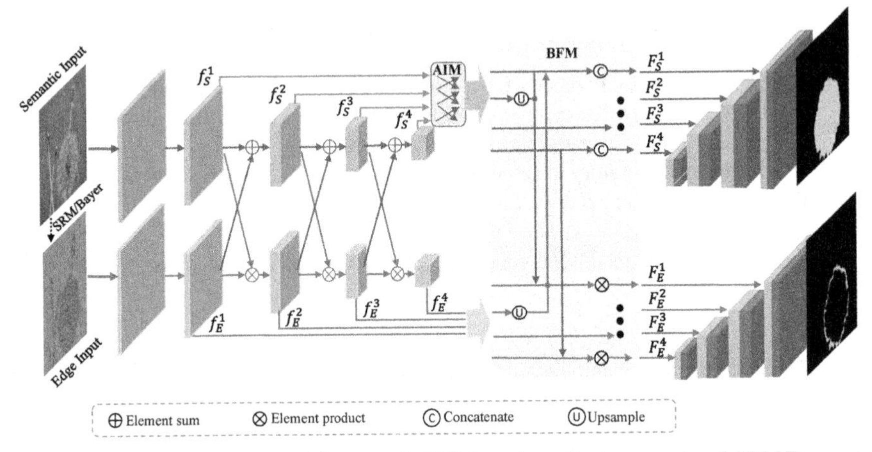

Fig. 1. Overview of the proposed framework SEINet, where the tampered and SRM/Bayar pre-processed image are the input of semantic and edge stream, respectively. The Cross Interaction (CI) extract the manipulated information and fused by Bidirectional Fusion Module (BFM) under the supervision of semantic and edge Ground-Truth.

Unlike the statistical properties of patches, the pixel-based methods usually require preprocessing to extract manipulating traces. Bayar [9] proposed a constrained convolution kernel to restrict the image content and enhance the tampering trace information, which had been applied in convolutional networks to extract features for median filter detection [10] and camera models [11]. In splicing detection, a FCN-based framework was proposed to detect and localize splicing regions [12]. ManTra-net regarded manipulation detection as anomaly detection problem and proposed a network to distinguish hundreds of operation types [13]. RGB-N [14] and D-CNNs [15] used dual streams to extract features, where RGB-N extracted color and noise features, and D-CNNs focused on spatial and frequency features, respectively. LBP features and clustering algorithm can recognize the structural relevance [16, 17]. SPAN utilized a self-attentive hierarchical structure, which can help to locate regions [18].

The above methods just consider semantic and frequency features in manipulation detection, which ignored the edges of the manipulated area. We propose an end-to-end framework named Semantic-Edge Interaction Network (SEINet), which utilize dual streams to extract edge and semantic features under the supervision of edge and image mask respectively. Besides, the Cross Interaction (CI) extract the information of the semantic and edge streams, which are fused by Bidirectional Fusion Module (BFM), as is shown in Fig. 1. Our contributions are summarized as follows:

(1) We consider four interaction patterns and prove that the Cross Interaction (CI) pattern extract more comprehensive features in manipulation operation.

(2) We propose the Bidirectional Fusion Module (BFM) that makes further use of the relationships between semantic and edge information to optimize the manipulation features.
(3) By combining the interaction in semantic and edge branches, a novel manipulation localization method SEINet is proposed, which is more effective to locate manipulated regions, and robust to postprocessing operations.

The rest of this article is organized as follows. The Sect. 2 briefly describes the related work. In Sect. 3, we elaborate the structure of the proposed network. The ablation experiments and comparison experiments are analyzed in Sect. 4, and finally we draw conclusions in Sect. 5.

2 Related Work

Multi-level Feature Extraction: Full convolutional networks is able to perform pixel-level classification. U-net was developed from a fully convolutional network using a symmetric structure, sampled on the corresponding stage, and the recovered feature map incorporating more shallow information, also allowing the fusion of features at different scales [19]. DMVN [20] and AR-Net [21] focused on multi-level feature extraction, which used jump structure in the last three layers to fuse features. The saliency detection network MINet exploits the feature similarity of neighboring layers in feature extraction and uses the information from the preceding and following layers to optimize the features in this layer [22], but multiple reuse of features from the same layer is prone to information redundancy. In this paper, we propose a cross extraction strategy, which makes use of the complementary information of adjacent layers to avoid feature redundancy.

Dual-Stream Frameworks: Zhou applied the steganalysis feature of the face classification stream and patch stream to analyze the degree of similarity in the embedding space for facial manipulation detection [23]. The Busternet fused features from the similarity and detection branches before classifying [24]. The RGB-N applied the target detection framework Faster-rcnn to extract the color and noise features of the image separately, and finally fused the results using bilinear pooling [14]. DCNNs considered the spatial and frequency domains and transformed the results to fully connected layer for classification [15]. However, the above methods ignore the correlation between the two branches, where we focus on the correlation of semantic and edge, and consider four interaction patterns for dual-stream feature extraction.

Edge Information: The edge information can be used to optimize the segmentation results. The splicing detecting network MFCN, incorporates parallel edge branch, uses ground-truth mask and edge mask as supervision to learn image content features and edge features, it determines the result through an intersection operation. Nevertheless, the regions localized in this way are slightly coarse, and taking the intersection may lose some of the correct regions when one branch contains the correct localization results. The saliency detection network SCRN is designed to optimize both object feature detection and edge detection, with the saliency features at each level spliced with the edge map of all levels before decoding [25]. JSENet designs shared encoders on a 3D semantic segmentation task, using a semantic decoder and an edge decoder to jointly optimize the features [26]. In contrast to MFCN which takes the intersection directly, when extracting manipulated features, we mutually optimize the semantic and edge information from different layers and forward propagate to extract higher level features.

3 Method

Step-by-step instructions of proposed SEINet are provided in this section, as is shown in Fig. 1. The VGG16 is applied as the backbone in semantic and edge branches, features are extracted through cross interaction pattern, and semantic information is aggregated in multiple levels to enhance the representation of features. In addition, a bidirectional fusion module is optimized to obtain pixel-level prediction results.

3.1 Cross Interaction Pattern

The mainstream frameworks focus on the semantic level information when locating the manipulated regions. Manipulation localization is essentially a pixel binary classification problem, where the edge features can strengthen the feature discrimination ability of the network to obtain more precise localization results.

The input of edge branch is the image after SRM [27] and Bayar [9] preprocessing, which can suppress image content and enhance tampering features. For fusing the feature maps in semantic and edge branches, we consider four interaction patterns, namely None Interaction (NI), Top-Bottom Interaction (TBI), Bottom-Top Interaction (BTI), and Cross Interaction (CI), as is shown in Fig. 2. The relevant experiments are analyzed in Sect. 4.2, where the experiments prove that the CI works better. Inspired by the correlation of the semantic and edge, the semantic branch is interacted by summation so the area is more complete and the edge is interacted by multiplication so the edge is more precise. The feature maps in semantic stream f_S^i and the edge stream f_E^i are expressed in Eq. (1),

$$f_S^i = \begin{cases} E^i\left(f_S^{i-1}\right) & i = 1, 2 \\ E^i\left(f_S^{i-1} + f_E^{i-1}\right) & i = 3, 4, 5 \end{cases} \quad f_E^i = \begin{cases} E^i\left(f_S^{i-1}\right) & i = 1, 2 \\ E^i\left(f_E^{i-1} * f_S^{i-1}\right) & i = 3, 4, 5 \end{cases} \tag{1}$$

where f_S^0 and f_E^0 denote the feature maps after convolution and pooling operation of the semantic and edge inputs, $E^i(\cdot)$ denotes the convolution and pooling operation of i-th layer.

3.2 Aggregate Interaction Module

In the semantic branch, different convolution layers correspond to the corresponding levels of features. The Aggregate Interactive Module(AIM) enhances the representation of individual resolution features, and the information from adjacent convolutional layers can provide an effective complement to this layer. In SEINet, the aggregation operation

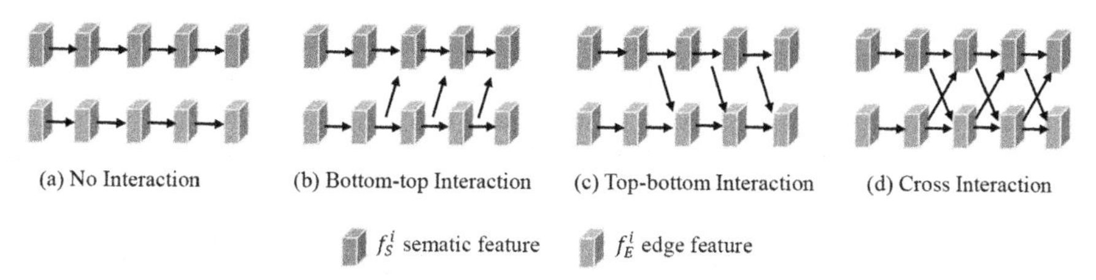

Fig. 2. Illustration of four interaction patterns, which are None Interaction(NI), Top-Bottom-Interaction(TBI), Bottom-Top Interaction(BTI), and Cross Interaction(CI).

$Agg(\cdot)$ is applied in the last four layers of the semantic branch. The feature maps after AIM is recorded as f_A^i, where the formula is described as follows.

$$f_A^i = f_S^i + Agg^i(f_S^i) \tag{2}$$

Since the dimension of front and next layers is different, the aggregation operation is merged by the downsampling of the front layer and upsampling of the next layer. To ensure that information from neighboring layers just plays a complementary role, the weights of this layer are emphasized using residual joins, as is shown in Fig. 3.

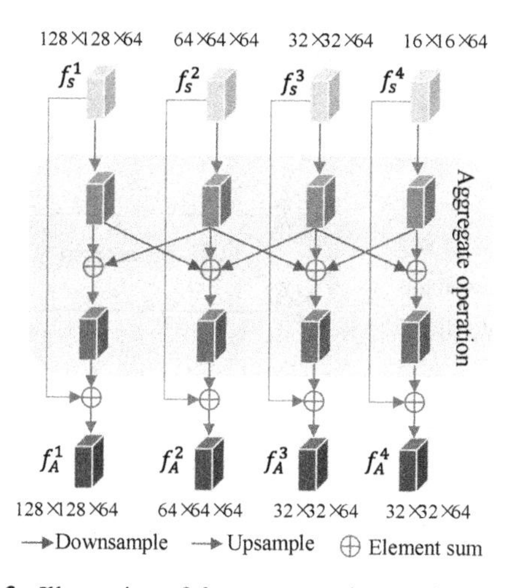

Fig. 3. Illustration of the aggregate interactive module

3.3 Bidirectional Fusion Module

Edge information could guide manipulation localization, so edge features and semantic information can be optimized by each other. The semantic branch learn content information and the edge branch learn attribute information, which combine to explore tampering traces from different perspectives and make the detection results more comprehensive. The features obtained from semantic and edge branches are denoted as f_A^i

and f_E^i, $i = (1, 2, 3, 4)$, and a convolution of kernel size 1 is used to fix the dimension of the features. The bidirectional fusion module consists of two steps, as is shown in Fig. 4. The first step use the edge features of the corresponding and adjacent layers to concatenate together with the semantic features. For example, the edge features in $i + 1$th layer are upsampled to the same size as the semantic feature of i-th layer. For removing useless information, the second step that attention mechanism SK kernel $att(\cdot)$ is utilized to distinguish the importance of different channels [28]. Besides, the residual join is to sum up the input features and the fused features. The output of BFM in semantic F_S^i and edge F_E^i features can be computed as follows,

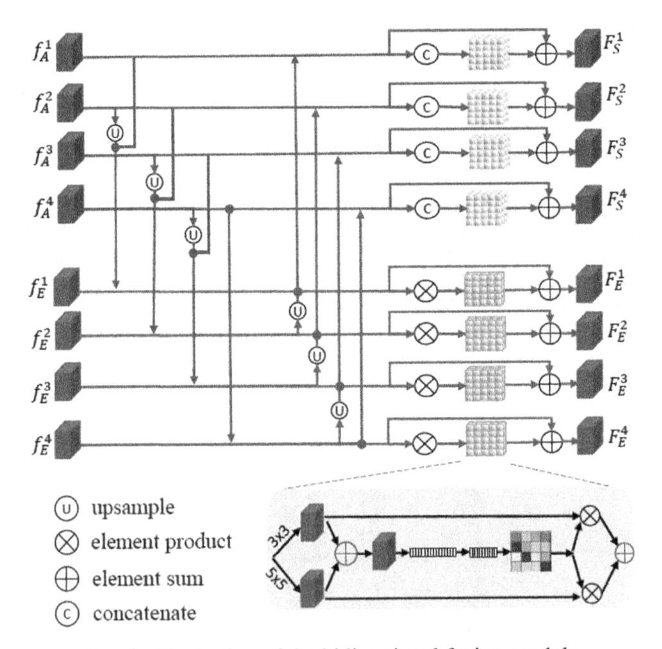

Fig. 4. Illustration of the bidirectional fusion module.

$$\begin{cases} F_S^i = f_A^i + att(Con(f_E^i, Up(f_E^{i+1}), f_A^i)) \\ F_E^i = f_E^i + att(Pro(f_A^i, Up(f_A^{i+1}), f_E^i)) \end{cases} \tag{3}$$

where $Con(\cdot)$ and $Pro(\cdot)$ are concatenate and element product operation, the $Up(\cdot)$ denotes the upsample operation.

3.4 Training Loss

The Binary Cross Entropy (BCE) loss in semantic stream L_{bce}^S and edge stream L_{bce}^E are applied to measure the difference between the predicted and ground-truth values, which are computed as Eq. (4):

$$L_{bce}^S = -\sum\nolimits_{(i,j)} [G^S(i,j)\log(P(i,j)) + (1 - G^S(i,j))\log(1 - P(i,j))] \tag{4}$$

where $G^S(i,j)$ is the ground-truth of semantic branch, $P(i,j)$ is the probability value of pixel predicted to be manipulated. The binary cross-entropy loss L_{bce}^E in edge branch is similar with L_{bce}^S, where the $G^S(i,j)$ is replaced by the ground-truth of edge branch $G^E(i,j)$. The CEL loss L_{CEL} is applied in semantic stream, which associates all pixels for global constraints and accelerate model convergence. The L_{CEL} is defined as follows,

$$L_{CEL} = \frac{\sum[G^S(i,j) - P(i,j)G^S(i,j)] + \sum[P(i,j) - P(i,j)G^S(i,j)]}{\sum P(i,j) + \sum G^S(i,j)} \tag{5}$$

The total loss L_{loss} is computed as follows,

$$L_{loss} = L_{bce}^S + L_{bce}^E + L_{CEL} \tag{6}$$

4 Experiments

4.1 Datasets and Implementation Details

For a fair comparison, the synthetic dataset that include 10k tampered images is set as training dataset. The ground-truth of edge branch is the binary mask, which is obtained by the Sobel edge extraction algorithm. For testing, we use two standard datasets NIST16 [29] and CASIA [30] and randomly choose 300 images from synthetic dataset.

NIST16 is a challenging dataset that contains splicing, copy-move and removal manipulations. The dataset provides the ground-truth, where 404 images were for fine-tuning and 160 images were testing.

CASIA contains copy-move and splicing images of multiple targets, and the tampered regions are carefully selected, with a post-processing operation such as filtering or blurring. We use CASIA2.0 for training and CASIA1.0 for testing.

The module is initialized by kaiming_uniform [31] and the training images is resized to 256 × 256. The batch-size is set to 4, optimizer is SGD with a momentum of 0.9 and the initial learning rate is 1.0e−3. All the experiments are conducted on a machine with Intel Xeon W2123 CPU @ 3.60 GHz, 32 GB RAM, and a single GPU (GTX 1080Ti).

4.2 Evaluation Metrics

To evaluate the effectiveness of SEINet, $F1$ and AUC score (Area Under the receiver operating characteristic Curve) are measured in pixel level. The $F1$ score focus on reducing model misses, while the AUC focus on reducing model false positives. The $F1$, True Positive Rate (TPR) and False Positive Rate (FPR) are calculated as follows,

$$F1 = \frac{2 \times precision \times recall}{precision + recall} \tag{7}$$

$$TPR = \frac{TP}{TP + FP} \tag{8}$$

$$FPR = \frac{FP}{FP + TN} \tag{9}$$

where $precision = TP/TP + FP$, $recall = TP/TP + FP$, TP and FN denote the number of manipulated pixels that predicted correctly and incorrectly, TN and FP denote the number of genuine pixels that predicted correctly and falsely.

4.3 Ablation Studies

In order to verify the effectiveness of the four interaction patterns and bidirectional fusion module, the various frameworks are set as follows,

- **Base(img)**: The framework of single stream with the inputting of image.
- **Base(pre)**: The framework of single stream with the inputting of image after preprocessing.
- **BTI:** The framework of dual streams with features extraction through top-bottom interaction, as shown in Fig. 2(b).
- **TBI:** The framework of dual streams with features extraction through bottom-top interaction, as shown in Fig. 2(c).
- **CI:** The framework of dual streams with features extraction through cross interaction, as shown in Fig. 2(d).
- **SEINet:** The framework of dual streams with features extraction through cross interaction and bidirectional fusion module.

The $F1$ on Synthetic, NIST16 and CASIA datasets are shown in Table 1, where the greatest performance are masked in bold. Visual comparisons are provided in Fig. 5, Base(pre) is able to detect regions that are not detected by the Base(img), which facilitates the network to capture the cues after learning manipulated traces. Besides, the CI has a suppressing effect on the detection noise, and the results of SEINet reveal accurate location with unabridged edge.

Table 1. Training and testing split (number of images) on different datasets

Inputs	Framework	$F1$		
		Synthetic	NIST16	CASIA
Img	Base(img)	0.929	0.550	0.443
SRM/Bayar	Base(pre)	0.932	0.554	0.447
Img+ SRM/Bayar	BTI	0.941	0.660	0.466
Img+ SRM/Bayar	TBI	0.948	0.649	0.464
Img+ SRM/Bayar	CI	0.951	0.703	0.481
Img+ SRM/Bayar	SEINet	**0.957**	**0.891**	**0.488**

4.4 Robustness Analysis

To verify the robustness of the postprocessing attacks on the SEINet, we apply three attack operations. The mean is zero and the variances are 0.01 and 0.05 in Gaussian noise. The window size is 3×3 and variances are 0.5 and 1 in the blur. And JPEG compression factors are 90, 60, 30. The results are shown in Table 2, which demonstrates that the SEINet has a great robustness except the Gaussian noise. Besides, to compare

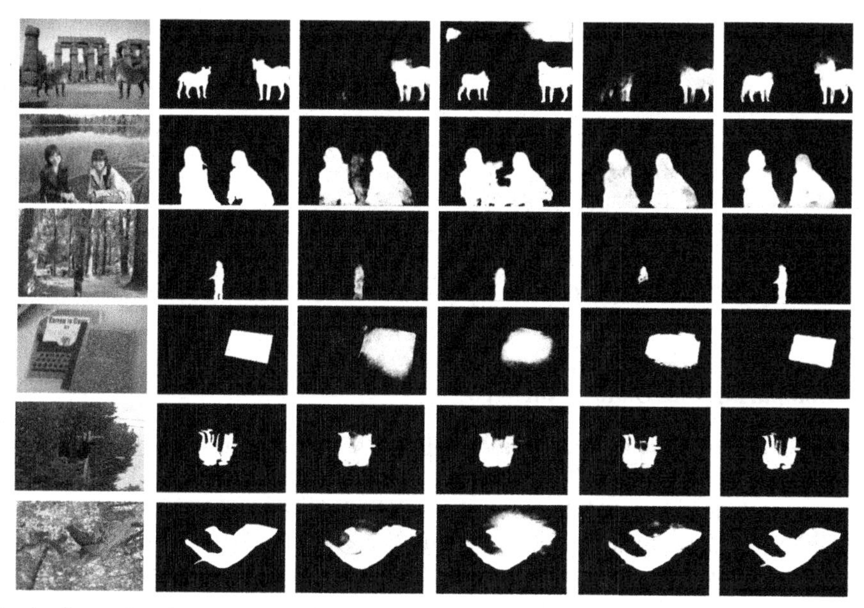

Fig. 5. Exemplar visualization of the ablation studies. From left to right: the first two columns are tampered images and corresponding ground truth, the third to sixth columns are the results of Base(img), Base(pre), CI and CESNet.

with other methods, we use the setting which is same with RGB-N [14] and evaluated on NIST dataset. The comparison result is shown in Table 3, where the greatest performance are masked in bold and other result data is from RGB-N. It shows that the robustness of the proposed SEINet is more stable than other methods.

Table 2. Robustness validation of SEINet on different postprocessing attacks

Postprocessing attacks	$F1$	AUC
None	0.9568	0.9982
Noise(0,0.01)	0.8419	0.9579
Noise(0,0.05)	0.5662	0.7375
Blur(3 × 3,0.5)	0.9563	0.9982
Blur(3 × 3,1)	0.9553	0.9983
JPEG(90)	0.9568	0.9982
JPEG(60)	0.9568	0.9982
JPEG(30)	0.9568	0.9982

Table 3. $F1$ score comparison on NIST16 dataset for JPEG and resizing of different methods

JPEG/Resizing	100/1	70/0.7	50/0.5
NOI	0.285/0.285	0.142/0.147	0.140/0.155
ELA	0.236/0.236	0.119/0.141	0.114/0.114
CFA	0.174/0.174	0.152/0.134	0.139/0.141
RGB-N	0.722/0.722	0.677/0.689	0.677/0.681
SEINet	**0.891/0.891**	**0.891/0.891**	**0.888/0.890**

4.5 Comparing with State-of-the-Art

In this section, we compare the proposed SEINet with the mainstream algorithms on NIST and CASIA datasets. The traditional methods NOI [32], ELA [33], CFA [1], and the convolutional network-based methods RGB-N [14], LSTM-En [7], ManTra-net [13], SPAN [18] are evaluated through $F1$ and AUC. The $F1$ and AUC scores of different methods are shown in Table 4, where the results are from published paper.

Our proposed SEINet almost has the greatest performance even though the training datasets ManTra-Net and SPAN included 180k images. The visual experimental results on NIST16 and CASIA2.0 are shown in Fig. 6, where SEINet can accurately locate the manipulating region in most cases. Then we analyze the comparison with other methods on different types of manipulated images in NIST dataset, the proposed SEINet achieves better results than ManTra-Net and SPAN on both removal and splicing, the quantitative results are shown in Table 5.

Table 4. $F1$ and AUC Score comparison of state-of-the arts methods on NIST and CASIA datasets.'–' denotes the results are not evaluated in the published paper.

Methods	$F1$		AUC	
	NIST	CASIA	NIST	CASIA
ELA	0.236	0.214	42.9	61.3
NOI	0.285	0.263	48.7	61.2
CFA	0.174	0.207	50.1	52.2
RGB-N	0.722	0.408	93.7	79.5
LSTM-En	–	0.391	79.3	76.2
ManTra-Net	–	–	79.5	81.7
SPAN	0.582	0.382	96.1	**83.8**
GSCNet	0.837	0.471	91.6	83.3
SEINet	**0.891**	**0.488**	**98.0**	80.1

Table 5. Pixel level $F1$ and AUC Score comparison on different manipulation types with state-of-the arts methods on NIST16 dataset.

Methods	Splicing		Removal		Copy-move	
	$F1$	AUC	$F1$	AUC	$F1$	AUC
ManTra-Net	0.386	85.89	0.149	65.52	0.150	79.8
SPAN	0.829	**99.15**	0.500	90.95	0.405	90.94
SEINet	**0.907**	96.57	**0.887**	**99.60**	**0.845**	**99.31**

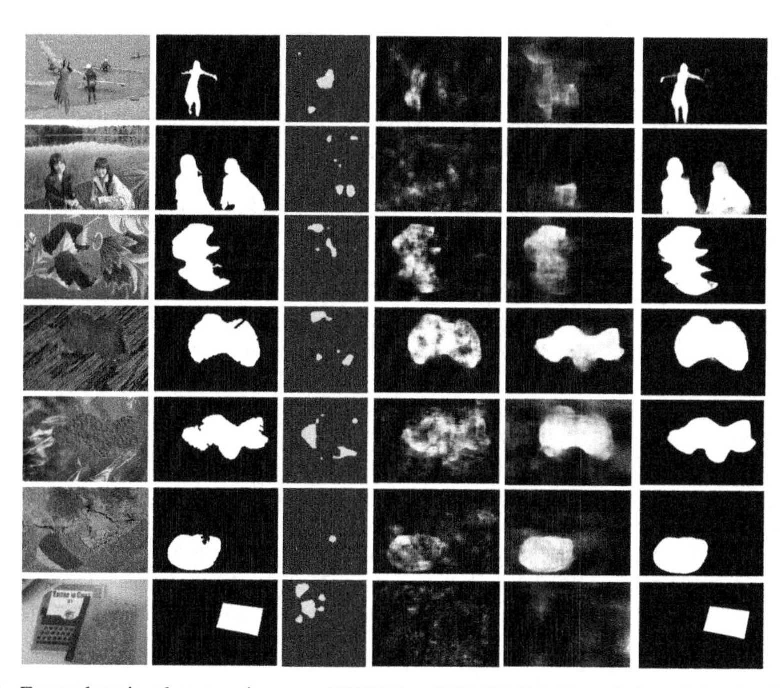

Fig. 6. Exemplar visual comparisons on NIST16 and CASIA2.0. From left to right: the first two columns are tampered images and corresponding ground truth, the third to sixth columns are results based on LSTM-En, ManTra-Net, SPAN and SEINet.

5 Conclusion

We propose a manipulation detection network SEINet, which can detect multiple image manipulation operations. The model establishes relationships between the semantics and edges of tampered regions, where the cross interaction pattern explores multi-level features and bidirectional fusion module optimizes the semantic and edge features with the decoder of U-Net. Our extensive experiments show that proposed SEINet is robust to post-processing operations and outperforms existing state-of-the-art methods. However, the SEINet is not sufficiently adaptable to new samples, more framework generalization will be considered in our future work.

Acknowledgement. This work was supported in part by the National Natural Science Foundation of China (Grant 61806071 and Grant 6210071784), the Natural Science Foundation of Hebei Province, China (Grant F2021202030, Grant F2019202381 and Grant F2019202464), Guangdong Basic and Applied Basic Research Foundation (Grant 2019B151502001), Shenzhen R&D Program (Grant JCYJ20200109105008228), the Sci-tech Research Projects of Higher Education of Hebei Province, China (Grant QN2020185), National Natural Science Foundation of China (Grant 62102129) and the Alibaba Group through Alibaba Innovative Research (AIR) Program.

References

1. Ferrara, P., Bianchi, T., De Rosa, A., Piva, A.: Image forgery localization via fine-grained analysis of CFA artifacts. IEEE Trans. Inf. Forensics Secur. **7**(5), 1566–1577 (2012)
2. Mayer, O., Stamm, M.: Improved forgery detection with lateral chromatic aberration. In: 2016 IEEE International Conference on Acoustics, Speech and Signal Processing, pp. 2024–2028 (2016)
3. Korus, P., Huang, J.: Multi-scale analysis strategies in PRNU-based tampering localization. IEEE Trans. Inf. Forensics Secur. **12**(4), 809–824 (2016)
4. Goljan, M., Fridrich, J., Kirchner, M.: Image manipulation detection using sensor linear pattern. Electron. Imaging. **2018**(7), 119–121 (2018)
5. Cozzolino, D., Verdoliva, L.: Camera-based image forgery localization using convolutional neural networks. In: 2018 26th European Signal Processing Conference, pp. 1372–1376 (2018)
6. Cozzolino, D., Verdoliva, L.: Noiseprint: a CNN-based camera model fingerprint. IEEE Trans. Inf. Forensics Secur. **15**, 144–159 (2019)
7. Bappy, J.H., Simons, C., Nataraj, L., Manjunath, B.S., Roy-Chowdhury, A.K.: Hybrid LSTM and encoder–decoder architecture for detection of image forgeries. IEEE Trans. Image Process. **28**(7), 3286–3300 (2019)
8. Shi, Z., Shen, X., Chen, H., Lyu, Y.: Global semantic consistency network for image manipulation detection. IEEE Signal Process. Lett. **27**, 1755–1759 (2020)
9. Bayar, B, Stamm, M.C.: A deep learning approach to universal image manipulation detection using a new convolutional layer. In: Proceedings of the 4th ACM Workshop on Information Hiding and Multimedia Security, pp. 5–10 (2016)
10. Castillo Camacho, I., Wang, K.: A simple and effective initialization of CNN for forensics of image processing operations. In: Proceedings of the ACM Workshop on Information Hiding and Multimedia Security, pp. 107–112 (2019)
11. Ghosh, A., Zhong, Z., Boult, T.E., Singh, M.: SpliceRadar: a learned method for blind image forensics. In: IEEE Conference on Computer Vision and Pattern Recognition Workshops, pp. 72–79 (2019)
12. Salloum, R., Ren, Y., Kuo, C.C.J.: Image splicing localization using a multi-task fully convolutional network (MFCN). J. Vis. Commun. Image Represent. **51**, 201–209 (2018)
13. Wu, Y., AbdAlmageed, W., Natarajan, P.: ManTra-net: manipulation tracing network for detection and localization of image forgeries with anomalous features. In: Proceedings of the IEEE Conference on Computer Vision and Pattern Recognition, pp. 9543–9552 (2019)
14. Zhou, P., Han, X., Morariu, V.I., Davis, L.S.: Learning rich features for image manipulation detection. In: Proceedings of the IEEE Conference on Computer Vision and Pattern Recognition, pp. 1053–1061 (2018)
15. Shi, Z., Shen, X., Kang, H., Lv, Y.: Image manipulation detection and localization based on the dual-domain convolutional neural networks. IEEE Access. **6**, 76437–76453 (2018)

16. Remya Revi, K., Wilscy, M.: Image forgery detection using deep textural features from local binary pattern map. J. Intell. Fuzzy Syst. **38**(5), 6391–6401 (2020)
17. Pham, N.T., Lee, J.W., Park, C.S.: Structural correlation based method for image forgery classification and localization. Appl. Sci. **10**(13), 4458 (2020)
18. Hu, X., Zhang, Z., Jiang, Z., Chaudhuri, S., Yang, Z., Nevatia, R.: SPAN: spatial pyramid attention network for image manipulation localization. In: Vedaldi, A., Bischof, H., Brox, T., Frahm, J.-M. (eds.) ECCV 2020. LNCS, vol. 12366, pp. 312–328. Springer, Cham (2020). https://doi.org/10.1007/978-3-030-58589-1_19
19. Ronneberger, O., Fischer, P., Brox, T.: U-net: Convolutional networks for biomedical image segmentation. In: Navab, N., Hornegger, J., Wells, W.M., Frangi, A.F. (eds.) MICCAI 2015. LNCS, vol. 9351, pp. 234–241. Springer, Cham (2015). https://doi.org/10.1007/978-3-319-24574-4_28
20. Liu, Y., Zhu, X., Zhao, X., Cao, Y.: Adversarial learning for constrained image splicing detection and localization based on atrous convolution. IEEE Trans. Inf. Forensics Secur. **14**(10), 2551–2566 (2019)
21. Zhu, Y., Chen, C., Yan, G., Guo, Y., Dong, Y.: AR-Net: adaptive attention and residual refinement network for copy-move forgery detection. IEEE Trans. Ind. Inf. **16**(10), 6714–6723 (2020)
22. Pang, Y., Zhao, X., Zhang, L., Lu, H.: Multi-scale interactive network for salient object detection. In: Proceedings of the IEEE/CVF Conference on Computer Vision and Pattern Recognition, pp. 9413–9422 (2020)
23. Zhou, P., Han, X., Morariu, V.I., Davis, L.S.: Two-stream neural networks for tampered face detection. In: 2017 IEEE Conference on Computer Vision and Pattern Recognition Workshops, pp. 1831–1839 (2017)
24. Yue, W., Abd-Almageed, W., Natarajan, P.: Busternet: detecting copy-move image forgery with source/target localization. In: Ferrari, V., Hebert, M., Sminchisescu, C., Weiss, Y. (eds.) ECCV 2018. LNCS, vol. 11210, pp. 170–186. Springer, Cham (2018). https://doi.org/10.1007/978-3-030-01231-1_11
25. Wu, Z., Su, L., Huang, Q.: Stacked cross refinement network for edge-aware salient object detection. In: Proceedings of the IEEE/CVF International Conference on Computer Vision, pp. 7264–7273 (2019)
26. Hu, Z., Zhen, M., Bai, X., Fu, H., Tai, C.L.: Jsenet: Joint semantic segmentation and edge detection network for 3d point clouds. arXiv preprint arXiv:2007.06888 (2020)
27. Fridrich, J., Kodovsky, J.: Rich models for steganalysis of digital images. IEEE Trans. Inf. Forensics Secur. **7**(3), 868–882 (2012)
28. Li, X., Wang, W., Hu, X., Yang, J.: Selective kernel networks. In: Proceedings of the IEEE/CVF Conference on Computer Vision and Pattern Recognition, pp. 510–519 (2019)
29. Nist nimble 2016 datasets Homepage. https://www.nist.gov/itl/iad/mig/. Accessed 2016
30. Dong, J., Wang, W., Tan, T.: Casia image tampering detection evaluation database. In: 2013 IEEE China Summit and International Conference on Signal and Information Processing, pp. 422–426 (2013)
31. He, K., Zhang, X., Ren, S., et al.: Delving deep into rectifiers: surpassing human-level performance on imagenet classification. In: Proceedings of the IEEE International Conference on Computer Vision, pp. 1026–1034 (2015)
32. Mahdian, B., Saic, S.: Using noise inconsistencies for blind image forensics. Image Vis. Comput. **27**(10), 1497–1503 (2009)
33. Krawetz, N., Solutions, H.F.: A picture's worth. hacker factor. Solutions **6**(2), 2 (2007)

Video-Based Reconstruction of Smooth 3D Human Body Motion

Han Zhang, Jianming Wang, and Hui Liu(✉)

School of Computer Science and Technology, TianGong University, Tianjin 300387, China

Abstract. We describe a method to directly recover a 3D human mesh from videos. The existing methods often show that the movement is not smooth and the motion jitter in a certain frame. The contribution of our method is to judge whether the difference between two frames exceeds the threshold range by adding constraint loss, and then optimize it. It effectively limits the changes of pose and shape parameters in the video sequence, and solves the jitter and mutation issues of the human model. Using an adversarial learning framework, the generator outputs the predicted human body parameters, and the discriminator to distinguish real human actions from those produced by our generator. Such adversarial training can produce kinematically plausible motion results. We use GRU network to effectively learn the temporal information which are hidden in the sequence. This is conducive to the continuity and smoothness of the human movement. We conduct some experiments to analyze the importance of constraint loss and demonstrate the effectiveness of our method on the challenging 3D pose estimation datasets.

Keywords: Reconstruction of 3D human body · Constraint loss · Adversarial learn

1 Introduction

The reconstruction of 3D human body is helpful for us to observe and understand the movement of human body better. Most of the current works use an end-to-end framework, to predict human parameters directly from the network and then output the human body mesh. The estimation methods of 3D human body model from a single image have achieved great performances [8,10]. Although the work of human body regression from a single image also has many application scenarios, the human body is usually dynamic. The video-based body model is a great significance to the study of human motion, and some great progresses have been made in this research work.

Most of the previous methods are two-stage [16], in which 2D joints are first estimated and then 3D model parameters are estimated from these joints. Such

This work was supported by The Tianjin Science and Technology Program (19PTZWH Z00020).

H. Ma et al. (Eds.): PRCV 2021, LNCS 13020, pp. 42–53, 2021.
https://doi.org/10.1007/978-3-030-88007-1_4

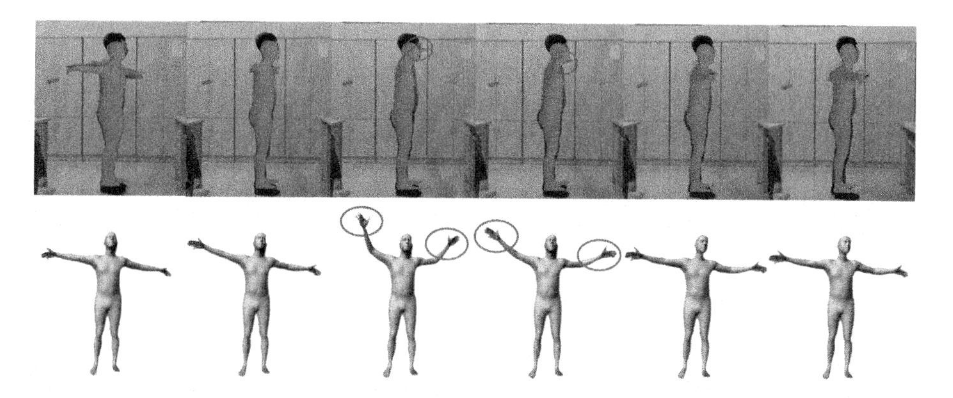

Fig. 1. In a video sequence, if there are some changes in a certain frame, the motion of the model result will be discontinuous, and jitter will occur. The top row shows the predicted body mesh in camera view, and the bottom row shows the predicted mesh from an alternate view point.

methods have the inherent ambiguity in 2D-to-3D estimation. The 2D joints are sparse, whereas the human body is defined by a surface in 3D space. The problem of depth ambiguity often exists in the process of restoring 3D joint. For example, multiple 3D models can explain the same 2D projections [25].

To solve this problem, many approaches use the generative adversarial networks GANs [12,19], which shows significant performance advantages over the two-stage. So we also experiment on this network. Input the video sequence with 2D ground-truth into the Generator to predict the SMPL human body parameters: pose and shape, and then input these predicted parameters into the SMPL model to generate the human body mesh. At the same time, the dataset with 3D ground-truth was used to train a Discriminator, which was mainly used to judge whether the generated human body model was normal rather than a monster (the joint rotation angle of the pose is abnormal, and the shape is too fat or thin). The model parameters predicted by the Generator are identified by the Discriminator. Therefore, the Discriminator needs a large 3D dataset for training, and we chose the AMASS [19].

Current studies still lack sufficient training datasets, especially in-the-wild datasets with some complex movements, and 3D ground-truth labels are more difficult to obtain. This makes previous studies combine 3D datasets with 2D ground-truth for training. Many methods have used 2D in-the-wild datasets like MPII [3] to learn richer poses. For 3D pose estimation, it may be sufficient. However, for 3D human mesh reconstruction, 2D pose is not enough to recover the complex 3D shape and pose details of human body. Moreover, the indoor environment conditions are limited, which leads to the limitation of the number of people and the complexity of human motion in the dataset. So that the model is not robust, the predicted results are not accurate, the complexity and variability of human motion cannot be restored. When a 3D human body is recovered from a video, the problem of model jitter as shown in Fig. 1 is often encountered. This makes the model unsmooth in pose and shape, and it often has some problems such as jitter, mutation and poor continuity of movement. Therefore, we add constraints to deal with noise and ensure the smoothness of the model.

Our main contribution was to add constraints to Pose and Shape. The idea is that given a video sequence, the image features are extracted by CNN, and then the parameters of SMPL are predicted by temporal encoder and body parameter regressor [12]. By calculating the constraint loss, the difference between two adjacent frames is constrained, in turn to ensure the continuity of action and reduce the mutation of pose and shape. Then, a motion discriminator takes predicted poses along with the poses sampled from the AMASS dataset and outputs a real/fake label for each sequence. The whole model is supervised by adversarial loss and regression loss to minimize the error between prediction and ground-truth, pose and shape.

2 Related Work

2.1 3D Human Mesh from Single Images

Most people define human pose estimation as the problem of locating the 3D joints of the human body from images. The human movement is often more complex and changeable, but this notion of "pose" is overly simplistic. Methods of human model reconstruction are mainly divided into two categories: two-stage and direct estimation. The two-stage method firstly uses a 2D pose detector to predict the 2D joint position [23,26], and then predicts the 3D joint position from the 2D joint by regressor [21,22] or model fitting. The two-stage method has good robustness to the motion localization of the model, but ignores other details of the human body. The direct estimation method estimates SMPL parameters from the image in an end-to-end way. In order to limit the ambiguity inherent in the 2D-to-3D estimation process, many methods use a variety of priors [25], most of which make some assumptions about limb length or proportion [7]. For example, the work of HMR [12] uses GANs, the generator predicts SMPL parameters, and the discriminator identify their authenticity. In this method, training only uses 2D label, which is superior to all two-stage methods.

2.2 3D Human Mesh from Video

By using deep networks, various networks have been developed to take the advantage of temporal information. For example, the self-attention temporal network proposed by Yu Sun et al. [24] guides learning of motion dynamics in video through an unsupervised adversarial training strategy. Angjoo et al. [13] proposed the temporal encoder of image features to learn the representation of 3D human dynamics on the time window centering on frame T, as well as the change of posture in adjacent frames. They also showed that web videos with 2D key-point lables by detectors could reduce the need for 3D ground-truth.

2.3 GANs for Modeling

GANS [4,9] has an important impact on image-based model and compound. It mainly consists of a generator and a discriminator. The Generator can use random numbers to generate meaningful data. The Discriminator can learn how to determine which is a real data and which is a generated data, and then transfer the learning experience back to the Generator, allowing it to generate data that looks more like real data from random numbers. They are trained at the same time and compete with each other, constantly optimizing the accuracy of the model. The study [1] shows that the combination of video sequence and adversarial training can predict the human motion sequence of future frames based on the current frame. Similarly, we use a cyclic architecture to predict the posture of the input data, and a motion discriminator is trained with 3D motion capture data [19] to competitively encode the pose and shape parameters.

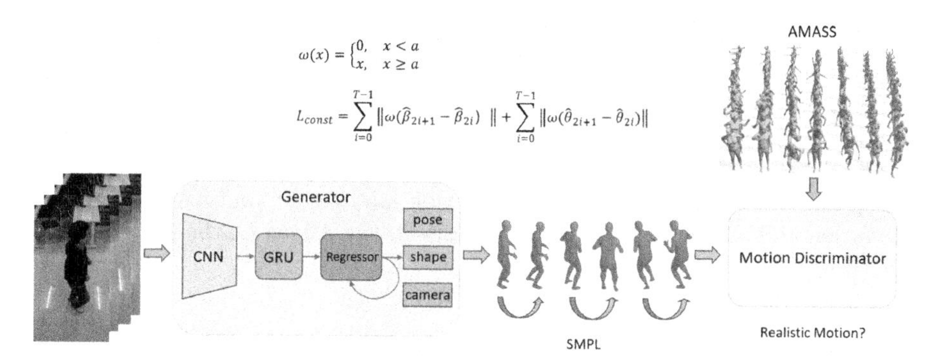

Fig. 2. We estimate the parameters of the SMPL body model for each frame in the video sequence using a temporal generation network. Constraint loss is added to reduce the possibility of a frame motion mutation. By learning the hidden temporal information in the video sequence, the motion discriminator can determine whether the mesh is consistent with the real human movement.

3 Approach

Our goals are 1) to learn 3D representation of the human body from the video and 2) to get a smooth 3D human model prediction. Using the GANs network, the generator first uses a pretrained CNN to extract the features of each frame from the video, and then trains a GRU (Gated Recurrent Units) temporal encoder. It outputs potential variables containing past and future frame information, iterates on the average SMPL parameters, and predicts the SMPL parameters. Finally, it inputs the SMPL parameters from feature regressor and a large number of datasets into the discriminator to judge, whether the generated model results are true or not. Figure 2 illustrates this framework.

3.1 3D Body Representation

We use the Skinned Multi-Person Linear (SMPL) model to output the body mesh. SMPL [18] is a generative model that represents different human bodies through pose, shape and proportion. Since the joints will rotate at different angles when the human body makes movements, and the muscles will also deform accordingly, the SMPL model establishes an effective mapping $\mathrm{M}(\beta,\theta):\mathbb{R}^{|\beta|\times|\theta|}\mapsto\mathbb{R}^{3\times 6890}$ for the relationship between pose and shape. The shape parameter $\beta\in\mathbb{R}^{10}$ is the linear combination weights of 10 basic shape. The pose parameter $\theta\in\mathbb{R}^{3\times 23}$ represents relative 3D rotation of 23 joints in axis-angle representation. SMPL is a differentiable function that outputs a triangulated mesh with $N=6890$ vertices, which is obtained by shaping the template body vertices conditioned on β and θ, then articulating the bones according to the joint rotations θ via forward kinematics, and finally deforming the surface with linear blend skinning.

The 3D keypoints used for reprojection error, $X(\theta,\beta)\in\mathbb{R}^3$, are obtained by linear regression from the final mesh vertices. The mapping relationship between 3D space and 2D image plane is established by using a weak-perspective camera model, which is convenient for supervising 3D mesh and 2D pose label. Finally, a set of parameters representing the 3D reconstruction of the human body is expressed as an 85 dimensional vector $\Theta=\{\theta,\beta,R,t,s\}$, where $R\in\mathbb{R}^{3\times 3}$ is an axis-angle representation of global rotation, and $t\in\mathbb{R}^2$ and $s\in\mathbb{R}$ represent translation and scale in the image plane respectively. Given Θ, the projection of $X(\theta,\beta)$ is:

$$\hat{\mathbf{x}}=s\Pi(RX(\theta,\beta))+t \tag{1}$$

where Π is an orthographic projection.

3.2 Temporal Encoder

Because the motion between frames of the video is continuous, the loop architecture is used to establish the information connection between the current frame and the future frame of human body posture. When the occlusion and truncation of the human body are encountered, this association can be used to solve and constrain the attitude estimation by using the past information.

The temporal encoder, as a generator, consists of CNN and GRU. First, by inputting a video $V=\{I_t\}_{t=1}^{T}$ of length T, people in each frame is detected. Then the features of each frame are extracted through a convolutional neural network. The feature extractor f outputs one vector $f_i\in\mathbb{R}^{2048}$ for each frame $f(I_1),\ldots,f(I_T)$. These vectors are sent to a Gated Recurrent Unit(GRU) layer and the latent feature vectors g_i for each frame $g(f_1),\ldots,g(f_T)$, which are generated based on the previous frames. Finally, g_i is inputted into the regressor T with iterative feedback. The mean pose $\bar{\Theta}$ is taken as the initial. In each iteration of k, the feature g_i is taken as the input together with the current Θ_k, and Θ is adjusted through calculating the loss of parameter projection and ground-truth. During the regressor training, we chose to use a loop fitting method similar to [15], and used shape consistency to propose smooth terms in the objective function as in [5].

3.3 Constraint Loss

The continuity of motion has always been an important research object in the video-based reconstruction of 3D Human Body. Our goal is to separate the useful signal from the noise through certain calculation methods and programs, so as to improve the anti-interference and analysis accuracy of the signal. In the direction of the filter, the main purpose of our constraint loss is to use linear equations to filter out the noise in the predicted SMPL parameters, the part of the parameters that cause the jitter and mutation of the model results, and focus on obtaining refined and accurate predictions.

In order to prevent the generated jitter and mutation of posture, we added a constraint loss. We divided the two adjacent frames into a group and calculated the difference between their predicted pose and shape parameters respectively. We give this difference a threshold range a, and when the difference between two frames is less than this range, the constraint loss is not calculated, otherwise it is calculated and optimized. Jitter and mutation changes are often caused by large differences in parameters between two adjacent frames. The main purpose of constraint loss is to constrain the range of change. We control for the scope of this variation to make the movements of our predicted human body smoother.

In general, the proposed temporal encoder loss is composed of 2D(x), 3D(X), Pose(θ), Shape(β) losses and constraint loss. This is combined with an adversarial discriminator loss. Specifically, the total loss of the generator is:

$$L_{\mathcal{G}} = L_{3D} + L_{2D} + L_{SMPL} + L_{const} + L_{adv} \tag{2}$$

where each term is calculated as:

$$L_{2D} = \sum_{t=1}^{T} \|x_t - \hat{x}_t\|_2 \tag{3}$$

$$L_{3D} = \sum_{t=1}^{T} \left\|X_t - \hat{X}_t\right\|_2 \tag{4}$$

$$L_{SMPL} = \|\beta - \hat{\beta}\|_2 + \sum_{t=1}^{T} \left\|\theta_t - \hat{\theta}_i\right\|_2 \tag{5}$$

$$\omega(x) = \begin{cases} 0, x < a \\ x, x \geq a \end{cases}, \quad L_{const} = \sum_{i=0}^{T-1} \left\|\omega\left(\widehat{\beta}_{2i+1} - \widehat{\beta}_{2i}\right)\right\| + \sum_{i=0}^{T-1} \left\|\omega\left(\hat{\theta}_{2i+1} - \hat{\theta}_{2i}\right)\right\| \tag{6}$$

where L_{adv} is the discriminator adversarial loss. L_{2D} and L_{3D} are the joint reprojection error for predicted values and annotation respectively. Additional direct 3D supervision can be used when paired ground-truth 3D data is available. The most common form of 3D annotation is the 3D joints. Supervision in terms of SMPL parameters $[\beta, \theta]$ may be obtained through MoSh [17] when raw 3D MoCap marker data is available.

3.4 Motion Discriminator

Some methods use a body discriminator to predict the body mesh of a single image, however the discriminator and reprojection loss force the generator to produce a real body pose. This has limitations in alignment with the 2D joint position: not sufficient in sequence to explain the continuity of motion. Multiple inaccurate postures may be recognized as valid when temporal continuity of movement is ignored. We used a motion discriminator to determine whether the generated sequence of poses was consistent with the real one.

The generator output $\hat{\Theta}$ is used as the input of the discriminator. After a multi-layer GRU model, the information that needs to be remembered in the long sequence is retained, and the hidden information h_i are estimated at each time step i, which helps to learn the coherent information of the action. Then use self-attention to aggregate these hidden states $[h_i, \cdots, h_T]$. Finally, a linear layer is used to predict a value $\in [0, 1]$, indicating the probability that Θ belongs to a reasonable human motion. The discriminator is trained by the ground-truth position, so it can learn reasonable skeletal proportion and rotation limits for each joint. This can prevent the network from producing totally unconstrained human bodies. The adversarial loss term that is back propagated to generator is:

$$L_{adv} = \mathbb{E}_{\Theta \sim p_G}\left[\left(\mathcal{D}(\hat{\boldsymbol{\Theta}}) - 1\right)^2\right] \tag{7}$$

The loss of the discriminator is:

$$L_{\mathcal{D}} = \mathbb{E}_{\Theta \sim p_R}\left[(\mathcal{D}(\boldsymbol{\Theta}) - 1)^2\right] + \mathbb{E}_{\Theta \sim p_G}\left[\mathcal{D}(\hat{\boldsymbol{\Theta}})^2\right] \tag{8}$$

Where p_R is the real motion sequence from the AMASS dataset, and p_G is the generated motion sequence by the generator.

Recurring networks update their hidden states as they process inputs continuously, and the final hidden states contain brief information in the sequence. The self-attention [6] is mainly used to amplify the contribution of the most important frames in the sequence. The formulas are:

$$\phi_i = \phi\left(h_i\right), \quad a_i = \frac{e^{\phi_i}}{\sum_{t=1}^{N} e^{\phi_t}}, \quad r = \sum_{i=1}^{N} a_i h_i \tag{9}$$

The weights a_i are learned by the multilayer perceptron (MLP) ϕ, and then they are normalized using softmax to form a probability distribution.

4 Experiments

We begin with describing the details of the training process and the datasets which we used. Next, we compare our results with other methods. Finally, we also conducted some ablation studies on constraint loss.

4.1 Implement Details

Our generator mainly used video datasets with 2D ground-truth, such as Pennaction [27] and Posetrack [2]. We also used InstaVariety [13], which are the pseudo ground-truth datasets annotated using a 2D keypoint detector. For the discriminator, we used AMASS [19] for adversarial training to obtain real samples. We used 3DPW [20] dataset to conduct ablation experiments to ensure the performance of the model in an in-the-wild setting.

We use a pre-trained ResNet-50 network [11] as an image encoder to output a single frame image feature $f_i \in \mathbb{R}^{2048}$. $T = 16$ is used as the sequence length, and the mini-batch size is 32. For the temporal encoder, we use a two-layer GRU with a hidden size of 1024. The regressor has 2 fully-connected layers with 1024 neurons per layer, and the last layer outputs SMPL parameters $\hat{\Theta} \in \mathbb{R}^{85}$ including pose, shape, and camera parameters. The output of the generator is taken as the fake sample, and the data sequence with ground-truth is taken as the real sample, which are inputted into the discriminator together. For self-attention, we used two MLP layers, 1024 neurons in each layer and *tanh* activated learning attention weights. The final linear layer predicts a false/true probability for each sample. We also used the Adam optimizer, and the learning rates for generator and discriminator are 5×10^{-5} and 1×10^{-4}, respectively. The value range of constraint loss threshold a is derived from the difference between the joint coordinates of a frame with action mutation and its adjacent frames. The approximate range is obtained by averaging these differences. We believe that the more accurate the value of threshold a is, the better the effect of action constraint will be. It can be set according to the situation. In the experiment, we set $a = 0.001$.

Table 1. We compared our method with others on 3DPW and MPI-INF-3DHP datasets, and our results are improved in both evaluation metrics and experimental results. "–" shows the results that is not available.

Model	3DPW				MPI-INF-3DHP	
	PA-MPJPE ↓	MPJPE ↓	PVE ↓	Accel ↓	PA-MPJPE ↓	MPJPE ↓
Kolotouros et al. [15]	59.2	96.9	116.4	29.8	67.5	105.2
Kanazawa et al. [13]	72.6	116.5	139.3	**15.2**	–	–
HMR [12]	73.6	120.1	142.7	34.3	–	89.8
VIBE [14]	51.9	82.9	**99.1**	23.4	64.6	**96.6**
Ours	**51.5**	**82.4**	99.0	23.1	**63.6**	98.1

4.2 Comparison to Other Methods

Our goal is to recover the 3D mesh of the human body from the video. In Table 1, we compare our model to the previous methods. We validate on the challenging in-the-wild datasets, which are 3DPW and MPIINF-3DHP. Our method has an effective improvement, because our model encourages temporal pose and shape consistency, while adding constraints between frames. These results confirm our hypothesis that constraining the pose and shape between two frames can improve the estimation from the video.

Procrustes aligned mean per joint position error (PA-MPJPE), mean per joint position error (MPJPE), Acceleration Error (Accel) and Per Vertex Error (PVE) are reported and compared with both frame-based and temporal methods.

The method most similar to ours is VIBE [14], which also uses GRU to learn motion dynamics for time optimization. The results show that our method performs well in generalization, and the constraint loss not only limits the abrupt change of the action to a certain extent, but also preserves the subtle change of the predicted model action well for the computational constraint of two frames in a group.

4.3 Ablation Experiments

In adding constraint losses, we considered adding a constraint module after the generator predicted the SMPL parameters. This constraint module mainly directly adds constraint operations to the results, constrains the continuous sequence, and tries its best to ensure that the variation range of predicted parameters is not too large. If it's too large, we make the parameters of the next frame equal to the previous frame one. However, the experiment results were not well. Although the model showed some improvement in the performance of the validation set, it did not perform well in the metrics, as shown in Table 2. After analyzing the experimental results and the reasons, we believe that this method

Table 2. We did an ablation experiment on how to add constraints. They are adding constraint module, constraint loss of a sequence and constraint loss of two frames as a group. The table shows their performance on the 3DPW and MPI-INF-3DHP datasets.

Model	3DPW				MPI-INF-3DHP	
	PA-MPJPE ↓	MPJPE ↓	PVE ↓	Accel ↓	PA-MPJPE ↓	MPJPE ↓
VIBE [14]	51.9	82.9	**99.1**	23.4	64.6	**96.6**
Constraint Module	56.1	86.5	102.8	23.3	67.1	100.1
Constraint Loss(sequence)	52.2	83.1	101.1	**22.3**	64.2	99.7
Constraint Loss(group)	**51.5**	**82.4**	99.0	23.1	**63.6**	98.1

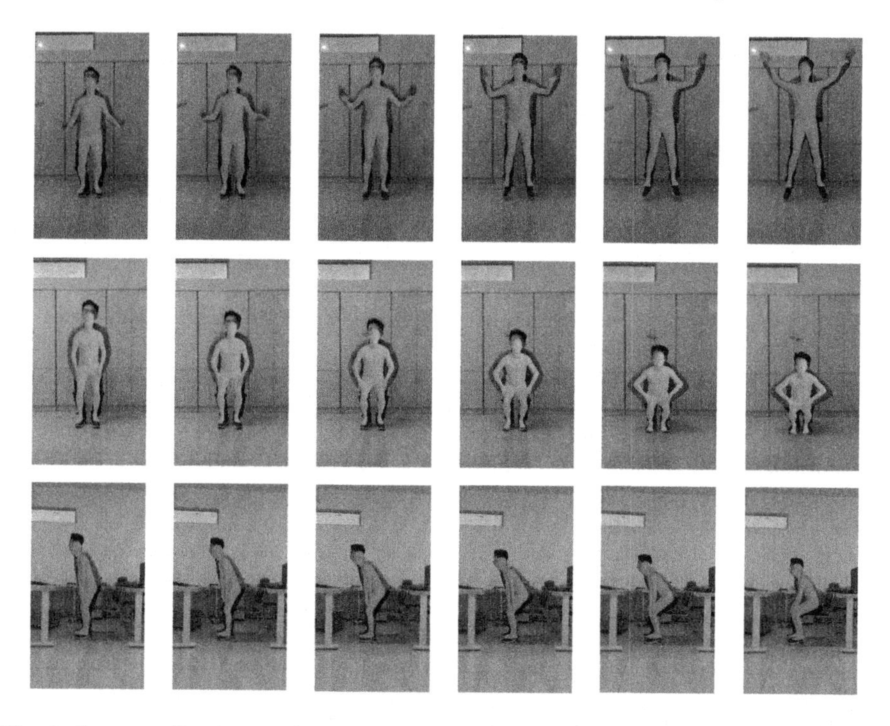

Fig. 3. Some qualitative results of our work. In a continuous video frame, the continuity and smoothness of action are improved.

is inconsistent with the variation of human movement. Forced parameter constraints will lead to unobvious changes in the predicted results and inaccurate movements of the generated human mesh.

At the same time, when we think about the constraint loss, we also consider whether the constraint loss is calculated for a sequence or for a group. So we also ran a set of experiments. For the constraint loss of a sequence frame, there is the problem of inaccuracy action as mentioned above. Therefore, we finally decided to take two frames as a group to calculate the constraint loss. It not only preserves the variation of the movement, but also restricts the variation range to improve the smoothness. The experiment results are shown in Table 2 (Fig.3).

5 Conclusion

Estimating human movement from videos is crucial to understanding human behavior. We have extended some static methods and added improvements: (1) We use GANs to predict SMPL parameters and improve the performance of the model through adversarial learning. (2) Constraint loss is added to constrain the range of the predicted adjacent frame parameters to reduce the jitter of the reconstructed human motion. (3) Use GRU network to learn the temporal hidden information in the sequence. Our model performs well on the in-the-wild datasets and has an effective improvement. Future work will explore the issue of occlusion using motion information tracking, as well as ambiguity in multiplayer model reconstruction. In addition, the detail restoration of hands and feet is inaccurate, which will also be the direction of our further study.

References

1. Aksan, E., Kaufmann, M., Hilliges, O.: Structured prediction helps 3D human motion modelling. In: Proceedings of the IEEE/CVF International Conference on Computer Vision, pp. 7144–7153 (2019)
2. Andriluka, M., et al.: PoseTrack: a benchmark for human pose estimation and tracking. In: Proceedings of the IEEE Conference on Computer Vision and Pattern Recognition, pp. 5167–5176 (2018)
3. Andriluka, M., Pishchulin, L., Gehler, P., Schiele, B.: 2D human pose estimation: new benchmark and state of the art analysis. In: Proceedings of the IEEE Conference on Computer Vision and Pattern Recognition, pp. 3686–3693 (2014)
4. Arjovsky, M., Chintala, S., Bottou, L.: Wasserstein generative adversarial networks. In: International Conference on Machine Learning, pp. 214–223. PMLR (2017)
5. Arnab, A., Doersch, C., Zisserman, A.: Exploiting temporal context for 3D human pose estimation in the wild. In: Proceedings of the IEEE/CVF Conference on Computer Vision and Pattern Recognition, pp. 3395–3404 (2019)
6. Bahdanau, D., Cho, K., Bengio, Y.: Neural machine translation by jointly learning to align and translate. arXiv preprint arXiv:1409.0473 (2014)
7. Barrón, C., Kakadiaris, I.A.: Estimating anthropometry and pose from a single uncalibrated image. Comput. Vis. Image Underst. **81**(3), 269–284 (2001)
8. Bogo, F., Kanazawa, A., Lassner, C., Gehler, P., Romero, J., Black, M.J.: Keep It SMPL: automatic estimation of 3D human pose and shape from a single image. In: Leibe, B., Matas, J., Sebe, N., Welling, M. (eds.) ECCV 2016. LNCS, vol. 9909, pp. 561–578. Springer, Cham (2016). https://doi.org/10.1007/978-3-319-46454-1_34
9. Goodfellow, I.J., et al.: Generative adversarial networks. arXiv preprint arXiv:1406.2661 (2014)
10. Guler, R.A., Kokkinos, I.: HoloPose: holistic 3D human reconstruction in-the-wild. In: Proceedings of the IEEE/CVF Conference on Computer Vision and Pattern Recognition, pp. 10884–10894 (2019)
11. He, K., Zhang, X., Ren, S., Sun, J.: Identity mappings in deep residual networks. In: Leibe, B., Matas, J., Sebe, N., Welling, M. (eds.) ECCV 2016. LNCS, vol. 9908, pp. 630–645. Springer, Cham (2016). https://doi.org/10.1007/978-3-319-46493-0_38
12. Kanazawa, A., Black, M.J., Jacobs, D.W., Malik, J.: End-to-end recovery of human shape and pose. In: Proceedings of the IEEE Conference on Computer Vision and Pattern Recognition, pp. 7122–7131 (2018)
13. Kanazawa, A., Zhang, J.Y., Felsen, P., Malik, J.: Learning 3D human dynamics from video. In: Proceedings of the IEEE/CVF Conference on Computer Vision and Pattern Recognition, pp. 5614–5623 (2019)
14. Kocabas, M., Athanasiou, N., Black, M.J.: Vibe: video inference for human body pose and shape estimation. In: Proceedings of the IEEE/CVF Conference on Computer Vision and Pattern Recognition, pp. 5253–5263 (2020)
15. Kolotouros, N., Pavlakos, G., Black, M.J., Daniilidis, K.: Learning to reconstruct 3D human pose and shape via model-fitting in the loop. In: Proceedings of the IEEE/CVF International Conference on Computer Vision, dpp. 2252–2261 (2019)

16. Lassner, C., Romero, J., Kiefel, M., Bogo, F., Black, M.J., Gehler, P.V.: Unite the people: closing the loop between 3D and 2D human representations. In: Proceedings of the IEEE Conference on Computer Vision and Pattern Recognition, pp. 6050–6059 (2017)
17. Loper, M., Mahmood, N., Black, M.J.: Mosh: motion and shape capture from sparse markers. ACM Trans. Graph. (TOG) **33**(6), 1–13 (2014)
18. Loper, M., Mahmood, N., Romero, J., Pons-Moll, G., Black, M.J.: SMPL: a skinned multi-person linear model. ACM Trans. Graph. (TOG) **34**(6), 1–16 (2015)
19. Mahmood, N., Ghorbani, N., Troje, N.F., Pons-Moll, G., Black, M.J.: AMASS: archive of motion capture as surface shapes. In: Proceedings of the IEEE/CVF International Conference on Computer Vision, pp. 5442–5451 (2019)
20. von Marcard, T., Henschel, R., Black, M.J., Rosenhahn, B., Pons-Moll, G.: Recovering accurate 3D human pose in the wild using IMUs and a moving camera. In: Ferrari, V., Hebert, M., Sminchisescu, C., Weiss, Y. (eds.) ECCV 2018. LNCS, vol. 11214, pp. 614–631. Springer, Cham (2018). https://doi.org/10.1007/978-3-030-01249-6_37
21. Martinez, J., Hossain, R., Romero, J., Little, J.J.: A simple yet effective baseline for 3D human pose estimation. In: Proceedings of the IEEE International Conference on Computer Vision, pp. 2640–2649 (2017)
22. Moreno-Noguer, F.: 3D human pose estimation from a single image via distance matrix regression. In: Proceedings of the IEEE Conference on Computer Vision and Pattern Recognition, pp. 2823–2832 (2017)
23. Newell, A., Yang, K., Deng, J.: Stacked hourglass networks for human pose estimation. In: Leibe, B., Matas, J., Sebe, N., Welling, M. (eds.) ECCV 2016. LNCS, vol. 9912, pp. 483–499. Springer, Cham (2016). https://doi.org/10.1007/978-3-319-46484-8_29
24. Sun, Y., Ye, Y., Liu, W., Gao, W., Fu, Y., Mei, T.: Human mesh recovery from monocular images via a skeleton-disentangled representation. In: Proceedings of the IEEE/CVF International Conference on Computer Vision, pp. 5349–5358 (2019)
25. Taylor, C.J.: Reconstruction of articulated objects from point correspondences in a single uncalibrated image. Comput. Vis. Image Underst. **80**(3), 349–363 (2000)
26. Wei, S.E., Ramakrishna, V., Kanade, T., Sheikh, Y.: Convolutional pose machines. In: Proceedings of the IEEE Conference on Computer Vision and Pattern Recognition, pp. 4724–4732 (2016)
27. Zhang, W., Zhu, M., Derpanis, K.G.: From actemes to action: a strongly-supervised representation for detailed action understanding. In: Proceedings of the IEEE International Conference on Computer Vision, pp. 2248–2255 (2013)

A Unified Modular Framework with Deep Graph Convolutional Networks for Multi-label Image Recognition

Qifan Lin[1,2], Zhaoliang Chen[1,2], Shiping Wang[1,2], and Wenzhong Guo[1,2(✉)]

[1] College of Mathematics and Computer Science, Fuzhou University, Fuzhou, Fujian, China
guowenzhong@fzu.edu.cn

[2] Fujian Provincial Key Laboratory of Network Computing and Intelligent Information Processing, Fuzhou University, Fuzhou, Fujian, China

Abstract. With the rapid development of handheld photographic devices, a large number of unlabeled images have been uploaded to the Internet. In order to retrieve these images, image recognition techniques have become particularly important. As there is often more than one object in a picture, multi-label image annotation techniques are of practical interest. To enhance its performance by fully exploiting the interrelationships between labels, we propose a unified modular framework with deep graph convolutional networks (MDGCN). It consists of two modules for extracting image features and label semantic respectively, after which the features are fused to obtain the final recognition results. With classical multi-label soft-margin loss, our model can be trained in an end-to-end schema. It is important to note that a deep graph convolutional network is used in our framework to learn semantic associations. Moreover, a special normalization method is employed to strengthen its own connection and avoid features from disappearing in the deep graph network propagation. The results of experiments on two multi-label image classification benchmark datasets show that our framework has advanced performance compared to the state-of-the-art methods.

Keywords: Multi-label image recognition · Convolutional neural networks · Graph convolutional networks · Feature extraction

1 Introduction

Every time we press the shutter, we get a picture. Nowadays, anyone can easily takes a picture using a mobile phone, which has led to an explosion in the number of images circulating on the Internet. How a computer performs image recognition automatically is a significant task for computer vision researchers. Since there is often more than one subject in the images we take, the task of multi-label image recognition is of great practical importance. And it has been

H. Ma et al. (Eds.): PRCV 2021, LNCS 13020, pp. 54–65, 2021.
https://doi.org/10.1007/978-3-030-88007-1_5

used in many areas, for example, in medical image recognition analysis [14] to help doctors improve diagnostic efficiency and in the image retrieval field [16] to effectively reduce the amount of manually annotated data.

The basis of image recognition is image feature extraction. In single-label image classification, deep convolutional neural networks have shown advanced performance. Once ResNet [10] was proposed in 2016, it has been gradually replacing the VGG [17] in many practical scenarios. In the following year, Kaiming He's team proposed an upgraded version of the deep residual network called RexNeXt [22]. It can be used for image feature extraction. By simply appending a linear layer of the network extracting feature dimensions to the categories with threshold control, it can perform multi-label image classification tasks.

The feature extraction network utilizes the images from the training set, but does not fully exploit the semantic features of the given labels. Therefore, CNN-RNN [19], SRN [24], RNN-Attention [20], ML-GCN [4] and other frameworks that combine deep neural networks with labeled semantic information have been proposed one after another.

How to depict the relationship between labels is an important issue, and it is worth noting that graph networks have this advantage. Graph convolutional network (GCN) was proposed by Kipf and Welling [11] in 2017. It was initially used in semi-supervised image classification and achieved advanced performance. A number of researchers have invested in studying GCN and have proposed many methods based on it, such as Cluster-GCN [5] and AdaGCN [18].

In this paper, we introduce a unified modular framework with deep graph convolutional networks. It contains two main modules, one module adopts convolutional neural networks to extract image features, and the other module employs deep graph convolutional networks to extract label semantic features. The final multi-label image classification results will be obtained by fusing the features of the two aforementioned modules. The main contributions of this paper can be concluded as follows:

- An end-to-end trainable framework is proposed for multi-label image recognition tasks, where both image representation and label semantic features can be extracted and fully used.
- A deep graph convolutional network is employed to extract semantic features, where the adjacency matrix is regularized so that the contribution of nearby nodes is greater than that of the more distant nodes.
- Experiments are carried out on two multi-label image benchmark datasets, MS-COCO and VOC 2007, and the experimental results show the competitive performance of our approach compared to the state-of-the-art methods.

2 Related Work

In the past fifteen years, the performance of image classification has witnessed rapid progress. The reason for this is due to the establishment of large-scale hand-labeled datasets such as ImageNet [6], Microsoft COCO [13] and PASCAL VOC [7], and the fast development of deep convolutional networks such as ResNet

[10]. Many studies have been devoted to extending deep convolutional networks to the field of multi-label image recognition.

A straightforward way to solve the multi-label image classification problem is to train multiple binary classifiers. However, this is not realistic as the number of binary classifiers required for it is exponential, requiring 2^{20} binary classifiers for a dataset of just twenty classes. It can also be performed by training a classifier for each class and controlling the classification results using thresholding or ranking [9]. However, many researchers believe that what limits its performance improvement is not considering the dependencies between labels.

In order to exploit label dependencies, some studies have been carried out. Wang et al. [19] used recurrent neural networks to learn a joint low-dimensional image-label embedding to model the semantic relevance between images and labels. Zhu et al. [24] proposed the Spatial Regularization Net to generate attention maps for all labels and capture the underlying relations between them via learnable convolutions. Wang et al. [20] designed a special recurrent memorized-attention module, composed of a spatial transformer and an LSTM network, to take full advantage of the label distributions.

Different from these methods mentioned above, the researchers found that the graph structure could be more effective in modeling label co-occurrence relationships. Chen et al. [4] proposed a framework for combining multi-label image recognition with graph convolutional networks, using the graph network to map label representations into a series of inter-dependent object classifiers. Chen et al. [3] proposed a graph neural network semantic extraction structure called SSGRL, which extracts specific semantic relationships between labels through multiple graph neural network layers.

Unlike previous work, our framework MDGCN focuses on using deep graph convolutional networks to obtain better semantic features. In addition, a normalization method is used to consider that the information should be enhanced for nodes nearby rather than distant nodes.

3 Proposed Method

Given the image set I, the label set L, multi-label image recognition is supposed to predict the image labels $\hat{y}$. In this paper, we propose our MDGCN framework, as shown in Fig. 1. We design an image feature extraction module to perform representation learning in image set. In label set L, a label semantic extraction module is designed to fully explore the label semantic information and co-occurrence relationship between labels. Then, the final classification result $\hat{y}$ is obtained by feature fusion.

3.1 Image Feature Extraction Module

We consider an RGB image as a tensor with the dimension of $w \times h \times 3$. Original input would result in a model with too large parameters and slow down the training process, attributed to which we designed the image feature extraction

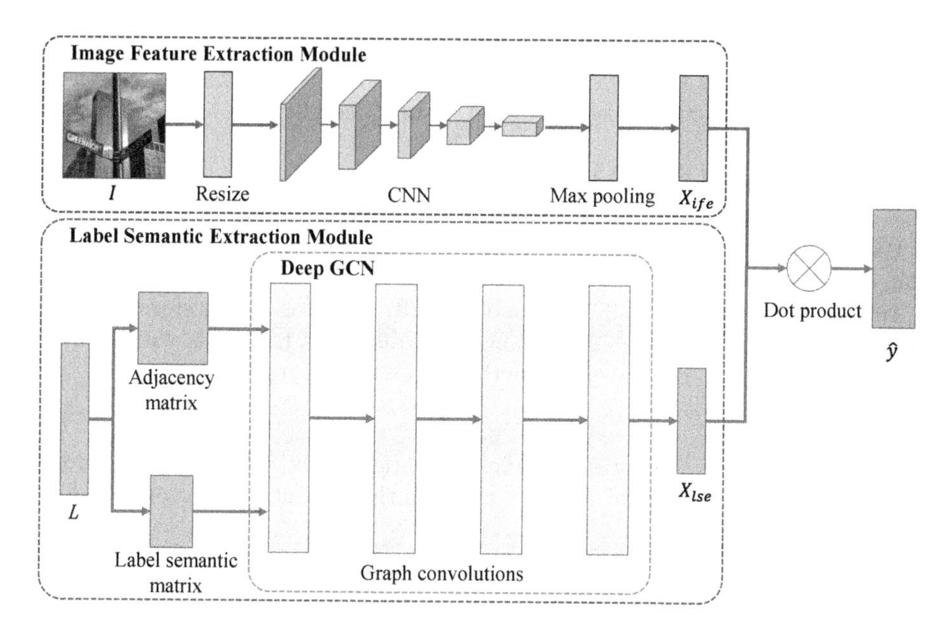

Fig. 1. Overall framework of our MDGCN model for multi-label image recognition.

module (IFEM) to perform feature extraction on the original input. In this module, we use the convolutional neural networks (CNN) which is commonly used in representation learning. It can be summarized as

$$F_{image} = f_{cnn}(f_{resize}(I)), \tag{1}$$

where $I \in \mathbb{R}^{h \times w \times 3}$ represents the set of input images, $f_{resize}(\cdot)$ denotes a size scaling function, and F_{image} is the features obtained after the convolutional layers of CNN. We can adopt any kind of CNN in this part. To adapt its output dimension to the entire framework, we employ a max pooling function. And it enables the model to have better generalization ability. The formula is expressed as follow

$$X_{ife} = f_{mp}(F_{image}), \tag{2}$$

where $f_{mp}(\cdot)$ denotes a max pooling function, and $X_{ife} \in \mathbb{R}^{n \times s}$ indicates the image features obtained by the image feature extraction module. The input image set I is transformed into a feature representation with dimension s, according to Eq. (1) and Eq. (2).

3.2 Label Semantic Extraction Module

Our intention in designing the label semantic extraction module (LSEM) is to make full use of the given label information to achieve better multi-label image classification results. The first question we think about is what information we can get from the labels of the training set and how we can make full use of this information. Previous work has given us some inspirations. We can obtain the meaning of words themselves by learning word embedding and calculate the label co-occurrence probability through statistical methods.

Semantic Vector of LSEM. The training set labels, such as dog, aircraft, cars, are often treated as completely different category markers in classification methods. However, the semantic meaning of the label is overlooked. We consider that aircraft and cars are both vehicles, so the distance between them should be smaller than that between dogs and aircraft. In order to transform semantic labels into vectors with smaller Euclidean distances for labels with similar semantics, we use a word-to-vector method. The definition is as follows

$$X_{lsv} = f_{wsr}(L_t), \tag{3}$$

where $X_{lsv} \in \mathbb{R}^{C \times X_{gci}}$ denotes the label semantic matrix, which is obtained by passing the training set label $L_t \in \mathbb{R}^C$ through the semantic-to-vector transformation function $f_{wsr}(\cdot)$.

Correlation Matrix of LSEM. How the dependencies between labels are described is another key point for LSEM to be effective. As shown in Fig. 2, we employ conditional probabilities to construct label co-occurrence probability. For example, $P(C_a \mid C_b)$ denotes the co-occurrence probability of class a, given the co-occurrence of class b. Therefore, $P(C_a \mid C_b)$ is not equal to $P(C_b \mid C_a)$, which constitutes a directed graph.

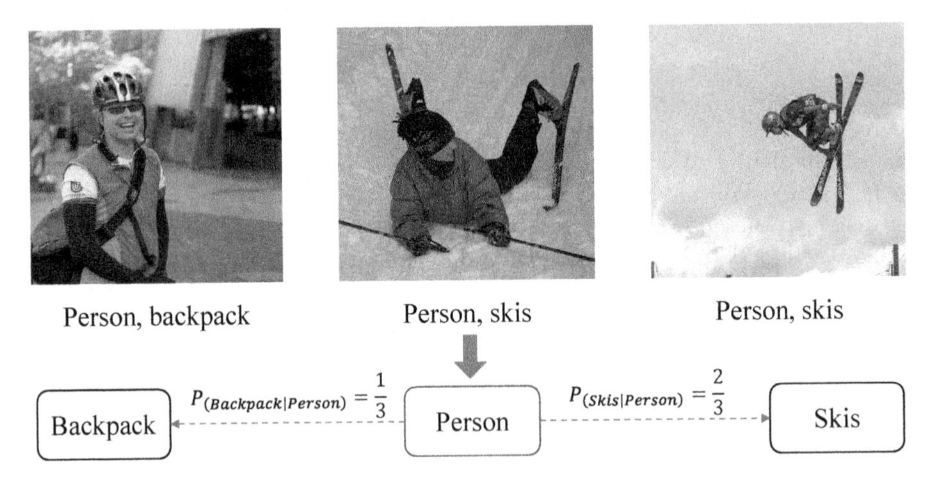

Fig. 2. Schematic diagram of correlation matrix construction. It shows the co-occurrence probability between "person" and the other two labels.

We denote the label co-occurrence matrix by P. Therefore, we need to count the number of times that the label appears together. The formula for calculating the co-occurrence probability P of a label is as follows

$$P_i = T_i / N_i, \tag{4}$$

where $T \in \mathbb{R}^{C \times C}$ indicates the number of times that the label co-occurs, and N_i is the number of appearances of the i-th tag in the training set.

Deep Graph Convolutional Networks. The original recursive formula for a graph convolutional network is expressed as follows

$$X^{(l+1)} = f\left(A, X^{(l)}\right), \tag{5}$$

where $X^{(l)}$ denotes the l-th layer of GCN, and A is the adjacency matrix used to describe neighborhood relationships.

As the input matrix increases, the performance of the original two-layer GCN exhibits some limitations. Common approaches are to expand the parameters of each layer of the network or to deepen the network. Therefore, we have designed a deep graph convolutional network (DGCN).

We define a DGCN as $f(A, X^{(1)})$ where A substitutes the adjacency matrix and the input to first layer is $X^{(1)} = X_{lsv}$. We improve the original $\tilde{A} = (A + I)$ by adding an identity and normalizing, that is,

$$\tilde{A} = (D + I)^{-1}(A + I), \tag{6}$$

where $A = P$ and $D_{ii} = \sum_j A_{ij}$, then we can define the recursive formula of DGCN as follows

$$X^{(l+1)} = \delta\left(\left(\tilde{A} + \text{diag}(\tilde{A})\right) X^{(l)} W^{(l)}\right), \tag{7}$$

where $\delta(\cdot)$ is an activation function. Using Eq. (7), we can get the final output of the label semantic extraction module X_{lse}.

3.3 Prediction Results and Training Scheme

After the two modules described above, we can obtain the final prediction results $\hat{y}$ according to the following equation:

$$\hat{y} = X_{ife} \otimes X_{lse}, \tag{8}$$

where $\otimes$ denotes dot product, X_{ife} is the output of image feature extraction module, and X_{lse} is the output of label semantic extraction module.

In this work, MDGCN is trained with a multi-label soft-margin loss, which creates a criterion that optimizes a multi-label one-versus-all loss between predicted result $\hat{y}$ and the ground truth label y based on max-entropy. The loss function equation is defined as follows

$$\mathcal{L} = -\frac{1}{C} \sum_{c=1}^{C} y_c \log\left(\sigma\left(\hat{y}_c\right)\right) + \left(1 - y_c\right) \log\left(1 - \sigma\left(\hat{y}_c\right)\right), \tag{9}$$

where C is the number of classes, and $\sigma(\cdot)$ denotes a sigmoid function. Algorithm 1 illustrates an epoch of the proposed MDGCN in detail.

Algorithm 1. MDGCN

Input: Image set I, label set L, and depth of the graph convolutional network l_{max}.
Output: The trained model parameters Θ.
1: In IFEM, X_{ife} is generated by Equations (1), (2) and the set of images I;
2: In LSEM, X_{lsv} and P are generated from L by Equations (3), (4);
3: $X^{(1)} = X_{lsv}$, $A = P$;
4: $\tilde{A} = (D+I)^{-1}(A+I)$;
5: **for** $l = 1 \rightarrow l_{max}$ **do**
6: $\quad X^{(l+1)} = \delta\left(\left(\tilde{A} + \text{diag}(\tilde{A})\right) X^{(l)} W^{(l)}\right)$;
7: **end for**
8: $X_{lse} = X^{l_{max}}$;
9: The prediction results $\hat{y}$ are obtained by the dot product of X_{ife} and X_{lse};
10: Calculate the loss $\mathcal{L}$ according to Equation (9);
11: Update Θ with back propagation and the loss $\mathcal{L}$;
12: **return** Model parameters Θ.

4 Experiments

In this section, first, we present the evaluation metrics for multi-label image recognition. Then, we describe the implementation details of our experiment. And we report the results of our experiments on MS-COCO [13] and VOC 2007 [7]. Finally, we show the results of the ablation experiments, as well as the visualization of the adjacency matrix.

4.1 Evaluation Metrics

For a fair comparison with existing methods, we follow the conventional evaluation metrics as [3,19]. We adopt the average precision (AP) to reflect the performance of each category and mean average precision (mAP) to reflect the overall performance. Top-3 evaluation metric is also used in our experiments. In addition, we use traditional evaluation metrics in the image classification which are defined in Eq. (10). It includes overall precision, recall, F1-measure (OP, OR, OF1) and average per-class precision, recall, F1-measure (CP, CR, CF1),

$$\begin{aligned} \text{OP} &= \frac{\sum_i N_i^{cp}}{\sum_i N_i^{p}}, & \text{OR} &= \frac{\sum_i N_i^{cp}}{\sum_i N_i^{g}}, & \text{OF1} &= \frac{2 \times \text{OP} \times \text{OR}}{\text{OP} + \text{OR}}, \\ \text{CP} &= \frac{1}{C}\sum_i \frac{N_i^{cp}}{N_i^{p}}, & \text{CR} &= \frac{1}{C}\sum_i \frac{N_i^{cp}}{N_i^{g}}, & \text{CF1} &= \frac{2 \times \text{CP} \times \text{CR}}{\text{CP} + \text{CR}}, \end{aligned} \tag{10}$$

where C is the category number of labels. For the i-th label, N_i^{cp} denotes the number of correct predictions among the predicted labels, N_i^{p} denotes the number of labels that predicted by the model, N_i^{g} denotes the number of ground truth labels provided by the dataset.

4.2 Implementation Details

In our experiments, the parameter settings we use will be elaborated as follows. The input image is cropped at random centers for data augmentation. Then, each input image is converted to a tensor of dimension $\mathbb{R}^{512\times512\times3}$ and fed into the CNN. We used a pre-trained RexNeXt-101 to accelerate the model training and added a max pooling to transform the dimension to 2048. In LSEM, unlike the traditional two-layer GCN, we employ a four-layer GCN with input and output dimensions of 300 for the first three layers, and 300 for the fourth layer input, and 2048 for the output dimension. And we adopt GloVe [15] in word-to-vector transformation. We adopt LeakyReLU [12] with the negative slope of 0.18 as the non-linear activation function. Dropout layers with hyperparameter 0.15 are also added to prevent the model from falling into an over-fitting state. In the optimization process, we used a stochastic gradient descent method.

The initial learning rate is set to 0.1. And it changes to one-tenth of the original learning rate every 20 rounds, with a lower limit of 0.001. We train 300 epochs in total and adopt an early stopping strategy to preserve the best results and reduce the impact of over-fitting. We use PyTorch to implement our framework. Our experimental server is configured with Ubuntu 16.04, an Intel Xeon E5-2620 CPU, 128 G of RAM and a Tesla P100 GPU.

4.3 Experimental Results

In order to effectively compare our framework with current state-of-the-art algorithms, we have conducted comparative experiments on mainstream benchmark datasets including MS-COCO and VOC 2007. Figure 3 presents some sample images from both datasets.

Datasets	Microsoft COCO		PASCAL VOC 2007	
Images				
Ground truths	Person, baseball bat, baseball glove, sports ball	Bicycle, dog, person	Person, horse	Boat, car, person
Prediction results	Person, baseball bat, sports ball	Dog, bicycle, person	Person, horse	Car, person, boat

Fig. 3. Some example images from MS-COCO and VOC 2007 datasets.

Table 1. Comparisons with state-of-the-art methods on the MS-COCO dataset. The best results are marked in bold.

Methods	mAP	Top-3						All					
		CP	CR	CF1	OP	OR	OF1	CP	CR	CF1	OP	OR	OF1
CNN-RNN [19]	61.2	66.0	55.6	60.4	69.2	66.4	67.8	–	–	–	–	–	–
RNN-Attention [20]	–	79.1	58.7	67.4	84.0	63.0	72.0	–	–	–	–	–	–
Order-Free RNN [1]	–	71.6	54.8	62.1	74.2	62.2	67.7	–	–	–	–	–	–
SRN [24]	77.1	85.2	58.8	67.4	87.4	62.5	72.9	**81.6**	65.4	71.2	82.7	69.9	75.8
Multi-Evidence [8]	–	84.5	62.2	70.6	89.1	64.3	74.7	80.4	70.2	74.9	85.2	72.5	78.4
ML-GCN (Binary) [4]	80.3	84.9	61.3	71.2	88.8	65.2	75.2	81.1	70.1	75.2	83.8	74.2	78.7
MDGCN	**81.7**	**85.0**	**64.4**	**73.3**	**90.2**	**66.7**	**76.7**	81.5	**73.1**	**77.1**	**84.7**	**76.5**	**80.4**

Experimental Results on MS-COCO. Microsoft COCO (MS-COCO) [13] dataset is a large-scale object detection, image segmentation and image description generation dataset. MS-COCO version we used in our experiments is 2014, which contains 82,783 images in the training set and 40,504 images in the validation set, which used as a test set in this experiment. It is divided into 80 classes. It is therefore a challenging dataset for multi-label image recognition tasks.

The results of the comparison experiments on the COCO dataset are shown in Table 1. It can be seen that our algorithm achieves comparable performance and, on average, leads on the performance metrics mAP, CF1 and OF1. Compared to the latest algorithms our method has an advantage in precision, this is because we use a deep graph convolutional network that captures semantic information with higher accuracy.

Experimental Results on VOC 2007. PASCAL Visual Object Classes Challenge (VOC 2007) [7] dataset is now widely used for multi-label image classification tasks. It contains a total of 9963 images, of which 5011 images from the original divided training set plus the validation set are used as the training set in the experiment, and 4952 images from the test set are used as the test set in the experiment. It has 20 classes with an average of 2.9 labels per image.

The results of our experiment are presented in Table 2. The results indicate that our method achieves advanced performance on sixteen of the twenty classes, and comparable performance on the remaining four. The mAP metric reflects the overall accuracy of the model, and in terms of this metric, our model has a performance of 94.8, ahead of ML-GCN's 94.0. This shows that our model has an advantage in the use of feature extraction for most classes. And in our experiments, the computational time of our method is relatively similar to that of ML-GCN, although we employ a more complex mechanism.

4.4 Ablation Studies

To explore the effectiveness of our MDGCN, we conducted ablation experiments on the picture feature extraction network part. The experimental results are shown in Table 3. It can be seen that the performance of ResNeXt for direct

Table 2. Comparisons of AP and mAP with state-of-the-art methods on the VOC 2007 dataset. The best results are highlighted in bold.

Methods	Aero	Bike	Bird	Boat	Bottle	Bus	Car	Cat	Chair	Cow	Table	Dog	Horse	Motor	Person	Plant	Sheep	Sofa	Train	Tv	mAP
VeryDeep [17]	98.9	95.0	96.8	95.4	69.7	90.4	93.5	96.0	74.2	86.6	**87.8**	96.0	96.3	93.1	97.2	70.0	92.1	80.3	98.1	87.0	89.7
CNN-RNN [19]	96.7	83.1	94.2	92.8	61.2	82.1	89.1	94.2	64.2	83.6	70.0	92.4	91.7	84.2	93.7	59.8	93.2	75.3	**99.7**	78.6	84.0
FeV+LV [23]	97.9	97.0	96.6	94.6	73.6	93.9	96.5	95.5	73.7	90.3	82.8	95.4	97.7	95.9	98.6	77.6	88.7	78.0	98.3	89.0	90.6
HCP [21]	98.6	97.1	98.0	95.6	75.3	94.7	95.8	97.3	73.1	90.2	80.0	97.3	96.1	94.9	96.3	78.3	94.7	76.2	97.9	91.5	90.9
RNN-Attention [20]	98.6	97.4	96.3	96.2	75.2	92.4	96.5	97.1	76.5	92.0	87.7	96.8	97.5	93.8	98.5	81.6	93.7	82.8	98.6	89.3	91.9
Atten-Reinforce [2]	98.6	97.1	97.1	95.5	75.6	92.8	96.8	97.3	78.3	92.2	87.6	96.9	96.5	93.6	98.5	81.6	93.1	83.2	98.5	89.3	92.0
SSGRL [3]	99.5	97.1	97.6	97.8	82.6	94.8	96.7	98.1	78.0	97.0	85.6	97.8	98.3	96.4	98.8	84.9	96.5	79.8	98.4	92.8	93.4
ML-GCN [4]	99.5	**98.5**	98.6	98.1	80.8	94.6	97.2	98.2	82.3	95.7	86.4	98.2	98.4	96.7	99.0	84.7	**96.7**	84.3	98.9	93.7	94.0
MDGCN	**99.9**	97.4	**98.8**	**98.3**	**83.5**	**96.5**	**98.0**	**98.7**	**83.2**	**97.7**	86.9	**98.9**	**98.8**	**97.1**	**99.2**	**85.8**	95.8	**86.9**	99.0	**95.7**	**94.8**

multi-label image classification with the same data augmentation technique approach is also higher than that of ResNet. And, our framework, incorporating semantic features, thus outperforms ResNeXt on two datasets.

We tested the effect of DGCN with different network depths. The experimental results in Table 4 show that the best performance is achieved when the depth of DGCN is 4. It also illustrates that although the regularization method has been used to avoid deepening the network with weakened neighbor node connections, a DGCN is not always better.

Table 3. The impact of choosing different CNN on performance.

Datasets	Methods	mAP	CF1	OF1
MS-COCO	ResNet-101 [10]	77.3	72.8	76.8
MS-COCO	RexNeXt-101 [22]	78.2	72.7	76.7
MS-COCO	MDGCN	**81.7**	**77.1**	**80.4**
VOC 2007	ResNet-101 [10]	89.9	83.1	84.5
VOC 2007	RexNeXt-101 [22]	91.6	84.7	86.7
VOC 2007	MDGCN	**94.8**	**88.9**	**89.9**

Table 4. Comparisons with different depths of DGCN.

Depths	MS-COCO			VOC 2007		
	mAP	CF1	OF1	mAP	CF1	OF1
2	81.0	75.9	79.2	94.2	88.0	88.2
3	81.6	76.9	80.3	94.7	88.5	89.7
4	**81.7**	**77.1**	**80.4**	**94.8**	**88.9**	**89.9**
5	81.5	76.6	80.1	94.1	87.8	89.2
6	81.2	76.2	79.3	93.9	87.5	88.1

4.5 Adjacency Matrix Visualization

The adjacency matrix is very important for the label semantic extraction module, therefore we present it visually in Fig. 4. From the visualization results, the original A reflects the co-occurrence probability, from which we can see that the 15-th class of VOC 2007 and the 50-th class of MS-COCO are "person", and have a high co-occurrence probability with other labels. It is worth noting that compared to the traditional GCN's A', we construct an A' with the feature that nearby nodes have more contribution compared to distant nodes.

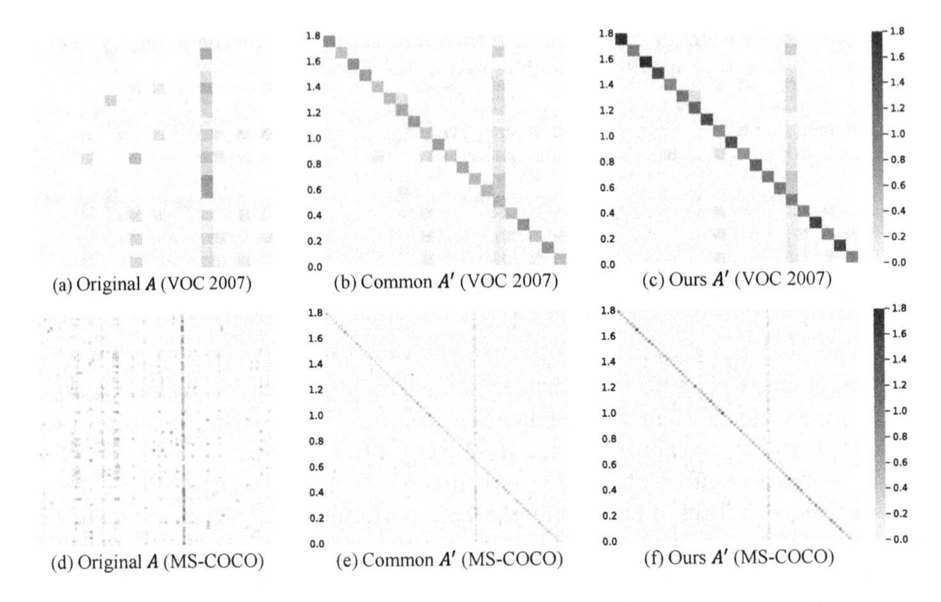

Fig. 4. Visualization of the label co-occurrence matrix A and the matrix A' obtained from it. Original A indicates that we construct it according to Eq. (4), common A' represents that it is constructed according to the original GCN, and ours A' denotes the adjacency matrix constructed by our method.

5 Conclusion

We introduced a new end-to-end multi-label image recognition framework called MDGCN. In order to make full use of the available label information, our framework was designed based on deep graph convolutional networks. In addition, we employed the method of enhancing the diagonal elements of the adjacency matrix to avoid the weakening of its own information during the propagation of DGCN. We conducted validation experiments and we achieved comparable performance. So we believe that our proposed framework can be used for more complex and larger multi-label image recognition tasks.

Acknowledgments. This work is in part supported by the National Natural Science Foundation of China (Grant No. U1705262), the Natural Science Foundation of Fujian Province (Grant Nos. 2020J01130193 and 2018J07005).

References

1. Chen, S., Chen, Y., Yeh, C., Wang, Y.F.: Order-free RNN with visual attention for multi-label classification. In: AAAI, pp. 6714–6721 (2018)
2. Chen, T., Wang, Z., Li, G., Lin, L.: Recurrent attentional reinforcement learning for multi-label image recognition. In: AAAI, pp. 6730–6737 (2018)
3. Chen, T., Xu, M., Hui, X., Wu, H., Lin, L.: Learning semantic-specific graph representation for multi-label image recognition. In: ICCV, pp. 522–531 (2019)

4. Chen, Z., Wei, X., Wang, P., Guo, Y.: Multi-label image recognition with graph convolutional networks. In: CVPR, pp. 5177–5186 (2019)
5. Chiang, W., Liu, X., Si, S., Li, Y., Bengio, S., Hsieh, C.: Cluster-GCN: an efficient algorithm for training deep and large graph convolutional networks. In: ACM SIGKDD, pp. 257–266 (2019)
6. Deng, J., Dong, W., Socher, R., Li, L., Li, K., Li, F.: ImageNet: a large-scale hierarchical image database. In: CVPR, pp. 248–255 (2009)
7. Everingham, M., Gool, L.V., Williams, C.K.I., Winn, J.M., Zisserman, A.: The pascal visual object classes (VOC) challenge. IJCV **88**(2), 303–338 (2010)
8. Ge, W., Yang, S., Yu, Y.: Multi-evidence filtering and fusion for multi-label classification, object detection and semantic segmentation based on weakly supervised learning. In: CVPR, pp. 1277–1286 (2018)
9. Gong, Y., Jia, Y., Leung, T., Toshev, A., Ioffe, S.: Deep convolutional ranking for multilabel image annotation. In: ICLR (2014)
10. He, K., Zhang, X., Ren, S., Sun, J.: Deep residual learning for image recognition. In: CVPR, pp. 770–778 (2016)
11. Kipf, T.N., Welling, M.: Semi-supervised classification with graph convolutional networks. In: ICLR (2017)
12. Maas, A.L., Hannun, A.Y., Ng, A.Y.: Rectifier nonlinearities improve neural network acoustic models. In: ICML, pp. 1–6 (2013)
13. Lin, T.-Y., et al.: Microsoft COCO: common objects in context. In: Fleet, D., Pajdla, T., Schiele, B., Tuytelaars, T. (eds.) ECCV 2014. LNCS, vol. 8693, pp. 740–755. Springer, Cham (2014). https://doi.org/10.1007/978-3-319-10602-1_48
14. Ma, X., et al.: Understanding adversarial attacks on deep learning based medical image analysis systems. Pattern Recogn. **110**, 107332 (2021)
15. Pennington, J., Socher, R., Manning, C.D.: Glove: global vectors for word representation. In: Proceedings of the 2014 Conference on Empirical Methods in Natural Language Processing, pp. 1532–1543 (2014)
16. Qian, S., Xue, D., Zhang, H., Fang, Q., Xu, C.: Dual adversarial graph neural networks for multi-label cross-modal retrieval. In: AAAI, pp. 2440–2448 (2021)
17. Simonyan, K., Zisserman, A.: Very deep convolutional networks for large-scale image recognition. In: ICLR (2015)
18. Sun, K., Lin, Z., Zhu, Z.: AdaGCN: adaboosting graph convolutional networks into deep models. In: ICLR (2021)
19. Wang, J., Yang, Y., Mao, J., Huang, Z., Huang, C., Xu, W.: CNN-RNN: a unified framework for multi-label image classification. In: CVPR, pp. 2285–2294 (2016)
20. Wang, Z., Chen, T., Li, G., Xu, R., Lin, L.: Multi-label image recognition by recurrently discovering attentional regions. In: ICCV, pp. 464–472 (2017)
21. Wei, Y., et al.: HCP: a flexible CNN framework for multi-label image classification. IEEE TPAMI **38**(9), 1901–1907 (2016)
22. Xie, S., Girshick, R.B., Dollár, P., Tu, Z., He, K.: Aggregated residual transformations for deep neural networks. In: CVPR, pp. 5987–5995 (2017)
23. Yang, H., Zhou, J.T., Zhang, Y., Gao, B., Wu, J., Cai, J.: Exploit bounding box annotations for multi-label object recognition. In: CVPR, pp. 280–288 (2016)
24. Zhu, F., Li, H., Ouyang, W., Yu, N., Wang, X.: Learning spatial regularization with image-level supervisions for multi-label image classification. In: CVPR, pp. 2027–2036 (2017)

3D Correspondence Grouping with Compatibility Features

Jiaqi Yang[1,2], Jiahao Chen[1,2], Zhiqiang Huang[1,2], Zhiguo Cao[3(✉)], and Yanning Zhang[1,2]

[1] School of Computer Science, Northwestern Polytechnical University, Xi'an 710129, China
[2] The National Engineering Laboratory for Integrated Aero-Space-Ground-Ocean Big Data Application Technology, Xi'an 710129, China
[3] School of Artificial Intelligence and Automation, Huazhong University of Science and Technology, Wuhan 430074, China
zgcao@hust.edu.cn

Abstract. We present a simple yet effective method for 3D correspondence grouping. The objective is to accurately classify initial correspondences obtained by matching local geometric descriptors into inliers and outliers. Although the spatial distribution of correspondences is irregular, inliers are expected to be geometrically compatible with each other. Based on such observation, we propose a novel feature representation for 3D correspondences, dubbed compatibility feature (CF), to describe the consistencies within inliers and inconsistencies within outliers. CF consists of top-ranked compatibility scores of a candidate to other correspondences, which purely relies on robust and rotation-invariant geometric constraints. We then formulate the grouping problem as a classification problem for CF features, which is accomplished via a simple multilayer perceptron (MLP) network. Comparisons with nine state-of-the-art methods on four benchmarks demonstrate that: 1) CF is distinctive, robust, and rotation-invariant; 2) our CF-based method achieves the best overall performance and holds good generalization ability.

Keywords: 3D point cloud · Compatability feature · 3D correspondence grouping

1 Introduction

3D correspondence grouping (a.k.a. 3D correspondence selection or 3D mismatch removal) is essential to a number tasks based on point-to-point correspondences, such as 3D point cloud registration, 3D object recognition, and 3D reconstruction. The aim is to classify initial feature correspondences between two 3D point clouds obtained by

This work was supported in part by the National Natural Science Foundation of China (NFSC) under Grant 62002295, the Natural Science Basic Research Plan in Shaanxi Province of China under Grant No. 2020JQ-210, the Ningbo Natural Science Foundation Project under Grant 202003N4058, and the Fundamental Research Funds for Central Universities under Grant D5000200078.

H. Ma et al. (Eds.): PRCV 2021, LNCS 13020, pp. 66–78, 2021.
https://doi.org/10.1007/978-3-030-88007-1_6

matching local geometric descriptors into inliers and outliers. Due to a number of factors, e.g., repetitive patterns, keypoint localization errors, and data nuisances including noise, limited overlap, clutter and occlusion, heavy outliers are generated in the initial correspondence set [19]. Thus, it is very challenging to mine the consistency of scarce inliers and find those inliers.

Existing 3D correspondence grouping methods can be divided into two categories: group-based and individual-based. Group-based methods [2,3,6,14,18] assume that inliers constitute a cluster in a particular domain and struggle to recover such cluster. By contrast, individual-based ones [1,7,10,20] usually first assign confidence scores to correspondences based on feature or geometrics constraints, and then select top-scored correspondences independently. However, as revealed by a recent evaluation [19], existing methods in both categories **1)** generalize poorly across datasets with different application scenarios and data modalities, and **2)** deliver limited precision performance which is critical to applications (e.g., 3D registration and 3D object recognition) relying on sparse and accurate correspondences.

To overcome above limitations, we present a new feature presentation to describe 3D correspondences dubbed compatibility feature (CF) along with a CF-based 3D correspondence grouping method. CF consists of top-ranked compatibility scores of a candidate to other correspondences. CF is supposed to hold strong discriminative power because *inliers are geometrically compatible with each other whereas outliers are unlikely to be compatible with either outliers or inliers* due to their unordered spatial distributions. This results in clear distinctions between CF features of inliers and outliers. Since the correspondence grouping problem can be viewed as a *binary classification problem*, we train a simple multilayer perceptron (MLP) network as a robust classifier to distinguish inliers and outliers. Although there have been some "end-to-end" learning-based 2D correspondence selection methods, our method follows a "geometry + learning" fashion due to the following reasons. **First,** even for 2D images with pixel coordinate values being in a small range, training "end-to-end" networks still requires a huge amount number of labeled image pairs [12]. By contrast, the coordinates of 3D points can be arbitrary in a 3D space, greatly increasing the challenges of training data preparation and training. **Second,** pixel/point coordinates are sensitive to rotation. Although augmenting training data can sometimes alleviates this problem, the network is still not fully rotation-invariant in nature. By contrast, CF features are extracted with rotation-invariant geometric constraints and are robust to arbitrary 3D rotations. **Third,** most of existing "end-to-end" methods are still not practical on real-world data because of the generalization issue. In a nutshell, this paper has the following contributions:

- A compatibility feature (CF) representation is proposed to describe 3D feature correspondences. CF captures the key differences between inliers and outliers regarding pairwise geometrical compatibility, which is distinctive, robust, and rotation-invariant.
- A 3D correspondence grouping method based on CF is proposed. Our method has the "geometry + learning" peculiarity that fully leverages the advantages of geometry and learning methods. Comprehensive experiments and comparisons with all methods assessed in a recent evaluation [19] on datasets with different application contexts and data modalities verify that our method has good generalization abilities and achieves outstanding precision performance.

2 Related Work

This section briefly reviews group-based and individual-based methods for 3D correspondence grouping. Methods in both categories are geometric-only ones. Because our method includes a learning-based classier, we also discuss some learning-based techniques for correspondence problems typically in 2D domain.

2.1 3D Correspondence Grouping

Group-Based Methods. Random sampling consensus [3] is arguably the most commonly used method for 3D correspondence grouping and transformation estimation. It iteratively estimates a model from correspondences and verifies its rationality; correspondences coherent with the best estimated model are served as inliers. The variants of RANSAC [4] generally follow such pipeline. Some methods try to find the main cluster within initial correspondences by analyzing the affinity matrix computed for correspondences. For instance, game theory matching (GTM) [14] and spectral technique [6] perform dynamic evolution and spectral analysis on the affinity matrix to determine the inlier cluster, respectively. Geometric consistency (GC) [5] performs inlier cluster selection more straightforwardly. In particular, GC forms a cluster for each correspondence by ensuring correspondences in the cluster are compatible with the query correspondence; the cluster with the maximum element count is served as the inlier cluster. Different from above iterative methods, 3D Hough voting (3DHV) [18] is a one-shot method, which first transforms correspondences to 3D points in a 3D Hough space and then finds the cluster in Hough space.

As demonstrated in a recent evaluation [19], *group-based methods often miss isolated inliers and are sensitive to low inlier ratios.*

Individual-Based Methods. In early studies, some individual-based methods group correspondences based on feature distances only [4,10], which are straightforward but over-reliant on the distinctiveness of descriptors. To achieve more robust grouping, several voting-based methods have been proposed such as search of inliers (SI) [1] and consistency voting (CV) [20]. The common peculiarity of these methods is that one or more voting sets are first defined and then all voters will cast a vote to each correspondence based on some pre-defined rules.

Compared with group-based methods, individual-based ones assign scores to correspondences independently and thus can more reliably recall isolated inliers. However, existing individual-based methods still exhibit limited precision performance. *We note that the proposed method is individual-based as well, but is highly selective with outstanding precision performance.*

2.2 Learning for Correspondence Grouping

Most of existing 3D correspondence grouping methods are still geometric-based ones [19]. In 2D domains, there exist a few mismatch removal methods based on deep learning [8,12,21]. Yi et al. [12] presented the first attempt to find inliers with an "end-to-end" network. To mine local information, Ma et al. [8] and Zhao et al. [21] associated

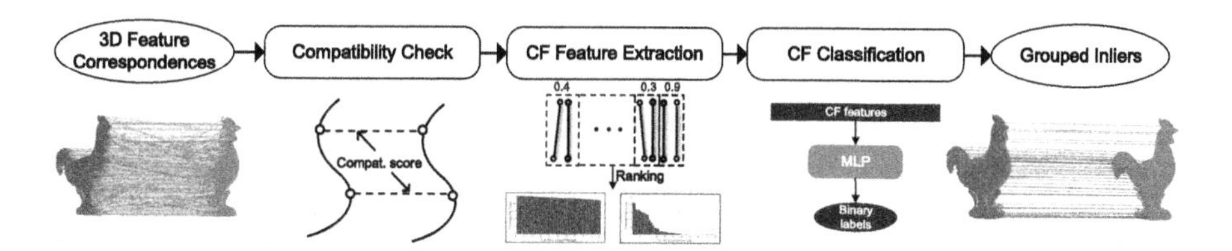

Fig. 1. Pipeline of the proposed method. *Compatibility check*: computing the compatibility scores of a correspondence with others; *CF feature extraction*: parameterizing each correspondence by a distinctive CF feature; *CF classification*: classifying CF features as inliers and outliers with an MLP network.

spatial and compatibility-specific neighbors to each correspondence for classifier training, respectively.

Nonetheless, most of existing learning-based image correspondence grouping methods suffer from the following limitations: **1)** the requirement of a large amount of training matching pairs; **2)** the sensitivity to rotations due to the input of coordinate information; **3)** redundant network architectures. By contrast, *our method properly interprets the roles of geometric and learning techniques, and can effectively overcome these limitations.*

3 Methodology

The pipeline of our method is presented in Fig. 1. It consists of three main steps, including compatibility check, CF feature extraction, and CF classification. To improve readability, we introduce the following notations. Let $\mathbf{P}^s \in \mathbb{R}^3$ and $\mathbf{P}^t \in \mathbb{R}^3$ be the source point cloud and the target point cloud, respectively. A feature correspondence set $\mathbf{C} \in \mathbb{R}^6$ can be generated by matching local geometric descriptors for $\mathbf{P}^s$ and $\mathbf{P}^t$. The aim of our method is to assign a binary label (inlier or outlier) to each element $\mathbf{c} = (\mathbf{p}^s, \mathbf{p}^t)$ in $\mathbf{C}$, where $\mathbf{p}^s \in \mathbf{P}^s$ and $\mathbf{p}^t \in \mathbf{P}^t$.

3.1 Compatibility Check

In order to distinguish inliers and outliers, we should fully mine the consistency information within inliers. As depicted in Fig. 2, an important observation is that inliers are geometrically compatible with each other, while outliers are unlikely to be compatible with either outliers or inliers, because the spatial distribution of outliers are unordered. Following this cue, we are motivated to define a metric to check the compatibility between two correspondences.

In the context of 3D point cloud matching, we consider distance and angle constraints [1,20] that are invariant to rotations for compatibility metric definition. Let $\mathbf{n}$ be the normal of $\mathbf{p}$, the distance and angle constraints for two correspondences $(\mathbf{c}_i, \mathbf{c}_j)$ are respectively defined as:

$$s_{dist}(\mathbf{c}_i, \mathbf{c}_j) = \left| ||\mathbf{p}_i^s - \mathbf{p}_j^s|| - ||\mathbf{p}_i^t - \mathbf{p}_j^t|| \right|, \tag{1}$$

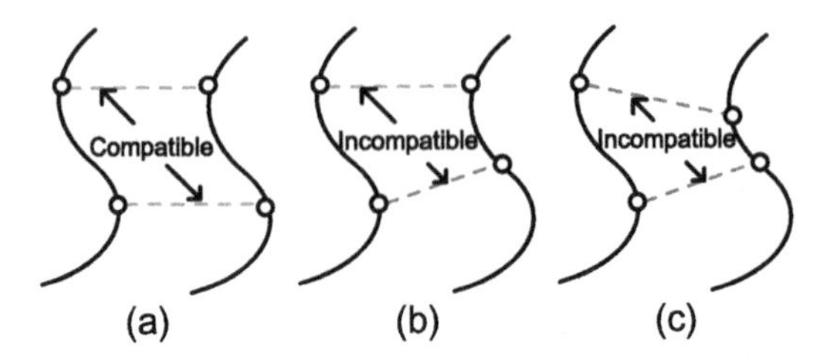

Fig. 2. Illustration of the statement that (a) inliers are compatible with each other, while outliers are usually incompatible with either (b) inliers or (c) outliers. Green and red dashed lines denote inliers and outliers, respectively. (Color figure online)

and

$$s_{ang}(\mathbf{c}_i, \mathbf{c}_j) = \left|\operatorname{acos}(\mathbf{n}_i^s \cdot \mathbf{n}_j^s) - \operatorname{acos}(\mathbf{n}_i^t \cdot \mathbf{n}_j^t)\right|. \tag{2}$$

We note that s_{dist} and s_{ang} are calculated based on linear operation on relative distances and angles, thus being rotation-invariant. Both constraints are complementary to each other (Sect. 4.2). By integrating the two constraints, we define the compatibility metric as:

$$S(\mathbf{c}_i, \mathbf{c}_j) = \exp(-\frac{s_{dist}(\mathbf{c}_i, \mathbf{c}_j)^2}{2\alpha_{dist}^2} - \frac{s_{ang}(c_i, c_j)^2}{2\alpha_{ang}^2}), \tag{3}$$

where α_{dist} and α_{ang} represent a distance parameter and an angle parameter, respectively. One can see that $S(\mathbf{c}_i, \mathbf{c}_j) \in [0, 1]$ and $S(\mathbf{c}_i, \mathbf{c}_j)$ equals 1 only if both constraints are fully satisfied.

3.2 CF Feature Extraction

With a compatibility metric, a naive way for correspondence grouping is to first assess the greatest compatibility score of each correspondence to others and then set a threshold to filter those with low scores. This is not robust and the distinctiveness of a single compatibility score is limited, as demonstrated in [2]. Instead, we consider top-k compatibility scores and render them as a feature vector. Remarkably, most prior works focus on assign scores to correspondences, and the main difference among them is the scoring functions. Our methods differs from those ones as we exact feature vectors for correspondences.

Specifically, the calculation of CF features consists of three steps: **1)** compute the compatibility scores of $\mathbf{c}$ to other correspondences in $\mathbf{C}$ based on Eq. 3, obtaining a score set $F = \{S(\mathbf{c}, \mathbf{c}_1), \cdots, S(\mathbf{c}, \mathbf{c}_{D-1})\}$ (D being the cardinality of $\mathbf{C}$); **2)** sort elements in F by a descending order, resulting in $\mathbf{F} = \left[S(\mathbf{c}, \mathbf{c}_1') \cdots S(\mathbf{c}, \mathbf{c}_{D-1}')\right]$; **3)** compute the N-dimensional CF feature $\mathbf{f}(\mathbf{c})$ of $\mathbf{c}$ as the concatenation of the former N elements in $\mathbf{F}$, i.e., $\mathbf{f}(\mathbf{c}) = \left[\mathbf{F}(1) \cdots \mathbf{F}(N)\right]$.

Assume that: **1)** an ideal compatibility scoring metric is defined, which assigns '1' to correspondence pairs composed by inliers and '0' to those with at least one outlier, and **2)** a proper N is defined, we can obtain CF features with all elements being '1' and '0' for inliers and outliers, respectively. Hence, *from the theoretic perspective, our*

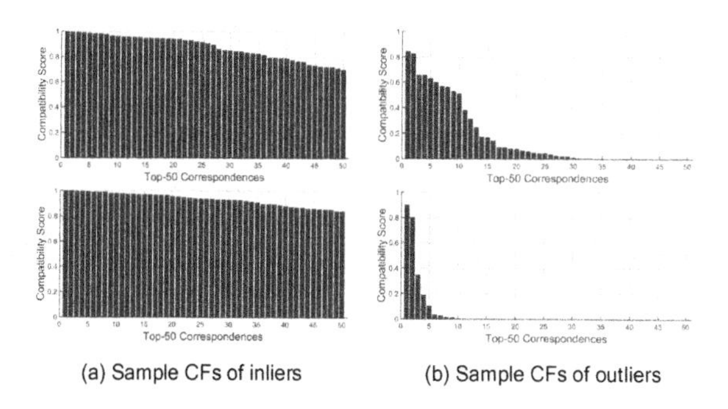

Fig. 3. Sample CF features of (a) inliers and (b) outliers. We find that with the metric defined in Eq. 3 and a proper dimensionality N (50 in the figure), the generated CF features are quite distinctive and intuitively classifiable.

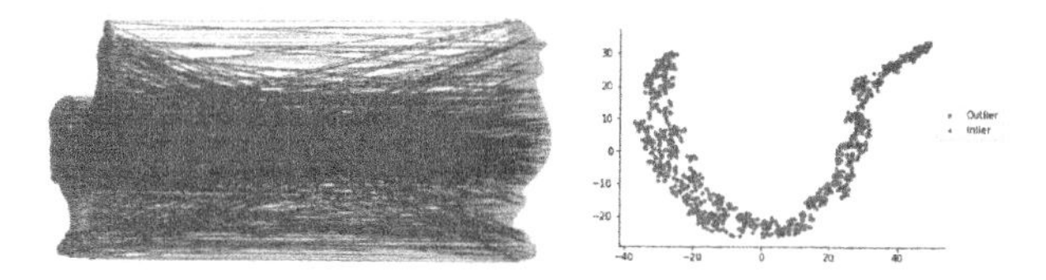

Fig. 4. Classifying CF features in cases with low inlier ratios appears to be a non-linear classification problem. Left: feature correspondences between two 3D point clouds, where green lines and red lines represent inliers and outliers, respectively. Right: the CF features of all correspondences are projected in a 2D space with t-SNE [9]. (Color figure online)

proposed CF can be ultra distinctive. At present, robust compatibility metric definition for 3D correspondences is still an open issue [20] and estimating a proper N appears to be a chicken-and-egg problem, resulting in *noise* in CF features. However, with the metric defined in Eq. 3 and an empirically determined N (based on experiments in Sect. 4.2), *our CF features, in real case, still hold strong distinctiveness,* as shown in Fig. 3.

3.3 CF Classification

Finally, the 3D correspondence grouping problem boils down to a binary feature classification problem. In recent years, deep learning has achieved remarkable success in classification tasks. In addition, we find that classifying CF features in cases with low inlier ratios sometimes appears to be a non-linear classification problem. As shown in Fig. 4, the CF features of inliers and outliers cannot be linearly separated. Thus, we are motivated to employ a deep-learning classifier.

In particular, the MLP network is suffice to our task because spatially 1-D CF feature vectors are inputs to the network. *This makes the network ultra lightweight* as compared with other networks for image correspondence problem [12,17,21], which is also demonstrated to be quite effective as will be verified in the experiments. The

Table 1. Experimental datasets and their properties.

Dataset	Scenario	Nuisances	Modality	# Matching pairs	Avg. inlier ratio
U3M [11]	Registration	Limited overlap, self-occlusion	LiDAR	496	0.1480
BMR [16]	Registration	Limited overlap, self-occlusion, real noise	Kinect	485	0.0563
U3OR [10]	Object recognition	Clutter, occlusion	LiDAR	188	0.0809
BoD5 [16]	Object recognition	Clutter, occlusion, real noise, holes	Kinect	43	0.1575

employed MLP network has 6 layers with 50, 128, 128, 64, 32, and 2 neurons, respectively. We note that the training samples of inliers and outliers are imbalanced for 3D correspondence grouping problem, and we employ the focal loss to train our network.

4 Experiments

4.1 Experimental Setup

Datasets. Four datasets are considered in our experiments, including UWA 3D modeling (U3M) [11], Bologna Mesh Registration (BMR) [16], UWA 3D object recognition (U3OR) [10], and Bologna Dataset5 (BoD5) [16]. The main properties of experimental datasets are summarized in Table 1. These datasets have **1)** different application scenarios, **2)** a variety of nuisances, and **3)** different data modalities, which can ensure a comprehensive evaluation. For each dataset, we use correspondence data generated by 75% matching pairs for training and the remaining for testing. Note that we will also test the generalization performance of our method without training a model for each dataset.

Metrics. Precision (P), Recall (R), and F-score (F) are popular metrics for evaluating the performance of correspondence grouping [19,21]. A correspondence $\mathbf{c} = (\mathbf{p}^s, \mathbf{p}^t)$ is judged as correct if $||\mathbf{R}_{gt}\mathbf{p}^s + \mathbf{t}_{gt} - \mathbf{p}^t|| < d_{inlier}$, where d_{inlier} is a distance threshold; $\mathbf{R}_{gt}$ and $\mathbf{t}_{gt}$ denote the ground-truth rotation matrix and translation vector, respectively. We set d_{inlier} to 5 pr as in [19]. The unit 'pr' denotes the point cloud resolution, i.e., the average shortest distance among neighboring points in the point cloud. Thus, precision is defined as $\mathrm{P} = \frac{|\mathbf{C}_{inlier}|}{|\mathbf{C}_{group}|}$, and recall is defined as $\mathrm{R} = \frac{|\mathbf{C}_{inlier}|}{|\mathbf{C}^{gt}_{inlier}|}$, where $\mathbf{C}_{group}$, $\mathbf{C}_{inlier}$, and $\mathbf{C}^{gt}_{inlier}$ represent the grouped inlier set by a grouping method, the true inlier subset in the grouped inlier set, and the true inlier subset in the raw correspondence set, respectively. F-score is given by $\mathrm{F} = \frac{\mathrm{PR}}{\mathrm{P+R}}$.

We note that 3D correspondence grouping methods are typically applied to rigid registration tasks, e.g., point cloud registration and 3D object recognition, which require sparse and accurate correspondences. *Thus, the precision performance is more critical to these practical applications.*

Implementation Details. For our method, the compatibility check and CF feature exaction modules are implemented in the point cloud library (PCL) [15], and the MLP

Table 2. Performance of our method when varying the dimensionality of CF features.

	10	20	50	100	200
P	0.8031	0.7625	0.7483	0.7386	0.7468
R	0.4754	0.5364	0.5308	0.5114	0.4870
F	0.5973	0.6298	0.6211	0.6044	0.5896
# Epochs	77	44	7	15	9

classifier is trained in PyTorch with a GTX1050 GPU. The network is optimized via stochastic gradient descent (SGD) with a learning rate of 0.02. All evaluated methods in [19] are compared in our experiments, including similarity score (SS), nearest neighbor similarity ratio (NNSR) [7], spectral technique (ST) [6], random sampling consensus (RANSAC) [3], geometric consistency (GC) [2], 3D Hough voting (3DHV) [18], game theory matching (GTM) [14], search of inliers (SI) [1], and consistency voting (CV) [20].

To generate 3D feature correspondences between point clouds, we match the signatures of histograms of orientations (SHOT) descriptors extracted on keypoints detected by the Harris 3D (H3D) detector. The initial number of correspondences used for training in the U3M dataset is around 490k. It has been verified in [19] that H3D+SHOT can generate correspondences with *different spatial distributions, different scales, and different inlier ratios*, enabling a thorough evaluation.

4.2 Method Analysis

The following experiments were conducted on the U3M dataset (the largest scale one) to analyze the rationality, peculiarities, and parameters of our method.

Dimensionality of CF Features. The dimensionality N of CF features is a key parameter of the proposed method. We test the performance of our method with N being 10, 20, 50, 100, and 200, respectively. The results are shown in Table 2.

The results indicate that $N = 20$ and $N = 50$ achieve the best and the second best performance, respectively. Thus, a proper N is needed to maximize the distinctiveness between the CF features of inliers and outliers. In addition, we find that the network converges much faster with $N = 50$ than other settings, thus we set N to 50 by default.

Varying Compatibility Metrics. A critical factor to the proposed CF features is the definition of compatibility metrics. In our defined compatibility metric (Eq. 3), both distance and angle constraints are considered. Here, we test the effect when using looser constraints, i.e., solely using either distance constraint or angle constraint, as shown in Table 3.

It is interesting to see that using a slightly looser constraint (distance only) can achieve better F-score performance than using both constraints. However, when the constraint is too loose (angle only), the network does not converge as the generated CF features are ambiguous. Because using both constraints achieves the best precision performance, which is preferred in most application scenarios, so we consider both constraints to define the compatibility metric.

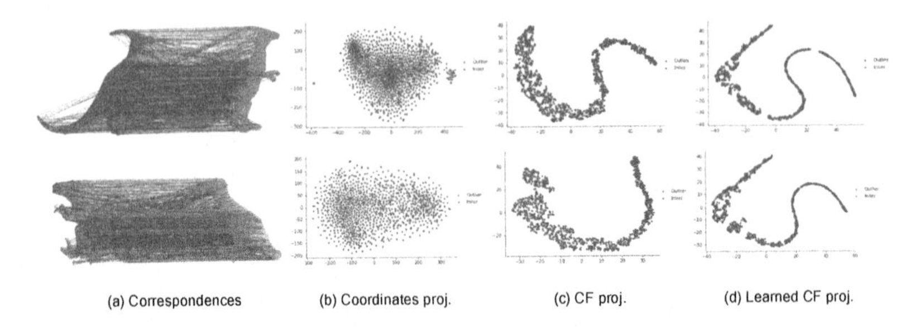

Fig. 5. Sample results of (a) 3D feature correspondences, and 2D projections (by t-SNE [9]) of (b) correspondence coordinates, (c) CF features, and (d) the features of the second last layer of MLP.

PointNet vs. MLP. As similar to some 2D correspondence methods [12,17], directly setting the coordinates of correspondences as the input to networks can be another way for grouping. We tested the performance of using coordinate information for learning with PointNet [13] on testing data with and without arbitrary rotations. The results are reported in Table 4.

Table 3. The effect of using compatibility metrics with different geometric constraints (NC: not converge).

	Distance	Angle	Both
P	0.6443	NC	0.7483
R	0.6885	NC	0.5308
F	0.6657	NC	0.6211

Table 4. Comparison of PointNet [13] and our method on testing data with and without arbitrary $SO(3)$ rotations, which input with point coordinates and CF features, respectively.

	PointNet	PointNet ($SO(3)$)	Ours	Ours ($SO(3)$)
P	0.3888	0.1290	0.7483	0.7483
R	0.0355	0.0018	0.5308	0.5308
F	0.0651	0.0035	0.6211	0.6211

Two observations can be made from the table. **1)** PointNet with coordinates being the input achieves significantly worse performance than our MLP architecture with CF features being input. This is because the range of 3D real-world coordinate information is too large, which makes the network very difficult to mine the patterns within training data. **2)** Coordinates are sensitive to rotations, making the performance of PointNet even worse when the testing data undergoing rotations. By contrast, because our CF features consist of compatibility scores computed based on rotation-invariant constraints, this makes CF and the CF-based learning network rotation-invariant as well.

To further support our statement, we visualize some exemplar results of feature correspondences, projections of correspondence coordinates, CF features, and the features of the second last layer of MLP in Fig. 5. Obviously, one can hardly mine consistencies within inliers from the coordinate information. By contrast, CF features hold strong distinctiveness. In addition, learned CF features by MLP can further enhance the distinctiveness (the clusters of inliers and outliers in Fig. 5(d) are tighter than these in Fig. 5(c)).

Table 5. Comparison of the proposed method with nine state-of-the-art methods in terms of precision, recall, and F-score performance on four experimental datasets (bold: the best; underlined: the second best).

	SS	NNSR [7]	ST [6]	RANSAC [3]	GC [2]	3DHV [18]	GTM [14]	SI [1]	CV [20]	CF (Ours)
(a) *U3M dataset*										
P	0.0374	0.1289	0.3984	<u>0.5442</u>	0.2920	0.1960	0.5285	0.0380	0.1092	**0.7483**
R	0.3819	0.4084	0.5833	0.8493	0.7499	0.6999	0.5987	**0.9996**	0.9839	0.5308
F	0.0681	0.1960	0.4734	**0.6634**	0.4203	0.3062	0.5614	0.0733	0.1966	<u>0.6211</u>
(b) *BMR dataset*										
P	0.0243	0.0606	0.2993	0.3737	0.1458	0.1492	<u>0.3946</u>	0.0350	0.0700	**0.8575**
R	0.3405	0.0967	0.3734	0.8178	<u>0.5740</u>	0.5049	0.3626	0.5522	**0.9438**	0.1529
F	0.0454	0.0745	0.3323	**0.5129**	0.2325	0.2304	<u>0.3779</u>	0.0658	0.1303	0.2596
(c) *BoD5 dataset*										
P	0.0474	0.1635	0.5660	<u>0.5961</u>	0.5207	0.3927	**0.7022**	0.0748	0.3593	0.5699
R	0.2024	0.1136	0.4086	0.8747	0.7559	<u>0.8890</u>	0.4556	0.7337	**0.9869**	0.4151
F	0.0768	0.1341	0.4746	**0.7090**	<u>0.6166</u>	0.5448	0.5527	0.1359	0.5268	0.4804
(d) *U3OR dataset*										
P	0.0171	0.0724	0.1119	<u>0.5812</u>	0.1918	0.1190	0.4907	0.0143	0.0523	**0.8641**
R	0.4111	0.5296	0.1670	0.2442	0.6302	0.3537	0.5224	**1.0000**	<u>0.9461</u>	0.3196
F	0.0328	0.1274	0.1340	0.3438	0.2941	0.1781	**0.5061**	0.0282	0.0991	<u>0.4666</u>

4.3 Comparative Results and Visualization

Start-of-the-Art Comparison. All evaluated methods in a recent evaluation [19] are compared with the proposed method on four experimental datasets. All methods are tested on the same testing data. The results are shown in Table 5.

The following observations can be made from the table. **1)** Our method achieves the best precision performance on the U3M, BMR, and U3OR dataset. Moreover, the gap between our method and the second best one is significant on the BMR and U3OR datasets. On the BoD5 dataset, our method is surpassed by GTM and RANSAC. However, this dataset is less challenging than the other three ones (Table 1). This indicates that our method can achieve superior precision performance especially on data with low inlier ratios. We also note that only 33 pairs of data are leveraged to train our network on the BoD5 dataset. **2)** In terms of the recall performance, SI and CV, as two typical individual-based methods, achieve top-ranked performance. Unfortunately, their precision performance is quite limited. This could result in inaccurate and time-consuming rigid registration results due to heavy outliers in the grouped inlier set. *We note that a looser geometric constraint can be used if a balance is needed between precision and recall (as verified in Table* 3), indicating that our method is flexible. **3)** Although the proposed method is an individual-based one, it is quite selective with superior precision performance. Notably, GTM is the most selective method as evaluated by [19], while our method generally outperforms it by a large margin regarding precision.

Generalization Performance. We use the model trained on the initial U3M dataset to predict inliers on the following datasets: BMR dataset, variants of U3M datasets

Table 6. Generalization performance of the proposed method (the model is trained on the original U3M dataset).

	BMR	U3M+noise	U3M+simplification	ISS+FPFH
P	0.6928	0.7407	0.7088	0.7409
R	0.3241	0.4111	0.3247	0.4342
F	0.4416	0.5287	0.4454	0.5475

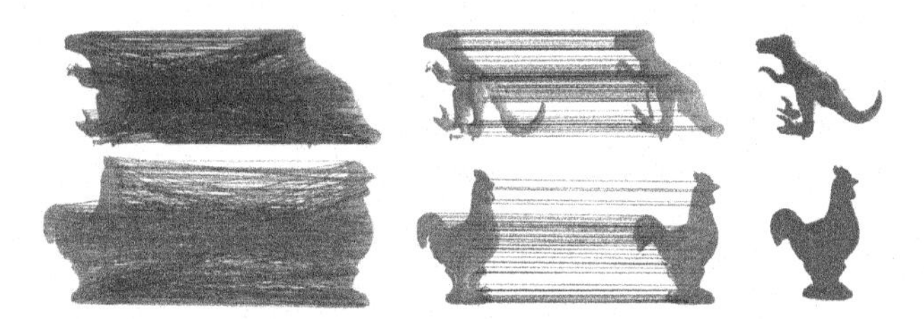

Fig. 6. Sample visualization results. From left to right: initial correspondences with colors obtained by projecting CF features to a 3D RGB space, grouped correspondences by our method, and the registration result with the grouped correspondences using PCL [15].

with 0.3 pr Gaussian noise, $\frac{1}{8}$ random data decimation, and "ISS detector + FPFH descriptor", respectively. The results are shown in Table 6. One can see that the model trained on the U3M dataset also achieves decent performance when changing the testing dataset, injecting additional nuisances, and changing "detector-descriptor" combinations. This is potentially because the eventual impact caused by above test conditions is the variation of inlier ratios, while our CF features can effectively mine the hidden consistencies of inliers and inconsistencies of outliers in different inlier ratio cases.

Visualization. Finally, we give some visualization results of our method in Fig. 6. Two observations can be made. First, the colors of correspondences obtained by projecting CF features to a 3D RGB space can reflect the consistency of inliers. Second, the grouped correspondences by our method are quite consistent and can achieve accurate 3D registration results.

5 Conclusions

We presented a CF feature representation to describe 3D feature correspondence, along with a CF-based 3D correspondence grouping method. CF captures the main distinctiveness between inliers and outliers regarding pairwise geometrical compatibility, which is rotation-invariant as well. With CF features, a lightweight MLP network is able to classify them and achieve outstanding performance. Experiments on four standard datasets with various applications, paired with comparisons with the state-of-the-arts, have demonstrated the overall superiority of our method.

References

1. Buch, A.G., Yang, Y., Krüger, N., Petersen, H.G.: In search of inliers: 3D correspondence by local and global voting. In: Proceedings of the IEEE Conference on Computer Vision and Pattern Recognition, pp. 2075–2082. IEEE (2014)
2. Chen, H., Bhanu, B.: 3D free-form object recognition in range images using local surface patches. Pattern Recogn. Lett. **28**(10), 1252–1262 (2007)
3. Fischler, M.A., Bolles, R.C.: Random sample consensus: a paradigm for model fitting with applications to image analysis and automated cartography. Commun. ACM **24**(6), 381–395 (1981)
4. Guo, Y., Sohel, F., Bennamoun, M., Lu, M., Wan, J.: Rotational projection statistics for 3D local surface description and object recognition. Int. J. Comput. Vis. **105**(1), 63–86 (2013)
5. Johnson, A.E., Hebert, M.: Surface matching for object recognition in complex three-dimensional scenes. Image Vis. Comput. **16**(9), 635–651 (1998)
6. Leordeanu, M., Hebert, M.: A spectral technique for correspondence problems using pair-wise constraints. In: Proceedings of the International Conference on Computer Vision, vol. 2, pp. 1482–1489. IEEE (2005)
7. Lowe, D.G.: Distinctive image features from scale-invariant keypoints. Int. J. Comput. Vis. **60**(2), 91–110 (2004)
8. Ma, J., Jiang, X., Jiang, J., Zhao, J., Guo, X.: LMR: learning a two-class classifier for mismatch removal. IEEE Trans. Image Process. **28**(8), 4045–4059 (2019)
9. Maaten, L.V.D., Hinton, G.: Visualizing data using t-SNE. J. Mach. Learn. Res. **9**(1), 2579–2605 (2008)
10. Mian, A.S., Bennamoun, M., Owens, R.: Three-dimensional model-based object recognition and segmentation in cluttered scenes. IEEE Trans. Pattern Anal. Mach. Intell. **28**(10), 1584–1601 (2006)
11. Mian, A.S., Bennamoun, M., Owens, R.A.: A novel representation and feature matching algorithm for automatic pairwise registration of range images. Int. J. Comput. Vis. **66**(1), 19–40 (2006)
12. Moo Yi, K., Trulls, E., Ono, Y., Lepetit, V., Salzmann, M., Fua, P.: Learning to find good correspondences. In: Proceedings of the IEEE Conference on Computer Vision and Pattern Recognition, pp. 2666–2674 (2018)
13. Qi, C.R., Su, H., Mo, K., Guibas, L.J.: PointNet: deep learning on point sets for 3D classification and segmentation. In: Proceedings of the IEEE Conference on Computer Vision and Pattern Recognition, pp. 652–660 (2017)
14. Rodolà, E., Albarelli, A., Bergamasco, F., Torsello, A.: A scale independent selection process for 3D object recognition in cluttered scenes. Int. J. Comput. Vis. **102**(1–3), 129–145 (2013)
15. Rusu, R.B., Cousins, S.: 3D is here: Point Cloud Library (PCL). In: Proceedings of the IEEE International Conference on Robotics and Automation, pp. 1–4 (2011)
16. Salti, S., Tombari, F., Di Stefano, L.: SHOT: unique signatures of histograms for surface and texture description. Comput. Vis. Image Underst. **125**, 251–264 (2014)
17. Sun, W., Jiang, W., Trulls, E., Tagliasacchi, A., Yi, K.M.: ACNe: attentive context normalization for robust permutation-equivariant learning. In: Proceedings of the IEEE/CVF Conference on Computer Vision and Pattern Recognition, pp. 11286–11295 (2020)
18. Tombari, F., Di Stefano, L.: Object recognition in 3D scenes with occlusions and clutter by Hough voting. In: Proceeding of the Pacific-Rim Symposium on Image and Video Technology, pp. 349–355. IEEE (2010)
19. Yang, J., Xian, K., Wang, P., Zhang, Y.: A performance evaluation of correspondence grouping methods for 3D rigid data matching. IEEE Trans. Pattern Anal. Mach. Intell. (2019). https://doi.org/10.1109/TPAMI.2019.2960234

20. Yang, J., Xiao, Y., Cao, Z., Yang, W.: Ranking 3D feature correspondences via consistency voting. Pattern Recogn. Lett. **117**, 1–8 (2019)
21. Zhao, C., Cao, Z., Li, C., Li, X., Yang, J.: NM-Net: mining reliable neighbors for robust feature correspondences. In: Proceedings of the IEEE Conference on Computer Vision and Pattern Recognition, pp. 215–224 (2019)

Contour-Aware Panoptic Segmentation Network

Yue Xu[1,2,3], Dongchen Zhu[1,3](✉), Guanghui Zhang[1,3], Wenjun Shi[1,3], Jiamao Li[1,3], and Xiaolin Zhang[1,2,3]

[1] Shanghai Institute of Microsystem and Information Technology, Chinese Academy of Sciences, Shanghai 200050, China
{dchzhu,zhanggh,wjs,jmli,xlzhang}@mail.sim.ac.cn
[2] School of Information Science and Technology, ShanghaiTech University, Shanghai 201210, China
xuyue@shanghaitech.edu.cn
[3] University of Chinese Academy of Sciences, Beijing 100049, China

Abstract. Panoptic segmentation incorporating semantic segmentation and instance segmentation plays an important role in scene understanding. Although current research has made remarkable progress, it has great potential for improvement in the contour region. In this paper, we propose a Contour-Aware Panoptic Segmentation Network (CAPSNet) with a panoptic contour branch and a new structure loss. The panoptic contour branch is designed to enhance structural cues perception by guiding feature extraction. The novel structure loss is presented for associating the consistency between panoptic segmentation and contour. Experimental results on the Cityscapes dataset show that CAPSNet achieves 60.0 in PQ metric, improving by 0.9% over baseline, and the proposed panoptic contour branch and structure loss are able to realize promising boosts on panoptic segmentation task.

Keywords: Panoptic segmentation · Contour-aware · Structure loss

1 Introduction

Scene understanding is one of the essential parts of computer vision, which mainly includes instance segmentation determining pixel-level instance annotations, and semantic segmentation assigning a semantic label to each pixel in an image. He *et al.* [13] unify the two tasks as panoptic segmentation task and provide corresponding evaluation metrics. The task requires assigning a semantic label and an instance id to each pixel, which is a rich and complete scene segmentation.

De *et al.* [8] and Arnab *et al.* [1] process the two tasks separately and then fuse them to obtain the results of the panoptic segmentation by post-processing. With the in-depth development of research, end-to-end panoptic segmentation networks have been proposed, such as Panoptic FPN [12], UPSNet [24], and AUNet [16]. Most of them implement each task separately by using a shared backbone

H. Ma et al. (Eds.): PRCV 2021, LNCS 13020, pp. 79–90, 2021.
https://doi.org/10.1007/978-3-030-88007-1_7

and adopting different branches on top of the shared features. Although these approaches have achieved remarkable improvement in forecast accuracy, they still ignore the basic and important structure in the image. Especially, the lack of structural information generally causes misclassification on different objects with high similarity and complex skeletonized regions, which affects the overall segmentation accuracy.

As illustrated by Zhou *et al.* [29], 18% of the cells in area V1 and more than 50% of the cells in V2 and V4 on cerebral cortex are responsible for coding according to the direction of the boundary. That is to say, contour perception plays a vital role in helping humans distinguish different objects when understanding the surrounding environment. Just as shown in Fig. 1, it is easy to identify objects such as people, cars, signal lights, trees and skateboards from Fig. 1(c). This also shows that structural information in panoptic contour can effectively assist people in discriminating the targets.

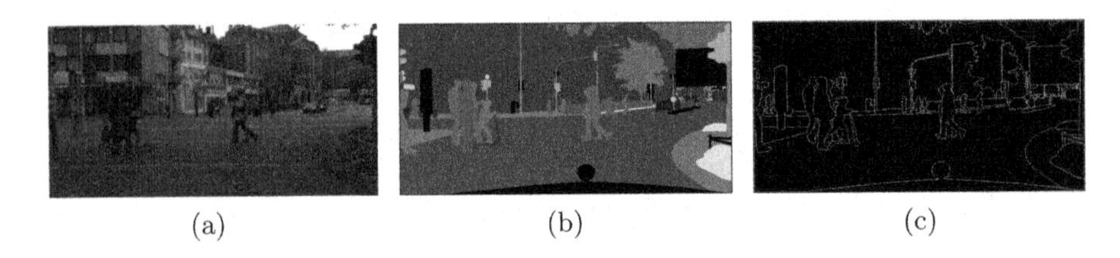

(a) (b) (c)

Fig. 1. (a) image; (b) panoptic segmentation; (c) panoptic contour(boundary of panoptic segmentation). It is not difficult to restore the information of (b) from (c).

To enhance structural perception, we propose a Contour-Aware Panoptic Segmentation Network (CAPSNet) inspired by the above biological mechanism of human beings in scene understanding. Specifically, a panoptic contour branch is added to the top of the shared backbone, which focuses on guiding feature extraction to make the features contain structural information. In consequence, the panoptic segmentation accuracy is boosted. Besides, a novel structure loss function is proposed to explicitly supervise the contours of the panoptic segmentation result. Through the learning of consistency between panoptic segmentation and contour, the performance of the network in the contour regions is improved.

In summary, the contributions of this work are as follows:

- To obtain features with strong structural information, a panoptic contour branch is introduced to panoptic segmentation network to leverage contour knowledge.
- A novel structure loss function is developed to consider the structural constraint between panoptic segmentation and contour to boost network performance.
- Experimental results on the Cityscapes dataset show that the proposed network exploiting contour cues is effective.

2 Related Work

Panoptic Segmentation. The current panoptic segmentation can be divided into two-stage panoptic segmentation network and one-stage panoptic segmentation network. Among them, the two-stage panoptic segmentation networks are mainly evolved from the basis of Mask R-CNN [10], such as [8,12,16,18,24]. AUNet [16] adds two attentions as additional connections across foreground and background branches to provide object-level and pixel level cues, respectively. UPSNet [24] introduces a parameter-free panoptic head solving the panoptic segmentation by pixel-wise classification. Above methods introduce semantic branches on Mask R-CNN. They take the advantage of Mask R-CNN's high accuracy in instance segmentation, and adopt semantic segmentation networks such as PSPNet [27], FCN [19] to improve panoptic segmentation results. A similar framework is explored in this paper, which compensates for the insensitivity to structural information of the above methods by introducing panoptic contour branch.

Owing to the development of one-stage instance segmentation and Transformer [21] applications in computer vision, numerous work on one-stage panoptic segmentation emerges. For example, FPSNet [9], EPSNet [3], DeeperLab [25], Axial-DeepLab [22], Panoptic-DeepLab [4] and DETR [30] are the representatives of one-stage panoptic segmentation networks. They do not require generating proposals, which improve the efficiency of panoptic segmentation.

Multi-task Learning. Multi-task learning [2] utilizes additional auxiliary tasks to boost the accuracy of main task and reduce the computation cost through sharing feature extraction. Several algorithms [20,26,28] explore the combination of contour detection and semantic segmentation. RPCNet [28] presents a joint multi-task learning network for semantic segmentation and boundary detection, which couples two tasks via iterative pyramid context module (PCM). Gated-SCNN [20] proposes a novel two-stream (i.e., shape stream and classical stream) architecture in parallel for semantic segmentation and connect the intermediate layers of the two streams by a new type of gates. Inspired by those work, we introduce panoptic contour detection task into our panoptic segmentation to achieve higher accuracy.

3 Approach

Figure 2 illustrates the proposed CAPSNet network in detail, which adopts a residual network (ResNet) [11] as the backbone and combines a feature pyramid network (FPN) [17]. As shown in Fig. 2, the input is an image and the outputs are panoptic segmentation and panoptic contour results. The network consists of two branches: panoptic contour branch generating panoptic contour and panoptic segmentation branch predicting panoptic labels, which are elaborated in Sect. 3.1 and Sect. 3.2, respectively. Besides, we present the designed structure loss function and total loss of our algorithm in Sect. 3.3.

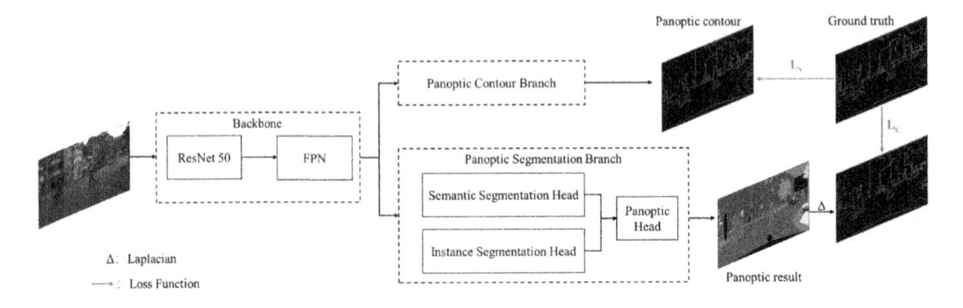

Fig. 2. Framework of CAPSNet.

3.1 Panoptic Contour Branch

As described in Fig. 1 above, contour information is capable of helping distinguish different objects more completely and accurately. Therefore, we propose the panoptic contour branch as an auxiliary task for panoptic segmentation network. Figure 3 shows the architecture of the panoptic contour branch, which will be elaborated in the following.

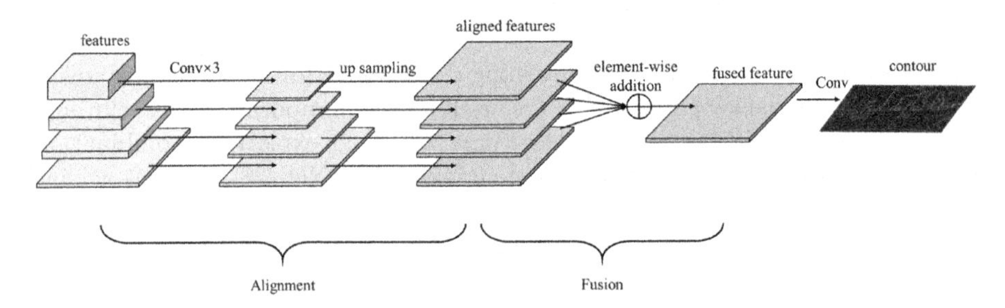

Fig. 3. The architecture of our panoptic contour branch.

The main goal of the branch is to generate panoptic contours and to enforce the structural information(i.e., low-level contour cue) in the feature extraction with supervision of contour annotations. The branch can be summarized as:

$$\begin{aligned} C = \gamma(\Phi(F_i)),\ i \in 1, 2, ..., m \\ \Phi(F_i) = \sum_i \phi(F_i) \end{aligned} \tag{1}$$

where Φ and ϕ denote the fusion and alignment of different scales' features respectively. For ϕ, it refers to the stacked three composite convolution operations consisting of convolution with 3×3 kernel and ReLU function, and upsampling to align features of different sizes by interpolating, corresponding to the Alignment in Fig. 3. The Φ refers to fusion process through element-wise addition. This operation can enhance contour information at different scales, taking into account targets of different sizes. γ denotes a convolution operation and sigmoid function. F_i represents input features of different scales. C is the output contour map.

3.2 Panoptic Segmentation Branch

This branch which consists of semantic segmentation head, instance segmentation head, and panoptic head, outputs panoptic predicted labels.

For semantic segmentation head, the deformable convolution [6] is employed, which is able to enlarge the receptive field. To be specific, we stack two 3×3 deformable convolution layers following FPN features. Similar to PSPNet [27], we first scale the features of different sizes to the same scale to achieve fusion of different granularity. Then, the semantic presentations are obtained through concatenating these scaled features. Eventually, we utilize an 1×1 convolution to perform semantic prediction on the semantic presentation.

Following the Mask R-CNN [10], our instance segmentation head produces three outputs (*i.e.*, bounding box regression, classification, and segmentation mask). The purpose of this head is to provide pixel-level annotations for each instance.

For panoptic segmentation head, we utilize the similar architecture with UPSNet [24], which fuses semantic segmentation and instance segmentation results and outputs pixel-wise panoptic annotations. Different from it, we remove the unknown prediction layer. The main purpose of this layer is to eliminate predictions for pixels that are easily ambiguous. Although the model with this layer achieves low false positive and high panoptic quality (PQ) indicator, it leads to some pixels such as contour pixels that have no significative labels. Whereas the proposed network without the unknown prediction layer realizes the prediction of all pixels, and achieves good segmentation results in contour regions.

3.3 Structure Loss Function

As there exists the structure constraint between panoptic segmentation and panoptic contour detection, we develop a novel structure loss function to enhance contour consistency for panoptic segmentation task.

As shown in Fig. 1(c), the panoptic contour is derived as the boundary of panoptic segmentation (Fig. 1(b)). The difference between panoptic prediction result and the ground truth of panoptic contour could be constructed as structure loss term to explicitly impose the structural information constraint on the results. Furthermore, the structure loss term is differentiable due to the pixel-wise operation, which is capable of achieving back propagation during training.

Specifically, the loss term focuses on minimizing the difference between ground truth of panoptic contour and the predicted result's boundary, which is obtained via Laplacian operator. Considering the distribution imbalance of contour and non-contour points, inspired by HED [23], we utilize the class-balanced sigmoid cross entropy loss, which can be expressed as:

$$L_s = -\alpha \hat{C} log(sigmoid(C)) - (1-\alpha)(1-\hat{C}) log(1 - sigmoid(C)),$$
$$\alpha = \frac{C_-}{C} \tag{2}$$

where C denotes the ground truth of panoptic contour, and $\hat{C}$ denotes the predicted panoptic contour. α is a balance factor. C_- denotes the contour ground truth label sets. At the same time, we also apply the loss term to the supervision of the panoptic contour branch, denoted as L_c.

Following Mask R-CNN [10], PSPNet [27] and UPSNet [24], our network also employs the RPN proposal loss L_{rpn} (bounding box regression and classification), instance segmentation loss L_{ins} (bounding box classification, bounding box regression, and mask segmentation), the semantic segmentation loss L_{seg} (cross entropy loss), and the panoptic loss L_{pano} (cross entropy loss).

To sum up, the total loss L_{total} is formulated as below:

$$L_{total} = L_{rpn} + L_{ins} + L_{seg} + L_{pano} + \lambda_c L_c + \lambda_s L_s, \tag{3}$$

where λ_c and λ_s are loss weights to control the balance between panoptic contour branch loss and structure loss.

4 Experiments

In this section, we evaluate our approach on the Cityscapes dataset [5]. The UPSNet without unknown prediction is used as our baseline. We compare the proposed CAPSNet with the baseline and similar models based on Mask-RCNN including UPSNet [24], TASCNet [15], AUNet [16], Panoptic FPN [12] and OCFusion [14]. Furthermore, the extensive ablative analysis is provided to verify the effectiveness of the panoptic contour branch and structure loss.

4.1 Dataset

Cityscapes [5]**:** This dataset contains 5000 images, of which 2975 for training, 500 for validation. Notice that the panoptic contour ground truth utilized to supervise the proposed panoptic contour branch is obtained by calculating the gradient of the panoptic ground truth label. Therefore, in our experiments, we exploit training set for training and validation set for testing.

4.2 Evaluation Metrics

Following He *et al.* [13], the Panoptic Quality(PQ) is adopted as summary metric:

$$PQ = \underbrace{\frac{\sum_{(p,g)\in TP} IoU(p,g)}{|TP|}}_{segmentation\ quality(SQ)} \times \underbrace{\frac{|TP|}{|TP| + \frac{1}{2}|FP| + \frac{1}{2}|FN|}}_{recognition\ quality(RQ)}, \tag{4}$$

which is composed of two parts, segmentation quality and recognition quality. The p and g in the formula denote matched predicted and ground truth respectively, and TP, FP, FN represent true positives, false positives, and false negatives, respectively. When this IoU is greater than 0.5, it is regarded as a positive detection. Note that PQ^{th} and PQ^{st} denote the PQ averaged over *thing* classes and *stuff* classes.

4.3 Implementation Details

All of experiments are implemented based on PyTorch platform and trained using 4 TITAN V GPUs. For our experiments, models are trained in a batch size of 1 per GPU, learning rate of 0.02 and weight decay of $1e^{-4}$ for 48k steps in total and decay the learning rate by a factor of 0.1 at step 36k. All models employ ResNet-50 pre-trained on ImageNet [7] as our backbone. Both the weight parameters of λ_c and λ_s are set to 0.5.

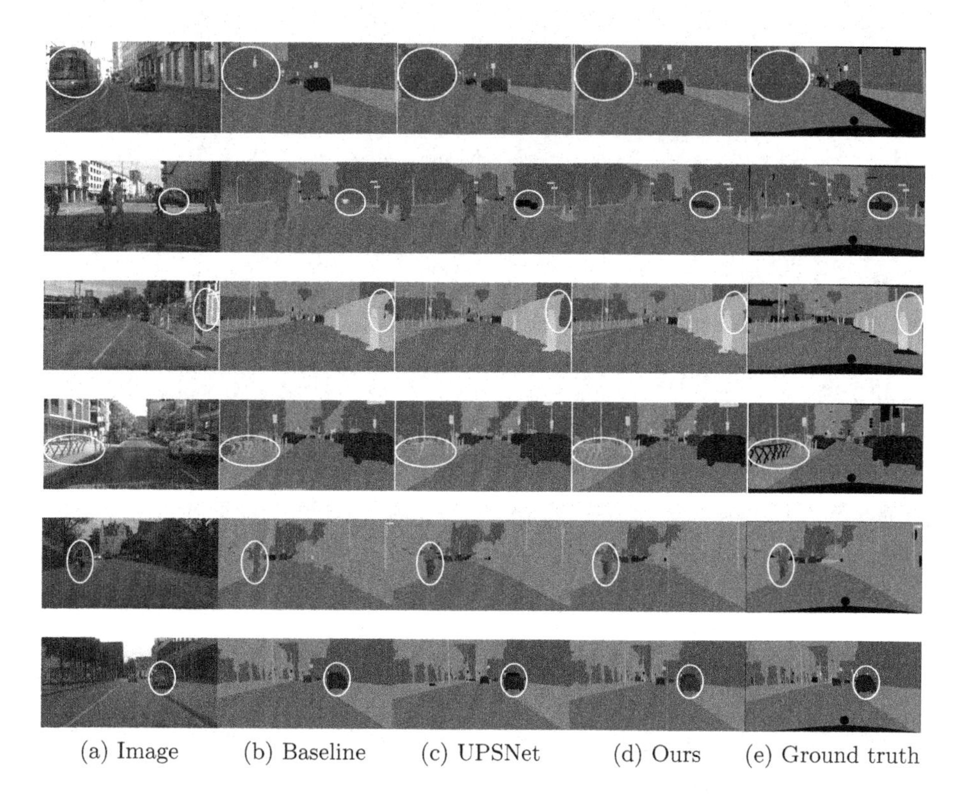

Fig. 4. Visual examples of panoptic segmentation on CityScapes.

Table 1. Performance on Cityscapes validation set.

Methods	PQ	PQ^{th}	PQ^{st}	SQ	SQ^{th}	SQ^{st}	RQ	RQ^{th}	RQ^{st}
TASCNet [15]	59.3	**56.3**	61.5	–	–	–	–	–	–
AUNet [16]	56.4	52.7	59.0	–	–	–	–	–	–
Panoptic FPN [12]	58.1	42.5	62.5	–	–	–	–	–	–
UPSNet [24]	59.3	54.6	62.7	79.7	79.3	80.1	73.0	68.7	76.2
OCFusion [14]	41.3	49.4	29.0	–	–	–	–	–	–
Baseline	59.1	54.1	62.7	80.1	79.9	**80.3**	72.4	67.5	76.0
Ours	**60.0**	55.7	**63.1**	**80.2**	**80.2**	80.2	**73.5**	**69.3**	**76.6**

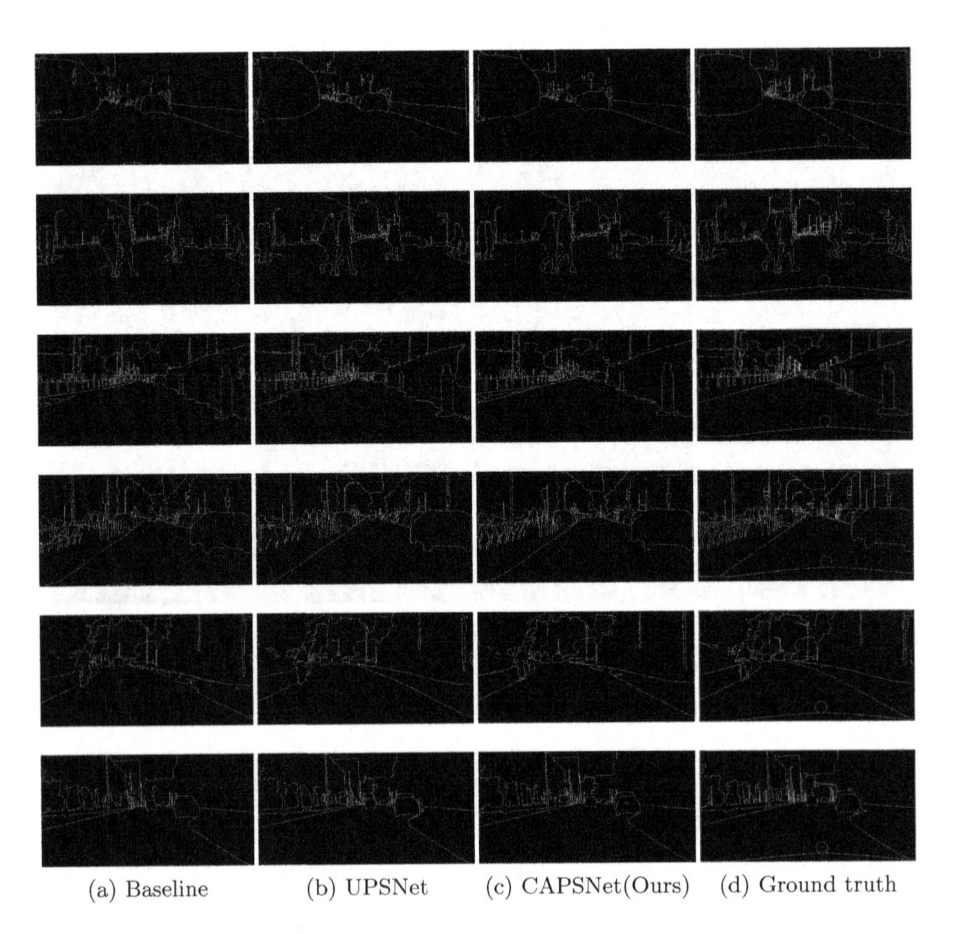

Fig. 5. Visual examples of panoptic contour on CityScapes.

4.4 Comparisons with Other Methods

The comparison of our model with other two-stage methods on Cityscapes *val* set is reported in the Table 1. In terms of PQ, our CAPSNet outperforms the baseline by 0.9%. Comparing the *thing* and *stuff* columns of PQ, SQ, and RQ between our method and baseline, it can be observed that our approach achieves higher gain on *thing* classes than *stuff* classes. For example, the proposed model realizes 1.6% in PQ^{th} and 0.4% in PQ^{st}, 0.8% in RQ^{th} and 0.6% in RQ^{st} improvements, respectively. The main reason is that the contours of *thing* classes are more regular and distinct than *stuff* classes. This is consistent with our original intention of designing contour perception.

Figure 4 presents some visual examples of our algorithm. The first two rows show that the misclassification caused by the similarity of cars or buses with buildings due to the perspective and lighting is alleviated. The third and fourth rows show that our approach effectively improves the performance in the wall and skeletonized regions, presenting a better structure. The fifth row implies that a more refined contour on the bicycle is achieved and the sixth row indicates that our method is able to distinguish two similar overlapping objects thanks for contour knowledge.

In addition, the contours of our predicted results are illustrated in Fig. 5. Note that the image of each row in this figure refers to the image of the corresponding row in the above Fig. 4. It is easy to observe that the contours of the predicted results are more consistent with ground truth, and the qualitative performance boost is in line with the above quantitative analysis that contour perception improves the distinction of objects, especially for *thing* classes.

4.5 Ablative Analysis

To demonstrate the effectiveness of each component in our network, we conduct related ablation experiments. Table 2 shows the quantitative ablative analysis, where empty cells mean the corresponding components are not adopted. The second row (panoptic contour branch on ResNet) and third row are the results of the model adding panoptic contour branch after ResNet and FPN, respectively. The results indicate that our panoptic contour branch is able to improve the PQ indicator by 0.5% on ResNet and 0.4% on FPN respectively due to the reason that feature extraction guided by contour information is more sensitive to structural information. Figure 6 exhibits the feature maps of UPSNet and our method. It shows that UPSNet focuses on assigning a high response value to the region, whereas our method expertise to capture contour or structural information.

Comparing the first row and fourth row, the model with structure loss performs better than baseline by 0.5% in PQ metric, which proves the effectiveness of designed loss. This is mainly due to the loss function that explicitly supervises the segmentation results through incorporating corresponding relationship between panoptic segmentation and contour.

Table 2. Ablation study on Cityscapes dataset.

Panoptic contour branch		Structure loss	PQ	PQ^{th}	PQ^{st}	SQ	SQ^{th}	SQ^{st}	RQ	RQ^{th}	RQ^{st}
ResNet	FPN										
			59.1	54.1	62.7	80.1	79.9	**80.3**	72.4	67.5	76.0
✓			59.6	54.6	**63.2**	80.1	80.0	80.2	73.1	68.0	**76.8**
	✓		59.5	55.6	62.4	79.7	79.6	79.7	73.3	**69.6**	75.9
		✓	59.6	55.3	62.7	79.8	79.7	79.9	73.3	69.2	76.2
✓		✓	**60.0**	**55.7**	63.1	**80.2**	**80.2**	80.2	**73.5**	69.3	76.6

The values of PQ^{th} and PQ^{st}, RQ^{th} and RQ^{st} show that the improvement on *things* is more significant than on *stuff*, which indicates that our loss is more effective for the more regular contours of *things*.

However, the exploitation of this loss degrades the accuracy of SQ. The main reasons are that SQ evaluates the result of semantic segmentation and loss resorts to explicit and strong constraint benefiting instance segmentation, which will inevitably affect the segmentation between different overlapping objects with same category. As is shown in the sixth row, the best results are achieved in indicators such as PQ, SQ, and RQ when the panoptic contour branch and structure loss are utilized simultaneously.

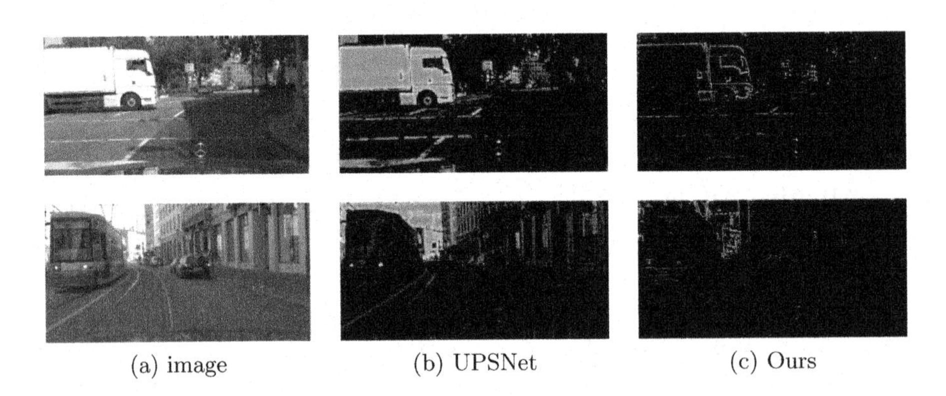

(a) image (b) UPSNet (c) Ours

Fig. 6. Feature maps of UPSNet and ours.

5 Conclusion

In this paper, we put forward the CAPSNet which incorporates contour cues for panoptic segmentation task. It enforces the structural information by introducing a panoptic contour branch and a structure loss function. The panoptic contour branch embraces structural information during feature extraction, and the structure loss function utilizes the constraint between panoptic segmentation and contour. Finally, experimental results on Cityscapes dataset demonstrate the effectiveness of our network. In future work, we would like to explore the unsupervised contour-aware panoptic segmentation network.

Acknowledgments. This work was supported by Shanghai Sailing Program (19YF1456000), Shanghai Municipal Science and Technology Major Project (ZHANGJIANG LAB) under Grant (2018SHZDZX01), Shanghai Post-doctoral Excellence Program (2020483) and Youth Innovation Promotion Association, Chinese Academy of Sciences (2018270, 2021233).

We would also like to thank Jiafei Song and Yuan Yuan for their support during the development of this work.

References

1. Arnab, A., Torr, P.H.: Pixelwise instance segmentation with a dynamically instantiated network. In: Proceedings of the IEEE Conference on Computer Vision and Pattern Recognition, pp. 441–450 (2017)
2. Caruana, R.: Multitask learning. Mach. Learn. **28**(1), 41–75 (1997)
3. Chang, C.Y., Chang, S.E., Hsiao, P.Y., Fu, L.C.: EPSNet: efficient panoptic segmentation network with cross-layer attention fusion. In: Proceedings of the Asian Conference on Computer Vision (2020)
4. Cheng, B., et al.: Panoptic-DeepLab. arXiv preprint arXiv:1910.04751 (2019)
5. Cordts, M., et al.: The cityscapes dataset for semantic urban scene understanding. In: Proceedings of the IEEE Conference on Computer Vision and Pattern Recognition, pp. 3213–3223 (2016)
6. Dai, J., et al.: Deformable convolutional networks. In: Proceedings of the IEEE International Conference on Computer Vision, pp. 764–773 (2017)
7. Deng, J., Dong, W., Socher, R., Li, L.J., Li, K., Fei-Fei, L.: ImageNet: a large-scale hierarchical image database. In: 2009 IEEE Conference on Computer Vision and Pattern Recognition, pp. 248–255. IEEE (2009)
8. de Geus, D., Meletis, P., Dubbelman, G.: Panoptic segmentation with a joint semantic and instance segmentation network. arXiv preprint arXiv:1809.02110 (2018)
9. de Geus, D., Meletis, P., Dubbelman, G.: Fast panoptic segmentation network. IEEE Rob. Autom. Lett. **5**(2), 1742–1749 (2020)
10. He, K., Gkioxari, G., Dollár, P., Girshick, R.: Mask R-CNN. In: Proceedings of the IEEE International Conference on Computer Vision, pp. 2961–2969 (2017)
11. He, K., Zhang, X., Ren, S., Sun, J.: Deep residual learning for image recognition. In: Proceedings of the IEEE Conference on Computer Vision and Pattern Recognition, pp. 770–778 (2016)
12. Kirillov, A., Girshick, R., He, K., Dollár, P.: Panoptic feature pyramid networks. In: Proceedings of the IEEE/CVF Conference on Computer Vision and Pattern Recognition, pp. 6399–6408 (2019)
13. Kirillov, A., He, K., Girshick, R., Rother, C., Dollár, P.: Panoptic segmentation. In: Proceedings of the IEEE/CVF Conference on Computer Vision and Pattern Recognition, pp. 9404–9413 (2019)
14. Lazarow, J., Lee, K., Shi, K., Tu, Z.: Learning instance occlusion for panoptic segmentation. In: Proceedings of the IEEE/CVF Conference on Computer Vision and Pattern Recognition, pp. 10720–10729 (2020)

15. Li, J., Raventos, A., Bhargava, A., Tagawa, T., Gaidon, A.: Learning to fuse things and stuff. arXiv preprint arXiv:1812.01192 (2018)
16. Li, Y., et al.: Attention-guided unified network for panoptic segmentation. In: Proceedings of the IEEE Conference on Computer Vision and Pattern Recognition. pp. 7026–7035 (2019)
17. Lin, T.Y., Dollár, P., Girshick, R., He, K., Hariharan, B., Belongie, S.: Feature pyramid networks for object detection. In: Proceedings of the IEEE Conference on Computer Vision and Pattern Recognition, pp. 2117–2125 (2017)
18. Liu, H., et al.: An end-to-end network for panoptic segmentation. In: Proceedings of the IEEE/CVF Conference on Computer Vision and Pattern Recognition, pp. 6172–6181 (2019)
19. Long, J., Shelhamer, E., Darrell, T.: Fully convolutional networks for semantic segmentation. In: Proceedings of the IEEE Conference on Computer Vision and Pattern Recognition, pp. 3431–3440 (2015)
20. Takikawa, T., Acuna, D., Jampani, V., Fidler, S.: Gated-SCNN: gated shape CNNs for semantic segmentation. In: Proceedings of the IEEE/CVF International Conference on Computer Vision, pp. 5229–5238 (2019)
21. Vaswani, A., et al.: Attention is all you need. arXiv preprint arXiv:1706.03762 (2017)
22. Wang, H., Zhu, Y., Green, B., Adam, H., Yuille, A., Chen, L.-C.: Axial-DeepLab: stand-alone axial-attention for panoptic segmentation. In: Vedaldi, A., Bischof, H., Brox, T., Frahm, J.-M. (eds.) ECCV 2020. LNCS, vol. 12349, pp. 108–126. Springer, Cham (2020). https://doi.org/10.1007/978-3-030-58548-8_7
23. Xie, S., Tu, Z.: Holistically-nested edge detection. In: Proceedings of the IEEE International Conference on Computer Vision, pp. 1395–1403 (2015)
24. Xiong, Y., et al.: UPSNet: a unified panoptic segmentation network. In: Proceedings of the IEEE/CVF Conference on Computer Vision and Pattern Recognition, pp. 8818–8826 (2019)
25. Yang, T.J., et al.: DeeperLab: single-shot image parser. arXiv preprint arXiv:1902.05093 (2019)
26. Yu, Z., Feng, C., Liu, M.Y., Ramalingam, S.: CASENet: deep category-aware semantic edge detection. In: Proceedings of the IEEE Conference on Computer Vision and Pattern Recognition, pp. 5964–5973 (2017)
27. Zhao, H., Shi, J., Qi, X., Wang, X., Jia, J.: Pyramid scene parsing network. In: Proceedings of the IEEE Conference on Computer Vision and Pattern Recognition, pp. 2881–2890 (2017)
28. Zhen, M., et al.: Joint semantic segmentation and boundary detection using iterative pyramid contexts. In: Proceedings of the IEEE/CVF Conference on Computer Vision and Pattern Recognition, pp. 13666–13675 (2020)
29. Zhou, H., Friedman, H.S., Von Der Heydt, R.: Coding of border ownership in monkey visual cortex. J. Neurosci. **20**(17), 6594–6611 (2000)
30. Zhu, X., Su, W., Lu, L., Li, B., Wang, X., Dai, J.: Deformable DETR: deformable transformers for end-to-end object detection. arXiv preprint arXiv:2010.04159 (2020)

VGG-CAE: Unsupervised Visual Place Recognition Using VGG16-Based Convolutional Autoencoder

Zhenyu Xu[1,2], Qieshi Zhang[1,2,3](✉), Fusheng Hao[1,3], Ziliang Ren[1,3], Yuhang Kang[1,3], and Jun Cheng[1,2,3]

[1] Shenzhen Institute of Advanced Technology, Chinese Academy of Sciences, Shenzhen, China
qs.zhang@siat.ac.cn
[2] Shenzhen College of Advanced Technology, University of Chinese Academy of Sciences, Beijing, China
[3] The Chinese University of Hong Kong, Hong Kong, China

Abstract. Visual Place Recognition (VPR) is a challenging task in Visual Simultaneous Localization and Mapping (VSLAM), which expects to find out paired images corresponding to the same place in different conditions. Although most methods based on Convolutional Neural Network (CNN) perform well, they require a large number of annotated images for supervised training, which is time and energy consuming. Thus, to train the CNN in an unsupervised way and achieve better performance, we propose a new place recognition method in this paper. We design a VGG16-based Convolutional Autoencoder (VGG-CAE), which uses the features outputted by VGG16 as the label of images. In this case, VGG-CAE learns the latent representation from the label of images and improves the robustness against appearance and viewpoint variation. When deploying VGG-CAE, features are extracted from query images and reference images with post-processing, the Cosine similarities of features are calculated respectively and a matrix for feature matching is formed accordingly. To verify the performance of our method, we conducted experiments with several public datasets, showing our method achieves competitive results comparing to existing approaches.

Keywords: Visual place recognition · Convolutional autoencoder · VGG-16

1 Introduction

Visual Place Recognition (VPR) plays a critical role in the localization of mobile robots [18] by matching query images to reference images indicating the same place [20,23,36]. State-of-the-art VPR technology performs well when the query image is significantly different from the reference image caused by viewpoint and appearance variation. Initially, methods based on handcrafted feature [11,22,30]

H. Ma et al. (Eds.): PRCV 2021, LNCS 13020, pp. 91–102, 2021.
https://doi.org/10.1007/978-3-030-88007-1_8

achieve decent performance, combining with sequence information and Bag-of-Visual-Words (BoVW) [5,10,12,26,27,29]. However, they are not robust enough when the environment vastly changes.

Recent methods based on CNNs are more robust than the previous ones [6,8,15,34]. There exist two kinds of approaches to use CNN-based features: (a) use Siamese network or Triple network to fine-tune a pre-trained model to output similarity score of paired images [14,16,28]. (b) feed a labeled image directly into the CNN to obtain a special representation for matching [7,8,21]. However, the robustness of these models is at the cost of a large amount of labeled training images and requires images from the same place under different conditions, which is laborious and hard to define the ground truth. Therefore, it is important to maintain the good performance of the model while avoiding the requirement of annotated images [24,38].

To address this issue mentioned above, the Autoencoder (AE) architecture [13] was introduced. To learn abstractive representations, researchers use a Convolutional Autoencoder (CAE) with extra optimization constraints [25]. The AE and CAE models, albeit performing well in the case of extreme changing conditions, still have unsatisfactory recognition accuracy compared to the features of traditional deep networks such as VGG16 [32].

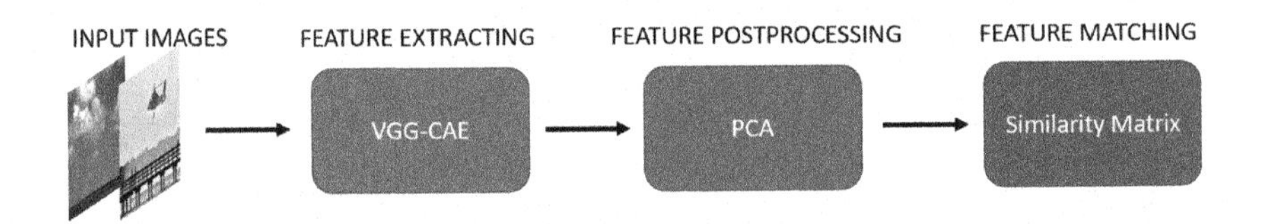

Fig. 1. Our brief architecture for recognition

To address the challenges mentioned above, we integrate VGG and CAE as a novel model, VGG-CAE, by exploiting their merits, that VGG16 extracts abstractive features of images, and CAE further learns the compact representations. Specifically, prior works use hand-crafted features to label the images but still unable to capture inherent patterns of places. In contrast, herein, the deep learning model VGG-16 is used to extract abstractive features that are considered as the labels of the images. We further employ the CAE model simultaneously to learn a more compact feature. To achieve this, the l_2 loss function is exploited to ensure the extracted compact features of CAE comparatively represent the extracted features of VGG16. As such, our proposed model is able to train without annotated images, capture the robust representation of images, and achieve the comparative performance of the VGG16 model. Our proposed model does not require labels and instead exploits the intermediate features of CAE to replace images. Therefore, the VGG-CAE realizes the unsupervised visual place recognition under complex environmental conditions.

Our main contributions are as follows:

- We propose a novel model, VGG-CAE, that could extract compact features from features of VGG16 in an unsupervised way.
- We adopt principal component analysis to post-process the features of VGG-CAE, improving the recognition performance in a large margin.
- On different datasets with challenging appearance and viewpoint changes, our method is compared with the state-of-the-art methods, achieving decent performance.

2 Realted Work

2.1 Handcraft-Based Methods

Commonly, the handcraft-based methods are applied to the technology like the Fast Appearance-based Mapping (FAB-MAP) algorithm [10] and BoVW [12]. These techniques achieved excellent robustness in slight viewpoint and illumination changes while suffering from drastic environmental changes. To make up for this shortcoming, a sequence-based method was introduced. Milford used sequence matching [26], and it works well under drastic appearance changes. However, this method directly compares the values of down-sampled images, which makes it incompetent in large viewpoint changes. Recently, a novel technique [35] combines sequence information and CoHOG [37] achieves viewpoint-invariance.

2.2 CNN-Based Methods

Recently, CNN-based methods have achieved better results in environmental changes mentioned above. It uses CNN to extract features from images and the features as descriptors for image matching. Chen introduced this kind of method firstly [8]. They used the features extracted by Overfeat network [31] as a global descriptor and got a better place recognition effect than hand-crafted features. In order to improve the precision in VPR, multiple methods were proposed with optimization constraints. Sünderhauf divided an image into multiple salient areas and extracts several features as a descriptor and named those local areas as landmarks [34]. However, the original image contains many landmarks with low information content. Sem [15], a landmark with more semantic information, was proposed to make the detected landmarks more distinguishable. When matching images, geometry-preserving landmark pairs provide spatial constraints and remove incorrect matches. The methods above produce a large-dimensional descriptor requiring a long time for computing.

To address the time-consuming issue, Kenshimov proposed a filter to reduce some irrelevant layers in the feature maps while improving cross-seasonal place recognition [16]. Besides, another method applied pre-training CNN, Siamese and triplet network was proposed to adapt the drastic appearance changes caused by the turn of seasons [28]. Considering the methods above, Hausler tries to combine the filter with the Siamese network and triplet network [14], and the result showed that the method reduces a large number of the feature map, while maintaining the same effect as the original. Instead of using a generic CNN, Training a specific network for place recognition was introduced in [7] for the first time. Similarly, Lopez-Antequera proposed a method using a triplet network to accelerate computation without any dimensionality reduction techniques [21]. These networks are capable of extracting features, however, they require a large number of labeled images for training.

2.3 AE-Based Methods

To solve the problem mentioned above, Gao introduced the Autoencoder (AE) architecture in [13]. The internal representation learned by the AE is compact and competent to represent the original image as a feature. To learn a composite internal representation, the CAE architecture was introduced in [25]. They warp the input images by randomized projective transformation and use Histograms of Oriented Gradients (HOG) as an extra optimization constraint, making their CAE reconstruct a HOG descriptor instead of an image.

In this paper, we combine the CAE architecture with a deep-learning network VGG16 [32], utilizing the features of VGG16 as the label for training. Our model can achieve an equal level of precision to VGG16, while greatly reducing the extraction time and improving the robustness to environment changes.

3 VGG16-Based Convolutional Autoencoder

This section provides the details of our VGG-CAE network architecture. We employ the features of VGG16-fc2 as the label of images while training and utilize the coding ability of CAE to learn compact features. Through computing the Cosine similarity among the images, we find the best match for each query image. Results show that the VGG-CAE features could achieve a comparative performance to the one achieved by VGG16-fc2.

3.1 Model Architecture

Our model architecture is shown in Fig. 2. The training images are scaled to RGB images, I_m, size $224 \times 224 \times 3$. Then they are extracted by VGG16 as the labels of the training images, V_1. At the same time, I_m are fed into our CAE to get the vector, V_2. To ensure the V_2 comparatively represents the V_1, we use l_2 loss function to reduce the difference between V_1 and V_2. Note that VGG16 has the weights trained on ImageNet [3], and its second fully-connected layer, VGG16-fc2, is used to produce V_1.

Our CAE includes two parts, an encoder and a decoder. The encoder comprises four convolution and max-pooling paired layers. The decoder comprises two fully-connected layers, which are similar to upsampling layers and transposed convolutional layers. By using fully-connected layers, we can shape the

V_2 similar to V_1 easily without computing the layer size specially. As the number of convolutional network layers increases, to avoid overfitting, we add the l_2 regularization to all convolutional layers.

In this case, the encoder has the ability to compress the I_m into the internal representation that contains abstractive and compact information to reconstruct the V_2. Note that when V_1 and V_2 are similar enough, the internal representation is a compressed product of V_1. Therefore, our model exploits the internal representation, instead of V_2, as a compact feature for matching.

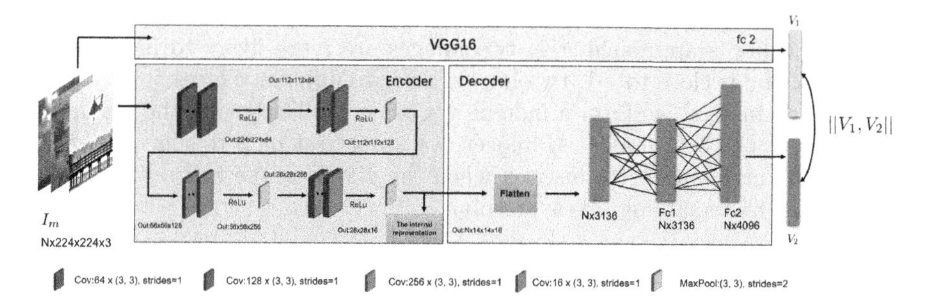

Fig. 2. Model architecture

3.2 Training

Due to the unsupervised characteristics of our model, training does not require a large number of labeled images, so we focus on the choice of training images types. Considering that the training dataset should include various places that makes the model learn environmental changes, we choose Places365-Standard [39] as the training dataset. The dataset contains 1.8 million images from 365 different scenes and each scene is provided with 5000 images. Consequently, the latent representations in the same scene are extracted by our model with a strong generalization ability.

We use Keras [2] to train the network for its efficiency. The Adam optimizer adopted a fixed learning rate of 1×10^{-4}, and the training process took about 8 epochs. Also, we use EarlyStopping to avoid overfitting, which stops training when the absolute difference of validation loss and training loss is less than 0.3.

3.3 Matching

After training, we use VGG-CAE for place recognition as shown in Fig. 3. When performing feature matching, we remove the decoder and only use the encoder, which is lightweight with a computation resource of the only 88MB, about 30% of the VGG16 when deploying. References images and query images are scaled to size $N \times 224 \times 224 \times 3$, mapping to $N \times 3136$ vectors, R_i and Q_i. By observing the vectors, we found that they contain many zero values, which implies that these vectors are redundant for recognition. So we use Principal Component Analysis (PCA) to map R_i and Q_i to $N \times 128$ vectors, which also improves the matching performance. Then, the similarity between Q_i and R_i is computed.

To compare whether two images refer to the same place, the traditional method for computing similarity between vectors mostly is Cosine similarity, that is defined as

$$c = \frac{\sum_{i=1}^{n}(Q_i \times R_i)}{\sqrt{\sum_{i=1}^{n}(Q_i)^2} \times \sqrt{\sum_{i=1}^{n}(R_i)^2}} = \frac{Q_i \bullet R_i}{\| Q_i \| \| R_i \|}. \quad (1)$$

When the value is approaching 1, two images are more likely to indicate the same place. If c is close to -1, two images refer to different places.

To match images, we form a matrix of Cosine similarity, and the maximum of each row is the best match. Moreover, we use a heat map to show the result directly. The matching performs well when the pixels on the main diagonal are green, while the remaining pixels are blue.

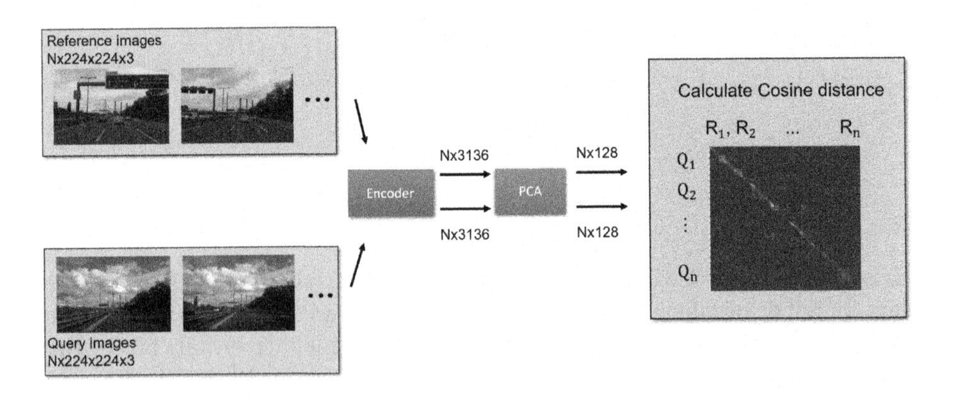

Fig. 3. Workflow of place recognition (Color figure online)

4 Experiments

4.1 Datasets

We select some common datasets listed in Table 1 as benchmarks to comprehensively evaluate the performance of our model. Figure 4 shows several typical examples of evaluated images.

Gardens Point Dataset: The Gardens Point dataset [1] contains moderate viewpoint and strong illumination variations, recorded on the Gardens Point Campus of the Queensland University of Technology. The dataset includes three sequences for the same route: two during the day and one during the night. Two day routes one was recorded on the left side, while the other on the right side. For night one, it provides large appearance variations recorded on the right side. To evaluate the appearance and viewpoint robustness of our method, we choose the route recorded in the daytime along the left side as query images and the other one from the night right side as reference images.

Berlin Dataset: The Berlin A100 and the Berlin Kudamm datasets are firstly introduced in [9]. Recording from the perspectives of bike, car, and bus, they contain more significant viewpoint changes than the garden point dataset. Additionally, each paired image contains the same landmarks with some dynamic objects, which brings the challenge to recognize places with occlusion and perceptual aliasing. Therefore, We choose them to evaluate the robustness against severe viewpoint and appearance.

Table 1. Test dataset description

Dataset	Environment	Appearance	Viewpoint	Qurey	Reference
Gardens Point [1]	Campus	Severe	Moderate	200 (day-left)	200 (night-right)
Berlin A100 [9]	Urban	Moderate	Severe	81 (car)	85 (car)
Berlin Kudamm [9]	Urban	Moderate	Severe	222 (bike)	201 (bus)

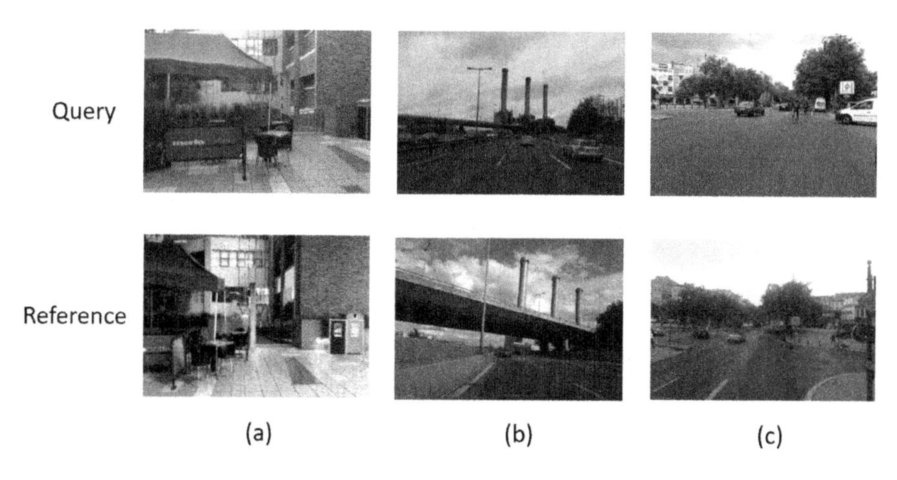

Fig. 4. Examples of dataset images: (a) Gardens Point, (b) BerlinA100 and (c) Berlin Kudamm dataset.

4.2 State-of-the-Art Approaches

Our method exploits CNN features for further encoding and obtains an image representation, which is different from existing methods. Therefore, we choose various methods to compare with as follows:

- **HOG** [11]: a handcraft-based feature is robust to illumination changes. This comparison is to prove that the VGG-CAE features are more robust than the ones extracted by the HOG in illumination changes.
- **FAB-MAP** [10]: a feature-based method was implemented by [17].
- **Seq-SLAM** [26]: a sequenced-based method was implemented by [17], which makes full use of sequence information.
- **Sum pool** [4]: a famous method encodes deep features to produce global descriptors.
- **Cross pool** [19]: a method expriots the convolutional layer activations as the image representation by using cross-convolutional-layer pooling.
- **SPP HybirdNet** [7]: a specific model for place recognition was implemented by [17], which employed SPP on conv5 of the model.
- **Cross-Region-VLAD** [17]: a novel method based on Regions Of Interest (ROI), which is available significant viewpoint and appearance changes.
- **VGG16** [32]: research shows that in the deep neural networks, the features extracted by the highest layer are robust to viewpoint variation [33]. Additionally, considering the label of images, we choose 4096-dimensional vectors of VGG16-fc2 to compare with our methods.

4.3 Ground Truth

Since sequential images of the same place are continuous, we add a frame tolerance to define the ground truth in the experiment. The frame tolerance is described by

$$| frame_{query} - frame_{search} | < 4. \tag{2}$$

In our experiment, a match can be retrieved once the Cosine similarity is greater than a threshold and a match is correct if the sequence index of the retrieved image is within 3.

4.4 Comparison and Discussion

We choose the Area Under the precision-recall Curve (AUC) for the evaluation. Precision is defined as the number of correct matches over the total number of matches retrieved by the place recognition algorithm and recall is defined as the number of correct matches over the total number of correct matches and the matches that were not retrieved but should have been. By observing the AUC value, we can better understand the generalization ability of the model. The higher the AUC is, the better the model performs. The performance of various methods and the related AUC values are visually shown in Fig. 5.

Gardens Point Dataset: Since the dataset contains strong temporal correlation, Seq-SLAM takes the highest value by utilizing the sequence information. However, note that our method just extracts features without any image sequence information and outperforms others, which indicates our method is available for these drastically changing conditions. The Cross-Region-VLAD method achieves close performance to ours, outperforming others.

Berlin A100 Dataset: This dataset mainly presents severe viewpoint changes and dynamic objects. The VGG-CAE showed the highest AUC, followed by the Cross-Region-VLAD, indicating the significance of using CAE architecture for improving the robustness of CNN-based features. While SPP HybirdNet suffers in this case. This may be caused by strong dynamic object occlusion, which affects HybridNet's ability to recognize salient buildings.

Berlin Kudamm Dataset: This dataset is collected from two routes: bike and bus, causing severe viewpoint changes along with different degrees of illumination changes and dynamic occlusion. It is the most challenging dataset mentioned in this paper. However, our method achieved the best results under this condition, which was 0.12 AUC more than the second-best results. This confirms that our model can extract more abstract features to represent images without being affected by changes in viewpoint and dynamic objects.

On the whole, even though the CNN-based methods achieve good results, the output features still have a certain degree of redundancy, which may be the reason why PCA has significantly improved our features. This may give the inspiration to check whether the output of the deep model is a sparse feature. If features are sparse, there exists a lot of redundant information in features, and adopting dimensionality reduction methods can be used to not only reduce storage but also improve the performance of place recognition.

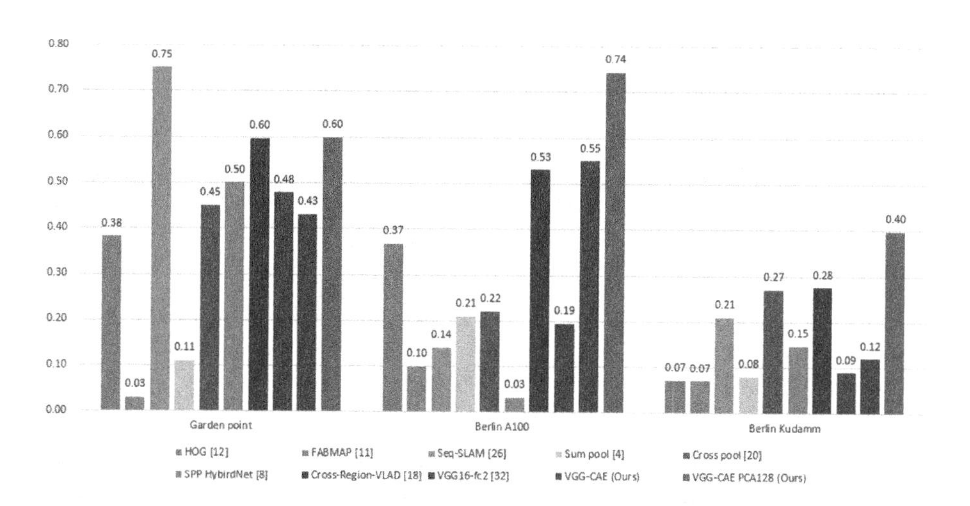

Fig. 5. The AUC of state-of-art methods

5 Conclusion

In this paper, to improve the precision of AE and CAE models, we proposed a novel model, VGG-CAE, that is available with unsupervised learning for VPR.

The abstractive features are extracted by deep network VGG16 as the labels of the input images and the CAE is trained to learn a more compact feature. Additionally, our model does not require labeled images and exploits CAE to replace images with the intermediate features for matching. Therefore, our VGG-CAE can perform robustly in extreme viewpoint and appearance variations.

Acknowledgments. This work was supported by the National Natural Science Foundation of China (nos. U1913202, U1813205, U1713213, 61772508), CAS Key Technology Talent Program, Shenzhen Technology Project (nos. JCYJ20180507182610734, JSGG20191129094012321)

References

1. Glover, A.: Gardens point walking (2014). https://wiki.qut.edu.au/display/raq/day+and+night+with+lateral+pose+change+datasets
2. Chollet, F., et al.: Keras (2015). https://keras.io/
3. ImageNet, an image database organized according to the wordnet hierarchy. http://www.image-net.org/
4. Babenko, A., Lempitsky, V.: Aggregating local deep features for image retrieval. In: IEEE International Conference on Computer Vision (ICCV), pp. 1269–1277 (2015)
5. Bampis, L., Amanatiadis, A., Gasteratos, A.: Fast loop-closure detection using visual-word-vectors from image sequences. Int. J. Robot. Res. (IJRR) **37**(1), 62–82 (2018)
6. Camara, L.G., Gäbert, C., Přeučil, L.: Highly robust visual place recognition through spatial matching of CNN features. In: IEEE International Conference on Robotics and Automation (ICRA), pp. 3748–3755 (2020)
7. Chen, Z., et al.: Deep learning features at scale for visual place recognition. In: 2017 IEEE International Conference on Robotics and Automation (ICRA), pp. 3223–3230 (2017)
8. Chen, Z., Lam, O., Jacobson, A., Milford, M.: Convolutional neural network-based place recognition. arXiv preprint arXiv:1411.1509 (2014)
9. Chen, Z., Maffra, F., Sa, I., Chli, M.: Only look once, mining distinctive landmarks from convnet for visual place recognition. In: IEEE/RSJ International Conference on Intelligent Robots and Systems (IROS), pp. 9–16 (2017)
10. Cummins, M., Newman, P.: FAB-MAP: probabilistic localization and mapping in the space of appearance. Int. J. Robot. Res. (IJRR) **27**(6), 647–665 (2008)
11. Dalal, N., Triggs, B.: Histograms of oriented gradients for human detection. In: IEEE Computer Society Conference on Computer Vision and Pattern Recognition (CVPR), vol. 1, pp. 886–893 (2005)
12. Gálvez-López, D., Tardos, J.D.: Bags of binary words for fast place recognition in image sequences. IEEE Trans. Robot. (TRO) **28**(5), 1188–1197 (2012)
13. Gao, X., Zhang, T.: Unsupervised learning to detect loops using deep neural networks for visual SLAM system. Auton. Robot. **41**(1), 1–18 (2015). https://doi.org/10.1007/s10514-015-9516-2
14. Hausler, S., Jacobson, A., Milford, M.: Feature map filtering: improving visual place recognition with convolutional calibration. arXiv preprint arXiv:1810.12465 (2018)

15. Hou, Y., Zhang, H., Zhou, S., Zou, H.: Use of roadway scene semantic information and geometry-preserving landmark pairs to improve visual place recognition in changing environments. IEEE Access **5**, 7702–7713 (2017)
16. Kenshimov, C., Bampis, L., Amirgaliyev, B., Arslanov, M., Gasteratos, A.: Deep learning features exception for cross-season visual place recognition. Pattern Recognit. Lett. (PRL) **100**, 124–130 (2017)
17. Khaliq, A., Ehsan, S., Chen, Z., Milford, M., McDonald-Maier, K.: A holistic visual place recognition approach using lightweight CNNs for significant viewpoint and appearance changes. IEEE Trans. Robot. (TRO) **36**(2), 561–569 (2019)
18. Labbe, M., Michaud, F.: Online global loop closure detection for large-scale multi-session graph-based SLAM. In: IEEE International Conference on Intelligent Robots and Systems (IROS), pp. 2661–2666 (2014)
19. Liu, L., Shen, C., van den Hengel, A.: The treasure beneath convolutional layers: cross-convolutional-layer pooling for image classification. In: IEEE Conference on Computer Vision and Pattern Recognition (CVPR), pp. 4749–4757 (2015)
20. Liu, Y., Xiang, R., Zhang, Q., Ren, Z., Cheng, J.: Loop closure detection based on improved hybrid deep learning architecture. In: IEEE International Conferences on Ubiquitous Computing & Communications (IUCC) and Data Science and Computational Intelligence (DSCI) and Smart Computing, Networking and Services (SmartCNS), pp. 312–317 (2019)
21. Lopez-Antequera, M., Gomez-Ojeda, R., Petkov, N., Gonzalez-Jimenez, J.: Appearance-invariant place recognition by discriminatively training a convolutional neural network. Pattern Recogn. Lett. (PRL) **92**, 89–95 (2017)
22. Lowe, D.G.: Distinctive image features from scale-invariant keypoints. Int. J. Comput. Vis. (IJCV) **60**(2), 91–110 (2004)
23. Lowry, S., et al.: Visual place recognition: a survey. IEEE Trans. Robot. (TRO) **32**(1), 1–19 (2015)
24. Maffra, F., Teixeira, L., Chen, Z., Chli, M.: Real-time wide-baseline place recognition using depth completion. IEEE Robot. Autom. Lett. (RAL) **4**(2), 1525–1532 (2019)
25. Merrill, N., Huang, G.: Lightweight unsupervised deep loop closure. arXiv preprint arXiv:1805.07703 (2018)
26. Milford, M.J., Wyeth, G.F.: SeqSLAM: visual route-based navigation for sunny summer days and stormy winter nights. In: IEEE International Conference on Robotics and Automation (ICRA), pp. 1643–1649 (2012)
27. Naseer, T., Ruhnke, M., Stachniss, C., Spinello, L., Burgard, W.: Robust visual SLAM across seasons. In: IEEE International Conference on Intelligent Robots and Systems (IROS), pp. 2529–2535 (2015)
28. Olid, D., Fácil, J.M., Civera, J.: Single-view place recognition under seasonal changes. arXiv preprint arXiv:1808.06516 (2018)
29. Pepperell, E., Corke, P.I., Milford, M.J.: All-environment visual place recognition with smart. In: IEEE International Conference on Robotics and Automation (ICRA), pp. 1612–1618 (2014)
30. Rublee, E., Rabaud, V., Konolige, K., Bradski, G.: ORB: an efficient alternative to SIFT or SURF. In: IEEE International Conference on Computer Vision (ICCV), pp. 2564–2571 (2011)
31. Sermanet, P., Eigen, D., Zhang, X., Mathieu, M., Fergus, R., LeCun, Y.: Overfeat: integrated recognition, localization and detection using convolutional networks. arXiv preprint arXiv:1312.6229 (2013)

32. Simonyan, K., Zisserman, A.: Very deep convolutional networks for large-scale image recognition. In: Proceedings of the International Conference on Learning Representations, pp. 1–14 (2015)
33. Sünderhauf, N., Shirazi, S., Dayoub, F., Upcroft, B., Milford, M.: On the performance of convnet features for place recognition. In: IEEE International Conference on Intelligent Robots and Systems (IROS), pp. 4297–4304 (2015)
34. Sünderhauf, N., et al.: Place recognition with convnet landmarks: viewpoint-robust, condition-robust, training-free. Robot. Sci. Syst. (RSS) **XI**, 1–10 (2015)
35. Tomită, M.A., Zaffar, M., Milford, M., McDonald-Maier, K., Ehsan, S.: Convsequential-slam: a sequence-based, training-less visual place recognition technique for changing environments. arXiv preprint arXiv:2009.13454 (2020)
36. Xiang, R., Liu, Y., Zhang, Q., Cheng, J.: Spatial pyramid pooling based convolutional autoencoder network for loop closure detection. In: IEEE International Conference on Real-time Computing and Robotics (RCAR), pp. 714–719 (2019)
37. Zaffar, M., Ehsan, S., Milford, M., McDonald-Maier, K.: CoHOG: a light-weight, compute-efficient, and training-free visual place recognition technique for changing environments. IEEE Robot. Autom. Lett. **5**(2), 1835–1842 (2020)
38. Zaffar, M., Khaliq, A., Ehsan, S., Milford, M., Alexis, K., McDonald-Maier, K.: Are state-of-the-art visual place recognition techniques any good for aerial robotics? arXiv preprint arXiv:1904.07967 (2019)
39. Zhou, B., Lapedriza, A., Khosla, A., Oliva, A., Torralba, A.: Places: a 10 million image database for scene recognition. IEEE Trans. Pattern Anal. Mach. Intell. (TPAMI) **40**(6), 1452–1464 (2017)

Slice Sequential Network: A Lightweight Unsupervised Point Cloud Completion Network

Bofeng Chen, Jiaqi Fan, Ping Zhao, and Zhihua Wei(✉)

Tongji University, Shanghai, China
{1751386,1930795,zhaoping,zhihua_wei}@tongji.edu.cn

Abstract. The point cloud is usually sparse and incomplete in reality, and the missing region of the point cloud is coherent when it is blocked by other objects. To tackle this problem, we propose a novel light-weight unsupervised model, namely the Slice Sequential Network, for point cloud completion. Our method only generates the missing parts with high fidelity, while many previous methods output the entire point cloud and leave out some important details. Specifically, we slice the incomplete point cloud and force the model to exploit the information lying between slices. In addition, we design a new algorithm for extracting geometric information, which can extract multi-scale features of points of the point cloud to enhance the use of slices. The qualitative and quantitative experiments show that our method is more lightweight and has better performance than the existing state-of-the-art methods.

Keywords: 3D point cloud completion · Unsupervised learning

1 Introduction

3D vision is becoming increasingly popular nowadays, and significant progress has been made in 3D shape representation [11,16]. Among the various representations of 3D objects, point clouds can describe 3D objects in a delicate and straightforward way. However, due to the limitations of resolution and occlusion of other objects, the actual 3D point cloud is usually partially missing. Therefore, point cloud completion has become a hot spot in recent research, which is essential to many practical applications such as scene understanding [3] and robotic navigation [18].

Previous works usually take the incomplete point cloud as the input and output its complete form [2,4,7,13,32]. However, the overall structure of the input point cloud is not damaged in these works. When the missing ration increases, the performance can be significantly declined. There are some existing models that output only the missing part of the point cloud such as PF-Net [6], but it is

B. Chen and J. Fan—These authors contributed equally to this work and should be considered co-first authors.

H. Ma et al. (Eds.): PRCV 2021, LNCS 13020, pp. 103–114, 2021.
https://doi.org/10.1007/978-3-030-88007-1_9

too giant to be deployed in real time. By contrast, our lightweight network aims to finish the completion process in a relatively short time and achieve better completion results.

It makes sense that people usually observe objects in a certain order, which means that we do not have a leap frog observation. For instance, when we see an airplane, we first notice its front, then its wings, then its tail, or in the reverse order. That is to say, our attention usually do not jump from the front to the tail and skip the middle. Some previous researchers also put forward some feasible part learning methods by considering some typical features of an object, such as structural regularity and symmetry [19,28,30]. In our work, we imitate the way that people observe by constructing a novel sequential network. We input the incomplete point cloud in slices and progressively generate the missing slices. Our model is capable of aggregating the information of input slices, and predict slices consistent to the ground truth. More specifically, it can memorize previous slices and predict the next one which can be memorized later.

The advent of PointNet [5] makes it possible to learn global features of a point cloud in a convenient way, and its variants manage to make up for some of its weaknesses in different ways, like PointNet++ [1], DGCNN [9], etc. However, they have some intrinsic drawbacks, PointNet++ [1] is time-consuming while DGCNN [9] leaves out some geometrical information. The use of these traditional networks as encoders will reduce the quality of the generated parts in point cloud completion.

Therefore, we design a feature encoder that extracts multi-scale features of point clouds to enhance the use of slices. Inspired by PAN [10], we propose an aligned displacement vector to describe the neighboring region of the point. Since point clouds are usually not uniform, we use the nearest neighbor k to sample points in order to avoid redundant points, thereby improving the performance of local feature descriptors. We adopt several dilation rates to capture multi-scale local feature for fusion, which effectively improves the performance of the feature encoder.

The contributions of this paper can be summarized as follows:

(1) We present a novel point cloud completion network, the Slice Sequential Network, which is a sequential model that utilizes the information within the point cloud in a self-supervised way.
(2) We propose a new algorithm for extracting multi-scale geometric information. The global and local features extracted at the same time can make up for the shortcomings of the previous feature extraction methods and effectively improve the performance.
(3) Experiments demonstrate that our method is superior to existing methods in terms of both computation and performance.

2 Related Work

2.1 3D Learning

Aside from point cloud, there are other representations of 3D objects like meshes and voxels [16]. Voxel representations are easier and more regular to understand. However, voxel-based 3D convolution [11] is costly and thus not practical. In comparison, point cloud is simple but difficult to deal with. The innovative work PointNet [5] proposed a way to analyze the point cloud is widely used as the backbone to extract point features. Later, PointNet++ [1] is proposed and it extracts local feature by repeatedly using PointNet within a point cloud. However, within a sample region, the local information is still yet utilized. Then, DGCNN [9] uses graph convolution directly on the point and its k nearest neighbors. In spite of some potential weaknesses, a number of models for 3D learning [27,28] are proposed based on PointNet [5] series.

2.2 3D Completion

3D completion plays a key role in real-time applications like auto-driving, and various methods on 3D completion have been put forward. A variety of models have been put forward to do 3D completion in different backgrounds and achieve promising results respectively [21,23,25]. Recently a number of point cloud based approaches have been proposed to perform shape completion, such as PF-NET [6], which can generate the missing part with high fidelity, but it takes a lot of time because of its frequent Farthest Point Sampling [20] operations. [31] also uses two main branches to complete the completion task, which will lead to more memory overhead. In contrast, our architecture is simple but effective and it takes up less time and space than previous works.

3 Our Method

3.1 Overview

Our goal is to generate the missing region of the point cloud with fine details. And our model can be divided into the following parts: the Slicer, the Multi-scale point Encoder, the Sequential Predictor and the Shape Prediction Decoder. Figure 1 shows the overall architecture of our model.

3.2 Slicer

Before putting the incomplete point cloud into the network, we need to set a criteria to cut the incomplete point cloud into slices. Assume that the incomplete input point cloud has N points, and the number of point cloud slices is $M = \{1, 2, \cdots, m\}$. Since distinct slices store more information, there is no overlap between point cloud slices. Therefore, we use primary component analysis (PCA) to process point clouds to maximize the distinctiveness between point cloud slices

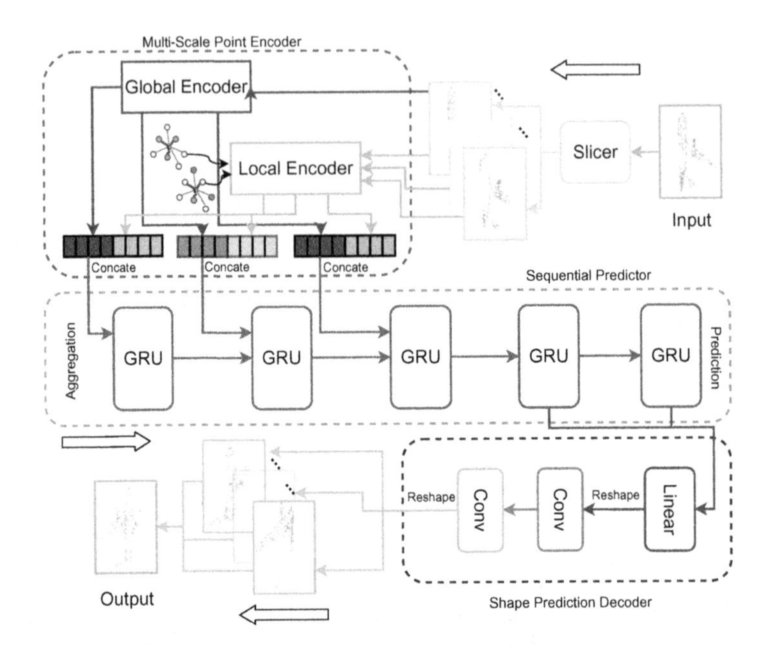

Fig. 1. Overall structure of our model. It mainly consists of the Slicer, the Multi-scale point Encoder, the Sequential Predictor and Shape Prediction Decoder.

while simplifying calculations. For a given point cloud X, let x_i denotes the i-th point of X. Let j denotes the j-th new dimension and u_j denotes the invertible matrix to get the new projection. The data projection formula is defined as follows:

$$
\begin{aligned}
J_j &= \frac{1}{N}\sum_{i=1}^{m}(x_i^T u_j - x_{center}^T u_j)^2 \\
&= \frac{1}{N} u_j^T X X^T u_j \\
&= u_j^T S u_j \qquad s.t. u_j^T u_j = 1
\end{aligned} \tag{1}
$$

where S is the covariance matrix. We use variance J_j to judge how distinct the slices are and our goal is to find the invertible matrix u_j to maximize the variance. We manage to transform 3D data into low-dimensional data, where the variance J_j reaches the maximum value.

3.3 Multi-scale Point Encoder

We propose a new multi-scale point encoder to extract global and local features in parallel. Specifically, the encoding process can be divided into several steps. First, we extract the global feature of the point cloud slice using PointNet [5].

Next, we use Atrous KNN (shown in Fig. 2) to depict the local feature, which is easy to compute. On this basis, we assign several atrous rates to get multi-scale features and use MLPs to encode them respectively. In this way, since we have considered the shape at different scales, we utilize the information of each point and its neighboring space and enhance the global features in a bottom-up manner. After that, we concatenate the global and multi-scale local feature and get it ready for the sequential predictor.

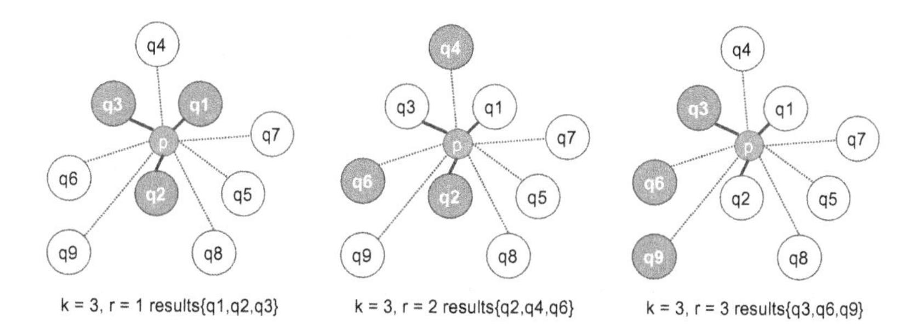

Fig. 2. The illustration of our Atrous KNN method, where k is the number of sampling points and r is the expansion rate. We use this method to obtain the displacement vector of the local feature between the center point and the sampling points.

3.4 Sequential Predictor

Our sequential predictor can predict missing slices based on current and prior input slices. It is capable of filtering and extracting the information. Usually, we use latent feature maps to describe high-level features that are difficult to interpret. Similarly, the relationship among different slices is generally unintuitive, so it is vital for our predictor to learn this relationship. To solve this problem, we use GRU, which works like a temporal-attention module, as the backbone of our decoder. For a given training sample, the input slices are $I_1, I_2, \cdots, I_m$ and the output slices are $O_1, O_2, \cdots, O_n$. There is $O_n = G(I_1, I_2, \cdots, I_m; O_1, O_2, \cdots, O_n)$, wherethe function G is determined by the gates' parameters of the GRU unit.

3.5 Shape Prediction Decoder

The shape prediction decoder can predict the overall shape and preserve geometrical characteristics of the input point cloud. To be exact, the decoder includes a full-connected layer, followed by two convolutional layers. It takes vector V from the sequential predictor as the input, and outputs the final point cloud. The fully-connected layer-based decoder only retains the final information, so we use several MLP layers to reorganize and arrange the feature. The visualized result demonstrates that our decoder can predict the overall shape and retain geometrical characteristics.

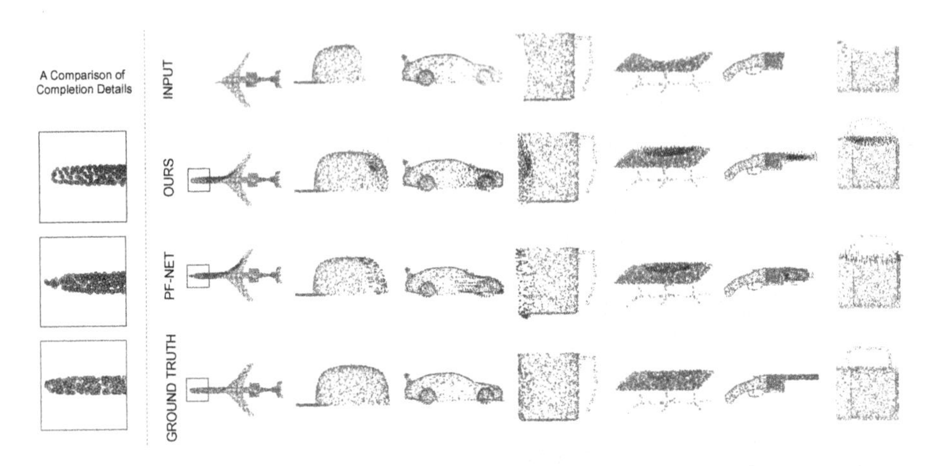

Fig. 3. The comparison between our network and PF-Net [6]. The number of points in the complete point cloud is 2048, and the ration of missing points is 25%. That is, the number of entered points is 1536, and the number of completed points is 512. The leftmost column is a comparison of the completion details of the airplane sample.

3.6 Loss Function

For 3D point cloud completion, there are two major evaluation metrics, Earth's Mover Distance and Chamfer Distance [14]. We adopt the latter one as our metrics because it uses fewer computational resources and enables the model to converge more easily.

$$CD(P,G) = \frac{1}{|P|}\sum_{p \in P} \min_{g \in G} \|p-g\|_2^2 + \frac{1}{|G|}\sum_{g \in G} \min_{p \in P} \|p-g\|_2^2 \tag{2}$$

where P denotes the predicted slices and G denotes the ground truths. CD measures the average closest point squared distance between two point clouds. Our loss function is the sum of CD between different predicted slices and ground truths. Given the number of predicted slices m, the loss function is defined as:

$$Loss = \sum_{i=1}^{m} \lambda_i CD(P_i, G_i) \tag{3}$$

where λ is a constant coefficient and is generally set to 1.

4 Experiments

4.1 Datasets and Implementation Details

In this work, we evaluate our method on ModelNet40 [19] and ShapeNet-Part [15]. For fair comparison, we choose 13 categories of different objects (Airplane, Bag, Cap, Car, Chair, Guitar, Lamp, Laptop, Motorbike, Mug, Pistol,

Skateboard, Table) in ShapeNet-Part [15], 14473 (11705 for training and 2768 for testing) objects are selected. The ground-truth point cloud data is created by sampling 2048 points on each shape. The incomplete point cloud data is generated by randomly selecting a viewpoint as a center in multiple viewpoints and removing points close to this center and we can freely control the amount of missing points. When comparing our method with other methods, we set the number of sample points to 5, dilation rate to $1, 2, 3$, the occlusion rate to 25%, that is, 512 points are missing. All our models are trained using the Adam [17] optimizer. We set initial learning rates for our model as 10^{-4} and batch size as 32. The learning rate is decayed by 0.5 after every 50 epochs. We train one single model for all categories. Figure 3 demonstrates the comparison of qualitative results among different models.

Table 1. Point cloud completion results of the missing point cloud.

Category	LGAN-AE [12]	PCN [2]	3D-Capsule [4]	PF-Net [6]	Ours
Airplane	4.487	6.303	4.077	**2.161**	2.357
Bag	11.110	7.565	9.430	7.697	**7.320**
Cap	13.576	11.255	15.779	10.090	**8.463**
Car	7.049	4.864	9.452	4.328	**4.325**
Chair	9.698	6.253	5.256	**3.898**	3.992
Guitar	1.374	2.108	1.287	0.885	**0.880**
Lamp	12.091	18.800	15.760	8.582	**7.967**
Laptop	9.062	4.492	3.862	2.234	**2.229**
Motorbike	6.950	6.884	11.325	3.981	**3.592**
Mug	11.874	7.181	10.323	6.376	**6.169**
Pistol	5.368	5.898	7.762	**2.177**	2.307
Skateboard	7.296	4.765	13.534	2.472	**2.468**
Table	5.142	4.955	7.027	**4.169**	4.365
Mean	7.998	7.021	8.837	4.543	**4.320**

4.2 Point Cloud Completion Results

We compare our method with several existing methods, including 3D point-Capsule Networks L-GAN-AE [12], PCN [2], PF-NET [6]. Table 1 shows the completion quality of the entire point cloud, by concatenating the incomplete point cloud and the generated missing region, we may compute the chamfer distance between the generated point cloud and ground-truth. Table 2 shows the completion quality of the missing region, and we directly compute the chamfer distance between the generated missing region and ground-truth. Quantitative results show our method outperforms PF-NET [6] in 9 out of 13 categories.

Table 2. Point cloud completion results of overall point cloud.

Category	LGAN-AE [12]	PCN [2]	3D-Capsule [4]	PF-Net [6]	Ours
Airplane	1.578	1.600	1.707	**0.501**	0.527
Bag	6.096	6.017	6.950	1.698	**1.675**
Cap	6.353	6.140	6.283	2.395	**1.993**
Car	3.919	4.062	4.416	1.023	**1.021**
Chair	3.014	3.130	3.241	0.914	**0.914**
Guitar	0.748	0.773	0.759	0.199	**0.191**
Lamp	5.099	4.760	5.183	1.677	**1.551**
Laptop	2.236	2.346	2.530	0.546	**0.538**
Motorbike	3.283	3.158	3.255	0.911	**0.816**
Mug	5.678	5.714	6.047	1.484	**1.472**
Pistol	2.080	1.926	2.175	**0.496**	0.521
Skateboard	1.907	2.022	2.159	**0.397**	0.583
Table	3.295	3.394	3.619	**0.929**	0.935
Mean	3.484	3.464	3.640	1.013	**0.982**

Further, our method has considerable advantages over other methods. According to Table 2, We obtain 4.91% relative improvement on the mean chamfer distance. From Fig. 3 we can observe that our method successfully generate missing point cloud with fine details.

4.3 Analysis of Encoder

To test our encoder, we evaluate the classification accuracy on ModelNet40 [19] and compare it with PointNet-MLP and PF-NET [6] encoder. Quantitative results are shown in Table 3. And it indicates that our encoder has a better performance on understanding semantic information.

4.4 Robustness to Occlusion

The occlusion may be serious in real-time applications, so it is necessary to test the robustness to occlusion of the network. In this subsection, we control the missing ratio of the point cloud, ranging from 25% to 75%, and make comparisons with PF-NET [6]. And it should be noted that data in Table 4 are obtained in our experiments. The results shows that our method is more robust to occlusion. Figure 4 also shows the comparison in different missing rations.

4.5 Comparison of Complexity

The training speed is an important aspect of evaluating the performance of the model. For a given point cloud composed of N points, PF-NET applies FPS to the entire point cloud, which requires $O(N^3)$ time. As for our method, the slicer takes $O(N)$ time. For the Atrous KNN module, it takes $O(N^2)$ time to calculate the distance between points, and it take $O(N^2 * lgK)$ time to select the K nearest points for each point of the point cloud. In short, our method is $O(N^2)$ in time complexity, while PF-NET is $O(N^3)$. We also evaluate the model parameters, as Table 5 shows. It is clear that previous models are too giant to be deployed in real time and our model is much more lightweight because we share some parameters in global encoder and local encoder. The performance of our model is better than that of the previous method, but the parameters are only 1/5 of those of the previous method.

Table 3. Classification accuracy result of models on ModelNet40 [19]

Method	PointNet [5]	PointNet++ [1]	PF-Net [6]	Ours
Acc. (%)	90.6	92.2	90.9	**92.2**

Table 4. Quantitative comparison for occluded point cloud under different occlusion rates. The evaluation metrics is CD(10^{-4}).

Method	Occlusion ratios		
	25%	50%	75%
PF-Net [1]	4.543	4.914	5.888
Ours	**4.320**	**4.698**	**5.774**

Table 5. Quantitative comparisons for model parameters

Method	PF-Net [6]	Ours
Parameters	76.7M	**15.7M**

5 Ablation Study

We denote our method with PointNet [5] encoder as the baseline and evaluates its performance on point cloud completion. Experiment is done on 2048 points of each shape in ShapeNet-Part [15]. The results in Table 6 indicates that our approach of extracting local feature is effective. With our local encoder, the relative improvement is 3.79%.

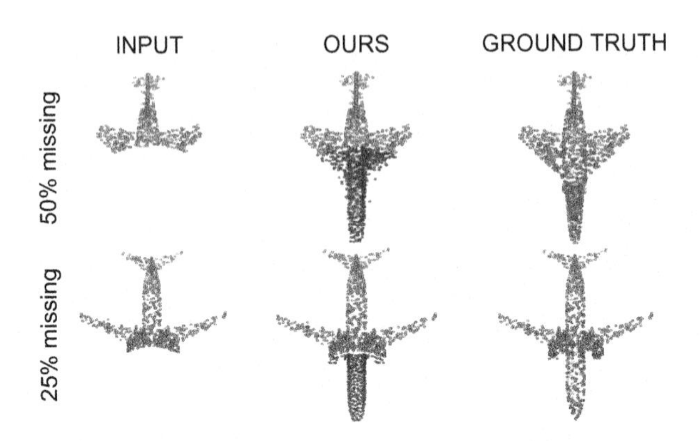

Fig. 4. Visualization results of different missing rations of input points. The upper line indicates the input point cloud is 50% missing and the bottom line indicates the input point cloud is 25% missing.

Table 6. Point cloud completion results for our model with and without local encoder. Ours* refers to our model without local encoder

Method	Ours*	Ours
CD(10^{-4})	4.490	**4.320**

6 Conclusion

In this paper, we propose a novel network that uses partial input to complete the point cloud, namely Slice Sequential Network. Our model is able to learn the relationship among different slices and retain the local features of incomplete point clouds. Besides, our network can generate missing parts with high fidelity at relatively low computational cost. Quantitative and qualitative experiments show that our method outperforms existing state-of-the-art methods.

References

1. Qi, C.R., Yi, L., Su, H., Guibas, L.J.: Pointnet++: deep hierarchical feature learning on point sets in a metric space. In: NeurIPS (2017)
2. Yuan, W., Khot, T., Held, D., Mertz, C., Hebert, M.: PCN: point completion network. In: 2018 International Conference on 3D Vision, pp. 728–737. IEEE (2018)
3. Dai, A., Ritchie, D., Bokeloh, M., Reed, S., Sturm, J., Nießner, M.: Scancomplete: large scale scene completion and semantic segmentation for 3D scans. In: Proceedings on Computer Vision and Pattern Recognition. IEEE (2018)
4. Zhao, Y., Birdal, T., Deng, H., Tombari, F.: 3D point capsule networks. In: Proceedings of the IEEE Conference on Computer Vision and Pattern Recognition. IEEE (2018)
5. Qi, C.R., Su, H., Mo, K., Guibas, L.J.: Pointnet: deep learning on point sets for 3D classification and segmentation. In: Proceedings of the IEEE Conference on Computer Vision and Pattern Recognition. IEEE (2017)

6. Huang, Z., Yu, Y., Xu, J., Ni, F., Le, X.: PF-net: point fractal network for 3D point cloud completion. In: Proceedings of the IEEE Conference on Computer Vision and Pattern Recognition. IEEE (2020)
7. Wang, X., Ang Jr, M.H., Lee, G.H.: Cascaded refinement network for point cloud completion. In: Proceedings of the IEEE Conference on Computer Vision and Pattern Recognition. IEEE (2020)
8. Han, Z., Wang, X., Liu, Y.S., Zwicker, M.: Multi-angle point cloud-VAE: unsupervised feature learning for 3D point clouds from multiple angles by joint self-reconstruction and half-to-half prediction. In: Proceedings of the IEEE Conference on Computer Vision and Pattern Recognition. IEEE (2019)
9. Wang, Y., Sun, Y., Liu, Z., Sarma, S.E., Bronstein, M.M., Solomon, J.M.: Dynamic graph CNN for learning on point clouds. ACM Trans. Graph. **38**(5), 1–12 (2018)
10. Pan, L., Wang, P., Chew, C.-M.: Point atrous convolution for point cloud analysis. IEEE Robot. Autom. Lett. **4**(4), 4035–4041 (2019)
11. Han, X., Li, Z., Huang, H., Kalogerakis, E., Yu, Y.: High-resolution shape completion using deep neural networks for global structure and local geometry inference. In: Proceedings of the IEEE International Conference on Computer Vision, pp. 85–93. IEEE (2017)
12. Achlioptas, P., Diamanti, O., Mitliagkas, I., Guibas, L.J.: Learning representations and generative models for 3D point clouds. In: International Conference on Machine Learning (2018)
13. Tchapmi, L.P., Kosaraju, V., Rezatofighi, H., Reid, I., Savarese, S.: Topnet: structural point cloud decoder. In: Proceedings of the IEEE Conference on Computer Vision and Pattern Recognition. IEEE (2019)
14. Fan, H., Su, H., Guibas, L.J.: A point set generation network for 3D object reconstruction from a single image. In: Proceedings of the IEEE Conference on Computer Vision and Pattern Recognition. IEEE (2017)
15. Yi, L., et al.: A scalable active framework for region annotation in 3D shape collections. ACM Trans. Graph. **35**(6), 210:1-210:12 (2016)
16. Park, J.J., Florence, P., Straub, J., Newcombe, R., Lovegrove, S.: DeepSDF: learning continuous signed distance functions for shape representation. In: Proceedings of the IEEE Conference on Computer Vision and Pattern Recognition, pp. 165–174. IEEE (2019)
17. Kingma, D.P., Ba, J.: Adam: a method for stochastic optimization. arXiv:1412.6980 (2014)
18. Mur-Artal, R., Montiel, J.M.M., Tardos, J.D.: ORB-SLAM: a versatile and accurate monocular slam system. IEEE Trans. Robot. **31**(5), 1147–1163 (2015)
19. Wu, Z., et al.: 3D shapenets: a deep representation for volumetric shapes. In: Proceedings of the IEEE Conference on Computer Vision and Pattern Recognition. IEEE (2015)
20. Birdal, T., Ilic, S.: A point sampling algorithm for 3D matching of irregular geometries. In: Proceedings of International Conference of Intelligent Robots and Systems (2017)
21. Girdhar, R., Fouhey, D.F., Rodriguez, M., Gupta, A.: Learning a predictable and generative vector representation for objects. In: Leibe, B., Matas, J., Sebe, N., Welling, M. (eds.) ECCV 2016. LNCS, vol. 9910, pp. 484–499. Springer, Cham (2016). https://doi.org/10.1007/978-3-319-46466-4_29
22. Shen, C.-H., Fu, H., Chen, K., Hu, S.-M.: Structure recovery by part assembly. ACM Trans. Graph. (TOG) **31**(6), 136 (2012)

23. Rock, J., Gupta, T., Thorsen, J., Gwak, J., Shin, D., Hoiem, D.: Completing 3D object shape from one depth image. In: Proceedings of the IEEE Conference on Computer Vision and Pattern Recognition, pp. 2484–2493. IEEE (2015)
24. Kim, V.G., Li, W., Mitra, N.J., Chaudhuri, S., DiVerdi, S., Funkhouser, T.: Learning part-based templates from large collections of 3D shapes. ACM Trans. Graph. **32**(3), 29 (2013)
25. Li, D., Shao, T., Hongzhi, W., Zhou, K.: Shape Completion from a single RGBD image. IEEE Trans. Vis. Comput. Graph. **23**(7), 1809–1822 (2017)
26. Han, Z., et al.: SeqViews2SeqLabels: learning 3D global features via aggregating sequential views by RNN with attention. IEEE Trans. Image Process. **28**(2), 658–672 (2018)
27. Han, Z., Liu, X., Liu, Y.S., Zwicker, M.: Parts4Feature: learning 3D global features from generally semantic parts from different views. In: International Joint Conference on Artificial Intelligence (2019)
28. Li, J., Chen, B.M., Lee, G.H.: SO-Net: self organizing network for point cloud analysis. In: Proceedings of the IEEE Conference on Computer Vision and Pattern Recognition, pp. 9397–9406 (2018)
29. Pauly, M., Mitra, N.J., Wallner, J., Pottmann, H., Guibas, L.J.: Discovering structural regularity in 3d geometry. ACM Trans. Graph. **27**(3), 43 (2008)
30. Mitra, N.J., Guibas, L.J., Pauly, M.: Partial and approximate symmetry detection for 3D geometry. ACM Trans. Graph. **25**(3), 560–568 (2006)
31. Luo, Z., et al.: Learning sequential slice representation with an attention-embedding network for 3D shape recognition and retrieval in MLS point clouds. ISPRS J. Photogramm. Remote. Sens. **161**, 147–163 (2020)
32. Dai, A., Qi, C.R., Niebner, M.: Shape completion using 3D-encoder-predictor CNNs and shape synthesis. In: Proceedings of the IEEE Conference on Computer Vision and Pattern Recognition. IEEE (2017)

From Digital Model to Reality Application: A Domain Adaptation Method for Rail Defect Detection

Wenkai Cui[1], Jianzhu Wang[1], Haomin Yu[1], Wenjuan Peng[1], Le Wang[2], Shengchun Wang[2], Peng Dai[2], and Qingyong Li[1](✉)

[1] Beijing Key Lab of Traffic Data Analysis and Mining, Beijing Jiaotong University, Beijing, China
liqy@bjtu.edu.cn

[2] Infrastructure Inspection Research Institute, China Academy of Railway Sciences LTD, Beijing, China

Abstract. Recently, vision-based rail defect detection has attracted much attention owing to its practical significance. However, it still faces some challenges, such as high false alarm rate and poor feature robustness. With the development of deep neural networks (DNNs), deep learning based models have shown the potential to solve the problems. Nevertheless, these models usually require a large number of training samples, while collecting and labeling sufficient defective rail images is somewhat impractical. On the one hand, the probability of defect occurrence is low. On the other hand, we are not able to annotate samples that include all types of defects. To this end, we propose to generate defective training images in the digital space. In order to bridge the gap between virtual and real defective samples, this paper presents a domain adaptation based model for rail defect detection. The proposed method is evaluated on a real-world dataset. Experimental results show that our proposed method is superior to five established baselines.

Keywords: Rail defect detection · Digital model · 3D point cloud · Object detection · Domain adaptation

1 Introduction

By the end of 2020, the railway running mileage of China has reached about 150,000 km, including about 30,000 km of high-speed railways. Obviously, with the rapid development of railways, there comes an urgent and higher requirement on railway safety. Among all the factors, various rail defects can directly

This work was supported in part by the National Natural Science Foundation of China under Grant U2034211, in part by the Fundamental Research Funds for the Central Universities under Grant 2020JBZD010, in part by the Beijing Natural Science Foundation under Grant L191016 and in part by the China Railway R&D Program under Grant P2020T001.

H. Ma et al. (Eds.): PRCV 2021, LNCS 13020, pp. 115–126, 2021.
https://doi.org/10.1007/978-3-030-88007-1_10

affect the operation security and reliability of the railway transportation system. Consequently, rail defect detection is of great practical significance.

In the early years, rail defects are mainly detected manually by railway workers, which can be influenced by various elements and can not meet the demands of modern railway inspection. To this end, many automatic rail defect detection methods have been proposed, such as ultrasonic inspection [4], eddy current inspection [19], magnetic flux leakage inspection [6] and vision-based inspection [13,14,27]. Compared to other non-destructive detection methods, vision-based inspection has more advantages including fast speed, simple deployment, low cost and high precision. Therefore, vision-based methods are widely used in real scenarios [10].

Traditional vision-based inspection methods rely on various hand-crafted features. Based on local normalization, Li et al. [13] proposed to enhance the contrast of the image and detect defects through the projection profile. Yu et al. [27] proposed a coarse-to-fine model to identify defects, which worked on three scales from coarse to fine: subimage level, region level and pixel level. However, due to the limitations of hand-crafted features [10,13,27], these methods are not able to obtain a satisfactory performance when dealing with complex visual appearances.

Compared with hand-crafted features, features extracted by DNNs are regarded to have stronger robustness and higher effectiveness [11]. As a result, DNN-based object detection methods have been successfully applied in many fields, such as steel surface defect detection [17] and remote sensing imagery object detection [9]. Although achieving great progress, these methods heavily rely on big datasets. In other words, if there is no specific and sufficient data, these methods will fail to achieve a good result. Also, rail defect detection can be considered as an object detection problem. In this scenario, the defective images are scarce, which will degrade the performance of general DNN-based object detection methods. Song et al. [26] and Wei et al. [25] tried to employ object detection methods to detect rail defects. However, due to the lack of defective samples, their experimental results were limited.

To deal with the aforementioned issues, we propose a domain adaptation based method for rail defect detection. For solving the shortage problem of defective samples, we manipulate rail 3D surface model and rail 3D point cloud model in the digital space to generate virtual defective images. Particularly, we can automatically obtain the ground truth of these virtual images since the defects are generated by ourselves. To overcome the distribution mismatch problem between virtual and real images, we present a model named DA-YOLO, which can achieve domain adaptation from virtual digital space to real physical space. We evaluate our method on a real-world dataset. The experimental results show that the proposed method outperforms five baselines, demonstrating its superiority for rail inspection tasks. The contributions of this work can be summarized as follows:

- We propose to generate defective samples in the digital space instead of using common data augmentations. This enables us to obtain training samples in

absence of defective images. What is more, we are able to control the diversity and complexity of the defects while obtaining the ground truth automatically.
- We put forward the concept of dummy-target domain, which makes it possible to customize a domain adaptation method for rail defect detection.
- We design a model named DA-YOLO, which integrates two alignment modules from the image and feature levels and significantly improves the detection performance.

2 Preliminaries

Digital Twin. A digital twin (DT) is a digital representation of a physical object or system [22]. Based on the digital representation, various problems can be analyzed. In our work, by constructing a large deal of defective rail digital models, we are able to overcome the shortage problem of defective samples. Meanwhile, because it needs a lot of time and efforts to manually construct digital models, a point cloud method is adopted to automatically generate digital models.

Domain Adaptation. Domain adaptation belongs to transductive transfer learning and aims to solve the domain shift problem [18]. Domain shift refers to that the testing set has a different distribution from the training set. In domain adaptation, the testing set and training set correspond to the target domain and source domain, respectively. Its basic idea is to minimize the distance between two domains' distributions. In the early days, the distance is measured in a statistical manner [8,16,24]. Motivated by some theoretical results [1,2], many methods have begun to use domain classifier to measure the distance [7,20,23].

Usually, the domain adaptation methods take a lot of samples from testing set to minimize the distance. However, it is hard to take sufficient samples from testing set in our scenario due to the fact that real defective images are scarce. In view of this, we come up with the concept of the dummy-target domain. The details of dummy-target domain will be elaborated in Sect. 3.2.

YOLO-V3. YOLO-V3 [21] is a real-time one-stage object detection model. It predicts the locations, size and classes of the objects directly. Recently, some new YOLO models have been proposed [3], which mainly try to find the best combination from a variety of object detection components. But they are redundant for our problem. Therefore, we still refer to YOLO-V3 to design our DA-YOLO.

3 Method

In this section, we introduce the DT-based data generation scheme and the DA-YOLO model. Meanwhile, to clearly illustrate our model, we also detail the dummy-target domain.

3.1 DT-Based Virtual Data Generation

As illustrated in the top half of Fig. 1, the data generation scheme can be divided into four steps, which are rail modeling, defect generation, augmentation and images and labels acquiring. In the first process, we construct a rail 3D surface model in the digital space. In the second process, we manually add various defects on the rail model to get some defective rail models. In the third process, we further expand the diversity and quantity of defective rail models. In the last process, we acquire the surface images of defective rail models and their corresponding labels.

Particularly, we adopt a point cloud method to achieve the augmentation process. The specific details are illustrated in the bottom half of Fig. 1, including defect extraction, defect augmentation and defect filling. Formally, we first convert the defective rail surface models into 3D point cloud models and extract defects from the point cloud models. Then various transformations are performed on the defects to get some new type of defects. Lastly, the augmented defects are filled to random position of point cloud models to generate some new defective rail 3D models.

However, the new defective rail 3D models have jaggies at the transition position between the defect and the background, as illustrated in Fig. 1(b). This is because the rail is curved, and the curvature information is discrepant between the position of extracting defect and filling defect. To address this issue, we propose the curvature elimination and curvature addition operations as shown in Fig. 1(a). Specifically, after extracting the defect, we first eliminate defect's curvature information in the extracted position. Next, before filling the defect into point cloud model, the curvature information in the filled position is added to the defect. Finally, the jaggies can be eliminated ideally, as illustrated in Fig. 1(c).

3.2 Dummy-Target Domain

In this section, the dummy-target domain will be introduced in detail. First, we use $P_S(B,I)$ and $P_T(B,I)$ to denote the joint distribution of source domain and target domain, where B, I, S and T denote bounding-box of defect, image, source domain and target domain, respectively. In our scenario, the source and target domain correspond to the virtual and real defective rail images, respectively. In DNNs, only when the $P_S(B,I)$ and $P_T(B,I)$ are similar, can the model trained on the source domain achieve a good performance on the target domain. However, as illustrated in Fig. 2(a) and 2(d), there is a large gap between the two domains in our scenario. That is:

$$P_S(B,I) \nsim P_T(B,I) \tag{1}$$

where $\nsim$ denotes dissimilarity. Using the Bayers' formula, formula (1) can be rewritten as:

$$P_S(B|I)\,P_S(I) \nsim P_T(B|I)\,P_T(I). \tag{2}$$

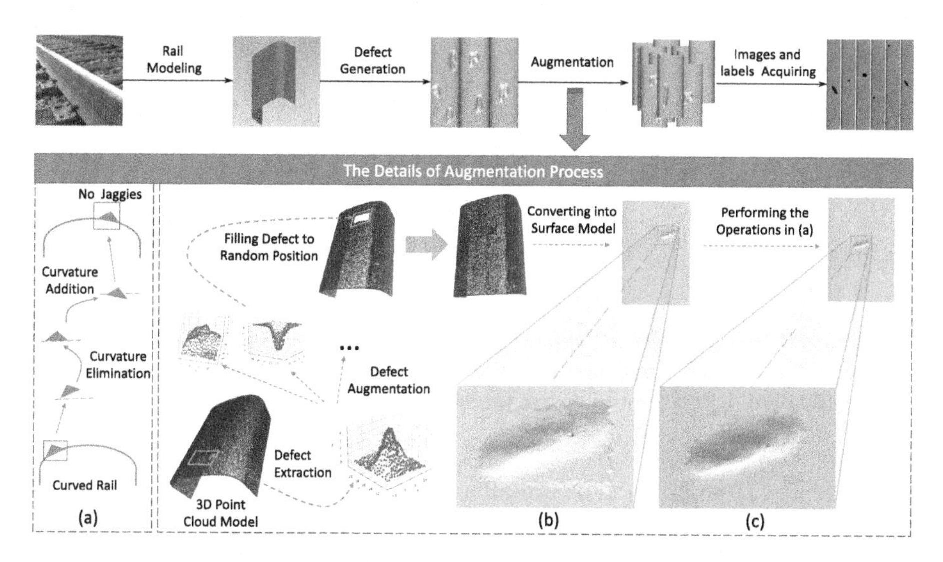

Fig. 1. The architecture of the proposed data generation scheme. (a) The curvature elimination and curvature addition operations. (b) Defective rail model with jaggies. (c) Defective rail model without jaggies.

According to the covariate shift assumption of domain adaptation [5], the domain discrepancy is caused by the difference in the marginal distribution:

$$P_S(I) \nsim P_T(I), P_S(B|I) \sim P_T(B|I) \tag{3}$$

where $\sim$ denotes similarity. So our goal is to make $P_S(I) \sim P_T(I)$. However, as mentioned in Sect. 2, it is difficult to implement this goal because target domain samples are scarce in our scenario. To this end, we propose the dummy-target domain, which is defined as a domain very similar to the target domain:

$$P_D(I) \sim P_T(I) \tag{4}$$

where D and $P_D(I)$ denote the dummy-target domain and its distribution, respectively. Obviously, we can achieve $P_S(I) \sim P_T(I)$ by making $P_S(I) \sim P_D(I)$.

In our scenario, the real normal images are sufficient and very similar to the samples of target domain as shown in Fig. 2(c) and 2(d). Therefore, they can be used as samples of the dummy-target domain.

3.3 DA-YOLO

In this section, we detail our proposed DA-YOLO model derived from YOLO-V3 [21]. Based on the analysis in Sect. 3.2, the performance of model will be enhanced by making $P_S(I) \sim P_D(I)$. Therefore, compared with the raw YOLO-V3, our model integrates the image alignment and feature alignment modules

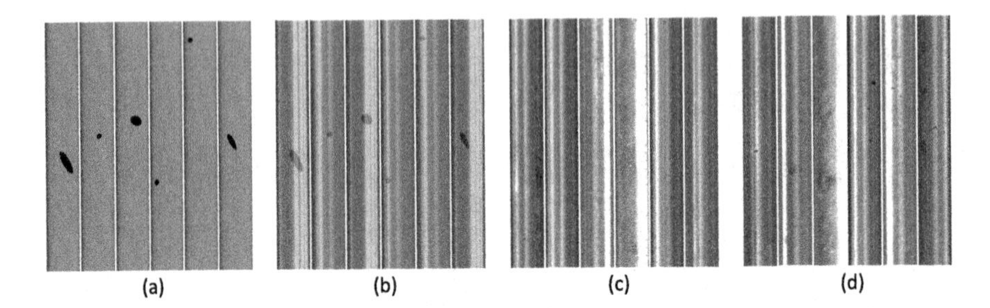

Fig. 2. (a) Virtual defective rail images: original source domain. (b) Aligned virtual defective rail images: aligned source domain. (c) Real normal rail images: dummy-target domain. (d) Real defective rail images: target domain.

to achieve $P_S(I) \sim P_D(I)$. The overall architecture of DA-YOLO is illustrated in Fig. 3. It mainly consists of a feature extractor F, a detector G and a domain classifier H. The F and G are from the Darknet53 of YOLO-V3 and compose a complete object detection model. The H is designed to accomplish the feature alignment. The details are elaborated as follows.

Image Alignment. To let $P_S(I) \sim P_D(I)$, we first align the images that come from the source domain and dummy-target domain. For each image in the source domain, we randomly select an image from the dummy-target domain and blend them in a certain proportion. The process can be formulated by:

$$E = \{e|\mu s + (1 - \mu)d, s \in S, d \in D\} \tag{5}$$

where s and d denote images from source domain and dummy-target domain, respectively. e and E denote the aligned image and aligned source domain, respectively. μ is a trade-off parameter to balance the blending process.

After the image alignment, we obtain the aligned source domain, which is more similar to dummy-target domain than original source domain. Images in original source domain, aligned source domain and target domain can be found in Fig. 2(a), 2(b) and 2(d), respectively.

Feature Alignment. To make $P_S(I) \sim P_D(I)$ further, we align the features extracted from the source domain and dummy-target domain. Motivated by Ben-David et al.'s theory [1], we define the distance between two domains' feature distributions as follows:

$$dis(S, D) = 2\left(1 - \min_{H}(err_S(H(x)) + err_D(H(x)))\right) \tag{6}$$

where x denotes a feature vector extracted by F (i.e. $x = F(I)$). H is our classifier for predicting which domain the x comes from. err_S and err_D are the predicting error rates of H on the source domain and dummy-target domain

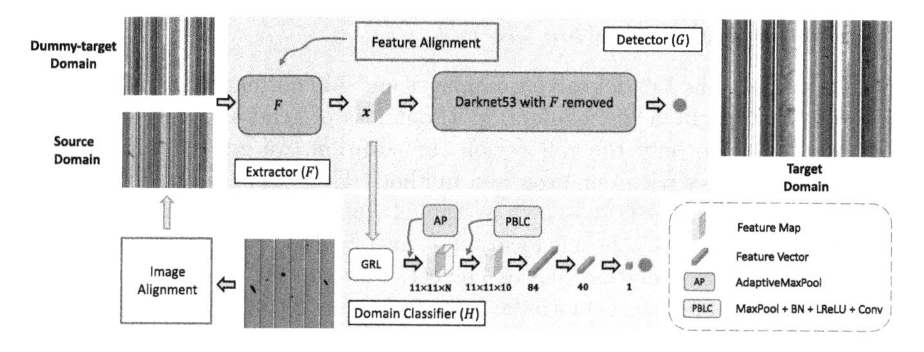

Fig. 3. Overview of the proposed DA-YOLO model. Conv, BN, LReLU and FC stand for Convolutional layer, BatchNormalization, Leaky rectified linear unit and Fully connected layer, respectively.

samples. Therefore, to make $P_S(I) \sim P_D(I)$ further, F is required to minimize the $dis(S, D)$:

$$\begin{aligned} &\min_F dis(S, D) \\ \Leftrightarrow &\min_F \left(1 - \min_H (err_S(H(x)) + err_D(H(x)))\right) \\ \Leftrightarrow &\max_F \min_H (err_S(H(x)) + err_D(H(x))). \end{aligned} \tag{7}$$

Formula (7) indicates that the F and H can be optimized in an adversarial method. In our work, we use the gradient reverse layer (GRL) [7] to achieve the adversarial process.

Loss Function. The loss function is composed of two components. The first is detection loss L_{det}, which is the same as the loss function used in YOLO-V3. The second is the feature alignment loss L_{align}, which depends on the predicting error rate of binary classifier H. Thus, we use the cross entropy loss to define L_{align}:

$$L_{align} = -\sum [C_i Log(p_i) + (1 - C_i) Log(1 - p_i)] \tag{8}$$

where $C_i \in \{0, 1\}$ denotes the domain label of i-th image, and 0 represents this image is from the source domain. p_i is the output of H. With the help of GRL module, we make H minimize the L_{align} and F maximize the L_{align} in the training process. Ultimately, the loss of the proposed network is a summation of two components. In training process, we minimize the L_{det} and L_{align} alternately.

4 Experiment

In this section, we evaluate our method on a real-world dataset. All experiments are conducted on a 64-bit Ubuntu 16.04 computer with 18 Intel 2.68 GHz CPUs, 256 GB memory and 8 NVIDIA TITAN Xp GPUs.

4.1 Dataset and Evaluation Metrics

The dataset contains 143 defective rail images and 740 normal rail images, which are both captured by a line camera. Without loss of generality, all images are cropped to contain only the rail region. In addition, we generate 757 virtual defective rail images with our proposed method. These images are divided into training set and testing set as shown in Table 1. Intersection over Union (IoU) is used to determine the acceptable location deviation, and we regard the detection results with IoU greater than 0.3 as correct. Precision(P), Recall(R) and F1-score($F1$) are adopted to evaluate the proposed method's performance, and they are defined as follows:

$$P = \frac{TP}{TP + FP}, R = \frac{TP}{TP + FN}, F1 = \frac{2 \times P \times R}{P + R} \tag{9}$$

where TP, FN and FP denote the number of rail defects that are accurately detected, missed and wrongly detected, respectively.

Table 1. Division of dataset. VDR, RNR and RDR stand for virtual defective rail, real normal rail and real defective rail, respectively.

Dataset	# VDR images	# RNR images	# RDR images	# Defects
Training set	757	740	0	757
Testing set	0	0	143	176

4.2 Experiment Settings

Experiment Design. First, to demonstrate the effectiveness of proposed method, we compare our approach with three traditional methods (i.e. LN+DLBP [13], MLC+PEME [14] and CTFM [27]) and two DNN-based object detection methods (i.e. YOLO-V3 [21] and SSD [15]). Note, detection results of MLC+PEME and CTFM are binary images rather than images with bounding-boxes. In order to quantitatively evaluate their performance, we consider the results as true positives if the segmentation areas are within the ground truth bounding-boxes. Second, we conduct a set of ablation experiments to fully investigate the effectiveness of each proposed module in DA-YOLO model. (1) Baseline network; (2) Network with only image alignment; (3) Network with only feature alignment; (4) DA-YOLO. Finally, considering that our dummy-target domain has no high-level semantic information, we conduct a set of controlled experiments to explore whether aligning fewer layers would gain better performance. We randomly select to align 5 layers, 9 layers and 37 layers in the feature alignment module. Aligning 5 layers represents that we select the first 5 layers of Darknet53 [21] as the feature extractor F.

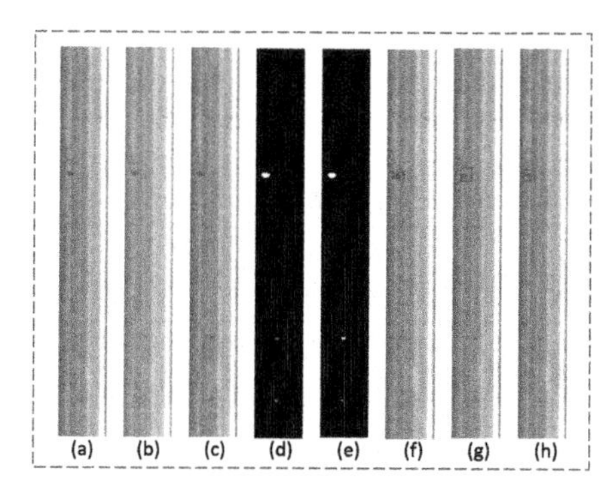

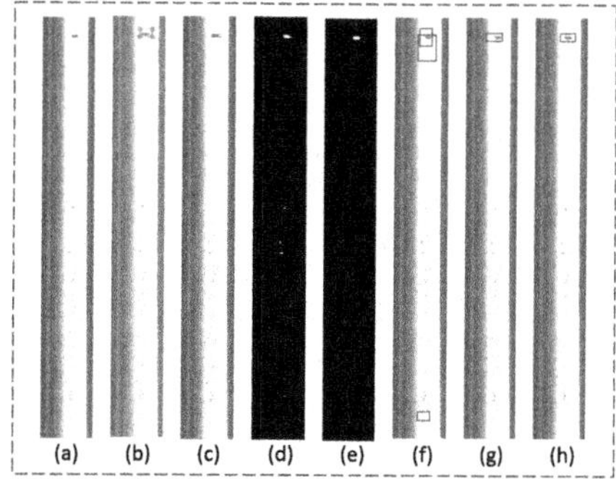

Fig. 4. Sample defect images and detection results of different methods. (a) Original image. (b) Ground truth. (c) LN+DLBP. (d) MLC+PEME. (e) CTFM. (f) YOLO-V3. (g) SSD. (h) Our method.

Implementation Details. In DA-YOLO, we use Adam optimizer [12] with $\beta_1 = 0.9$ and $\beta_2 = 0.99$. Learning rate lr and batch size are set to 10^{-3} and 8, respectively. Parameters of GRL are set to the defaults in [7]. Also, the trade-off parameter μ in equation (5) is set to 0.5 in all experiments. In YOLO-V3, we set the parameters same as those in DA-YOLO. In SSD, we use amdegroot's implementation[1] and set the lr and batch size to 5×10^{-3} and 8. In LN+DLBP, the upper bound of defect proportion dp is set to 0.3. In MLC+PEME, the contribution controlling parameter γ is set to 3. In CTFM, the three weighting parameters ($w1$, $w2$ and $w3$) of the weighted geometric mean are set to 0.7, 0.6 and 1.7, respectively.

4.3 Experimental Results

Detection results of different methods are visualized in Fig. 4, and corresponding quantitative results are listed in Table 2. From Fig. 4, we can clearly see our method has a better performance than other methods. From Table 2, we have the following detailed results. Firstly, LN+DLBP, MLC+PEME and CTFM can identify most defects but yield high false detection. Compared with the three methods, DA-YOLO is superior and achieves at least +13.1% improvement in F1-score, which proves that the features extracted by our model are more robust than the traditional methods. Secondly, compared with SSD and YOLO-V3, DA-YOLO achieves at least +19.3% improvement in F1-score. This is because our model integrates the image alignment and feature alignment modules to overcome the domain gap between virtual training images and real test images. Lastly, it can be seen that our model has an outstanding performance in precision, which indicates that our method can effectively reduce the pressure of manual reinspection in real scenarios.

[1] [2021-05-30] https://github.com/adafruit/Adafruit_SSD1306.

Table 2. Experimental results of various models. Correct, Wrong and Missed denote the number of defects that are correctly detected, wrongly detected and missed, respectively.

Model	# Correct	# Wrong	# Missed	Precision	Recall	F1-score
LN+DLBP [13]	124	410	52	0.232	0.705	0.349
MLC+PEME [14]	156	423	20	0.269	0.886	0.413
CTFM [27]	158	118	18	0.572	**0.898**	0.699
YOLO-V3 [21]	91	159	85	0.364	0.517	0.427
SSD [15]	129	100	47	0.563	0.733	0.637
Ours	147	31	29	**0.826**	0.835	**0.830**

The results of ablation experiments are shown in Table 3, from which we have the following observations. First, compared to the baseline, the image alignment and feature alignment modules can respectively bring +35.2% and 36.2% improvement in F1-score. These results prove that each of the proposed modules contributes to improving the model's performance. Second, by combining the two modules, we can get a further improvement of 4.1% in F1-score, which demonstrates that the two modules are able to promote each other.

Table 3. Ablation studies on image alignment and feature alignment modules.

Method	# Correct	# Wrong	# Missed	Precision	Recall	F1-score
Baseline	91	159	85	0.364	0.517	0.427
Image alignment	160	75	16	0.681	**0.909**	0.779
Feature alignment	142	42	34	0.772	0.807	0.789
Both of them	147	31	29	**0.826**	0.835	**0.830**

Table 4. The influence of aligning different number of layers.

The number of aligned layers	#Correct	#Wrong	#Missed	Precision	Recall	F1-score
5	147	31	29	0.826	**0.835**	**0.830**
19	127	25	49	**0.836**	0.722	0.775
37	133	51	43	0.723	0.756	0.739

From Table 4, we can find that the performance of DA-YOLO is decreasing with the increasing number of aligned layers. This is because the higher layers extract more high-level features, which mainly include information about the defect detection task [11]. However, the samples of dummy-target domain do not contain defects, which leads to aligning features at deeper layers useless or counterproductive.

5 Conclusion

In this paper, we propose a domain adaptation based method for rail defect detection, which can alleviate the reliance on real defective rail images. Based on DT technology, we are able to generate various and sufficient virtual defective images, whose ground truth can be simultaneously obtained in the generation process. To bridge the gap between virtual and real defective samples, we present DA-YOLO, which integrates image alignment and feature alignment modules to achieve domain adaptation from digital space to physical space. Our method is almost no manpower consumption and superior to other popular methods.

In future research, we will extend the current work in two aspects. First, the training process is not stable enough due to the adversarial architecture so we expect to find a way to overcome the shortage. Second, our method is transferable and can be used to solve similar problems in other scenarios. Therefore, expanding the application range of our method is also the aim in our next stage.

References

1. Ben-David, S., Blitzer, J., Crammer, K., Kulesza, A., Pereira, F., Vaughan, J.W.: A theory of learning from different domains. Mach. Learn. **79**(1), 151–175 (2010)
2. Ben-David, S., Blitzer, J., Crammer, K., Pereira, F., et al.: Analysis of representations for domain adaptation. In: Advances in Neural Information Processing Systems, vol. 19, p. 137. MIT 1998 (2007)
3. Bochkovskiy, A., Wang, C.Y., Liao, H.Y.M.: YOLOv4: optimal speed and accuracy of object detection. arXiv preprint arXiv:2004.10934 (2020)
4. Bombarda, D., Vitetta, G.M., Ferrante, G.: Rail diagnostics based on ultrasonic guided waves: an overview. Appl. Sci. **11**(3), 1071 (2021)
5. Chen, Y., Li, W., Sakaridis, C., Dai, D., Van Gool, L.: Domain adaptive Faster R-CNN for object detection in the wild. In: IEEE Conference on Computer Vision and Pattern Recognition, pp. 3339–3348 (2018)
6. Chen, Z., Xuan, J., Wang, P., Wang, H., Tian, G.: Simulation on high speed rail magnetic flux leakage inspection. In: IEEE International Instrumentation and Measurement Technology Conference, pp. 1–5. IEEE (2011)
7. Ganin, Y., Lempitsky, V.: Unsupervised domain adaptation by backpropagation. In: International Conference on Machine Learning, pp. 1180–1189. PMLR (2015)
8. Ghifary, M., Kleijn, W.B., Zhang, M.: Domain adaptive neural networks for object recognition. In: Pham, D.-N., Park, S.-B. (eds.) PRICAI 2014. LNCS (LNAI), vol. 8862, pp. 898–904. Springer, Cham (2014). https://doi.org/10.1007/978-3-319-13560-1_76
9. Guo, Q., Liu, Y., Yin, H., Li, Y., Li, C.: Multi-scale dense object detection in remote sensing imagery based on keypoints. In: Peng, Y., et al. (eds.) PRCV 2020. LNCS, vol. 12305, pp. 104–116. Springer, Cham (2020). https://doi.org/10.1007/978-3-030-60633-6_9
10. Hajizadeh, S., Núnez, A., Tax, D.M.: Semi-supervised rail defect detection from imbalanced image data. IFAC-PapersOnLine **49**(3), 78–83 (2016)
11. He, K., Zhang, X., Ren, S., Sun, J.: Deep residual learning for image recognition. In: IEEE Conference on Computer Vision and Pattern Recognition, pp. 770–778 (2016)

12. Kingma, D.P., Ba, J.: Adam: a method for stochastic optimization. arXiv preprint arXiv:1412.6980 (2014)
13. Li, Q., Ren, S.: A real-time visual inspection system for discrete surface defects of rail heads. IEEE Trans. Instrum. Measure. **61**(8), 2189–2199 (2012)
14. Li, Q., Ren, S.: A visual detection system for rail surface defects. IEEE Trans. Syst. Man Cybern. Part C (Appl. Rev.) **42**(6), 1531–1542 (2012)
15. Liu, W., et al.: SSD: single shot multibox detector. In: Leibe, B., Matas, J., Sebe, N., Welling, M. (eds.) ECCV 2016. LNCS, vol. 9905, pp. 21–37. Springer, Cham (2016). https://doi.org/10.1007/978-3-319-46448-0_2
16. Long, M., Cao, Y., Wang, J., Jordan, M.: Learning transferable features with deep adaptation networks. In: International Conference on Machine Learning, pp. 97–105. PMLR (2015)
17. Luo, Q., Fang, X., Liu, L., Yang, C., Sun, Y.: Automated visual defect detection for flat steel surface: a survey. IEEE Trans. Instrum. Measure. **69**(3), 626–644 (2020)
18. Pan, S.J., Yang, Q.: A survey on transfer learning. IEEE Trans. Knowl. Data Eng. **22**(10), 1345–1359 (2009)
19. Papaelias, M.P., Lugg, M., Roberts, C., Davis, C.: High-speed inspection of rails using ACFM techniques. NDT & E Int. **42**(4), 328–335 (2009)
20. Pei, Z., Cao, Z., Long, M., Wang, J.: Multi-adversarial domain adaptation. In: AAAI Conference on Artificial Intelligence, vol. 32 (2018)
21. Redmon, J., Farhadi, A.: YOLOv3: an incremental improvement. arXiv preprint arXiv:1804.02767 (2018)
22. Tao, F., Zhang, H., Liu, A., Nee, A.Y.: Digital twin in industry: state-of-the-art. IEEE Trans. Ind. Inform. **15**(4), 2405–2415 (2018)
23. Tzeng, E., Hoffman, J., Saenko, K., Darrell, T.: Adversarial discriminative domain adaptation. In: IEEE Conference on Computer Vision and Pattern Recognition, pp. 7167–7176 (2017)
24. Tzeng, E., Hoffman, J., Zhang, N., Saenko, K., Darrell, T.: Deep domain confusion: maximizing for domain invariance. arXiv preprint arXiv:1412.3474 (2014)
25. Wei, X., Wei, D., Suo, D., Jia, L., Li, Y.: Multi-target defect identification for railway track line based on image processing and improved YOLOv3 model. IEEE Access **8**, 61973–61988 (2020)
26. Yanan, S., Hui, Z., Li, L., Hang, Z.: Rail surface defect detection method based on YOLOv3 deep learning networks. In: 2018 Chinese Automation Congress (CAC), pp. 1563–1568. IEEE (2018)
27. Yu, H., et al.: A coarse-to-fine model for rail surface defect detection. IEEE Trans. Instrum. Measure. **68**(3), 656–666 (2018)

FMixAugment for Semi-supervised Learning with Consistency Regularization

Huibin Lin[1,2], Shiping Wang[1,2], Zhanghui Liu[1,2], Shunxin Xiao[1,2], Shide Du[1,2], and Wenzhong Guo[1,2(✉)]

[1] College of Mathematics and Computer Science, Fuzhou University, Fujian, China
guowenzhong@fzu.edu.cn
[2] Fujian Provincial Key Laboratory of Network Computing and Intelligent Information Processing, Fuzhou University, Fujian, China

Abstract. Consistency regularization has witnessed tremendous success in the area of semi-supervised deep learning for image classification, which leverages data augmentation on unlabeled examples to encourage the model outputting the invariant predicted class distribution as before augmented. These methods have been made considerable progress in this area, but most of them are at the cost of utilizing more complex models. In this work, we propose a simple and efficient method FMixAugment, which combines the proposed MixAugment with Fourier space-based data masking and applies it on unlabeled examples to generate a strongly-augmented version. Our approach first generates a hard pseudo-label by employing a weakly-augmented version and minimizes the cross-entropy between it and the strongly-augmented version. Furthermore, to improve the robustness and uncertainty measurement of the model, we also enforce consistency constraints between the mixed augmented version and the weakly-augmented version. Ultimately, we introduce a dynamic growth of the confidence threshold for pseudo-labels. Extensive experiments are tested on CIFAR-10/100, SVHN, and STL-10 datasets, which indicate that our method outperforms the previous state-of-the-art methods. Specifically, with 40 labeled examples on CIFAR-10, we achieve 90.21% accuracy, and exceed 95% accuracy with 1000 labeled examples on STL-10.

Keywords: Semi-supervised learning · Image classification · Consistency regularization · Data augmentation

1 Introduction

Deep neural networks have achieved immense success in most computer vision applications in recent years, including image classification [1], image segmentation [2], object detection [3], scene labeling [4]. However, if there are not enough labeled training examples, it usually leads to overfitting of the model,

H. Lin—Student.

The original version of this chapter was revised: a missing co-author has been added. The correction to this chapter is available at
https://doi.org/10.1007/978-3-030-88007-1_55

H. Ma et al. (Eds.): PRCV 2021, LNCS 13020, pp. 127–139, 2021.
https://doi.org/10.1007/978-3-030-88007-1_11

which causes the generalization ability to plummet. Moreover, it is both time-consuming and expensive to obtain label information because it usually involves expert knowledge, such as the application in medical imaging diagnosis. Semi-supervised learning (SSL) is a learning task, which training model on a small amount of labeled data with a large amount of unlabeled data. SSL alleviates the scarcity of labeled examples and improves the performance of the model by utilizing unlabeled samples. Since it is usually relatively simple and inexpensive to obtain unlabeled data, any performance improvement that SSL brings is typically low-cost.

The research in the area of SSL has made significant progress in recent years. There are various methods used in SSL, of which consistency regularization is a relatively mature technology. The idea of consistency regularization is to make slight perturbations for input examples or networks, then the predicted results output by the model should be consistent. This framework is essentially a weak variant of the smoothness assumption. Intuitively, it ensures that the model to be robust for any predicted data with minor changes.

Under the framework of consistency regularization, the main difference of such methods lies in how to inject the noise and measure consistency. The primarily used approaches for injecting noise are as follows: First, exploiting image augmentation technology to perturb the examples. Second, improving the robustness of model perturbations by utilizing different networks. As for the method of measuring consistency, the mean square error, KL divergence, and cross-entropy are usually used for the purpose of the cost function.

In this work, we improve the current state-of-the-art methods and apply them to our approach to enhance the performance of SSL. Firstly, inspired by AugMix [5] and FMix [6], we combine the improved AugMix method (MixAugment) with our FMask and propose FMixAugment, which is utilized to produce a severely distorted version of a given image, called a strongly-augmented version. Since most of the current masking methods usually crop a rectangular area, while such techniques produce completely horizontal and vertical masks that are unlikely to be a salient feature, it is easy for the model to memorize the augmentation. Specifically, the approach we proposed used for masking is called FMask, which obtains a binary mask by imposing a magnitude to the low-frequency image sampled from Fourier space. FMask preserves the low-frequency region of the input image in the continuous part, which maximizes the edge shape space whilst maintaining local features, thereby effectively augmented the dataset. Secondly, for the labeled set, a weakly-augmented image is fed into the model to obtain a predicted label distribution, and match it with a one-hot label via minimizing cross-entropy. As for the unlabeled samples, we predict the distribution of a weakly-augmented version and convert it into a hard pseudo-label (one-hot label) when the maximum value of the distribution of the weakly-augmented version is greater than the threshold. A cross-entropy loss is minimized between the hard pseudo-label and the strongly-augmented version after processed by FMixAugment. At the same time, in order to speed up the convergence and robustness of the model, we also propose a new regularization term, which utilizes the

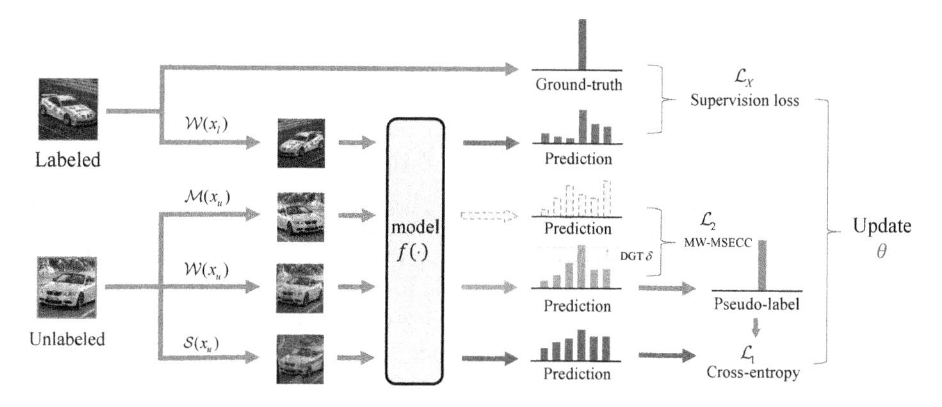

Fig. 1. The framework of FMixAugment method, where $\mathcal{W}(x)$ refers to a weakly-augmented version, $\mathcal{M}(x_u)$ and $\mathcal{S}(x_u)$ are mixed version and strongly-augmented version, respectively. The model parameters θ are updated via calculating the gradient relative to $\mathcal{L}$, where $\mathcal{L}$ is the combination of the three loss functions in the figure.

mean square error (MSE) loss to constrain the consistency of predicted class distribution between the image with only undergo MixAugment operations and the weakly-augmented version. This regularization term is referred to as MW-MSECC for short. Finally, our research found that setting a fixed threshold for pseudo-labeled examples is not conducive to the model's early convergence. On this basis, we propose a dynamic growth threshold (DGT) method that can make the model converge more quickly. A diagram of the proposed FMixAugment for consistency regularization is illustrated in Fig. 1. The threshold δ setting uses the proposed DGT method and the supervision loss utilizes cross-entropy as mentioned above. FMixAugment learns a model $f(\cdot)$ via updating the model parameters θ. To conclude, the main contributions of this paper are as follows:

- A new augmentation method is proposed to improve the robustness and uncertainty estimation of the model when encountering invisible data distribution shifts, and integrate it into a consistency framework.
- In order to improve the stability and efficiency of the model, the proposed MW-MSECC regularization term constrains the consistency between MixAugment and the weakly-augmented on unlabeled samples.
- A DGT approach is proposed to restraint the confidence-based thresholding and retain the pseudo-label with high confidence, which further improves the convergence speed and confidence calibration of the model.

2 Related Work

In order to lay the foundation for FMixAugment, we mainly focus on current mainstream technologies from the following two aspects, which are closely relevant to the proposed approach.

Consistency Regularization. In recent years, consistency regularization is one of the popular research directions in SSL. This idea is to encourage the model output prediction unchanged when fed unlabeled data is perturbed. This method was first proposed in [7] and became popular in Π-Model [8]. Mean Teacher [9] leveraged the exponential moving average of the weights during each iteration of training to obtain a teacher model. Virtual Adversarial Training (VAT) [10] added the adversarial perturbation on the unlabeled example as a term of regularization. ICT [11] enforced the model's predictions for unlabeled samples at interpolation points to be consistent with the predicted interpolations. UDA [12] applied advanced data augmentation to generate various noisy unlabeled samples. MixMatch [13] performed label guessing on augmented unlabeled samples using sharpen technology and utilized MixUp [14] to mix labeled and unlabeled data. ReMixMacth [15] was an improvement of MixMatch. FixMatch [16] proposed a new simple and effective algorithm by combining pseudo-labeling and consistency regularization. Such a category of approach is also used in [17,18].

Data Augmentation. Existing research has shown that data augmentation plays a vital role in SSL and greatly improves generalization performance. Mixed sample data augmentation (MSDA) has witnessed great attention in recent years. For MSDA, there are many triumphant methods, such as MixUp [14], CutMix [19] and FMix [6]. Some other data augmentation strategy based on masking approach or interpolation method has also attracted the attention of researchers in this field. Cutout [20] applied the idea of random erasing, and the mask enables the network to improve the robustness in the situation of occlusion. In particular, AugMix [5] is a simple and effective data augmentation technology, which leveraged convex combination to mix the results of multiple augmentation chains to increase the diversity and quality of images.

3 Methods

FMixAugment is a method that utilizes image augmentation, pseudo-labeling, and combines consistency regularization technology. Before introducing our approach, it is necessary to declare a set of notations in this paper. Suppose $x \in \mathbb{R}^d$ and $y \in \{1, \cdots, C\}$ respectively denote an input feature and an output label, where d is the input feature dimension and C is the number of label classes. We are eager to learn a model f to predict the label y of the input samples x based on a predicted label distribution $f(y \mid x, \theta)$, where θ represents the model's parameter. $\mathrm{H}(q, p)$ is the cross-entropy between distributions p and q.

3.1 FMixAugment: MixAugment Combined with FMask

The essential motivation of our approach is that we improve the robustness and uncertainty estimation of the model by augmenting the image whilst considering the consistency constrain. First and foremost, we introduce MixAugment, which is a weak variant of AugMix combined with RandAugment. Formally,

MixAugment begins by sampling k augmentation chains, and each augmentation chain consists of one or two randomly selected augmentation operations. In the experiment, we set $k = 2$ by default. The weight coefficients w_i of the k augmentation chains are sampled by applying Dirichlet(β, β) distribution. After combining the augmentation chains, we mix the original image and the image combined with the augmented chains proportionally by randomly sampling from Beta(β, β) distribution, and call it the mixed augmented version. β is set to 1 in our experiment. Ultimately, we perform random augmentation operations on the image that has undergone the above operations.

FMask allows us to generate various masks, which is vital for inducing the robustness of the model. Since horizontal or vertical masks are not conducive to capturing edge information of images in the field of mask robustness, diverse masks can eliminate the model's memory of sample fixation. FMask first samples low-frequency images from Fourier-based space and converts them into gray-scale masks. In particular, suppose $Z = \mathbb{V}^{w\times h}$ denotes a complex random variable, where $\mathbb{V}$ is its corresponding value domain, w and h indicate the width and height of the image. In addition, we make the complex random variable Z obey Gaussian distribution $P_{\mathcal{R}} = \mathcal{N}(\mathbf{0}, \mathbb{1}_{w\times h})$ and $P_{\mathcal{I}} = \mathcal{N}(\mathbf{0}, \mathbb{1}_{w\times h})$, where $\mathcal{R}$ and $\mathcal{I}$ represent the real and imaginary parts of Z, respectively. Subsequently, in order to obtain a low-pass filter, we apply the sample frequency with decay power τ corresponding to Z, which can be demonstrated as

$$\text{filter}(Z,\tau)[i,j] = \frac{Z[i,j]}{\text{freq }[i,j]^{\tau}}, \tag{1}$$

where freq $[i,j]$ represents the fast Fourier transform sample frequency of pixels at the position i,j'th bin for the size of $w\times h$ images. It is worth noting that a higher value of τ corresponds to an increase in the attenuation of high-frequency information, namely, higher τ values result in smoother masks. Next, a gray-scale image can be obtained by inverting the discrete Fourier transform and taking the real-valued part, which can be expressed as

$$\mathcal{G} = \mathcal{R}\left(\mathcal{F}^{-1}(\text{ filter }(Z,\tau))\right). \tag{2}$$

Here, the gray-scale image is normalized into the scale of $[0,1]$. Further, we convert the gray-scale image into a binary mask by selecting the first n largest pixel value and set it to 1, and the rest to 0. Specifically, n is ηwh and η is a hyperparameter of magnitude, which is utilized to remain the image unmasked area, and we set it to 0.8 in our experiments. The binary mask is shown as

$$\text{binary\_mask}[i,j] = \begin{cases} 1, & \text{if } \mathcal{G}[i,j] \in \text{top}(\eta wh, \mathcal{G}); \\ 0, & \text{otherwise .} \end{cases} \tag{3}$$

After obtaining the binary mask, we also need to obtain a gray color mask. The approach of the gray color mask is similar to the binary mask, with the difference that the first n elements are set to 0, and the rest are set to gray color value 0.5. Finally, we convert the above operations into FMask, as shown below.

$$\text{FMask}\ (x) = x * \text{binary\_mask} + \text{gray\_mask}, \tag{4}$$

where x is an image of the input unlabeled sample. Figure 2 depicts an illustration of the main steps of our proposed augmentation operation of FMixAugment. Specifically, x_{mix} and $x_{FMixAug}$ respectively denote mixed augmented version and strongly-augmented version in the figure.

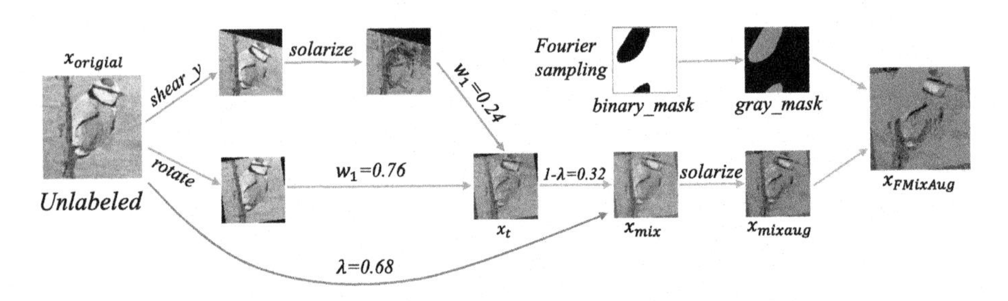

Fig. 2. A cascade of visualization augmentation operations of FMixAugment. The weights w_i and λ are sampled from a Dirichlet(β, β) and a Beta(β, β) distribution, respectively. The combination of augmented chains helps us explore the semantically equivalent input space around the image and improve the uncertainty of the model.

3.2 Improved Consistency Regularization

After the input image is subjected to the strongly-augmented method FMix-Augment, we combine the method with unlabeled samples to constrain the consistency. In this task, we are given labeled data domain $\mathcal{D}_l$ and unlabeled data domain $\mathcal{D}_u$. We randomly sample batch size B examples with their labels $\mathcal{X} = \{(x_l, y_l)_{l=1}^{B}\}$ on labeled data domain. The supervision loss of the classifier on $\mathcal{D}_l$ is defined as

$$\mathcal{L}_X = \frac{1}{B} \sum_{l=1}^{B} \text{H}(\hat{f}_l, f(y \mid \mathcal{W}(x_l); \theta)), \tag{5}$$

where $\mathcal{W}(\cdot)$ is a weakly-augmented version and $\hat{f}_l$ is a one-hot label corresponding to label y. For unlabeled data domain $\mathcal{D}_u$, we randomly sample a batch of unlabeled data $\mathcal{U} = \{(x_u)_{u=1}^{\xi B}\}$, where ξ is the unlabeled data ratio. A weakly-augmented operation is performed on unlabeled data and the predicted class distribution is obtained by the deep network: $f_u = f\,(y \mid \mathcal{W}(x_u)\,; \hat{\theta})$, where $\hat{\theta}$ is a fixed copy of the current model parameter $\hat{\theta}$, suggesting $\hat{\theta}$ does not back-propagated to update the gradient. Then, a pseudo-labeled example is generated by controlling the threshold δ. Thus, the loss function between the weakly-augmented and the strongly-augmented image is calculated as

$$\mathcal{L}_1 = \text{H}\left(\hat{f}_u, f(y \mid \mathcal{S}(x_u); \theta)\right), \tag{6}$$

where $\hat{f}_u$ is the hard pseudo-label, that is, $\hat{f}_u = \arg\max(f_u)$. $\mathcal{S}(x_u)$ refers to the strongly-augmented version that has been processed by FMixAugment. Further, to improve the stability and efficiency of the model, we also perform consistency operation between the mixed augmented version and the weakly-augmented version of unlabeled data. Specifically, the mixed augmented version here refers to perform MixAugment without random augmentation again, which is denoted as $\mathcal{M}(x_u)$. The loss function of MW-MSECC can be expressed as

$$\mathcal{L}_2 = \frac{1}{C} \left\| f_u, f(y \mid \mathcal{M}(x_u); \theta) \right\|_2^2, \tag{7}$$

The consistency cost MW-MSECC is defined as the expected distance of predicted class distribution between the mixed augmented version and the strongly-augmented version. The whole unsupervised loss function is given by

$$\mathcal{L}_U = \frac{1}{\xi B} \sum_{u=1}^{\xi B} \mathbb{1}\left(\max(f_u{}^{(Sharpen)}) \geq \delta\right)(\lambda_1 \mathcal{L}_1 + \lambda_2 \mathcal{L}_2),$$
$$(f_u^{(Sharpen)})_i = \frac{\exp(f_u^{(i)}/T)}{\sum_{i=1}^{C} \exp(f_u^{(i)}/T)}, \tag{8}$$

where $\mathbb{1}(\cdot)$ is an indicator function, λ_1 and λ_2 are hyperparameters of two regularization terms respectively. When judging the confidence of the weakly-augmented unlabeled samples, we also employ sharpen technology, where $f_u^{(i)}$ is the i-th entry of the predicted label distribution f_u and T is sharpening temperature. Lastly, the full objective function is $\mathcal{L} = \mathcal{L}_X + \mathcal{L}_U$.

3.3 Dynamic Growth Threshold

For semi-supervised deep learning, directly predicting the results of unlabeled examples usually makes the model fail to converge due to the scarcity of the proportion of labeled data, which leads to poor generalization performance. To some extent, it is necessary to use the training model to predict the labels on unlabeled examples to generates pseudo-labels when training a model on unlabeled data. Therefore, most studies choose to generate pseudo-labels by exploiting the high confidence level of the threshold δ. It can be roughly regarded as the data item is penalized only when the maximum value of the model's predicted distribution for unlabeled data is greater than the confidence level of the threshold δ. For example, a fixed threshold is set, as reported in UDA and FixMatch.

However, most of the methods mentioned above leverage a fixed-size threshold to generate pseudo-labeled samples. Unlike the above works, in our research, we propose a dynamic growth threshold (DGT) approach to solve the problem of slow convergence caused by a fixed-size threshold. This method is based on gradually increasing the fixed value we set to change the threshold dynamically. The motivation of this idea is substantially simple: we found that in the early stages of model training, especially for the framework based on pseudo-labeling, the value of the threshold during training is crucial to achieving fast and efficient convergence. Since the model does not predict the pseudo-labeled samples accurately enough in the initial training phase, the model converges slowly when the fixed threshold value is set too high. Therefore, as the number of training iterations increases, we will gradually increase the threshold size to ensure that the model improves its generalization ability. In our experiments, we gradually raise the threshold δ from a fixed value to a confidence level close to 1. After introducing the proposed approach, we demonstrate the detailed procedure of FMixAugment algorithm in Algorithm 1.

Algorithm 1. FMixAugment algorithm pseudocode.

1: **Input:** Labeled batch $\mathcal{X} = \{(x_l, y_l)_{l=1}^{B}\}$, unlabeled batch $\mathcal{U} = \{(x_u)_{u=1}^{\xi B}\}$, sharpening temperature T, threshold δ, unsupervised loss weight λ_1, λ_2.
2: **function** FMIXAUGMENT($x_u, k = 2, \beta = 1$)
3: Sample weight $(w_1, w_2 \dots w_k) \sim$ Dirichlet(β, β) and weight $\lambda \sim$ Beta(β, β);
4: Initialize x_t to a zero with the same dimensions as x_u;
5: **for** $i = 1$ to k **do**
6: Randomly sample operations $\mathrm{op}_1, \mathrm{op}_2 \sim \mathcal{O}$;
7: Different depths compose augmentation chain, i.e., op_1, $\mathrm{op}_{12} = \mathrm{op}_2 \circ \mathrm{op}_1$;
8: $x_t + = w_i \cdot$ chain (x_u);
9: **end for**
10: Interpolated mixed sample $x_{mix} = \lambda x_u + (1 - \lambda) x_t$;
11: Randomly Sample operations $\mathrm{op}_3 \sim \mathcal{O}$ and $x_{mixaug} = op_3(x_{mix})$;
12: Apply Eq. (4) to generate FMask: $x_{FMixAug} = \mathrm{FMask}(x_{mixaug})$;
13: **return** x_{mix}, $x_{FMixAug}$.
14: **end function**
15: $\mathcal{M}(x_u)$, $\mathcal{S}(x_u) =$ FMIXAUGMENT($x_u, k = 2, \beta = 1$);
16: Update the threshold δ using DGT method;
17: Substitute $\mathcal{M}(x_u)$ and $\mathcal{S}(x_u)$ into Eq. (6) and Eq. (7), and compute $\mathcal{L}$;
18: Update the parameters θ of the model via taking the gradients with respect to $\mathcal{L}$;
19: **return** θ.

4 Experiments

In this section, we will evaluate FMixAugment on a variety of image recognition tasks. In order to compare the performance of different methods more fairly, we follow the standard SSL evaluation protocol as suggested by [21], which compares the performance between different approaches under the same network architecture. Four publicly available datasets are used to verify the effectiveness, which are CIFAR-10[1], SVHN[2], CIFAR-100[3], and STL-10[4], respectively.

[1] https://www.cs.toronto.edu/~kriz/cifar-10-matlab.tar.gz.
[2] http://ufldl.stanford.edu/housenumbers/{}_32x32.mat.
[3] https://www.cs.toronto.edu/~kriz/cifar-100-matlab.tar.gz.
[4] http://ai.stanford.edu/~acoates/stl10/stl10_binary.tar.gz.

4.1 Implementation Details

In our experiments, the optimizer we employed SGD with Nesterov momentum 0.9. For the scheduler of the learning rate, which is tuned as 0.03 and then using the cosine annealing learning rate decay. The weight decay coefficient we used for CIFAR-10, SVHN and STL-10 is set to 0.0005, and the weight for CIFAR-100 to 0.001. The test model is composed of an exponential moving average of train model following [9], where the decay coefficient is fixed as 0.999. The batch size $B = 64$ for labeled data. As for unlabeled data, the batch size $\xi B = 448$, namely, $\xi = 7$. The total number of training epochs is 1024, and each epoch includes 1024 steps. Next, we will introduce some hyperparameters used in the experiments.

The confidence threshold δ used to predict pseudo-labels is set to 0.93 as its initial value, then linearly ramp up its weight from initial value 0.93 to its final value of 0.9999 during the training epochs. Note that the sharpening temperature T for the threshold of pseudo-labeling is fixed as 0.5, except in the case of labeled samples that are extremely scarce where we set it to 1, such as CIFAR-10 with 40 labels, CIFAR-100 with 250 labels, SVHN with 40 labels. For FMask, we set the η to 0.8, representing the area of the gray color mask as 20% of the total image area, and the τ is fixed as 3. Finally, for hyperparameters λ_1 and λ_2 are set to 1 and 0.05, respectively, except in CIFAR-100 where we set λ_2 to 0.01.

4.2 Experimental Results

In this section, we will present the experimental results by adopting FMixAugment on CIFAR-10/100, SVHN, and STL-10 datasets, which are prevalent semi-supervised image classification tasks. In most of the following experiments, we report the mean and standard deviation of the accuracy with 5 run trials, where the best and second best performances are highlighted in bold and underlined. In addition, to compare the experiments more comprehensively, we run these algorithms with different numbers of labels. Table 1 compares our method against current state-of-the-art techniques on the SVHN and CIFAR-10.

An examination of the data displayed on the left side of Table 1 indicates that our FMixAugment is superior to other state-of-the-art SSL approaches in most cases. Specifically, our method is the first accuracy rate to exceed 90% when CIFAR-10 has only 40 labeled samples under the most commonly used architecture Wide ResNet-28-2. This side-effect suggests that our method is likely to be more applicable to the case where there are extremely few labeled samples. Further, the accuracy rates obtained by FMixAugment are competitive with other robust SSL methods for SVHN on the left side.

From Table 2, we can see that our proposed FMixAugment outperforms most of the baselines under a WRN-28-8 structure with 400, 2500, and 10K labeled training data on CIFAR-100. Moreover, our method outperforms all of the baselines under a WRN-37-2 network with 250 and 1000 labeled training data on STL-10. In conclusion, our method shows better test accuracy rates in a fair comparison on these datasets.

Table 1. Accuracy rates (%) for CIFAR-10, SVHN on 5 different trials, where the best and second best performance are highlighted in bold and underlined respectively (mean and standard deviation).

Method	CIFAR-10			SVHN		
	40 labels	250 labels	4000 labels	40 labels	250 labels	1000 labels
Π-Model	17.43 ± 0.89	46.98 ± 2.05	82.59 ± 0.37	19.19 ± 1.87	81.04 ± 1.92	92.46 ± 0.36
VAT	29.04 ± 1.43	63.97 ± 2.82	88.95 ± 2.28	81.94 ± 4.29	94.65 ± 0.24	95.49 ± 0.15
Mean Teacher	35.07 ± 2.49	67.68 ± 2.30	89.64 ± 0.25	82.57 ± 3.39	96.43 ± 0.11	96.58 ± 0.07
ICT	36.40 ± 6.73	58.34 ± 2.54	92.34 ± 0.17	44.33 ± 13.54	94.10 ± 0.59	96.47 ± 0.07
MixMatch	52.46 ± 11.50	88.95 ± 0.86	93.58 ± 0.10	57.45 ± 14.53	96.02 ± 0.23	96.50 ± 0.28
ReMixMatch	80.90 ± 9.64	94.56 ± 0.05	95.28 ± 0.13	**96.66 ± 0.20**	97.08 ± 0.48	97.35 ± 0.08
UDA	70.95 ± 5.93	91.24 ± 0.90	94.71 ± 0.25	47.37 ± 20.51	97.24 ± 0.17	97.45 ± 0.99
FixMatch (RA)	86.19 ± 3.37	<u>94.93 ± 0.65</u>	<u>95.74 ± 0.05</u>	96.04 ± 2.17	**97.52 ± 0.38**	<u>97.72 ± 0.11</u>
FixMatch (CTA)	<u>88.61 ± 3.35</u>	**94.93 ± 0.33**	95.69 ± 0.15	92.35 ± 7.65	97.36 ± 0.64	97.64 ± 0.19
FMixAugment	**90.21 ± 1.36**	94.88 ± 0.21	**95.83 ± 0.14**	<u>96.23 ± 0.85</u>	<u>97.48 ± 0.21</u>	**97.79 ± 0.18**

Table 2. Accuracy rates (%) for CIFAR-100, STL-100 on 5 different trials, where the best and second best performance are highlighted in bold and underlined respectively (mean and standard deviation).

Method	CIFAR-100			STL-10	
	400 labels	2500 labels	10000 labels	250 labels	1000 labels
Π-Model	9.61 ± 1.37	42.75 ± 0.48	62.12 ± 0.11	40.30 ± 1.33	73.77 ± 0.82
VAT	14.26 ± 1.88	47.07 ± 0.94	65.65 ± 0.14	41.70 ± 1.14	76.57 ± 0.41
Mean Teacher	13.98 ± 0.70	46.09 ± 0.57	64.17 ± 0.24	47.60 ± 0.82	78.57 ± 2.39
ICT	16.74 ± 1.61	49.82 ± 0.10	66.49 ± 0.69	51.13 ± 2.96	76.40 ± 0.23
MixMatch	32.39 ± 1.32	60.06 ± 0.37	71.69 ± 0.33	64.57 ± 3.49	89.59 ± 0.61
ReMixMatch	**55.72** ± 0.26	<u>72.57 ± 0.31</u>	76.97 ± 0.56	68.87 ± 1.14	94.77 ± 0.45
UDA	49.72 ± 0.88	66.87 ± 0.22	75.50 ± 0.25	77.29 ± 1.11	92.34 ± 0.56
FixMatch (RA)	51.15 ± 1.75	71.71 ± 0.11	<u>77.40 ± 0.12</u>	91.17 ± 0.82	92.02 ± 1.50
FixMatch (CTA)	50.05 ± 0.31	71.36 ± 0.24	76.82 ± 0.11	92.36 ± 0.69	<u>94.83 ± 0.63</u>
FMixAugment	<u>53.69 ± 0.45</u>	**73.07 ± 0.25**	**78.55 ± 0.34**	**94.06 ± 0.41**	**95.26 ± 0.36**

4.3 Ablation Study

In this subsection, we will provide several ablation studies to better validate the efficiency of our proposed FMixAugment method. To recap, the loss function used in our algorithm has three components, the supervision loss $\mathcal{L}_X$, the consistency constraint loss $\mathcal{L}_1$, and the MW-MSECC loss term $\mathcal{L}_2$. Among these three parts, we are more interested in the strongly-augmented method used in $\mathcal{L}_1$ and the MW-MSECC loss term $\mathcal{L}_2$. Therefore, in the ablation learning, we remove our strongly-augmented version $\mathcal{S}(x_u)$ of FMixAugment and replace it with RandAugment, namely, the strongly-augmented version $\mathcal{S}(x_u)$ and the mixed augmented version $\mathcal{M}(x_u)$ used in $\mathcal{L}_1$ and $\mathcal{L}_2$, respectively, are both replaced with RandAugment. Figure 3 shows the test accuracy and corresponding training loss dynamics plots of the FMixAugment algorithm when different components are used on the CIFAR-10 dataset.

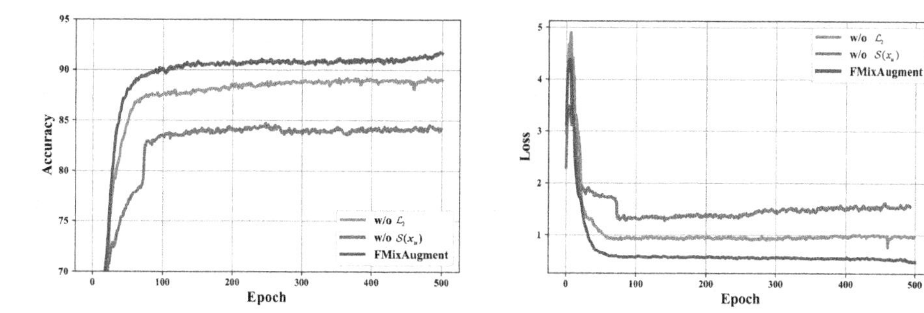

Fig. 3. Ablation study on the MW-MSECC $\mathcal{L}_2$ and the strongly-augmented version $\mathcal{S}(x_u)$, where 'w/o' is short for 'without'. The left plot is the test accuracy on CIFAR-10 dataset, and the right plot is the corresponding dynamics loss on CIFAR-10 dataset. (Color figure online)

In both two figures, the blue line indicates the full objective function used for training with the proposed method FMixAugment. The orange line represents that only $\mathcal{L}_1$ and $\mathcal{L}_X$ are applied. The green line expresses without using the strongly-augmented version FMixAugment. We can see from Fig. 3 that each component contributes to the performance of our algorithm, especially for the introduced FMixAugment strongly-augmented approach. For the augmentation methods we proposed, the performance of the algorithm can be improved to a great extent. For the without used FMixAugment, it will increase the loss at about 300 to 500 epochs. However, FMixAugment helps us to accelerate the convergence of the model to achieve better performance. Further, by comparing the orange and blue lines, we can find that the MW-MSECC can further accelerate the convergence and improve the efficiency of the model, thus demonstrating the necessity of the MW-MSECC regularization term.

5 Conclusion and Future Work

In this paper, we have proposed an efficient and robust algorithm for semi-supervised image recognition, FMixAugment, which combines MixAugment and FMask to generate a strongly-augmented image. In order to improve the confidence of the weakly-augmented images, we also propose the MW-MSECC regularization term for consistency constraints and DGT settings. Specifically, our proposed approach can significantly improve performance compared with other methods. In particular, our method has a clear advantage over an extremely scarce number of labeled samples. In future work, we also continue to explore a new augmentation method and the loss function of the model to achieve better performance on fewer samples. In addition, we hope that our approach would be extended to other domains and conduct further research, such as applying to semi-supervised multi-label image classification.

Acknowledgments. This work is in part supported by the National Natural Science Foundation of China (Grant No. U1705262), the Natural Science Foundation of Fujian Province (Grant Nos. 2020J01130193 and 2018J07005).

References

1. Xu, N., Liu, Y.-P., Geng, X.: Partial multi-label learning with label distribution. In: Proceedings of the Thirty-Fourth AAAI Conference on Artificial Intelligence, vol. 34, pp. 6510–6517 (2020)
2. Chen, L.-Z., Lin, Z., Wang, Z., Yang, Y.-L., Cheng, M.-M.: Spatial information guided convolution for real-time RGBD semantic segmentation. IEEE Trans. Image Process. **30**, 2313–2324 (2021)
3. Yang, X., Yan, J.: Arbitrary-oriented object detection with circular smooth label. In: Vedaldi, A., Bischof, H., Brox, T., Frahm, J.-M. (eds.) ECCV 2020. LNCS, vol. 12353, pp. 677–694. Springer, Cham (2020). https://doi.org/10.1007/978-3-030-58598-3_40
4. Wang, Q., Gao, J., Yuan, Y.: A joint convolutional neural networks and context transfer for street scenes labeling. IEEE Trans. Intell. Transp. Syst. **19**(5), 1457–1470 (2017)
5. Hendrycks, D., Mu, N., Cubuk, E.D., Zoph, B., Gilmer, J., Lakshminarayanan, B.: Augmix: a simple data processing method to improve robustness and uncertainty. In: Proceedings of the Eighth International Conference on Learning Representations (2020)
6. Harris, E., Marcu, A., Painter, M., Niranjan, M., Prugel-Bennett, A., Hare, J.: Fmix: enhancing mixed sample data augmentation, arXiv preprint arXiv:2002.12047, vol. 2, no. 3, p. 4 (2020)
7. Bachman, P., Alsharif, O., Precup, D.: Learning with pseudo-ensembles. In: Proceedings of the Twenty-Eighth Conference on Advances in Neural Information Processing Systems, pp. 3365–3373 (2014)
8. Rasmus, A., Valpola, H., Honkala, M., Berglund, M., Raiko, T.: Semi-supervised learning with ladder networks. In: Proceedings of the Twenty-Ninth Conference on Advances in Neural Information Processing Systems, pp. 3546–3554 (2015)
9. Tarvainen, A., Valpola, H.: Mean teachers are better role models: weight-averaged consistency targets improve semi-supervised deep learning results. In: Proceedings of the Thirty-First Conference on Advances in Neural Information Processing Systems, pp. 1195–1204 (2017)
10. Miyato, T., Maeda, S., Koyama, M., Ishii, S.: Virtual adversarial training: a regularization method for supervised and semi-supervised learning. IEEE Trans. Pattern Anal. Mach. Intell. **41**(8), 1979–1993 (2019)
11. Verma, V., Lamb, A., Kannala, J., Bengio, Y., Lopez-Paz, D.: Interpolation consistency training for semi-supervised learning. In: Proceedings of the Twenty-Eighth International Joint Conference on Artificial Intelligence, pp. 3635–3641 (2019)
12. Xie, Q., Dai, Z., Hovy, E.H., Luong, T., Le, Q.: Unsupervised data augmentation for consistency training. In: Proceedings of the Thirty-Fourth Conference on Advances in Neural Information Processing Systems (2020)
13. Berthelot, D., Carlini, N., Goodfellow, I., Papernot, N., Oliver, A., Raffel, C.A.: Mixmatch: a holistic approach to semi-supervised learning. In: Proceedings of the Thirty-Third Conference on Advances in Neural Information Processing Systems, pp. 5049–5059 (2019)

14. Zhang, H., Cisse, M., Dauphin, Y.N., Lopez-Paz, D.: mixup: beyond empirical risk minimization. In: Proceedings of the Sixth International Conference on Learning Representations (2018)
15. Berthelot, D., et al.: Remixmatch: semi-supervised learning with distribution matching and augmentation anchoring. In: Proceedings of the Eighth International Conference on Learning Representations (2020)
16. Sohn, K., et al.: Fixmatch: simplifying semi-supervised learning with consistency and confidence. In: Proceedings of the Thirty-Fourth Conference on Advances in Neural Information Processing Systems (2020)
17. Xie, Q., Luong, M.-T., Hovy, E., Le, Q.V.: Self-training with noisy student improves imagenet classification. In: Proceedings of the IEEE Conference on Computer Vision and Pattern Recognition, pp. 10687–10698 (2020)
18. Cui, S., Wang, S., Zhuo, J., Li, L., Huang, Q., Tian, Q.: Towards discriminability and diversity: batch nuclear-norm maximization under label insufficient situations. In: Proceedings of the IEEE Conference on Computer Vision and Pattern Recognition, pp. 3941–3950 (2020)
19. Yun, S., Han, D., Chun, S., Oh, S.J., Yoo, Y., Choe, J.: Cutmix: regularization strategy to train strong classifiers with localizable features. In: Proceedings of the IEEE International Conference on Computer Vision, pp. 6022–6031 (2019)
20. DeVries, T., Taylor, G.W.: Improved regularization of convolutional neural networks with cutout, arXiv preprint arXiv:1708.04552 (2017)
21. Oliver, A., Odena, A., Raffel, C.A., Cubuk, E.D., Goodfellow, I.: Realistic evaluation of deep semi-supervised learning algorithms. In: Proceedings of the Thirty-Second Conference on Advances in Neural Information Processing Systems, pp. 3235–3246 (2018)

IDANet: Iterative D-LinkNets with Attention for Road Extraction from High-Resolution Satellite Imagery

Benzhu Xu, Shengshuai Bao, Liping Zheng, Gaofeng Zhang, and Wenming Wu(✉)

School of Computer Science and Information Engineering,
Hefei University of Technology, Hefei 230009, China
wwming@hfut.edu.cn
http://ci.hfut.edu.cn/

Abstract. Road information plays a fundamental role in many application fields, while satellite images are able to capture a large area of the ground with high resolution. Therefore, extracting roads has become a hot research topic in the field of remote sensing. In this paper, we propose a novel semantic segmentation model, named IDANet, which adopts iterative D-LinkNets with attention modules for road extraction from high-resolution satellite images. Our road extraction model is built on D-LinkNet, an effective network which adopts encoder-decoder structure, dilated convolution, and pretrained encoder for road extraction task. The attention mechanism can be used to achieve a better fusion of features from different levels. To this end, a modified D-LinkNet with attention is proposed for more effective feature extraction. With this network as the basic refinement module, we further adopt an iterative architecture to maximize the network performance, where the output of the previous network serves as the input of the next network to refine the road segmentation and obtain enhanced results. The evaluation demonstrates the superior performance of our proposed model. Specifically, the performance of our model exceeds the original D-LinkNet by 2.2% of the IoU on the testing dataset of DeepGlobe for road extraction.

Keywords: Road extraction · Semantic segmentation · Convolutional neural network · Attention mechanism · Iterative architecture

1 Introduction

Accurate and real-time road information update is of great significance in many applications, such as urban planning and construction, vehicle navigation, natural disaster analysis, etc. Benefiting from the development of remote sensing

This work was supported in part by the National Natural Science Foundation of China (No. 61972128, 61906058) and the Natural Science Foundation of Anhui Province, China (No. 1808085MF176, 1908085MF210) and the Fundamental Research Funds for the Central Universities, China (No. PA2021KCPY0050).

H. Ma et al. (Eds.): PRCV 2021, LNCS 13020, pp. 140–152, 2021.
https://doi.org/10.1007/978-3-030-88007-1_12

technology, continuous ground observations via remote sensing satellites have been achieved, and a large number of high-resolution satellite images can be easily obtained, providing a reliable and abundant data source with rich spatial structure and geometric texture for extracting various ground targets.

Traditional methods aim to detect roads by carefully designing multi-level and multi-scale features to distinguish between the road and non-road. However, it is challenging to choose effective features, especially in complex heterogeneous regions where surroundings such as buildings, vegetation are all contextual features that also affect the original features of the road. Recently, with the rise of deep learning, convolutional neural networks (CNN) have shown considerable development in feature extraction and have been applied in many computer vision tasks, including image recognition [11] and semantic segmentation [14]. The "era of deep learning" also popularizes the use of deep neural networks in the field of remote sensing, which has enabled effective automatic road extraction.

CNN-based approaches have been proposed for extracting roads, and remarkable improvements have been made. Most of them consider road extraction as semantic segmentation. However, this problem is far from solved. On the one hand, ground information provided by satellite images has increased dramatically with altitude, which reduces the differences of ground targets. On the other hand, roads often have differences in width, color, and structure. Surroundings usually cause shadow and occlusion issues. The above-mentioned has undoubtedly increased the difficulty of road extraction with inherent complexity and variability. CNN-based methods extract road parts with some important parts missed and poor connectivity. The road is sparser compared to other ground targets, and the imbalance of categories increases the difficulty of semantic segmentation. Key features can not be well extracted from various feature information.

To this end, we propose a novel semantic segmentation model named IDANet, which adopts iterative D-LinkNets [20] with attention modules [19] for road extraction with much better accuracy, connectivity, and completeness than previous methods. We equip the original D-LinkNet with more advanced dilated convolution modules, which can further enhance the ability of feature learning. To fix the issue of category imbalance, a powerful loss function with focal loss [13] is adopted. We also introduce recent popular attention modules into our model for more effective feature extraction. Furthermore, we use an iterative architecture to obtain enhanced results. The output of the previous network serves as the input of the next network to refine the results of semantic segmentation. The main contributions of our work are summarized as follows:

- A improved encoder-decoder network with attention is introduced as the basic module of our model, which can significantly improve the results of semantic segmentation.
- A novel dilated convolution module which has fewer parameters but better performance than that in original D-LinkNet.
- A novel iterative framework is proposed for road extraction from high-resolution satellite images, which can further refine and enhance the results of road extraction.

2 Related Work

Our work relates to a line of research on semantic segmentation in the field of deep learning. We refer the reader to the recent survey [8] on semantic segmentation in deep learning for discussions on a variety of methods. CNN has achieved unprecedented success in many computer vision tasks, which also provides various powerful tools for semantic segmentation. Fully convolutional networks (FCN) [14] has made milestone progress in image segmentation, which extends the classification at the image level into the pixel level with small storage and high segmentation efficiency. Since then, various FCN-based methods [2,21] continuously refresh the record of semantic segmentation. U-Net [16] adopts a symmetric U-shaped structure, which lays the foundation for the following design of the segmentation network. To increase the receptive field without information loss due to decreased resolution, DeepLab [5,6] is based on dilated convolution and has shown strong abilities to increase the segmentation accuracy.

Inspired by the research of image segmentation, CNN-based methods provide new chances for road extraction. Excellent works [1] have been proposed for road extraction. Mattyus et al. [15] develop a variant of FCN using ResNet as an encoder with a fully deconvolutional decoder to estimate road topology. Zhang et al. [20] introduce a deep residual U-Net for road extraction. Skip connections are used to obtain improved performance by information propagation. Bastani et al. [3] propose RoadTracer to extract road network using an iterative search process guided by a patch-based CNN decision function. Based on this, Lian and Huang [12] develop a road network tracking algorithm for road extraction. Zhou et al. [22] propose D-LinkNet for road semantic segmentation, which contains an encoder–decoder structure with a dilated convolution part in the center. D-LinkNet achieves obvious improvement in road extraction but retains several issues concerning road connectivity and recognition. Based on this, Huang et al. [10] propose D-CrossLinkNet by adding cross-resolution connections in D-LinkNet. Our proposed model also makes full use of the outstanding extraction capability of the D-LinkNet architecture. Long-distance spatial information learning is very important in road extraction. Except for dilated convolution, attention mechanism [9,18,19] can also achieve good results in global information learning. Therefore, we introduce recent popular attention modules [19] into our model for more effective feature extraction, which could also reduce the loss of short-distance spatial features caused by dilated convolution.

3 Methodology

3.1 Overview

We propose a novel semantic segmentation model IDANet, which adopts iterative D-LinkNets with attention modules for road extraction from high-resolution satellite images. The whole network is designed in an iterative architecture, as shown in Fig. 1, which can strengthen the learning of semantic segmentation by

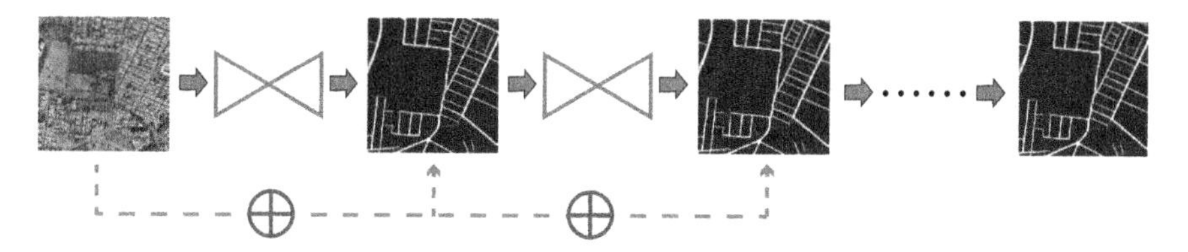

Fig. 1. Iterative architecture. IDANet adopts iterative D-LinkNets with attention modules for road extraction.

fusing the original input and the intermediate result generated in each iteration. IDANet uses D-LinkNet as the basic iteration module, but some effective modifications have been made. First of all, to enhance the effect of the dilated convolution, we modify the dilation module in the original D-LinkNet to make better use of feature information at different levels. We also introduce attention modules into our model for more effective feature extraction. Finally, a powerful loss function with focal loss is adopted to fix the issue of category imbalance. The iteration module is repeated to achieve self-correction and further improve the segmentation output. Therefore, IDANet can extract road information with much better accuracy, connectivity, and completeness than previous methods.

3.2 Basic Iteration Module

D-LinkNet is a classical encoder-decoder network that receives high-resolution images as input. The encoder part and decoder part of our model remain the same as the original D-LinkNet, so our model can also work with high-resolution images. The encoder part reduces the resolution of the feature map through the pooling layers, if an image of size 1024×1024 goes through the encoder part, the output feature map will be of size 32×32. The decoder part uses several transposed convolution layers to do upsampling, restoring the resolution of the feature map from 32×32 to 1024×1024, as shown in Fig. 2.

Dilation Module. Having a large receptive field is important for road extraction, as roads in most satellite images span the whole image with some natural properties such as narrowness, connectivity, complexity. The highlight of D-LinkNet is its focus on increasing the receptive field of feature points by embedding dilation convolutions in the network without decreasing the resolution of feature maps. In the original D-LinkNet, different dilated convolutions are stacked both in cascade mode and parallel mode. This will inevitably lead to the loss of short-distance spatial information due to the "holes" introduced in the computation of dilated convolution. A better approach is to use the original input to further strengthen the output of the dilated convolution to achieve the supplement of information features and decrease the information loss in the computation of dilated convolution. Therefore, we only keep the backbone dilation network where the dilation rates of the stacked dilated convolution layers are 1, 2, 4, 8, respectively, and discard other parallel branch networks. Inspired

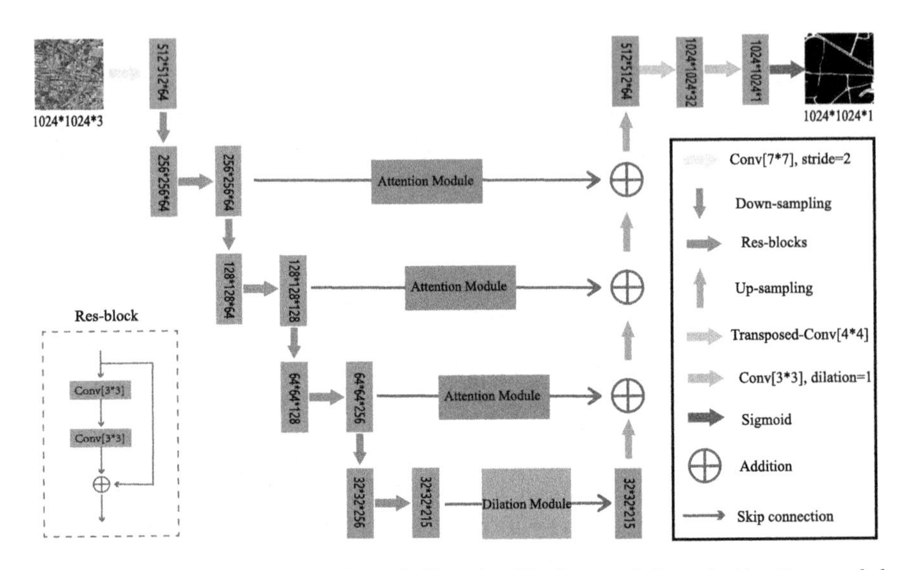

Fig. 2. The basic iteration module including the dilation module and attention module.

by the identity mapping, we also add identity mapping between different dilated convolution layers, as shown in Fig. 3(a).

Attention Module. Upsampling can supplement some lost image information, but it is certainly incomplete. Therefore, the feature map from upsampling of the decoder part and which from the corresponding layer of the encoder part are concatenated together by skip connection which bypasses the input of each encoder layer to the corresponding decoder. This operation can bring in the features of lower convolution layers which contain rich low-level spatial information. To achieve better fusion between these two feature maps, we introduce attention modules into our model. We choose recent popular attention architecture CBAM (Convolutional Block Attention Module) [19] as our attention module, as shown in Fig. 3(b). CBAM can achieve good results in global and long-distance spatial information learning by combining the channel attention module and spatial attention module. Specifically, the channel attention module squeezes the spatial dimension of the input feature map and focuses on which channels are meaningful. The spatial attention module generates a spatial attention map by utilizing the inter-spatial relationship of features and focuses on where is an informative part given an input image. Then these two attention maps are multiplied to the input feature map for adaptive feature refinement.

Loss Function. D-LinkNet uses BCE (Binary Cross Entropy) and dice coefficient loss as loss function. Binary Cross-Entropy performs pixel-level classification and defined as:

$$L_{BCE}(y, \hat{y}) = -(ylog(\hat{y}) + (1 - y)log(1 - \hat{y})) \tag{1}$$

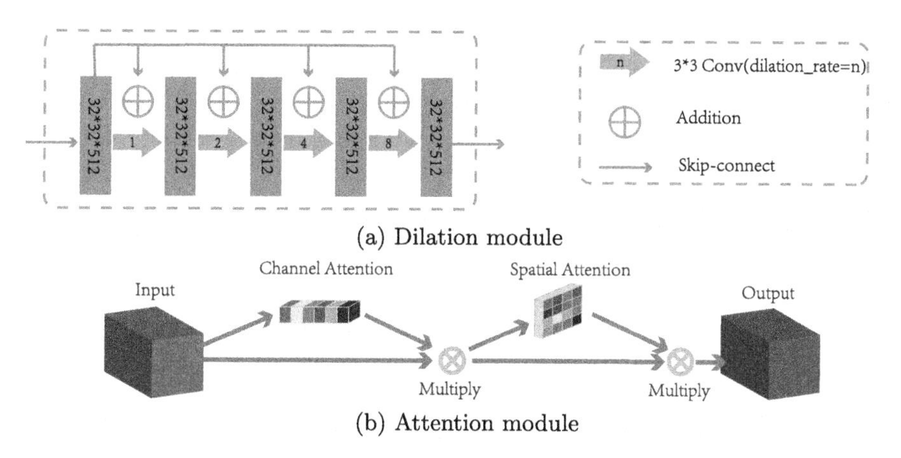

Fig. 3. Dilation module(a) and attention module(b) used in IDANet.

Here, $\hat{y}$ is the predicted value. Dice coefficient loss is widely used to calculate the similarity between two images and defined as:

$$DL_(y, \hat{p}) = 1 - \frac{2y\hat{p} + 1}{y + \hat{p} + 1} \tag{2}$$

Here, 1 is added to ensure that the function is well defined when $y = \hat{p} = 0$.

However, we have a category imbalance problem, since most of the areas in satellite images are non-road pixels (more than 90% of the area pixels are non-road pixels), which also causes that it is difficult to train sparse samples and seriously affect the training effect. To this end, the focal loss [13] is introduced. It can not only alleviate the imbalance of categories but also down-weight the contribution of easy-training examples and enable the model to focus more on learning hard-training samples. Focal Loss is defined as:

$$FL_(p_t) = -\alpha_t(1 - p_t)^\gamma log(p_t) \tag{3}$$

$$p_t = \begin{cases} p, & \text{if y = 1} \\ 1 - p, & otherwise \end{cases} \tag{4}$$

Here, $\gamma > 0$ and $0 \leq \alpha_t \leq 1$. Since the real goal of image segmentation is to maximize IoU metrics, and dice coefficient is calculated based on IoU, dice coefficient loss is especially suitable for the optimization of IoU. Therefore, we adopt focal loss and dice coefficient loss as our loss function:

$$L = \alpha_f FL + \alpha_d DL \tag{5}$$

Here, α_f and α_d are weights for focal loss and dice coefficient loss, respectively.

3.3 Iterative Architecture

Using the above modified D-LinkNet as the basic module, we adopt an iterative architecture to refine the road segmentation and obtain enhanced results, as shown in Fig. 1. From the perspective of data processing, our iterative algorithm can be regarded to enhance the post-processing of output results. In our iterative architecture, the output of the previous network serves as the input of the next network to refine the road segmentation and obtain enhanced results. The input image of our model is expressed as I, the input of the ith iteration is expressed as I_i, and the output of the ith iteration is expressed as O_i. Each iteration step can be defined as follows:

$$D\text{-}LinkNet(I_{i+1}) \longrightarrow O_{i+1}, i = 1, ..., n \tag{6}$$

I_{i+1} is the splicing result of I and O_i along the channel in the tth basic iterative module, which is defined as follows:

$$I_{i+1} = concat(I, O_i), i = 1, ..., n \tag{7}$$

where n is the number of basic iteration modules. The iterative architecture integrates information through repeated iterative enhancement learning of the splicing results of D-LinkNet output results and original images, which can further refine and enhance the results of road extraction. The proposed iterative architecture is very useful and efficient. The training time for a single iteration module is decreased with the iteration increasing, while impressive performance can be obtained. It contributes about 0.5% at IoU in our experiment.

4 Experiment

4.1 Datasets

We have performed our experiments on two diverse datasets: 1) the DeepGlobe dataset [7] and 2) the Beijing-Shanghai dataset [17], as shown in Fig. 4. The DeepGlobe dataset consists of 6226 annotated satellite images with an image resolution of 1024 × 1024. Among them, 5226 images are used for training, and the left is used as the testing set. There are 348 satellite images (298 Beijing maps and 50 Shanghai maps) in the Beijing-Shanghai dataset. Each image has a size of 1024×1024. During the experiments, 278 images are used for training, and the left is used as the testing set. For these benchmark datasets, the road labels on the image are manually marked. Specifically, roads are labeled as foreground, and other objects are labeled as background.

4.2 Implementation Details

We use PyTorch to implement and train our networks. All models are trained and tested on an NVIDIA RTX 2080TI with 11 GB memory. Limited by the hardware, we can only train IDANet step by step, which means that the results of the previous network serve as the training data for the next network. During the training process, we find that only the first iteration network takes about 200 epochs to converge, and the following iteration modules have faster convergence

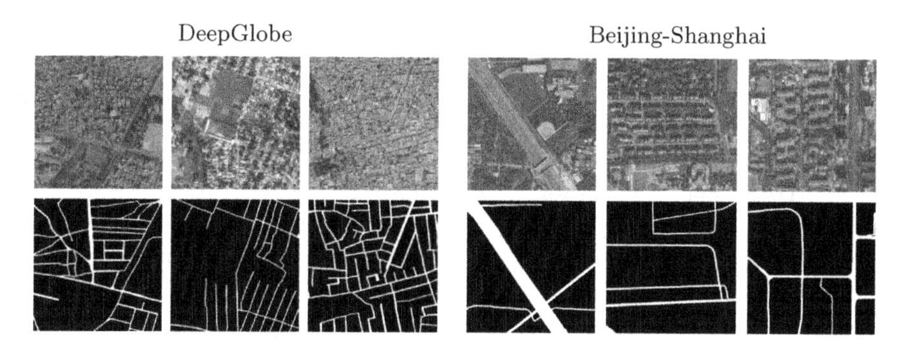

Fig. 4. Visualization of two benchmark datasets.

with less than 100 epochs. The number of the iteration network is an important architecture parameter that can affect the final segmentation results. The number of basic iteration modules is set to $n = 3$, which is an experimental value and enables to obtain the final stable segmentation results. For more discussion and analysis of the iteration number, please refer to Sect. 5.3. In our experiment, the loss weights for focal loss and dice coefficient loss are empirically set for different datasets to achieve the best results. Specifically, $\alpha_f = 30$ & $\alpha_d = 1$ for the DeepGlobe dataset and $\alpha_f = 20$ & $\alpha_d = 1$ for the Beijing-Shanghai dataset in our experiments. We also implement data augmentation similar to D-LinkNet, which can make full use of the limited amount of training data.

5 Results

Accuracy, *Recall*, and *IoU* (Intersection over Union) are commonly used as the evaluation indicators for semantic segmentation. Specifically, *Accuracy* is the ratio of the number of correctly predicted samples to the total number of predicted samples. *Recall* is the ratio of the number of correctly predicted positive samples to the total number of positive samples. *IoU* refers to the ratio between the intersection of the road pixels predicted and the true road pixels and the result of their union. In our experiments, road pixels are labeled as foreground, and other pixels are labeled as background, so we also adopt *Accuracy*, *Recall*, and *IoU* to evaluate the segmentation results at the pixel level.

5.1 Comparison of Road Segmentation Methods

To evaluate our model, we select U-Net [16], LinkNet [4], D-LinkNet [22], and D-CrossLinkNet [10] as competitors. Specifically, U-Net is trained with 7 pooling layers, and LinkNet is trained with pretrained encoder but without dilated convolution in the center part. We also compare our proposed model with the non-iterative version of IDANet, namely BaseNet. We have trained these models on two benchmark datasets. *Accuracy*, *Recall*, and *IoU* of each method on

Table 1. Comparison results on the testing dataset of different models

Dataset	Method	*Accuracy*	*Recall*	*IoU*
DeepGlobe	U-Net	0.9652	0.5637	0.5202
	LinkNet	0.9779	0.7948	0.6681
	D-LinkNet	0.9752	0.8005	0.6700
	D-CrossLinkNet	0.9749	0.8011	0.6710
	BaseNet	0.9802	0.8066	0.6876
	IDANet	**0.9813**	**0.8201**	**0.6921**
Beijing-Shanghai	Unet	0.9359	0.7131	0.5478
	LinkNet	0.9404	0.7458	0.5800
	D-LinkNet	0.9429	0.7282	0.5835
	D-CrossLinkNet	**0.9431**	0.7294	0.5841
	BaseNet	0.9419	0.7502	0.5891
	IDANet	0.9416	**0.7710**	**0.5948**

testing datasets are calculated. The results of these models on different datasets are shown in Table 1.

For the DeepGlobe dataset, it can be observed that, compared with U-Net, LinkNet, D-LinkNet, and D-CrossLinkNet, our model (both BaseNet and IDANet) achieves the best performance in all of evaluation metrics. Both our BaseNet and IDANet exceed D-LinknNet and D-CrossLinkNet considerably. Taking the results of D-LinkNet as a baseline, the *Accuracy* improves 0.61%, the *Recall* improves 1.96%, and the *IoU* improves 2.21%. The experimental results on the DeepGlobe dataset are shown Fig. 5.

In terms of the Beijing-Shanghai dataset where the amount of data is much smaller than the DeepGlobe dataset, the results are much poor for all indicators. Even so, IDANet achieves the best performance in both *Recall* and *IoU*. The *Recall* and *IoU* of IDANet are 4.28% and 1.13% higher than the results of D-LinkNet, respectively. For the results of *Accuracy*, IDANet is slightly lower than D-LinkNet's 0.9429 and D-CrossLinkNet's 0.9416. This small difference in *Accuracy* can be attributed to insufficient training data. Furthermore, *Recall* and *Accuracy* are mutually restricted in general. In the case of small data sets, the pursuit of high *Recall* will lead to lower *Accuracy*, which is a normal phenomenon. This result can be understood that our method can further improve the *Recall* and *IoU* while maintaining the *Accuracy* compared with other models.

5.2 Ablation Experiment

This section aims to further certify the effectiveness and universality of the modules introduced in IDANet, including the dilation module, attention module,

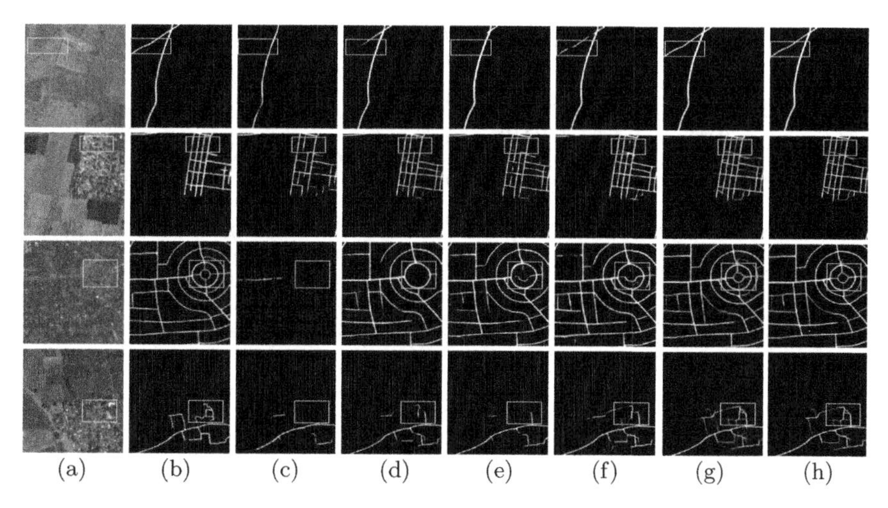

Fig. 5. Visualization results on the DeepGlobe dataset. From left to right: (a) Satellite image, (b) Ground truth, (c) U-Net, (d) LinkNet, (e) D-LinkNet, (f) D-CrossLinkNet, (g) BaseNet, (h) IDANet.

and loss function. Therefore, we have done an ablation experiment on the DeepGlobe dataset. We only perform one iteration for IDANet as well as other models derived from IDANet. We first remove the dilation and attention module from IDANet, denoted as IDANet-D and IDANet-A, respectively. Then we remove both the dilation and attention module from IDANet, denoted as IDANet-DA. We denote the network where BCE loss and dice coefficient loss are used to train our IDANet as IDANet+diceBCE. The results of the ablation experiment are shown in Table 2. For the DeepGlobe dataset, IDANet outperforms other derived versions in terms of the *Accuracy* and *IoU* in the ablation experiment. For the results of *Recall*, the performance of IDANet is slightly lower than IDANet-Dilation. As discussed in Sect. 5.1, *Recall* and *Accuracy* are mutually restricted, and high *Accuracy* results in lower *Recall*, which means that the network is more capable of correcting errors. Compared with IDANet+diceBCE, our loss function can alleviate the imbalance of training samples and focus on learning hard-training samples, which can effectively improve the segmentation results. Experiments show that the dilation module, attention module, and loss function are effective for the semantic segmentation of road extraction.

5.3 The Influence of Network Iteration

In this section, we evaluate the effects of different iterations of IDANet on the performance of road extraction for the DeepGlobe dataset. The results are described in Fig. 6. When the iteration number n is increased, the *IoU* of IDANet is also increasing until the iteration number reaches $n = 3$. After that, the performance of IDANet changes slightly to converge. The results prove that the performance of the model is robust to parameter $n = 3$ for the DeepGlobe dataset and achieves a tradeoff between accuracy and efficiency.

Table 2. Ablation experiment

Method	Accuracy	Recall	IoU
IDANet	**0.9802**	0.8066	**0.6876**
IDANet-D	0.9791	**0.8086**	0.6845
IDANet-A	0.9788	0.8053	0.6842
IDANet-DA	0.9764	0.7999	0.6792
IDANet+diceBCE	0.9758	0.8007	0.6769

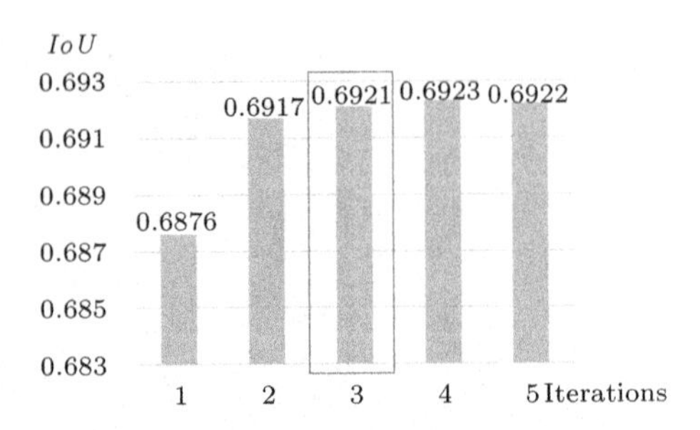

Fig. 6. Different iterations of IDANet

6 Conclusion

This paper aims to improve the accuracy of road extraction from high-resolution satellite images by a novel encoder-decoder network, IDANet, which adopts an iterative architecture to refine the road segmentation and obtain enhanced results. For more effective feature learning, a modified D-LinkNet with attention is proposed as the basic iteration module. We evaluate IDANet on two benchmark datasets. Experimental results show that our model can extract road information from high-resolution satellite images with much better accuracy, connectivity, and completeness than previous methods.

Nevertheless, there is still a big gap between the results we have obtained and the expected. The main issues of current results are the missed identification and wrong recognition. The overall segmentation accuracy is still unsatisfactory. Therefore, our future work will address two aspects. On the one hand, data is always the foundation of deep learning. We plan to adopt more advanced data augmentation techniques and more effective data post-processing methods to enable supervised learning. On the other hand, network is always the key to deep learning. We plan to do more research on the design of refinement networks to realize global and local improvement of semantic segmentation.

References

1. Abdollahi, A., Pradhan, B., Shukla, N., Chakraborty, S., Alamri, A.: Deep learning approaches applied to remote sensing datasets for road extraction: a state-of-the-art review. Remote Sens. **12**(9), 1444 (2020)
2. Badrinarayanan, V., Kendall, A., Cipolla, R.: Segnet: a deep convolutional encoder-decoder architecture for image segmentation. IEEE Trans. Pattern Anal. Mach. Intell. **39**(12), 2481–2495 (2017)
3. Bastani, F., et al.: Roadtracer: automatic extraction of road networks from aerial images. In: Proceedings of the IEEE Conference on Computer Vision and Pattern Recognition, pp. 4720–4728 (2018)

4. Chaurasia, A., Culurciello, E.: Linknet: exploiting encoder representations for efficient semantic segmentation. In: 2017 IEEE Visual Communications and Image Processing (VCIP), pp. 1–4. IEEE (2017)
5. Chen, L.C., Papandreou, G., Kokkinos, I., Murphy, K., Yuille, A.L.: Deeplab: semantic image segmentation with deep convolutional nets, atrous convolution, and fully connected CRFs. IEEE Trans. Pattern Anal. Mach. Intell. **40**(4), 834–848 (2017)
6. Chen, L.C., Zhu, Y., Papandreou, G., Schroff, F., Adam, H.: Encoder-decoder with atrous separable convolution for semantic image segmentation. In: Proceedings of the European Conference on Computer Vision (ECCV), pp. 801–818 (2018)
7. Demir, I., et al.: Deepglobe 2018: a challenge to parse the earth through satellite images. In: Proceedings of the IEEE Conference on Computer Vision and Pattern Recognition Workshops, pp. 172–181 (2018)
8. Hao, S., Zhou, Y., Guo, Y.: A brief survey on semantic segmentation with deep learning. Neurocomputing **406**, 302–321 (2020)
9. Hu, J., Shen, L., Sun, G.: Squeeze-and-excitation networks. In: Proceedings of the IEEE Conference on Computer Vision and Pattern Recognition, pp. 7132–7141 (2018)
10. Huang, K., Shi, J., Zhang, G., Xu, B., Zheng, L.: D-CrossLinkNet for automatic road extraction from aerial imagery. In: Peng, Y., et al. (eds.) PRCV 2020. LNCS, vol. 12305, pp. 315–327. Springer, Cham (2020). https://doi.org/10.1007/978-3-030-60633-6_26
11. Krizhevsky, A., Sutskever, I., Hinton, G.E.: Imagenet classification with deep convolutional neural networks. Adv. Neural. Inf. Process. Syst. **25**, 1097–1105 (2012)
12. Lian, R., Huang, L.: Deepwindow: sliding window based on deep learning for road extraction from remote sensing images. IEEE J. Sel. Top Appl. Earth Obs. Remote Sens. **13**, 1905–1916 (2020)
13. Lin, T.Y., Goyal, P., Girshick, R., He, K., Dollár, P.: Focal loss for dense object detection. In: Proceedings of the IEEE International Conference on Computer Vision, pp. 2980–2988 (2017)
14. Long, J., Shelhamer, E., Darrell, T.: Fully convolutional networks for semantic segmentation. In: Proceedings of the IEEE Conference on Computer Vision and Pattern Recognition, pp. 3431–3440 (2015)
15. Máttyus, G., Luo, W., Urtasun, R.: Deeproadmapper: extracting road topology from aerial images. In: Proceedings of the IEEE International Conference on Computer Vision, pp. 3438–3446 (2017)
16. Ronneberger, O., Fischer, P., Brox, T.: U-Net: convolutional networks for biomedical image segmentation. In: Navab, N., Hornegger, J., Wells, W.M., Frangi, A.F. (eds.) MICCAI 2015. LNCS, vol. 9351, pp. 234–241. Springer, Cham (2015). https://doi.org/10.1007/978-3-319-24574-4_28
17. Sun, T., Di, Z., Che, P., Liu, C., Wang, Y.: Leveraging crowdsourced GPS data for road extraction from aerial imagery. In: Proceedings of the IEEE/CVF Conference on Computer Vision and Pattern Recognition, pp. 7509–7518 (2019)
18. Wang, F., et al.: Residual attention network for image classification. In: Proceedings of the IEEE Conference on Computer Vision and Pattern Recognition, pp. 3156–3164 (2017)
19. Woo, S., Park, J., Lee, J.Y., Kweon, I.S.: Cbam: convolutional block attention module. In: Proceedings of the European Conference on Computer Vision (ECCV), pp. 3–19 (2018)
20. Zhang, Z., Liu, Q., Wang, Y.: Road extraction by deep residual u-net. IEEE Geosci. Remote Sens. Lett. **15**(5), 749–753 (2018)

21. Zhao, H., Shi, J., Qi, X., Wang, X., Jia, J.: Pyramid scene parsing network. In: Proceedings of the IEEE Conference on Computer Vision and Pattern Recognition, pp. 2881–2890 (2017)
22. Zhou, L., Zhang, C., Wu, M.: D-linknet: Linknet with pretrained encoder and dilated convolution for high resolution satellite imagery road extraction. In: Proceedings of the IEEE Conference on Computer Vision and Pattern Recognition Workshops, pp. 182–186 (2018)

Disentangling Deep Network for Reconstructing 3D Object Shapes from Single 2D Images

Yang Yang[1], Junwei Han[1], Dingwen Zhang[1](✉), and De Cheng[2]

[1] School of Automation, Northwestern Polytechnical University, Xi'an 710072, Shaanxi, People's Republic of China
tp030ny@mail.nwpu.edu.cn

[2] The School of Telecommunication Engineering, Xidian University, Xi'an 710071, Shaanxi, People's Republic of China
dcheng@xidian.edu.cn

Abstract. Recovering 3D shapes of deformable objects from single 2D images is an extremely challenging and ill-posed problem. Most existing approaches are based on structure-from-motion or graph inference, where a 3D shape is solved by fitting 2D keypoints/mask instead of directly using the vital cue in the original 2D image. These methods usually require multiple views of an object instance and rely on accurate labeling, detection, and matching of 2D keypoints/mask across multiple images. To overcome these limitations, we make effort to reconstruct 3D deformable object shapes directly from the given unconstrained 2D images. In training, instead of using multiple images per object instance, our approach relaxes the constraint to use images from the same object category (with one 2D image per object instance). The key is to disentangle the category-specific representation of the 3D shape identity and the instance-specific representation of the 3D shape displacement from the 2D training images. In testing, the 3D shape of an object can be reconstructed from the given image by deforming the 3D shape identity according to the 3D shape displacement. To achieve this goal, we propose a novel convolutional encoder-decoder network—the Disentangling Deep Network (DisDN). To demonstrate the effectiveness of the proposed approach, we implement comprehensive experiments on the challenging PASCAL VOC benchmark and use different 3D shape ground-truth in training and testing to avoiding overfitting. The obtained experimental results show that DisDN outperforms other state-of-the-art and baseline methods.

Keywords: 3D shape reconstruction · Disentangling deep network · Point cloud

1 Introduction

3D shape reconstruction of an object from a single image is a hotspot issue in the field of computer vision, and it is widely used in augmented reality, automatic driving and other scenes. However, this problem is full of challenges because of the lack of reliable shape prior information in a single image to ensure the completeness and authenticity of the recovered 3D shape. In particular, given an image of the object, the 2D observation

H. Ma et al. (Eds.): PRCV 2021, LNCS 13020, pp. 153–166, 2021.
https://doi.org/10.1007/978-3-030-88007-1_13

only describes the visible part of the object, while the self-occluded part is unknown. Without knowing the global shape prior information, there are many possible 3D shapes to explain this 2D observation.

To solve the ill-posed problem of reconstructing the complete 3D shape of an object from a single image, the deformation based 3D shape reconstruction method has attracted much attention in computer vision. This method estimates the deformation parameters of the instance from the input image and deforms a 3D template shape (or mean shape) to make it consistent with the 2D observation from image. The 3D template shape (or mean shape) provides the effective shape prior information for reconstructing 3D shape from a single image. The deformation based 3D reconstruction method can not only estimate the complete 3D shape of the object but also effectively constraint the reconstruction process to infer the geometry structure of the invisible part of the object. However, the traditional deformation parameter estimation needs to solve complex optimization problems such as viewpoint estimation, 3D shape to 2D plane projection, and 2D shape fitting. These problems increase the difficulty and time-consuming of 3D reconstruction.

Recently, deep neural networks have been widely used for addressing many computer vision tasks in an end-to-end manner. The research of 3D shape reconstruction of a single image based on the deep neural network has become a hot topic in the field of computer vision. In order to learn the complete 3D structure information of the object, the traditional deep network based method is often required the constrained dataset for training the 3D reconstruction network model, i.e., the training dataset contains multiple view 2D observations of an object. However, for the real image dataset, obtaining the corresponding multi-perspective observations of objects and independent fine-detail annotations is extremely expensive.

To address the aforementioned issues, in this paper, we make effort to achieve the end-to-end single image 3D shape reconstruction by learning the shape prior from the unconstrained image dataset. Intuitively, for a category of objects, they obviously share a common attribute with each other and have their own exclusive attributes. The common attribute of a category of objects contains the generic information of the objects and is class-invariant. The exclusive attributes contain the specific information of each object in the category. The abstracted common information can be concretized into the spatial 3D shape of the category of objects, which serves as a priori to provide help for the 3D reconstruction of a single image. The abstracted exclusive information of each object can be concretized into individuals' differences for helping the 3D reconstruction model to recover their specific 3D shapes. In general, the common information and exclusive information in the feature space of the object are entangled together, and the traditional feature learning method cannot separate the information and make it a priori of constrained 3D reconstruction. Inspired by the works of disentangled representation learning [9,18,19], in this paper, we disentangle the low dimensional feature of the object into two parts: a category-specific representation containing the shared common information between a category of objects and an instance-specific representation containing exclusive information of the object instance. The former representation describes the common geometric properties shared within the category, the later representation describes the shape deformation specific to each object instance. Thus,

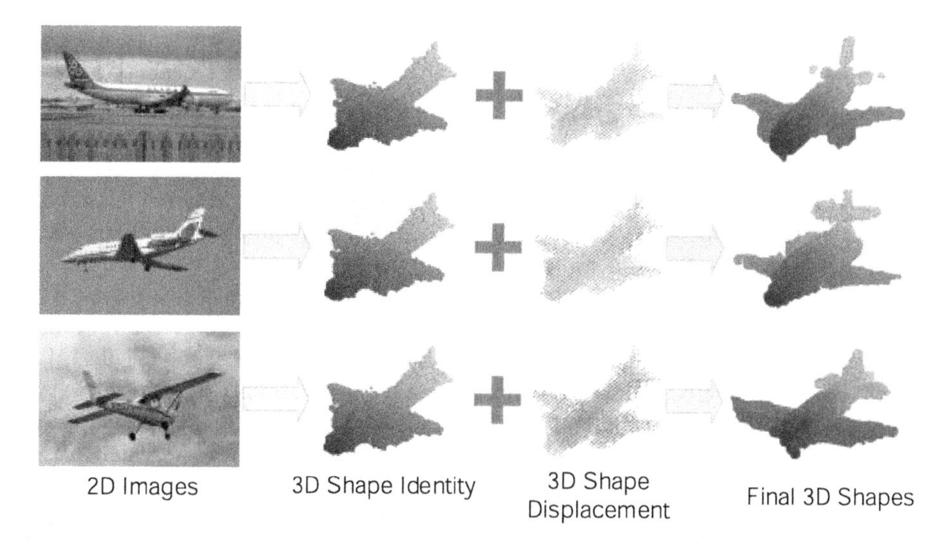

Fig. 1. 3D shape reconstruction by disentangling the 3D shape identity and 3D shape displacement from 2D images.

we can concretize those representations into the 3D shape identity and 3D shape displacement to represent the common shape of a category of objects and the differences respectively, and then use a simple fusion strategy to achieve 3D reconstruction. The 3D shape identity and 3D shape displacement are shown in Fig. 1.

In building our deep neural network, specifically, we cast the 3D shape reconstruction problem as the problem of 1) disentangling the latent identity representation and the latent deformation representation from the given 2D image 2) and then generating the real 3D shape identity and 3D shape displacement from such obtained latent representations. Combining the shape identity and instance-specific displacement, an instance-specific 3D shape can be easily reconstructed. The deep neural network we built is thus named as the Disentangling Deep Network(DisDN).

To sum up, three major contributions are presented in this paper.

- We make the effort to propose an end-to-end framework for reconstructing the 3D deformable object shapes from single 2D images.
- We cast the 3D shape reconstruction problem as the problem of disentangling the deep representations into the 3D shape identity and 3D shape displacement.
- The experimental results demonstrate the rationality of network components and superior performance of DisDN.

2 Related Works

Among the past years, a number of traditional 3D shape reconstruction methods [12,14,17,20,26] have achieved decent performance given images of one object captured at multiple views. Later methods [8,11,22,23,28,30] relief such limitation by reconstructing 3D shapes for category-specific images where there is only one image per each object instance. Zia *et al.* [30] proposed 3D geometric object category representations for object recognition in a single image. Tulsiani *et al.* [22] learned category-specific the 3D mean shape and a set of the deformation basis, which were then used to reconstruct 3D shape from a test image. Note that most approaches relying on 2D keypoints and segmentation masks in training images. Zhang *et al.* [28] achieved the same goal with a weakly-supervised learning approach which only requires category-specific images and 2D keypoints.

With the advance of deep networks, recent works [2,5,7,24] attempt to estimate 3D shape directly from 2D images. Choy *et al.* [2] realized 3D object reconstruction from wild images by using the long short term memory (LSTM) model. Similar with [2], Girdhar *et al.* [5] rendered multi-view images for each object to train a 'TL-Embedding' network, where 3D voxels were predicted from the embedding representation that corresponds to the input 2D image. Specifically, the 64-D embedding representation is obtained by feeding both a 2D RGB image and a 3D voxel into an auto-encoder network. Wu *et al.* [24] employed the variational auto-encoder generative adversarial network (VAE-GAN) to learn a disentangled generative and discriminative representations for 3D objects. They reconstructed 3D object shapes in forms of 3D voxels, which requires to use the time-consuming operation of 3D convolution.

Different from the above previous works, our proposed DisDN combines the advantage of both the traditional and the deep learning-based 3D object reconstruction methods. Firstly, learning by the deep convolutional network, DisDN can capture the desired hidden patterns (in a data-driven way) to build 3D shapes from the collection of 2D observations without the manual annotations on key points and segmentation masks. Secondly, inspired by the traditional category-specific 3D shape reconstruction methods, we learn the network to disentangle the representation of 3D shape identity and displacement, which provides the global prior to predict the general appearance of the unseen object part and thus helps overcome the problem of self-occlusion of the object in the 2D image.

3 Disentangling Deep Network

3.1 Network Architecture

Denote $\{(\mathbf{X}_1, \mathbf{S}_1), ..., (\mathbf{X}_N, \mathbf{S}_N)\}$ the training dataset of N entries, in which $\mathbf{X}_n$ denotes the nth image, $n \in [1, N]$, $\mathbf{S}_n$ is the 3D shape ground truth in forms of point cloud. We also denote the variables $\hat{\mathbf{S}}$, $\hat{\mathbf{S}}_I$ and $\hat{\mathbf{S}}_D$ as the predicted 3D shape, the predicted 3D shape identity and the predicted 3D shape displacement, respectively. Given a 2D image $\mathbf{X}_n$, our DisDN aims at estimating a 3D shape identity $\hat{\mathbf{S}}_{I,n}$ and a 3D shape displacement $\hat{\mathbf{S}}_{D,n}$ that correspond to the included object of interest. Finally, the predicted 3D shape reconstructed from this image is represented as $\hat{\mathbf{S}}_n = \hat{\mathbf{S}}_{I,n} + \hat{\mathbf{S}}_{D,n}$. The 3D shape is expressed in the form of point clouds with size 2048×3.

Figure 2 shows the network architecture of DisDN. Specifically, DisDN consists of three components: a disentangling encoder network, a 3D shape displacement generative network, and a 3D shape identity generative network. The disentangling encoder

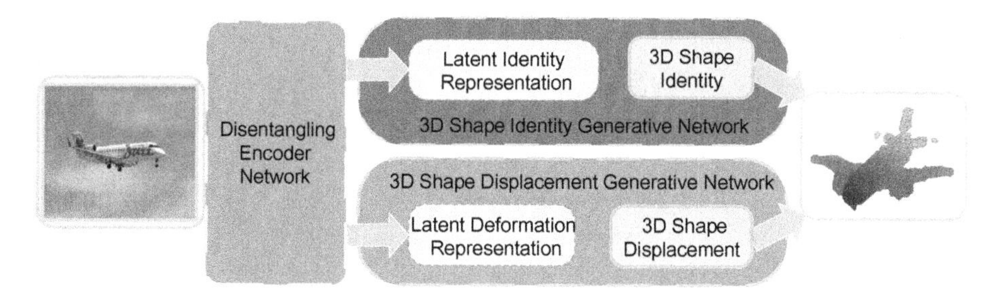

Fig. 2. Illustrator of the Disentangling Deep Network (DisDN). The disentangling encoder network maps the input image to low-dimensional latent representations shared by the two streams of the decoder network. The final 3D object shape is obtained by combining the predicted 3D shape identity and 3D shape displacement, i.e., the point-wise summation of the predicted 3D shape identity vector and the 3D shape displacement vector.

network maps the input images into two different latent representations: (1) the latent deformation representation $\mathbf{z}_D$ (formed by the encoded mean μ and variance Σ), and (2) the latent identity representation $\mathbf{z}_I$, as shown in Fig. 3(A). Then the latent deformation representation $\mathbf{z}_D$ is decoded by the 3D shape displacement generative network to generate the 3D shape displacement $\hat{\mathbf{S}}_D$, as shown in Fig. 3(B). The 3D shape identity generative network decodes the latent identity representation $\mathbf{z}_I$ to generate a 3D shape identity $\hat{\mathbf{S}}_I$, as shown in Fig. 3(C).

3.2 Learning Objective Functions

For learning DisDN, we introduce the encoding losses to avoid building trivial latent representations and the generation losses to guide the network to obtain the desired 3D shape identity and displacement for each given 2D image. We alternatively train DisDN to optimize two objective functions: (1) Fixing the 3D displacement generative network $G_D(\cdot)$, the 3D identity generative network $G_I(\cdot)$ is learned over three losses: identity representation consistency loss from 2D images (L_{I-RC}), the 3D shape identity reconstruction loss (L_{I-UD}) and the 3D shape identity consistency loss L_{3D-I}.

$$L_{identity} = L_{I-RC} + L_{I-UD} + L_{3D-I}.$$

(2) Fixing the 3D shape identity generative network $G_I(\cdot)$, optimizing the 3D displacement generative network $G_D(\cdot)$ by three losses : the 3D shape displacement generation loss (L_{D-UD}), the KL-divergence loss (L_{D-KL}) and the 3D shape smoothness loss L_{D-SM}.

$$L_{deform} = L_{D-KL} + L_{D-UD} + L_{D-SM}.$$

Details of the learning objectives are provided below.

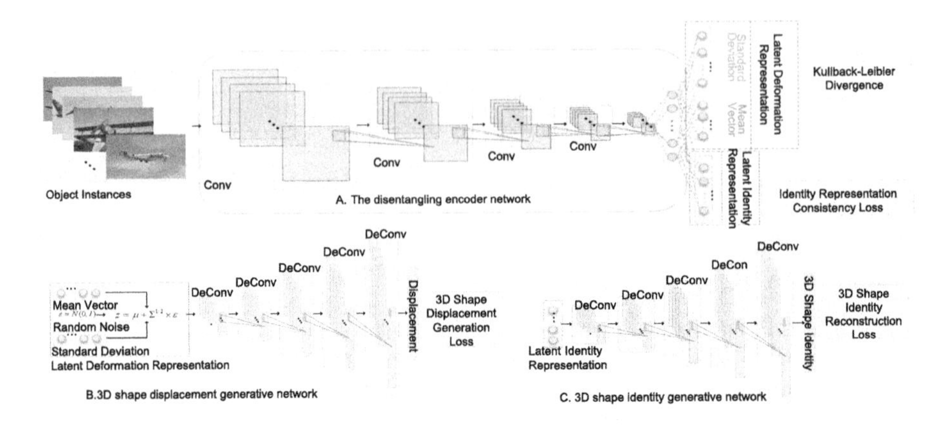

Fig. 3. Illustrator of the network blocks for DisDN. The "Conv" denotes 2D-convolution, "Deconv" denotes the 1D-deconvolution.

Encoding Losses: For training the encoder $q(\mathbf{z}_{D,n}|\mathbf{X}_n)$, the VAE regularizes it by imposing a prior over the latent distribution $p(\varepsilon)$. Following [16], We employ the KL-divergence loss to constrain the latent deformation representation z_D as:

$$L_{D-KL} = \frac{1}{N}\sum_{n=1}^{N} D_{KL}(q(\mathbf{z}_{D,n}|\mathbf{X}_n)\,\|p(\varepsilon)), \tag{1}$$

$$\mathbf{z}_{D,n} = \boldsymbol{\mu}_n + \boldsymbol{\Sigma}_n^{1/2} \times \varepsilon; \varepsilon \sim N(0, I). \tag{2}$$

where $\mathbf{X}_n$ denotes the nth training image, $\mathbf{z}_{D,n}$ denotes its latent deformation representation which is decided by the mean μ_n, the variance $\boldsymbol{\Sigma}_n$ and a random noise ε obeying the Normal Gaussian distribution $N(0, I)$. $q(\mathbf{z}_{D,n}\,|\mathbf{x}_n)$ and $p(\varepsilon)$ represents the variational distribution and the probability distributions of $\mathbf{z}_{D,n}$, respectively. The KL-divergence $D_{KL}(q(\mathbf{z}_{D,n}\,|\mathbf{X}_n)\,\|p(\varepsilon))$ encourages the variational distribution $q(\mathbf{z}_{D,n}\,|\mathbf{X}_n)$ towards to the prior distribution $p(\varepsilon)$.

We design the identity representation consistency loss Eq. (3) to constrain the encoder network to obtain the latent representation $\mathbf{z}_{I,n}$ which represent the mean shape information of such object category.

$$L_{I-RC} = \frac{\sum\limits_{n=1}^{N}\sum\limits_{l=1}^{L}\left(z_I^{n,l} - \frac{1}{N}\sum\limits_{n=1}^{N}\left(z_I^{n,l}\right)\right)^2}{\sum\limits_{l=1}^{L}\sum\limits_{n=1}^{N}\left(z_I^{n,l} - \frac{1}{L}\sum\limits_{l=1}^{L}\left(z_I^{n,l}\right)\right)^2}. \tag{3}$$

where $z_I^{n,l}$ denotes the lth point in the nth latent identity representation $\mathbf{z}_{I,n}$. As can be seen, Eq. (3) both consideres the consistency of category-specific objects and the discrepancy between its elements, where the numerator term encourages the latent identity representations of object instances to be as similar as possible while the denominator term maxims the variance of each raw in latent identity representation to prevent the elements being limited to a small range.

Generation Losses: We define two loss functions (i.e. the 3D shape identity loss and the 3D shape displacement loss) to train our two-steam generative network, which actually optimize two target tasks jointly: 1) Updating the parameters in the 3D shape identity generative network to explore the category-specific mean shape. 2) Updating the parameters in the 3D shape displacement generative network to explore the 3D shape deformation with respect to each single object instance. We follow [15] to use the Chamfer distance (CD) to build the loss functions for 3D point cloud.

When calculating the 3D shape identity loss, we fix the 3D shape displacement generative network to obtain the 3D shape displacement $\mathbf{S}_{D,n}$ and formulate the corresponding loss function as follows:

$$L_{I-UD} = \sum_{n=1}^{N} d_{CD}(\hat{\mathbf{S}}_{I,n}, (\mathbf{S}_n - \mathbf{S}_{D,n})). \tag{4}$$

where $\hat{\mathbf{S}}_{I,n}$ denotes the generated 3D Identity of image $\mathbf{X}_n$, $\mathbf{S}_n$ presents its corresponding 3D shape ground truth, and $\mathbf{S}_{D,n}$ is the generated 3D displacement by fixing the parameters of the 3D displacement generation network.

In addition, We design the 3D shape identity consistency loss Eq. (5) to keep the generated 3D shape identities are consistent among object instances.

$$L_{3D-I} = \frac{1}{\Phi} \sum_{n=1}^{N} \sum_{k=1}^{K} \sum_{l=1}^{L} \left(\hat{s}_{(n,l,k)} - \frac{1}{N} \sum_{n=1}^{N} \left(\hat{s}_{(n,l,k)} \right) \right)^2 . \tag{5}$$

where $\hat{s}_{(n,l,k)}$ denotes variable at the lth row and the kth column in n^{th} predicted 3D shape identity $\hat{\mathbf{S}}_{I,n}$, and $\Phi = N \times K \times L$. Given input training images $\mathbf{X}_n$, the 3D identity constraint only consider minimizing the variance between each predicted 3D shape identity $\hat{\mathbf{S}}_{I,n}$.

The 3D shape displacement network is trained by fixing the 3D shape identity generative network and $\mathbf{S}_{I,n}$ by using the 3D shape displacement loss:

$$L_{D-UD} = \sum_{n=1}^{N} d_{CD}(\hat{\mathbf{S}}_{D,n}, (\mathbf{S}_n - \mathbf{S}_{I,n})). \tag{6}$$

where, $\hat{\mathbf{S}}_{D,n}$ denotes the predicted 3D shape displacement, the $\mathbf{S}_n$ presents the 3D shape ground truth, the $\mathbf{S}_{I,n}$ present the generated 3D shape identity by fixing the 3D shape identity generative network.

When the 3D shape identity and the 3D shape displacements are available, the 3D shape can be formulated as $\hat{\mathbf{S}}_n = \hat{\mathbf{S}}_{D,n} + \hat{\mathbf{S}}_{I,n}$. We also introduce the 3D shape displacement smoothness loss by imposing a quadratic penalty between every point and its neighbors in final 3D shape $\hat{\mathbf{S}}_n$:

$$L_{D-SM} = \frac{1}{\Psi} \sum_{n=1}^{N} \sum_{l=1}^{L} \sum_{o \in O(l)} \left(\|\hat{\mathbf{s}}_{n,l} - \hat{\mathbf{s}}_{n,o}\| - \hat{\delta}_n \right)^2 . \tag{7}$$

where $\hat{\mathbf{s}}_{n,l}$ is the lth 3D point in the reconstructed 3D shape $\hat{\mathbf{S}}_n$, the O(l) denotes the neighbors of the point $\hat{\mathbf{s}}_{n,l}$ and $\hat{\delta}_n$ denotes the mean squared displacement among the neighbors of the point $\hat{\mathbf{s}}_{n,l}$. This item achieves the smooth processing of the 3D shape by limiting the abnormal points in the local neighborhood of the generated point cloud.

3.3 Training Strategy

We first initialize the network parameters randomly and initialize the 3D shape identity as the category-specific convex hull [22] obtained by using the visual hull [23] method. According to [23], the visual hull method is a sampling process which attempts to intersect an object's projection cone with the cones of minimal subsets of other similar objects among those pictured from certain vantage points. We could also pre-train the network with a randomly initialized 3D shape, but an acceptable 3D mean shape could strengthen the performance for DisDN. The training instances (category-specific images) and its corresponding silhouettes are obtained by cropping and resizing the category-specific images to 64×64. These images and silhouettes are provided by PASCAL dataset [3]. When the silhouettes of category-specific objects is available, the goal of visual hull is to find a binary map by minimization:

$$E(L) = \sum_{v \in V} l_v \bar{C}(v). \tag{8}$$

where V is a set of voxels, $l_v = 1$ if voxel v is inside the shape, and $l_v = 0$ otherwise, and $C(\cdot)$ denotes the signed distance function to the camera cone of instance.

By minimizing Eq. (8), we can obtain the 3D shape $\bar{V}$. We convert such 3D shape $\bar{V}$ into 3D point cloud from $\bar{S}$ which is conveniently used in our DisDN framework. After that, we can produce the pre-training data S^n_{id} for the 3D displacement generative network by computing the difference $S^n_{id} = S_n - \bar{S}$ between ground-truth 3D shape S_n and the initialized 3D shape identity $\bar{S}$. The pre-training process simultaneously trains the disentangling encoder network, the 3D identity generative network and the 3D displacement generative network. We use the pre-training data $\bar{S}$ and $\{S^n_{id}\}$ to provide the objective losses for both generative networks. The 3D shape identity consistency loss and the both smoothness loss do not involve pre-training optimization.

After the initialization phase, we learn the disentangling encoder network and the 3D shape displacement generative network of DisDN by optimizing $L_{D-KL}+L_{D-UD}$. Then, we use the learned network parameters to generate the predicted 3D shape displacement $\hat{\mathbf{S}}_D$ and use it to fine-tune the disentangling encoder network and the 3D shape identity generative network of DisDN by optimizing $L_{L_I-RC} + L_{I-UD}$. This alternative optimization learning process is performed continuously until convergence. Figure 4 visualizes the evolution of 3D shape identity during the training process of DisDN. At the beginning of training process, the output of 3D shape identity generative networks is a convex hull. At the end of training process, DisDN estimated a fine-detailed 3D shape identity of the "Aeroplane" category.

Fig. 4. The evolution of 3D shape identity during the training process of our DisDN. The 3D shape identity starts at a initialized convex hull over all training 3D shapes of the same object category (e.g., aeroplane). Our DisDN learns a high quality category-specific shape identity through training.

4 Experiments

DisDN was evaluated in three folds: (1) Ablation analysis. (2) 3D Shape reconstruction performance against state-of-the-art methods. (3) Effects of 3D shape identity in 3D shape reconstruction.

4.1 Implementation Details

Dataset: We employ the real-images database PASCAL VOC 2012 [3] for training and testing the proposed DisDN. For fairly testing 3D reconstruction methods on the real-images database, we use the same "train\test" splits with previous approaches [22, 28], where 80% and 20% images from the dataset are used for training and testing, respectively.

However, the dataset Pascal3D+ [25] just provides approximate 3D shape ground-truth for each real-image using a small set of CAD shapes (e.g. 7 shapes for "Aeroplane"), and the same set of approximate 3D shape ground-truths are shared across training and testing objects. Thus, the 3D shape ground-truth provided by the dataset Pascal3D+ [25] cannot be directly used to train the 3D reconstruction network. We employ the ShapeNet [25] to provide the 3D shape ground-truth for training DisDN instead of using the 3D annotations from Pascal3D+. This can avoid over-fitting to the small set of 3D models when training DisDN.

Data Processing: Training DisDN requires the 3D point cloud models for supervising. To this end, we use a random sampling strategy to collect the vertex points on the surface of the 3D CAD models. These points are then arranged by column to constitute the point cloud. Each point cloud contains 2048 points. Similar to the previous point cloud prediction methods [4], for each model, we normalize the radius of bounding hemisphere to unit 1 and aligned their ground plane.

Testing DisDN requires the volumetric form to represent the 3D shape, we follow the post-processing method in work [4] to use a local method to convert the point cloud into a volumetric representation with $32 \times 32 \times 32$.

Metrics: We evaluate DisDN and related approaches using two popular metrics. (1) Mesh Error(Tulsiani *et al.* [22]): The mesh error is computed as the Hausdorff distance normalized by the diagonal of the tightest 3D bounding box size of the ground truth model. The lower Mesh Error represents the better reconstruction results. (2)

Table 1. VIoU of 3D shapes reconstructed by DisDN and the baseline methods. The higher VIoU denote the better reconstruction results.

Categories	aero	bike	boat	bus	car	chair	mbike	sofa	train	tv	mean
DisDN-w/o-KL	0.581	0.447	0.560	0.631	0.698	0.285	0.687	0.404	0.550	0.543	0.539
DisDN-w/o-I	0.575	0.463	0.567	0.625	0.675	0.296	0.678	0.372	0.541	0.535	0.533
DisDN-w-RPt	0.601	0.481	0.578	0.697	0.714	0.304	0.707	0.414	0.560	0.559	0.562
VAE [13]	0.552	0.436	0.546	0.624	0.687	0.289	0.435	0.325	0.432	0.539	0.487
DisDN(Ours)	**0.620**	**0.486**	**0.593**	**0.712**	**0.721**	**0.305**	**0.713**	**0.436**	**0.575**	**0.569**	**0.573**

Voxel Intersection-over-Union (VIoU) (Choy *et al.* [2]): We convert the output shape of DisDN to the 3D voxel ($32 \times 32 \times 32$) and compute VIoU with respect to the 3D voxel of the original 3D CAD shape models. The VIoU metric is computed as:

$$VIoU = \frac{\sum \left(\delta\left(p_{(i,j,k)} = 1\right) * \delta\left(y_{(i,j,k)}\right)\right)}{\sum \left(\delta\left(p_{(i,j,k)} = 1\right) + \delta\left(y_{(i,j,k)}\right)\right)}. \tag{9}$$

where $\delta(\cdot)$ denotes an indicator function. The Higher VIoU denotes the better reconstruction results.

4.2 Ablation Analysis

We evaluated four baselines with our DisDN on the real-images Pascal3D+ [25] for analyzing the effectiveness of modules in the proposed DisDN. The first baseline (DisDN-w/o-KL) consists of the latent identity representation encoding network and 3D shape identity generative network with the identity representation consistency loss L_{I-RC}. The deformation-network is replaced by a simple encoder-decoder network. The second baseline (DisDN-w/o-I) only restricts the displacement latent representation with KL-divergence D_{KL} and replaces the 3D shape identity generative parts by an encoder-decoder network. The third one (DisDN-w-RPt) uses all of those loss functions in our proposed method but initialize the 3D shape identity with a random shape.

From Table 1, we have three observations: 1) Compared with VAE and other baseline deep models, our proposed network architecture can better reconstruct 3D shape from real-world images. 2) To a certain extent, the 3D shape displacement and 3D shape identity are of equal importance to promoting DisDN's performance (see the comparisons of DisDN-w/o-KL, DisDN-w/o-I and Ours). 3) The proposed visual hull-based initialization process can effectively improve the learning performance of DisDN, but it is not as significant as the influence brought by the designed network architecture (see the comparison among the three baseline methods and the final DisDN).

4.3 3D Reconstruction

We compare our DisDN with state-of-the-art methods including Shin *et al.* [21], Johnston *et al.* [10], Choy [2] and Kar *et al.* [11] on 3D shape reconstruction. From the Table 2, we observe that DisDN outperforms the state-of-the-art approaches in almost

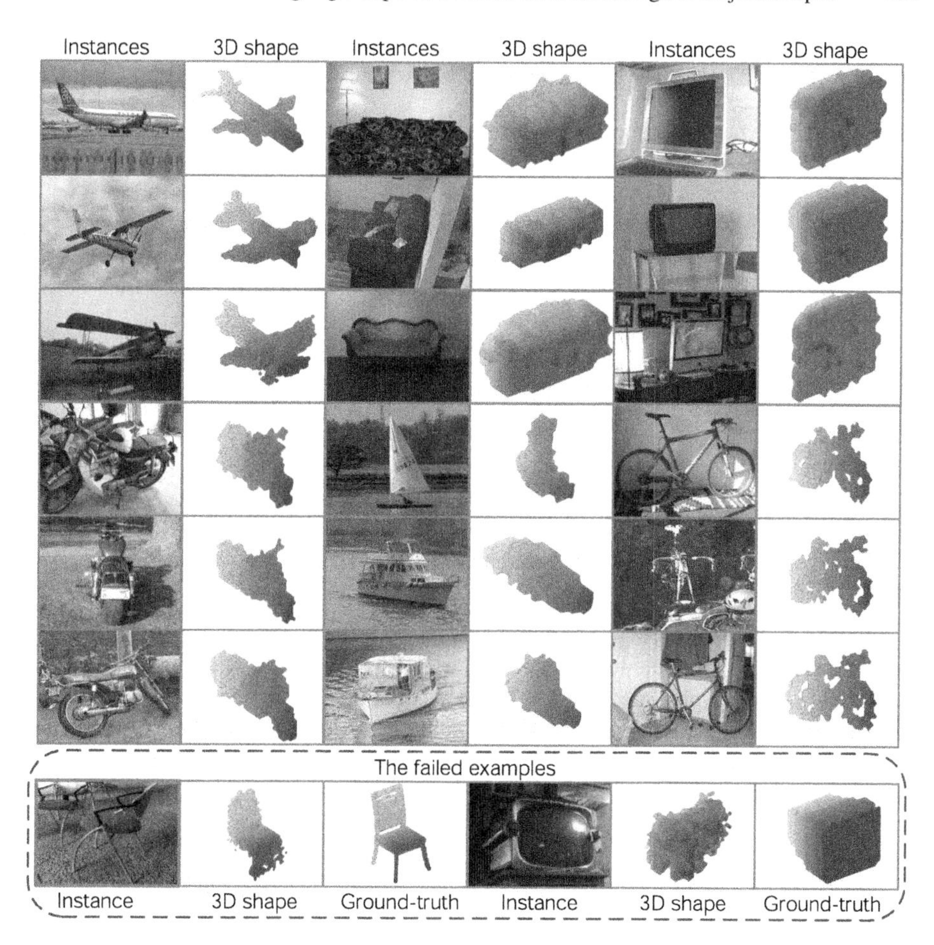

Fig. 5. Examples of 3D shapes reconstructed by DisDN. **Top:** successful cases. **Bottom:** failure cases. Images are best viewed in color. (Color figure online)

all categories as well as the mean score. We show the 3D reconstruction result and some failed instances in Fig. 5. As shown in Fig. 5, 3D shape reconstruction by using our DisDN has some difficulties to recovering trivial parts, such as the thin legs of chairs. Additionally, our DisDN easily fails to inconspicuous appearances of some 'tv' 'monitors'. Nevertheless, our DisDN demonstrates a competitive 3D shape reconstruction at the aspect of quantitative analysis.

4.4 Effects of 3D Shape Identity

In this experiment, we demonstrate the effectiveness of the 3D shape identities learned by our approach. Specifically, we compare DisDN with Tulsiani *et al.* [22] and Zhang *et al.* [28], by embedding the 3D shape identities into the framework of [22] to perform category-specific 3D shape reconstruction. In the inference process, only the 3D

Table 2. Comparison of VIoU between 3D shape reconstructed by DisDN and state-of-the-art methods on Pascal3D+[25]. The higher VIoU denotes the better reconstruction results.

Categories	aero	bike	boat	bus	car	chair	mbike	sofa	train	tv	mean
Kar *et al.* [11]	0.298	0.144	0.188	0.501	0.472	0.234	0.361	0.149	0.249	0.492	0.309
Shin *et al.* [21]	0.362	0.362	0.331	0.497	0.566	0.362	0.487	**0.555**	0.301	0.383	0.421
Johnston *et al.* [10]	0.555	0.489	0.523	0.776	0.622	0.249	0.656	0.462	0.574	0.549	0.547
Choy *et al.* [2]	0.544	0.499	0.560	**0.816**	0.699	0.280	0.649	0.332	**0.672**	**0.574**	0.563
DisDN(Ours)	**0.620**	**0.486**	**0.593**	0.712	**0.721**	**0.305**	**0.713**	**0.436**	0.575	0.569	**0.573**

Table 3. Comparison of Mesh Error between 3D shape identities estimated by DisDN and the weakly supervised baseline methods. The lower Mesh Error denotes the better reconstruction results.

Categories	aero	bike	boat	bus	car	chair	mbike	sofa	train	tv	mean
Zhang *et al.* [28]	1.870	3.000	4.150	2.960	2.240	2.320	2.220	5.830	8.010	8.310	4.090
Tulsiani *et al.* [22]	1.720	1.780	3.010	1.900	1.770	2.180	1.880	2.130	2.390	3.280	2.200
DisDN(Ours)	**1.640**	**1.474**	**2.035**	**1.828**	**1.684**	**2.174**	**1.793**	**1.814**	**1.871**	**2.806**	**1.932**

displacement will be updated by matching the 3D-2D projection with object's masks. Since the framework proposed by [22,28] only generates 3D meshes, we report Mesh Error in this experiment. All the compared methods need to use additional manually labeled keypoints in testing. As shown in Table 3, our DisDN constantly outperformed the state-of-the-art methods Tulsiani *et al.* [22] and Zhang *et al.* [28].

5 Conclusion

In this paper, we propose a novel disentangling deep network (DisDN) to reconstruct the 3D shape of objects from single view images. In this way, the 3D shape identity of the object can be automatically mined to represent the prior information of the class. We also propose a 3D shape displacement to represent the difference between each object in one class. Therefore, 3D shape deformation can be formulated as a simple additive operation of the 3D identity and the 3D displacement. Experiments show that our model is able to yield complete 3D shape reconstruction from a single view image. On the challenging PASCAL VOC benchmark, DisDN outperformed state-of-the-art methods on both reconstructing 3D shape models from single wild images and generating 3D shape identity for object categories. In our future research, we will further explore the possibility of 3D identity as a prior in improving the performance of weakly supervised target detection [6,27,29] and few-shot image classification [1] tasks.

Acknowledgements. This work was supported by the Key-Area Research and Development Program of Guangdong Province(2019B010110001), the National Science Foundation of China (Grant Nos. 61876140, 61806167, 61936007 and U1801265), and the research funds for interdisciplinary subject, NWPU.

References

1. Cheng, G., Li, R., Lang, C., Han, J.: Task-wise attention guided part complementary learning for few-shot image classification. Sci. China Inf. Sci. **64**(2), 1–14 (2021). https://doi.org/10.1007/s11432-020-3156-7
2. Choy, C.B., Xu, D., Gwak, J., Chen, K., Savarese, S.: 3D–R2N2: a unified approach for single and multi-view 3D object reconstruction. In: Proceedings of the European Conference on Computer Vision, pp. 628–644 (2016)
3. Everingham, M., Van Gool, L., Williams, C.K.I., Winn, J., Zisserman, A.: The pascal visual object classes challenge. Int. J. Comput. Vision **88**(2), 303–338 (2010)
4. Fan, H., Su, H., Guibas, L.: A point set generation network for 3D object reconstruction from a single image. In: IEEE Conference on Computer Vision and Pattern Recognition, vol. 38 (2017)
5. Girdhar, R., Fouhey, D.F., Rodriguez, M., Gupta, A.: Learning a predictable and generative vector representation for objects. In: Proceedings of the European Conference on Computer Vision, pp. 484–499 (2016)
6. Gong, C., Yang, J., You, J.J., Sugiyama, M.: Centroid estimation with guaranteed efficiency: a general framework for weakly supervised learning. IEEE Trans. Pattern Anal. Mach. Intell. (2020)
7. Gwak, J., Choy, C.B., Garg, A., Chandraker, M., Savarese, S.: Weakly supervised generative adversarial networks for 3D reconstruction. arXiv preprint arXiv:1705.10904 (2017)
8. Han, J., Yang, Y., Zhang, D., Huang, D., Xu, D., De La Torre, F.: Weakly-supervised learning of category-specific 3D object shapes. IEEE Trans. Pattern Anal. Mach. Intell. **43**(4), 1423–1437 (2019)
9. Higgins, I., et al.: Towards a definition of disentangled representations. arXiv preprint arXiv:1812.02230 (2018)
10. Johnston, A., Garg, R., Carneiro, G., Reid, I., van den Hengel, A.: Scaling CNNs for high resolution volumetric reconstruction from a single image. In: Proceedings of the IEEE International Conference on Computer Vision Workshops, pp. 939–948 (2017)
11. Kar, A., Tulsiani, S., Carreira, J., Malik, J.: Category-specific object reconstruction from a single image. In: IEEE Conference on Computer Vision and Pattern Recognition, pp. 1966–1974 (2015)
12. Kim, Y.M., Theobalt, C., Diebel, J., Kosecka, J., Miscusik, B., Thrun, S.: Multi-view image and ToF sensor fusion for dense 3D reconstruction. In: Proceedings of the IEEE International Conference on Computer Vision Workshops, pp. 1542–1549 (2009)
13. Kingma, D.P., Welling, M.: Auto-encoding variational bayes. arXiv preprint arXiv:1312.6114 (2013)
14. Kolev, K., Klodt, M., Brox, T., Cremers, D.: Continuous global optimization in multiview 3D reconstruction. Int. J. Comput. Vision **84**(1), 80–96 (2009)
15. Kurenkov, A., et al.: Deformnet: free-form deformation network for 3D shape reconstruction from a single image. arXiv preprint arXiv:1708.04672 (2017)
16. Larsen, A.B.L., Sønderby, S.K., Larochelle, H., Winther, O.: Autoencoding beyond pixels using a learned similarity metric. arXiv preprint arXiv:1512.09300 (2015)
17. Owens, A., Xiao, J., Torralba, A., Freeman, W.: Shape anchors for data-driven multi-view reconstruction. In: Proceedings of the IEEE International Conference on Computer Vision, pp. 33–40 (2013)
18. Reed, S., Sohn, K., Zhang, Y., Lee, H.: Learning to disentangle factors of variation with manifold interaction. In: Proceedings of the International Conference on Machine Learning, pp. 1431–1439 (2014)

19. Sanchez, E.H., Serrurier, M., Ortner, M.: Learning disentangled representations via mutual information estimation. In: Proceedings of the European Conference on Computer Vision, pp. 205–221 (2020)
20. Seitz, S.M., Curless, B., Diebel, J., Scharstein, D., Szeliski, R.: A comparison and evaluation of multi-view stereo reconstruction algorithms. In: Proceedings of the IEEE Conference on Computer Vision and Pattern Recognition, vol. 1, pp. 519–528 (2006)
21. Shin, D., Fowlkes, C.C., Hoiem, D.: Pixels, voxels, and views: a study of shape representations for single view 3D object shape prediction. In: Proceedings of the IEEE Conference on Computer Vision and Pattern Recognition, pp. 3061–3069 (2018)
22. Tulsiani, S., Kar, A., Carreira, J., Malik, J.: Learning category-specific deformable 3D models for object reconstruction. IEEE Trans. Pattern Anal. Mach. Intell. **39**(4), 719–731 (2017)
23. Vicente, S., Carreira, J., Agapito, L., Batista, J.: Reconstructing pascal VOC. In: IEEE Conference on Computer Vision and Pattern Recognition, pp. 41–48 (2014)
24. Wu, J., Zhang, C., Xue, T., Freeman, B., Tenenbaum, J.: Learning a probabilistic latent space of object shapes via 3D generative-adversarial modeling. In: Advances in Neural Information Processing Systems, pp. 82–90 (2016)
25. Xiang, Y., Mottaghi, R., Savarese, S.: Beyond pascal: a benchmark for 3D object detection in the wild. In: Proceedings of the IEEE Winter Conference on Applications of Computer Vision, pp. 75–82 (2014)
26. Yingze Bao, S., Chandraker, M., Lin, Y., Savarese, S.: Dense object reconstruction with semantic priors. In: Proceedings of the IEEE Conference on Computer Vision and Pattern Recognition, pp. 1264–1271 (2013)
27. Zhang, D., Han, J., Cheng, G., Yang, M.H.: Weakly supervised object localization and detection: a survey. IEEE Trans. Pattern Anal. Mach. Intell. (2021)
28. Zhang, D., Han, J., Yang, Y., Huang, D.: Learning category-specific 3D shape models from weakly labeled 2D images. In: Proceedings of the IEEE Conference on Computer Vision and Pattern Recognition, pp. 4573–4581 (2017)
29. Zhang, D., Zeng, W., Yao, J., Han, J.: Weakly supervised object detection using proposal-and semantic-level relationships. IEEE Trans. Pattern Anal. Mach. Intell. (2020)
30. Zia, M.Z., Stark, M., Schiele, B., Schindler, K.: Detailed 3D representations for object recognition and modeling. IEEE Trans. Pattern Anal. Mach. Intell. **35**(11), 2608–2623 (2013)

AnchorConv: Anchor Convolution for Point Clouds Analysis

Youngsun Pan, Andy J. Ma(✉), and Yiqi Lin

School of Computer Science and Engineering, Sun Yat-sen University, Guangzhou, China
majh8@mail.sysu.edu.cn

Abstract. Point cloud, especially in 3D, is a common and important data structure. Recently, many point convolution methods have been proposed for point cloud processing, in which features of each point is updated by aggregating those of its neighbor points. Though existing methods can achieve satisfactory performance for point cloud analysis on several tasks, the performance may degrade when the distribution is non-uniform. In this paper, we present a new point convolution method namely Anchor Convolution (AnchorConv) for analysis of irregular and unordered point clouds. Inspired by standard grid convolution for images, each point in a point cloud is updated by fusing information from uniformly distributed anchors instead of non-uniformly distributed neighbor points. Features of each anchor are estimated by aggregating features of its neighborhood and distance relative to the anchor. Since the estimation biases and variances of different anchors are not the same, anchors are reweighted to obtain better feature representation of the center point. Experiments on ModelNet40, ShapeNetPart, S3DIS, and Semantic3D datasets show that the proposed AnchorConv outperforms state-of-the-art methods for classification and segmentation of 3D point clouds.

1 Introduction

As the usage of 3D scanning sensors (e.g. LIDAR) increases rapidly in a wide range of applications including autonomous driving, face recognition, 3D point cloud data processing has become an active research topic. Point clouds contain not only feature information of all the points $\mathcal{F} \in \mathbb{R}^{N \times D}$ but also structural information of points $\mathcal{P} \in \mathbb{R}^{N \times d}$, where N is the number of points, D is the dimension of the features and d denotes the dimension of position information ($d = 3$ for 3D point clouds). Different from 2D imagery with trivial position information (i.e., regularly distributed pixels/features), the structural information in 3D point clouds provide important cues to reduce ambiguities for many vision tasks.

Lots of methods have been proposed to analyze such irregular and unordered point cloud data. In [10,15], point clouds are converted to regularly distributed volume data by assigning each point to a certain volume so that standard grid convolution operation can be employed. Methods based on Shared multilayer perceptrons (MLPs) [12,14,25] and graph convolution networks (GCN) [2,7,11,16] can process point cloud data with less computation overhead. Nevertheless, the structural information $\mathcal{P}$ has not been fully ultilized in these methods, so that the performance may be suboptimal on complex tasks like object detection and semantic segmentation.

H. Ma et al. (Eds.): PRCV 2021, LNCS 13020, pp. 167–179, 2021.
https://doi.org/10.1007/978-3-030-88007-1_14

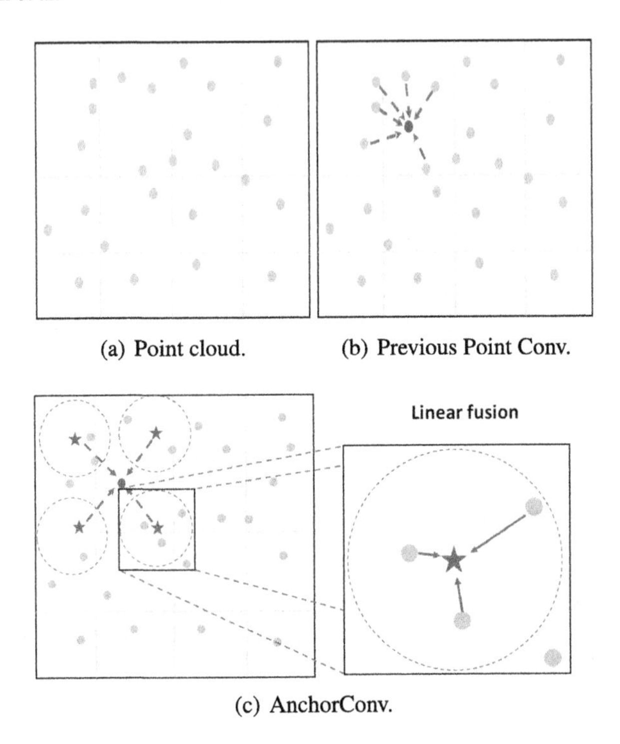

Fig. 1. Motivation: The ideas of existing point convolution methods and the proposed AnchorConv are illustrated on 2D point cloud. (a) An example 2D point cloud. (b) Previous point convolution methods: Input points (blue points) are aggregated to update center points (red points). However, the information from low-density regions may be neglected in the aggregating step (e.g. the bottom right area of the center point). (c) AnchorConv: Anchors are uniformly distributed around the center point to collect information from regions nearby (red dash-line circles) by linear fusion. Center point features are updated by aggregating these anchors. (Color figure online)

More recently, in [5,8,18,20,22], inspired by standard grid convolution, point convolutions are performed by using the relative position information for better feature update. While these methods update features of each point by fusing information from the neighbor points, the problem of non-uniform distribution may result in the update with a bias towards the high-density region. As shown in Fig. 1(b), when updating features of the center point using existing methods, the low-density regions may hardly contribute to the feature update.

To address this problem, a novel point convolution method namely AnchorConv is proposed for point cloud analysis in this paper. The proposed AnchorConv updates features of the center point by aggregating features of uniformly distributed anchors rather than features of non-uniformly distributed neighbor points. As a result, the negative influence by the non-uniform density of points can be reduced. The main idea of the proposed method is shown in Fig. 1(c). In this figure, each center point (colored in red) is updated by uniformly distributed anchors (red stars) that collect information from regions nearby. We design a linear fusion strategy to estimate anchor features with linear correlation function and normalization function. The linear correlation function ensures that points closer to each anchor measured by e.g. Euclidean distance plays a more important role in determining features of the anchor. With the anchor normalization function, the information from regions of different anchors can be balanced, such that the non-uniform distribution of points is converted to uniform distribution of anchors. Therefore, our method is advantageous over the state-of-the-art due to the insensitivity to non-uniform distribution of points.

Since features of each anchor are estimated by neighbor points in a region, the estimation accuracy may differ for different anchors. Consequently, anchors are reweighted to reduce the importance of less reliable anchors with larger estimation error for better feature representation of the center point. In this paper, two reweighting methods are introduced. The first reweighting method is named as bias-variance reweighting. Based on the analysis of bias and variance for anchor features estimation, the weight of each anchor is determined by the confidence of anchor features, i.e., the correlation summation between points and the anchor. The second method namely adaptive reweighting is derived based on the powerful representation capacity of multilayer perceptrons (MLPs). In this method, we leverages MLPs to automatically learn the weight of each anchor.

In a word, the main contributions of this work are summarized as follows:

- We propose a novel Anchor Convolution approach for Point Cloud Analysis, which reduce the negative influence of non-uniform distribution of data points. In the proposed method, features of center points are updated by uniformly distributed anchors rather than non-uniformly distributed neighbor points.
- Since the estimation errors of anchor features differ for different anchors, two reweighting schemes, i.e., bias-variance reweighting and adaptive reweighting have been proposed to assign larger/smaller weights to more/less reliable anchors for better feature representation of center points.
- Experiments show that our method outperforms state-of-the-art point cloud classification and semantic segmentation methods on several benchmarks. Ablation studies verify that the anchor normalization makes the proposed AnchorConv robust to different densities of point clouds and the proposed anchor reweighting methods can further improve the performance.

2 Related Work

In this section, we will briefly review previous deep learning methods developed for point cloud data analysis.

Grid Convolution Networks. Many grid convolution networks have been developed and achieved great success in grid data based applications. In [13, 17, 24], point clouds are projected into a sequence of 2D images which can be processed by normal grid convolution. These methods have been applied to classification and retrieval tasks and obtained convincing performances. For more complex tasks such as scene segmentation, point clouds are projected into volumetric grids [10] by 3D quantization so that 3D convolution can be employed. Nevertheless, these methods are computationally expensive with high memory. However, this kind of approach may still be not suitable for large-scale point cloud data due to the computational complexity.

Point MLP Networks. PointNet [12] is one of the first research works by using shared multi-layer perceptrons (MLP) for point cloud deep learning. After MLP, a global max pooling layer is inserted to obtain global features of point clouds in the PointNet. Though the PointNet has achieved competitive accuracy for classification, it does not consider local structure information in the point clouds, which may degrade the performance for point cloud segmentation or detection. To solve this problem, PointNet++ [14] improves PointNet by incorporating hierarchical architectures to aggregate local neighborhood information. But, local structural information may still be underutilized in this kind of approach.

Graph Convolution Networks. GCN are designed to process graph structure data which is composed of nodes and edges. In most GCN based methods [7,11,16], the nodes of the graph are defined by the points and the edges of the graph are defined by the relative distance of two nodes. Similar to shared point MLP networks, the relative positions of points containing important structure information are not fully utilized in graph convolutions.

Point Convolution Networks. Point convolution networks have been proposed more recently, in which convolution kernels are defined for point clouds and the relative position information can be better utilized. The Pointwise CNN [5] sets the kernel shape as voxel bins, and each bins process its internal points with corresponding weights. PointConv [20] uses MLP to generate the weight matrix for every neighbor point based on the relative position between the center point and the neighbor point. KPConv [18] assigns kernel points around the center point, and convolution weights of each neighbor point are decided by relative position between itself and every kernel point.

In the feature aggregating step, almost all previous methods update center point features with neighbor points which didn't consider the non-uniform sampling problem. Different from existing methods, AnchorConv updates the center point features with normalized features of uniformly distributed anchors rather than neighbor points which is the major difference between our method and existing works. By this means, the proposed AnchorConv can reduce the sensitivity to the changes in point cloud density.

3 Proposed Method

In this section, we will introduce the proposed AnchorConv in Sect. 3.1. Then, anchor reweighting modules include the bias-variance weighting and adaptive weighting are presented in Sect. 3.2. Finally, the network architecture is described in Sect. 3.3.

3.1 AnchorConv

As shown in Fig. 2(a), the standard grid convolution approach is applicable to only grid data such as images. In the convolution operations, convolution kernels which can be considered as anchors are uniformly placed around each center point (pixel). In the

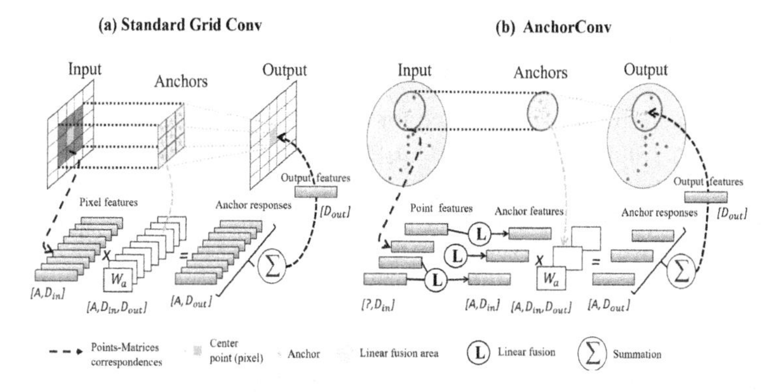

Fig. 2. (a): Illustration of example standard grid convolution. (b): Framework of the proposed AnchorConv for irregular Point Cloud data. A is the number of anchors, and ? represents a varying value. D_{in} and D_{out} are the number of input and output channels, respectively. For regularly distributed grid data like images, each anchor corresponds to one pixel. In the proposed AnchorConv, input points cannot aligned with the anchors, so we propose to compute features of each anchor by linear fusion, and then multiplied by the corresponding weight matrix.

cases of standard grid convolution, it can be observed that each anchor corresponds to exactly one point (pixel). Nevertheless, for point clouds data, there is not a trivial one-to-one matching between points and anchors due to irregularity. As shown in Fig. 1(c) and Fig. 2(b), we propose to place uniformly distributed anchors around the center points of point cloud data similar to standard grid convolution. Each anchor collects information from a small region surrounding it to obtain feature representation of the anchor. As a result, standard grid convolution can be viewed as a special case of the proposed AnchorConv, if the point cloud data is a set of regularly and uniformly distributed points.

Let x be the center point to be updated and denote the set of uniformly distributed anchors around x as $\mathcal{A}$. For the feature representation function f of the point cloud, features of the center point are updated by summing feature vectors of its anchors, i.e.,

$$(f * g)(x) = \sum_{i \in \mathcal{A}} F_i \tag{1}$$

where $*$ denotes the convolution operation, g is the convolution function, F_i is anchor responses of the ith anchor before anchor fusion (see in Fig. 2(b)). The features of the ith anchor F_i is computed as follows,

$$F_i = h\Big(\{I(y, a_i) g(f(y)) | y \in \mathcal{R}_i\}\Big) W_i \tag{2}$$

where a_i is the ith anchor, h denotes the normalization function, $W_i \in \mathbb{R}^{D_{in} \times D_{out}}$ is the weight matrix of the ith anchor, and I is the correlation function between y and a_i. In Eq. (2), $\mathcal{R}_i$ represents the set of all the points in the small region of the ith anchor a_i is (red dotted circle in Fig. 1(c)). In 3D point clouds, this kind of region is a ball with radius r_a. The correlation function I is decided by Euclidean distance between the neighbor point in the small ball and the anchor, which is given as follow:

$$I(y, a_i) = max(0, 1 - \frac{||y - a_i||}{r_a}) \tag{3}$$

where r_a is the radius of the small region for each anchor and $||\cdot||$ denotes Euclidean distance. With Eq. (3), when y is closer to a_i, the output of $I(y, a_i)$ becomes higher.

The normalization function h is a novel and important design in the proposed AnchorConv. Due to the non-uniform points, each anchor region contains different numbers of points (see Fig. 1(c)). Without the normalization step, the center point may tend to focus more on information from anchor regions with more points during aggregation and neglect the information from regions with fewer points. The normalization function can balance the information of different regions by assigning a uniform weight of each non-empty anchor which is independent of density. The normalization operation for the ith anchor can be formulated as:

$$h\Big(\{I(y, a_i)g(f(y)) | y \in \mathcal{R}_i\}\Big) = \sum_{y \in \mathcal{R}_i} \sigma_{yi} I(y, a_i) g(f(y)) \tag{4}$$

$$\sigma_{yi} = \frac{I(y, a_i)}{\sum_{p \in \mathcal{R}_i} I(p, a_i) + \theta} \tag{5}$$

where θ is a small positive constant to ensure non-zero denominator. The linear fusion operation (see Fig. 1(c) and Fig. 2(b)) is given by the combination of the correlation function I and normalization function h.

3.2 Anchor Reweighting Module

As explained above, features of the anchors obtained by the linear fusion operation has the desired property of density invariance. However, the bias and variance for estimating anchor features inevitably differ with different anchors. It is necessary to utilize gates for dynamicaly reweighting the anchors during aggregation, i.e., giving lower weight to anchors which have larger bias and variance. Therefore, the anchor feature summation step in Eq. (1) can be modified by anchor reweighting as:

$$(f * g)(x) = \sum_{i \in \mathcal{A}} \beta_i F_i \tag{6}$$

where β_i is the weight of the ith anchor. Two anchor reweighting methods are introduced in the following two subsection, respectively.

Bias-Variance Reweighting: Let F_i^t be the ground-truth feature vector of the ith anchor. Because there is usually no data point sampling at the exact position of the ith anchor, the ground-truth F_i^t is not available. Denote the feature vector of y in the small region of the ith anchor (i.e. $y \in \mathcal{R}_i$) as F_y. Then, F_y can considered as an approximation to the ground-truth F_i^t, i.e.,

$$F_y = F_i^t + \varepsilon_y$$

where ε_y represents the bias vector. We assume that the feature space is continuous and differentiable. The norm of the bias vector $||\varepsilon_y||$ can be The bounded as follows,

$$||\varepsilon_y|| \leq ||y - a_i||G_i = r_a(1 - I(y, a_i))G_i \tag{7}$$

where $I(y, a_i)$ is the weight of y for the ith anchor in Eq. (3), and G_i is the maximum norm of the gradient in the small region of the ith anchor. According to (7), $1 - I(y, a_i)$ (or $I(y, a_i)$) is (inversely) proportional to the upper bound of the bias $||\varepsilon_y||$.

On the other hand, if points in an anchor region are independent and F_i^t is estimated by averaging F_y over $\mathcal{R}_i$, the variance of the estimation is given by,

$$\sigma_{N_i}^2 = D(\frac{1}{N_i} \sum_{y \in \mathcal{R}_i} \varepsilon_y) = \frac{\sigma_1^2}{N_i} \tag{8}$$

where D is the variance function, σ_1^2 denotes the variance of the single point estimate, $\sigma_{N_i}^2$ is the variance of the average estimation, and N_i is the number of points in $\mathcal{R}_i$. By Eq. (8), the variance becomes smaller when the number of sampling points increases. Therefore, the variance is inversely proportional to the number of points N_i.

With Eqs. (7) and (8), we propose the bias-variance coefficient η_i to determine the weight β_i for each anchor, i.e.,

$$\beta_i = \begin{cases} \frac{\eta_i}{\tau} & \eta_i < \tau \\ 1 & \eta_i \geq \tau \end{cases} \tag{9}$$

where τ is a threshold hyperparameter and $\eta_i = \sum_{y \in \mathcal{R}_i} I(y, a_i)$.

Adaptive Reweighting: Multilayer perceptrons (MLP) is a powerful technique for representation learning. Therefore, we propose the adaptive reweighting module by utilizing MLP to decide the weight of each anchor and reweights them. For the adaptive reweighting, the β_i in Eq. (6) can be defined as:

$$\beta_i = Sigmoid([F_i || g(f(x))]\vec{v_i})$$

where F_i is the feature vector of the ith anchor (see Eq. (1)), $g(f(x))$ is the input feature vector of the center point, $||$ denotes the concatenation operation, and $\vec{v_i}$ is a vector.

3.3 Network Architectures

Subsampling Layer: Followed by [18], we use grid subsampling method, which chose a support points of each grid cells to represent grids' features.

AnchorConv Layer: The proposed AnchorConv layer takes full advantage of both the structural information $\mathcal{P} \in \mathbb{R}^{N \times 3}$ and features $\mathcal{F} \in \mathbb{R}^{N \times D_{in}}$. As explained before, anchors are uniformly placed in a ball with a radius r_s. The effective receptive field of an AnchorConv layer mainly depends on radius r_s. Intuitively, the deeper the convolution layer, the larger the receptive field, so each subsampling layer will double the radius r_s.

Segmentation Network: We use the encoder-decoder structure for segmentation tasks. In order to make a fair comparison with previous methods, our encoder contains 8 residual blocks and 4 subsampling layers. In the decoder part, we use the nearest upsampling layer to get the final pointwise features. As in previous methods, skip links are used to directly pass the features from encoder layers to corresponding decoder layers. Those two features are concatenated and followed by a shard fully connected (FC) layer.

Classification Network: The encoder of the classification network is the same as the segmentation network but followed by a global max pooling layer rather than decoder. We use two FC layers and a softmax layer to output prediction results.

Table 1. Quantitative results for 3D Shape Classification and Segmentation tasks. Our method is compared with the previous methods by using overall accuracy (OA) as the evaluation metrics on Modelnet40, class average IoU (mcIoU) and instance average IoU (mIoU) on S3DIS and ShapeNetPart. $AnchorConv_{BVR}$ and $AnchorConv_{AR}$ are AnchorConv with the bias-variance and adaptive reweighting module, respectively.

Methods	ModelNet40	ShapeNet Part		S3DIS area 5
	OA	mcIoU	mIoU	mIoU
PointNet [12] (CVPR17)	–	–	–	41.1
PointNet++ [14] (NIPS17)	90.7	81.9	85.1	–
PointCNN [8] (NIPS18)	92.2	84.6	86.1	57.3
PointConv [20] (CVPR19)	92.5	82.8	85.7	50.4
DynPoints [9] (ICCV19)	91.9	–	86.1	60.0
KPConv [18] (ICCV19)	92.4	84.7	86.2	65.6
RandLA-Net [4] (CVPR20)	–	–	–	62.5
$AnchorConv_{BVR}$	92.6	85.0	86.5	**67.8**
$AnchorConv_{AR}$	**92.9**	**85.1**	**86.7**	**67.8**

4 Experiments

To evaluate the proposed AnchorConv network, we conduct experiments for classification and segmentation tasks on four publicly available datasets including **ModelNet40** [21], **ShapeNet Part** [23], **S3DIS** [1], and **Semantic3D** [3].

In the experiments, our method is implemented with Tensorflow and optimized by SGD optimizer with 0.9 momenta. The initial learning rate is set as 0.01 and decreases by 10% after every 40,000 liters. Anchors are distributed on each vertex of regular icosahedron and one more at the center (12+1 anchors). Since the small region of anchors must be changed with the distribution of anchors, we set the anchor region radius r_a (see Eq. (3)) according to the anchor distribution radius r_s: $r_a = 0.6 \times r_s$. Meanwhile the anchor distribution radius r_s of the jth layer is set as $5.0 \times r_j$, where r_j is the subsampling grid size of the jth layer. The threshold τ in Eq. (9) is set to 0.5. The channel number of the first block is set to 64 and doubled after every subsampling layer.

4.1 Classification on ModelNet40

ModelNet40 contains 12,311 3D models from 40 categories. We use the official split which uses 9,843 instances to train and 2,468 3D models for testing. For benchmarking, we use the data provided by PointNet++ [14] without normals. In this experiment, we set the pooling grid sizes r_0 of the first subsampling layer as 2.0 cm and the batch size is equal to 16. As shown in Table 1, the proposed AnchorConv achieves state-of-the-art performance for the 3D classification task with out normals information.

Table 2. Quantitative results of the proposed AnchorConv compared with state-of-the-art approaches on Semantic3D (reduced-8) [3]. The best results are in bold and the second best results are underlined.

	mIoU	OA	Man-made	Natural	High veg	Low veg	Buildings	Scape	Art	Cars
SPG [6]	73.2	94.0	97.4	<u>92.6</u>	<u>87.9</u>	44.0	83.2	31.0	63.5	76.2
KPConv [18]	74.6	92.9	90.9	82.2	84.2	47.9	94.9	40.0	**77.3**	79.7
RGNet [19]	74.7	94.5	<u>97.5</u>	93.0	**88.1**	48.1	94.6	36.2	72.0	68.0
RandLA-Net [4]	77.4	<u>94.8</u>	95.6	91.4	86.6	51.5	**95.7**	**51.5**	69.8	76.8
$AnchorConv_{BVR}$	**77.8**	94.6	97.1	91.1	<u>87.9</u>	<u>53.2</u>	94.8	41.5	<u>75.7</u>	81.1
$AnchorConv_{AR}$	<u>77.7</u>	**94.9**	**97.6**	**93.0**	85.7	53.1	95.1	42.1	72.1	<u>82.7</u>

4.2 ShapeNet Part Segmentation

ShapeNet dataset has 16,881 instances and 50 categories from 16 classes. The input is the shapes represented by a point cloud, and the goal is to predict a part category label to each point in the input point clouds. We follow the experiment setup in [18]. As the same as the above experiment, the first pooling grid side r_0 is set to 2.0 cm. Point intersection-over-union (IoU) is the most commonly used metric to evaluate the segmentation performance. As shown in Table 1, with only taking point clouds as input, the proposed AnchorConv with AR module obtains a class average IoU (mcIOU) of 85.1% and an instance average IoU (mIoU) of 86.7%.

Table 3. Ablation Study on area 5 of S3DIS. BVR and AR denote bias-variance reweighting and adaptive reweighting, respectively. h denotes the normalization operation (see Eq. (1)).

BVR module	AR module	h	mIoU
–	–	–	65.5
–	–	✓	66.7
✓	–	✓	**67.8**
–	✓	✓	**67.8**
✓	✓	✓	67.6

4.3 3D Segmentation of Indoor Scene

3D Scene Segmentation is a challenging task. S3DIS contains six large-scale indoor areas with a total of 273 million points for 13 classes from 3 different buildings. We use Area-5 as test scene while the other 5 areas are used to train models. We follow [18] to sample small subclouds contained in spheres during training and testing. In this experiment, we set r_0 as 4.0 cm and the batch size is set to 8. As shown in Table 1, the point convolution method KPConv can obtain better performance than other existing methods on the 3D scene point segmentation task. Both $AnchorConv_{BVR}$ and $AnchorConv_{AR}$ achieve 67.8 mIOU on this dataset, which is higher than the state-of-the-art results by 2.2 mIoU for 3D Scene Segmentation.

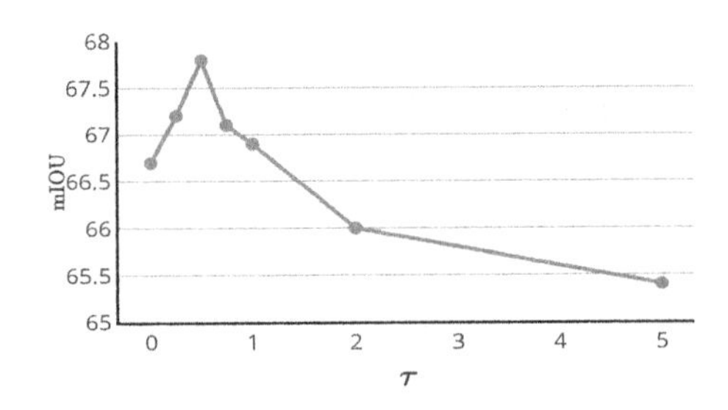

Fig. 3. Evaluation of the threshold τ: $AnchorConv_{BVR}$ results on S3DIS with different τ.

4.4 3D Segmentation of Outdoor Scene

The Semantic3D reduce-8 dataset [3] consists of 15 point clouds for training and 4 for online testing. In this experiment, we set the pooling grid sizes r_0 as 6.0 cm. The raw points belong to 8 classes with 3D coordinates information, RGB information, and intensity information. We use only the 3D coordinates and color information for training and testing. The Mean Intersection-over-Union (mIoU), Overall Accuracy (OA) of all classes, and IoU of each class are used for performance measurement of methods. As shown in Table 2, both the proposed $AnchorConv_{BVR}$ and $AnchorConv_{AR}$ achieve competitive or better performance compared with state-of-the-art methods.

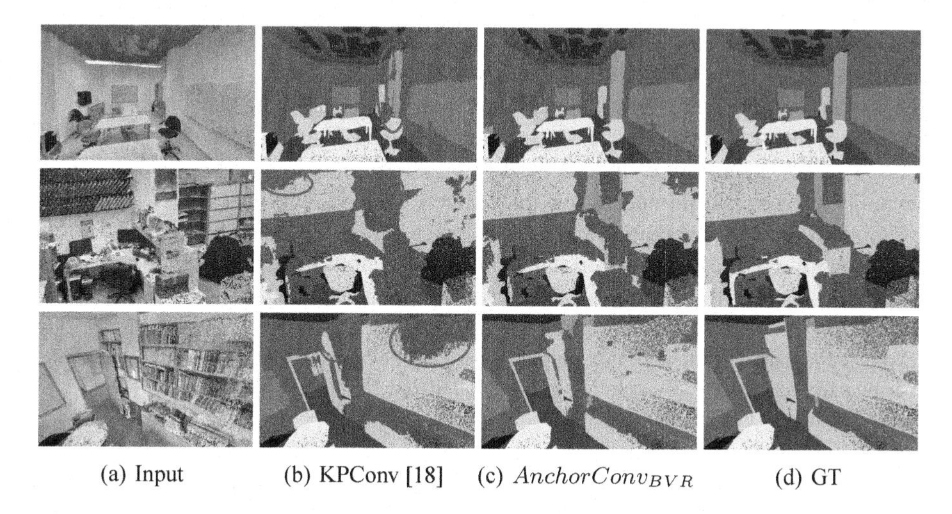

(a) Input (b) KPConv [18] (c) $AnchorConv_{BVR}$ (d) GT

Fig. 4. Qualitative results of the proposed AnchorConv with bias-variance reweighting module and the state-of-the-art KPConv [18] on area 5 of S3DIS. Best view in color. (Color figure online)

4.5 Ablation Study

In order to explore the impact of anchor reweighting modules, we implement extensive ablation experiments on S3DIS (area 5 as test set).

Table 3 presents the results of all ablated networks. It is noteworthy that, without normalization function h, the performance of our method drops by 1.2 mIoU, which indicates that anchors' features captured by our proposed linear fusion are indeed more robust to density. What's more, AnchorConv can achieve 67.8 mIoU with BVR module or AR module, 1.1 mIoU better than our model without reweighting module. BVR module or AR module can bring significant improvement to our method, but using them together will not bring greater gains.

Figure 3 presents the results of $AnchorConv_{BVR}$ with different τ. As τ increases, mIoU generally ascend in first and descend at last, and the best result is achieved when τ is equal to 0.5. It should be noted that when τ is too large, $AnchorConv_{BVR}$ will no longer be robust to the density of points. For example, if τ equals to 5.0, the result of $AnchorConv_{BVR}$ (65.4 mIoU) is similar to the model without h (65.5 mIoU).

4.6 Qualitative Results

Example semantic segmentation results are visualized on S3DIS in Fig. 4. From these results, we can see that the proposed AnchorConv network can obtain better results on complex 3D scene segmentation tasks than the state-of-the-art approach (KPConv).

5 Conclusion

In this work, we propose a novel point convolution method namely AnchorConv that can process irregular and unordered data. AnchorConv uniformly places anchors around the center point for feature updating. Each anchor processes features obtained from the containing small region by the proposed linear fusion. Due to the novel design which updates the center point by aggregating feature representation of the anchors rather than neighbor points, our method is less sensitive to changes in point cloud density. Since features of the anchor are estimated by the neighboring points, the bias and variance inevitably differ with different anchors due to the irregularity of point clouds. We further develop two anchor reweighting modules as gates to filter features of anchors before aggregating them. Extensive experiments on four benchmark datasets demonstrate that our method is better than the state-of-the-art for point cloud analysis.

Acknowledgements. This work was supported partially by NSFC (No. 61906218), Guangdong Basic and Applied Basic Research Foundation (No. 2020A1515011497) and Science and Technology Program of Guangzhou (No. 202002030371).

References

1. Armeni, I., et al.: 3D semantic parsing of large-scale indoor spaces. In: CVPR, pp. 1534–1543 (2016)
2. Bronstein, M.M., Bruna, J., LeCun, Y., Szlam, A., Vandergheynst, P.: Geometric deep learning: going beyond Euclidean data. IEEE Sig. Process. Mag. **34**(4), 18–42 (2017)
3. Hackel, T., Savinov, N., Ladicky, L., Wegner, J.D., Schindler, K., Pollefeys, M.: SEMANTIC3D.NET: A new large-scale point cloud classification benchmark. In: ISPRS, vol. IV-1-W1, pp. 91–98 (2017)
4. Hu, Q., et al.: RandLA-Net: efficient semantic segmentation of large-scale point clouds. In: CVPR (2020)
5. Hua, B.S., Tran, M.K., Yeung, S.K.: Pointwise convolutional neural networks. In: CVPR, pp. 984–993 (2018)
6. Landrieu, L., Simonovsky, M.: Large-scale point cloud semantic segmentation with superpoint graphs. In: CVPR, pp. 4558–4567 (2018)
7. Li, G., Muller, M., Thabet, A., Ghanem, B.: DeepGCNs: can GCNs go as deep as CNNs? In: ICCV, pp. 9267–9276 (2019)
8. Li, Y., Bu, R., Sun, M., Wu, W., Di, X., Chen, B.: PointCNN: convolution on x-transformed points. In: NIPS, pp. 820–830 (2018)
9. Liu, J., Ni, B., Li, C., Yang, J., Tian, Q.: Dynamic points agglomeration for hierarchical point sets learning. In: ICCV, October 2019
10. Maturana, D., Scherer, S.: VoxNet: a 3D convolutional neural network for real-time object recognition. In: IROS, pp. 922–928. IEEE (2015)
11. Monti, F., Boscaini, D., Masci, J., Rodola, E., Svoboda, J., Bronstein, M.M.: Geometric deep learning on graphs and manifolds using mixture model CNNs. In: CVPR, pp. 5115–5124 (2017)
12. Qi, C.R., Su, H., Mo, K., Guibas, L.J.: PointNet: deep learning on point sets for 3D classification and segmentation. In: CVPR, pp. 652–660 (2017)
13. Qi, C.R., Su, H., Nießner, M., Dai, A., Yan, M., Guibas, L.J.: Volumetric and multi-view CNNs for object classification on 3D data. In: CVPR, pp. 5648–5656 (2016)
14. Qi, C.R., Yi, L., Su, H., Guibas, L.J.: PointNet++: deep hierarchical feature learning on point sets in a metric space. In: NIPS, pp. 5099–5108 (2017)
15. Riegler, G., Osman Ulusoy, A., Geiger, A.: OctNet: Learning deep 3D representations at high resolutions. In: ICCV, pp. 3577–3586 (2017)

16. Simonovsky, M., Komodakis, N.: Dynamic edge-conditioned filters in convolutional neural networks on graphs. In: CVPR, pp. 3693–3702 (2017)
17. Su, H., Maji, S., Kalogerakis, E., Learned-Miller, E.: Multi-view convolutional neural networks for 3D shape recognition. In: ICCV, pp. 945–953 (2015)
18. Thomas, H., Qi, C.R., Deschaud, J.E., Marcotegui, B., Goulette, F., Guibas, L.J.: KPConv: flexible and deformable convolution for point clouds. In: ICCV, pp. 6411–6420 (2019)
19. Truong, G., Gilani, S.Z., Islam, S.M.S., Suter, D.: Fast point cloud registration using semantic segmentation. In: DICTA, pp. 1–8. IEEE (2019)
20. Wu, W., Qi, Z., Fuxin, L.: PointConv: deep convolutional networks on 3D point clouds. In: CVPR, pp. 9621–9630 (2019)
21. Wu, Z., et al.: 3D shapeNets: a deep representation for volumetric shapes. In: CVPR, pp. 1912–1920 (2015)
22. Xu, Y., Fan, T., Xu, M., Zeng, L., Qiao, Yu.: SpiderCNN: deep learning on point sets with parameterized convolutional filters. In: Ferrari, V., Hebert, M., Sminchisescu, C., Weiss, Y. (eds.) ECCV 2018. LNCS, vol. 11212, pp. 90–105. Springer, Cham (2018). https://doi.org/10.1007/978-3-030-01237-3_6
23. Yi, L., et al.: A scalable active framework for region annotation in 3D shape collections. ACM Trans. Graph. (TOG) **35**(6), 1–12 (2016)
24. Yue, X., Wu, B., Seshia, S.A., Keutzer, K., Sangiovanni-Vincentelli, A.L.: A lidar point cloud generator: from a virtual world to autonomous driving. In: ICMR, pp. 458–464 (2018)
25. Zaheer, M., Kottur, S., Ravanbakhsh, S., Poczos, B., Salakhutdinov, R.R., Smola, A.J.: Deep sets. In: NIPS, pp. 3391–3401 (2017)

IFR: Iterative Fusion Based Recognizer for Low Quality Scene Text Recognition

Zhiwei Jia, Shugong Xu(✉), Shiyi Mu, Yue Tao, Shan Cao, and Zhiyong Chen

Shanghai Institute for Advanced Communication and Data Science,
Shanghai University, Shanghai 200444, China
{zhiwei.jia,shugong,mushiyi,yue_tao,cshan,bicbrv_g}@shu.edu.cn

Abstract. Although recent works based on deep learning have made progress in improving recognition accuracy on scene text recognition, how to handle low-quality text images in end-to-end deep networks remains a research challenge. In this paper, we propose an **I**terative **F**usion based **R**ecognizer (**IFR**) for low quality scene text recognition, taking advantage of refined text images input and robust feature representation. IFR contains two branches which focus on scene text recognition and low quality scene text image recovery respectively. We utilize an iterative collaboration between two branches, which can effectively alleviate the impact of low quality input. A feature fusion module is proposed to strengthen the feature representation of the two branches, where the features from the **R**ecognizer are **F**used with image **R**estoration branch, referred to as **RRF**. Without changing the recognition network structure, extensive quantitative and qualitative experimental results show that the proposed method significantly outperforms the baseline methods in boosting the recognition accuracy of benchmark datasets and low resolution images in TextZoom dataset.

Keywords: Scene text recognition · Iterative collaboration · Feature fusion

1 Introduction

In recent years, scene text recognition (STR) has attracted much attention of the computer vision community. STR aims to recognize the text in the scene images, which is an important part of the downstream task, such as license plate recognition [26], receipts key information extraction [22], etc. Recent works based on deep learning have succeeded in improving recognition accuracy on clear text images. Benefiting from the development of sequence-to-sequence learning, STR methods can be roughly divided into two major techniques [11], Connectionist Temporal Classification [19,21] and Attention mechanism [9,13,20,24].

Previous works focus on texts in natural [1,19] and curve scenes [9,13,20], which prove the outstanding performance on clear images. However, as shown

Z. Jia—Student

H. Ma et al. (Eds.): PRCV 2021, LNCS 13020, pp. 180–191, 2021.
https://doi.org/10.1007/978-3-030-88007-1_15

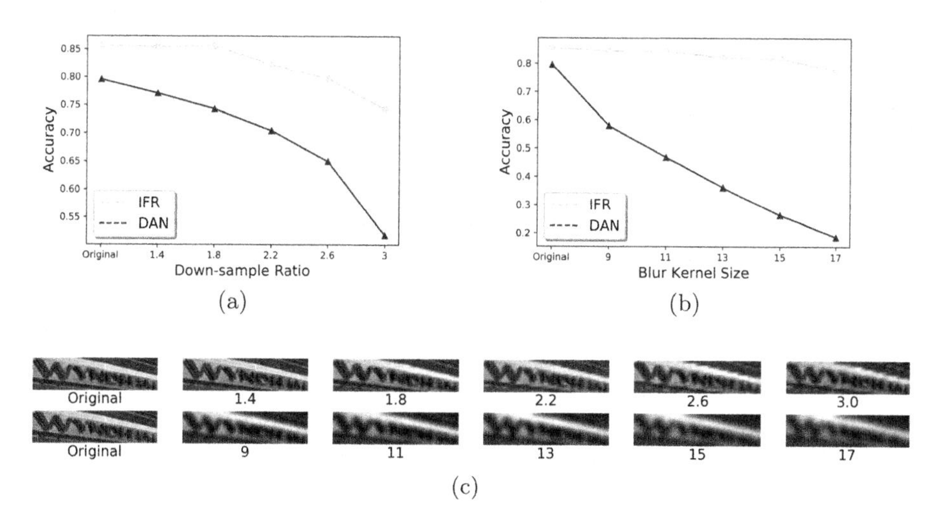

Fig. 1. Accuracy comparison on SVTP [17] dataset. (a-b) Compared to state-of-the-art DAN [24] (in blue), our IFR model (in red) is shown to achieve robust results for different image degradation methods. (c) Visualization results of different degrees of down-sample ratio(up) and blur kernel size (down). (Color figure online)

in Fig. 1, both of the methods are facing a problem that their performances drop sharply when text image quality is poor, including low-resolution, blurred, low contrast, noisy, etc. Therefore, how to generalize the recognizer to both high and low quality text images is still a research challenge. The biggest problem is that it is hard to extract robust semantic information due to the lack of sufficient text region visual details. Recent works have noticed this problem. TSRN [25] (Fig. 2a) introduced super-resolution (SR) methods as a pre-processing procedure before recognizer and show good performance on the scene text SR dataset TextZoom, which is the first real paired scene text image super-resolution dataset. The limitation is that TSRN can only handle the problem of low-resolution text recognition. A more reasonable way is image restoration(IR) [27], which is the task of recovering a clean image from its degraded version. However, compared to single image restoration, text image restoration only considers text-level features instead of complex scenes. So it is necessary to share features between STR and IR. In this way, the feature of the recognizer backbone can be strong prior knowledge for restoration. Although Plut-Net [15] (Fig. 2b) proposed an end-to-end trainable scene text recognizer together with a pluggable super-resolution unit, the super-resolved images from the IR branch are not utilized to enhance the recognition accuracy. As the IR branch restores higher quality images, they can be used as new inputs into the recognition branch to get more accurate results.

Inspired by these works, we propose an iterative fusion based recognizer as shown in Fig. 2c. IFR contains two networks focusing on scene text recognition and low-quality scene text image restoration, respectively. Benefit from the IR

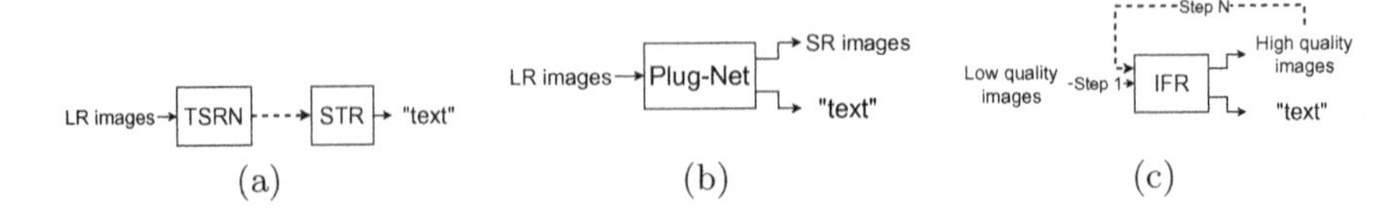

Fig. 2. (a) Pipeline of TSRN. The two network are trained separately and the dotted line represent inference. (b) Pipeline of Plug-Net which containing two branches. (c) Our iterative fusion recognizer IFR.

branch, IFR uses refined text images as new inputs for better recognition results in the iterative collaboration manner, which utilizes image restoration knowledge in scene text recognition. In each step, previous outputs of IR branch are fed into the STR branch in the following step. Different from previous methods, the text image restoration and recognition processes facilitate each other progressively. A fusion module RRF is designed to combine the image shallow feature and the STR backbone feature. In this way, the IR branch can get more semantic information to refine the detail of text region instead of background and noise. In return, feature sharing enhances the feature representation of different quality text images. We also claim that our proposed RRF and collaboration manner can boost the performance on standard benchmarks and low-resolution dataset TextZoom without modifying the recognizer. Additionally, only exploiting synthetic augmented training data can achieve the comparable results on TextZoom.

The contributions of this work are as follow:

* Firstly, an iterative fusion based recognizer IFR are proposed for scene text recognition which contains two branches focusing on scene text recognition and low-quality scene text image recovery, respectively. An iterative collaboration are then proposed between two branches, which can effectively alleviate the impact of low-quality input.
* Secondly, a fusion module RRF is designed to strengthen the feature representation of the two tasks. Benefited from this module, the recognition branch can enhance the feature extraction for recognizing low-quality text images.
* Thirdly, without modifying the recognizer, quantitative and qualitative experimental results show the proposed method outperforms the baseline methods in boosting the recognition accuracy of benchmark datasets and LR images in TextZoom dataset.

2 Related Works

Text Recognition. Early work used a bottom-up approach which relied on low-level features, such as histogram of oriented gradients descriptors [23], connected components [16], etc. Recently, deep learning based methods have achieved remarkable progress in various computer vision tasks including scene text recognition. STR methods can roughly divided into two major techniques [11]:

Connectionist Temporal Classification [2] and Attention mechanism. CRNN [19] integrates feature extraction, sequence modeling, and transcription into a unified framework. CTC is used to translate the per-frame predictions into the final label sequence. As the attention mechanism was wildly used in improving the performance of natural language process systems, an increasing number of recognition approaches based on the attention mechanism have achieved significant improvements. ASTER [20] proposed a text recognition system which combined a Spatial Transformer Network (STN) [6] and an one dimension attention-based sequence recognition network. DAN [24] propose a decoupled attention network, which decouples the alignment operation from using historical decoding results to solve the attention drift problem. In this work, we choose DAN as our baseline recognizer to boost the recognition accuracy of the low-quality images.

Low-Quality Scene Text Image Recognition. Recent works have noticed the text image degradation problem. TSRN introduced the first real paired scene text SR dataset TextZoom and proposed a new text SR method as a pre-processing procedure before recognition methods. Plug-net proposed an end-to-end trainable scene text recognizer together with a pluggable super-resolution unit for a more robust feature representation.

These works have achieved notable success by super resolution module, but they treat the two networks as an independent task. While recursive networks and feature fusion promote the development of image restoration [27], a few methods have employed their generative power in low-quality text recognition. To our best knowledge, this is the first attempt to integrate text recognition and image restoration iteratively into a single end-to-end trainable network.

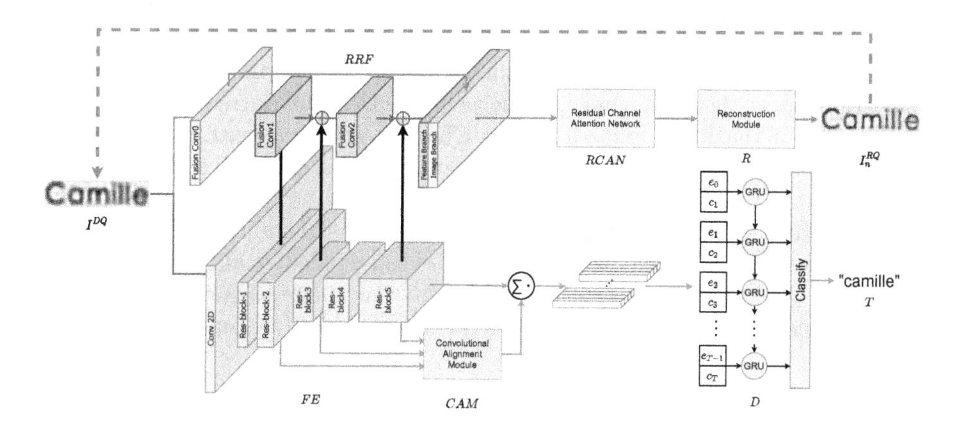

Fig. 3. Overall framework of IFR.

3 Method

In this section, we present our proposed method IFR in detail. As shown in Fig. 3, we aim to recognize the text of the input degraded quality text images $\mathbf{I}^{DQ}$ and get the high-quality result $\mathbf{I}^{HQ}$. We design a deep iterative collaboration network that estimates high-quality text images and recognizes texts iteratively and progressively between the recognizer and IR process. In order to enhance the collaboration, we design a novel fusion module RRF that integrates two sources of information effectively.

3.1 Iterative Collaboration

The previous methods are very sensitive to image quality as low-quality images may lose enough information for recognition. Therefore, our method alleviates this issue by an iterative fusion IFR as shown in Fig. 3. In this framework, text image recovery and recognition are performed simultaneously and recursively. We can obtain more accurate recognition result if the text image is restored well. Both processes can enhance each other and achieve better performance progressively. Finally, we can get accurate recognition results and HQ text images with enough steps.

The recognition branch uses DAN as the baseline, which includes a CNN-based feature encoder FE, a convolutional alignment module CAM, and a decoupled text decoder D. The CAM takes multi-scale visual features from the feature encoder as input and generates attention maps with a fully convolutional network [10] in channel-wise manner. Compare with other attention decoders, DAN avoids the accumulation and propagation of decoding errors. The IR branch contains a fusion module RRF, a residual channel attention network RCAN [28] and a reconstruction module R. RCAN contains two residual groups, each of which has two residual channel attention blocks. Residual groups and long skip connection allow the main parts of the network to focus on more informative components of the low-quality features. Channel attention extracts the channel statistic among channels to further enhance the discriminative ability of the network.

For the first step, the recognition branch extracts the multi-scale image features and decodes the text result by using the degraded quality text images, denoted as $\{\mathbf{f}_1^T\}$ and $\mathbf{T}_1$, respectively. Besides, the image features are fed into the fusion module with the input image. Therefore, the first step can be formulated by:

$$\{\mathbf{f}_1^T\} = FE(\mathbf{I}^{DQ}), \tag{1}$$

$$\mathbf{T}_1 = D(CAM(\{\mathbf{f}_1^T\}), \{\mathbf{f}_1^T\}), \tag{2}$$

$$\mathbf{f}_1^I = RRF(\mathbf{I}^{DQ}, \{\mathbf{f}_1^T\}), \tag{3}$$

$$\mathbf{I}_1^{RQ} = R(RCAN(\mathbf{f}_1^I)), \tag{4}$$

For the n_{th} step where $n = 2, ..., N$, the difference is the input in feature encoder FE and fusion module RRF is the reconstructed picture $\mathbf{I}_{n-1}^{RQ}$ from the previous step $n-1$, as follows:

$$\{\mathbf{f}_n^T\} = FE(\mathbf{I}_{n-1}^{RQ}), \tag{5}$$

$$\mathbf{T}_n = D(CAM(\{\mathbf{f}_n^T\}), \{\mathbf{f}_n^T\}), \tag{6}$$

$$\mathbf{f}_n^I = RRF(\mathbf{I}_{n-1}^{RQ}, \{\mathbf{f}_n^T\}), \tag{7}$$

$$\mathbf{I}_n^{RQ} = R(RCAN(\mathbf{f}_n^I)), \tag{8}$$

After N iterations, we get the recognition results $\{\mathbf{T}_n\}_{n=1}^N$ and reconstruction result $\{\mathbf{I}_n^{RQ}\}_{n=1}^N$. In order to optimize the output of each iteration, we calculate the loss for each step of the output. In this way, the output results are gradually optimized through supervision.

3.2 Fusion Module RRF

Previous works considered text image restoration as independent tasks to improve the image quality or multi-task to obtain a more robust feature representation. Our method exploits the merits of both designs. In our method, a fusion module is designed to strengthen the feature representation of the two tasks, where the features from the recognizer are fused with IR branch. As shown in Fig. 3, the fusion module contains a 3×3 convolutional layer for the image shallow feature extraction and a cascade convolutional layer to generate the multi-scale fusion features. The inputs are degraded quality text images $\mathbf{I}^{DQ}$ at the first step or $\mathbf{I}_n^{RQ}$ from the last step, and visual features $\{\mathbf{f}_n^T\}$ from the feature encoder. These multi-scale features are first encoded by cascade convolutional layers then concatenate with image shallow feature as output. The proposed RRF has several merits. First, the STR branch can extract more robust features for recognizing low-quality text images. Second, the multi-scale features of the FE help to enrich the semantic features of the IR branch.

3.3 Loss Functions

Here, IFR is trained end-to-end using the cross-entropy loss L_{rec} and the pixel-wise loss L_{pixel}. The L_{rec} is defined as follow:

$$L_{rec} = -\frac{1}{2}\sum_{i=1}^{T}(\log p_{LR}(y_t|\mathbf{I}) + \log p_{RL}(y_t|\mathbf{I})) \tag{9}$$

where y_t is the ground-truth text represented by a character sequence. p_{LR} and p_{RL} are bidirectional decoder distributions, respectively. The pixel-wise loss function is defined as follow:

$$L_{pixel} = \frac{1}{N * W * H}\sum_{n=1}^{N}\sum_{i=1}^{W}\sum_{j=1}^{H}||\mathbf{I}^{HQ} - \mathbf{I}_n^{RQ}|| \tag{10}$$

W and H refer to the width and height of the input image. The model is optimized by minimizing the following overall objective function:

$$L = L_{rec} + \lambda L_{pixel} \tag{11}$$

Note that the gradients can be back-propagated to both the STR and IR in a recursive manner. The STR can be supervised by not only the recognition loss but also by the revision of image restoration loss through the fusion module.

3.4 Paired Training Data Generate

For fair comparison, our model is trained on the MJSynth MJ [4,5] and SynthText ST [3]. As we all know, the smaller the domain gap between the training dataset and the real scene, the better the performance of the test dataset. TSRN experiment shows that fine-tune ASTER on TextZoom training set can improve the accuracy of the TextZoom test set but harm the performance of other high-quality benchmarks. In our work, data augmentation is a key method to generate paired high and low-quality training text images. Owing to the good quality of the two synthetic datasets, we use random data augmentation, Gaussian kernel and down-up sampling, to generate paired training data. Inspired by Plug-Net, these methods randomly generate paired data. The degraded text images $\mathbf{I}^{DQ}$ are the input at the first step and $\mathbf{I}^{HQ}$ provide ground-truth supervisory signals useful for the progressive image restoration at each step. Different from other augment methods, The degraded images contain both clear and different degrees low-quality text images as the random strategy, which enables IR branch to learn not only "how" but also "when" to restore a text image.

4 Experiments

4.1 Datasets and Implementation Details

Eight standard benchmarks include ICDAR 2003 (IC03) [12], ICDAR 2013 (IC13) [8], ICDAR 2015 (IC15) [7], IIIT5K (IIIT) [14], Street View Text (SVT) [23], Street View Text-Perspective (SVTP) [17], CUTE80 (CUTE) [18] and TextZoom are as the testing datasets. Details of these datasets can be found in the previous works [24]. In addition, TextZoom dataset is divided into three testing subsets [25] by difficulty and a training set.

We adopt an opensource implementation of DAN and reproduced bidirectional decoding according to [24]. The STR and IR model dimension C are set to 512 and 64 throughout respectively. Balanced factor λ is set to 10. Ground-truths are directly resized to 32 $\times$ 128 greyscale images. Then the augment functions are used to generate the paired degraded data randomly. The range of blur kernel size is 9 to 17. The down-up sample ratio is in the range of 1 to 3. The model is trained by Adadelta optimizer. The initial learning rate is 1 and is decayed to 0.1 and 0.01 respectively after 4 and 5 epochs. Recognition results are evaluated with accuracy. IR results are evaluated with PSNR and SSIM. They are computed on the greyscale space. Our experiments are implemented on Pytorch with NVIDIA RTX 2080Ti GPUs.

4.2 Ablation Study

Effectiveness of Iterative Learning. In our work, we use the execution manner of iterative collaboration between STR and IR branches. To better show the influence of the proposed iterative learning scheme, we evaluate the recognition accuracy and quality of the IR outputs on different steps. The performance on TextZoom subsets is given in Table 1, where the testing steps are set to 1 to 4. As we can see from the results, iterating the IFR three times can achieve a significant improvement especially on the second step. Specifically, there are little gains on the third and fourth steps. As mentioned above, we use PSNR and SSIM as image recovery measurement metrics. From the first to fourth step, SSIM and PSNR get better progressively on most of the dataset. Therefore, the comparison proves that our method is able to achieve progressively better image quality and recognition accuracy simultaneously.

Table 1. Ablation study of iterative steps on TextZoom subsets. Step means the iteration times.

Step	Hard			Medium			Easy		
	Accuracy	SSIM	PSNR	Accuracy	SSIM	PSNR	Accuracy	SSIM	PSNR
1	45.87	0.6867	19.75	62.03	0.6504	**19.50**	78.43	0.8127	23.23
2	52.51	0.6940	19.82	69.31	0.6536	19.49	83.15	0.8280	23.57
3	52.77	0.7005	**19.87**	69.56	0.6562	19.45	83.36	0.8389	23.84
4	**52.85**	**0.7037**	19.86	**69.64**	**0.6568**	19.38	**83.43**	**0.8445**	**23.97**

We further explore the difference of iteration steps between training and testing. Increase the number of iteration steps at the training stage will increase the memory usage and training time. The same as the testing stage. The average accuracy on SVT–P in Fig. 4 suggests that: 1) training without applying iterative collaboration, the reconstructed images will harm the accuracy in STR branch when testing; 2) iterating during the training phase is helpful, as it provides recovery training images for STR branch; 3) the accuracy improves significantly in the second step and slowly improves in the next few steps. Therefore testing with a big iteration step is unnecessary.

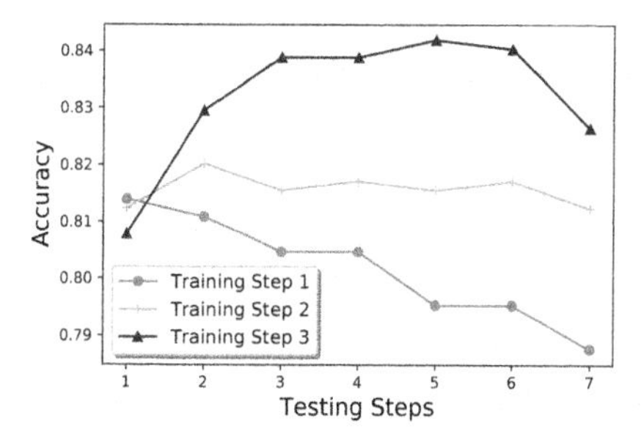

Fig. 4. Accuracy of iterating steps in training and testing phase on SVTP.

Table 2. Ablation study of training data augmention and fusion module RRF. DAN_{RP} represents the result we reproduced. Subscript numbers indicate the testing steps.

Method	Step	Aug	RRF	IIIT	SVT	IC03	IC13	IC15	SVTP	CUTE	Hard	Medium	Easy
				3000	647	867	1015	2077	645	288	1175	1209	1442
DAN_{RP}	1	×	×	93.9	88.7	94.5	92.6	74.2	79.5	**82.3**	33.2	45.2	62.3
DAN_{RP}	1	✓	×	93.9	**89.2**	94.7	**93.7**	**76.5**	**82.1**	80.6	36.3	50.0	70.3
IFR_1	1	✓	✓	**94.5**	**89.2**	**94.8**	93.4	76.0	81.3	80.6	**37.9**	**50.8**	**71.2**
IFR_2	2	✓	×	94.0	89.8	94.7	93.6	76.0	81.7	**82.3**	**40.0**	50.8	74.5
IFR_2	2	✓	✓	**94.2**	**90.3**	**95.5**	**93.9**	**77.7**	**82.3**	81.9	39.6	**53.3**	**76.4**

Effectiveness of Fusion Module. Firstly, we discuss the performance from the training data. Different from the baseline method DAN, our work uses degraded images as input. For a fair comparison, we train the recognition branch DAN with original and augmentation datasets. Table 2 shows the result of two recognition models in seven text recognition benchmark datasets and three TextZoom low-resolution datasets. Reproduced DAN is slightly different from the open-source model by using data augmentation, bidirectional decoding, etc. With the help of the data augmentation, the recognition accuracy in TextZoom has improved from 33.2%, 45.2%, 62.3% to 36.3%, 50.0%, 70.3%. To some extent, data augmentation improves the generalization in text recognition tasks.

We further implement an ablation study to measure the effectiveness of the fusion module. We discuss the performance from two aspects: training with or without fusion module; different fusion stages of the STR backbone. When we don't use the iterative strategy, IFR_1 without fusion module is equivalent to an independent recognition branch. As depicted in Table 2, training with fusion module is useful which boosts the accuracy on most of the benchmarks, especially on TextZoom subsets. When training the IFR_2 with one iteration, it achieves competitive advantage in accuracy. Besides, IFR_2 shows better performance especially on challenging datasets such as SVTP and TextZoom. Besides, we further explore the fusion stage of the STR backbone. The average accuracy on benchmark and TextZoom is in Table 3 suggests that the deeper the feature depth, the higher the accuracy on test datasets.

From the fusion module ablation study we can conclude: 1) learning with gradient from the IR branch can make the feature extractor more robust on image quality. 2) by further equipping fusion module with iterative training, the image quality problem can be alleviated, which is recommended to deal with challenging datasets such as TextZoom and SVTP.

Table 3. Ablation study of fusion stage of the STR backbone.

Method	Fusion Stage	Benchmark	TextZoom
IFR_1	1,2,3	87.60	53.99
IFR_1	2,3,4	87.69	53.95
IFR_1	1,3,5	**87.80**	**54.31**

Table 4. Comparison with SOTA methods. 'A' means using data augmentation when training. 'T' means the TextZoom training dataset. Top accuracy for each benchmark is shown in **bold**.

Method	Training data	IIIT	SVT	IC03	IC13	IC15	IC15	SVTP	CUTE
		3000	647	867	1015	1811	2077	645	288
CRNN (2015) [19]	MJ	78.2	80.8	–	86.7	–	–	–	–
FAN (2017) [1]	MJ+ST	87.4	85.9	94.2	93.3	–	70.6	–	–
ASTER (2019) [20]	MJ+ST	93.4	89.5	–	91.8	76.1	–	78.5	79.5
SAR (2019) [9]	MJ+ST	91.5	84.5	–	91.0	–	69.2	76.4	83.3
MORAN (2019) [13]	MJ+ST	91.2	88.3	95.0	92.4	–	68.8	76.1	77.4
PlugNet (2020) [15]	MJ+ST	94.4	**92.3**	95.7	**95.0**	82.2	–	84.3	**85.0**
DAN (2020) [24]	MJ+ST	94.3	89.2	95.0	93.0	–	74.5	80.0	84.4
IFR_3	MJ+ST+A	94.6	91.7	95.6	94.2	–	78.5	83.9	82.6
IFR_3	MJ+ST+A+T	**94.9**	92.0	**96.0**	94.8	–	**80.2**	**85.4**	82.3

4.3 Comparisons with State-of-the-Arts

We compare our proposed IFR with state-of-the-art STR methods. Table 4 shows the recognition results among 7 widely used benchmarks. It is noteworthy that IFR with two iterations outperforms baseline method DAN by a large margin without changing the STR structure. By the progressive collaboration between the STR and IR processes, IFR can help the STR network obtain more robust feature maps and generate higher-quality text images at the same time. Especially in two low-quality text datasets as SVT and SVTP, our method shows a much robust performance. So, the iterative collaboration and RRF may also be useful for other STR networks. When the real low-quality dataset TextZoom adds to training data, the accuracy will be further improved.

Table 5 shows the results on low-resolution dataset TextZoom. Compared with TSRN, we achieve comparable results with only the synthetic training dataset. Meanwhile, the proposed method achieves better performance when training on both synthetic and TextZoom datasets.

Table 5. Comparison with SOTA methods on the TextZoom dataset. In each sub-set, the left column only contains 36 alphanumeric characters while the other contains 93 classes, in corresponding to 10 digits, 52 case sensitive letters, 31 punctuation characters. We use the model published on the TSRN to reproduce the results on the 36-class subsets.

Methods	Training dataset	Hard		Medium		Easy	
		1175	1343	1209	1411	1442	1619
TSRN	Synthetic	–	33.00	–	45.30	–	67.50
	TextZoom	41.45	40.10	58.56	56.30	73.79	75.10
IFR_3	Synthetic	41.87	–	55.91	–	77.12	–
	Synthetic+TextZoom	**53.19**	–	**69.73**	–	**83.01**	–

5 Conclusion

In this paper, we have proposed IFR which explores iterative collaboration for utilizing image restoration knowledge in scene text recognition. The IFR can extract robust feature representation by fusion module RRF and provide refined text images as input for better recognition results by iterative collaboration manner. Quantitative and qualitative results on standard benchmarks and low-resolution dataset TextZoom have demonstrated the superiority of IFR, especially on low-quality images. We also claim that exploiting synthetic augmented data can achieve comparable results.

References

1. Cheng, Z., Bai, F., Xu, Y., Zheng, G., Pu, S., Zhou, S.: Focusing attention: towards accurate text recognition in natural images. In: Proceedings of the IEEE International Conference on Computer Vision, pp. 5076–5084 (2017)
2. Graves, A., Fernández, S., Gomez, F., Schmidhuber, J.: Connectionist temporal classification: labelling unsegmented sequence data with recurrent neural networks. In: Proceedings of the 23rd international Conference on Machine Learning, pp. 369–376 (2006)
3. Gupta, A., Vedaldi, A., Zisserman, A.: Synthetic data for text localisation in natural images. In: Proceedings of the IEEE Conference on Computer Vision and Pattern Recognition, pp. 2315–2324 (2016)
4. Jaderberg, M., Simonyan, K., Vedaldi, A., Zisserman, A.: Synthetic data and artificial neural networks for natural scene text recognition. In: NIPS Deep Learning Workshop (2014)
5. Jaderberg, M., Simonyan, K., Vedaldi, A., Zisserman, A.: Reading text in the wild with convolutional neural networks. Int. J. Comput. Vis. **116**(1), 1–20 (2016)
6. Jaderberg, M., Simonyan, K., Zisserman, A., Kavukcuoglu, K.: Spatial transformer networks. In: Advances in Neural Information Processing Systems, vol. 28 (2015)
7. Karatzas, D., et al.: ICDAR 2015 competition on robust reading. In: 13th International Conference on Document Analysis and Recognition), pp. 1156–1160 (2015)
8. Karatzas, D., et al.: ICDAR 2013 robust reading competition. In: 2013 12th International Conference on Document Analysis and Recognition, pp. 1484–1493 (2013)
9. Li, H., Wang, P., Shen, C., Zhang, G.: Show, attend and read: a simple and strong baseline for irregular text recognition. In: AAAI Conference on Artificial Intelligence (2019)
10. Long, J., Shelhamer, E., Darrell, T.: Fully convolutional networks for semantic segmentation. In: Proceedings of the IEEE Conference on Computer Vision and Pattern Recognition, pp. 3431–3440 (2015)
11. Long, S., He, X., Yao, C.: Scene text detection and recognition: the deep learning era. Int. J. Comput. Vis. **129**(1), 161–184 (2021)
12. Lucas, S.M., et al.: ICDAR 2003 robust reading competitions: entries, results, and future directions. Int. J. Doc. Anal. Recogn. (IJDAR) **7**(2–3), 105–122 (2005)
13. Luo, C., Jin, L., Sun, Z.: Moran: a multi-object rectified attention network for scene text recognition. Pattern Recogn. **90**, 109–118 (2019)
14. Mishra, A., Alahari, K., Jawahar, C.: Scene text recognition using higher order language priors. In: British Machine Vision Conference (BMVC) (2012)

15. Mou, Y., et al.: PlugNet: degradation aware scene text recognition supervised by a pluggable super-resolution unit. In: The 16th European Conference on Computer Vision (ECCV 2020), pp. 1–17 (2020)
16. Neumann, L., Matas, J.: Real-time scene text localization and recognition. In: 2012 IEEE Conference on Computer Vision and Pattern Recognition, pp. 3538–3545. IEEE (2012)
17. Quy Phan, T., Shivakumara, P., Tian, S., Lim Tan, C.: Recognizing text with perspective distortion in natural scenes. In: Proceedings of the IEEE International Conference on Computer Vision, pp. 569–576 (2013)
18. Risnumawan, A., Shivakumara, P., Chan, C.S., Tan, C.L.: A robust arbitrary text detection system for natural scene images. Expert Syst. Appl. **41**(18), 8027–8048 (2014)
19. Shi, B., Bai, X., Yao, C.: An end-to-end trainable neural network for image-based sequence recognition and its application to scene text recognition. IEEE Trans. Pattern Anal. Mach. Intell. **39**(11), 2298–2304 (2016)
20. Shi, B., Yang, M., Wang, X., Lyu, P., Yao, C., Bai, X.: Aster: an attentional scene text recognizer with flexible rectification. IEEE Trans. Pattern Anal. Mach. Intell. **41**(9), 2035–2048 (2018)
21. Wan, Z., Xie, F., Liu, Y., Bai, X., Yao, C.: 2D-CTC for scene text recognition. arXiv preprint arXiv:1907.09705 (2019)
22. Wang, J., et al.: Towards robust visual information extraction in real world: new dataset and novel solution. In: Proceedings of the AAAI Conference on Artificial Intelligence (2021)
23. Wang, K., Babenko, B., Belongie, S.: End-to-end scene text recognition. In: 2011 International Conference on Computer Vision. pp. 1457–1464. IEEE (2011)
24. Wang, T., et al.: Decoupled attention network for text recognition. In: AAAI Conference on Artificial Intelligence (2020)
25. Wang, W., et al.: Scene text image super-resolution in the wild. In: Vedaldi, A., Bischof, H., Brox, T., Frahm, J.-M. (eds.) ECCV 2020. LNCS, vol. 12355, pp. 650–666. Springer, Cham (2020). https://doi.org/10.1007/978-3-030-58607-2_38
26. Wu, C., Xu, S., Song, G., Zhang, S.: How many labeled license plates are needed? In: Lai, J.-H., et al. (eds.) PRCV 2018. LNCS, vol. 11259, pp. 334–346. Springer, Cham (2018). https://doi.org/10.1007/978-3-030-03341-5_28
27. Zamir, S.W., et al.: Multi-stage progressive image restoration. arXiv preprint arXiv:2102.02808 (2021)
28. Zhang, Y., Li, K., Li, K., Wang, L., Zhong, B., Fu, Y.: Image super-resolution using very deep residual channel attention networks. In: Proceedings of the European Conference on Computer Vision (ECCV), pp. 286–301 (2018)

Immersive Traditional Chinese Portrait Painting: Research on Style Transfer and Face Replacement

Jiayue Li(✉), Qing Wang, Shiji Li, Qiang Zhong, and Qian Zhou

College of Information and Electrical Engineering, China Agricultural University, Beijing 100083, China
{s20193081370,wangqingait}@cau.edu.cn

Abstract. Traditional Chinese portrait is popular all over the world because of its unique oriental charm. However, how to use neural network to express the aesthetic and feelings in instantiated Chinese portrait effectively is still a challenging problem. This paper proposes a Photo to Chinese Portrait method (P-CP) providing immersive traditional Chinese portrait painting experience. Our method can produce two groups intriguing pictures. One is Chinese Portrait Style Picture, the other is Immersive Chinese Portrait Picture. We pay attention to neural style transfer for traditional Chinese portrait for the first time, and have trained a fast feedforward generative network to extract the corresponding style. The generative network principle is explored to guide the style transfer adjustment in detail. Face replacement is added to form a more appealing stylized effect. We also solve the problems of color and light violation, and unnatural seam. We hope this work offers a deeper and immersive conversation between modern society and antiques, and provides a useful step towards related interdisciplinary areas.

Keywords: Style transfer · Face replacement · Traditional Chinese portrait

1 Introduction

As an ancient form of artistic expression, Chinese painting, with its unique painting tools and creative characteristics, has become a typical symbol of Chinese culture and even Eastern civilization. Making the ancient paintings have more interesting and vivid expression, can effectively shorten the distance between the audience and history.

Image style transfer, namely image stylization, is a class of computer image processing technology that retains the content framework of ordinary photos and adds a specific artistic style. The idea of applying deep learning to image style transfer originally came from the research of Gatys et al. [1–3]. Gatys and others use convolutional neural network to automatically extract image features, and calculate texture, so as to present the style features of an image. By matching the image content features with the weighted style features, we can get a great artistic image output. The processed image with artistic style caters to people's curiosity psychology and the pursuit of beauty, that has been widely praised and independent social communication. This kind of image style transfer

H. Ma et al. (Eds.): PRCV 2021, LNCS 13020, pp. 192–203, 2021.
https://doi.org/10.1007/978-3-030-88007-1_16

also has practical applications in short video creation, live video, film special effects and other fields, and is commonly noticed in application software Prisma and Tiktok. Most of style transfer effects are oil painting, cartoon and sketch, but have little cultural characteristics, especially effects related to traditional Chinese painting.

Our first task is to combine traditional Chinese portrait with fast neural style transfer. As far as I am concerned, only after a long period of practice can Chinese painting artists create realistic traditional Chinese style paintings. However, using neural network, we can quickly generate an adjustable and impressive Chinese style painting picture. How to accomplish it is the technical difficulty we need to overcome. At the same time, it is also a challenge to pay attention to the style transfer of portrait painting, because the human visual system is extremely sensitive to the most subtle changes on the face, and the use of common style transfer techniques will normally deform the whole facial structure. Figure 1 shows an example. When using Gatys neural style transfer, it takes about 11 h for 600 iterations with CPU. Although it adds exquisite pattern elements, its output does not keep the facial information of the content image well, under the limited computing power. Our P-CP method requires 3–5 h training process with GPU; however, it can generate a target Chinese portrait style image in seconds through the pretrain network.

(a) Gatys (b) Modified Gatys (c) DeepArt (d) Ours

Fig. 1. Comparison of classic methods of neural style transfer. (a) Gatys method; (b) Gatys method with the same content and style ratio as ours; (c) Style transfer experience website [4]; (e) Our P-CP method.

Inspired by the cardboard cutout for group photo in amusement parks and museums, this paper takes Chinese traditional portrait painting as an example, and also studies the neural style transfer for face replacement scenes. Traditional face replacement experience is often directly superimposed or customized color and matting through image processing software. Instead, we use facial feature points to calculate mask and affine matrix, then apply color correction and edge blur to complete the face replacement from photo to portrait painting, which provides another gratifying output application.

In general, our P-CP model provides immersive traditional Chinese portrait painting experience, shows a more attractive style, and is faithful to the content image. Our method provides more than one output options, and has unique advantages in Museum scenes. It can be widely used in online cultural communication, interesting social interaction, short video experience, online marketing. The main contributions of this paper are summarized as follows.

- We pay attention to the research of portrait in the style transfer of traditional Chinese painting for the first time. Since portrait is the most eye-catching part of the application scene of style transfer, this problem not only fills the corresponding vacancy, but also brings new challenges.
- We demonstrate the P-CP method, which produce two groups intriguing pictures. One is Chinese Portrait Style Picture, the other is Immersive Chinese Portrait Picture
- We verify the basic principle of neural network to extract color, texture and other image features, while adjusting the network to fulfill our expectations for output.
- We have verified the possibility of applying Chinese portrait style transfer in museum human-computer interaction, using artificial intelligence to replace user's face to ancient paintings, which provides a novel choice for related application.

2 Related Work

2.1 Neural Style Transfer

Gaty's [1–3] input is a white noise image. By calculating the style loss and content loss, the image is updated iteratively. The local and global style features are weighted to produce high quality images. However, due to the long iterative optimization process caused by numerous CNN parameters, the computation and time cost is relatively high.

Johnson [5] use VGG to extract perceptual features, and then use a deep residual image transform network to optimize the perceptual loss. Using the pretrained network, the generation of images only needs one single feed forward transmission, which is three orders faster that Gaty's method. Instance Normalization (IN) [6] design can improve the quality of the output image, which structure is also the cornerstone of the multi-style network. IN is to normalize the channel in each sample, which is more in line with the original intention of Gram matrix.

Recently, more researchers begin to work on multi-style. [7] provides a network that can easily add new style incremental training, and region-specific style fusion function. [8] present a method that can predict the conditional instance normalization parameters directly from a style image for multi-style transfer networks. [9] proposed an adaptive instance normalization (AdaIN) layer that aligns the mean and variance of the content features with those of the style features, and enables arbitrary style transfer in real-time.

The application and progress of the traditional Chinese painting study along with neural network. Before the potential of neural network in image processing is fully explored, researchers mainly concentrate on soft brushes interactive model [10], and technique to animate or produce a traditional Chinese style painting [11–13]. With neural network, [14] propose a multiscale deep neural network that is able to transform sketches into Chinese paintings. [15] devise different strategies to transfer not only the ink tone but also the painting skills from fine brushwork painting and freehand brushwork painting. [16] present a method focusing on abstract style transfer. [17] bring out the model generating end to end Chinese landscape paintings, without conditional input. There are still many interdisciplinary areas worthy of further study.

2.2 Face Replacement

Face feature is a crucial part of human identification. Therefore, researches on human face emerge in endlessly, such as reenactment, replacement, editing and synthesis [18].

Face replacement, as a branch of human face related research, is the key technical feature of this paper. Replacement refers to replacing the target identity with the content of the source identity, so that the target identity has partial characteristics of the source identity. It is commonly used in the fashion industry, memes making and selfie filter applications. Another healthy use of face changing is to anonymize identity in a fuzzy or pixelated way on public platforms and occasions.

The method based on shape fitting is the most intuitive face replacement algorithm, that is, by detecting the key points to calculate the deformation between the two face shapes, and then adding image fusion and other post-processing. Currently, this method is relatively stable, and it is also the method used in our research. Early face replacement algorithms pay more attention to facial expression migration, among which the method of multi-image or video sequence based on 3D model and dense optical rheological transformation is relatively popular [19, 20]. Pix2Pix [21] and CycleGAN [22] are designed to solve the problem of face-to-face image translation, so many related derivative models are also followed. The state-of-the-art face replacement algorithm is to find the optimal solution of data, speed, availability and quality.

Researchers have noticed the use of Generative deep learning algorithms for unethical and malicious applications. However, we still hope that technological progress will be used in legitimate and ethical application scenarios.

3 The P-CP Method

3.1 Network Architecture

With the theoretical basis of Chinese portrait style transfer and face replacement, we combine the advantages of extracting texture and color from style transfer with the novel experience of face replacement. We propose the P-CP method (see Fig. 2).

3.2 Neural Style Transfer Network

We use Tensorflow to train this feedforward fast style transfer network. Image transform network plays the role of condition generator. Loss network is a pretrained VGG-19 with fixed parameters. Instead of per pixel loss, we calculate the loss function by forward propagation, extract advanced features from the pretrained network, update the parameters of the previous network by back propagation, and get a feedforward network that can be used for real-time image transfer task.

The overall network structure design is consistent with Johnson [5]. Image transform network (see Fig. 3) consists of 3 down sampling layers, followed by 5 residual blocks for learning identification function [23], 2 up sampling layers, 1 down sampling layer to limit the channel number to 3, and the result is mapped to (0, 225) by tanh function. It takes 3×3 for all convolutional layers, except for the first and last layers which use 9×9. When VGG-19 is used to extract the content features of images, the higher-level network

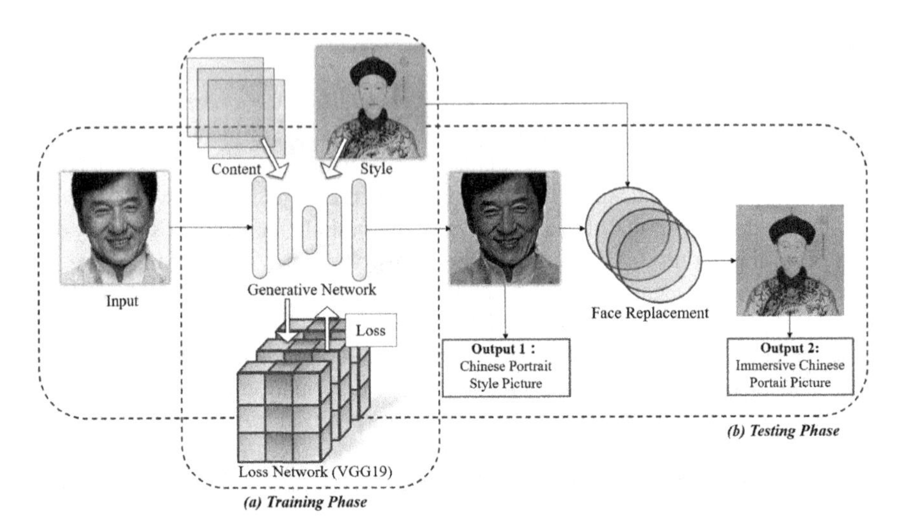

Fig. 2. P-CP Architecture. Our P-CP architecture involves the training phase and the testing phase. The purpose of training phase is to generate a target generative network. Input the specific style image and content image training dataset, pass them through the generative network and loss network, and calculate the loss function of the latter network to train the former, then we finally get the Chinese portrait style transfer model. The purpose of the testing phase is to generate Chinese Portrait Style Picture with the help of the pretrained model, and generate the Immersive Chinese Portrait Picture with the help of face replacement functional structure. Both of the output pictures are of ornamental and practical value.

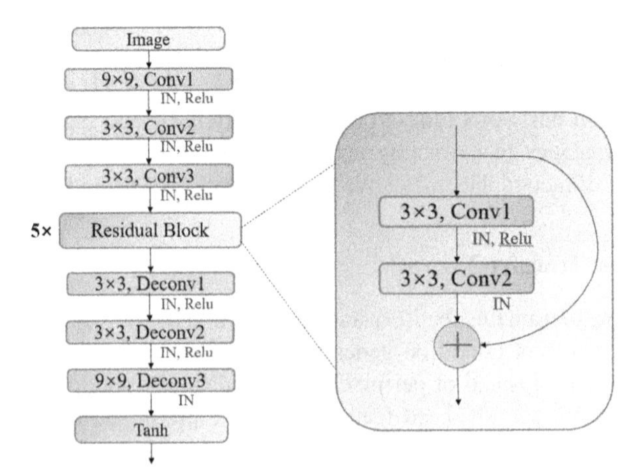

Fig. 3. Generative Network Architecture

retains the content and overall spatial structure of the image, but details such as color and texture are lost. When extracting the style features of images, due to the addition of Gram matrix, higher-level network can extract a wider range of styles, and lower-level network can extract more detailed styles, so we usually choose the combination of different levels of style reconstruction.

We optimize and improve the Johnson transform network, and use a slightly different loss function. Among the Generative Network, Batch Normalization (BN) is replaced by Ulyanov's Instance Normalization (IN) [6], and the scaling and offset of output tanh layer is slightly modified. We choose VGG-19 with deeper layers instead of VGG-16, and choose different layers according to the target requirements.

The training process of Generative Network is the optimization process of minimizing the total loss function. Given content images I_c and one style image I_s, our goal is to generate the stylized image I through training. The total loss function is composed of content loss L_C, style loss L_S and total variation L_{TV}, and λ_S, λ_C, λ_{TV} are their weight respectively.

$$L_T(I, I_c, I_s) = \lambda_C L_C(I, I_c) + \lambda_S L_S(I, I_s) + \lambda_{TV} L_{TV}(I) \tag{1}$$

The pretrained network ϕ can get the $l-$th feature map $F_l(I) = \phi^l(I)$. The size of $F_l(I)$ is $C_l(I) \times H_l(I) \times W_l(I)$, where $C_l(I)$ stands for the channel number of the $l-$th layer, $H_l(I)$ is the height of $l-$th feature map and $W_l(I)$ is the width. The content loss is the mean-squared distance between the feature map of I_c and I at certain layer:

$$L_C(I, I_c) = \frac{1}{C_l H_l W_l} \sum\nolimits_{i,j}^{L} \left[F_{ij}^l(I) - F_{ij}^l(I_c)\right]^2 \tag{2}$$

The style loss L_S is the mean-squared distance between the correlations Gram matrices of I_s and I at several appointed layers: Among them, Gram matrix $G_{ij}^l(I)$ is used to measure the correlation between different features in feature map.

$$G_{ij}^l(I) = \frac{1}{C_l H_l W_l} \sum\nolimits_{h=1}^{H_l} \sum\nolimits_{w=1}^{W_l} \phi_i^l(I) \phi_j^l(I) \tag{3}$$

$$L_s(I, I_s) = \sum\nolimits_{l=0}^{L} \left\{ \frac{1}{C_l^2 H_l^2 W_l^2} \sum\nolimits_{i,j} \left[G_{ij}^l(I) - G_{ij}^l(I_s)\right]^2 \right\} \tag{4}$$

Total Variation (TV) regularizer [6] is a strategy to remove defects while retaining important details such as edges, which is widely used in denoising, super-resolution and other problems due to its edge preserving ability. By calculating the sum of the mean square error between adjacent pixels, L_{TV} is minimized to encourage the spatial smoothness in the output image.

$$L_{TV}(I) = \sum\nolimits_{i,j} \left[\left(I_{i,j+1} - I_{ij}\right)^2 + \left(I_{i+1,j} - I_{ij}\right)^2 \right] \tag{5}$$

Denote the parameters of the feed-forward network as y, the training objective is as follows:

$$\hat{y} = \arg\min_y E_I[L_T(I, I_c, I_s)] \tag{6}$$

where E is the expectation.

3.3 Face Replacement

In order to make the partial portrait of the first photo (Fig. 4 A) superimposed on the ancient portrait painting (Fig. 4 B). We apply open-source Toolkit Dlib's object detector to locate human faces, and its human face pose prediction function to identify the locations of important facial landmarks [26]. According to the position information of feature points, we draw the region mask of facial features, including eyes, nose and mouth, then calculate the feature transformation matrix of two input images. Finally, according to the position information, the facial content of the image A extracted by the mask is transformed to the image B using Ordinary Procrustes Analysis. The flow chart is shown in the figure.

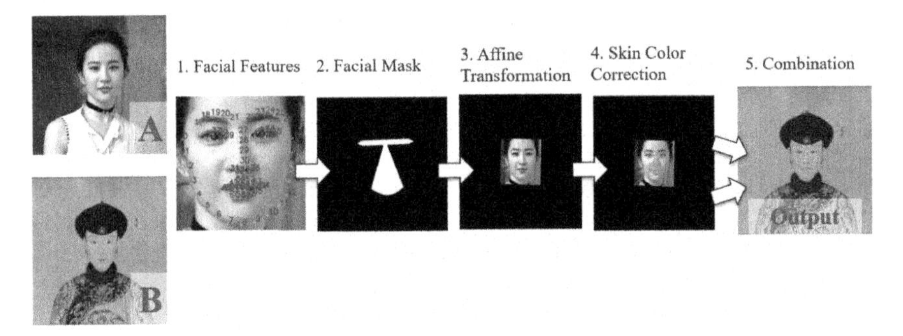

Fig. 4. Face replacement flow chart

Because the color and illumination information of the two input images are not close, we need to correct the local color of image A before superimposing. Here we choose local histogram matching and edge Gaussian blur for primary processing. We will introduce the use of style transfer in the next chapter. Local histogram matching matches the histogram of the image region to another image, and changes the RGB color distribution, so as to adjust the color and illumination information to the specified style.

4 Experiment

Our experiments are carried out on Windows 10, with support of Tensorflow-GPU 2.1.0, CUDA 10.1.243 and cuDNN 7.6.5.32. The optimization pipeline is implemented by Tensorflow. CUDA is the framework responsible for managing the allocation of computing units. cuDNN is used to accelerate VGG network. All our experiments are conducted on a PC with an Intel Core i5 processor and NVIDIA GeForce GTX 1070.

In particular, picture(b) and picture(d) from Fig. 5 are from the Palace Museum and the Cleveland Museum of Art official website, and others are from Internet. All images are for academic use.

4.1 Comparison of Different Traditional Chinese Painting Styles

We first selected a variety of representative traditional Chinese paintings, including paintings of different periods, different creative styles and different nationalities, trained the network and obtained five representative results (see Fig. 5).

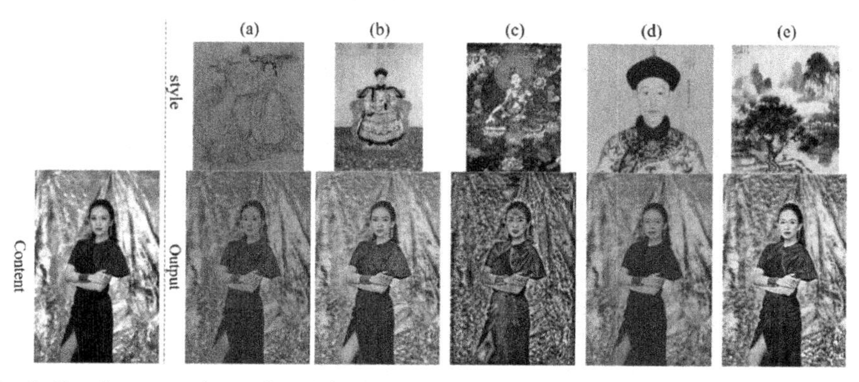

Fig. 5. Results comparison using different traditional Chinese style painting. (a) Part of Born of Gautama Buddha (1024 × 1157 pixels, 72dpi); (b) Portrait of the Qianlong Emperor in Court Rode (1890 × 2685 pixels, 300dpi); (c) Classic Thang-ga (825 × 110 pixels, 72dpi); (d) Part of Mind Picture of a Well-Governed and Tranquil Reign (2048 × 2048 pixels, 300dpi); (e) Xu Beihong's Landscape Paintings (604 × 690 pixels, 72dpi)

Different traditional paintings have different output styles. For the painting (a) focusing on lines, its output is more faithful to the original color of traditional canvas (Xuan paper or silken cloth). For Thangka (c), which uses rich color blocks, its output retains the color diversity. For freehand brushwork painting (e), its output shows the typical halo dyeing style of Chinese ink painting. For fine brushwork painting (b & d), the delicacy and exquisite feeling brought by painting skills is well preserved, and the traditional style can be added on the foundation of retaining the characters' facial features. The standard of fine brushwork painting for line is neat, delicate and rigorous, and its color is required to be gorgeous and unified. Therefore, most fine brushwork is elegant and has strong Chinese national aesthetic color and characteristics.

Based on the demand for face detail protection and aesthetic pursuit, we decide on the Qing Dynasty portraits of Qianlong emperor and his twelve consorts as our standard style sample, which is also titled as Mind Picture of a Well-Governed and Tranquil Reign. Further details optimization and face replacement experiments are carried out as follows.

4.2 Image Detail Exploration and Optimization

In order to make our network more universal in portrait and specific in traditional Chinese painting style, we test the effect of using different layers of VGG as content extraction (Fig. 6 a), and using different style weights and content weights (Fig. 6 b) in the loss function. Exploring and summarizing the neural network principle, we chose the appropriate parameters for the certain scene. The selection criteria are: 1. It has the obvious characteristics of Chinese portrait painting; 2. The facial details of the characters in the output pictures are relatively complete.

We use the MS-COCO2014 dataset, batch size 8, epoch iteration 2. We take Relu1_2, Relu2_2, Relu3_3 and Relu4_3 weighted as style, Adam optimizer, learning rate 0.001. The network training takes about 4 h, after that, it takes about 10 s to finish one feed forward calculation for a 2048 × 2048 pixels image.

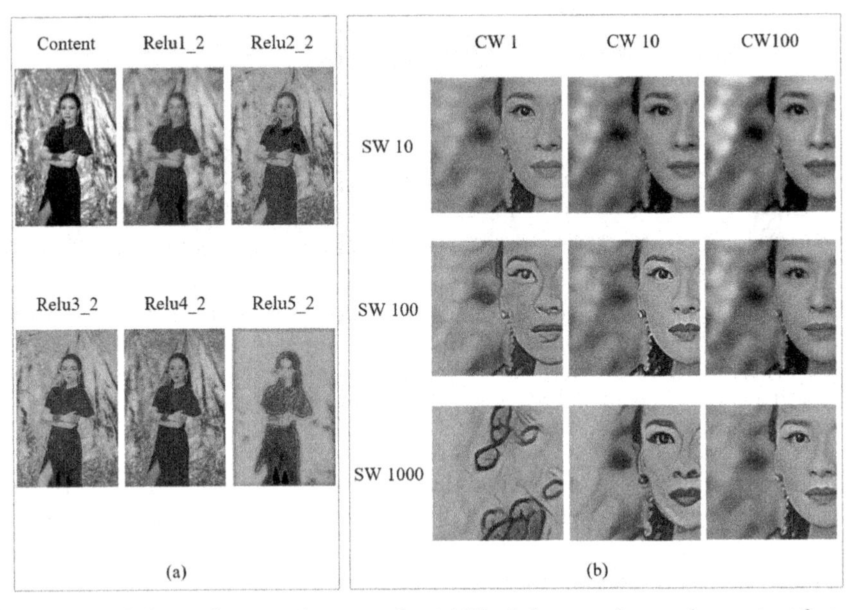

Fig. 6. Network fine tuning experiment results. (a) The influence of extracting content features from different layers of VGG; (b) The influence of different style weight (SW, λ_S) and content weight (CW, λ_C).

It shows that the higher-level layers extract more abstract features. Relu4_ 2 reaches the ideal state. In the weight adjustment test, the output with the same SW and CW ratio has quite similar effect. When the style and content ratio is 10:1, the picture expression is relatively stable. When CW is 100, the output shows obvious deconvolution checkerboard pattern [25], and the larger the SW is, the more obvious the effect is.

In addition, we test the same style image of different sizes to train the network. Using larger pixel images, the training effect is more delicate and natural, closer to the original painting expression, and the probability of line distortion and noise insertion is less.

4.3 Improvement of Face Replacement with Style Transfer

In the implementation of facial feature replacement, if the color of the facial region is not corrected (Fig. 7 a), it is obvious that the output is inconsistent. If we only do

Gaussian distribution matching (Fig. 7 b), it can still be observed obvious seam and over-exposure. However, after using the specific style transfer (Fig. 7 c), the shortcomings of the above operation are well overcome. On the image with 3177 × 3400 pixels and 300 dpi resolution, the sense of disobedience caused by seams is obviously weakened, the color overflow is greatly reduced, and the unity and harmony of the painting texture are maintained. In addition, it takes about 1 min to process images of that size.

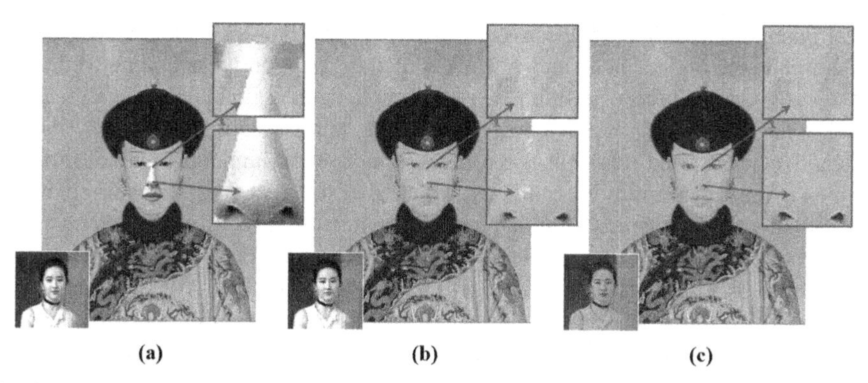

Fig. 7. Results Comparison without color correction (a), with color correction (b), and with our style transfer pretreatment (c)

Fig. 8. Stability verification test

We tested the experimental effect of P-CP on different gender, different skin color, even black and white photos, and found that the experimental effect is quite universal. See Fig. 8.

For a single image, the processing and optimization of style transfer implementation in color change and Chinese painting texture addition is subtle and difficult to detect, but if this method is applied to high-definition video processing, the effect will be much more obvious. Imagine the traditional Chinese portrait painting and real-time face replacement experience, any change of user's expression will be displayed on the display screen in real time, then the ability of style transfer to deal with the details and defects will be greatly reflected. That is also the focus of our following work.

5 Conclusion

We have presented the P-CP method, which can provide authentic traditional Chinese portal painting experience. We have discussed the mechanism of image neural style transfer and face replacement, and gave a detailed description for neural network style transfer applied in Chinese portrait face replacement details. Crucially, the method proposed in this paper can not only convert the photos into Chinese Portrait Style Picture (around 10 s in our condition), but also generate effective and Immersive Chinese Portrait Picture (around 1 min in our condition), which also solves the problems of color and light violation, and unnatural seam in practical application. In addition, we compare the influence of different style and content weights on the output image, adjust the network and loss function settings, and finally get the ideal output results. We note, however, when the face in picture is not facing the camera or there are some obstructions, the face replacement cannot achieve good results. What's more, the whole generation cycle needs to be shortened to the real-time level in the following study.

References

1. Gatys, L.A., Ecker, A.S., Bethge, M.: Texture synthesis using convolutional neural networks. arXiv preprint arXiv:1505.07376 (2015)
2. Gatys, L.A., Ecker, A.S., Bethge, M.: A neural algorithm of artistic style. J. Vis. (2015)
3. Gatys, L.A., Ecker, A.S., Bethge, M.: Image style transfer using convolutional neural networks. In: Conference on Computer Vision and Pattern Recognition (CVPR). IEEE (2016)
4. DeepArt Homepage. https://deepart.io/#. Accessed 1 May 2021
5. Johnson, J., Alahi, A., Fei-Fei, L.: Perceptual losses for real-time style transfer and super-resolution. In: Leibe, B., Matas, J., Sebe, N., Welling, M. (eds.) ECCV 2016. LNCS, vol. 9906, pp. 694–711. Springer, Cham (2016). https://doi.org/10.1007/978-3-319-46475-6_43
6. Ulyanov, D., Vedaldi, A., Lempitsky, V.: Instance normalization: the missing ingredient for fast stylization. arXiv preprint arXiv:1701.02096 (2016)
7. Chen, D., Yuan, L., Liao, J., Yu, N., Hua, G.: StyleBank: an explicit representation for neural image style transfer. In: IEEE Conference on Computer Vision and Pattern Recognition, pp. 2770–2779 (2017)
8. Ghiasi, G., Lee, H., Kudlur, M., Dumoulin, V., Shlens, J.: Exploring the structure of a real-time, arbitrary neural artistic stylization network. In: British Machine Vision Conference 2017, London, United Kingdom. BMVA Press (2017)
9. Huang, X., Belongie, S.: Arbitrary style transfer in real-time with adaptive instance normalization. In: International Conference on Learning Representations, ICLR, Toulon, France (2019)
10. Lee, J.: Simulating oriental black-ink painting. Comput. Graph. Appl. **19**(3), 74–81 (1999)
11. Way, D.L., Lin, Y.R., Shih, Z.C.: The synthesis of trees in Chinese landscape painting using silhouette and texture strokes. J. WSCG, 499–506 (2002)
12. Xu, S., Xu, Y., Kang, S.B., Salesin, D.H., Pan, Y., Shum, H.: Animating chinese paintings through stroke-based decomposition. ACM Trans. Graph. **25**(2), 239–267 (2006)

13. Zhang, S., Chen, T., Zhang, Y., Hu, S., Ralph, M.: Video-based running water animation in Chinese painting style. Sci. China Ser. F-Inf. Sci. **52**(2), 162–171 (2009)
14. Lin, D., Wang, Y., Guangluan, X., Li, J., Kun, F.: Transform a simple sketch to a Chinese painting by a multiscale deep neural network. Algorithms **11**(1), 4 (2018)
15. Sheng, J., Song, C., Wang, J., Han, Y.: Convolutional neural network style transfer towards Chinese paintings. IEEE Access **7**, 163719–163728 (2019)
16. Li, B., Xiong, C., Tianfu, W., Zhou, Y., Zhang, L., Chu, R.: Neural abstract style transfer for Chinese traditional painting. In: Jawahar, C.V., Li, H., Mori, G., Schindler, K. (eds.) ACCV 2018. LNCS, vol. 11362, pp. 212–227. Springer, Cham (2019). https://doi.org/10.1007/978-3-030-20890-5_14
17. Xue, A.: End-to-End Chinese landscape painting creation using generative adversarial networks. In: IEEE Winter Conference on Applications of Computer Vision (2021)
18. Mirsky, Y., Lee, W.: The creation and detection of deepfakes: a survey. ACM Comput. Surv. **54**(1), 1–41 (2021)
19. Suwajanakorn, S., Seitz, S.M., Kemelmacher-Shlizerman, I.: What makes tom hanks look like tom hanks. In: Proceedings of the IEEE International Conference on Computer Vision, pp. 3952–3960. Institute of Electrical and Electronics Engineers Inc., Santiago (2015)
20. Thies, J., Zollhofer, M., Stamminger, M., Theobalt, C., Niessner, M.: Face2Face: real-Time face capture and reenactment of RGB videos. Commun. ACM **62**(1), 96–104 (2019)
21. Isola, P., Zhu, J., Zhou, T., Efros, A.A.: Image-to-image translation with conditional adversarial networks. In: IEEE Conference on Computer Vision and Pattern Recognition, pp. 5967–5976 (2017)
22. Zhu, J., Park, T., Isola, P., Efros, A.A.: Unpaired image-to-image translation using cycle-consistent adversarial networks. In: IEEE International Conference on Computer Vision, pp. 2242–2251 (2017)
23. He, K., Zhang, X., Ren, S., Sun, J.: Deep residual learning for image recognition. In: Proceedings of the IEEE Computer Society Conference on Computer Vision and Pattern Recognition, Las Vegas, NV, United States, pp 770–778. IEEE Computer Society (2016)
24. Rudin, L.I., Osher, S., Fatemi, E.: Nonlinear total variation based noise removal algorithms. Phys. D-Nonlinear Phenom. **60**(1–4), 259–268 (1992)
25. Mahendran, A., Vedaldi, A.: Visualizing deep convolutional neural networks using natural pre-images. Int. J. Comput. Vis. **120**(3), 233–255 (2016)
26. Kazemi, V., Sullivan, J.: One millisecond face alignment with an ensemble of regression trees. In: IEEE Conference on Computer Vision and Pattern Recognition, pp. 1867–1874 (2014)

Multi-camera Extrinsic Auto-calibration Using Pedestrians in Occluded Environments

Junzhi Guan(✉), Hujun Geng, Feng Gao, Chenyang Li, and Zeyong Zhang

CETC Key Laboratory of Aerospace Information Applications, Shijiazhuang 050000, China

Abstract. In this paper, we propose a novel extrinsic calibration method for camera networks based on a pedestrian who walks on a horizontal surface. Unlike existing methods which require both the feet and head of the person to be visible in all views, our method only assumes that the upper body of the person is visible, which is more realistic in occluded environments. Firstly, we propose a method to calculate the normal of the plane containing all head positions of a single pedestrian. We then propose an easy and accurate method to estimate the 3D positions of a head w.r.t. to each local camera coordinate system. We apply orthogonal procrustes analysis on the 3D head positions to compute relative extrinsic parameters connecting the coordinate systems of cameras in a pairwise fashion. Finally, we refine the extrinsic calibration matrices using a method which minimizes the reprojection error. Experimental results show that the proposed method provides an accurate estimation of the extrinsic parameters.

Keywords: Extrinsic calibration · Camera network · Pedestrians · Orthogonal procrustes · Normal

1 Introduction

Calibration of camera networks is an essential step for further developing a visual surveillance system or a visual tracking system. The parameters of a camera to be calibrated are divided into two classes: extrinsic and intrinsic. The extrinsic parameters denote the coordinate system transforms from 3D world coordinates to 3D camera coordinates. The intrinsic parameters define the imaging geometry and the optical characteristics of each camera individually. While intrinsic parameters usually need to be estimated once for a given camera (unless the camera has a variable focal length), the extrinsic parameters must be recomputed whenever cameras are moved or reoriented (on purpose or accidentally). Also, intrinsic calibration can be performed in the lab, before deploying the cameras, whereas extrinsic calibration is only possible after the deployment. Especially in the case of temporary installations, e.g., to capture public events, the available time for calibration at deployment time is often limited and an efficient and reliable procedure is needed. In this paper, we assume that the intrinsic parameters are known and we focus on extrinsic calibration.

H. Ma et al. (Eds.): PRCV 2021, LNCS 13020, pp. 204–215, 2021.
https://doi.org/10.1007/978-3-030-88007-1_17

Classical methods [7,20] require highly accurate tailor-made 3D calibration objects to achieve a high calibration accuracy. Given sufficient point correspondences between 3D world points on the calibration object and corresponding image points, calibration of a single camera can be done very efficiently with these methods. However, these approaches require an expensive calibration apparatus and an elaborate setup. Zhang [23] proposed a calibration algorithm which uses a planar grid pattern as the calibration object. This method is mainly for intrinsic calibration. Patterns required for this method [23] are easy and cheap to manufacture, which makes it flexible for calibration of a single camera. However, for calibration of a camera network, it is necessary for all involving cameras to simultaneously observe a number of points. It is hardly possible to achieve this with these methods [7,20] if one camera is mounted in the front of a room while another in the back. One way to solve this problem is to divide the cameras into smaller groups such that all cameras within the same group can see the calibration pattern. Each group of cameras is then calibrated separately, and finally all camera groups are registered into a common reference coordinate system. However, it will make the calibration tedious and time-consuming.

All the aforementioned calibration methods depend on placing the calibration object in the scene before or during the operational phase of camera networks. For camera networks which mainly deal with the analysis of humans, pedestrians can be used as calibration objects for self-calibrating camera networks. Various calibration methods [16] have been proposed using pedestrians as the calibration object. The common assumption made by these methods is that the head and feet of the person are visible in all views. However, the feet of the person are prone to occlusion (e.g., by tables, by chairs or other furniture) especially for indoor visual surveillance system. In that case, only the upper body of the person is visible. Also all positions of the head of a single pedestrian lie in a plane, which is a degenerate case for all methods. To the best of our knowledge, no method exists that can do extrinsic calibration for this specific but quite normal scenario.

To address the aforementioned problem, we propose a method for calibration based on pedestrians while solving this problem. Our method only assumes that the head and center line of the person are visible. The first contribution of this paper is that we propose a robust method to calculate the normal of the plane composed by all head positions w.r.t. each local camera coordinate system. The second contribution is that we propose an easy and accurate method to estimate 3D positions of the head w.r.t. each camera coordinate system, which entitles us to obtain the extrinsic parameters by computing the 3D rigid body transformation which optimally aligns two sets of points for which the correspondence is known [5]. Arun *et al.* [1] have shown that extrinsic parameters can still be uniquely found even when all 3D points (all head positions) are coplanar if the 3D positions are known.

We will evaluate the proposed method in a camera network in Sect. 5. The camera network monitor occluded environments, which makes the feet of pedestrians hardly visible. Thus most of the existing state-of-the-art methods are not applicable. We will show that the proposed method provides accurate calibration for camera networks with different configurations.

The remainder of this paper is organized as follows. Section 2 gives a survey of works in the literature for extrinsic calibration. Section 3 explains how to estimate 3D positions of the head w.r.t. the camera coordinate system, and how to do the actual extrinsic calibration based on the aforementioned 3D positions. The routines for refinement are presented in Sect. 4. Section 5 shows the experimental results. Finally, Sect. 6 concludes the paper.

2 Related Work

In this section, we present an overview of state-of-the-art approaches for extrinsic calibration based on pedestrians. Various calibration methods [13,16] have been proposed using pedestrians as the calibration objects.

A drawback of the aforementioned methods is that they rely on estimating vanishing points [3], which is extremely sensitive to noise, as reported by Micuisik *et al.* [17]. Moreover, the common assumption made by these methods is that the head and feet of the person are visible in all views. However, the feet of the person are prone to occlusion (e.g., by tables, chairs or other furniture) especially in indoor environments. In that case, only the upper body of the person is visible, and all heads positions will lie in a plane, which will fail all the aforementioned methods.

Guan [9] proposed another extrinsic calibration approach for camera networks by analyzing tracks of pedestrians. First of all, they extract the center lines of walking persons by detecting their heads and feet in the camera images. An easy and accurate method was proposed to estimate the 3D positions of the head and feet w.r.t. a local camera coordinate system from these center lines. Then, they apply the orthogonal procrustes approach to compute relative extrinsic parameters connecting the coordinate systems of cameras in a pairwise fashion. This method also assumes that the head and feet of the person are visible in all views, which make it not applicable for occluded environments.

For extrinsic calibration of cameras with known intrinsic parameters, epipolar geometry also plays an import role. Many methods have been proposed for calibration using epipolar geometry [4,14,19]. Longuet-Higgins [15] showed how an essential matrix relating a pair of calibrated views can be estimated from eight or more point correspondences by solving a linear equation, and also how the essential matrix can be decomposed to give relative camera orientation and position. Hödlmoser *et al.* [12] use head and feet of the person as the corresponding points to estimate the essential matrix. Then they decompose the essential matrix to obtain the camera rotation and translation parameters. However, decomposing the essential matrix, multiple triangulations are needed for the chirality check [11,22], which make the method more prone to erroneous correspondences between heads (or feet) in different camera views. Moreover, the method using the essential matrix will also fail when the feet of pedestrians are not visible in all cameras. In that case, all head positions lie in a plane, which is a degenerate case for estimating essential matrix [10]. Only the homography matrix between two views can be obtained since only the image position were explored, but it is not possible to obtain the unique extrinsic parameters by decomposing the homography matrix [6,21]. This problem can be possibly solved if the scene contains two or more pedestrians with different height. However, the essential matrix-based methods will become unstable if the height difference of different persons is much smaller compared to the distance between the camera and the pedestrians, which is quite common in most scenarios.

In contrast, we propose a method to estimate the unit vector which is parallel to the person's center line. With the estimated unite vector, we then propose an approach to estimate the 3D positions of the head of the person w.r.t the local camera coordinate. Knowing the 3D positions entitle us the ability to estimate extrinsic parameters by computing the 3D rigid body transformation which optimally aligns two sets of points for which the correspondence is known [1,5]. Our method still works when only the upper body is visible as we can estimate the 3D positions w.r.t. a single camera coordinate system. Arun *et al.* [1] have shown that extrinsic parameters can still be uniquely found even when all 3D points are coplanar if the 3D positions are known.

3 Calibration Based on 3D Positions

3.1 3D Head Positions in Local Camera Coordinates

Extraction of Image Positions of the Head and Center Line. To determine the image positions of the head and center line of a walking person, we propose to detect the person's silhouette in a first step. Since we are not mainly focusing on solving the problem of background subtraction and tracking, we just apply a well known background subtraction method [24] to obtain a rough foreground blob of the pedestrian. We then fit a line and a bounding box to person's blob. The intersections between the line and the upper edge of the bounding box are taken as $\vec{u}_\mathrm{h}$.

Figure 1 shows the detected head and center line positions using the aforementioned method.

Normal Vector Estimation. Suppose a person moves to N different positions while keeping a fixed posture. Suppose that all cameras see the head of the person. At each time t, we first calculate $\tilde{\vec{u}}_\mathrm{h}^{(k)}(t)$ and the center line using technique described in Sect. 3.1 for camera k. We denote the intersections between the center line and the top and bottom boarder of the image as $\tilde{\vec{u}}_\mathrm{t}^{(k)}(t)$ and $\tilde{\vec{u}}_\mathrm{b}^{(k)}(t)$. Let $\tilde{\vec{x}}_\mathrm{t}^{(k)}(t)$ and $\tilde{\vec{x}}_\mathrm{b}^{(k)}(t)$ be the normalized homogeneous image coordinates of the top and bottom points detected at time t in camera k. Furthermore, let $Z_\mathrm{b}^{(k)}(t)$ and $Z_\mathrm{t}^{(k)}(t)$ be the corresponding unknown $Z^{(k)}$ coordinates. Finally, assuming the person is walking upright and let $\vec{r}_\mathrm{t}^{(k)}(t)$ and $\vec{r}_\mathrm{b}^{(k)}(t)$ be the 3D camera coordinates of the top and bottom points, define $h^{(k)}(t)$ as the real 3D distance between

Fig. 1. Head and center line detection in a image. The red circles represent detected head position using bounding box fitting and line fitting. The line represents the center line.

top and bottom points within local camera coordinate. By defining $\vec{e}_z^{(k)}$ the unit vector which is parallel to the person's center line within camera k, we have $\vec{r}_\mathrm{t}^{(k)}(t) = \vec{r}_\mathrm{b}^{(k)}(t) + h^{(k)}(t)\vec{e}_z^{(k)}$ since the vectors $\vec{r}_\mathrm{t}^{(k)}(t)$, $\vec{r}_\mathrm{b}^{(k)}(t)$ and $h^{(k)}(t)\vec{e}_z^{(k)}$ compose a triangle. Figure 2 shows the triangle composed by these 3 vectors. With $\vec{r}_\mathrm{t}^{(k)}(t) = Z_\mathrm{t}^{(k)}(t)\tilde{\vec{x}}_\mathrm{t}^{(k)}(t)$ and $\vec{r}_\mathrm{b}^{(k)}(t) = Z_\mathrm{b}^{(k)}(t)\tilde{\vec{x}}_\mathrm{b}^{(k)}(t)$, we have

$$Z_\mathrm{t}^{(k)}(t)\tilde{\vec{x}}_\mathrm{t}^{(k)}(t) - Z_\mathrm{b}^{(k)}(t)\tilde{\vec{x}}_\mathrm{b}^{(k)}(t) = h^{(k)}(t)\vec{e}_z^{(k)}, \tag{1}$$

where $\vec{e}_z^{(k)} \triangleq R^{(k)}\vec{e}_z$, in which $\vec{e}_z$ is the unit vector in the world coordinates system.

Now define $\vec{m}^{(k)}(t) = \tilde{\vec{x}}_\mathrm{t}^{(k)}(t) \times \tilde{\vec{x}}_\mathrm{b}^{(k)}(t)$ with $\times$ representing cross product, which is a vector orthogonal to $\tilde{\vec{x}}_\mathrm{t}^{(k)}(t)$ and $\tilde{\vec{x}}_\mathrm{b}^{(k)}(t)$. It is also the normal of the plane spanned by two vectors $\tilde{\vec{x}}_\mathrm{t}^{(k)}(t)$ and $\tilde{\vec{x}}_\mathrm{b}^{(k)}(t)$. It is obvious that $h^{(k)}(t)\vec{e}_z^{(k)}$ is also on that plane, which leads to

$$\left(\vec{m}^{(k)}(t)\right)^T h^{(k)}(t)\vec{e}_z^{(k)} = 0, \tag{2}$$

canceling $h^{(k)}(t)$ leads to

$$\left(\vec{m}^{(k)}(t)\right)^T \vec{e}_z^{(k)} = 0. \tag{3}$$

Let $M^{(k)}$ be the matrix with rows $\left(\vec{m}^{(k)}(t)\right)^T$, then we have

$$M^{(k)}\vec{e}_z^{(k)} = 0. \tag{4}$$

Therefore $\vec{e}_z^{(k)}$ must be in the null space of the matrix $M^{(k)}$. Hence $\vec{e}_z^{(k)}$ can be determined by SVD(Singular Value Decomposition) decomposition of $M^{(k)}$: $\vec{e}_z^{(k)}$ is the singular vector which is corresponding to the lowest singular value. $\vec{e}_z^{(k)}$ is also the normal of the plane composed by all heads' positions when the person walks upright on a horizontal surface.

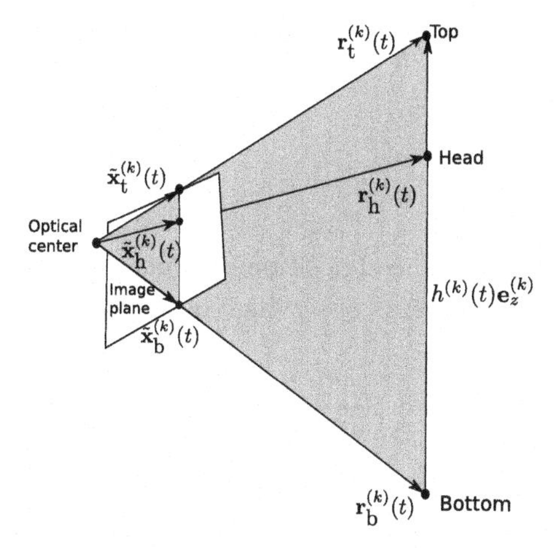

Fig. 2. Co-planarity of $\vec{r}_{\mathrm{t}}^{(k)}(t)$, $\vec{r}_{\mathrm{b}}^{(k)}(t)$ and $h^{(k)}(t)\vec{e}_z^{(k)}$. $\vec{e}_z^{(k)}$ remains the same while walking. $h^{(k)}(t)$ changes when the pedestrian is walking.

Robust Estimation Using RANSAC. The technique described in Sect. 3.1 can provide a good estimation of $\vec{e}_z^{(k)}$ when there are no outliers in the estimated center lines of the person. However, in practice the center lines estimation may not be accurate due to noise in the image. So we propose to combine RANSAC [8] with the method proposed in Sect. 3.1, from which we deal with outliers in the dataset and make the estimation of normal vector robust.

3D Position Estimation. In camera coordinates, the head lies in the plane

$$(\vec{e}_z^{(k)})^T \vec{r}_{\mathrm{h}}^{(k)}(t) = d^{(k)}, \tag{5}$$

with $d^{(k)}$ is the Euclidean distance between the center of camera k and the plane. Substitute $\vec{r}_{\mathrm{h}}^{(k)}(t) = Z_{\mathrm{h}}^{(k)}(t)\tilde{\vec{x}}_{\mathrm{h}}^{(k)}(t)$ in Eq. (5), we get

$$(\vec{e}_z^{(k)})^T Z_{\mathrm{h}}^{(k)}(t)\tilde{\vec{x}}_{\mathrm{h}}^{(k)}(t) = d^{(k)}, \tag{6}$$

from which we can obtain

$$Z_{\mathrm{h}}^{(k)}(t) = \frac{d^{(k)}}{(\vec{e}_z^{(k)})^T \tilde{\vec{x}}_{\mathrm{h}}^{(k)}(t)}. \tag{7}$$

The equation shows that the full camera coordinates (including depth) of the head positions can be expressed in terms of only one unknown but constant number (for each specific camera): $d^{(k)}$.

In order to find the correct ratio between $d^{(k)}$ and $d^{(1)}$ for camera k and camera 1, we consider two different head positions: $\vec{r}_{\mathrm{h}}^{(k)}(t)$ and $\vec{r}_{\mathrm{h}}^{(k)}(t')$. The Euclidean distance between these two head positions is

$$e_{t,t'}^{(k)} = \|\vec{r}_{\text{h}}^{(k)}(t) - \vec{r}_{\text{h}}^{(k)}(t')\| = d^{(k)} q_{t,t'}^{(k)}, \tag{8}$$

in which

$$q_{t,t'}^{(k)} = \left\| \frac{\tilde{\vec{x}}_{\text{h}}^{(k)}(t)}{(\vec{e}_z^{(k)})^T \tilde{\vec{x}}_{\text{h}}^{(k)}(t)} - \frac{\tilde{\vec{x}}_{\text{h}}^{(k)}(t')}{(\vec{e}_z^{(k)})^T \tilde{\vec{x}}_{\text{h}}^{(k)}(t')} \right\|. \tag{9}$$

The distance between two head positions is a physical distance and obviously does not depend on the cameras observing the scene, so we have $e_{t,t'}^{(k)} = e_{t,t'}^{(1)}$, which leads to

$$\frac{d^{(k)}}{d^{(1)}} = \frac{q_{t,t'}^{(1)}}{q_{t,t'}^{(k)}}. \tag{10}$$

Since we choose the coordinate system of the first camera as the world coordinate system, we set $d^{(1)} = 1$, and find $d^{(k)}$ for all the other cameras using Eq. (10). In order to robustly find $d^{(k)}$, we use different t, t' combinations with image distance of the head larger then a certain threshold. Then we compute the mean value of $q_{t,t'}^{(1)}/q_{t,t'}^{(k)}$. Once $d^{(k)}$ is found, we can then obtain $\vec{r}_{\text{h}}^{(k)}(t)$.

3.2 Registration of 3D Point Sets

We compute the extrinsic parameters of all the other cameras w.r.t. the first camera. From equations in Sect. 3.1, we obtain $\vec{r}_{\text{h}}^{(k)}(t)$ (3D positions of the head) of a person in each camera. So we can compute the extrinsic parameters by optimally aligning two sets of 3D points . We use the method [18] which is based on orthogonal procrustes analysis. The calibration proceeds as follows:

1. Obtain 3D positions of the head $\vec{r}_{\text{h}}^{(k)}(t)$ and $\vec{r}_{\text{h}}^{(1)}(t)$ while the person is walking in the scene. we have $\vec{r}_{\text{h}}^{(k)}(t) = R^{(k)} \vec{r}_{\text{h}}^{(1)}(t) + \vec{c}^{(k)}$, in which $R^{(k)}$ and $\vec{c}^{(k)}$ represent the rotation matrix and translation vector of camera k respectively.
2. Calculate the centroid of $\vec{r}_{\text{h}}^{(1)}(t)$ and $\vec{r}_{\text{h}}^{(k)}(t)$ using

$$\overline{\vec{r}^{(1)}} = \frac{1}{N} \sum_{t=1}^{N} \vec{r}_{\text{h}}^{(1)}(t), \ \overline{\vec{r}^{(k)}} = \frac{1}{N} \sum_{t=1}^{N} \vec{r}_{\text{h}}^{(k)}(t). \tag{11}$$

3. Define $H^{(1)}$ (a $3 \times t$ matrix) as the matrix with columns $\vec{r}_{\text{h}}^{(1)}(t) - \overline{\vec{r}^{(1)}}$, $t = 1, 2 \ldots N$, and $H^{(k)}$ as the matrix with columns $\vec{r}_{\text{h}}^{(k)}(t) - \overline{\vec{r}^{(k)}}$. Decompose $H^{(1)} {H^{(k)}}^T$ using singular value decomposition as $H^{(1)} {H^{(k)}}^T = U_k S_k V_k^T$.
4. Calculate the rotation matrix as

$$R^{(k)} = \begin{cases} V_k U_k^T & \text{if } \det(V_k U_k^T) = 1 \\ V_k' U_k^T & \text{if } \det(V_k U_k^T) = -1 \end{cases}, \tag{12}$$

where V_k' is obtained by changing the sign of the last column of matrix V_k.

5. Calculate $\vec{c}^{(k)}$ using

$$\vec{c}^{(k)} = \overline{\vec{r}^{(k)}} - R^{(k)}\overline{\vec{r}^{(1)}}. \tag{13}$$

6. RANSAC implementation of the above procedures as proposed in our previous method [9].

4 Refinement

The approach proposed in Sect. 3 is elegant because the extrinsic parameters can be computed in closed form. However, it often provides suboptimal performance in noisy real-world conditions since measurement noise is not explicitly considered or modelled.

Under the assumption that the observed 2D head positions are corrupted by zero-mean Gaussian noise, we propose to do the refinement using maximum likelihood estimation. The optimization step minimizes the total reprojection error over all the extrinsic parameters. The objective function is the discrepancy between the observed positions of the head in the image and their image reprojections computed using the estimated extrinsic calibration matrices, which is defined as

$$\sum_{i=1}^{n}\sum_{k=1}^{K} d(\vec{\hat{u}}_{ik}^{r}, \vec{u}_{ik}), \tag{14}$$

where $\vec{u}_{ik}$ denotes the observed i^{th} image position of the head in camera k, and

$$\vec{\hat{u}}_{ik}^{r} = p(A^{(k)}, R^{(k)}, \vec{c}^{(k)}, \vec{u}_{ik}) \tag{15}$$

represents the corresponding image position after reprojection, $A^{(k)}$ represents the intrinsic matrix. We apply an optimization method which was developed by Bouguet [2]. It is optimized by an iterative gradient descent procedure.

5 Experiments and Results

We evaluated the proposed method in one camera network. We used the performance measures proposed in the method [9], which include the triangulation error, the projection error, the reprojection error.

We calibrated a multi-camera people tracking system composed of three side view cameras. The cameras were mounted at a height of about 2 m in three corners of a room and have a resolution of 640 by 480 pixel. These cameras were intrinsically calibrated using the method of Zhang [23]. There is all kinds of furniture in the monitored area. Thus the feet of the person are hardly even visible in all views. We use a single pedestrian as the calibration object. Some of the performance criteria require ground truth 3D position of test samples expressed in the user specified world coordinate system. For this purpose, we need to convert the global camera coordinate system to a user specified one. Thus, we measured the positions of 8 non-coplanar markers using a tape measure. To test the accuracy of calibration, we also captured 49 more test samples. In this case, the markers were placed in different positions than for the alignment samples.

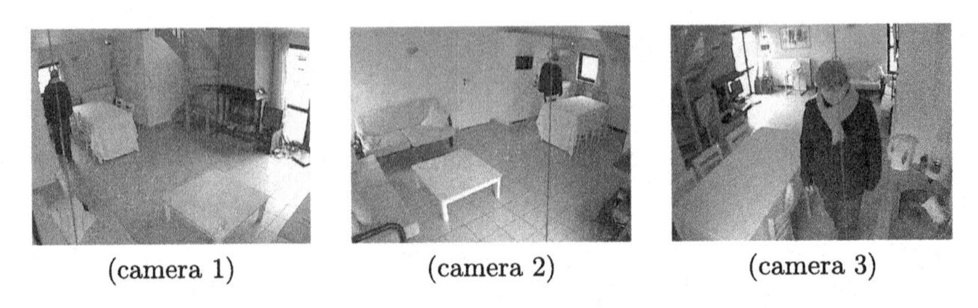

Fig. 3. Detected head and center line positions for each camera in the camera network. Red circles represent detected head positions, red lines represent detected center line positions.

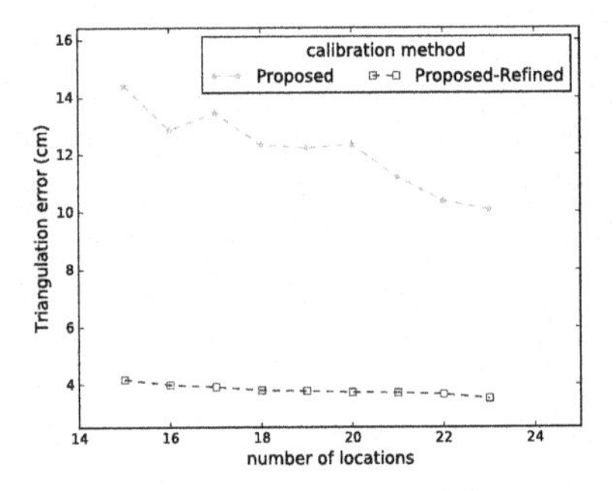

Fig. 4. Triangulation error of our method (Proposed) and the method after refinement (Proposed-Refined), using different numbers of locations.

Figure 3 shows the detected head and center lines of the person in each scene. We detected the head and center lines when the person stood at 26 different locations in the common view of all cameras.

We evaluated the proposed method using different observation numbers of the person. We repeat the calibration procedure 100 times for each N (number of locations), for various sets of locations, randomly selected from the available ones.

Figure 4, Fig. 5, Fig. 6 show the performance in terms of the triangulation error, the projection error and the reprojection error respectively.

It can be observed from these three figures that the accuracy of the proposed method improves by increasing the number (from 15 to 23) of locations of the person. The refinement also improves the accuracy a lot. Thus the proposed method provides good performance of the extrinsic parameters.

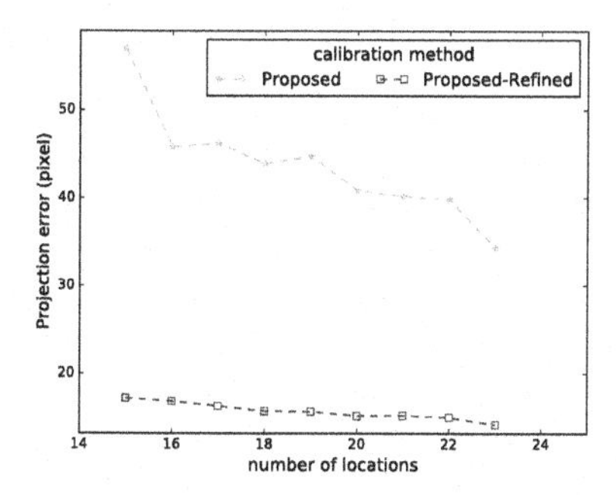

Fig. 5. Projection error of our method (Proposed) and the method after refinement (Proposed-Refined), using different numbers of locations.

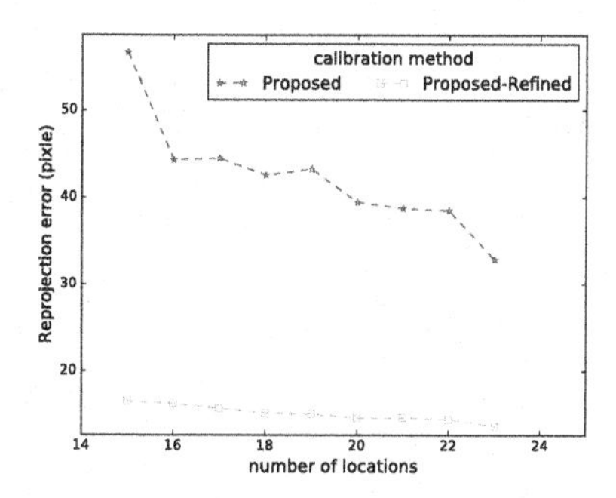

Fig. 6. Reprojection error of our method (Proposed) and the method after refinement (Proposed-Refined), using different numbers of locations.

6 Conclusion

For indoor camera networks which are intended for people surveillance, the feet of a pedestrian are usually occluded by the furniture, which is the degenerate case for most of the current state-of-the-art calibration methods due to co-planarity of all heads' positions of a single pedestrian. To the best of our knowledge, no work exists that deals with camera network calibration for this specific scenario. We proposed a method for extrinsic calibration of camera networks using a walking human as the calibration object, assuming only the head and center line of the person are visible. Thus, the proposed method can be very useful for many of the existing multi-camera visual surveillance systems that observe people indoors.

We first presented a simple and robust method to compute the unit vector which is parallel to the center line of the person w.r.t. each camera coordinate system. Based on the estimated unit vector, we proposed a method for the extrinsic calibration. Firstly, we proposed an easy and accurate method to estimate 3D positions of the head w.r.t. each camera coordinate system, then the extrinsic parameters were estimated using the method of Arun *et al.* [1] in a pairwise fashion. A classical optimization routine was also applied to refine the proposed method.

We tested the proposed method on one real data sequences of multiple cameras observing pedestrians in occluded environments. The experimental results demonstrated the practicality and the feasibility of the proposed method.

We assumed that all cameras share a common volume in our experiments. In theory, only pairwise overlap is required, as we do calibration in a pairwise fashion. For alignment with world coordinate system and determination of the scale, we propose to measure the positions of at least 3 marker w.r.t. the user specified world coordinate system. The scale could also be obtained by assuming known walking speed of the person, which will make the method completely automatic. We will explore this in our future work.

References

1. Arun, K., Huang, T., Blostein, S.: Least-squares fitting of two 3-D point sets. IEEE Trans. Pattern Anal. Mach. Intell. **9**(5), 698–700 (1987). https://doi.org/10.1109/TPAMI.1987.4767965
2. Bouguet, J.Y.: Camera calibration toolbox for matlab. http://www.vision.caltech.edu/bouguetj/calib_doc/. Visited 19 Feb 2015
3. Caprile, B., Torre, V.: Using vanishing points for camera calibration. Int. J. Comput. Vis. **4**(2), 127–140 (1990)
4. Chen, X., Davis, J., Slusallek, P.: Wide area camera calibration using virtual calibration objects. In: IEEE Conference on Computer Vision and Pattern Recognition, SC, USA, 13–15 June, 2000, vol. 2, pp. 520–527 (2000). https://doi.org/10.1109/CVPR.2000.854901
5. Eggert, D.W., Lorusso, A., Fisher, R.B.: Estimating 3-D rigid body transformations: a comparison of four major algorithms. Mach. Vision Appl. **9**(5–6), 272–290 (1997)
6. Faugeras, O., Lustman, F.: Motion and structure from motion in a piecewise planar environment. Int. J. Pattern Recognit Artif Intell. **02**(03), 485–508 (1988). https://doi.org/10.1142/S0218001488000285
7. Faugeras, O.D., Toscani, G.: The calibration problem for stereo. In: IEEE Conference on Computer Vision and Pattern Recognition, Miami, FL, USA, 22–26 June 1986, pp. 15–20 (1986)
8. Fischler, M.A., Bolles, R.C.: Random sample consensus: a paradigm for model fitting with applications to image analysis and automated cartography. Commun. ACM **24**(6), 381–395 (1981)
9. Guan, J., et al.: Extrinsic calibration of camera networks based on pedestrians. Sensors **16**(5), 654 (2016)
10. Hartley, R., Zisserman, A.: Multiple View Geometry in Computer Vision, 2nd edn. Cambridge University Press, New York (2003)
11. Hartley, R.I.: Chirality. Int. J. Comput. Vis. **26**(1), 41–61 (1998)
12. Hödlmoser, M., Kampel, M.: Multiple camera self-calibration and 3D reconstruction using pedestrians. In: Bebis, G., et al. (eds.) ISVC 2010. LNCS, vol. 6454, pp. 1–10. Springer, Heidelberg (2010). https://doi.org/10.1007/978-3-642-17274-8_1 http://dl.acm.org/citation.cfm?id=1940006.1940008
13. Krahnstoever, N., Mendonca, P.R.S.: Bayesian autocalibration for surveillance. In: IEEE International Conference on Computer Vision, Beijing, China, 17–21 October 2005, vol. 2, pp. 1858–1865 (2005). https://doi.org/10.1109/ICCV.2005.44

14. Kurillo, G., Li, Z., Bajcsy, R.: Wide-area external multi-camera calibration using vision graphs and virtual calibration object. In: ACM/IEEE International Conference on Distributed Smart Cameras, California, USA, 7–11 September 2008, pp. 1–9 (2008). https://doi.org/10.1109/ICDSC.2008.4635695
15. Longuet-Higgins, H.C.: A computer algorithm for reconstructing a scene from two projections. Nature **293**, 133–135 (1981)
16. Lv, F., Zhao, T., Nevatia, R.: Camera calibration from video of a walking human. IEEE Trans. Pattern Anal. Mach. Intell. **28**(9), 1513–1518 (2006). https://doi.org/10.1109/TPAMI.2006.178
17. Micusik, B., Pajdla, T.: Simultaneous surveillance camera calibration and foot-head homology estimation from human detections. In: IEEE Conference on Computer Vision and Pattern Recognition, San Francisco, USA, 13–18 June 2010, pp. 1562–1569 (2010). https://doi.org/10.1109/CVPR.2010.5539786
18. Schönemann, P.: A generalized solution of the orthogonal procrustes problem. Psychometrika **31**(1), 1–10 (1966)
19. Svoboda, T., Martinec, D., Pajdla, T.: A convenient multi-camera self-calibration for virtual environments. PRESENCE Teleoperators Virtual Environ. **14**(4), 407–422 (2005)
20. Tsai, R.Y.: A versatile camera calibration technique for high-accuracy 3D machine vision metrology using off-the-shelf TV cameras and lenses. IEEE J. Rob. Autom. **3**(4), 323–344 (1987)
21. Weng, J., Ahuja, N., Huang, T.S.: Motion and structure from point correspondences with error estimation: planar surfaces. IEEE Trans. Sig. Process. **39**(12), 2691–2717 (1991). https://doi.org/10.1109/78.107418
22. Werner, T., Pajdla, T.: Cheirality in Epipolar geometry. In: IEEE International Conference on Computer Vision, Vancouver, Canada, 07–14 July 2001, vol. 1, pp. 548–553 (2001). https://doi.org/10.1109/ICCV.2001.937564
23. Zhang, Z.: A flexible new technique for camera calibration. IEEE Trans. Pattern Anal. Mach. Intell. **22**(11), 1330–1334 (2000)
24. Zivkovic, Z.: Improved adaptive gaussian mixture model for background subtraction. In: IEEE International Conference on Pattern Recognition, Cambridge, UK, 23–26 August 2004, vol. 2, pp. 28–31 (2004). https://doi.org/10.1109/ICPR.2004.1333992

Dual-Layer Barcodes

Kang Fu, Jun Jia, and Guangtao Zhai(✉)

The Institute of Image Communication and Network Engineering,
Shanghai Jiao Tong University, Shang-hai 200240, China
{fuk20-20,jiajun0302,zhaiguangtao}@sjtu.edu.cn

Abstract. With the emergence of mobile networks and smartphones, Quick Response Code (QR Code) has been widely used in many scenes in life, e.g. mobile payments, advertisement, and product traceability. However, when the encoded information of QR Codes is leaked, the QR Codes can be easily copied, which increases the risk of mobile payment and the difficulty of product traceability. To solve these problems, this paper proposes a novel approach to expand the information channels of QR Code based on invisible data hiding. The proposed architecture consists of an information encoder to hide messages into QR Codes while maintaining the original appearances of the QR codes and an information decoder to extract the hidden messages. To make the hidden messages detectable by smartphones, we use a series of noise layers to process the encoder output between the end-to-end training of the encoder and the decoder. The noise layers simulate the general distortions caused by camera imaging, e.g. noise, blur, JPEG compression, and light reflection. Experimental results show that the proposed method can achieve a high decoding accuracy of the hidden messages without affecting the decoding rate of the QR Codes used as the containers.

Keywords: QR Code · Information hiding · Adversarial training · Convolutional networks

1 Introduction

In the age of the Internet of Things, people want to get information quickly from offline to online, which leads to the invention of a variety of offline communication media, such as Radio Frequency Identification (RFID), 1D barcode, and 2D barcode. Among these, Quick Response Code (QR Code) is the most popular 2D barcode that is widely applied in our life. Compared to other kinds of barcodes, QR Code has comparable information capacity and stronger ability of error-correcting while the fine design of the localization rule makes QR Code more robustness to geometric deformation. Thus, QR Code becomes popular in mobile payments, advertisements, social software, and product traceability.

However, QR Codes can be easily copied when the encoded information of them is leaked, which limits the applications of QR Codes in many scenes. Imagine a typical scene of product traceability, where the packaging of a product

H. Ma et al. (Eds.): PRCV 2021, LNCS 13020, pp. 216–227, 2021.
https://doi.org/10.1007/978-3-030-88007-1_18

carries a QR code that links to the product information and the official website of the product. In this case, forgers can easily get the information encoded in the QR Code to make similar packaging to disguise their fake products. A similar problem may occur in mobile payment when the QR Codes used for payment are forged. The forgers use their payment QR code to replace the merchant's, thereby stealing the merchant's income. To solve the above problems, this paper proposes an information hiding method to hide invisible information into QR Codes, making the hidden information detectable by the cameras of smartphones without affecting the decoding of the QR Codes used as the containers. With the help of our method, the ownership information is encoded in the QR Codes as an additional invisible information channel, which decreases the risk of mobile payment and the difficulty of product traceability. In addition, our method can be applied in the information interaction of advertisement and product packaging.

Compared to the traditional technology of property protection like digital watermarking, our challenges are from two aspects: (1) the hidden information can be detectable by smartphones with cameras, (2) the hidden information cannot affect the decoding of the QR Code used as the containers. For the first challenge, traditional digital watermarking only consider the robustness to digital signal processing, e.g. noise and compression. However, the distortion caused by camera shooting is much more complicated than general digital signal processing. Inspired by the recent research on deep information hiding, e.g. HiDDeN [27], StegaStamp [17], RIHOOP [7] and Two-Layer QR Codes [26], we propose an end-to-end architecture consisting of an information encoder to hide messages into QR Codes and an information decoder to extract the hidden messages, which is shown in Fig. 1. To make the hidden messages detectable by cameras, we establish a series of differentiable noise layers between the encoder and the decoder. The differentiable noise layers process the output of the encoder by complicated distortion including noise, blur, JPEG compression, and light reflection, which can simulate the distortion introduced by camera imaging during training.

The second challenge requires that the encoded information inherent in the QR code is not destroyed by the embedding of invisible messages. A standard QR Code consists of finder patterns, alignment patterns, timing patterns, error-correcting codes, data codes, and version information, which represent the location information and the encoded information of the QR Code in the form of images. Thus, we simplify the second challenge to embed messages into QR Codes while maintaining the original appearances of the QR codes. We combine a variety of loss functions for image generation to train our information encoder, making the hidden messages invisible to human eyes. Experimental results show that our approach can achieve high decoding accuracy for the hidden messages under different shooting conditions without affecting the decoding of the QR Coders. The main contributions of this paper are summarized below:

- We propose a novel method to hide invisible messages into QR Code without affecting the decoding of the QR code. To the best of our knowledge, our

method is the first time to make the hidden messages encoded in the QR Code detectable in the scenes of camera-print and camera-display.
- We design a series of noise layers to process the output of the encoder, which can simulate the distortion caused by camera shooting. The distorted QR Code is then fed to the decoder thereby increasing the decoding accuracy in the scenes of camera-display and camera-printing.
- We add some image quality evaluation indicators to the loss function, so that the visual quality of the image is maintained when the message is embedded.

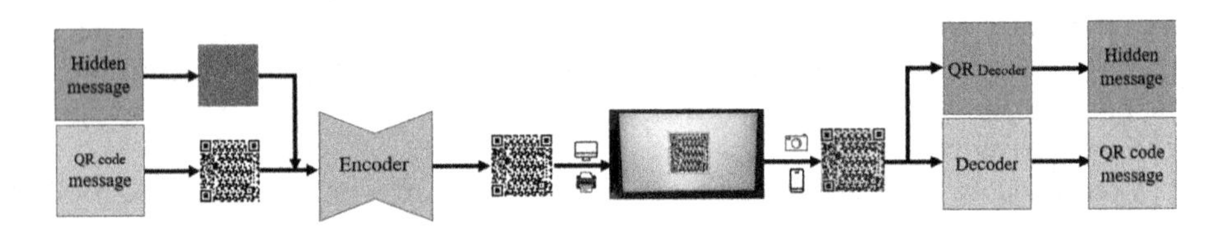

Fig. 1. A general pipeline of our structure to hide messages into QR Codes. First, we concatenate a hidden message and the image of a QR Code as the input of our encoder. The encoder network outputs a QR Code with the hidden message. Then, this QR Code is transmitted and captured by cameras. Finally, we use the decoder to recover the hidden message after detecting the QR Code. The hidden message does not affect the decoding of the QR Code that is as the containers.

2 Related Work

2.1 Steganography

Steganography is a technology that hides secret information in normal digital carriers such as images, digital audio, and video signals. The embedding of additional data does not change the visual and auditory effects of the carrier signal, nor does it change the size and format of the carrier file, so that the hidden information can be transmitted in an unknown manner.

To achieve the goal of being undetected by steganalysis, many different methods of steganography have been proposed. Traditional steganography can be divided into two parts: those in the spatial domain and those in the transform Domain [9]. The simplest and most classic spatial steganography algorithm: Least Significant Bit (LSB) algorithm is to compare the binary least significant bit of the carrier image with the bit of the secret information. If they are the same, the value of the corresponding bit of the carrier image remains unchanged; otherwise, the value of the least significant bit of the carrier image is replaced by the bit of the secret information [13,20]. The LSB method can keep the carrier image unchanged in terms of visual effects through some methods. However, this usually systematically changes the statistical information of the image which can be detected [24]. There is a classic algorithm Jsteg steganography in the transform domain. The Jsteg algorithm embeds secret information into the Least significant bit of the quantized DCT coefficient [12]. Generally speaking, transform domain steganography is more complicated but safer than spatial domain. In recent years, more and more people have begun to use deep learning methods for steganography. Many of them have good results, For example, SGAN [21], SSGAN [15], and ASDL-GAN [18], Hayes et al. [6] are new models for generating image-like containers based on Convolutional Neural Network (CNN). They generate more steganalysis-secure message embedding than the previous algorithm. None of the models proposed above consider the influence of noise in the transmission, so the robustness in screen and print tests is not very good. HiDDeN [27] considers the noise generated by JPGE compression in network transmission. At the same time, Light Field Messaging (LFM) [23] uses a data set containing many actual captured images to obtain the noise introduced during the shooting process and improve the robustness of the algorithm.

2.2 Watermarking

Watermarking is also an important research direction of information hiding technology. The main difference between it and steganography is that watermarking is more robust than steganography and requires less information capacity. Therefore, watermarks are often more suitable for scenarios such as anti-counterfeiting and copyright protection. Similar to steganography, traditional methods of watermarking are also divided into spatial domain and transform domain.

For example, [8,10] is the method in the spatial domain. [8]proposed spatial perceptual mask is based on the cover image prediction error sequence and matches very well with the properties of the human visual system. [5,19] is the method in the transform domain. [19] propose local daisy feature transform (LDFT), Then, the binary space partitioning (BSP) tree is used to partition the geometrically invariant LDFT space, finally, watermarking sequence is embedded bit by bit into each leaf node of the BSP tree. Other work focuses on modeling the physical transmission process of the watermark. [11,16] mainly solve the printer-camera transform problem. [4] Commits to solving the special distortions caused by the screen-shooting process, including lens distortion, light source distortion and moire distortion. Inspired by the above work, [17] proposed a method that automatically learns how to hide and transmit data in a way that is robust to many different combinations of printers/displays, cameras, lighting, and viewpoints. [7] improves the imperceptibility of the watermark based on some image quality evaluation methods, and at the same time increases the robustness of the watermark. We refer to their algorithm to implement our subject watermarking algorithm.

2.3 Barcode

Barcode is a graphic identifier that separates multiple wide black bars and blanks according to certain coding rules. These black bars and blanks can indicate the relevant information of the good, such as the country of production, manufacturer, product name, production date, Book classification number, category, and date. The traditional barcode is one-dimensional, which can increase the speed of information entry and reduce the error rate, but it has some shortcomings: The data capacity is small and can only contain letters and numbers; after the barcode is damaged, it cannot be read and The space utilization rate is low, so it is usually used for supermarket sales checkout.

To solve this problem, QR Code was invented. The advantages of QR code are large information capacity, wide coding range, low cost, easy to make, and can store a large amount of data without the database itself. At the same time, the fault-tolerant mechanism of the QR code ensures that the picture part can be correctly identified after it is damaged. The fault tolerance rate can be as high as 30%. Many mobile applications can locate and recognize QR codes, which has led to the widespread use of QR codes in social networks, express delivery, and mobile payment. These applications generally use open source QR code positioning and detection libraries such as ZBar [2], ZXing [3], and OpenCV [1]. Among them, Opencv is most popular tool that is applied to detect QR Codes by many social apps such as WeChat. We refer to Opencv to implement our QR code location detection.

3 Method

This section outlines the proposed method. The architecture of our method was shown in Fig. 2.

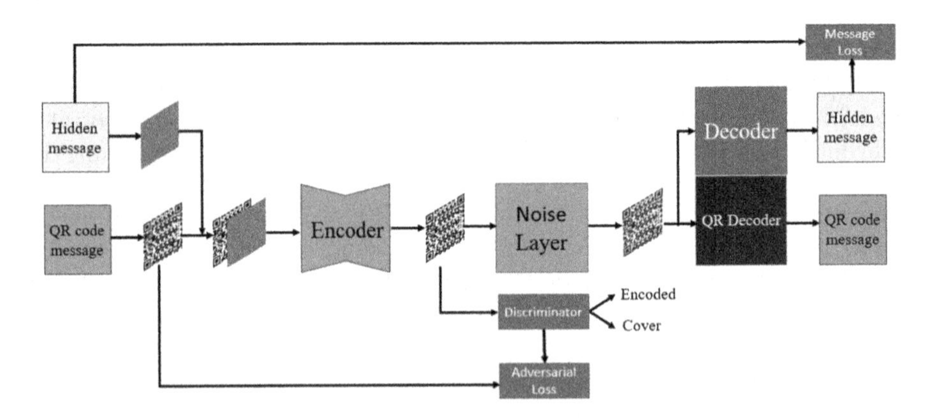

Fig. 2. The architecture of our method. This architecture consists of 5 modules: (1) encoder, (2) decoder, (3) decoder, (4) noise layer, (5) discriminator.

3.1 Encoder

The input of the entire system is two strings, one of which is the QR code message, and the other is the hidden message. We first use the QR code message to generate a QR code image (the size is M * N * 3). At the same time, we preprocess the hidden message through the dense layer to convert its size to M * N * 1 because this preprocessing applied to the hidden message helps the system to converge [7]. Then concatenate it with the QR code picture to generate an M * N * 4 image and input it into the decoder.

We use U-Net [14] style architecture as our encoder. It can convert the input image into an M * N * 3 QR code image with a hidden message. This image can be considered as the addition of the original QR code image and the residual image. In order to make the output image of the encoder visually similar to the original image as much as possible and make the message is embedded in an imperceptible place. We choose to add a JND (Just Noticeable Distortion) [25] loss function when training the encoder.

3.2 Decoder

The output image of the encoder is input to the decoder after passing through the noise layer. The noise layer can enhance the robustness to resist small angle changes introduced when capturing and correcting the encoded image and the chromatic aberration caused by the printer camera.

3.3 Noise Layer

To enhance the robustness of our network against distortion caused during transmission, we assume that there is a camera that captures the output of the encoder (either on the screen or printed out). The noise layer adds perspective warp, motion and defocus blur, Color Manipulation, noise, JPEG compression distortion simulations.

Perspective Warp. To simulate the geometric deformation caused by shooting from different directions, we randomly generate homography to simulate the effect of the QR code image taken when the camera is not parallel to the image. We randomly perturb the four corner locations of the marker uniformly within a fixed range then solve for the homography that maps the original corners to their new locations.

Motion and Defocus Blur. Camera movement and incorrect autofocus may cause blur. We sample a random angle and generate a straight line blur kernel with a width between 3 and 7 pixels to simulate motion blur, and use a Gaussian blur kernel with its standard deviation randomly sampled between 1 and 3 pixels to simulate the diagonal blur.

Color Manipulation. Because the color gamut of printers and monitors is limited, and the exposure settings of cameras, white balance, and color correction matrix are used to modify their output, the final image and the encoder output image have a certain degree of distortion. We use a series of random color transformations to simulate it. We used hue shift, desaturation, brightness, and contrast adjustment.

JPEG Compression. Because camera images are usually stored in a lossy format (such as JPEG). These lossy formats compress the image by calculating the discrete cosine transform of each 8×8 block in the image and quantizing the resulting coefficients by rounding to the nearest integer. So we use discrete cosine transform to simulate lossy compression.

Noise. During the image transmission process, a lot of noise includes photon noise, dark noise, and shot noise will be introduced. We assume that the Gaussian noise model is used to solve the imaging noise.

Light. In actual shooting, the resulting image will largely affect the received ambient light, so we used five to ten randomly distant light sources and one to three Nearby light sources to simulate realistic near perpendicular light.

3.4 Discriminator

We use the discriminator of WGAN-style to supervise the training of our encoder to improve the encoded image's quality.

4 Experiments and Analysis

4.1 Dataset and Experimental Setting

To make the hidden message more invisible and robust to the QR code, we chose 1500 QR code images for training. The version of the QR Codes is 5 and the boxsize is 10. We chose 50 QR code images for testing. In experiments, we use a SANC-M2433PA as the monitors. To capture test photos, we select three mobile devices equipped with cameras: an iPad Air3, a Xiaomi 11, and an iPhone 6S. The training environments are an NVIDIA 2080 GPU for single image training. We use python (3.6), TensorFlow (1.13.1) and cuDNN (10.1).

4.2 Implementation Details

Encoder. We train the encoder to embed the hidden message in the QR code while minimizing the visual difference between the input image and the image with the hidden message embedded. We use U-Net style architecture as our encoder. We present examples of encoded images in Fig. 3.

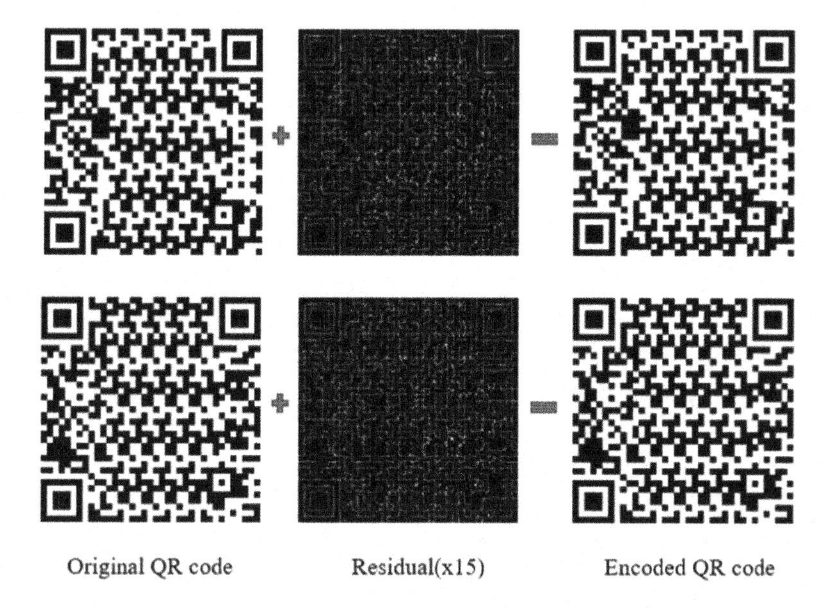

Fig. 3. The encoder will generate a residual based on the input hidden message and QR code and then add it to the original QR code to get the encoded QR code. For us to see the residuals clearly, we multiply each pixel by fifteen times.

Decoder. The decoder consists of seven convolutional layers and a dense layer and finally produces a final output with the same length as the hidden message. The decoder network is supervised using cross-entropy loss.

Our encoder (Table 1) and decoder (Table 2) are similar to the StegaStamp's [17] architecture.

4.3 Metrics

The bit accuracy rate refers to the correct ratio of the bit message we received after decoding the image compared to the original message. We use bit accuracy as our test metric. $BitAccuracy = \frac{rightbits}{allbits} * 100\%$. When the bit accuracy is bigger than 95%, the hidden message can be decoded (Table 3).

In order to test the invisibility of the hidden message, we use PSNR, MSE, and SSIM [22] as our metrics. The formulas of PSNR, MSE, and SSIM are as follows:

$$MSE = \frac{1}{mn} \sum_{i=0}^{m-1} \sum_{j=0}^{n-1} ||I(i,j) - K(i,j)||^2 \tag{1}$$

$$PSNR = 10 \log_{10}^{\frac{Maxvalue^2}{MSE}} = 10 \log_{10}^{\frac{2^{bits}-1}{MSE}} \tag{2}$$

$$SSIM = (x,y) = \frac{(2\mu_x \mu_y + c_1)(\sigma_{xy} + c_2)}{(\mu_x^2 + \mu_y^2 + c_1)(\sigma_x^2 + \sigma_y^2 + c_2)} \tag{3}$$

Table 1. The encoder network architecture. "conv" is convolutional layers and **k** is kernel size, **s** is stride, **chns** is number of input and output channels for each layer, **in** and **out** are accumulated stride for the input and output of each layer, **input** denotes the input of each layer with + meaning concatenation and "upsample" performing 2× nearest neighbor upsampling. A ReLU is applied after each layer except the last.

Layer	k	s	chns	in	out	input
Inputs			6			image + secret
conv1	3	1	6/32	1	1	inputs
conv2	3	2	32/32	1	2	conv1
conv3	3	2	32/64	2	4	conv2
conv4	3	2	64/128	4	8	conv3
conv5	3	2	128/256	8	16	conv4
up6	2	1	256/128	16	8	upsample (conv5)
conv6	3	1	256/128	8	8	conv4 + up6
up7	2	1	128/64	8	4	upsample (conv6)
conv7	3	1	128/64	4	4	conv3 + up7
up8	2	1	64/32	4	2	upsample (conv7)
conv8	3	1	64/32	2	2	conv2 + up8
up9	2	1	32/32	2	1	upsample (conv8)
conv9	3	1	70/32	1	1	conv1 + up9 + inputs
conv10	3	1	32/32	1	1	conv9
Residual	1	1	32/3	1	1	conv10

Table 2. The decoder network architecture. We indicate fully connected layers with the prefix "fc.".

Layer	k	s	chns	in	out	input
conv1	3	2	3/32	1	2	image_warped
conv2	3	1	32/32	2	2	conv1
conv3	3	2	32/64	2	4	conv2
conv4	3	1	64/64	4	4	conv3
conv5	3	2	64/64	4	8	conv4
conv6	3	2	64/128	8	16	conv5
conv7	3	2	128/128	16	32	conv6
fc0			20000			flatten (conv7)
fc1			20000/512			fc0
Secret			512/100			fc1

Table 3. Bit accuracy of the recovery message with different horizontal and vertical angles.

Angle	left 30°	left 20°	left 10°	0°	right 10°	right 20°	right 30°
Bit accuracy	95%	99%	99%	100%	99%	99%	96%
Angle	up 30°	up 20°	up 10°	0°	down 10°	down 20°	down 30°
Bit accuracy	97%	99%	99%	100%	99%	99%	96%

We randomly generated 100 QR codes, then added hidden message to them, and tested psnr, mse, ssim and the average decoding accuracy of zbar, zxing and oprncv for encoded QR codes (Tables 4 and 5).

Table 4. The average PSNR, MSE and SSIM values of the QR codes generated by our method.

Metirc	PSNR	MSE	SSIM
Score	42.65 (db)	3.53	0.99

Table 5. The average decoding accuracy of ZBar, ZXing and OpenCV.

Decoder	ZBar	ZXing	OpenCV
Decoding Rate	100%	100%	100%

5 Discussion

Our proposed method still has two limitations. First of all, image distortion caused by lighting is the biggest problem affecting decoding accuracy. Although we have added lighting simulation to the noise layer, this is usually only effective for one strong light source. When there are multiple strong light sources, the hidden message cannot be decoded. We believe that increasing the redundancy of the hidden message may increase the hidden message's resistance to the light source. Secondly, when the QR code on the shooting screen has moire stripes and the color gradation of the printed QR code is too large, the decoding accuracy will also be greatly affected, which will cause the variance of the decoding accuracy to be large. LFM [23] uses a large number of photos captured by different cameras and screen pairs to train the transformation function, but this method may overfit a specific screen and camera, so the effect is not very good. We think we can continue to adjust the noise layer to solve this problem.

6 Conclusion

This paper proposes an end-to-end method to hide the hidden message into the QR code. At the same time, to ensure the robustness of the encoder image, we have established a noise layer between the encoder and the decoder to enhance the encoded image. The noise layer involves a variety of distortions, such as perspective distortion, noise, and JPEG compression. We have added some image quality evaluation indicators to the loss function to maintain the visual quality of the image when the message is embedded. The experimental results show that our method has strong robustness in various practical situations. In the future, we plan to study how to simulate more complex actual distortion (for example, a combination of multiple barrel distortion and pincushion distortion) for the noise layer. We hope that someone can further optimize the neural network information hiding problem based on our work.

References

1. Opencv-wechatqrcode. https://docs.opencv.org/4.5.2
2. Zbar bar code reader. http://zbar.sourceforge.net
3. Zxing ("zebra crossing") barcode scanning library for java, android. https://github.com/zxing/zxing
4. Fang, H., Zhang, W., Zhou, H., Cui, H., Yu, N.: Screen-shooting resilient watermarking. IEEE Trans. Inf. Forensics Secur. **14**(6), 1403–1418 (2018)
5. Fang, H., Zhou, H., Ma, Z., Zhang, W., Yu, N.: A robust image watermarking scheme in DCT domain based on adaptive texture direction quantization. Multimedia Tools Appl. **78**(7), 8075–8089 (2019)
6. Hayes, J., Danezis, G.: Generating steganographic images via adversarial training. Adv. Neural. Inf. Process. Syst. **30**, 1954–1963 (2017)
7. Jia, J., et al.: RIHOOP: robust invisible hyperlinks in offline and online photographs. IEEE Trans. Cybern. (2020)
8. Karybali, I.G., Berberidis, K.: Efficient spatial image watermarking via new perceptual masking and blind detection schemes. IEEE Trans. Inf. Forensics Secur. **1**(2), 256–274 (2006)
9. Morkel, T., Eloff, J.H., Olivier, M.S.: An overview of image steganography. In: ISSA, vol. 1 (2005)
10. Mukherjee, D.P., Maitra, S., Acton, S.T.: Spatial domain digital watermarking of multimedia objects for buyer authentication. IEEE Trans. Multimedia **6**(1), 1–15 (2004)
11. Pramila, A., Keskinarkaus, A., Seppänen, T.: Increasing the capturing angle in print-cam robust watermarking. J. Syst. Softw. **135**, 205–215 (2018)
12. Qiao, T., Retraint, F., Cogranne, R., Zitzmann, C.: Steganalysis of JSteg algorithm using hypothesis testing theory. EURASIP J. Inf. Secur. **2015**(1), 1–16 (2015)
13. Qin, J., Xiang, X., Wang, M.X.: A review on detection of LSB matching steganography. Inf. Technol. J. **9**(8), 1725–1738 (2010)
14. Ronneberger, O., Fischer, P., Brox, T.: U-net: convolutional networks for biomedical image segmentation. In: Navab, N., Hornegger, J., Wells, W.M., Frangi, A.F. (eds.) MICCAI 2015. LNCS, vol. 9351, pp. 234–241. Springer, Cham (2015). https://doi.org/10.1007/978-3-319-24574-4_28

15. Shi, H., Dong, J., Wang, W., Qian, Y., Zhang, X.: SSGAN: secure steganography based on generative adversarial networks. In: Zeng, B., Huang, Q., El Saddik, A., Li, H., Jiang, S., Fan, X. (eds.) PCM 2017. LNCS, vol. 10735, pp. 534–544. Springer, Cham (2018). https://doi.org/10.1007/978-3-319-77380-3_51
16. Sitawarin, C., Bhagoji, A.N., Mosenia, A., Chiang, M., Mittal, P.: DARTS: deceiving autonomous cars with toxic signs. arXiv preprint arXiv:1802.06430 (2018)
17. Tancik, M., Mildenhall, B., Ng, R.: StegaStamp: invisible hyperlinks in physical photographs. In: Proceedings of the IEEE/CVF Conference on Computer Vision and Pattern Recognition, pp. 2117–2126 (2020)
18. Tang, W., Tan, S., Li, B., Huang, J.: Automatic steganographic distortion learning using a generative adversarial network. IEEE Signal Process. Lett. **24**(10), 1547–1551 (2017)
19. Tian, H., Zhao, Y., Ni, R., Qin, L., Li, X.: LDFT-based watermarking resilient to local desynchronization attacks. IEEE Trans. Cybern. **43**(6), 2190–2201 (2013)
20. Van Schyndel, R.G., Tirkel, A.Z., Osborne, C.F.: A digital watermark. In: Proceedings of 1st International Conference on Image Processing, vol. 2, pp. 86–90. IEEE (1994)
21. Volkhonskiy, D., Nazarov, I., Burnaev, E.: Steganographic generative adversarial networks. In: Twelfth International Conference on Machine Vision (ICMV 2019), vol. 11433, p. 114333. International Society for Optics and Photonics (2020)
22. Wang, Z., Bovik, A.C., Sheikh, H.R., Simoncelli, E.P.: Image quality assessment: from error visibility to structural similarity. IEEE Trans. Image Process. **13**(4), 600–612 (2004)
23. Wengrowski, E., Dana, K.: Light field messaging with deep photographic steganography. In: Proceedings of the IEEE/CVF Conference on Computer Vision and Pattern Recognition, pp. 1515–1524 (2019)
24. Wolfgang, R.B., Delp, E.J.: A watermark for digital images. In: Proceedings of 3rd IEEE International Conference on Image Processing, vol. 3, pp. 219–222. IEEE (1996)
25. Yang, X., Ling, W., Lu, Z., Ong, E.P., Yao, S.: Just noticeable distortion model and its applications in video coding. Sig. Process. Image Commun. **20**(7), 662–680 (2005)
26. Yuan, T., Wang, Y., Xu, K., Martin, R.R., Hu, S.M.: Two-layer QR codes. IEEE Trans. Image Process. **28**(9), 4413–4428 (2019). https://doi.org/10.1109/TIP.2019.2908490
27. Zhu, J., Kaplan, R., Johnson, J., Fei-Fei, L.: HiDDeN: hiding data with deep networks. In: Ferrari, V., Hebert, M., Sminchisescu, C., Weiss, Y. (eds.) ECCV 2018. LNCS, vol. 11219, pp. 682–697. Springer, Cham (2018). https://doi.org/10.1007/978-3-030-01267-0_40

Graph Matching Based Robust Line Segment Correspondence for Active Camera Relocalization

Mengyu Pan[1,2], Chen Meng[1,2], Fei-Peng Tian[1,2], and Wei Feng[1,2(✉)]

[1] College of Intelligence and Computing, Tianjin University, Tianjin, China
{panmengyu,chmeng,tianfeipeng,wfeng}@tju.edu.cn

[2] Key Research Center for Surface Monitoring and Analysis of Cultural Relics, SACH, Beijing, China

Abstract. Active camera relocalization (ACR) is a new challenge problem that focuses on the in-situ perception of the same scene by actively adjusting camera motion and relocate the camera to a target pose. It can support many real-world applications such as high-value objects (e.g., cultural heritage) monitoring, urban and natural environments evolvement studying, robot navigation etc. Advanced ACR methods need a large amount of accurate keypoint correspondences for camera pose estimation. However, on textureless regions or scenes suffering large illumination change, most feature point detection and description methods fail to discover sufficient and reliable keypoint matches to support estimating camera pose. To address it, we present the *Graph Matching based Line Segment Correspondence* for ACR. By exploring the stable spatial geometry relationship between line segments, we formulate the line segment correspondence as a graph matching problem and obtain reliable matching results without utilizing image appearance information. This design is more robust to large illumination differences. We apply the line segment correspondences on ACR workflow to produces a new image that is consistent with the reference one. Extensive experimental results demonstrate the effectiveness of our approach, which achieves more accurate results in challenging scenes.

Keywords: Active camera relocaliztion · Active vision · Line segment matching

1 Introduction

Active camera relocation (ACR) focuses on actively adjusting camera motion and relocate the camera to a target pose physically to obtain the in-situ perception of the same scene at different times. It can support many real-world applications such as high-value scene monitoring [4,23], urban and natural environments evolvement studying [6,19] and robot navigation [24] etc. Via active camera

This work was supported by the Natural Science Foundation of China (No. 62072334).

H. Ma et al. (Eds.): PRCV 2021, LNCS 13020, pp. 228–238, 2021.
https://doi.org/10.1007/978-3-030-88007-1_19

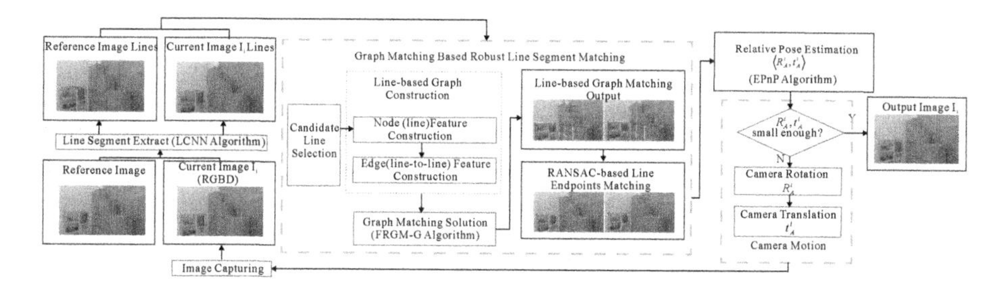

Fig. 1. Working flow of the proposed method. See text for details.

relocation, we can recover the camera pose and collect a current image consistent with the historical image, thus the scene change that occurred between two different times can be easily captured.

However, most existing methods concerns only register the camera pose into a known 3D scene as SfM or SLAM approaches. In this paper, we focus on physically and iteratively adjust the camera pose to the target one. Early methods, i.e., computational rephotography [1,20], build a visual tool as guidance to help users to relocate camera pose, whose accuracy is very low (>0.5 m). Recently advanced ACR methods [12,14] follow a framework of feature points detection and matching, pose estimation, camera pose adjustments in an iteratively way. ACR-B [14] uses a bisection approaching method to deal with the unknown translation scale and unknown hand-eye calibration problem with a monocular camera. The authors prove the convergence of such bisection approaching strategy and achieve a good high relocation accuracy. ACR-D [12] introduce RGBD camera to solve the unknown translation scale problem and speed up the camera relocalization process. As a result, the ACR-D needs much fewer iterations to achieve convergence.

A direct drawback of existing ACR methods is they all need many accurate keypoint correspondences to support camera pose estimation. However many man-made scenes like buildings are textureless and it may suffer large illumination changes on outdoor scenes at different times. Most keypoint detection and description methods like SIFT fail to discover sufficient and reliable keypoint matches due to it depends on image appearance and texture information. In this paper, we propose to use line segment matching to utilize scene structural information instead of appearance information. However, existing line matching methods build line descriptors [15] from surrounding image regions (e.g., MSLD [17] and LBD [22]) or utilize keypoint matching to facilitate line correspondence [3,8], which still require appearance information. In long interval scene monitoring, the environment can have large illumination changes, resulting in unconsistent image appearance and texture at two different times. Such line matching methods may fail to find reliable correspondences as well, making it not suitable for pose estimation for ACR.

Build line segment descriptors from only geometry structural information is a very challenging problem due to the lack of distinguishability. To conquer this problem, we study the relationship between the spatial location of structure lines and transform the line matching to a graph matching problem. For graph matching optimization, we propose a novel representation to build the unary and pair-wise terms and improved the optimization process. Because this matching progress only uses spatial structure relationship of line segments, our method can work well even in the extreme illumination difference scenes or low-texture regions.

We compare different line matching methods to prove the efficiency of ours and evaluate the proposed method in virtual and real scenes. Extensive experiments show that our method gets comparable results in normal scenes and outperforms the state-of-the-art ACR methods in hard scenes.

2 Method

2.1 System Overview

We want to recover the camera pose to its historical state that produces the input reference RGBD image, thus obtain a new image of the same scene at the same 3D pose. The main challenge is we cannot discover sufficient and reliable keypoint matches under scenes with low texture or significant lighting change to support camera pose estimation. To conquer this problem, we propose to utilize line segmentation matching rather than keypoint matching to facilitate pose estimation. Figure 1 shows the working flow of the proposed method. After initializing the camera pose and capture a current image, we first use LCNN [7] to detect the line segments for both current and reference images and remove unreliable or short line segments to generate two groups of line candidates. Then the line candidates are fed into a graph matching based line segment matching module to produce reliable line matching results. We extract point matchings from line segments matching and use RANSAC algorithm [5] to filter unreliable matching outliers. After that, we adopt EPnP algorithm [11] to produce a relative pose between the current and reference images, and execute mechanical motion to reduce the difference between the current and target (reference) camera pose. By iteratively adjusting camera pose via the above processes, we can finally relocate the camera to the target pose and capture a new relocalization image.

2.2 Robust Line Segment Matching

Existing line matching methods designing handcraft line descriptors [17,22] from image appearance or take advantage of matched points [3] to facilitate line correspondence searching. As a consequence, it cannot get a good matching performance under scenes with low texture or significant lighting change. Instead, we propose to utilize graph matching for robust line segment matching by building line structure information and adjacent relations. It can avoid the disadvantage of using image appearance to construct line similarities/dissimilarities. In the following, we first formulate the line matching as a graph matching problem, then explain the method to construct node and edge similarities as well as how to solve it.

Problem Formulation. Given a group of line segments $\mathcal{L}$, we build its corresponding graph by treating line segments as graph nodes and construct edges $\mathcal{E}$ from adjacency relations between two lines (see Sect. 2.2 for details). It results in a graph $\mathcal{G} = \{\mathcal{L}, \mathcal{E}\}$, where the edges $\mathcal{E}$ are generally represented by a adjacency weight matrix. We thus can build two graphs $\mathcal{G} = \{\mathcal{L}, \mathcal{E}\}$ and $\mathcal{G}' = \{\mathcal{L}', \mathcal{E}'\}$ of node size M and N according to the detected line segments from current and reference image, respectively. In the following, we use index m, n to represent nodes from two different graphs, i.e., $\mathcal{L}_m \in \mathcal{L}$ and $\mathcal{L}'_n \in \mathcal{L}'$ and use indexes i, j to indicate nodes from the same graph.

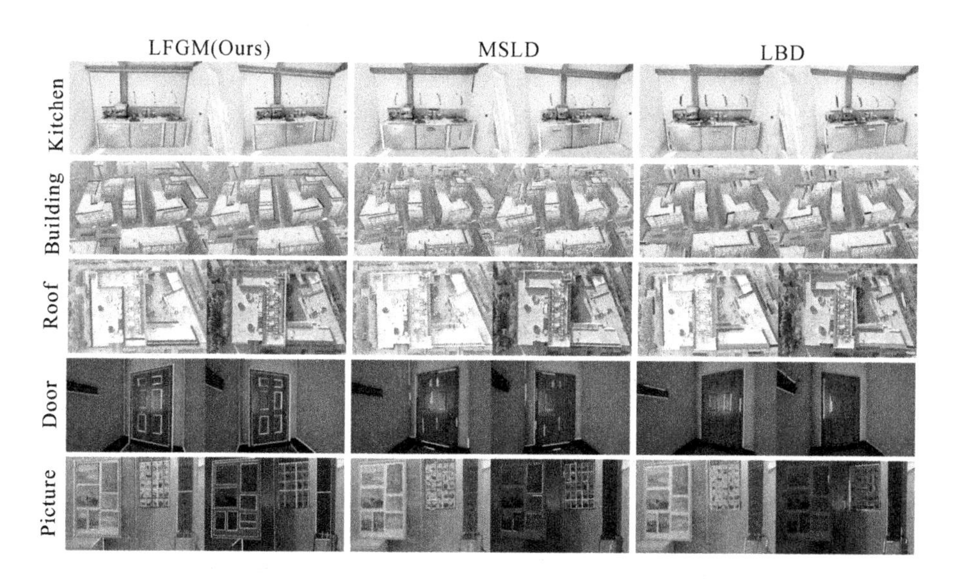

Fig. 2. Line matching visualization results.

Solving the line segment matching is thus transformed to a graph matching problem. The objective is to find an optimal binary node correspondences $\mathbf{C} \in \{0, 1\}^{M \times N}$ such that $\mathbf{C}_{m,n} = 1$ means two lines $\mathcal{L}_m$ and $\mathcal{L}'_n$ are matched, $\mathbf{C}_{m,n} = 0$ otherwise. This is a typical Quadratic Assignment Problem (QAP). To get the optimal solution, graph matching methods generally maximize or minimize an objective function with unary and pairwise constraints. In this paper, we adopt Koopmans-Beckmann's QAP [9,21] as our objective function, that is:

$$\max_{\mathbf{C}} -\text{tr}(\mathbf{U}^T\mathbf{C}) + \lambda\text{tr}(\mathcal{E}\mathbf{C}\mathcal{E}'\mathbf{C}^T),$$
$$s.t.\ \forall m, \sum_n \mathbf{C}_{m,n} \leq 1;\ \forall n, \sum_m \mathbf{C}_{m,n} \leq 1, \tag{1}$$

where tr(·) is the trace of a matrix, $\mathbf{U} \in \mathbb{R}^{M\times N}$ is the node dissimilarity. Equation (1) contains a unary term and a pair-wise term, which measures the node similarities and edge similarities, respectively. The constraint in Eq. (1) limits each node is supposed to have at most one correspondence. If the number of nodes in the two graphs is different, the Hungarian algorithm is used to solve the maximum number of nodes matching, then delete some nodes and transform the two pictures to the same number of nodes. If $\mathbf{C}$ is a orthogonal matrix, the Eq. (1) can be transformed into the following form [16]:

$$\min_{\mathbf{C}} \langle \mathbf{C}, \mathbf{U} \rangle_{\mathcal{F}} + \frac{\lambda}{2} \left\| \mathcal{E}_1 - \mathbf{C}\mathcal{E}_2\mathbf{C}^\top \right\|_{\mathcal{F}}^2 \tag{2}$$

where $\langle \cdot, \cdot \rangle_{\mathcal{F}}$ is the Frobenius dot-product and $\|\cdot\|_{\mathcal{F}}$ is the Frobenius matrix norm, $\langle \mathbf{C}, \mathbf{U} \rangle_{\mathcal{F}} = \sum_{ij} \mathbf{C}_{ij}\mathbf{U}_{ij}$, $\|\mathbf{P}\|_{\mathcal{F}} = \langle \mathbf{P}, \mathbf{P} \rangle_{\mathcal{F}}$.

Node and Edge Similarities Construction. The way to construct node and edge similarities or dissimilarities is critical for solving the QAP problem in Eq. (1). To build the node dissimilarities $\mathbf{U}$, we need to measure the difference between each line segment $\mathcal{L}_m$ in the current image and line segment $\mathcal{L}_n$ in the reference image. However, directly adopting line attribute differences (e.g., line slope error, center distance etc.) as node dissimilarities may result in unsatisfactory matching performance. This is because the matched line in two different views may have large attribute differences. Inspired by the idea of shape context [2], we propose to construct line shape context histograms for each view and use histogram differences combined with attribute differences to measure node dissimilarities. Specifically, we first represent a line segment (node) by a quintuple $\langle \rho, \theta, c^x, c^y, L \rangle$, where ρ and θ are the polar coordinates of the corresponding 2D line, among them, ρ is the distance from the straight line of the line segment to the origin of the coordinate, and θ is determined by the angle between the line segment and the positive horizontal axis. c^x and c^y are the center point coordinate in Euclidean space, L is the line segment length. We use Δ to represent the difference, e.g., $\Delta\rho_{ij} = \rho_i - \rho_j$. Then we build two shape context histograms h^{p} and h^{c} based on polar and center coordinates difference distributions respectively. For an arbitrary node indexing i, the k-th bin of histograms h_i^{p} is:

$$\#\{j \neq i, (\Delta\rho_{ij}, \Delta\theta_{ij}) \in bin(k)\}. \tag{3}$$

Similarly, the center context histogram h_i^{c} is

$$\#\{j \neq i, (\|(\Delta c_{ij}^x, \Delta c_{ij}^y)\|_2, \arctan\frac{\Delta c_{ij}^y}{\Delta c_{ij}^x}) \in bin(k)\}. \tag{4}$$

In shape context histogram construction, we normalize differences to range (0, 1) to achieve stable matching. We measure two histogram difference $\chi^2(h_m, h_n)$ by Chi-Square test. The final node dissimilarity $\mathbf{U}_{m,n}$ for a node $\mathcal{L}_m \in \mathcal{L}$ and a node $\mathcal{L}'_n \in \mathcal{L}'$ is:

$$\mathbf{U}_{m,n} = |L_m - L_n| + \alpha\chi^2(h^p_m, h^p_n) + \beta\chi^2(h^c_m, h^c_n) \tag{5}$$

where α and β are weights to balance three components.

The $\mathbf{U}_{m,n}$ measure the node dissimilarity between two graphs $\mathcal{G}$ and $\mathcal{G}'$. To construct the adjacency weight matrix $\mathcal{E}$ inside each graph, we simply normalize each element in quintuple $\langle \rho, \theta, c^x, c^y, L\rangle$ to $\langle \tilde{\rho}, \tilde{\theta}, \tilde{c}^x, \tilde{c}^y, \tilde{L}\rangle$ and construct the edge weight $\mathcal{E}_{ij}$ of two adjacent nodes $\mathcal{L}_i$ and $\mathcal{L}_j$ by

$$\mathcal{E}_{ij} = \|(\Delta c^x_{ij}, \Delta c^y_{ij})\|_2. \tag{6}$$

In this paper, we build the adjacent weights of two nodes based on the intersect relation. That is, if two line segments $\mathcal{L}_i$ and $\mathcal{L}_j$ intersect with each other after extended a fixed length on both endpoints, we set their adjacent weight $\mathcal{E}_{ij}$ by Eq. (6), otherwise $\mathcal{E}_{ij} = 0$.

Graph Matching Optimization. We solve the graph matching problem based on the FRGM algorithm [16]. It represents each graph by a linear function to reduce the space and time complexities in graph matching. We have made three improvements on the original FRGM algorithm [16]. First, we treat line segments as graph nodes and build novel nodes and edge similarities based on the combination of shape context and line segment attributes. Second, by only considering the unary term in Eq. (1), we use Hungarian algorithm [10,13] to obtain the matching initialization and feed the initial matching to FRGM algorithm and speed up the optimization process. Third, in FRGM optimization, we build adjacency weight matrix $\mathcal{E}$ on both complete graphs (i.e., each node is connected with all other nodes) and non-complete graphs (i.e., nodes connected only when two lines intersect) and alternately optimize on two types of graphs to get reliable matching. Using the graph matching algorithm described above, we can obtain line segment matching without utilizing image appearance information. Thus the line matching is robust to significant light condition change and can also work well on low-texture scenes.

2.3 Active Camera Relocation

Given the line segment correspondences from graph matching, we want to estimate the relative pose estimation from line matches of current and target images, then actively and physically relocate the camera from initial pose $\mathbf{P}^0_{\mathrm{A}} \in \mathrm{SE}(3)$ to target pose $\mathbf{P}^{\mathrm{ref}}_{\mathrm{A}} = \mathbf{I} \in \mathrm{SE}(3)$. To achieve it, we need to execute an optimal mechanical motion $\mathbf{M}^*_{\mathrm{B}}$ to carry the camera to target pose. That is,

$$\mathbf{M}^*_{\mathrm{B}} = \operatorname*{arg\,min}_{\mathbf{M}\in\mathrm{SE}(3)} \|\mathbf{X}\mathbf{M}\mathbf{X}^{-1}\mathbf{P}^0_{\mathrm{A}} - \mathbf{P}^{\mathrm{ref}}_{\mathrm{A}}\|^2_{\mathrm{F}}, \tag{7}$$

Table 1. Correct line matching ratio in virtual and real scenes

Virtual	Ours	LBD	MSLD	Real	Ours	LBD	MSLD
Door 1	**1**	0.69	0.38	Picture	**0.94**	0.88	0.63
Door 2	**0.92**	0.81	0.67	Window	**1**	0.92	0.62
Kitchen	**0.92**	0.38	0.6	Screen	**0.82**	0.8	0.2
Bedroom	**0.93**	0.81	0.76	Roof	**0.9**	0.83	0.67
Sofa	**0.94**	**0.94**	0.73	Building	0.96	0.95	**0.97**

where $\mathbf{X} \in \mathrm{SE}(3)$ is the relative pose between camera and motion platform. Due to the unknown $\mathbf{X}$ and inaccurate pose estimation $\mathbf{P}_{\mathrm{A}}^{0}$, we can asymptotically approach optimal platform motion $\mathbf{M}_{\mathrm{B}}^{*}$ by a series of motions $\mathbf{M}_{\mathrm{B}}^{i}$. Formally,

$$\mathbf{M}_{\mathrm{B}}^{*} = \prod_{i=n}^{0} \mathbf{M}_{\mathrm{B}}^{i} = \mathbf{M}_{\mathrm{B}}^{n}\mathbf{M}_{\mathrm{B}}^{n-1} \cdots \mathbf{M}_{\mathrm{B}}^{i} \cdots \mathbf{M}_{\mathrm{B}}^{1}\mathbf{M}_{\mathrm{B}}^{0}. \tag{8}$$

Existing methods [12,14] has proved when $\theta_{\mathrm{X}} \leq \frac{\pi}{4}$ (θ_{X} is the rotation angle of $\mathbf{X}$ in axis-angle representation), we can ignore the impact of $\mathbf{X}$ and execute platform motion $\mathbf{M}_{\mathrm{B}}^{i} = {\mathbf{P}_{\mathrm{A}}^{i}}^{-1}$ iteratively to relocate the camera to target pose while guarantee the convergence, where $\mathbf{P}_{\mathrm{A}}^{i}$ is the camera pose relative to target pose $\mathbf{P}_{\mathrm{A}}^{\mathrm{ref}}$ in i-th step.

To realize the active camera relocation, we need to estimate the camera pose $\mathbf{P}_{\mathrm{A}}^{i}$ at each step i. Different from keypoints matching, 2D lines correspondences do not have the epipolar constraint. To address it, we extract two endpoints from each line segment as well as cross points from every two intersected lines. Thus we can obtain the endpoints or cross points matching if the line segments or both intersected lines are matched. With a series of matched points and the aligned current depth image, we build 2D-3D point correspondences and feed them into the EPnP algorithm [11] to generate relative pose $\mathbf{P}_{\mathrm{A}}^{i}$. Thus we can adjust platform motion $\mathbf{M}_{\mathrm{B}}^{i}$ *by* ${\mathbf{P}_{\mathrm{A}}^{i}}^{-1}$ iteratively to recover the camera pose to the target one and capture a new image at the relocalized pose.

3 Experiments

3.1 Experimental Setup

We evaluate our method in both virtual and real scenes. The virtual scenes are construct by UnrealCV [18]. These scenes can be divided into normal scenes and hard scenes. The normal scenes have many feature points and little illumination difference. Previous ACR methods can work well in normal scenes. The hard scenes have many textureless regions or large illumination differences. We set the image resolution to 1280×720. Feature-point displacement flow (FDF) and average feature point displacement (AFD) [14] is used to quantitatively evaluate the relocalization accuracy, the lower FDF/AFD value is better.

Table 2. AFD and iteration number for normal/hard scenes

Scenes	ACR-L (ours)		ACR-D		ACR-B	
	AFD	Iter num	AFD	Iter num	AFD	Iter num
Door	**0.462**	8	0.978	10	0.914	19
Sofa	**0.375**	8	0.955	7	0.496	8
Screen	0.502	11	1.350	20	**0.472**	9
Screen (dark)	**1.1395**	8	–	–	–	–
Shelf	0.412	5	**0.331**	4	0.451	5
Shelf (dark)	**0.812**	20	–	–	–	–
Table	**0.214**	6	0.490	10	1.258	20
Table (dark)	**0.216**	7	–	–	–	–

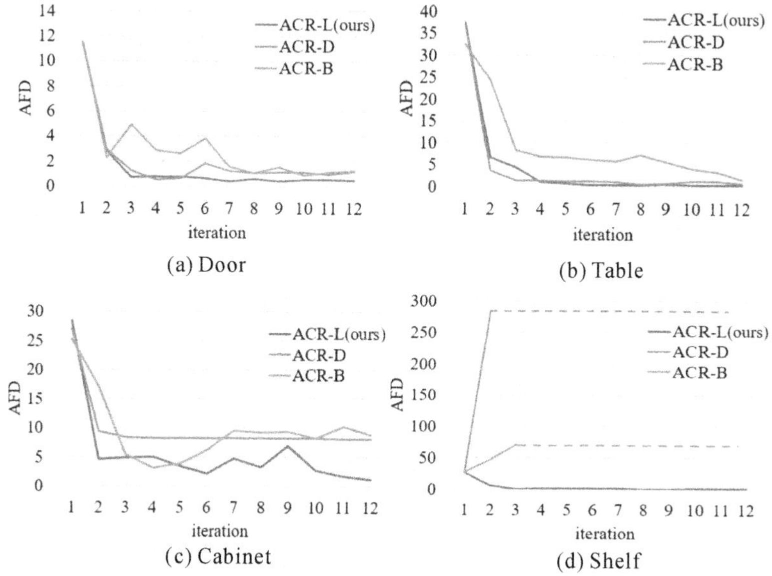

(a) Door

(b) Table

(c) Cabinet

(d) Shelf

Fig. 3. AFD curves in the process of ACR.

3.2 Analysis of Line Segment Matching

We compare our line segment matching method with LBD based method and MSLD based method. From Table 1 we can see that the right matching ratio outperforms the LBD/MSLD based method. This is because that our line segment the matching method does not rely on appearance feature, which is robust to illumination changes. The visualization of matching result is shown in Fig. 2.

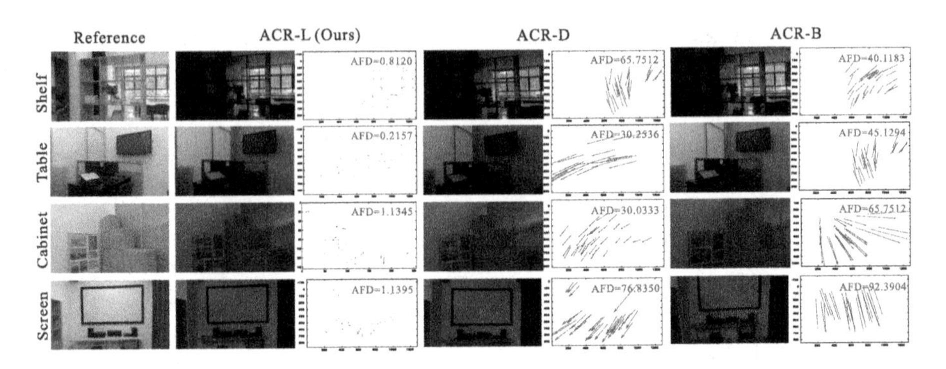

Fig. 4. Comparison of ACR methods and in hard scenes.

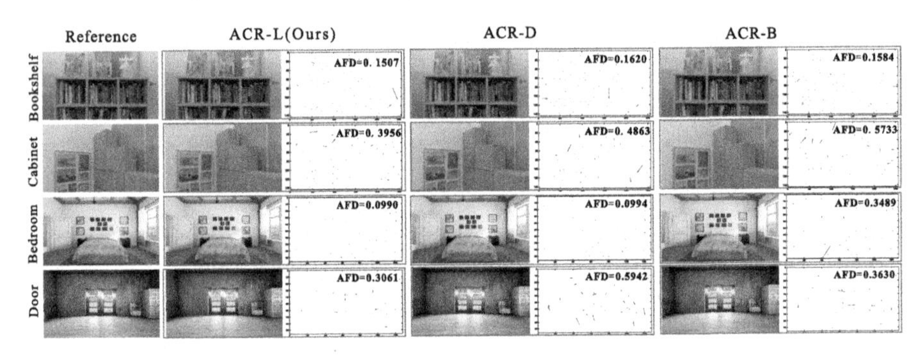

Fig. 5. Comparison of ACR methods and in normal scenes.

3.3 Analysis of Relocalization Accuracy and Convergence Speed

We compare our method (ACR-L) with ACR-B, ACR-D in normal scenes to analyze the relocalization accuracy and convergence speed. Figure 3 (a–c) show 3 example scenes of the iteration progress, which present the descending trend and speed of every ACR method. Table 2 (scenes without dark label) shows the final relocalization result and iteration number of all normal scenes. Among them (a) is a low texture scene, (b) is a general scene, (c) and (d) are scenes with significant lighting differences. Compare to the feature point-based methods (ACR-D and ACR-B), our method gets comparable relocalization accuracy and convergence in normal scenes and achieves significantly better performance in difficult scenes. This is because of the accurate line segment detection and matching results of our method.

3.4 Analysis of Robustness in Hard Scenes

To study the robustness of the ACR method in hard scenes, we select hard scenes which contain many textureless regions or large illumination differences. The AFD curves are shown in Fig. 3 (d). ACR-B and ACR-D fail to reach the reference pose and the AFD curves are not converged. The result of other hard scenes in Table 2 (scenes with dark label) also proved the failure of the feature point-based method in hard scenes. The descriptor of our line segment will not be affected by lighting differences. Furthermore, the structure line segment is more stable than the feature point, which will not have significant change with the weather and seasons. The visualization relocalization result is shown in Fig. 4 and Fig. 5.

4 Conclusion

In this paper, we have proposed a graph matching based line segment matching method for ACR. We utilize the stable spatial relationship between line segments to obtain correct line segment correspondence for camera pose estimation. The line segment based method is robust in man-made textureless regions and the line descriptor without appearance feature can work well in large illumination scenes. The line segment matching based method can be easily used with the feature point based method for stronger generality.

References

1. Bae, S., Agarwala, A., Durand, F.: Computational rephotography. ACM Trans. Graph. **29**(3), 1–15 (2010)
2. Belongie, S., Malik, J., Puzicha, J.: Shape context: a new descriptor for shape matching and object recognition. NIPS **13**, 831–837 (2000)
3. Fan, B., Wu, F., Hu, Z.: Line matching leveraged by point correspondences. In: CVPR, pp. 390–397 (2010)
4. Feng, W., Tian, F.P., Zhang, Q., Zhang, N., Wan, L., Sun, J.: Fine-grained change detection of misaligned scenes with varied illuminations. In: ICCV, pp. 1260–1268 (2015)
5. Fischler, M.A., Bolles, R.C.: Random sample consensus: a paradigm for model fitting with applications to image analysis and automated cartography. Commun. ACM **24**(6), 381–395 (1981)
6. Guggenheim, D., Gore, A., Bender, L., Burns, S.Z., David, L., Documentary, G.W.: An Inconvenient Truth. Paramount Classics Hollywood, CA (2006)
7. Huang, K., Wang, Y., Zhou, Z., Ding, T., Gao, S., Ma, Y.: Learning to parse wireframes in images of man-made environments. In: CVPR, pp. 626–635 (2018)
8. Jia, Q., Gao, X., Fan, X., Luo, Z., Li, H., Chen, Z.: Novel coplanar line-points invariants for robust line matching across views. In: Leibe, B., Matas, J., Sebe, N., Welling, M. (eds.) ECCV 2016. LNCS, vol. 9912, pp. 599–611. Springer, Cham (2016). https://doi.org/10.1007/978-3-319-46484-8_36
9. Koopmans, T.C., Beckmann, M.: Assignment problems and the location of economic activities. Econometrica: J. Econ. Soc., 53–76 (1957)
10. Kuhn, H.W.: The Hungarian method for the assignment problem. Nav. Res. Logist. Q. **2**(1–2), 83–97 (1955)
11. Lepetit, V., Moreno-Noguer, F., Fua, P.: EPnP: efficient perspective-n-point camera pose estimation. IJCV **81**(2), 155–166 (2009)

12. Miao, D., Tian, F.P., Feng, W.: Active camera relocalization with RGBD camera from a single 2D image. In: ICASSP, pp. 1847–1851. IEEE (2018)
13. Munkres, J.: Algorithms for the assignment and transportation problems. J. Soc. Ind. Appl. Math. **5**(1), 32–38 (1957)
14. Tian, F.P., et al.: Active camera relocalization from a single reference image without hand-eye calibration. IEEE TPAMI **41**(12), 2791–2806 (2019)
15. Von Gioi, R.G., Jakubowicz, J., Morel, J.M., Randall, G.: LSD: a fast line segment detector with a false detection control. IEEE TPAMI **32**(4), 722–732 (2008)
16. Wang, F.D., Xue, N., Zhang, Y., Xia, G.S., Pelillo, M.: A functional representation for graph matching. IEEE TPAMI **42**(11), 2737–2754 (2019)
17. Wang, Z., Wu, F., Hu, Z.: MSLD: a robust descriptor for line matching. Pattern Recogn. **42**(5), 941–953 (2009)
18. Qiu, W., et al.: UnrealCV: virtual worlds for computer vision. In: ACM Multimedia Open Source Software Competition (2017)
19. Wells II, J.E.: Western landscapes, western images: a rephotography of US Highway 89. Ph.D. thesis, Kansas State University (2012)
20. West, R., Halley, A., Gordon, D., O'Neil-Dunne, J., Pless, R.: Collaborative rephotography. In: ACM SIGGRAPH Studio Talks (2013)
21. Zaslavskiy, M., Bach, F., Vert, J.P.: A path following algorithm for the graph matching problem. IEEE TPAMI **31**(12), 2227–2242 (2008)
22. Zhang, L., Koch, R.: An efficient and robust line segment matching approach based on LBD descriptor and pairwise geometric consistency. JVCIR **24**(7), 794–805 (2013)
23. Zhang, Q., Feng, W., Wan, L., Tian, F.P., Tan, P.: Active recurrence of lighting condition for fine-grained change detection. In: Proceedings of the 27th International Joint Conference on Artificial Intelligence, pp. 4972–4978 (2018)
24. Zhu, Y., et al.: Target-driven visual navigation in indoor scenes using deep reinforcement learning. In: ICRA, pp. 3357–3364. IEEE (2017)

Unsupervised Learning Framework for 3D Reconstruction from Face Sketch

Youjia Wang, Qing Yan, Wenli Zhou, and Fang Liu(✉)

Beijing University of Posts and Telecommunications, Beijing, China
{2015212858,qingyan,zwl,lindaliu}@bupt.edu.cn

Abstract. Increasingly attention has been paid to 3D understanding and reconstruction recently, while the inputs of most existing models are chromatic photos. 3D shape modelling from the monochromatic input, such as sketch, largely remains under-explored. One of the major challenges is the lack of paired training data, since it is costly to collect such a database with one-to-one mapping instances of two modalities, e.g., a 2D sketch and its corresponding 3D shape. In this work, we attempt to attack the problem of 3D face reconstruction using 2D sketch in an unsupervised setting. In particular, an end-to-end learning framework is proposed. There are two key modules of the network, the 2D translation network and the 3D reconstruction network. The 2D translation network is utilized to translate an input sketch face into a form of realistic chromatic 2D image. Then an unsupervised 3D reconstruction network is proposed to further transform the 2D image obtained in the previous step into a 3D face shape. In addition, because there is no existing sketch-3D face dataset available, two synthetic datasets are constructed based on BFM and CelebA, namely SynBFM and SynCelebA, to facilitate the evaluation. Extensive experiments conducted on these two synthetic datasets validate the effectiveness of our proposed approach.

Keywords: Sketch-based 3D reconstruction · Unsupervised learning

1 Introduction

While 3D reconstruction from 2D face photo input has been extensively studied, few sights have been put on the reconstruction from 2D face sketch input. 2D sketches render more freedom for painters to capture key facial features and are fairly easy to obtain. Nevertheless, the abstract contour information is not sufficient in the field such as forensics and medicine, particularly when researchers want to learn more information related to the 3D vision. Thus, 3D reconstruction from the 2D face sketch is a necessary step to facilitate the further study. However, the conversion from 2D face sketch to 3D face model commonly requires professional knowledge and complex procedures, making it difficult to realize. Based on this situation, a new method is required to simplify this process while

The first author of this paper is a graduate student.

H. Ma et al. (Eds.): PRCV 2021, LNCS 13020, pp. 239–250, 2021.
https://doi.org/10.1007/978-3-030-88007-1_20

ensuring the quality of output 3D face models. This is also the problem that our research focuses on.

The essence of this problem can be seen as an unsupervised cross-domain data generation. There are works on unsupervised cross-domain generation but constrained with the 2D domain [13,19,32]. To mend the gap between 2D input domain and 3D output domain, several methods [7,26] attempt to generate the output 3D face model with the help of the predefined 3D face model 3DMM [2]. However, these algorithms require ground truth 2D facial landmarks or 3D face models as supervision, which are costly to obtain and hard to train. Considering the limitations of the supervised methods, unsupervised reconstruction methods [9,23,29] without 2D or 3D supervision data were proposed. Although these algorithms can recover the fine details in the output 3D face models, the input images are chromatic face images instead of face sketches. On the other hand, despite that some works [27,30] contribute to recover 3D shapes from sketch input, they are subjected to the problem of coarse surface geometry and texture.

In this work, we proposed an end-to-end unsupervised learning framework to reconstruct 3D human faces from abstract sketch input. The framework is divided into two main components: the 2D translation module and the 3D reconstruction module. The 2D translation module realizes the sketch-to-image translation with the idea of generative adversarial network (GAN), specifically CycleGAN [32]. It is pre-trained on the CUFS-CUFSF dataset [28,31], acting as an image encoder to encode the input sketch into a chromatic image. Then the recovered image is fed into the 3D reconstruction module, which first disentangles the image into different photometric components. Then the components are input to a 3D-to-2D renderer [17] to generate the projected 2D image. The learning process of the 3D reconstruction module is to minimize the distance between the input image and the projected image with custom reconstruction losses. Since there is no such large-scale sketch-3D dataset to enable the training and testing process, a large-scale sketch-3D dataset is required to fix this issue. Given that existing face datasets BFM [25] and CelebA [20] include both 2D images and available 3D depth maps but lack 2D sketches, we figured out an idea to synthesize the 2D sketches from original 2D images through image-to-image translation methods. After comparing different methods, the superior method CycleGAN [32] is adopted, and two large-scale synthetic sketch-3D datasets are constructed, namely SynBFM and SynCelebA. To demonstrate the performance of our model, we compare our approach with the state-of-the-art models in both unsupervised and supervised 3D face construction. Our contributions can be summarized as:

(1) We proposed an end-to-end unsupervised learning framework to reconstruct 3D face models from single-view face sketches without the supervision of ground truth 3D face model or pre-defined 3D face template. The 2D translation network and 3D reconstruction network are cascaded and trained in an end-to-end manner in this learning framework.
(2) We constructed two large-scale synthetic sketch-3D datasets of human faces, specifically SynBFM and SynCeleA. Different face-to-sketch synthesis

methods are attempted and compared, then the one with the best performance is selected as the final method for sketch synthesis.

2 Related Work

2.1 Image-to-Image Translation

The concept of image-to-image translation is initially proposed by [14] in conditional GAN, which learns an image-to-image mapping, followed by [15] applying this method to different tasks. However, these models are trained to learn the mappings from paired samples, which are sometimes hard to retrieve. Recent studies like CycleGAN [32] and others [24] attempt to learn the mapping from unpaired samples by associating the source with the target data domains. Another line of studies like StyleGAN [16] and MUNIT [13] try to learn a disentangled representation, encouraging the output to share the same content features as the input but with different style features. Although all the image-to-image translation algorithms mentioned above are limited to the 2D space, in this work, they are very useful in recovering the style and background information from the sketch input in the 2D translation process, making it possible for high-quality 3D reconstruction. In addition, specific image-to-image translation methods are also employed in photo-to-sketch synthesis to construct the synthetic face sketch datasets.

2.2 3D Shape Reconstruction

The algorithm to realize 3D shape reconstruction can be split into three main branches: monocular reconstruction, Non-Rigid SfM and category-specific reconstruction. Monocular reconstruction methods [10] achieve the dense reconstruction without 2D keypoints, but it requires multi-view images or videos as supervisions. Non-Rigid SfM [3] realize the single-view reconstruction, but often requires 2D annotations. In addition, category-specific reconstruction utilizes category-specific information as prior knowledge for 3D shape reconstruction, such as 3D ground truth [25] or 2D silhouettes [11]. Specific in 3D face reconstruction area, some methods [4,9,18] manage to reconstruct the 3D face from a raw human face image, but predefined shape templates such as 3DMM [2] or BFM [25] are usually used as an implicit supervision. To tackle the problem on learning 3D shapes from monocular 2D images in an unsupervised approach, a recent representative study [29] attempt to disentangle the raw input image into different photometric components, training the model by minimizing the distance between the source image and the recovered image obtained from a differential rendering module [17]. The idea of component-disentangling is also shown to be effective for 3D reconstruction in similar works [6,12]. Our method is mainly inspired by this idea, but changes the model input from original chromatic natural images to monochromatic sketches. The modality degradation of the input sketch is the main challenge to solve in our study.

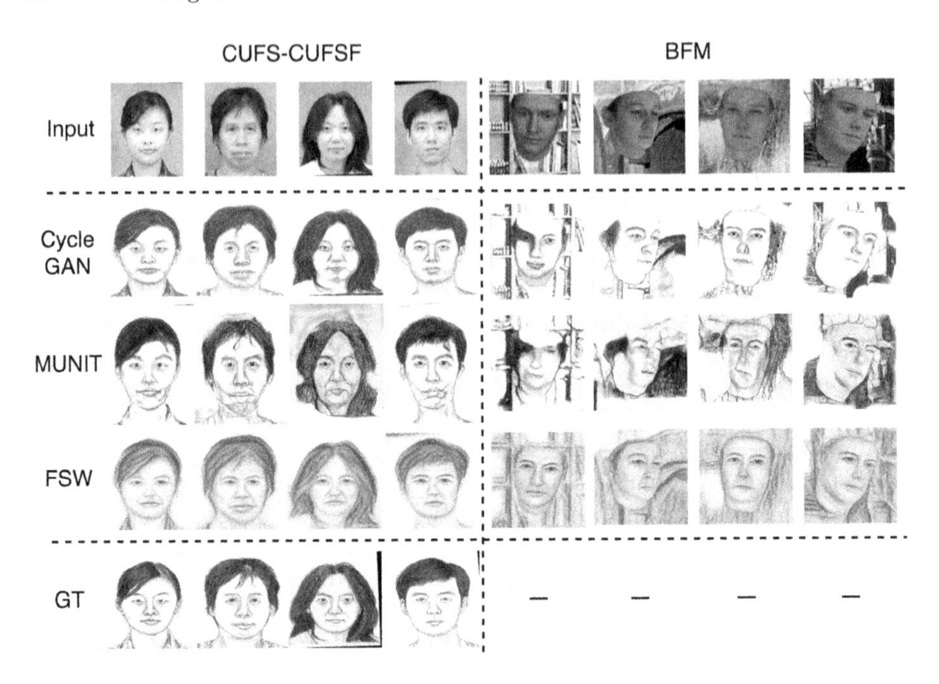

Fig. 1. Qualitative comparison on CUFS-CUFSF and BFM dataset

Table 1. Quantitative comparison between sketch synthesis methods on CUFS dataset. The up arrow means the higher the better, and vice versa.

Method	Metric	
	SSIM ↑	RMSE ↓
CycleGAN	**0.608**	**77.59**
MUNIT	0.607	82.17
FSW	0.554	85.87

3 Method

3.1 Dataset Construction

Since there is no large-scale sketch-3D dataset including both human face sketches and corresponding 3D depth maps to enable our training, we come up with an idea to synthesize the face sketch data from those large-scale face photo datasets such as BFM [25] and CelebA [20], whose 3D depth data is available. In this work, three remarkable work on image-to-sketch synthesis: CycleGAN [32], MUNIT [13] and Face-Sketch-Wild [5] are attempted. They are compared on the custom dataset which is made out of CUFS [28] and CUFSF [31], with 1175 paired face-sketch samples. Preprocessing techniques including gray-scaling and normalization are applied to mitigate the impact of the background colors, and various data augmentation skills like flipping and rotating are also applied to enlarge the training dataset to over 15,000 samples.

The qualitative generation results of the three models on CUFS-CUFSF fusion dataset and BFM dataset is shown in Fig. 1. As revealed in the qualitative results, the sketches generated by CycleGAN perform better in both facial features and details, and retain a higher resolution compared to the other two methods. We also estimate the Structural Similarity (SSIM) and Root Mean Square Error (RMSE) of different methods on CUFS dataset as shown in Table 1, and it is perceived that CycleGAN attains better results in both metrics. After both qualitative and quantitative comparisons, it can be confirmed that CycleGAN is a better solution to construct the large-scale synthetic face sketch dataset. After the whole process of image generation, two large-scale synthetic face-sketch datasets namely SynBFM and SynCelebA are obtained, which are prepared for training the model. Besides, a small-scale synthetic sketch-3D dataset namely SynFlo3D is obtained from the Florence3DFace dataset [1] following the same method, acting as a testset for evaluation.

3.2 Network Architecture

As demonstrated in Fig. 2, After the face sketch is input, the 2D translation network (Fig. 2-(2)) first utilizes a CycleGAN encoder (Fig. 2-(1)) to translate the input sketch $\mathbf{I}$ into a realistic photo-style representation $\mathbf{I}^{'}$, which will be fed into the 3D reconstruction network (Fig. 2-(3)). As for the CycleGAN encoder, it is pre-trained on the CUFS-CUFSF fusion face-sketch dataset, with the direction mapping from the sketch domain to the realistic image domain. With respect to the 3D reconstruction network, its objective is to learn a mapping function $\mathbf{\Phi}$ to decompose the input image $\mathbf{I}^{'} : \Omega \rightarrow \mathbb{R}^3$ into four components, which includes a depth map $\mathrm{D} : \Omega \rightarrow \mathbb{R}^+$, an albedo image $\mathrm{A} : \Omega \rightarrow \mathbb{R}^3$, a lighting direction $\mathrm{L} \rightarrow \mathbb{S}^2$ and a viewpoint $\omega \rightarrow \mathbb{R}^6$, so that the image $\hat{\mathbf{I}} \rightarrow \mathbb{R}^3$ rendered from the four components by the neural renderer [17] (Fig. 2-(4)) could progressively resemble the input image $\mathbf{I}^{'}$ during the training process. The reconstructed actual-view image $\hat{\mathbf{I}}$ is derived from the four photometric components through two functions, which are the lighting function Λ and reprojection function Π. The whole process can be expressed as:

$$\hat{\mathbf{I}} = \mathbf{\Pi}(\mathbf{\Lambda}(A, D, L), D, \omega) \tag{1}$$

In Eq. 1, the lighting function Λ generate the 3D object constrained with the albedo A, the depth map D and the lighting direction L viewed from a canonical viewpoint $\omega = 0$. The reprojection function Π introduces the viewpoint change and generates an image viewed and shaded from the viewpoint ω. The learning process encourages the reconstruction image $\hat{\mathbf{I}}$ to get closer to the input image $\mathbf{I}^{'}$ with different reconstruction losses during the training process, which will be explained in the next part.

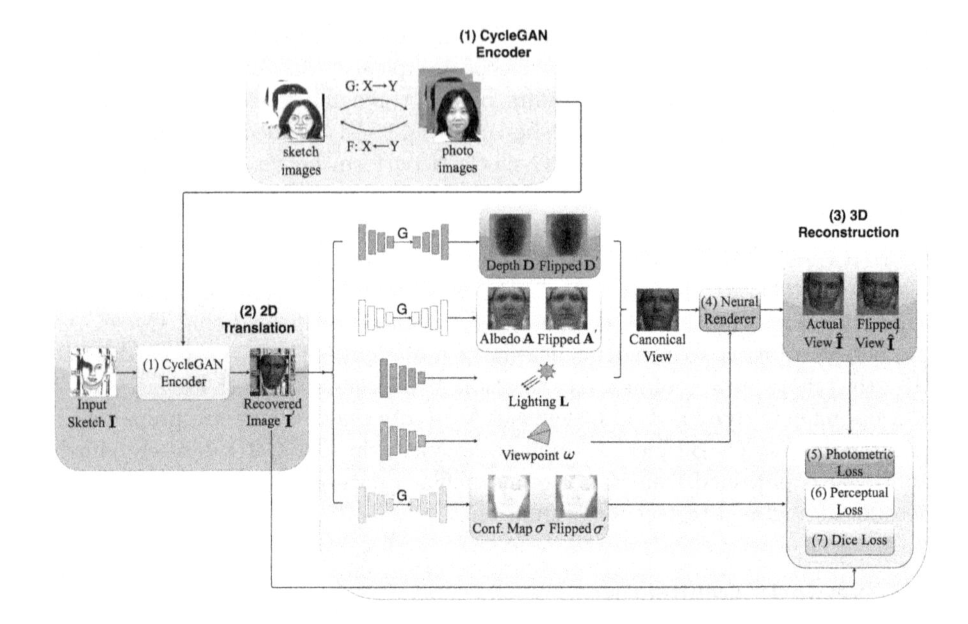

Fig. 2. Network architecture of our proposed model

3.3 Loss Functions

Three losses are applied in our network, which are photometric loss, perceptual loss and dice loss. The first loss is the photometric loss (Fig. 2-(5)). It measures the similarity between the image reconstructed from the 3D components and the input image, given by:

$$\mathcal{L}_{ph}(\hat{\mathbf{I}}, \mathbf{I}^{'}, \sigma) = -\frac{1}{\Omega}\sum_{uv\in\Omega} \ln[\frac{1}{\sigma_{uv}}\exp(-\frac{\sqrt{2}\|\hat{\mathbf{I}}_{uv} - \mathbf{I}^{'}_{uv}\|_1}{\sigma_{uv}})], \tag{2}$$

where $\|\hat{I}_{uv} - I^{'}_{uv}\|_1$ is the L_1 distance between the value of each pixel at the location (u, v). Meanwhile, a new parameter $\sigma \in \mathbb{R}_{+}^{W\times H}$ is introduced. It is also predicted by the network Φ, implying the confidence map which expresses the accidental uncertainty between the original image and the reconstructed image predicted by the model.

Considering the symmetry of human face objects as discussed in [29], we firstly assume that the two components depth D and albedo A of a face object are symmetric about a fixed axis. To achieve this, we need to minimize the distance between the original components and the flipped components, i.e. $D \approx flip(\hat{D})$ and $A \approx flip(\hat{A})$. Instead of training the two parameters separately making it difficult to balance the weights respectively, we utilize the flipped depth and albedo to reconstruct the whole flipped image, which is given by:

$$\hat{\mathbf{I}}^{'} = \mathbf{\Pi}(\mathbf{\Lambda}(A^{'}, D^{'}, L), D^{'}, \omega), A^{'} = flip(A), D^{'} = flip(D). \tag{3}$$

Similar to the photometric loss of the normal recovered image, the photometric loss of the flipping recovered image is defined as:

$$\mathcal{L}_{ph}(\hat{\mathbf{I}}^{'}, \mathbf{I}^{'}, \sigma^{'}) = -\frac{1}{\Omega} \sum_{uv \in \Omega} \ln[\frac{1}{\sigma^{'}_{uv}} \exp(-\frac{\sqrt{2}\|\hat{\mathbf{I}}^{'}_{uv} - \mathbf{I}^{'}_{uv}\|_1}{\sigma^{'}_{uv}})]. \quad (4)$$

Different from Eq. 2, a second confidence map $\sigma^{'}$ is introduced to learn the instance-specific asymmetric features on the face region, e.g. the hairstyle or the intensity of the reflected light. Thus, taking the flipping case into account, the overall photometric loss can be expressed as $\varepsilon_{ph}(\Phi; \mathbf{I}^{'}) = \lambda_1 \mathcal{L}_{ph}(\hat{\mathbf{I}}, \mathbf{I}^{'}, \sigma) + \lambda_2 \mathcal{L}_{ph}(\hat{\mathbf{I}}^{'}, \mathbf{I}^{'}, \sigma^{'})$, where λ_1 and λ_2 are the weighted factors of each case of the photometric loss, which can be tuned as hyperparameters.

The second loss is the perceptual loss (Fig. 2-(6)). Since the L_1 distance in Eq. 2 and Eq. 4 is a relatively coarse-grained measurement, causing failure in the recovery of geometric details, the perceptual loss is introduced as an enhancement to the original photometric loss. It utilizes a feature encoder e to predict a representation of the input image on kth layer of the encoder: $e^k(\mathbf{I}^{'}) \in \mathbb{R}^{C_k \times W_k \times H_k}$. Later, presuming that the perceptual loss would obey the normal distribution, the perceptual loss is defined as:

$$\mathcal{L}_{pe}(\hat{\mathbf{I}}, \mathbf{I}^{'}, \sigma^{(k)}) = -\frac{1}{\Omega} \sum_{uv \in \Omega_k} \ln[\frac{1}{\sigma^{(k)}_{uv}} \exp(-\frac{\|e^k_{uv}(\hat{\mathbf{I}}) - e^k_{uv}(\mathbf{I}^{'})\|_2}{2(\sigma^{(k)}_{uv})^2})]. \quad (5)$$

where $\|e^k_{uv}(\hat{\mathbf{I}}) - e^k_{uv}(\mathbf{I}^{'})\|_2$ represents the L_2 distance between the CNN-encoded representations and the original input at each pixel (u, v). The perceptual loss under the flipping case is also considered, with $\hat{\mathbf{I}}$ and $\sigma^{(k)}$ replaced with respective flipping version. With two predefined hyperparameters λ_3 and λ_4, the overall perceptual loss is expressed as $\varepsilon_{pe}(\Phi; \mathbf{I}^{'}) = \lambda_3 \mathcal{L}_{pe}(\hat{\mathbf{I}}, \mathbf{I}^{'}, \sigma) + \lambda_4 \mathcal{L}_{pe}(\hat{\mathbf{I}}^{'}, \mathbf{I}^{'}, \sigma^{'})$.

The third loss is the dice loss (Fig. 2-(7)). Dice loss is first proposed in V-Net [22]. It performs well in the field of the medical image segmentation by increasing the pixel-wise overlapping between the predicted image and the ground-truth image, improving the similarity of edges. Inspired by these characteristics, we employ the dice loss to our model, aiming to enhance the recovery of edge features of the reconstructed image. For our task, the dice coefficient of the dice loss can be rewritten as:

$$\mathbf{D}(\hat{\mathbf{I}}, \mathbf{I}^{'}) = \frac{2 \sum_{uv \in \Omega} \hat{\mathbf{I}}_{uv} \cdot \mathbf{I}^{'}_{uv}}{\sum_{uv \in \Omega} \hat{\mathbf{I}}^2_{uv} + \sum_{uv \in \Omega} \mathbf{I}^{'2}_{uv}} \quad (6)$$

Similarly, we use the dice loss for both the trivial case and the flipping case of the reconstructed image. The overall dice loss can be summarized as $\varepsilon_d(\Phi; \mathbf{I}^{'}) = \lambda_5 \mathcal{L}_d(\hat{\mathbf{I}}, \mathbf{I}^{'}) + \lambda_6 \mathcal{L}_d(\hat{\mathbf{I}}^{'}, \mathbf{I}^{'})$.

In summary, the final learning objective is the combination of the three losses above, given by $\varepsilon_{overall}(\Phi; \mathbf{I}^{'}) = \varepsilon_{ph}(\Phi; \mathbf{I}^{'}) + \varepsilon_{pe}(\Phi; \mathbf{I}^{'}) + \varepsilon_d(\Phi; \mathbf{I}^{'})$.

4 Experiments

4.1 Implementation Details

In the sketch-3D reconstruction network proposed, two main backbone networks are the 2D translation network and the 3D reconstruction network. With regards to the 2D translation network, CycleGAN is adopted to realize the sketch-to-image translation. It is trained on the CUFS-CUFSF dataset for 200 epochs. On the other hand, the 3D reconstruction network is implemented following the method part. The prediction networks of the photometric components and confidence maps are implemented with standard CNN encoder-decoder structures with group normalization. The model is trained for 30 epochs with the learning rate of 0.0001 on SynBFM dataset and SynCelebA dataset respectively. As for the hyperparameters of loss functions, we set $\lambda_1, \lambda_3, \lambda_5$ that are related to normal reconstruction losses to 1, set $\lambda_2, \lambda_4, \lambda_6$ that are related to flipping reconstruction losses to 0.5.

4.2 Quantitative Results and Ablation Study

As far as we learn, there is no such network to learn how to reconstruct 3D face models from 2D face sketches in an unsupervised manner, so we compare our model with the state-of-the-art model Unsup3D [29] in the field of unsupervised 3D shape reconstruction as well as its different variants. In addition, we also compare our model with supervised 3D face reconstruction methods: ONet [21], DECA [8], 3DDFA [33] and Deep3DFace [7]. The evaluation metrics for the baselines are mean hausdorff distance, scale-invariant depth error (SIDE) and mean angle deviation (MAD). Hausdorff distance originally measures how far two subsets of a metric space are from each other, and here we use it to measure the mean distance from the reconstructed 3D mesh to the ground-truth 3D mesh. The scale-invariant depth error (SIDE) depicted the difference on depth maps, given by $\mathbf{E}_{SIDE}(d_g, d_p) = (\frac{1}{WH}\sum_{uv}\Delta_{uv}^2 - (\frac{1}{WH}\sum_{uv}\Delta_{uv})^2)^{\frac{1}{2}}$, where $\Delta_{uv} = |\log d_{g,uv} - \log d_{p,uv}|$ represents the depth log-error at each pixel. The mean angle deviation (MAD) depicts the difference on surface normal, expressed as $\mathbf{E}_{MAD}(n_g, n_p) = \sum_{uv\in\Omega} \arccos\langle n_{g,uv}, n_{p,uv}\rangle$, where $n_{g,uv}$ and $n_{p,uv}$ represents the normal vectors at each point (u, v) of ground-truth depth and predicted depth respectively. We report the results of mean hausdorff distance on SynFlo3D dataset since only this dataset contains both 2D face sketches and 3D face shape ground truth, and report the results of SIDE and MAD on SynBFM dataset for those baselines that can generate depth maps. The complete experimental result is shown in Table 2. As shown in Table 2, in the comparison of mean hausdorff distance, our method outperforms all the unsupervised 3D reconstruction baselines as well as most of the supervised baselines (three out of four). Besides, in the comparison of SIDE and MAD related to the depth map, our method performs better than all the baselines that can produce depth prediction result to report these two metrics.

Table 2. Quantitative comparison with baseline models

Method	Hausdorff distance ($\times 10^{-1}$) @SynFlo3D ↓	SIDE ($\times 10^{-2}$) @SynBFM ↓	MAD @SynBFM ↓
Unsup3d	2.65 ± 0.26	2.31	41.55
Unsup3d w/ $zdim = 512$	2.61 ± 0.30	2.36	44.53
Unsup3d w/ UNet	2.61 ± 0.19	4.00	69.54
Unsup3d w/ TNet	2.62 ± 0.18	3.23	46.61
Unsup3d w/ ResNet	2.59 ± 0.23	2.66	43.25
ONet	2.49 ± 0.09	–	–
DECA	2.25 ± 0.15	–	–
3DDFA	$\mathbf{0.41 \pm 0.09}$	3.80	33.91
Deep3DFace	2.34 ± 0.12	–	–
Ours (proposed)	1.70 ± 0.14	**0.84**	**16.20**

Table 3. Ablation study on SynBFM sketch dataset

No.	Method	SIDE ($\times 10^{-2}$) ↓	MAD ↓
(1)	Ours full	0.839	**16.197**
(2)	w/o photom. loss	3.253	76.945
(3)	w/o perc. loss	2.813	54.892
(4)	w/o dice loss	**0.825**	16.207
(5)	w/o depth flip $\mathbf{D}'$	2.817	58.039
(6)	w/o albedo flip $\mathbf{A}'$	2.647	55.646
(7)	w/o view flip $\hat{\mathbf{I}}'$	2.670	53.768
(8)	w/o conf. map	3.321	71.542

Furthermore, we conduct the ablation experiments to study the effect of each component of our method in Table 3. Row (2) and Row (3) silences the photometric loss and the perceptual loss respectively, where we can see significant surges in both scale-invariant depth error (SIDE) and mean angle deviation (MAD). Row (4) silences the dice loss, a slight reduction could be witnessed in SIDE, but in turn the MAD metric rises compared to our proposed method. Row (5) removes the depth flipping, Row (6) removes the albedo flipping and Row (7) switches off the confidence map prediction, consequently in all these cases we receive deteriorations in both metrics.

4.3 Qualitative Results

Figure 3 demonstrates the 3D reconstruction performance of our model on the SynBFM dataset and SynCelebA dataset. As seen from the examples in Fig. 3, the global geometric features are well preserved as well as the fine details around

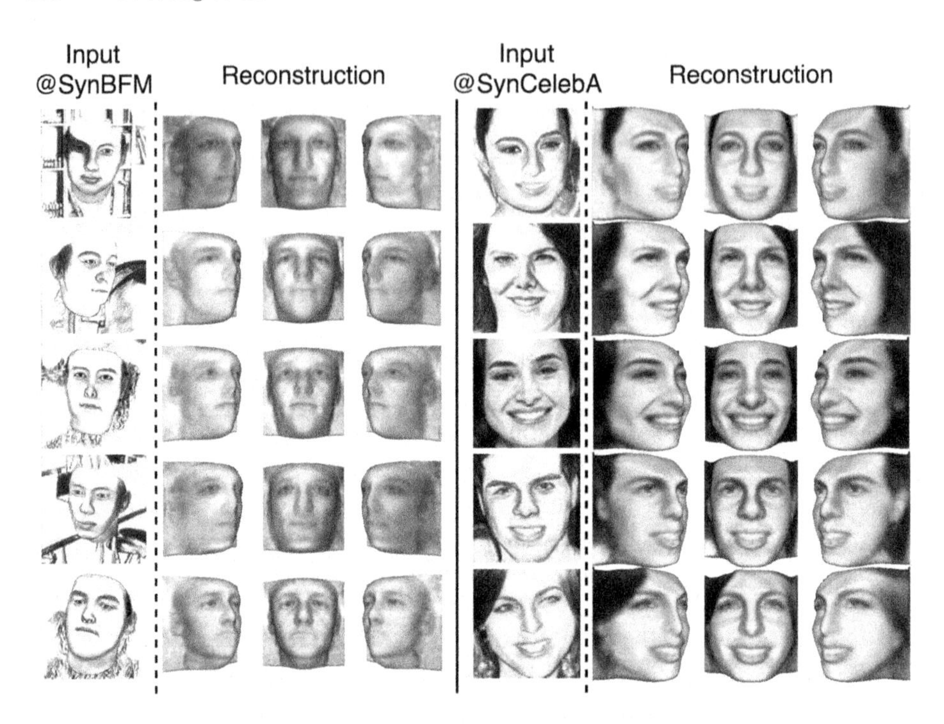

Fig. 3. Qualitative reconstruction results

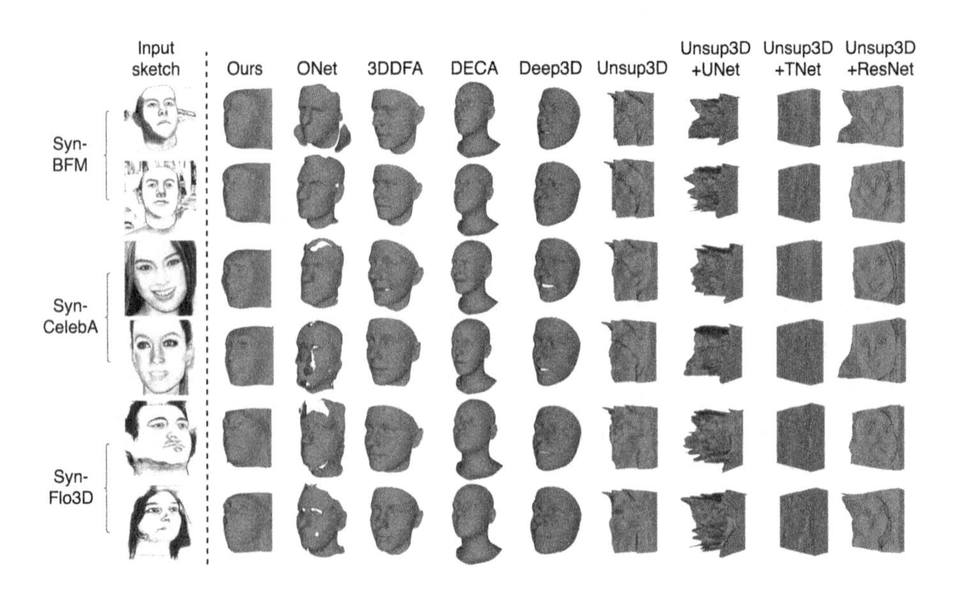

Fig. 4. Qualitative comparison with baseline models

local regions. The facial expression could also be well preserved even for big laughing faces. In Fig. 4, we compare the performances of our proposed model with different baseline models including Unsup3D and its variants as well as other supervised 3D reconstruction methods. As shown in Fig. 4, the 3D reconstruction results generated by our model achieve dramatically higher fidelity and resolution than Unsup3D and its variants, and approach or even outperform the results generated from supervised methods, in spite of the fact that there is no facial keypoints or presupposed face model such as 3DMM [2] used in the training or inference process of our method.

5 Conclusion

In this paper, we further explore the solution to the problem of unsupervised 3D reconstruction from 2D face sketches. We propose an end-to-end unsupervised learning framework composed of a 2D translation network and a 3D reconstruction network, optimized by custom-designed reconstruction losses to learn how to construct different photometric components for the 3D reconstruction. Due to the lack of large-scale sketch-3D dataset of human faces that is necessary for training and testing, we construct two large-scale synthetic face sketch-3D datasets, namely SynBFM and SynCelebA. As shown in the experimental results, our method achieves a competitive performance on this task and outperforms the state-of-the-art unsupervised 3D reconstruction Unsup3D and its variants, as well as most of the supervised 3D reconstruction methods.

References

1. Bagdanov, A.D., Del Bimbo, A., Masi, I.: The florence 2D/3D hybrid face dataset. In: J-HGBU (2011)
2. Blanz, V., Vetter, T.: A morphable model for the synthesis of 3D faces. In: Proceedings of the 26th Annual Conference on Computer Graphics and Interactive Techniques (1999)
3. Bregler, C., Hertzmann, A., Biermann, H.: Recovering non-rigid 3D shape from image streams. In: CVPR (2000)
4. Chen, A., Chen, Z., Zhang, G., Mitchell, K., Yu, J.: Photo-realistic facial details synthesis from single image. In: ICCV (2019)
5. Chen, C., Tan, X., Wong, K.K.: Face sketch synthesis with style transfer using pyramid column feature. In: WACV (2018)
6. Chen, R.T., Li, X., Grosse, R., Duvenaud, D.: Isolating sources of disentanglement in variational autoencoders. arXiv preprint arXiv:1802.04942 (2018)
7. Deng, Y., Yang, J., Xu, S., Chen, D., Jia, Y., Tong, X.: Accurate 3D face reconstruction with weakly-supervised learning: from single image to image set. In: CVPRW (2019)
8. Feng, Y., Feng, H., Black, M.J., Bolkart, T.: Learning an animatable detailed 3D face model from in-the-wild images. In: TOG (2021)
9. Gecer, B., Ploumpis, S., Kotsia, I., Zafeiriou, S.: GANFIT: generative adversarial network fitting for high fidelity 3D face reconstruction. In: CVPR (2019)

10. Godard, C., Mac Aodha, O., Brostow, G.J.: Unsupervised monocular depth estimation with left-right consistency. In: CVPR (2017)
11. Henzler, P., Mitra, N., Ritschel, T.: Escaping plato's cave using adversarial training: 3D shape from unstructured 2D image collections. In: ICCV (2019)
12. Hu, J., et al.: Information competing process for learning diversified representations. arXiv preprint arXiv:1906.01288 (2019)
13. Huang, X., Liu, M.Y., Belongie, S., Kautz, J.: Multimodal unsupervised image-to-image translation. In: ECCV (2018)
14. Isola, P., Zhu, J.Y., Zhou, T., Efros, A.A.: Image-to-image translation with conditional adversarial networks. In: CVPR (2017)
15. Karacan, L., Akata, Z., Erdem, A., Erdem, E.: Learning to generate images of outdoor scenes from attributes and semantic layouts. arXiv preprint arXiv:1612.00215 (2016)
16. Karras, T., Laine, S., Aila, T.: A style-based generator architecture for generative adversarial networks. In: CVPR (2019)
17. Kato, H., Ushiku, Y., Harada, T.: Neural 3D mesh renderer. In: CVPR (2018)
18. Lin, J., Yuan, Y., Shao, T., Zhou, K.: Towards high-fidelity 3D face reconstruction from in-the-wild images using graph convolutional networks. In: CVPR (2020)
19. Liu, M.Y., Breuel, T., Kautz, J.: Unsupervised image-to-image translation networks. In: NIPS (2017)
20. Liu, Z., Luo, P., Wang, X., Tang, X.: Deep learning face attributes in the wild. In: ICCV (2015)
21. Mescheder, L., Oechsle, M., Niemeyer, M., Nowozin, S., Geiger, A.: Occupancy networks: learning 3D reconstruction in function space. In: CVPR (2019)
22. Milletari, F., Navab, N., Ahmadi, S.: V-Net: fully convolutional neural networks for volumetric medical image segmentation. In: 3DV (2016)
23. Pan, X., Dai, B., Liu, Z., Loy, C.C., Luo, P.: Do 2D GANs know 3D shape? Unsupervised 3D shape reconstruction from 2D image GANs. arXiv preprint arXiv:2011.00844 (2020)
24. Park, T., Efros, A.A., Zhang, R., Zhu, J.-Y.: Contrastive learning for unpaired image-to-image translation. In: Vedaldi, A., Bischof, H., Brox, T., Frahm, J.-M. (eds.) ECCV 2020. LNCS, vol. 12354, pp. 319–345. Springer, Cham (2020). https://doi.org/10.1007/978-3-030-58545-7_19
25. Paysan, P., Knothe, R., Amberg, B., Romdhani, S., Vetter, T.: A 3D face model for pose and illumination invariant face recognition. In: AVSS (2009)
26. Sanyal, S., Bolkart, T., Feng, H., Black, M.J.: Learning to regress 3D face shape and expression from an image without 3D supervision. In: CVPR (2019)
27. Wang, L., Qian, C., Wang, J., Fang, Y.: Unsupervised learning of 3D model reconstruction from hand-drawn sketches. In: ACM Multimedia (2018)
28. Wang, X., Tang, X.: Face photo-sketch synthesis and recognition. TPAMI **31**(11), 1955–1967 (2008)
29. Wu, S., Rupprecht, C., Vedaldi, A.: Unsupervised learning of probably symmetric deformable 3D objects from images in the wild. In: CVPR (2020)
30. Xiang, N., et al.: Sketch-based modeling with a differentiable renderer. Comput. Anim. Virtual Worlds **31**(4–5), e1939 (2020)
31. Zhang, W., Wang, X., Tang, X.: Coupled information-theoretic encoding for face photo-sketch recognition. In: CVPR (2011)
32. Zhu, J.Y., Park, T., Isola, P., Efros, A.A.: Unpaired image-to-image translation using cycle-consistent adversarial networks. In: ICCV (2017)
33. Zhu, X., Liu, X., Lei, Z., Li, S.Z.: Face alignment in full pose range: a 3D total solution. TPAMI **41**(1), 78–92 (2017)

HEI-Human: A Hybrid Explicit and Implicit Method for Single-View 3D Clothed Human Reconstruction

Leyuan Liu[1,2], Jianchi Sun[1], Yunqi Gao[1], and Jingying Chen[1,2](✉)

[1] National Engineering Research Center for E-Learning, Central China Normal University, Wuhan 430079, China
{lyliu,chenjy}@ccnu.edu.cn, {sunjc0306,gaoyunqi}@mails.ccnu.edu.cn
[2] National Engineering Laboratory for Educational Big Data, Central China Normal University, Wuhan 430079, China

Abstract. Single-view 3D clothed human reconstruction is a challenging task, not only because of the need to infer the complex global topology of human body but also due to the requirement to recover delicate surface details. In this paper, a method named HEI-Human is proposed to hybridize an explicit model and an implicit model for 3D clothed human reconstruction. In the explicit model, the SMPL model is voxelized and then integrated into a 3D hourglass network to supervise the global geometric aligned features extraction. In the implicit model, 2D aligned features are first extracted by a 2D hourglass network, and then an implicit surface function is employed to construct the occupancy field of human body using the hybrid 2D and 3D aligned features. As the explicit model and implicit model are mutually beneficial, our HEI-Human method not only generates reconstructions with plausible global topology but also recovers rich and accurate surface details. The HEI-Human is evaluated on the current largest publicly available dataset, and the experimental results demonstrate that our method outperforms the state-of-the-art methods including DeepHuman, PIFu, and GeoPIFu.

Keywords: 3D human reconstruction · Hybrid model · Single-view

1 Introduction

3D human reconstruction has attracted more and more attention in the field of computer vision and computer graphics, since it has a wide range of potential applications such as virtual dressing [1] and game design [2]. To obtain accurate reconstructed 3D human models, conventional methods usually employ multi-view images [3] or stereo imaging sensors [4]. Recently, benefiting from the

This work was supported by the National Natural Science Foundation of China (62077026, 61937001), the Fundamental Research Funds for the Central Universities (CCNU19ZN004), and the Research Project of Graduate Teaching Reform of CCNU (2019JG01).

H. Ma et al. (Eds.): PRCV 2021, LNCS 13020, pp. 251–262, 2021.
https://doi.org/10.1007/978-3-030-88007-1_21

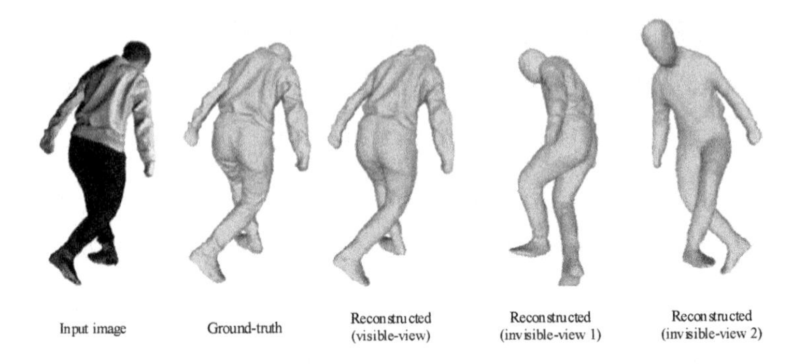

Fig. 1. 3D clothed human reconstruction results by our HEI-human method. Our method not only generates 3D human models with plausible global topology but also recovers rich and accurate surface details.

development of deep neural networks, promising progress [5–7] has been made in reconstructing 3D human models from a single RGB image.

Technically, single-view 3D human reconstruction is a very challenging task. To reconstruct satisfactory 3D human models, algorithms not only need to infer the plausible 3D global topology of the articulated human body from a monocular 2D image that lacks depth information but also are required to recover delicate surface details such as hairs and clothes wrinkles. Recent researches have tried to address these two challenges using different methods, which can be roughly divided into two categories: explicit methods and implicit methods. Explicit methods such as DeepHuman [5] and Tex2Shape [15] usually represent human body by a parametric model (e.g. SMPL [8]) and infer the 3D reconstructions explicitly. Due to the geometrical prior provided by the parametric model, explicit methods can infer the plausible 3D global topology of human body. However, constrained by the low resolution of the parametric model, such explicit methods often have difficulty in recovering surface details. Implicit methods like PIFu [6] and Geo-PIFu [7] employ implicit surface functions to estimate dense occupancy fields for reconstructing 3D human meshes. Benefiting from the dense sampling and interpolation strategies in the feature spaces, implicit methods are able to recover rather richer surface details than explicit methods. However, lacking of the guidance of global information, implicit methods tend to produce unreasonable artifacts on the reconstructed models.

To infer the reasonable global topology of 3D human body from a single RGB image while recovering rich surface details, in this paper, a method named HEI-Human is proposed to hybridize an explicit model and an implicit model for 3D clothed human reconstruction. In our explicit model, the SMPL model [8] is voxelized and then integrated into a 3D hourglass network [9] to supervise the 3D geometric aligned features extraction. In our implicit model, the 2D aligned features are first extracted by a 2D hourglass network and then an implicit surface function is employed to construct the occupancy field of the reconstructed human body using the hybrid 2D and 3D aligned features queried from interpolation feature spaces. The explicit model and implicit model in our HEI-human method are mutually beneficial. On one hand, as the implicit model is supervised by the topology prior provided by the explicit model, our method rarely produces unreasonable artifacts; On the other hand, the implicit model not only guarantees the recovery of surface details but also helps the explicit model to more accurate geometric aligned features by producing detailed reconstructions. As a result, our HEI-Human method not only generates reconstructions with plausible global topology but also recovers abundant and accurate surface details (As illustrated in Fig. 1). Experimental results on the DeepHuman dataset [5] demonstrates that our HEI-Human method achieves the state-of-the-art performance.

In summary, the main contributions of this paper are two-fold:

(1) We propose a framework to hybridize an explicit model and an implicit model for reconstructing 3D clothed human from a single RGB image. By squeezing the advantages of both the explicit model and the implicit model, this hybrid framework has the ability to generate 3D human reconstructions with plausible global topology and rich surface details.

(2) We design deep neural networks that mainly consist of 2D/3D hourglass structures to implement the hybrid framework for 3D clothed human reconstruction. Although very simple loss functions are used, our HEI-human method outperforms the current state-of-the-art methods including DeepHuman [5], PIFu [6], and GeoPIFu [7], which employ much more delicate and complex losses.

2 Related Work

Parametric Model Based 3D Human Reconstruction. A parametric model is an explicit model with three main representations: voxel, point cloud or mesh [10]. A parametric human model is a statistic template trained from many human models, which can be used to drive arbitrary human bodies with a limited number of parameters. The parametric human model allows for supervision and normalization during the 3D human reconstruction process, preventing the reconstruction results from varying significantly in comparison with the human model. Most parametric-based human reconstruction methods including HMR [11], SPIN [12], DaNet [13], and GCMR [14] focus on human shape and pose estimation. DeepHuman [5] uses the SMPL model to constrain the degrees of freedom in the output space. After obtaining the voxel occupation field, the surface normals and the computed depth values are employed to refine the details of the model surface, but this method has little effect due to the storage limitation. Tex2Shape [15] considers that the small resolution of the SMPL model can affect the details of the reconstruction, and adds normal maps and vector displacement maps to the SMPL model to enhance the details of the reconstruction. HMR uses the pose and shape parameters of the SMPL model to transform the 3D reconstruction problem into a parametric regression problem of the model but without surface reconstruction. What's more, model-based reconstruction can also lead to 3D reconstruction failure when the SMPL model is registering incorrectly [7].

Non-parametric Based 3D Human Reconstruction. Non-parametric human reconstruction does not require the 3D model as a priori hypothesis to obtain the 3D human model directly from a single RGB image. After the emergence of the 3D reconstruction task, some methods (such as Bodynet [16], TetraTSDF [17] and VRN [18]) reconstruct the 3D geometry of the human body via volumetric regression. Volumetric regression is limited by the resolution, which is a huge challenge for both network and memory of high-resolution volumetric regression. Secondly, volumetric regression ignores the details of the human surface. For solving these problems, the implicit function is taken into account for the 3D reconstruction task. In the field of 3D human reconstruction, the main work on implicit functions are PIFu [6], PIFuHD [19], ARCH [20], Geo-PIFu [7], SiCloPe [21] etc. The sampling strategy of the implicit surface function is an issue worth investigating. Dense and sparse sampling have a significant impact on the reconstruction results. The optimal sampling parameters by comparison experiments are given in PIFu. PIFuHD feeds higher resolution color images to obtain high quality reconstructions. ARCH proposes an opacity-aware distinguishable rendering in generating datasets to improve the implicit function representations in arbitrary poses. Geo-PIFu converts image 2D features to 3D latent features in feature extraction to constrain spatial degrees of freedom. SiCloPe uses an implicit representation of 2D silhouettes to describe complex human body. However, the implicit surface function lacks constraints on the global features of the human body model, which can cause some errors.

3 Methodology

3.1 Overview

Our HEI-human method is implemented by deep neural networks. As illustrated in Fig. 2, the networks are divided into two main parts: the explicit part (upper) and the implicit part (bottom). Given an input image ($\mathcal{I}$), our method feeds it to both the explicit part and the implicit part. The explicit part starts from the parametric SMPL model estimation and voxelization, then extracts latent 3D aligned features from the voxelized SMPL model ($\tilde{\mathcal{S}}$) by a 3D Hourglass network based encoder ($\mathcal{V}$), and finally reconstructs a coarse result ($\hat{\mathcal{S}}$) by a 3D convolutional network based decoder ($\mathcal{D}$). The implicit part first extracts latent a 2D aligned features using a 2D Hourglass network ($\mathcal{G}$), then feeds both the latent 2D and 3D aligned features into an implicit surface function (f) implemented by a multi-layer perception (MLP) for constructing an occupancy field. The occupancy field is constructed in a voxel-by-voxel manner. For each voxel in the occupancy field, the latent 2D and 3D aligned features with respect to the voxel are queried and hybridised to compute the probability of occupancy. Finally, the 3D human mesh is reconstructed from the refined occupancy fields ($\widehat{\mathcal{S}}$) via Marching Cubes [22].

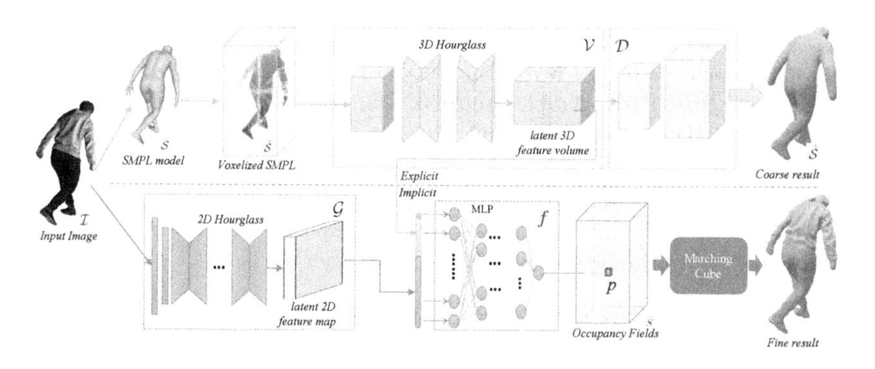

Fig. 2. Overview of our method. Our networks are divided into the explicit part (upper) and the implicit part (bottom). The explicit part extracts latent 3D aligned features from the voxelized SMPL model ($\tilde{\mathcal{S}}$) by a 3D Hourglass based encoder ($\mathcal{V}$). The implicit part first extracts the latent 2D aligned features using a 2D Hourglass network ($\mathcal{G}$), then hybridises both the latent 2D and 3D aligned features into an implicit surface function (f) to construct an occupancy field. Finally, the 3D human mesh is reconstructed from the occupancy filed via Marching Cubes.

3.2 Explicit Model

In our explicit model, a 3D human body is represented by a voxelized SMPL (Skinned Multi-Person Linear) [8]. The SMPL is a parametric model that represents a specific 3D human body by a shape vector α and a pose vector β:

$$\mathcal{T}(\alpha, \beta) = \overline{\mathcal{T}} + \mathcal{B}_s(\alpha) + \mathcal{B}_p(\beta) \tag{1}$$

$$\mathcal{S}(\alpha, \beta) = W(\mathcal{T}(\alpha, \beta), J(\alpha), \beta, \delta) \tag{2}$$

where $\overline{\mathcal{T}}$ is the mean model, $\mathcal{B}_s(\cdot)$ is a blend shape function, $\mathcal{B}_p(\cdot)$ is a pose-dependent blend shape function, $J(\cdot)$ is a joint prediction function, and $W(\cdot, \delta)$ is a skinning function with blend weights δ. In the public available SMPL model [8], $\overline{\mathcal{T}}$, $\mathcal{B}_s(\cdot)$, $\mathcal{B}_p(\cdot)$, $J(\cdot)$, and δ are given. The shape vector α and the pose vector β are estimated using HMR [11] and SMPLify [23]. For the sake of integrating the SMPL model into the deep learning network, the vertex-based SMPL model ($\mathcal{S}$) is voxelized into a voxel volume ($\tilde{\mathcal{S}}$):

$$\tilde{\mathcal{S}} = \mathcal{H}(\mathcal{S}) \tag{3}$$

where $\mathcal{H}(\cdot)$ denotes the voxelization algorithm [5]. As illustrated in Fig. 2, the voxelized SMPL model only generates a naked-like 3D human body with plausible pose and body shape to the input image.

A deep neural network with an "encoder-decoder" reconstruct is designed to reconstruct a clothed 3D human body ($\hat{\mathcal{S}}$) from the voxelized SMPL model ($\tilde{\mathcal{S}}$) explicitly:

$$\hat{\mathcal{S}} = \mathcal{D}_\varphi(\mathcal{V}_\mu(\tilde{\mathcal{S}})) \tag{4}$$

where $\mathcal{V}(\cdot)$ and $\mathcal{D}(\cdot)$ are respectively the encoder and decoder. The encoder $\mathcal{V}(\cdot)$, which is implemented by a 4-layer tandem 3D-Hourglass network with trainable weights μ, is utilized to extract latent 3D aligned features. The decoder $\mathcal{D}(\cdot)$ consists of two 3D convolutions with weights φ. It should note that only the latent 3D aligned features extracted by the encoder $\mathcal{V}$ are employed to infer the finally fine reconstruction by the implicit model, which will described in Subsect. 3.3. Although the decoder $\mathcal{D}$ also can generate a coarse result (i.e., $\hat{\mathcal{S}}$), it is only used for training the encoder.

3.3 Implicit Model

In our implicit model, the surface of a 3D human mesh is represented implicitly by the occupied/unoccupied decision boundary of a continuous occupancy field. An occupancy mapping function $f(\cdot)$ is employed to map each 3D point p in the occupancy field into an occupancy value o ($o \in [0, 1]$):

$$f(p) = o \tag{5}$$

An occupancy value $o > 0.5$ indicates point p is inside the mesh, while $o < 0.5$ means point p is outside the mesh. Thereby, the surface of a 3D human mesh is defined as a 0.5 level set of the continuous occupancy field.

Besides the input RGB image ($\mathcal{I}$), the latent feature volume ($\mathcal{V}$) produced by the explicit model is also utilized to learn the occupancy mapping function $f(\cdot)$. Thereupon, $f(\cdot)$ is formulated as:

$$f_\theta(\{\mathcal{G}_\omega(\mathcal{I}, \ddot{\mathcal{X}}(\pi(p))_m)_k\}_{k=1,\cdots K}^{m=1,\cdots M}, \{\mathcal{V}_\mu(\dddot{\mathcal{X}}(p)_n)_d\}_{d=1,\cdots D}^{n=1,\cdots N}, p_z) = o \tag{6}$$

where $\mathcal{G}(\mathcal{I}, \cdot)$ is a feature extraction function that generates latent feature maps from the input image $\mathcal{I}$, K and k are respectively the number of channels and the channel index of the feature map extracted by g, $\pi(p)$ represents the weak perspective transformation that projects the 3D query point p into the 2D feature map plane, D and d are the number of channels and the channel index of the latent feature volume ($\mathcal{V}$), $\ddot{\mathcal{X}}$ and $\dddot{\mathcal{X}}$ respectively denote the bi-linear and tri-linear interpolation functions, (M, m) and (N, n) are respectively the numbers and indexes of interpolations in the 2D and 3D feature spaces, and p_z represents the depth of p. Hence, our implicit occupancy mapping function with respect to the query point p totally has $(K \times M) + (D \times N) + 1$ input parameters, which fuse the 2D latent features extracted from the input image, the 3D latent features extracted by the explicit model, and the depth information of the query point.

As illustrated in Fig. 2, the functions $\mathcal{G}_\omega(\mathcal{I})$ and $f_\theta(\cdot)$ in Eq. (6) are respectively implemented by a 2D hourglass network with weights ω and a MLP with weight θ. Unlike the Geo-PIFu [7] that directly learns latent 2D and 3D aligned features from the input image, our method integrates the latent 3D aligned features learnt from the parametric SMPL model to regularize the implicit model. As a consequence, our method not only recovers rich surface details but also rarely produces unreasonable artifacts.

3.4 Loss Functions

The explicit part of our network is trained using an extended cross-entropy loss between the estimated voxel volume and ground-truth [24]:

$$\mathcal{L}_v(\ddot{\mathcal{S}}, \dot{\mathcal{S}}) = -\frac{1}{|\dot{\mathcal{S}}|}\sum_{x,y,z}\gamma\ddot{\mathcal{S}}_{x,y,z}\log\dot{\mathcal{S}}_{x,y,z} + (1-\gamma)(1-\ddot{\mathcal{S}}_{x,y,z})(1-\log\dot{\mathcal{S}}_{x,y,z}) \quad (7)$$

where $\ddot{\mathcal{S}}$ is the voxel volume voxelized from the ground-truth 3D human mesh, $\dot{\mathcal{S}} \in \{\hat{\mathcal{S}}, \widehat{\mathcal{S}}\}$ is the estimated (coarse or fine) voxel volume, (x, y, z) are the voxel indices for the width, height and depth axes, and γ is the weight to balance the losses of occupied/unoccupied voxels. The implicit part of our network is trained based on a set of query points randomly sampled from the occupancy field. The mean square error loss is adopted to measure the errors between the ground-truth and the predicted occupancy values:

$$\mathcal{L}_p = \frac{1}{|\mathcal{P}|}\sum_{p\in\mathcal{P}}(o_p - \hat{o}_p)^2 \quad (8)$$

where $\mathcal{P}$ is the set of sampled query points for a training sample, $|\mathcal{P}|$ denotes the number of sampled points, and o_p and $\hat{o}_p$ are respectively the ground-truth and the predicted occupancy values of the query point p.

4 Experiments

4.1 Dataset and Protocol

Dataset. The proposed 3D human reconstruction method is extensively evaluated on the DeepHuman dataset [5], which contains 6,795 data items of 202 subjects with various body shapes, poses, and clothes. The raw data of DeepHuman is captured by consumer-grade RGB-D cameras, and then processed into 3D human data items by an improved DoubleFusion algorithm [4]. Consequently, each data item in DeepHuman consists of a 3D textured surface mesh, a RGB image, and an aliened SMPL model. In our experiments, the aliened SMPL models are used.

Protocols. In our experiments, our method and the other competing 3D human reconstruction methods are all trained on the 5,436 training data items in the DeepHuman dataset, and evaluated on the rest 1,359 testing data items.

4.2 Training Details

A two-stage scheme is used to train our HEI-human model. At the first stage, the explicit model is trained only with the $\mathcal{L}_v(\ddot{\mathcal{S}}, \hat{\mathcal{S}})$ loss for 10 epochs. At the second stage, the whole network (explicit model + implicit model) is trained for another

10 epochs using $\lambda\mathcal{L}_v(\ddot{\mathcal{S}}, \widehat{\mathcal{S}}) + (1-\lambda)\mathcal{L}_q$. The weight parameter γ in Eq. (7) is set to 0.7, the number of sampled query points for each training sample in Eq. (8) is set as 5,000, and the hyper-parameter λ is set to 0.5. For both the two training stages, the RMSprop is adopted as the optimizer, the batch size is fixed to 8, and the learning rate is set as 0.0001 at the beginning and decayed by a factor of 10 after every 4 epochs. All the networks in our method are implemented by PyTorch [25], and the whole training task takes about 7 days on a computer with two NVIDIA GeForce 1080Ti GPUs.

4.3 Quantitative Results

Metrics. The Chamfer distance (ε_{cd}), point-to-surface distance (ε_{psd}), cosine distance (ε_{cos}), and L2-norm (ε_{l2}) are adopted as the metrics for evaluating different 3D human reconstruction methods. The Chamfer distance and point-to-surface distance focus more on the overall quality of model topology, while the cosine distance and L2-norm tend to evaluate local surface details. For all these four metrics, smaller values indicate better performance.

Table 1. Quantitative comparisons on the DeepHuman dataset.

Methods	ε_{cd}	ε_{psd}	ε_{cos}	ε_{l2}
DeepHuman [5]	11.928	11.246	0.2088	0.4647
PIFu [6]	2.6004	4.0174	0.0949	0.3048
Geo-PIFu (coarse) [7]	2.2907	2.6260	0.0874	0.3175
Geo-PIFu [7]	1.7794	1.9548	0.0717	0.2649
HEI-Human (Ours)	**0.1742**	**0.2297**	**0.0661**	**0.2540**

Results and Comparisons. The proposed HEI-Human is compared against three code available state-of-the-art 3D human reconstruction methods: the DeepHuman [5], PIFu [6], and Geo-PIFu [7]. The quantitative results achieved by these methods are presented in Table 1. In terms of global topology quality, our method outperforms the second-best method Geo-PIFu [7] (an implicit method) by a Chamfer distance (ε_{cd}) of 1.6052 and by a point-to-surface distance (ε_{psd}) of 1.4251; In terms of the local details, our method also surpasses Geo-PIFu by a cosine distance of 0.0056 and by a L2-norm of 0.0109. Although our method outperforms all the competing methods on the four metrics, it should be note that it is even not a completely fair comparison. Our method is only trained for 20 (10+10) epochs with a small batch-size of 8 due to the restriction in experimental condition, while all the three competing methods are trained for more than 40 epochs with a larger batch-size (Geo-PIFu [7] is trained for 45 epochs with a batch-size of 36).

4.4 Qualitative Results

Figure 3 shows the qualitative reconstruction results produced by DeepHuman [5], PIFu [6], Geo-PIFu [7], and our HEI-Human on the DeepHuman dataset. It can been seen that DeepHuman recovers rather less local surface details than the other three methods and generates "fattened" and "naked-like" human bodies. Although PIFu and Geo-PIFu can generate surface details of clothes and plausible global topology from the visible view, we can find many unreasonable artifacts (marked in red circles) on their results from the invisible view. Benefiting from the combination of the explicit voxelized SMPL model and the implicit surface function representation, our method not only generates better global topology from the views of both the visible and invisible sides but also recovers richer surface details (such as hair and clothes winkles) than the competing methods. However, the resolution of the 3D human model in the THuman dataset is low, and the details of the face are blurred. In Fig. 3, the facial details of all methods are not be recovered. 3D face reconstruction methods [26,27] can be employed to solve this problem.

Fig. 3. Qualitative 3D human reconstruction results on the DeepHuman dataset.

4.5 Ablation Studies

To further explain the effect of the explicit model and the implicit model, we conduct ablation studies. The implicit module (the bottom part in Fig. 2) and the explicit module (the upper part in Fig. 2) are respectively removed from our method and the remained networks are trained and tested with the same data and protocol described in Subsect. 4.1. Table 2 shows the results achieved by our method with different module configurations. Obviously, the Chamfer distance (ε_{cd}) and point-to-surface distance (ε_{psd}) produced by our method significantly increase after removing the explicit module, while the cosine distance (ε_{cos}) and L2-norm (ε_{l2}) grow larger after removing the implicit module. These quantitative changes have also been visually proved by the qualitative results illustrated in Fig. 4. On one hand, some topology artifacts and distortions (such as the arms and feet circled in red) can be found on the reconstructed human bodies from the invisible view after removing the explicit module. On the other hand, most of the local details on the surface of reconstructed 3D human bodies disappear after removing the implicit module. It can be inferred from these ablation experimental results that the explicit module is beneficial to improve the global regularities and the implicit module helps capture fine-scale surface details from the input image. By combining the explicit and implicit modules, our hybrid method (i.e., HEI-human) not only generates reconstructions with plausible global topology but also recovers abundant and accurate surface details.

Table 2. Quantitative results achieved by our method with different module configurations on the DeepHuman dataset.

Modules	ε_{cd}	ε_{psd}	ε_{cos}	ε_{l2}
Explicit + Implicit	**0.1742**	**0.2297**	**0.0661**	**0.2540**
Only implicit	2.6004	4.0174	0.0949	0.3048
Only explicit	0.6134	0.4997	0.0968	0.3211

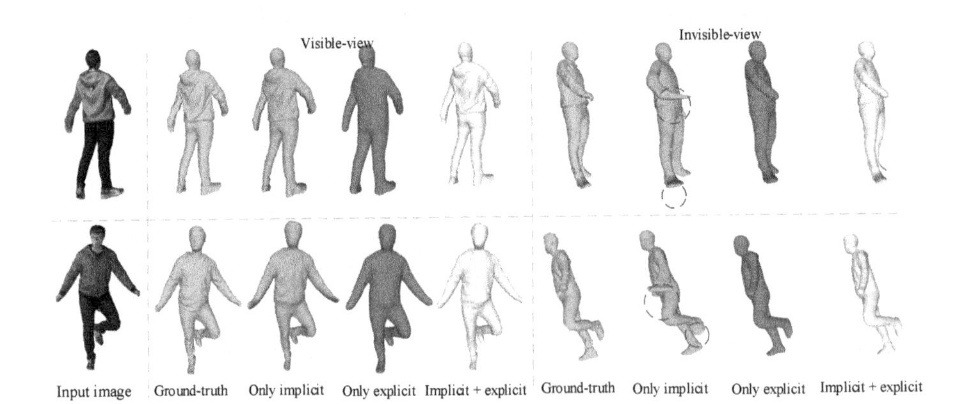

Fig. 4. Qualitative results achieved by our method with different module configurations.

5 Conclusions

As an inherently ill-pose problem, 3D human reconstruction is challenging not only because of the requirement to infer the complex global topology of human body but also due to the need to recover surface details. In order to outcome these challenges, a method named HEI-Human has been proposed to hybridize an explicit model and an implicit model for 3D clothed human reconstruction. Ablation studies have shown that the explicit model is beneficial for global topology while the implicit model mainly takes charge of recovering the surface details. As a consequence, our hybrid method recovers rich surface details and rarely produces unreasonable artifacts. Experimental results on the DeepHuman dataset demonstrate that our HEI-human outperforms the current state-of-the-art methods including DeepHuman [5], PIFu [6], and GeoPIFu [7].

References

1. Pons-Moll, G., Pujades, S., Hu, S., et al.: ClothCap: Seamless 4D clothing capture and retargetting. ACM Trans. Graph. (TOG) **36**(4), 1–15 (2017)
2. Cha, W., Price, T., Wei, Z., et al.: Towards fully mobile 3D face, body, and environment capture using only head-worn cameras. IEEE Trans. Visual. Comput. Graph. **24**(11), 2993–3004 (2018)
3. Ji, M., Gall, J., Zheng, H., et al.: Surfacenet: an end-to-end 3D neural network for multiview stereopsis. In: Proceedings of the IEEE International Conference on Computer Vision, pp. 2307–2315. IEEE, Italy (2017)
4. Yu, T., Zheng, Z., Guo, K., et al.: Doublefusion: real-time capture of human performances with inner body shapes from a single depth sensor. In: Proceedings of the IEEE Conference on Computer Vision and Pattern Recognition, pp. 7287–7296. IEEE, Salt Lake City (2018)
5. Zheng, Z., Yu, T., Wei, Y., et al.: Deephuman: 3D human reconstruction from a single image. In: Proceedings of the IEEE International Conference on Computer Vision, pp. 7739–7749. IEEE, Korea (2019)
6. Saito, S., Huang, Z., Natsume, R., et al.: PIFu: pixel-aligned implicit function for high-resolution clothed human digitization. In: Proceedings of the IEEE International Conference on Computer Vision, pp. 2304–2314. IEEE, Korea (2019)
7. He, T., Collomosse, J., Jin, H., et al.: Geo-PIFu: geometry and pixel aligned implicit functions for single-view human reconstruction. In: Advances in Neural Information Processing Systems, pp. 9276–9287, NIPS (2020)
8. Loper, M., Mahmood, N., Romero, J., et al.: SMPL: a skinned multi-person linear model. ACM Trans. Graph. (TOG) **34**(6), 1–16 (2015)
9. Newell, A., Yang, K., Deng, J.: Stacked hourglass networks for human pose estimation. In: Proceedings of the IEEE Conference on European Conference on Computer Vision, pp. 483–499. IEEE, Holland (2016)
10. Fan, H., Su, H., Guibas, L.J.: A point set generation network for 3D object reconstruction from a single image. In: Proceedings of the IEEE Conference on Computer Vision and Pattern Recognition, pp. 605–613. IEEE, U.S.A (2017)
11. Kanazawa, A., Black, J., Jacobs, W., et al.: End-to-end recovery of human shape and pose. In: Proceedings of the IEEE Conference on Computer Vision and Pattern Recognition, pp. 7122–7131. IEEE, U.S.A (2018)

12. Kolotouros, N., Pavlakos, G., Black, M.J., et al.: Learning to reconstruct 3D human pose and shape via model-fitting in the loop. In: Proceedings of the IEEE International Conference on Computer Vision, pp. 2252–2261. IEEE, Korea (2019)
13. Zhang, H., Cao, J., Lu, G., et al.: Learning 3D human shape and pose from dense body parts. IEEE Trans. Pattern Anal. Mach. Intell. (2020)
14. Kolotouros, N., Pavlakos, G., Daniilidis, K.: Convolutional mesh regression for single-image human shape reconstruction. In: Proceedings of the IEEE Conference on Computer Vision and Pattern Recognition, pp. 4501–4510. IEEE, U.S.A (2019)
15. Alldieck, T., Pons-Moll, G., Theobalt, C., et al.: Tex2shape: detailed full human body geometry from a single image. In: Proceedings of the IEEE International Conference on Computer Vision, pp. 2293–2303. IEEE, Korea (2019)
16. Varol, G., Ceylan, D., Russell, B., et al.: Bodynet: volumetric inference of 3D human body shapes. In: Proceedings of the European Conference on Computer Vision, pp. 20–36. IEEE, Germany (2018)
17. Onizuka, H., Hayirci, Z., Thomas, D., et al.: TetraTSDF: 3D human reconstruction from a single image with a tetrahedral outer shell. In: Proceedings of the IEEE International Conference on Computer Vision and Pattern Recognition, pp. 6011–6020. IEEE(2020)
18. Jackson, S., Manafas, C., Tzimiropoulos, G.: 3D human body reconstruction from a single image via volumetric regression. In: Proceedings of the European Conference on Computer Vision Workshops. IEEE, Germany (2018)
19. Saito, S., Simon, T., Saragih, J., et al.: PIFuHD: multi-level pixel-aligned implicit function for high-resolution 3D human digitization. In: Proceedings of the IEEE Conference on Computer Vision and Pattern Recognition, pp. 84–93. IEEE (2020)
20. Huang, Z., Xu, Y., Lassner, C., et al.: Arch: animatable reconstruction of clothed humans. In: Proceedings of the IEEE Conference on Computer Vision and Pattern Recognition, pp. 3093–3102. IEEE (2020)
21. Natsume, R., Saito, S., Huang, Z., et al.: Siclope: silhouette-based clothed people. In: Proceedings of the IEEE Conference on Computer Vision and Pattern Recognition, pp. 4480–4490. IEEE, U.S.A (2019)
22. Lorensen, E., Cline, E.: Marching cubes: a high resolution 3D surface construction algorithm. ACM SIGGRAPH Comput. Graph. **21**(4), 163–169 (1987)
23. Bogo, F., Kanazawa, A., Lassner, C., et al.: Keep it SMPL: automatic estimation of 3D human pose and shape from a single image. In: Proceedings of the IEEE International Conference on European Conference on Computer Vision, pp. 561–578. IEEE, Holland (2016)
24. Jackson, S., Bulat, A., Argyriou, V., et al.: Large pose 3D face reconstruction from a single image via direct volumetric CNN regression. In: Proceedings of the IEEE International Conference on Computer Vision, pp. 1031–1039. IEEE, Italy (2017)
25. Paszke, A., Gross, S., Massa, F., et al.: Pytorch: an imperative style, high-performance deep learning library. In: Advances in Neural Information Processing Systems, pp. 8024–8035. NIPS, Canada (2019)
26. Liu, L., Ke, Z., Huo, J., et al.: Head pose estimation through keypoints matching between reconstructed 3D face model and 2D image. Sensors **21**(5), 1841 (2021)
27. Liu, L., Zhang, L., Chen, J.: Progressive pose normalization generative adversarial network for frontal face synthesis and face recognition under large pose. In: Proceedings of the IEEE International Conference on Image Processing, pp. 4434–4438. IEEE, Taibei, China (2019)

A Point Cloud Generative Model via Tree-Structured Graph Convolutions for 3D Brain Shape Reconstruction

Bowen Hu[1,2], Baiying Lei[3], Yanyan Shen[1], Yong Liu[4], and Shuqiang Wang[1](✉)

[1] Shenzhen Institutes of Advanced Technology, Chinese Academy of Sciences, Shenzhen, Guangdong, China
sq.wang@siat.ac.cn
[2] University of Chinese Academy of Sciences, Beijing, China
[3] Shenzhen University, Shenzhen, Guangdong, China
[4] Renmin University of China, Beijing, China

Abstract. Fusing medical images and the corresponding 3D shape representation can provide complementary information and microstructure details to improve the operational performance and accuracy in brain surgery. However, compared to the substantial image data, it is almost impossible to obtain the intraoperative 3D shape information by using physical methods such as sensor scanning, especially in minimally invasive surgery and robot-guided surgery. In this paper, a general generative adversarial network (GAN) architecture based on graph convolutional networks is proposed to reconstruct the 3D point clouds (PCs) of brains by using one single 2D image, thus relieving the limitation of acquiring 3D shape data during surgery. Specifically, a tree-structured generative mechanism is constructed to use the latent vector effectively and transfer features between hidden layers accurately. With the proposed generative model, a spontaneous image-to-PC conversion is finished in real-time. Competitive qualitative and quantitative experimental results have been achieved on our model. In multiple evaluation methods, the proposed model outperforms another common point cloud generative model PointOutNet.

Keywords: 3D reconstruction · Generative adversarial network · Graph convolutional network · Point cloud

1 Introduction

With the continuous advancement of medical operation methods, minimally-invasive and robot-guided intervention technology have gradually been used in brain surgery in recent years, bringing patients smaller surgical wounds, shorter recovery time, and better treatment experience. However, the development also brings new requirements. Since doctors cannot directly observe lesions and surgical targets during operations, their experience is often not so efficient. Therefore, these technologies will not be fully mature before the surgical environment

H. Ma et al. (Eds.): PRCV 2021, LNCS 13020, pp. 263–274, 2021.
https://doi.org/10.1007/978-3-030-88007-1_22

and the real-time information acquisition ability improve. Recently, the application of intraoperative MRI (iMRI) has become more and more extensive, and some work used it to relieve the stricter visual restrictions in minimally invasive surgery [1,2]. But unlike the rich internal details of the brain, MRI cannot provide intuitive and visually acceptable information of the surface and the shape of target brains that is more important for surgery. Thus, it is an inevitable direction of development for these types of surgery to find some indirect 3D shape information acquisition methods that are accurate and controllable. Considering the limitations of the use of conventional scanners and the inconvenience of the acquisition of images in brain surgery, these methods should be based on algorithms, rather than physical medical equipment, and can reconstruct the 3D shape of the target from as little traditional information as possible.

There are several alternative representations in the field of 3D surface reconstruction, such as voxels [3], meshes [4], and point clouds [5,6]. A point cloud, which is represented as a set of points in 3D space, just uses N vertices to describe the overall shape of the target. As for comparisons, the voxel representation requires cubic space-complexity to describe the reconstruction result. And meshes, each of which will be treated as a domain to registration in the reconstruction, use a $N \times N$ dimensional adjacency matrix. Therefore, it is a reasonable proposal to choose point clouds as the reconstruction representation in surgical scenes that require less time consuming and flexibility.

In previous work, There are many kinds of research on the generative method of point cloud representation. In the PC-to-PC generation field, [7] proposed an auto-encoder (AE) model to fold and recover the point clouds. [8] proposed a variational auto-encoder (VAE) model to generate close point cloud results and then apply them to fracture detection and classification. [9–11] proposed different GAN architectures to learn the mapping from Gaussian distribution to multiple classes of point cloud representations, so as to reconstruct point clouds in an unsupervised manner. In the image-to-PC generation field, [12] proposed a new loss function named geometric adversarial loss to reconstruct point clouds representation that better fits the overall shape of images. In [13], a deep neural network model composed of an encoder and a predictor is proposed. This model named PointOutNet predicts a 3D point cloud shape from a single RGB image. On the basis of this work, [14] applied this model to the one-stage shape instantiation to reconstruct the right ventricle point cloud from a single 2D MRI image, thereby simplifying the two-stage method proposed by [15].

However, to the best of our knowledge, no effort has been devoted to the development of point cloud reconstruction of brains. Since Deep learning technology has been popularized in the medical prediction [16–18], and has been applied in many fields such as maturity recognition [19,20], disease analysis [21–25], data generation [26,27], there are many works that combine deep learning with 3D data for accurate reconstruction [28–30]. Generative adversarial network, as well as many of its variants [31–33], is a widely used generative model and is known

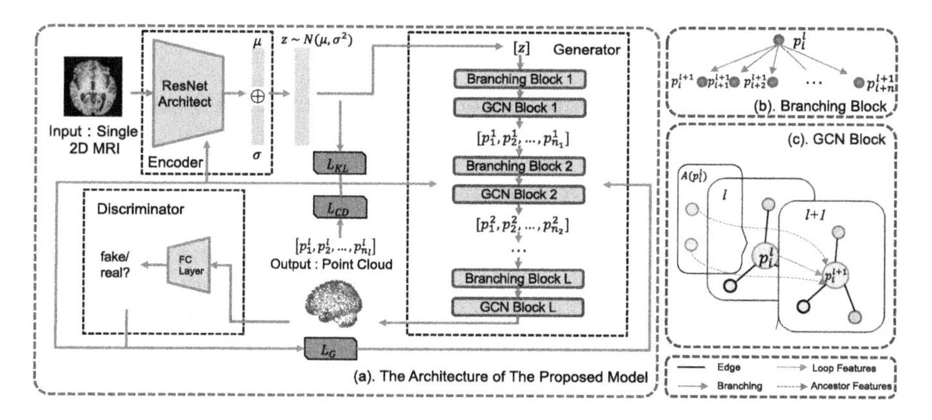

Fig. 1. The Architecture of The Proposed Model: (a) Pipeline of our model. Gray arrows represent how the information flows transfer, and brown arrows represent the backward propagation direction of the loss function. (b) How branching blocks in generator work. (c) Details of GCN blocks in generator.

for its good generation quality. Using it for image-point cloud conversion can preserve features of images to the greatest extent and improve the accuracy of point clouds. Variational method is another generative model [34–36], and can extract features to provide more reliable guidance for the generation of GAN. Meanwhile, graph convolutional networks (GCN) has achieved great success in solving problems based on graph structures [37–39], and has been proven to be very effective in the generation and analysis of point clouds [9,10]. To solve the problem that it is difficult for the rGAN model in [11] of generating realistic shapes with diversity, we consider construct novel GCN modules to replace the fully connected layer network in rGAN. In this paper, a new GAN architecture is proposed to convert a single 2D brain MRI image to the corresponding point cloud representation. Our model consists of an encoder, a novel generator that is composed of alternate GCN blocks and branching blocks, and a discriminator similar to WGAN-GP in [40]. A tree-structured graph convolutional generative mechanism is constructed to ensure that the generated point cloud guided by image features is as close to the target brain as possible.

The main contribution of this paper is to propose a general and novel GAN architecture to reconstruct accurate point clouds with high computational efficiency. This model is the first work to introduce point cloud generation into the fields of brain image analysis and brain shape reconstruction. We use the variational encoder method and graph convolutional algorithm to build and improve our GAN architect, and use different loss functions to ensure that all parts of the model achieve the best training effect.

2 Methods

2.1 Generative Architecture Based on Tree-Structured GCN

Given a specific brain subject, the point cloud of it can be expressed as a matrix $Y_{N\times 3}$ which denotes a set of N points and each row vector represents the 3D coordinate of a vertex. Preprocessed 2D images are required for the proposed generative model, which contains three networks, namely encoder, generator, and discriminator. The architecture of the model is shown in Fig. 1(a). Gray arrows represent how the information flows transfer, and brown arrows represent the backward propagation direction of the loss function. The encoder designed to have a similar architecture to ResNet in [41] takes such images $I_{H\times W}$ as inputs, where H is the height and W is the width of an image. It produces vectors $z \in \mathbb{R}^{96}$ from a Gaussian distribution with a specific mean μ and standard deviation σ as outputs, which are treated as a point set with only one single point by the generator. The generator uses a tree-structured graph network which has a series of GCN blocks and branching blocks to expand and adjust the initial point set. Then discriminator differentiates the output $Y_{N\times 3}$ (in this work, $N = 2048$) and real point cloud, and then enhances the generator to make generative point clouds closer to the ground truth. Specially, we use the model in the Wasserstein GAN with the gradient penalty method at the discriminator.

GCN Block. Some work has used graph convolutional networks for point cloud generation tasks. Inspired by [9,10], multi-layer improved GCNs are considered in our model to implement an efficient generator. Each layer consists of a GCN block and a branching block. The graph convolutions in GCN blocks (Fig. 1(c)) is defined as

$$p_i^{l+1} = \sigma\left(\boldsymbol{F}_K^l(p_i^l) + \sum_{q_j \in A(p_i^l)} U_j^l q_j + b^l\right), \tag{1}$$

where there are three main components: loop term S_i^{l+1}, ancestor term A_i^{l+1} and bias b^l. $\sigma(\cdot)$ is the activation function.

Loop term, whose expression is

$$S_i^{l+1} = \boldsymbol{F}_K^l(p_i^l), \tag{2}$$

is designed to transfer the features of points to the next layer. Instead of using a single parameter matrix W in conventional graph convolutional networks, the loop term uses a K-support fully connected layer $\boldsymbol{F}_K^l$ to represent a more accurate distribution. $\boldsymbol{F}_K^l$ has K nodes $(p_{i,1}^l, p_{i,2}^l, ..., p_{i,k}^l)$, and can ensure the fitting similarity in the big graph.

Ancestor term allows features to be propagated from the ancestors of a vertex to the corresponding next connected vertex. This term of a graph node in conventional GCN is usually named neighbors term and uses the information of its neighbors, rather than its ancestors. But in this work, point clouds are generated dynamically from a single vector, so the connectivity of the computational graph is unknown. Therefore, this item is modified to

$$A_i^{l+1} = \sum_{q_j \in A(p_i^l)} U_j^l q_j, \tag{3}$$

to ensure structural information is inherited and multiple types of point clouds can be generated. $A(p_i^l)$ is the set of ancestors of a specific point p_i^l. These ancestors map features spaces from different layers to p_i^l by using linear mapping matrix U_j^l and aggregate information to p_i^l.

Branching Block. Branching is an upsampling process, mapping a single point to more points. Different branching degrees $(d_1, d_2, ..., d_n)$ are used in different branching blocks (Fig. 1(b)). Given a point p_i^l, the result of the branching is d_l points. Therefore, after branching, the size of the point set becomes d_l times the upper layer. By controlling the degrees, we ensured that the reconstruction output in this work is accurate 2048 points.

The purpose of branching blocks is to build the tree structure of our generative model, thereby giving the generator the ability to transfer the feature from the root $z \in \mathbb{R}^{96}$ to points at the last layer and generate complex point cloud shapes, which can ensure the accuracy and effectiveness in brain shape reconstruction.

2.2 Training of the Proposed GAN Model

In addition to the generator, there are two networks, encoder and discriminator. Figure 1 shows the process that the encoder extracts the random vector z from the 2D MRI I and the generator generates the point cloud which was judged by discriminator. Because of the particular data format, conventional loss cannot train this generative network well. Chamfer distance is often used as loss function in point cloud generation. Given two point clouds Y and Y', Chamfer distance is defined as

$$\mathcal{L}_{CD} = \sum_{y' \in Y'} min_{y \in Y} ||y' - y||_2^2 + \sum_{y \in Y} min_{y' \in Y'} ||y - y'||_2^2, \tag{4}$$

where y and y' are points in Y and Y', respectively. Thus, we define the loss function of encoder and generator as

$$\mathcal{L}_{E-G} = \lambda_1 \mathcal{L}_{KL} + \lambda_2 \mathcal{L}_{CD} - \mathbb{E}_{z \sim \mathcal{Z}}[D(G(z))], \tag{5}$$

where $\mathcal{L}_{KL}$ is the Kullback-Leibler divergence, λ_1 and λ_2 are variable parameters and $\mathcal{Z}$ is a Gaussian distribution calculated by encoder.

Meanwhile, the loss which is similar with [40] is used on the discriminator. $\mathcal{L}_D$ is defined as

$$\mathcal{L}_D = \mathbb{E}_{z \sim \mathcal{Z}}[D(G(z))] - \mathbb{E}_{Y \sim \mathcal{R}}[D(Y))] + \lambda_{gp} \mathbb{E}_{\hat{x}}[(||\nabla_{\hat{x}} D(\hat{x})||_2 - 1)^2], \tag{6}$$

where $\hat{x}$ are sampled from line segments between real and fake point clouds, $\mathcal{R}$ represents the real point cloud distribution and λ_{gp} is a weighting parameter.

3 Experiments

Experimental Setting. The proposed model is trained on our in-house dataset with voxel-level segmentation converted into point clouds. The preprocessing was prepared under the professional guidance of doctors. The dataset consists of 317 brain MRIs with AD and 723 healthy brain MRIs. All bone structures of MRIs are removed and the remaining images are registered into format $91 \times 109 \times 91$ by software named FSL. 900 MRIs are randomly selected to construct the training set and the others are used for the test set. We choose some 2D slices of MRIs which were normalized to [0, 1] as the input of the model. 2000 epochs were trained for our generative model and the comparative model, PointOutNet. We use a CPU of Intel Core i9-7960X CPU @ 2.80 GHz $\times$ 32 and a GPU of Nvidia GeForce RTX 2080 Ti for our experiments. We set λ_1 and λ_{gp} to 0.1 and 10, respectively, and use Adam optimisers with an initial learning rate of 1×10^{-4}. Specific to the generator, we adjust training parameters dynamically by increasing λ_2 from 0.1 to 1.

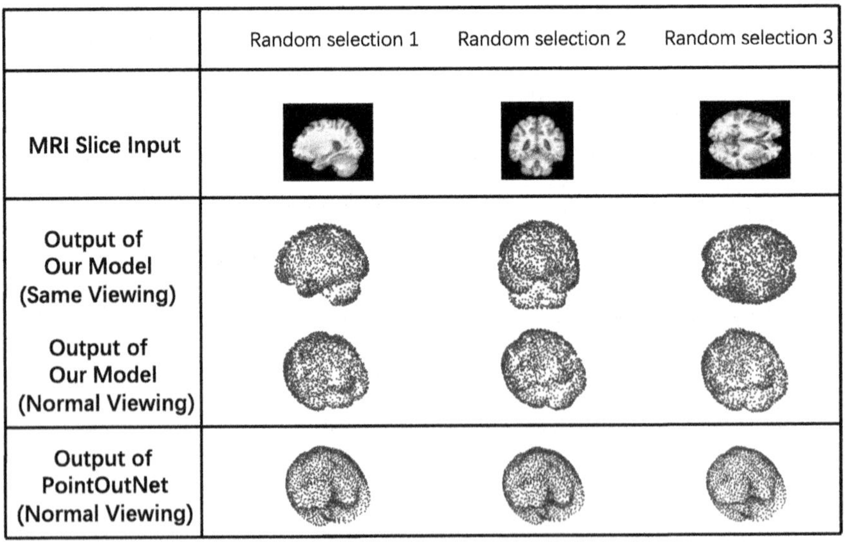

	Random selection 1	Random selection 2	Random selection 3
MRI Slice Input			
Output of Our Model (Same Viewing)			
Output of Our Model (Normal Viewing)			
Output of PointOutNet (Normal Viewing)			

Fig. 2. Intuitive illustrations of some angles of outputs with three views MRI inputs

Qualitative Evaluation. In order to investigate the performance of the proposed model, we conduct some qualitative evaluations and the results are in Fig. 2, which shows three views about the pairs of MRI inputs and point cloud

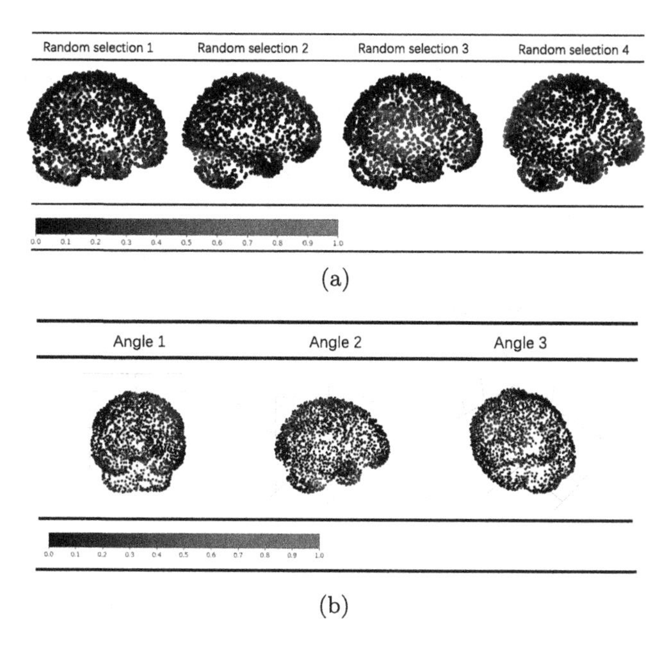

Fig. 3. Intuitive illustrations of (a) The Generated Point Clouds Whose Vertices is Colored by PC-to-PC Error and (b) A Generated Point Cloud Whose Vertices is Colored by PC-to-PC Error in different angles

outputs. In addition, we compared the output of our model with the output of PointOutNet under normal viewing. The point cloud outputs derived from PointOutNet have a highly similar appearance between each other, especially for the description of the overall structure of the brain. The structural features are disappearing and only some details are different. On the contrary, our model has the ability to aggregate point features and generate high-quality and detailed 3D point clouds from slices in multiple directions. Moreover, the outputs of PointOutNet are observed to be unevenly distributed in the 3D space, which is also avoided in our model.

Also, we show the point-by-point accuracy of our generated point cloud in Fig. 3. Figure 3(a) shows the reconstruction effect of different randomly selected objects under the same angle, while Fig. 3(b) shows the error analysis of the same object under different angles. Each vertex is colored by an index named $PC-to-PCerror_{\hat{y}}$ and introduced in [14]. Given two point clouds Y and Y', PC-to-PC error is defined as

$$PC-to-PCError_{\hat{y}} = min_{y \in Y}||y' - y||_2^2 \tag{7}$$

$$PC-to-PCError_{\hat{Y}} = \sum_{y' \in Y'} min_{y \in Y}||y' - y||_2^2 \tag{8}$$

where y and y' are points in Y and Y', respectively. The value on the axis is multiplied by a factor 10^{-4}. We can see that most of the vertices in the reconstruction results have a very small error.

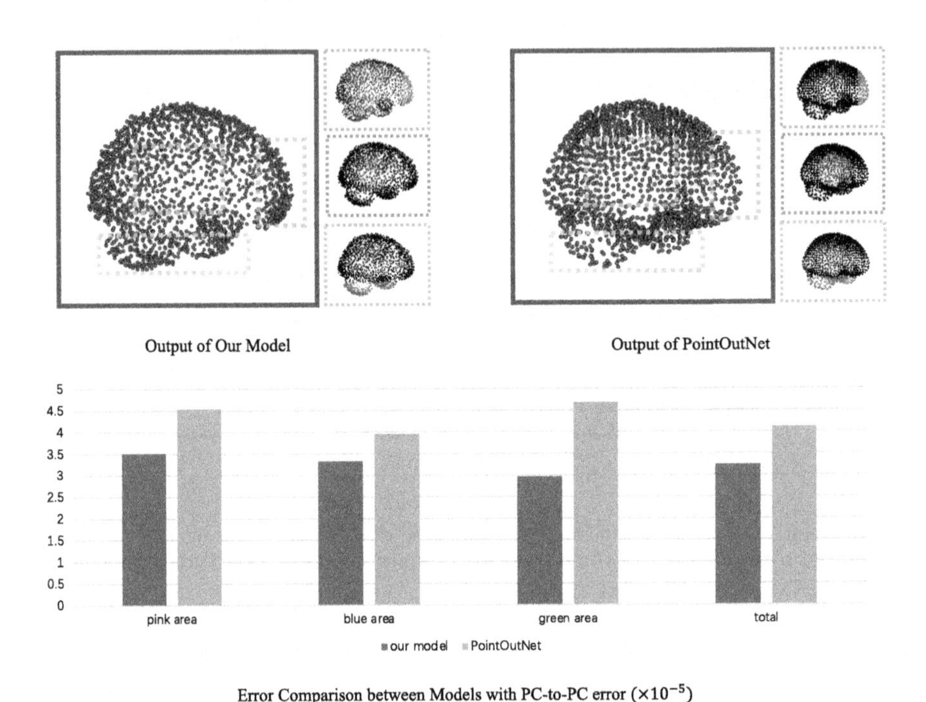

Fig. 4. The PC-to-PC Error of the different area and the Total Point Cloud (Lower is Better). (Color figure online)

Quantitative Measures. PC-to-PC error and Chamfer distance which are based on correlation and the positional relationship between vertices are employed to quantitatively measure the distance between different point clouds. Also, we apply the earth mover's distance (EMD) introduced in [11] to measure the overall distribution gap between the results of each model and the ground truth. The EMD is defined as

$$\mathcal{D}_{EMD} = min_{\phi:Y \rightarrow Y'} \sum_{x \in Y} ||x - \phi(x)||_2 \tag{9}$$

where ϕ is a bijection. In order to comprehensively compare the reconstruction results between our model and PointOutNet, we fairly selected 3 reconstruction key areas and marked them with pink, blue, and green. The relevant results are shown in Fig. 4. Considering that the number of generated points in each area may be different, we use the PC-to-PC error which is calculated for each point in

Table 1. The more specifically PC-to-PC error of the generated Point Cloud in Fig. 4.

Area	Pink Area	Blue Area	Green Area	Total
Our Model	**3.51**	**3.33**	**2.97**	**3.24**
PointOutNet	4.54	3.95	4.68	4.11

Table 2. Quantitative measures for test. In both metrics, the lower the better.

Metric	Our model	Our model without D	PointOutNet
CD	9.605	13.825	12.472
EMD($\times 10^{-1}$)	7.805	11.577	14.329

this evaluation. This indicator reveals that our method has better performance in the reconstruction of all the important areas and the overall point cloud. The more specific values of the related area are in Table 1. We also calculated the chamfer distance and earth mover's distance of our model and PointOutNet throughout the test. Specifically, we set up an additional model which removes the discriminator and uses Chamfer distance as a loss function to train the encoder and generator for a comparative experiment. We expect to compare this model with the proposed model to prove the advantages of the adversarial reconstruction of our model. Table 2 shows results quantitative results of all models using CD and EMD. As can be seen, our model outperforms the compared models with the lowest CD (9.605) and EMD (0.7805), indicating that the accuracy of our model is higher than that of PointOutNet in the test.

4 Conclusion

A shape reconstruction model with the GAN architect and the graph convolutional network is proposed in this paper to relieve the visual restrictions in minimally invasive surgery and robot-guided intervention. 3D point cloud is used as the representation. A novel tree-structured generative mechanism is constructed to transfer features between hidden layers accurately and generate point clouds reliably. To the best of our knowledge, this is the first work that achieves the 3D brain point cloud reconstruction model which uses brain medical images and reconstructs the brain shape. The proposed model has a very competitive performance compared to another generative model. In future work, this method will continue to be developed to solve the problem of visual limitation in minimally invasive surgery and other advanced medical techniques.

Acknowledgment. This work was supported by the National Natural Science Foundations of China under Grant 61872351, the International Science and Technology Cooperation Projects of Guangdong under Grant 2019A050510030, the Distinguished Young Scholars Fund of Guangdong under Grant 2021B1515020019, the Excellent Young Scholars of Shenzhen under Grant RCYX20200714114641211 and Shenzhen Key Basic Research Project under Grant JCYJ20200109115641762.

References

1. van Tonder, L., Burn, S., Iyer, A., et al.: Open resection of hypothalamic hamartomas for intractable epilepsy revisited, using intraoperative MRI. Child's Nerv. Syst. **34**(9), 1663–1673 (2018)
2. Thomas, J.G., Al-Holou, W.N., de Almeida Bastos, D.C., et al.: A novel use of the intraoperative MRI for metastatic spine tumors: laser interstitial thermal therapy for percutaneous treatment of epidural metastatic spine disease. Neurosurg. Clin. **28**(4), 513–524 (2017)
3. Wu, J., Zhang, C., Xue, T., et al.: Learning a probabilistic latent space of object shapes via 3D generative-adversarial modeling. In: 29th Conference and Workshop on Neural Information Processing Systems, pp. 82–90 (2016)
4. Wang, N., Zhang, Y., Li, Z., Fu, Y., Liu, W., Jiang, Y.-G.: Pixel2Mesh: generating 3D mesh models from single RGB images. In: Ferrari, V., Hebert, M., Sminchisescu, C., Weiss, Y. (eds.) ECCV 2018. LNCS, vol. 11215, pp. 55–71. Springer, Cham (2018). https://doi.org/10.1007/978-3-030-01252-6_4
5. Qi, C., Su, H., Mo, K., et al.: PointNet: deep learning on point sets for 3D classification and segmentation. In: Proceedings of the IEEE Conference on Computer Vision and Pattern Recognition, pp. 652–660 (2017)
6. Qi, C., Yi, L., Su, H., et al.: PointNet++: deep hierarchical feature learning on point sets in a metric space. arXiv preprint arXiv:1706.02413 (2017)
7. Yang, Y., Feng, C., Shen, Y., et al.: FoldingNet: point cloud auto-encoder via deep grid deformation. In: Proceedings of the IEEE Conference on Computer Vision and Pattern Recognition 2018, pp. 206–215 (2018)
8. Sekuboyina, A., Rempfler, M., Valentinitsch, A., Loeffler, M., Kirschke, J.S., Menze, B.H.: Probabilistic point cloud reconstructions for vertebral shape analysis. In: Shen, D., et al. (eds.) MICCAI 2019. LNCS, vol. 11769, pp. 375–383. Springer, Cham (2019). https://doi.org/10.1007/978-3-030-32226-7_42
9. Valsesia, D., Fracastoro, G., Magli, E.: Learning localized generative models for 3D point clouds via graph convolution. In: International Conference on Learning Representations 2019 (2019)
10. Shu, D. W., Park, S. W., Kwon, J.: 3D point cloud generative adversarial network based on tree structured graph convolutions. In: Proceedings of the IEEE/CVF International Conference on Computer Vision 2019, pp. 3859–3868 (2019)
11. Achlioptas, P., Diamanti, O., Mitliagkas, I., et al.: Learning representations and generative models for 3D point clouds. In: International Conference on Machine Learning 2018, pp. 40–49 (2018)
12. Jiang, L., Shi, S., Qi, X., Jia, J.: GAL: geometric adversarial loss for single-view 3D-object reconstruction. In: Ferrari, V., Hebert, M., Sminchisescu, C., Weiss, Y. (eds.) ECCV 2018. LNCS, vol. 11212, pp. 820–834. Springer, Cham (2018). https://doi.org/10.1007/978-3-030-01237-3_49
13. Fan, H., Su, H., Guibas, L.J.: A point set generation network for 3D object reconstruction from a single image. In: Proceedings of the IEEE Conference on Computer Vision and Pattern Recognition 2017, pp. 605–613 (2017)
14. Zhou, X.-Y., Wang, Z.-Y., Li, P., Zheng, J.-Q., Yang, G.-Z.: One-stage shape instantiation from a single 2D image to 3D point cloud. In: Shen, D., et al. (eds.) MICCAI 2019. LNCS, vol. 11767, pp. 30–38. Springer, Cham (2019). https://doi.org/10.1007/978-3-030-32251-9_4

15. Lee, S.-L., Chung, A., Lerotic, M., Hawkins, M.A., Tait, D., Yang, G.-Z.: Dynamic shape instantiation for intra-operative guidance. In: Jiang, T., Navab, N., Pluim, J.P.W., Viergever, M.A. (eds.) MICCAI 2010. LNCS, vol. 6361, pp. 69–76. Springer, Heidelberg (2010). https://doi.org/10.1007/978-3-642-15705-9_9
16. Yu, W., Lei, B., Ng, M, et al.: Tensorizing GAN with high-order pooling for Alzheimer's disease assessment. IEEE Trans. Neural Netw. Learn. Syst. (2021)
17. Hu, S., Yu, W., Chen, Z., et al.: Medical image reconstruction using generative adversarial network for alzheimer disease assessment with class-imbalance problem. In: 2020 IEEE 6th International Conference on Computer and Communications (ICCC), pp. 1323–1327 (2020)
18. Lei, B., Xia, Z., Jiang, F., et al.: Skin lesion segmentation via generative adversarial networks with dual discriminators. Med. Image Anal. **64**, 101716 (2020)
19. Wang, S., Shen, Y., Shi, C., et al.: Skeletal maturity recognition using a fully automated system with convolutional neural networks. IEEE Access **6**, 29979–29993 (2018)
20. Wang, S., Wang, H., Shen, Y., et al.: Automatic recognition of mild cognitive impairment and alzheimers disease using ensemble based 3D densely connected convolutional networks. In: 2018 17th IEEE International Conference on Machine Learning and Applications, pp. 517–523 (2018)
21. Wang, S., Shen, Y., Zeng, D., et al.: Bone age assessment using convolutional neural networks. In: 2018 International Conference on Artificial Intelligence and Big Data, pp. 175–178 (2018)
22. Wang, S., Hu, Y., Shen, Y., et al.: Classification of diffusion tensor metrics for the diagnosis of a myelopathic cord using machine learning. Int. J. Neural Syst. **28**(02), 1750036 (2018)
23. Wang, S., Li, X., Cui, J., et al.: Prediction of myelopathic level in cervical spondylotic myelopathy using diffusion tensor imaging. J. Magn. Reson. Imaging **41**(6), 1682–1688 (2015)
24. Zeng, D., Wang, S., Shen, Y., et al.: A GA-based feature selection and parameter optimization for support tucker machine. Procedia Comput. Sci. **111**, 17–23 (2017)
25. Wu, K., Shen, Y., Wang, S.: 3D convolutional neural network for regional precipitation nowcasting. J. Image Signal Process. **7**(4), 200–212 (2018)
26. Wang, S., Wang, X., Hu, Y., et al.: Diabetic retinopathy diagnosis using multichannel generative adversarial network with semisupervision. IEEE Trans. Autom. Sci. Eng. **18**(2), 574–585 (2020)
27. Hu, S., Shen, Y., Wang, S., et al.: Brain MR to PET synthesis via bidirectional generative adversarial network. In: International Conference on Medical Image Computing and Computer-Assisted Intervention, pp. 698–707 (2020)
28. Hackel, T., Wegner, J.D., Schindler, K.: Contour detection in unstructured 3D point clouds. In: Proceedings of the IEEE Conference on Computer Vision and Pattern Recognition, pp. 1610–1618 (2016)
29. Li, Y., Pirk, S., Su, H., et al.: FPNN: field probing neural networks for 3D data. In: 29th Conference and Workshop on Neural Information Processing Systems, pp. 307–315 (2016)
30. Zhou, Y., Tuzel, O.: VoxelNet: end-to-end learning for point cloud based 3D object detection. In: Proceedings of the IEEE Conference on Computer Vision and Pattern Recognition, pp. 4490–4499 (2018)
31. Goodfellow, I., Pouget-Abadie, J., Mirza, M., et al.: Generative adversarial nets. In: 27th Conference and Workshop on Neural Information Processing Systems, pp. 2672–2680 (2014)

32. Zhu, J., Park, T., Isola, P., et al.: Unpaired image-to-image translation using cycle-consistent adversarial networks. In: Proceedings of the IEEE International Conference on Computer Vision, pp. 2223–2232 (2017)
33. Radford, A., Metz, L., Chintala, S.: Unsupervised representation learning with deep convolutional generative adversarial networks. arXiv preprint arXiv:1511.06434 (2015)
34. Mo, L., Wang, S.: A variational approach to nonlinear two-point boundary value problems. Nonlin. Anal. Theory Methods Appl. **71**(12), e834–e838 (2009)
35. Wang, S.: A variational approach to nonlinear two-point boundary value problems. Comput. Math. Appl. **58**(11–12), 2452–2455 (2009)
36. Wang, S., He, J.: Variational iteration method for a nonlinear reaction-diffusion process. Int. J. Chem. React. Eng. **6**(1) (2008)
37. Kipf, T., Welling, M.: Semi-supervised classification with graph convolutional networks. arXiv preprint arXiv:1609.02907 (2016)
38. Schlichtkrull, M., Kipf, T., Bloem, P., et al.: Modeling relational data with graph convolutional networks. In: European Semantic Web Conference, pp. 593–607 (2018)
39. Wu, F., Souza, A., Zhang, T., et al.: Simplifying graph convolutional networks. In: International Conference on Machine Learning, pp. 6861–6871 (2019)
40. Gulrajani, I., Ahmed, F., Arjovsky, M., et al.: Improved training of wasserstein GANs. In: 30th Conference and Workshop on Neural Information Processing Systems, pp. 5767–5777 (2017)
41. He, K., Zhang, X., Ren, S., et al.: Deep residual learning for image recognition. In: Proceedings of the IEEE Conference on Computer Vision and Pattern Recognition 2016, pp. 770–778 (2016)

3D-SceneCaptioner: Visual Scene Captioning Network for Three-Dimensional Point Clouds

Qiang Yu[1,2], Xianbing Pan[3], Shiming Xiang[1,2], and Chunhong Pan[1(✉)]

[1] National Laboratory of Pattern Recognition, Institute of Automation, Chinese Academy of Sciences, Beijing 100190, China
{qiang.yu,smxiang,chpan}@nlpr.ia.ac.cn

[2] School of Artificial Intelligence, University of Chinese Academy of Sciences, Beijing 100049, China

[3] College of Mobile Telecommunications, Chongqing 401520, China

Abstract. Currently, image captioning has been widely studied with the development of deep neural networks. However, seldom work has been conducted to develop captioning models for three-dimensional (3D) visual data, for example, point clouds, which are now popularly employed for vision perception. Technically, most of these models first project the 3D shapes into multiple images, and then use the existing or similar framework for image captioning models to fulfill the task. Consequently, within such a technical framework, a large amount of useful information hidden in 3D vision is inevitably lost. In this paper, a captioning model for visual scenes directly based on point clouds is proposed. First, a deep model with densely connected point convolution is developed to extract visual features directly on point clouds, and the multi-task learning method is adopted to improve the visual features. Then, the visual features are converted into sentences through a caption generation module. As a whole, an end-to-end model is constructed for the task of 3D scene captioning. This model makes full use of the rich semantic information in point clouds, and generate more accurate captions. Since there do not exist large-scale datasets for this task, in this paper two new datasets are created on existing point cloud datasets by manually labeling captions. Comprehensive experiments conducted on three datasets (including one public benchmark) indicate the effectiveness of our model.

Keywords: Scene captioning · Three-dimensional vision · Point cloud.

1 Introduction

Visual scene captioning is an intuitive way to realize the scene understanding task, which aims at generating reasonable and compact description sentences (or

This research was supported by the National Key Research and Development Program of China under Grant No. 2018AAA0100400, and the National Natural Science Foundation of China under Grants 62071466, 62076242, and 61976208.

H. Ma et al. (Eds.): PRCV 2021, LNCS 13020, pp. 275–286, 2021.
https://doi.org/10.1007/978-3-030-88007-1_23

called captions) about the contents in the scene. Visual scene captioning methods can be widely used in robots, automatic navigation, self-driving and blind guiding. Thus, researches on visual scene captioning are of great significance to the development of intelligent technologies and their applications.

Point clouds are a kind of widely used representation of 3D visual data. Compared with images, point clouds contain more abundant information, including 3D coordinates, normals, geometries, and support multi-view observations. Therefore, many common problems in images, such as occlusion and data missing, can be easily solved by using point clouds.

Because of the rich information in point clouds and its great significance to visual scene understanding, researchers have proposed various point cloud processing methods in recent years [12,13,16,19,20]. For the scene captioning task based on point clouds, there are few related methods. Existing methods usually project 3D shapes to images, and then use image captioning methods to generate captions [6,7]. This kind of methods cannot make full use of the advantage of point clouds, so the quality of generated captions is limited.

In order to solve these problems, a visual scene captioning model directly based on point clouds is proposed in this paper. In this model, instead of projecting to multiple images, a feature extraction module with dense connections is designed to extract the high-level visual features of point clouds, and then a caption generation module is adopted to convert visual features to word sequence as captions. This model can take advantage of the rich information in point clouds to mine the complex semantic information in visual scenes, which is conducive to generating more accurate captions. At the same time, in order to reduce the difficulty in 3D visual feature learning, the Multi-Task Learning (MTL) method is adopted by adding a semantic segmentation module as an auxiliary task. By joint training of the semantic segmentation module and the caption generation module, the proposed model is able to learn for extracting effective visual features and accelerate the convergence.

In order to verify the capability of the proposed model, 3D point cloud captioning datasets are needed. However, there are only a few open point cloud captioning datasets, and they contain only small 3D shapes but no large-scale scene point clouds. Therefore, in this paper, two large-scale scene point cloud captioning datasets are created by manually labeling caption annotations on existing point cloud datasets. Extensive comparative experiments are conducted on these datasets to verify the performance of our model.

To sum up, the innovations of this paper are as follows:

1. A scene captioning model based on 3D point clouds is proposed, which consists of a feature extraction module and a caption generation module.
2. The Multi-Task Learning method is adopted with a semantic segmentation module to extract more effective high-level visual features.
3. Two large-scale point cloud captioning datasets are created, which can be used to train and evaluate scene captioning models based on point clouds.
4. Comprehensive experiments show that the proposed model can generate more accurate captions for point clouds than existing methods.

2 Related Work

2.1 Point Cloud Processing

Recently, with the rapid development of deep learning, lots of methods have been proposed to process point clouds. PointNet [12] is a pioneering work. It takes point clouds as input, learns point features independently through multiple Multi-Layer Perceptrons (MLPs), and extracts global features through a max-pooling layer. Based on PointNet, a hierarchical model PointNet++ [13] is then developed. By integrating sampling, grouping and PointNet based layers, it can capture fine geometric information and abstract local features from the neighbor of each point, thus greatly improving the ability of feature extraction.

The above-mentioned methods are mainly based on point-wise MLPs. In contrast, many other methods use 3D point cloud convolution operations similar to image convolution. Kernel Point Convolution (KPConv) [20] defines a set of learnable convolution kernel points to realize the convolution operation on the neighbor of a given point. In addition, the model can be used to construct a fully convolutional network for point cloud segmentation by up-sampling operations.

2.2 Scene Captioning

Most of the scene captioning methods are developed for images. The encoder-decoder structure is a popular framework in image captioning models [17,26]. For example, Neural Image Caption (NIC) [24] uses Convolutional Neural Network (CNN) to extract visual features, and takes the output of the last hidden layer as the initial input of Long-Short Term Memory (LSTM) to generate captions. Another typical kind of model is based on the attention mechanism, which dynamically focuses on partial contents of the input image when generating captions. For example, Neural Image Caption with Attention (NIC-A) [25] uses two different mechanisms, random hard attention and deterministic soft attention, to generate attention. Meshed-Memory Transformer [4] uses the self-attention mechanism to learn the multi-level representation of the relationships between image regions integrating prior knowledge.

For 3D vision, there exist seldom researches on 3D visual scene captioning. Y2Seq2Seq [7] adopts multi-view images as the representation of 3D shapes, which learns 3D shape features by combining the global features of each view, and uses Recurrent Neural Network (RNN) as encoder and decoder to model visual and text features together, so as to generate captions for 3D shapes. Although that model can generate reasonable captions, it cannot capture details of local parts of 3D shapes. To solve this problem, an improved version named ShapeCaptioner [6] is then developed. It first detects important parts in multi-view, and then generates more accurate captions by using RNN. However, as a result of projecting the 3D shapes into multiple images, those existing methods inevitably waste a large amount of useful information hidden in 3D vision, thus the quality of generated captions are limited.

3 Method

This section describes the proposed visual scene captioning model based on 3D point clouds, shown in Fig. 1, which is called 3D-SceneCaptioner.

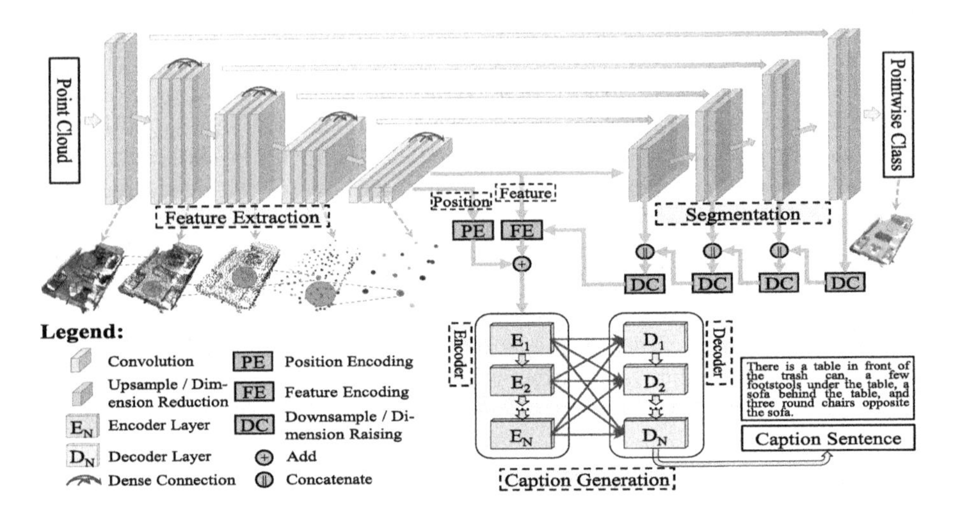

Fig. 1. The overall architecture of the proposed 3D-SceneCaptioner model.

3.1 Feature Extraction Module

The feature extraction module takes the 3D coordinates of all points in the point cloud as input, and extracts local features by multi-layer point convolution based on radius neighbor search.

Suppose the number of points in the point cloud is N. Let $\mathbf{x}_i \in \mathbb{R}^3$ be the coordinate of point i, and $\mathbf{f}_i \in \mathbb{R}^D$ be the corresponding D-dimensional feature vector. Thus, the coordinate set and feature set of all points are $\mathbf{P} = \{\mathbf{x}_i | 1 \leq i \leq N\} \in \mathbb{R}^{N\times 3}$ and $\mathbf{F} = \{\mathbf{f}_i | 1 \leq i \leq N\} \in \mathbb{R}^{N\times D}$, respectively. Then, the point convolution is defined as the weighted sum of point features:

$$g\left(\mathbf{x}|\mathbf{F}\right) = \sum\nolimits_{\mathbf{x}_i \in N_x} \text{kernel}\left(\mathbf{x}_i|\mathbf{x}\right)\mathbf{f}_i, \tag{1}$$

where $\mathbf{x}$ is the center point of the convolution kernel, N_x denotes the neighbor point set of $\mathbf{x}$, and kernel$(\cdot)$ represents the convolution kernel function. According to radius neighbor search, $N_x = \{\mathbf{x}_i \in \mathbf{P} | \, \|\mathbf{x}_i - \mathbf{x}\| \leq r\}$, where $r \in \mathbb{R}$ is the radius. Referring to the implementation in KPConv [20], a number of points in the neighbor space are selected uniformly as the carrier of the convolution kernel function kernel$(\cdot)$. These points are called kernel points, marked as K_x. Then, the convolution kernel function kernel$(\cdot)$ is defined as:

$$\text{kernel}\left(\mathbf{x}_i|\mathbf{x}\right) = \sum\nolimits_{\mathbf{x}_k \in K_x} \max\left(1 - \frac{\|\mathbf{x}_i - (\mathbf{x}_k + \Delta_k)\|}{\lambda}, 0\right)\mathbf{W}_k, \tag{2}$$

where $\mathbf{W}_k \in \mathbb{R}^{D_i \times D_o}$ is the feature mapping matrix, λ represents the influence range of convolution kernel points. When the distance between a neighbor point and a kernel point is longer than λ, the neighbor point does not participate in the calculation of this kernel point. Δ_k is the coordinate offset of the kernel point $\mathbf{x}_k$, which is a learnable parameter, and can make the point convolution adapt to extracting features for objects in different scales and shapes.

Based on the definition of the point convolution, the feature extraction module is designed as a series of down-sampling and convolution blocks. First, the input point cloud is down-sampled by grid-sampling. Denote the initial sampling cell size by d_0, and then the cell size grows twice each time, that is $d_i = 2d_{i-1}$. Next, higher dimension features are extracted through two point convolution layers. Then, every four point convolution layers with a down-sampling operation constitute a convolution block. To improve the capability of the model, dense connections are added between each two nonadjacent layers in each convolution block by concatenating feature maps. The output of the last convolution block is used as the extracted high-level features for subsequent modules. This module is demonstrated in the top left corner of Fig. 1.

3.2 Semantic Segmentation Module

In order to train the feature extraction module and make it reach the optimal state of extracting effective features, the semantic segmentation module is connected following the feature extraction module. The output of this module is point-wise semantic categories, which help to understand what kind of semantic contents exist and where are they in the scene.

Corresponding to the feature extraction module, the semantic segmentation module consists of four up-sampling layers, and each layer is followed by a feature dimension reduction layer. At the end, a softmax function is used to calculate semantic category probability of each point. The cross entropy loss function is adopted to guide the learning process of the feature extraction module through MTL. This module is demonstrated in the top right corner of Fig. 1.

3.3 Caption Generation Module

The caption generation module adopts an encoder-decoder architecture based on Transformer [21]. To deeply mine the semantic characteristics, the input features of this module consist of two parts: features from the feature extraction module and from the semantic segmentation module. The former is the output feature map of the last point convolution layer of the feature extraction module and the corresponding 3D coordinates of each feature. The latter is the fusion of the output feature maps of all feature dimension reduction layers of the semantic segmentation module, which helps to enhance the capability of the features. The number of features of the latter is the same as that of the former, and the corresponding 3D point coordinates are also the same. Let $\mathbf{X}_f \in \mathbb{R}^{N_f \times C_f}$ be the output features of the feature extraction module (where N_f is the number of feature points, and C_f is the feature dimension), $\mathbf{X}_p \in \mathbb{R}^{N_f \times 3}$ is the corresponding 3D coordinates, $\mathbf{X}_f^1 \in \mathbb{R}^{N_f^1 \times C_f^1}$, $\mathbf{X}_f^2 \in \mathbb{R}^{N_f^2 \times C_f^2}$, $\mathbf{X}_f^3 \in \mathbb{R}^{N_f^3 \times C_f^3}$ and $\mathbf{X}_f^4 \in \mathbb{R}^{N_f^4 \times C_f^4}$ represent the output feature map of each feature dimension reduction layer in the semantic segmentation module. Then, the input features of the caption generation module are calculated as follows:

$$\widehat{\mathbf{X}_f^4} = \text{Linear}^4\left(\text{Down}^4\left(\mathbf{X}_f^4\right)\right), \tag{3}$$

$$\widehat{\mathbf{X}_f^j} = \text{Linear}^j\left(\text{Down}^j\left(\left[\mathbf{X}_f^j, \widehat{\mathbf{X}_f^{j+1}}\right]\right)\right), j \in 3, 2, 1, \tag{4}$$

$$\widehat{\mathbf{X}_f} = \left[\mathbf{X}_f, \widehat{\mathbf{X}_f^1}\right], \tag{5}$$

$$\mathbf{PE} = \text{PE}\left(\mathbf{X}_p\right) \in \mathbb{R}^{N_f \times d_m}, \quad \mathbf{FE} = \text{FE}\left(\widehat{\mathbf{X}_f}\right) \in \mathbb{R}^{N_f \times d_m}, \tag{6}$$

$$\mathbf{X} = \mathbf{PE} + \mathbf{FE} \in \mathbb{R}^{N_f \times d_m}, \tag{7}$$

where Down^n and Linear^n represent the new down-sampling operation and feature dimension raising operation corresponding to the n-th feature dimension reduction layer in the semantic segmentation module, respectively, $[\cdot, \cdot]$ stands for the concatenation operation, $\text{PE}(\cdot)$ and $\text{FE}(\cdot)$ are the positional embedding function and the feature embedding function, respectively, which are implemented by single-layer linear mapping. Finally, the calculation result $\mathbf{X}$ is the input feature of the caption generation module.

Because the transformer model cannot effectively distinguish the position relationships among input features, it is necessary to integrate the corresponding position into the features. It should be noted that each feature needs to be coupled with a unique position index before calculating the position embedding. Different from the sequential indexing method in images, a multi-level position indexing method is designed based on octree for point clouds, as shown in Fig. 2. Let the maximum depth of the octree be d_o. According to 3D coordinates of each feature, the position index from 0 to 7 is determined at each level of the octree. Totally $d_o + 1$ position indices are used to calculate position embedding vectors for each feature by the position embedding function $\text{PE}(\cdot)$, and then the vectors are added as the final position embedding vector of the feature.

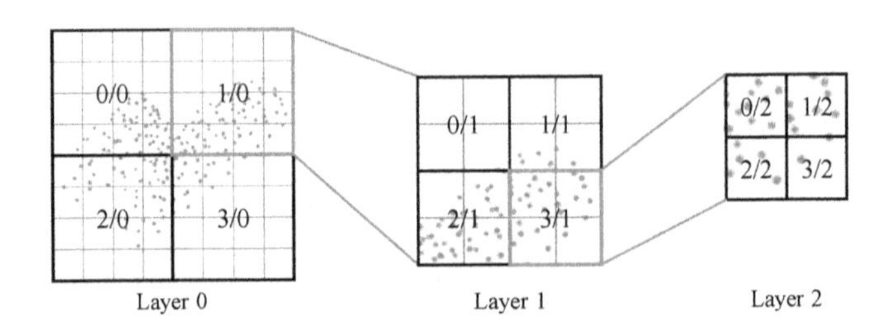

Fig. 2. Demonstration of the positional indexing method for point cloud features.

Inspired by Meshed-Memory Transformer [4], the caption generation module is composed of standard encoder and decoder based on self-attention mechanism. The encoder includes six identical encoder layers connected in sequence. Similarly, the decoder consists of six identical decoder layers. This module is demonstrated in the bottom right corner of Fig. 1.

4 Experiments

4.1 Datasets

To the best of our knowledge, currently 3D-Text [3] is the only related 3D captioning dataset. However, the size of the point clouds in this dataset is very small, which cannot fully verify the capability of our model. Therefore, two large-scale scene point cloud captioning datasets are created based on existing datasets by manually labeling captions: ScanNet-Caption dataset and NYUDepth-Caption dataset. The original datasets of the two datasets (ScanNet [5] and NYUDepth [10]) contain semantic segmentation annotations. Table 1 demonstrates the details.

Table 1. Details of large-scale 3D point cloud captioning datasets.

Dataset	Point cloud scale	Point cloud count		Caption count	
	Points per cloud	Total	Train/Val/Test	Total	Each
3D-Text	2715	15038	12,030/1504/1504	75,344	5
ScanNet-Caption (ours)	146,995	1100	900/100/100	5500	5
NYUDepth-Caption (ours)	285,200	1100	900/100/100	5500	5

4.2 Experimental Settings

The model proposed in this paper contains two outputs: the semantic segmentation categories used to guide the optimization of the feature extraction module, and the captions as the original goal. In the experiments, the MTL training method and the supervised learning method are used to jointly train the model on the above two outputs. First, the semantic segmentation module and the feature extraction module are trained together. The Stochastic Gradient Descent (SGD) with momentum is adopted to optimize the point-wise cross entropy loss function. The batch size is set to 16 and the momentum is 0.98. The learning rate increases linearly from 10^{-7} at beginning to 10^{-2} at the fifth epoch, and then decreases exponentially by 0.1 times every 40 epochs. Then, after 50 epochs, the semantic segmentation module is fixed and the caption generation module starts to be trained. Adam optimizer is used to optimize the word-wise cross entropy loss function, and the batch size is set to 50. The learning rate strategy is similar to that of semantic segmentation module, but increases from 10^{-7} to 10^{-3}. Next, after 100 epochs, the caption generation module is fixed, and the module of semantic segmentation starts to be trained, and so on. The learning rate of the same module decreases by 0.1 times each turn. After several turns, the whole model reaches the basic convergence state.

According to the general experience of image captioning model training, reinforcement learning training method can be used to improve the performance after reaching a stable state through cross entropy loss function training. In this paper, the CIDEr-D [22] value, which is closest to human evaluation criteria, is used as a reward. For each input point cloud, the reward value is calculated once after a complete caption is generated, and the model is optimized iteratively by reinforcement learning method [14] until the final convergence state is reached.

4.3 Comparative Analysis

The caption generation results are compared on 3D-Text dataset, ScanNet-Caption dataset and NYUDepth-Caption dataset respectively. For our proposed model, RGB colors are employed as the initial point features for the point convolution in the feature extraction module. All other methods are developed for multi-view projected images, and results of those methods are obtained by averaging the features of multiple images.

Table 2. Comparison between the proposed 3D-SceneCaptioner model and other methods on the 3D-Text dataset. "B-1" to "B-4" represent BLUE [11], "M" stands for METEOR [2], "R" stands for "ROUGE-L" [9], and "C" denotes "CIDEr-D" [22].

Method	B-1	B-2	B-3	B-4	M	R	C
NIC [24]	49.4	33.8	25.1	21.4	20.9	38.1	30.1
V2T [23]	67.6	43.0	26.0	15.0	21.0	45.0	27.0
GIF2T [18]	61.0	35.0	21.0	12.0	16.0	36.0	14.0
SLR [15]	40.0	17.0	08.0	04.0	11.0	24.0	05.0
Y2Seq2Seq [7]	80.0	65.0	54.0	46.0	30.0	56.0	72.0
ShapeCaptioner [6]	89.9	83.6	78.5	74.9	45.6	75.6	144.4
3D-SceneCaptioner (ours)	**92.1**	**91.5**	**90.2**	**88.9**	**52.4**	**87.8**	**153.2**

3D-Text Dataset. The results of caption generation on the 3D-Text dataset are shown in Table 2. The results show that the proposed model outperforms other methods on all metrics, which indicates the advantages of our model.

ScanNet-Caption Dataset and NYUDepth-Caption Dataset. the results of caption generation on ScanNet-Caption dataset and NYUDepth-Caption dataset are shown in Tables 3 and 4. It can be seen from the experimental results that our 3D-SceneCaptioner model also performs well on these two large-scale datasets, which indicates the strong feature extraction and caption generation ability of our model.

Table 3. Comparison between the proposed 3D-SceneCaptioner model and other methods on the ScanNet-Caption dataset.

Method	B-1	B-2	B-3	B-4	M	R	C
NIC [24]	41.6	23.9	14.2	09.0	18.0	39.6	22.6
SCST [14]	73.0	51.0	27.7	10.4	24.0	49.7	71.7
Up-Down [1]	73.7	52.0	29.0	11.5	24.4	50.3	75.1
MAD+SAP [8]	71.7	57.6	46.3	37.3	35.0	70.7	78.3
RUDet+MAD+SAP [27]	72.0	58.2	47.1	38.4	35.4	71.1	80.8
3D-SceneCaptioner (ours)	**79.5**	**69.7**	**62.1**	**56.2**	**42.2**	**78.7**	**128.4**

Table 4. Comparison between the proposed 3D-SceneCaptioner model and other methods on the NYUDepth-Caption dataset.

Method	B-1	B-2	B-3	B-4	M	R	C
NIC [24]	47.6	29.4	18.5	12.2	21.0	45.7	28.7
SCST [14]	73.8	52.2	29.2	11.7	24.5	50.4	75.7
Up-Down [1]	74.3	53.1	30.4	12.5	24.8	50.9	79.2
MAD+SAP [8]	72.7	59.2	48.4	39.9	35.9	71.7	84.5
RUDet+MAD+SAP [27]	74.0	61.1	50.8	42.8	36.9	72.9	92.1
3D-SceneCaptioner (ours)	**81.8**	**73.1**	**66.3**	**61.2**	**44.5**	**81.1**	**142.9**

Table 5. Ablation study results of the proposed 3D-SceneCaptioner model.

Model	MT	DF	SF	B-4	M	R	C
A				51.2	40.2	76.5	114.9
B	✓			54.3	41.3	77.7	122.8
C	✓	✓		55.5	41.8	78.3	125.9
D	✓	✓	✓	**56.2**	**42.2**	**78.7**	**128.4**

4.4 Ablation Study

In order to verify the effectiveness of the proposed innovations, ablation experiments are conducted on ScanNet-Caption dataset, and the experimental results are shown in Table 5, where "MT" (Multi-Task) represents the MTL method corresponding to the semantic segmentation module, "DF" (Density Feature) represents the dense connection in each convolution block in the feature extraction module, "SF" (Segmentation Feature) is the fusion feature returned by each feature dimension reduction layer in the semantic segmentation module. According to the CIDEr-D values in the experimental results, the most significant improvement on the model performance is made by the MTL method (increases by 4.5 points), and the other two innovations also improve the performance by 3.1 and 2.5 points.

GT: There is a table in the middle of the living room, two chairs in front of the table, some books on the table, and some pictures on the wall.
GE: There is a table in the living room, two chairs beside the table, some books, and some pictures on the wall.

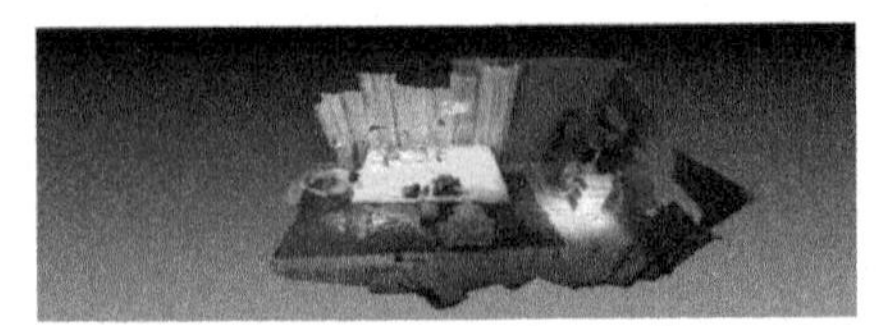

GT: There is a white table with some boxes, a tray standing next to the table, a hanger in the corner, and some white boxes next to the hanger.
GE: There is a white table with some boxes, a tray next to the table, and some boxes next to the hanger.

GT: There is a yellow table with some chairs around the table, and a small yellow table with some red sofas around the small yellow table.
GE: There is a yellow table with some chairs around, and a small yellow, and some red sofas.

GT: There is a bookcase in the corner, a white round table in the other corner, some chairs around the table, and some pictures on the wall.
GE: There is a bookcase in the corner, a white table in the corner, some chairs, and some pictures on the wall.

Fig. 3. Generated captions from the proposed 3D-SceneCaptioner on the ScanNet-Caption dataset ("GT" for ground truth, "GE" for generated caption).

GT: There is a big table in the room, seven chairs beside the table, a projector on the ceiling of the roof, and a white board on the wall in front of the table.
GE: There is a big table in the room, some chairs beside the table, a projector, and a white board on the wall.

GT: There are a few bottles on the dark brown table, a black chair next to the table, a drinking fountain next to the chair, and a brown door next to the drinking fountain.
GE: There are some bottles on the dark brown table, a black chair next to the table, a drinking fountain, and a brown door.

GT: There is a purple trash can on the ground, a table next to the trash can, some debris on the table, and a thick brown tube on the side of the table.
GE: There is a trash can on the ground, a table next to the trash can, and a thick brown tube next to the table.

GT: There are five red chairs on the side of the yellow table, a projection screen in front of the table, and a white board with some words on it.
GE: There are five red chairs beside the yellow table, a white board in front of the table.

Fig. 4. Generated captions from the proposed 3D-SceneCaptioner on the NYUDepth-Caption dataset ("GT" for ground truth, "GE" for generated caption).

4.5 Qualitative Results and Visualization

Figures 3 and 4 show examples of captions generated by the proposed model on two datasets. It is seen that the captions generated by the model can cover the main content of the scenes, which can accurately describe the colors and other characteristics of each object and the position relationships among them.

5 Conclusion

A visual scene captioning model based on 3D point clouds has been proposed in this paper, which is named as 3D-SceneCaptioner. Our model extracts high-level visual features from point clouds by point convolution with dense connections, and then generates captions by the caption generation module based on encoder-decoder architecture. At the same time, the semantic segmentation module is added to guide the training of the feature extraction through Multi-Task Learning. Experiments have shown that our model outperforms existing methods.

At present, the computational cost of the proposed model is not well considered. In the future, we would like to reduce the quantity of parameters, develop lightweight models, and apply them to mobile devices.

References

1. Anderson, P., et al.: Bottom-up and top-down attention for image captioning and visual question answering. In: IEEE Conference Computer Vision and Pattern Recognition, pp. 6077–6086 (2018)
2. Banerjee, S., Lavie, A.: METEOR: an automatic metric for MT evaluation with improved correlation with human judgments. In: Workshop on Intrinsic and Extrinsic Evaluation Measures for Machine Translation and/or Summarization, pp. 65–72 (2005)
3. Chen, K., Choy, C.B., Savva, M., Chang, A.X., Funkhouser, T.A., Savarese, S.: Text2shape: generating shapes from natural language by learning joint embeddings. In: Asian Conference on Computer Vision, pp. 100–116 (2018)
4. Cornia, M., Stefanini, M., Baraldi, L., Cucchiara, R.: Meshed-memory transformer for image captioning. In: IEEE Conference on Computer Vision and Pattern Recognition, pp. 10575–10584 (2020)
5. Dai, A., Chang, A.X., Savva, M., Halber, M., Funkhouser, T.A., Nießner, M.: Scannet: richly-annotated 3d reconstructions of indoor scenes. In: IEEE Conference on Computer Vision and Pattern Recognition, pp. 2432–2443 (2017)
6. Han, Z., Chen, C., Liu, Y., Zwicker, M.: Shapecaptioner: generative caption network for 3d shapes by learning a mapping from parts detected in multiple views to sentences. In: ACM International Conference on Multimedia, pp. 1018–1027 (2020)
7. Han, Z., Shang, M., Wang, X., Liu, Y., Zwicker, M.: Y2seq2seq: cross-modal representation learning for 3d shape and text by joint reconstruction and prediction of view and word sequences. In: Conference on Artificial Intelligence, pp. 126–133 (2019)
8. Huang, Y., Chen, J., Ouyang, W., Wan, W., Xue, Y.: Image captioning with end-to-end attribute detection and subsequent attributes prediction. IEEE Trans. Image Process. **29**(1), 4013–4026 (2020)
9. Lin, C.Y.: ROUGE: a package for automatic evaluation of summaries. In: Text Summarization Branches Out, pp. 1–8 (2004)
10. Silberman, N., Hoiem, D., Kohli, P., Fergus, R.: Indoor segmentation and support inference from RGBD images. In: Fitzgibbon, A., Lazebnik, S., Perona, P., Sato, Y., Schmid, C. (eds.) ECCV 2012. LNCS, vol. 7576, pp. 746–760. Springer, Heidelberg (2012). https://doi.org/10.1007/978-3-642-33715-4_54

11. Papineni, K., Roukos, S., Ward, T., Zhu, W.: BLEU: a method for automatic evaluation of machine translation. In: Meeting Association Computational Linguistics, pp. 311–318 (2002)
12. Qi, C.R., Su, H., Mo, K., Guibas, L.J.: Pointnet: deep learning on point sets for 3d classification and segmentation. In: IEEE Conference on Computer Vision and Pattern Recognition, pp. 77–85 (2017)
13. Qi, C.R., Yi, L., Su, H., Guibas, L.J.: Pointnet++: deep hierarchical feature learning on point sets in a metric space. In: Advances in Neural Information Processing Systems, pp. 5099–5108 (2017)
14. Rennie, S.J., Marcheret, E., Mroueh, Y., Ross, J., Goel, V.: Self-critical sequence training for image captioning. In: IEEE Conference Computer Vision and Pattern Recognition, pp. 1179–1195 (2017)
15. Shen, X., Tian, X., Xing, J., Rui, Y., Tao, D.: Sequence-to-sequence learning via shared latent representation. In: Conference on Artificial Intelligence, pp. 2395–2402 (2018)
16. Shi, S., Wang, X., Li, H.: Pointrcnn: 3d object proposal generation and detection from point cloud. In: IEEE Conference on Computer Vision and Pattern Recognition, pp. 770–779 (2019)
17. Song, J., Gao, L., Guo, Z., Liu, W., Zhang, D., Shen, H.T.: Hierarchical LSTM with adjusted temporal attention for video captioning. In: International Joint Conference on Artificial Intelligence, pp. 2737–2743 (2017)
18. Song, Y., Soleymani, M.: Cross-modal retrieval with implicit concept association, vol. 1, pp. 1–9. CoRR abs/1804.04318 (2018)
19. Su, H., Maji, S., Kalogerakis, E., Learned-Miller, E.G.: Multi-view convolutional neural networks for 3d shape recognition. In: IEEE International Conference on Computer Vision, pp. 945–953 (2015)
20. Thomas, H., Qi, C.R., Deschaud, J., Marcotegui, B., Goulette, F., Guibas, L.J.: Kpconv: Flexible and deformable convolution for point clouds. In: IEEE International Conference on Computer Vision, pp. 6410–6419 (2019)
21. Vaswani, A., et al.: Attention is all you need. In: Advances in Neural Information Processing Systems, pp. 5998–6008 (2017)
22. Vedantam, R., Zitnick, C., Parikh, D.: CIDEr: consensus-based image description evaluation. In: IEEE Conference Computer Vision and Pattern Recognition, pp. 4566–4575 (2015)
23. Venugopalan, S., Rohrbach, M., Donahue, J., Mooney, R.J., Darrell, T., Saenko, K.: Sequence to sequence - video to text. In: IEEE International Conference on Computer Vision, pp. 4534–4542 (2015)
24. Vinyals, O., Toshev, A., Bengio, S., Erhan, D.: Show and tell: a neural image caption generator. In: IEEE Conference on Computer Vision and Pattern Recognition, pp. 3156–3164 (2015)
25. Xu, K., et al.: Show, attend and tell: neural image caption generation with visual attention. In: International Conference on Machine Learning, pp. 2048–2057 (2015)
26. Yao, T., Pan, Y., Li, Y., Mei, T.: Hierarchy parsing for image captioning. In: IEEE International Conference on Computer Vision, pp. 2621–2629 (2019)
27. Yu, Q., Xiao, X., Zhang, C., Song, L., Pan, C.: Extracting effective image attributes with refined universal detection. Sensors **21**(1), 95 (2021)

Soccer Field Registration Based on Geometric Constraint and Deep Learning Method

Pengjie Li(✉), Jianwei Li, Shouxin Zong, and Kaiyu Zhang

School of Sports Engineering, Beijing Sports University, Beijing, China
lipj@bsu.edu.cn

Abstract. Registering a soccer field image into the unified standard template model image can provide preconditions for semantic analysis of soccer videos. In order to complete the task of soccer field registration, the soccer field marker lines need to be detected, which is a challenging problem. In order to solve the problem, we propose a soccer field registration framework based on geometric constraint and deep learning method. We construct a multi-task learning network to realize marker lines detection, homography matrix calculation and soccer field registration. Firstly, the input image is preprocessed to remove the background and occlusion areas and extract the marker lines to obtain the edge map of the soccer field; then, we extract the features on the edge map and on the standard template model image to calculate the homography matrix. In this paper, a two-stage deep training network is proposed. The first stage mainly completes the soccer field marker lines detection and the initial calculation of the homography matrix. The second stage mainly completes the optimization of the homography matrix using geometric constraint, which can provide more accurate homography matrix calculation. We propose to integrate the geometric constraint of the marker lines into the multi-task learning network, in which structural loss is constructed to import prior information such as the shape and direction of the marker lines of the soccer field. We evaluate our method on the World Cup dataset to show its performance against the state-of-the-art methods.

Keywords: Soccer field registration · Homography matrix · Geometric constraint · Multi-task learning

1 Introduction

With the development of computer vision, researchers began to use this technology to automatically analyze sports videos. Since these broadcast videos usually come from one or more cameras, in order to analyze the video effectively, we

Supported by the Open Projects Program of National Laboratory of Pattern Recognition (No.202100009), the Fundamental Research Funds for Central Universities (No. 2021TD006).

H. Ma et al. (Eds.): PRCV 2021, LNCS 13020, pp. 287–298, 2021.
https://doi.org/10.1007/978-3-030-88007-1_24

must uniformly register all the videos onto a template image. The registration process needs the calculation of homography matrix. If the process is completed manually, the task of marking the video with dozens of frames per second is large and takes a long time. Therefore, we need to make a sports field registration system to realize the analysis of the sport video automatically. After the sports field is registered onto a unified standard template model image, we can get some important semantic events, such as scoring, foul, serving, getting the position of players and ball in the uniform coordinate system, and automatically judging whether the ball or player is out of bounds etc., which also provides a prerequisite for the semantic analysis of sports videos. This paper mainly studies the soccer field registration. Firstly, we detect the soccer field marker lines in the video image and the soccer field lines in the template image. Then, we need to find the corresponding relationship between the soccer field lines in the two images and match them. The corresponding relationship is not a priori. The registration task is challenging due to the fact that the broadcast video images are taken from many different perspectives and from different distances, causing changes in the posture and scaling of the pitch in the images, and the lack of texture on the pitch in the soccer video images. In order to solve the problem, researchers have developed a number of methods based on geometric features [1–5], which estimate the homography matrix using geometric elements such as key points, lines or circles to complete field registration. These methods rely on Hough transform [6], local feature (SIFT, MSER) [7,8] or random sampling consensus (Ransac) [9], and require manual specification of color and geometric features. However, due to occlusion and illumination changes, marker lines are usually not detected. In serious cases, surrounding shadows and player movement often introduce irrelevant information that can impair detection performance, and the accuracy of these methods needs to be improved. In recent years, due to the success of deep learning in many research fields of computer vision [10–12], the application of registration of various field to standard template model image is not an exception [13–17]. Researchers predict the corresponding parameters directly based on the regression algorithm of convolutional neural network, and solve the problem of various field registration by using hierarchical strategy end-to-end deep learning framework. These methods do not consider the geometric constraints between the soccer field area and the marker lines. Moreover, the end-to-end training, if there is an error in the training network, it will be transferred to the whole network, resulting in performance degradation. In order to solve the above problems, we divide the training into two stages. The main contributions of this paper are as follows:

1) We propose a soccer field registration framework based on geometric constraints and deep learning method. By constructing a multi-task learning network, we can detect the marker lines, calculate the homography matrix and register the soccer field;
2) We propose a structural loss, which introduces prior information such as shape and direction of soccer field marker lines, and integrate geometric constraints into deep learning network;

3) We propose a two-stage deep learning training network. The first stage mainly completes the detection of soccer field marker line and the initial calculation of homography matrix, and the second stage mainly completes the optimization of homography matrix with geometric constraint, so the calculation of homography matrix accuracy will be improved.

2 Related Work

The researchers try to register the various fields based on the geometric features, and register the broadcast videos to the standard template model image [18] through a set of pre-calibrated reference images. These calibration reference images are used to estimate the relative pose of the salient region. In order to calculate the relative pose, these methods either assume that the image corresponds to a continuous frames [19] in the video, or use local features, for example, sift [7] and mser [8] to find the corresponding relationship between features. However, because local features are difficult to cope with long-term changes, these methods require that the calibrated image set contain some similar appearance to the current processed image. Although there is an alternative method of machine learning [20,21], they are online detected. The performance of pose estimation and registration is still limited. Mukhopadhyay et al. [6] used the data on the center circle of the sport field to align the image, and used the 6-D Hough transform to detect the ellipse. Because the angle of the image changes, the center circle is regarded as the ellipse in the image. The main disadvantages of these methods are high computational complexity and memory cost. Some researchers propose to use probability method to reduce the calculation cost, and some researchers have also proposed to detect lines tangent to ellipse [22]. However, these methods do not consider the detection of important edge points in the image, and are sensitive to outliers (i.e. points not belonging to ellipses), so their accuracy is unsatisfactory.

To overcome these limitations, recent researchers [13–17] focus on preprocessing the sport videos into images, which only contain the region of the sport field. They manually mark the soccer field, and then execute the registration of the sport field. Chen and little [14] further adopted the image conversion network in the hierarchical setting. Firstly, the sport field was subdivided, and then the sport field marker lines was extracted. The pose estimation is further optimized by distance transformation of edge images. Sharma et al. [16] directly extract the edges and lines of sports field instead of complex semantic segmentation. They use the generated database of isomorphic edge images to extract the pose, and then match to find the most similar edge images. The bottleneck of these methods is the dependence on the database. They can only deal with the existing pose or similar pose in the database correctly, which hinders their scalability. Homayounfar et al. [15] used deep learning network to segment video images, and then used branch and bound Markov random fields to optimize the pose. Although it is robust to some scene changes, its accuracy is still limited.

The key of sport field registration is to estimate homography matrix, which includes sparse feature-based method and dense feature-based method. Whether sparse or dense, these methods are mainly limited by the quality of local features or the sensitivity of the objective function used for optimization [23]. Researchers [24] proposed a method based on deep learning to calculate homography matrix estimation. The method is described as classification by discretizing the output of the regression network. Nguyen et al. [25] train the deep learning network unsupervised to estimate the corresponding homography matrix. However, the main focus of these methods is to improve the speed. Wei Jiang et al. [17] proposed to parameterize the homography matrix through the coordinates of four control points on the input image to register the input video image to the sport field template image. Citraro et al. [13] combined the accurate positioning and reliable identification of specific key points in the image by using the whole convolution neural network architecture. Although these methods are computational efficiency, they are highly dependent on the location of key points, and even small errors in point positioning will lead to performance degradation. Long Sha et al. [26] proposed a method to calculate the homography matrix based on the soccer field region matching. These methods [13–15] use end-to-end deep learning networks to optimize the homography matrix to solve the adjustment of dense constraints. These methods can not detect the site soccer field marker lines in complex scenes. In order to solve the above problems, we propose a multi-task learning framework to provide a unified method for soccer field marker line detection and the homography matrix calculation. In this paper, the geometric constraint between the field area and the marker line [27] is used to make the calculation of the homography matrix more accurate.

3 Homography Matrix

The homography matrix has eight degrees of freedom. It can be used to transform the soccer field image into the standard template image. The purpose of this paper is to complete the registration of soccer field image, which is realized by automatically calculating the homography matrix between the broadcast video image of the sports field and the standard template image. In order to complete the registration, the first step is to detect the image features such as corresponding lines or points. The correspondence between the video image and the standard template image is not a priori. Therefore, in order to calculate the homography matrix, all possible matching pairs need to be evaluated. After getting the homography matrix, the video image can be aligned to the standard template image, as shown in Fig. 1. According to the theoretical derivation [17], at least four groups of corresponding points are needed to calculate the homography matrix. However, due to the occlusion or not in the image, the calculation error will be caused. We use the corresponding relationship of line segments or arcs, which makes the calculation of homography matrix more robust. In this paper, we propose to combine geometric constraint with deep learning to detect the soccer field marker lines, so that the extraction of the marker lines is more accurate, which lays a foundation for the calculation of homography matrix.

 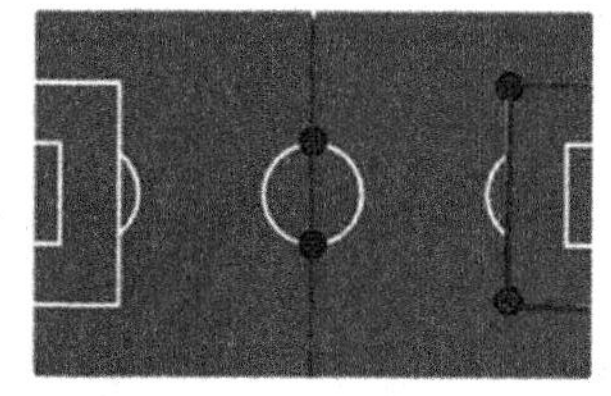 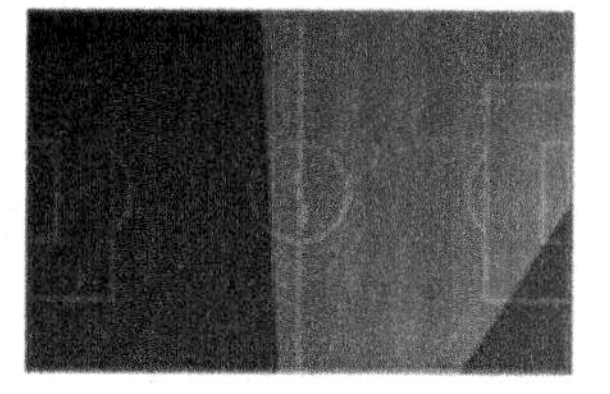

Fig. 1. Computing the homography and registering the soccer field.

4 Soccer Field Registration

In this paper, we propose a multi-task learning framework to provide a unified method for soccer field marker lines detection and the homography matrix calculation. The framework of deep learning network is based on the classification of full connection layer, and the features used are global features, which directly solve the problem of perceptual field. The problem of soccer field marker lines detection is defined as the collection of some rows and columns of boundary lines in the image. That is based on the position selection and classification in the two directions of rows and columns. In order to improve the performance of soccer field marker lines detection, we introduce the inherent geometric constraints into the multi-task learning framework. These geometric constraints are very important to find a consistent solution for soccer field segmentation and marker lines boundary detection. The method greatly enhances the robustness and accuracy of the soccer field marker lines location. Our algorithm framework is shown in Fig. 2, in which H denotes the homography matrix.

4.1 Soccer Field Marker Lines Detection

Before detecting the soccer field marker lines, we first preprocess the input image to remove the background areas (such as the stands, spectators, players, digital billboards, etc.) and extract the marked lines edge map of the soccer field area. Then, the features are calculated on the markers lines edge map.

These features are used to compute homography matrix from input image to the standard template image. Our algorithm is to detect the markers lines of soccer field in the direction of row and column. Because we don't conduct segmentation on every image pixel, in this way, the original $h \times w$ classification problem is simplified to $h + w$, so our method reduces the amount of calculation. Because of the direct position information in the horizontal row direction and vertical column direction, we can also add this information to the prior constraint of soccer field marker lines. In this paper, the problem of soccer field marker lines detection is transformed into the selection of row and column based on global image features. In the formula built in this paper, the marker lines are described as a series of row anchors (horizontal positions on a predefined row) and column anchors (vertical positions on a predefined column). To represent the location, the first step is to grid each row anchor and column anchor separately, and the

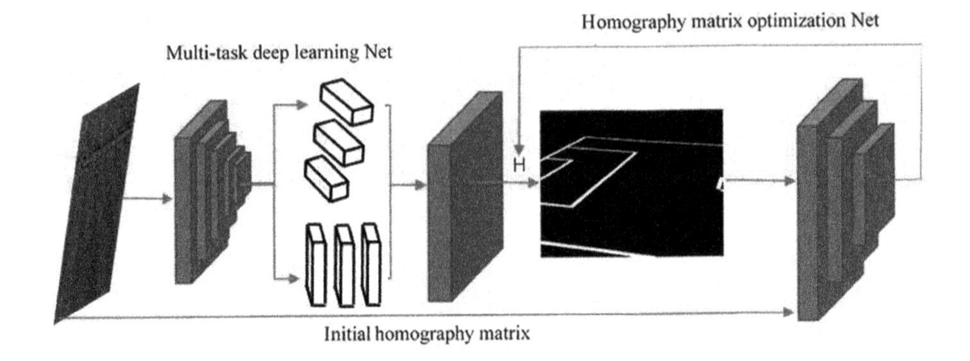

Fig. 2. The overall framework of soccer field registration method.

location is divided into many cells. In this way, the detection of marker lines can be described as the selection of specific cells on the predefined row anchor and column anchor [27].

4.2 Geometric Constraint Structure

In this paper, we use the structural loss of the geometric constraint of the soccer field marker lines to integrate the prior information such as the shape and direction into the deep learning network. The structure loss defines the marker lines detection as the collection of finding the position of the lines in some rows or columns in the image. This is the position selection and classification based on the row direction and column direction, with the purpose of calculating the position of the marker lines and the key points on the soccer field.

Suppose that the maximum number of marker lines in the row direction is C, the number of row anchors is D. The number of grids is ω, X is the image feature, and f_r^{ki} is the classifier used on the $k-th$ marker lines of the $i-th$ row anchor in the $r-th$ row direction, and the prediction probability of the landmark position in the row direction can be defined as common Equation (1). P_r^{ki} is a vector of $(\omega + 1)$ dimension, which represents the probability of selecting the $k-th$ marker line of the $i-th$ row anchor in the grid.

$$P_r^{ki} = f_r^{ki}(X) \tag{1}$$

where $1 \leq k \leq C, 1 \leq i \leq D$.

Assuming T_r^{ki} denotes the label of the correct position of the point on the marker line, the cross entropy loss function can be defined as formula (2).

$$L_{rls} = \sum_{k=1}^{C}\sum_{i=1}^{D} L_{CD}(P_r^{ki}, T_r^{ki}) \tag{2}$$

The task of marker lines detection is transformed into the selection of row and column based on global image features. In other words, the global feature is used to select the marker lines on each pre-defined row and column. Similarly, we can define the cross entropy loss function in the column direction. The geometric constraint structure loss function is defined as the sum of the cross entropy loss function in the row direction and the cross entropy loss function in the column direction. In this way, we can learn the geometric structure information more accurately. In the classical convolution neural network, each convolution layer receives input from the previous layer. After applying convolution and nonlinear activation, the output is passed to the next layers. Similarly, our algorithm regards the row or column of the feature map as a layer, and also uses convolution and nonlinear activation to realize the deep neural network in geometric space. This enables the spatial information to propagate on the neurons of the same layer and enhances the spatial information, which is particularly effective for identifying structured objects. This process allows information to be transferred between different neurons in the same layer, which can effectively enhance the geometric space information, learn spatial relations more effectively, and find continuous and strong prior structural targets smoothly.

4.3 Parameter Regression Model

In this work, we define the parameter set as 8-dimensional vector, and each vector corresponds to a parameter of homography matrix $H \in R^{(3\times 3)}$. We associate a predicted value $t = [t_1, t_2, t_3, t_4, t_5, t_6, t_7, t_8]$ corresponding to the homography matrix H, as shown in Eq. (3).

$$H = \begin{pmatrix} t_1 + 1 & t_2 & t_3 \\ t_4 & t_5 + 1 & t_6 \\ t_7 & t_8 & 1 \end{pmatrix} \quad (3)$$

According to the method in reference [26], we first normalized the coordinates of the input video image and the soccer field standard template image before calculating the homography matrix. We construct a small convolution neural network on top of the last convolution layer in the detection part of soccer field marker lines. We set the network architecture as $C_{16} - C_{32} - C_{64} - F_8$, where C_k represents convolution-batch normalization-activation function-maximum pooling with k filters, F_k represents the full connection layer with k neurons. Each convolution layer is a 3×3 filter, and the span of each maximum pooling layer is set to 2.

4.4 Deep Learning Training Network

In this section, we mainly introduce the implementation details of the training process. Our training is divided into two stages. The first stage is to detect the marker lines of soccer field and complete the calculation of initial homography matrix, which transforms the marker line of the input image soccer field into the edge map of the standard template image. In the second stage, we train the registration error network, which integrated the geometric constraint structure. In this paper, the disturbed homography matrix is used to project the image site marker lines to the standard template edge map, and the standard edge map is connected with the input image as the input data for training. In order to make the algorithm more robust, we use hierarchical method to add noise disturbs to the manually labeled homography matrix. Firstly, we add a global geometric constraint noise disturb. Then we multiply the input image by the perturbed homography matrix to get the projection image. By linking the input image and the projection image as the input of the registration error network [17] for the training model, so we will complete network training through minimizing Formula (4).

$$W_{err} = \left\| IoU(I, W(m, h_{per}^{gcs})) - T_{\psi}([I; W(m, h_{per}^{gcs})]) \right\|_2^2 \tag{4}$$

where I is the input image, m is the standard template image of the soccer field. h_{per}^{gcs} denotes the homography matrix with random noise perturbation, which was constrained by geometric structure. $W(m, h_{per}^{gcs})$ denotes the projection image obtained by mapping the input image to the template image using the homography matrix. $T_{\psi}([I; W(m, h_{per}^{gcs})])$ denotes the registration error calculated by combining the input and projected images.

5 Experiments and Results

In this section, we test the proposed method for soccer field registration. Firstly, we introduce data set, data augmentation, and performance evaluation indicators. Then, we compare the proposed method with the existing three methods (SIFT [7], PoseNet [28], OTLE [17]), and give the qualitative and quantitative results.

5.1 Data Augmentation

In this section, we will test the proposed method on the World Cup dataset [13]. The World Cup dataset [13] is composed of broadcast video images of soccer matches, recording 20 matches in Brazil's World Cup 2014. Due to the inherent structure and symmetry of the marker lines of soccer fields, the classification-based network may easily generate overfitting on the training set and show poor performance on the validation set. In order to solve this problem and gain stronger generalization ability, we use the data augmentation method composed of rotation, vertical and horizontal shift, as shown in Fig. 3. At the same time, in order to retain the soccer field structure, the marker lines are extended to the boundary of the image. We also use the unsupervised deep learning algorithm [16] to expand the size of the data set.

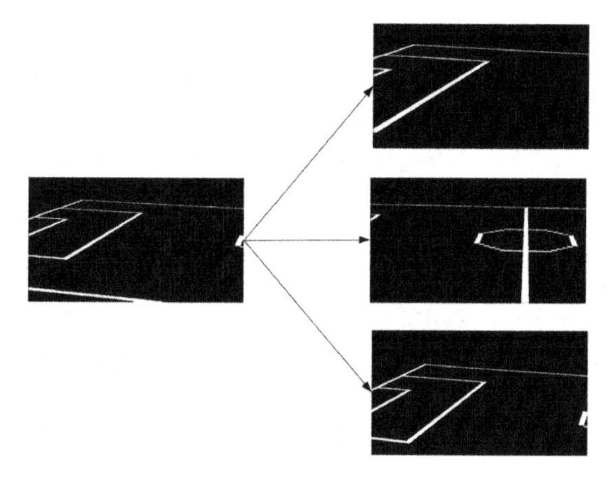

Fig. 3. Generating augmented data by adding three basic perturbations of Pan Tilt Zoom.

5.2 Performance Evaluation Indicator

We use IoU [13] as indicator to evaluate the performance of the algorithms. We compute the IoU on the top view and compare the intersection between the ground truth and the estimated homography. We measured the IoU not only on the whole soccer field IoU_{Whole} but also on the polygonal field area that appeared in the image IoU_{Part} . We report our results under both indicators for ease of comparison. For both indicators IoU_{Whole} and IoU_{Part}, we define two sub-indicators: mean, median [13].

5.3 Results

In this section, we compare the proposed algorithm with three existing algorithms on the World Cup dataset [13]. Our method can deal with soccer field registration in various scenarios. The qualitative results (partial results) are shown in Fig. 4. The quantitative results are summarized in Table 1. From Table 1, we can see that our method is better than only using the geometric features method [7] or only using the deep learning features method [28][17]. The method [17] also use two-stages deep learning training network, but this method did not import geometric constraint structure. Our method integrates geometric constraint into deep learning training network, so it has higher accuracy for soccer field registration.

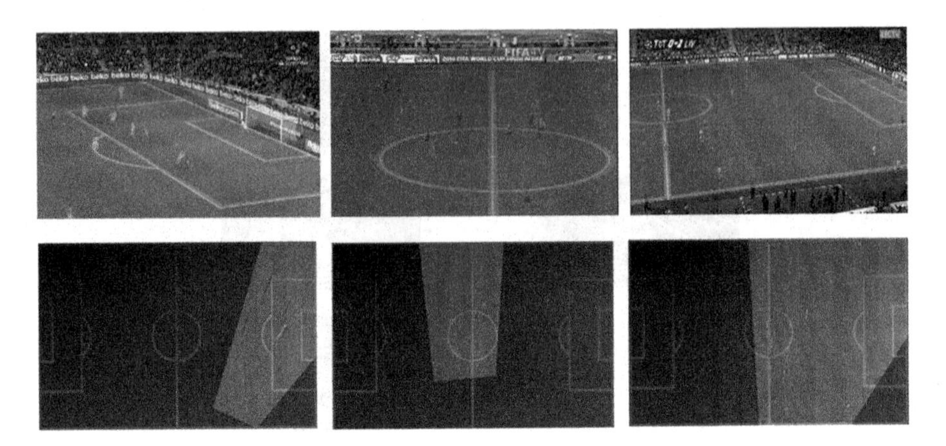

Fig. 4. Soccer field registration of our method. (Top) the field marker lines (yellow lines overlayed on the current view) generated by the predicted homography. (Bottom) current view overlayed on soccer field template image.

Table 1. Quantitative results on Soccer World Cup dataset.

Method name	IoU_{Whole}		IoU_{Part}	
	Mean	Median	Mean	Median
SIFT [7]	0.17	0.011	0.25	0.073
PoseNet [28]	0.528	0.559	0.59	0.622
OTLE [17]	0.898	0.929	0.951	0.967
Ours	0.921	0.943	0.962	0.974

6 Conclusion

In this paper, a framework for soccer field registration based on geometric constraints and deep learning method is proposed. Through the construction of multi-task learning network, we can detect the marker lines, calculate the homography matrix and register the soccer field. The training process is completed in two stages. The first stage mainly completes the detection of soccer field marker lines and the calculation of initial homography matrix, and the second stage mainly completes the optimization of homography matrix using geometric constraint structure. The second training network is independent of the first training network so that it does not overfit to the errors the initial training network makes. By optimizing through the registration error network using geometric constraints, accurate soccer registration results are obtained. Our method uses the geometric constraint relationship inherent in the soccer field marker lines. Our method uses the structural loss to integrate the prior information such as the shape and direction of the soccer field marker lines into the deep learning network, which makes the detection of the marker lines and the calculation of the homography matrix more accurate. Through experiments, we have shown that the proposed method can achieve state-of-the-art performance.

References

1. Cuevas, C., Quilon, D., Garcia, N.: Automatic soccer field of play registration. Pattern Recogn. **103**, 107278 (2020)
2. Bu, J., Lao, S., Bai, L.: Automatic line mark recognition and its application in camera calibration in soccer video. In: IEEE International Conference on Multimedia and Expo (ICME), pp. 1–6 (2011)
3. Dong, H., Prasad, D.K., Chen, I.-M.: Accurate detection of ellipses with false detection control at video rates using a gradient analysis. Pattern Recogn. **81**, 112–130 (2018)
4. Hess, R., Fern, A.: Improved video registration using non distinctive local image features. In: IEEE Conference on Computer Vision and Pattern Recognition (CVPR) (2007). https://doi.org/10.1109/CVPR.2007.382989
5. Lu, W.-L., Ting, J.-A., Little, J.J., Murphy, K.P.: Learning to track and identify players from broadcast sports videos. IEEE Trans. Pattern Anal. Mach. Intell. (TPAMI) **35**(7), 1704–1716 (2013)
6. Mukhopadhyay, P., Chaudhuri, B.B.: A survey of Hough transform. Pattern Recogn. **48**(3), 993–1010 (2015)
7. Lowe, D.: Distinctive image features from scale-invariant keypoints. Int. J. Comput. Vis. **20**(2), 91–110 (2004)
8. Pajdla, J.M.C.U.: Robust wide-baseline stereo from maximally stable extremal regions. Image Vis. Comput. **22**(10), 761–767 (2004)
9. Brachmann, E., et al.: DSAC-Differentiable RANSAC for camera localization. In: IEEE Conference on Computer Vision and Pattern Recognition (CVPR), pp. 2492–2500 (2017). https://doi.org/10.1109/CVPR.2017.267
10. Bochkovskiy, A., Wang, C.-Y., Liao, H.: YOLOv4: optimal speed and accuracy of object detection. In: IEEE Conference on Computer Vision and Pattern Recognition (CVPR) (2020)
11. Qin, Z., Wang, H., Li, X.: Ultra fast structure-aware deep lane detection. In: European Conference on Computer Vision (ECCV), pp. 1–16 (2020)
12. Pan, X., Shi, J., Luo, P., Wang, X., Tang, X.: Spatial as deep: spatial CNN for traffic scene understanding. In: AAAI Conference on Artificial Intelligence (2018)
13. Citraro, L.: Real-time camera pose estimation for sports fields. Mach. Vis. Appl. **31**(16), 1–12 (2020)
14. Chen, J., Little, J.J.: Sports camera calibration via synthetic data. In: IEEE Conference on Computer Vision and Pattern Recognition Workshops (2019)
15. Homayounfar, N., Fidler, S., Urtasun, R.: Sports field localization via deep structured models. In: IEEE Conference on Computer Vision and Pattern Recognition, pp. 4012–4020 (2017). https://doi.org/10.1109/CVPR.2017.427
16. Sharma, R.A., Bhat, B., Gandhi, V., Jawahar, C.V.: Automated top view registration of broadcast soccer videos. In: IEEE Winter Conference on Applications of Computer Vision, pp. 305–313, (2018). https://doi.org/10.1109/WACV.2018.00040
17. Wei, J., Camilo, J., Higuera, G., Angles, B., Javan, W.S.M., Yi, K.M.: Optimizing through learned errors for accurate sports field registration. In: IEEE Winter Conference on Applications of Computer Vision (WACV) (2020). https://doi.org/10.1109/WACV45572.2020.9093581

18. Gupta, A., Little, J.J., Woodham, R.: Using line and ellipse features for rectification of broadcast hockey video. In: Canadian Conference on Computer and Robot Vision, pp. 32–39 (2011). https://doi.org/10.1109/CRV.2011.12
19. Puwein, J., Ziegler, R., Vogel, J., Pollefeys, M.: Robust multi-view camera calibration for wide-baseline camera networks. In: IEEE Winter Conference on Applications of Computer Vision, pp. 321–328 (2011). https://doi.org/10.1109/WACV.2011.5711521
20. Detone, D., Malisiewicz, T., Rabinovich, A.: Superpoint: self-supervised interest point detection and description. In: IEEE Conference on Computer Vision and Pattern Recognition (CVPR) Workshop on Deep Learning for Visual SLAM (2018). https://doi.org/10.1109/CVPRW.2018.00060
21. Yi, K.M., Trulls, E., Lepetit, V., Fua, P.: LIFT: learned invariant feature transform. In: Leibe, B., Matas, J., Sebe, N., Welling, M. (eds.) ECCV 2016. LNCS, vol. 9910, pp. 467–483. Springer, Cham (2016). https://doi.org/10.1007/978-3-319-46466-4_28
22. Verdie, Y., Yi, K.M., Fua, P., Lepetit, V.: TILDE: a temporally invariant learned detector. In: IEEE Conference on Computer Vision and Pattern Recognition (CVPR), pp. 5279–5288 (2015). https://doi.org/10.1109/CVPR.2015.7299165
23. Yan, Q., Xu, Y., Yang, X., Nguyen, T.: HEASK: robust homography estimation based on appearance similarity and keypoint correspondences. Pattern Recogn. **47**(1), 368–387 (2014)
24. DeTone, D., Malisiewicz, T., Rabinovich, A.: Deep image homography estimation. In: RSS Workshop on Limits and Potentials of Deep Learning in Robotics (2016)
25. Nguyen, T., Chen, S.W., Shivakumar, S.S., Taylor, C.J., Kumar, V.: Unsupervised deep homography: a fast and robust homography estimation model. IEEE Robot. Autom. Lett. **3**(3), 2346–2353 (2018). https://doi.org/10.1109/LRA.2018.2809549
26. Sha, L., Hobbs, J., Felsen, P., Wei, X., Lucey, P., Ganguly, S.: End-to-end camera calibration for broadcast videos. In: IEEE Conference on Computer Vision and Pattern Recognition (CVPR), pp. 13624–13633 (2020). https://doi.org/10.1109/CVPR42600.2020.01364
27. Zhang, J., Xu, Y., Ni, B., Duan, Z.: Geometric constrained joint lane segmentation and lane boundary detection. In: European Conference on Computer Vision (ECCV), pp. 486–502 (2018)
28. Kendall, A., Grimes, M., Cipolla, R.: PoseNet: a convolutional network for real-time 6-DOF camera relocalization. In: International Conference on Computer Vision (ICCV), pp. 2938–2946 (2015)

Enhancing Latent Features for Unsupervised Video Anomaly Detection

Linmao Zhou[1,2], Hong Chang[1,2(✉)], Nan Kang[1,2], Xiangjun Zhao[3], and Bingpeng Ma[2]

[1] Key Laboratory of Intelligent Information Processing of Chinese Academy of Sciences (CAS), Institute of Computing Technology, Chinese Academy of Sciences, Beijing, China
{linmao.zhou,nan.kang}@vipl.ict.ac.cn, changhong@ict.ac.cn
[2] University of Chinese Academy of Sciences, Beijing, China
bpma@ucas.ac.cn
[3] Neusoft, Shenyang, China

Abstract. Recently, memory-augmented autoencoder has played a vital role in unsupervised video anomaly detection. The memory module is used for recording the prototypes of normal data and suppressing the feature representation capacity for anomalies. However, the feature representation capacity for normal data may also be suppressed due to the limited memory ability of the memory module. This may bring a serious problem to normal data being judged as anomaly. To this end, we propose a simple but effective method called Enhancing Latent Features, *i.e.* ELF, which integrates two modules Single Grid Feature Enhancement (SGFE) and Local Context Feature Enhancement (LCFE). They are performed sequentially to enhance the latent feature of normal instance based on single grid information and local context information, respectively. To capture the meaningful local context information, a learnable Local Sampling Location Prediction Network (LSLPN) is embedded into LCFE to predict the valuable sampling locations of local context information. Experimental results on standard benchmarks demonstrate the effectiveness of our method.

Keywords: Anomaly detection · Autoencoder · Feature enhancement

1 Introduction

Video anomaly detection is a very essential and challenging computer vision task, which has been widely applied to video surveillance [14]. Generally speaking, it is almost infeasible to collect all kinds of abnormal data and handle the problem with a general classification method. Therefore, it is widely accepted to consider anomaly detection as a special unsupervised learning problem, where only normal samples are available during training.

Recently, a promising approach to tackle this task is based on the powerful Autoencoder (AE) [2,9], which consists of an encoder to obtain a compressed

H. Ma et al. (Eds.): PRCV 2021, LNCS 13020, pp. 299–310, 2021.
https://doi.org/10.1007/978-3-030-88007-1_25

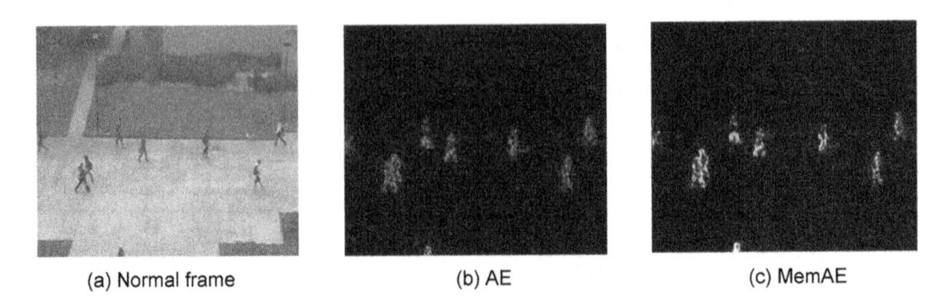

Fig. 1. Reconstruction error of AE and MemAE on a normal frame of UCSD Ped2 [15]. The highlight regions indicate the reconstruction error is large. For this normal frame, MemAE produces larger reconstruction errors than AE, thus the normal frame is more likely to be identified as an anomaly.

latent feature from the input and a decoder that can reconstruct the input data from the latent feature. Along the direction of reconstruction-based anomaly detection, many methods are proposed to improve original AE. Several types of regularized AEs [1,4] have been introduced to capture richer and more meaningful feature representation. Especially, LSA [1] proposes an additional autoregression density estimator to learn the distribution of the latent space by maximum likelihood principles.

Recently, DAGMM [26] and MemAE [6] point out that the assumption of reconstruction may not always hold in practice. Sometimes the AE "generalizes" so well that it can also reconstruct the abnormal inputs well, due to the very powerful representation capacity of convolutional neural networks (CNNs). Based on this discovery, MemAE and MNAD [17] adopt a memory module to record the main prototypes of normal data during the training stage. At test time, the memory module is also used for suppressing the latent feature of anomaly instance to strengthen the reconstruction error of it. However, the recorded prototypes in memory module cannot completely retrieve the latent feature of normal data because of the limited memory ability. Therefore, some normal data may sometimes produce large reconstruction errors and be recognized as anomalies, as shown in Fig. 1. This problem has also been observed in the existing literature MemAE [6]. Actually, the original latent feature extracted by the encoder is a complete feature. So it is possible to enhance the latent features of normal samples by transfer them from the original latent feature to the suppressed latent feature.

In this paper, to mitigate the drawback of the memory module, we propose an simple but effective method, *Enhancing Latent Features* (ELF), to enhance the suppressed latent feature based on the single grid information and local context information. Specifically, our method integrates two modules: *Single Grid Feature Enhancement* (SGFE) and *Local Context Feature Enhancement* (LCFE). In SGFE, to decide how much the information transferred between the original latent feature and the suppressed latent feature, we first calculate the aggregation weights for each feature grid based on the degree of difference between them. Then the aggregation weights are applied to the above two latent features to produce the enhanced latent feature. After that, LCFE captures the local context information to further enhance the latent feature. To ensure that LCFE can model the most useful local context features, we propose a *Local Sampling Location Prediction Network* (LSLPN) which is embedded into LCFE to predict the local sampling locations. During the training stage, the LSLPN is gradually learned and guided by the most meaningful sampling locations selected from the global context. In the test phase, the learned LSLPN is first used to predict the useful sampling locations. Then the local context information is modeled by the most relevant sampling locations. Consequently, the latent feature can be enhanced by the meaningful local context information.

ELF is quite simple to be implemented and trained end-to-end. Experimental results show our method significantly improves the performance of memory-augmented anomaly detectors [6,17] on the various datasets [13–15], especially on the most challenging ShanghaiTech dataset.

2 Related Work

Anomaly Detection. The unsupervised anomaly detection is to learn a normal profile given only the normal data examples as training samples. According to this definition, one-class classification methods [3] are very suitable for handling this task. The goal of the one-class classification is to learn a discriminative boundary surrounding the normal instances as much as possible. Besides, the unsupervised clustering methods [23] are also a reasonable choice for handling the task. For example, k-means method and Gaussian Mixture Models have played a vital role to model the low-dimensional data for identifying the anomalies. However, because of the curse of dimensionality, it is difficult to directly apply such methods to high dimensional data.

Autoencoder-Based: The core idea of this kind of method is that normal data instances can be better restructured from compressed latent representation than anomalies. The reconstruction errors [6,9] are adopted as a criterion for identifying anomalies under a given threshold. Many methods of improving AE aim to extract richer and more expressive latent representations. VAE [9,19] regularizes the latent representation by encoding the data instances to prevent overfitting and guarantee some good properties of learned latent space. Similarly, LSA [1] proposes a parametric density estimator learns the probability distribution underlying its latent representations through an autoregressive procedure. Recently, inspiring by memory-augmented networks [22], MemAE [6] and MNAD [17] adopt a memory module to improve the performance of AE.

Enhancing Feature Representation. Many deep learning approaches [25] enhance the feature representations by element-wise sum or concatenation operations. These operations are sub-optimal for the anomaly detection task because the anomaly features will be transferred as well, which is harmful to discovering anomalies. Compared to these methods, SGFE keeps the transfer of normal features and simultaneously reduces the transfer of anomaly features as much

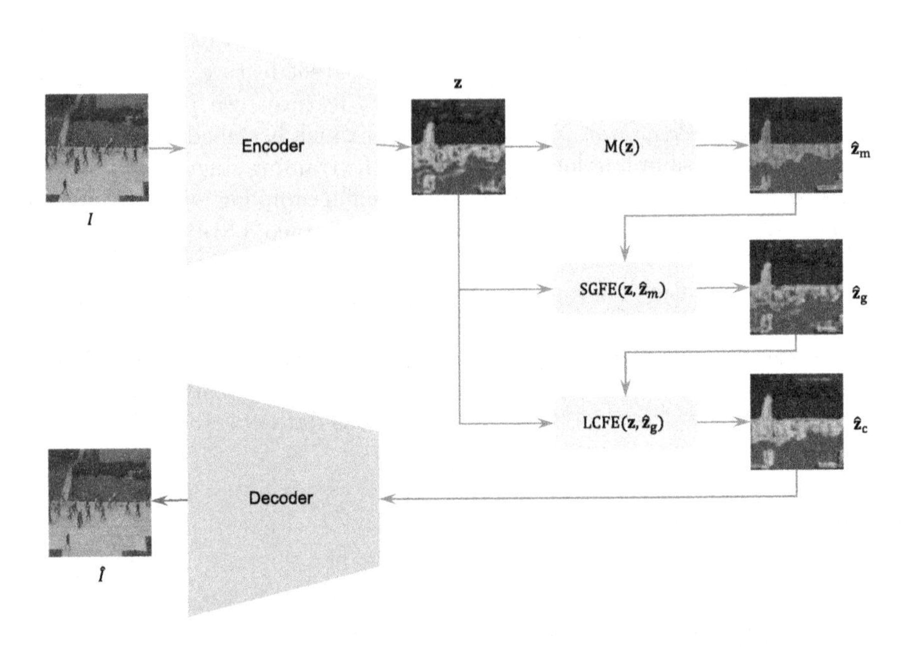

Fig. 2. Illustration of the overall pipepline of our method, where the suppressed latent features are gradually enhanced. $\mathbf{z}$, $\hat{\mathbf{z}}_m$, $\hat{\mathbf{z}}_g$, $\hat{\mathbf{z}}_c$ denote the latent feature extracted by encoder, the latent feature suppressed by $\mathrm{M}(\mathbf{z})$ and the latent feature enhanced by SGFE and LCFE, respectively.

as possible. From the perspective of modeling context, Non-local [20] is a popular way to capture the global context information to enhance feature representation, which has been widely applied to object detection [16] and person re-identification [7]. However, the global context information can introduce a large number of irrelevant features which leads to failure of anomaly detection. Different from Non-local, LCFE enhances the feature representation with the most relevant local context information.

3 Proposed Method

In this section, we first present the general framework of our method. Then, we detailed introduce the two proposed modules (SGFE and LCFE) that enhance the latent feature based on single-point information and local context information, respectively. Finally, we introduce the objective function used in our method and the way of calculating the abnormality score.

3.1 Overview

Our method can be applied to any memory-augmented anomaly detector [6,17]. The proposed modules SGFE and LCFE are embedded into them to enhance the latent feature. As shown in Fig. 2, given an input $\mathbf{I} \in \mathbb{R}^{H \times W \times C}$, the

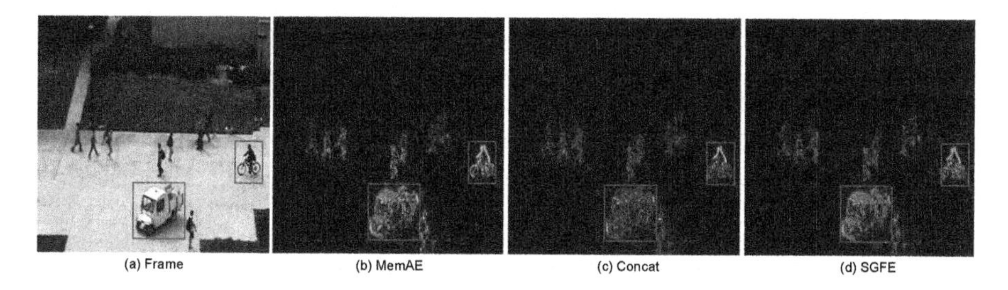

Fig. 3. Reconstruction error under different settings. (a) The red boxes enclose abnormal instances, and other areas are normal instances. (b) MemAE [6] increases the reconstruction errors of both abnormal instances and normal instances. (c) Concatenation operation lessens the reconstruction errors of both abnormal instances and normal instances. (d) SGFE reduces the reconstruction errors of normal instances but increase the reconstruction errors of abnormal instances.

encoder $E(\mathbf{I};\theta_e)$ first encodes the input to obtain the original latent feature $\mathbf{z} \in \mathbb{R}^{H' \times W' \times C'}$, where the H, W, C denote the height, width, the number of channels of the input image, H', W', C' is the height, width, the number of channels of the latent feature.

$$\mathbf{z} = E(\mathbf{I};\theta_e). \tag{1}$$

Then taken the latent feature $\mathbf{z}$ as a query, the memory module $M(\mathbf{z};\theta_m)$ retrieves the most relevant items in the memory to produce the suppressed latent feature $\hat{\mathbf{z}}_m$.

$$\hat{\mathbf{z}}_m = M(\mathbf{z};\theta_m). \tag{2}$$

Next, the suppressed latent feature $\hat{\mathbf{z}}_m$ is enhanced by SGFE and LCFE. Specifically, the SGFE first enhances the latent feature $\hat{\mathbf{z}}_m$ based on single grid information to produce $\hat{\mathbf{z}}_g$.

$$\hat{\mathbf{z}}_g = SGFE(\mathbf{z}, \hat{\mathbf{z}}_m). \tag{3}$$

Then the LCFE further enhances $\hat{\mathbf{z}}_g$ using the local context information to obtain $\hat{\mathbf{z}}_c$

$$\hat{\mathbf{z}}_c = LCFE(\mathbf{z}, \hat{\mathbf{z}}_g;\theta_c). \tag{4}$$

Finally, the enhanced latent feature $\hat{\mathbf{z}}_c$ is sent to the decoder $D(\hat{\mathbf{z}}_c;\theta_d)$ for reconstruction.

$$\hat{\mathbf{I}} = D(\hat{\mathbf{z}}_c;\theta_d). \tag{5}$$

The θ_e, θ_m, θ_c and θ_d denote the learnable parameters of encoder $E(\cdot)$, memory module $M(\cdot)$, $LCFE(\cdot,\cdot)$ and decoder $D(\cdot)$ respectively. The whole network is learned end-to-end.

3.2 Single Grid Feature Enhancement

Even for the latent feature of normal instance, the memory module $M(\cdot)$ cannot fully restore its latent feature in $\mathbf{z}$. In other words, the latent feature of normal instance may also be suppressed in $\hat{\mathbf{z}}_m$. This causes that the normal instance may also produce large reconstruction errors as shown Fig. 3 (b). A straightforward idea is to utilize the information of $\mathbf{z}$, $\hat{\mathbf{z}}_m$ to enhance latent feature by concatenation operation. Although this method can fully transfer the information of $\mathbf{z}$, it causes a problem that makes MemAE degenerate into AE. As shown in Fig. 3 (c), despite the reconstruction errors of normal instances decrease, the reconstruction errors of abnormal instances also greatly decrease as well.

To address the above problem, we propose an effective module SGFE to enhance the suppressed latent feature $\hat{\mathbf{z}}_m$ by the information transfer from $\mathbf{z}$ to $\hat{\mathbf{z}}_m$. Specifically, according to the degree of difference between $\mathbf{z}$ and $\hat{\mathbf{z}}_m$, we calculates the normalized aggregation weight $\mathbf{w} \in \mathbb{R}^{H' \times W' \times C'}$ of each feature grid for information transfer as follow:

$$\mathbf{w} = \sigma(\alpha|\mathbf{z} - \hat{\mathbf{z}}_m|). \tag{6}$$

The σ, α and $|\cdot|$ denote the sigmoid function, scaling factor and absolute function, respectively. The value of $\mathbf{w}$ determines how much the information of $\mathbf{z}$ is transferred to $\hat{\mathbf{z}}_m$. Specifically, we realize the information transfer from $\mathbf{z}$ to $\hat{\mathbf{z}}_m$ as follow:

$$\hat{\mathbf{z}}_g = \mathbf{w}\hat{\mathbf{z}}_m + (1 - \mathbf{w})\mathbf{z}, \tag{7}$$

by applying the aggregation weight $\mathbf{w}$.

As shown in Fig. 3 (d), the reconstruction errors of normal instances decrease but the reconstruction errors of abnormal instances still keep high. This indicates that SGFE is able to keep the transfer of normal features from $\mathbf{z}$ to $\hat{\mathbf{z}}_m$ and simultaneously reduces the transfer of anomaly features as much as possible. Besides, it is worth mentioning that this module dosen't add any extra parameters and all operations are element-wise that is very efficient.

3.3 Local Context Feature Enhancement

The SGFE only considers the single grid information, so leveraging the context information can further enhance the latent feature. In many computer vision tasks, the Non-local [20] operation is widely adopted to model the context information. However, the Non-local operation may pass a lot of irrelevant information because it models the global context by capturing relations between all pairs of pixels. Actually, the most relevant features are concentrated on only a few feature grids. So if the context information is modeled by the most relevant feature grids, the limitations of Non-local can be effectively mitigated.

To achieve this goal and address the above issue, we propose a LCFE to enhance latent features based on the local context information. The pipeline of LCFE is illustrated in Fig. 4 (left). In this module, the core step is to sample the most useful local context feature. The sampling operation is called *Local Context Sampling* (LCS). We first apply a tiny fully convolutional network called LSLPN to predict the offsets for each grid. Then $\hat{\mathbf{z}}_g$ and the sampling locations decoded from the offsets are passed to LCS to acquire local context information. Finally, the acquired local context is aggregated to $\hat{\mathbf{z}}_g$ to obtain $\hat{\mathbf{z}}_c$.

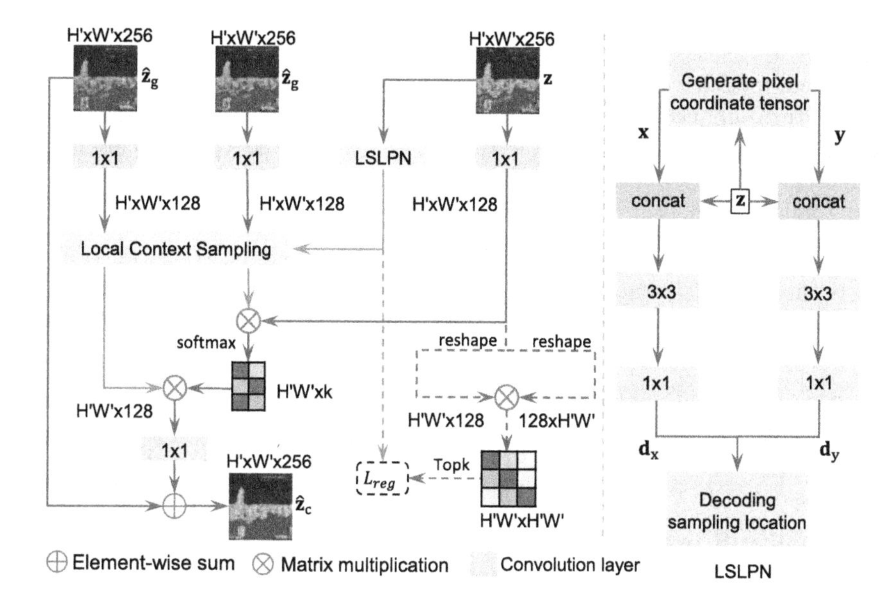

Fig. 4. Illustration of the detailed architecture of LCFE (left) and LSLPN (right). $\mathcal{L}_{reg}$, $\mathbf{x}$, $\mathbf{y}$, $\mathbf{d}_x$, $\mathbf{d}_y$ denote the regression loss, pixel coordinate of $\mathbf{z}$ and the predicted offset of sampling locations. The blue dashed line indicates that the ground-truth is generated for LSLPN during training. (Color figure online)

The pipeline of LSLPN is illustrated in Fig. 4 (right). In detail, LSLPN predicts offsets of sampling locations with respect to the location of each feature grid. This needs a model that is spatially variant, or in more precise words, position sensitive. However, the conventional convolutional operations are spatially invariant to some degree. Inspired by [21], we adopt the CoordConv [11] to introduce the spatial functionality. Specifically, we first build two tensors of same spatial size as input $\mathbf{z}$ that contain pixel coordinates of two directions ($\mathbf{x}$ and $\mathbf{y}$). Then they are normalized to $(-1, 1)$, concatenated to the input $\mathbf{z}$ and then feeded into the two convolutional branch, respectively. The two branch will predict two offset tensors ($\mathbf{d}_x$ and $\mathbf{d}_y$) with k channels, where k denotes the number of sampling location for each feature grid. Finally, we decode the offsets into the input tensor coordinate range.

Moreover, in order to ensure that LSLPN can be trained end-to-end with other modules, we design a reasonable "ground-truth" for it. Specifically, we first calculate the global relation matrix with shape of $(H' \times W') \times (H' \times W')$ that contains the entire relation between all pixels. Then we select the k most relevant positions for each position along the second dimension as the "ground-truth" for training LSLPN. Consequently, the shape of "ground-truth" is $(H' \times W') \times k$. Thanks to the use of this supervised information during training, LSLPN can generate the most useful sampling locations for the modeling context during inference. Thus LCFE can model the useful local context information based on well-learned sampling locations.

3.4 Training Objective and Abnormality Score

To train our model, we combine three constraints regarding frame reconstruction, latent feature enhancement and regression losses into our objective function. The total loss is formalized as $\mathcal{L} = \mathcal{L}_{rec} + \lambda_1 \mathcal{L}_{latent} + \lambda_2 \mathcal{L}_{reg}$. The λ_1 and λ_2 are hyper-parameters to balance total loss. The reconstruction loss is L2 distance between $\mathbf{I}$ and $\hat{\mathbf{I}}$ following [6,17].

Latent Feature Enhancement Loss. During the training stage, all training data is the normal instance, so we want the enhanced latent feature $\hat{\mathbf{z}}_c$ to be as close to $\mathbf{z}$ as possible. We adopt the L2 distance to measure the closeness between $\mathbf{z}$ and $\hat{\mathbf{z}}_c$ that is formalized as $\mathcal{L}_{latent} = \frac{1}{H'W'C'} \|\mathbf{z} - \hat{\mathbf{z}}_c\|_2$.

Regression Loss. To encourage LSLPN to predict the regression offset close to "ground-truth" as much as possible, the popular Smooth L1 loss [5] is adopted as $\mathcal{L}_{reg}$ to guide the learning of LSLPN. For a detailed definition of Smooth L1 loss, please refer to the literature Fast R-CNN [5].

Abnormality Score. Following [6,17], we normalize the reconstruction error of the sequences from the same video to calculate the abnormality score by the min-max normalization [12]. For finally comparison, we also apply a Gaussian filter to temporally smooth the frame-level anomaly score.

4 Experiments

In this section, we first briefly introduce the three popular benchmarks UCSD Ped2, CUHK Avenue, ShanghaiTech Campus and adopted evaluation metric. Then we present the implementation details. Finally, we evaluate our proposed method on the above three benchmarks to validate the effectiveness.

4.1 Experimental Setting

Dataset. We evaluate our proposed method on three publicly available anomaly detection datasets, including the UCSD Ped2 dataset [15], the CUHK Avenue dataset [13] and the ShanghaiTech Campus dataset [14]. The UCSD Ped2 dataset contains 16 training videos and 12 testing videos with 12 abnormal events, including riding a bike and driving a vehicle. The CUHK Avenue dataset contains 16 training videos and 21 testing ones with a total of 47 abnormal events, including throwing objects, loitering and running. The ShanghaiTech Campus dataset contains 330 training videos and 107 testing ones with 130 abnormal events which consists of 13 scenes and various anomaly types. It is the largest and the most challenging dataset among existing datasets for anomaly detection.

Evaluation Metric. Following [6,17], we adopt the popular evaluation metric Receiver Operation Characteristic (ROC) which is calculated by gradually changing the threshold of abnormality scores. Then the Area Under Curve (AUC) is cumulated to a scalar for performance evaluation. A higher value indicates better anomaly detection performance.

4.2 Implementation Details

Network Architecture. The network structure of encoder, memory module and decoder are followed from baseline methods [6,17] for fairly comparison. The proposed LSLPN of LCFE is a tiny fully convolutional network. The x-branch and y-branch are respectively passed to a convolutional layer with kernel size 3 to extract the useful feature representation which is used to predict the offsets using a convolutional layer with kernel size 1.

Training. Following [17], we resize each video frame to 256×256 and normalize it to the range of $(-1, 1)$. In order to make the training process more stable, we apply the tanh activation function to the output of the decoder to limit it to $(-1, 1)$. We use the Adam optimizer [8] with $\beta_1 = 0.9$ and $\beta_2 = 0.999$, over 4 GPUs with batch size 16. The total training epochs are 50, 15 and 10 for UCSD Ped2 [15], CUHK Avenue [13] and ShanghaiTech Campus [14] respectively and the learning rate is all set to 2e-4. The hyper-parameters of λ_1 and λ_2 are both set to 0.1 for balancing the different objective function. The scaling factor α in SGFE is set to 2. The number of sampling locations k is set to 5 for training LSLPN. All models are trained end-to-end using PyTorch [18], taking about 2, 8 and 25 h for UCSD Ped2, CUHK Avenue, and ShanghaiTech, respectively. The FPS is measured on a single GeForce GTX TITAN X GPU using batch size 1.

4.3 Ablation Study

To analyze the importance of each proposed module, we report the overall ablation study in Table 1(a). We first add SGFE and LCFE separately and then add them jointly to MemAE [6]. SGFE brings 0.9 points improvement on AUC compared with MemAE baseline, validating the fact that the suppressed latent feature $\hat{\mathbf{z}}_m$ is enhanced by the information transfer from $\mathbf{z}$ to $\hat{\mathbf{z}}_m$. LCFE improves the AUC from 94.1% to 94.8%. Compared with SGFE, this part of the improvement is a bit less. This result feels very unexpected. Logically speaking, LCFE should be more effective than SGFE as it consider local context informations. As shown in the last row of Table 1(a), LCFE can improve the AUC from 95.0% to 96.5% when SGFE is used. For LCFE, the improvement of 1.5% AUC with SGFE is much higher than 0.7% AUC without using SGFE. If the LCFE models the context on the latent feature $\hat{\mathbf{z}}_g$ (last row of Table 1(a)) where each grid feature has been enhanced by the SGFE, a more accurate context can be modeled. On the contrary, when LCFE is directly applied to the $\hat{\mathbf{z}}_m$ modeling context (third row in Table 1(a)), each grid feature is not enhanced, which makes it difficult to model an effective context. It's worth mentioning that the results of our LCFE (96.5% AUC) is notably better than Non-local [20] (95.5% AUC).

As mentioned in Sect. 3.2, we can achieve the information transfer from $\mathbf{z}$ to $\hat{\mathbf{z}}$ by the concatenation operation which has been adopted in existing work MNAD [17]. But this way of information transfer is sub-optimal in anomaly detection because all information in $\mathbf{z}$ is transferred to $\hat{\mathbf{z}}_m$. Especially, after the abnormal information is transferred, it may be harmful to the abnormal discovery. Different from concatenation operation, our proposed SGFE transfers

Table 1. Two modules (SGFE and LCFE) of our proposed method are applied to MemAE and MNAD. The results are evaluated on UCSD Ped2 dataset.

(a) Applying two modules on MemAE.

Settings	SGFE	LCFE	AUC	FPS
MemAE [6]			94.1	20.0
MemAE	✓		95.0	20.0
MemAE		✓	94.8	18.5
MemAE	✓	✓	**96.5**	18.4

(b) Applying two modules on MNAD.

Settings	SGFE	LCFE	AUC	FPS
MNAD [17]			90.2	34.5
MNAD	✓		93.0	35.4
MNAD	✓	✓	**94.3**	33.8

Table 2. AUC of different methods on UCSD Ped2 [15], CUHK Avenue [13] and ShanghaiTech [14] datasets. Numbers in bold indicate the best performance.

Method	UCSD Ped2	CUHK Avenue	ShanghaiTech	Year
FramePred [12]	95.4	85.1	72.8	2018
LSA [1]	95.4	–	72.8	2019
BMAN [10]	96.6	**90.0**	76.2	2019
VEC [24]	97.3	89.6	74.8	2020
MNAD [17]	90.2	82.8	69.8	2020
MemAE [6]	94.1	83.3	71.2	2019
ELF (ours)	**98.2**	88.6	**76.8**	–

the information from $\mathbf{z}$ to $\hat{\mathbf{z}}_m$ according to the aggregation weight which is calculated by the difference between $\mathbf{z}$ and $\hat{\mathbf{z}}_m$. It can prevent the abnormal information of $\mathbf{z}$ from being passed to $\hat{\mathbf{z}}_m$ to some extent. As shown in the second row of Table 1(b), when we replace the concatenation operation with SGFE in MNAD, it can achieve a significant improvement on AUC from 90.2% to 93.0%. Finally, applying LCFE to MNAD can further improve the performance.

Runtime. As shown in Table 1(a) and Table 1(b), ELE only brings about 1.6 FPS and 0.7 FPS time-consumption compared to baseline MemAE and MAND, respectively.

4.4 More Results on Three Benchmark Datasets

To show the competitive performance of ELF on different datasets, we compare our results on UCSD Ped2 [15], CUHK Avenue [13] and ShanghaiTech [14] datasets as shown in Table 2. Compared with the baseline method MemAE [6], our method achieves significant improvements on UCSD Ped2 4.1% AUC, CUHK Avenue 5.3% AUC and ShanghaiTech 5.6% AUC, respectively. Even comparing with methods [1,10,12,24] that use many additional techniques, *e.g.* optical flow, adversarial training, complex autoregression layer, bidirectional multi-scale aggregation, pre-trained object detector, the performance of the proposed ELF is still comparable. Note that the purpose of our experiment is not pursuing the

Fig. 5. Each frame shows the sampling locations ($5 \times 9 = 45$ red points in each frame) in 3×3 feature grids on (left to right) UCSD Ped2 [15], CUHK Avenue [13], and ShanghaiTech [14]. The green points denote the centers of 3×3 feature grids. (Color figure online)

highest AUC on certain applications but to demonstrate the advantages of the proposed improvement of memory-augmented AE.

4.5 Visualization

To verify that LSLPN can predict effective sampling locations for modeling the most useful local context information, we decode the regression offset predicted by LSLPN into the coordinates of the input frame and then visualize it. As shown in Fig. 5, most sampling locations are concatenated in the most similar parts of the local region, and only a few irrelevant feature are transferred.

5 Conclusion

In this work, to alleviate the weakness of the memory-augmented AE, we propose a simple but effective method, ELF, to enhance the latent features. Specifically, our method integrates two modules SGFE and LCFE which can be embedded into memory-augmented AE and executed sequentially. With this two modules, the latent features are gradually enhanced by the single grid features and the useful local context features. As a future work, we will pursue the possibility of applying ELF on other challenging application scenarios.

Acknowledgement. This work is partially supported by Natural Science Foundation of China (NSFC): 61976203 and the Open Project Fund from Shenzhen Institute of Artificial Intelligence and Robotics for Society, under Grant No. AC01202005015.

References

1. Abati, D., Porrello, A., Calderara, S., Cucchiara, R.: Latent space autoregression for novelty detection. In: CVPR (2019)
2. Bengio, Y., Lamblin, P., Popovici, D., Larochelle, H.: Greedy layer-wise training of deep networks. In: NeurIPS (2007)

3. Chalapathy, R., Menon, A.K., Chawla, S.: Anomaly detection using one-class neural networks. arXiv:1802.06360 (2018)
4. Doersch, C.: Tutorial on variational autoencoders. arXiv:1606.05908 (2016)
5. Girshick, R.: Fast R-CNN. In: ICCV (2015)
6. Gong, D., et al.: Memorizing normality to detect anomaly: memory-augmented deep autoencoder for unsupervised anomaly detection. In: ICCV (2019)
7. Hou, R., Ma, B., Chang, H., Gu, X., Shan, S., Chen, X.: Interaction-and-aggregation network for person re-identification. In: CVPR (2019)
8. Kingma, D.P., Ba, J.: Adam: a method for stochastic optimization. arXiv:1412.6980 (2014)
9. Kingma, D.P., Welling, M.: Auto-encoding variational bayes. arXiv:1312.6114 (2013)
10. Lee, S., Kim, H.G., Ro, Y.M.: BMAN: bidirectional multi-scale aggregation networks for abnormal event detection. IEEE Trans. Image Process. **29**, 2395–2408 (2019)
11. Liu, R., et al.: An intriguing failing of convolutional neural networks and the coordconv solution. In: NeurIPS (2018)
12. Liu, W., Luo, W., Lian, D., Gao, S.: Future frame prediction for anomaly detection-a new baseline. In: CVPR (2018)
13. Lu, C., Shi, J., Jia, J.: Abnormal event detection at 150 fps in matlab. In: ICCV (2013)
14. Luo, W., Liu, W., Gao, S.: A revisit of sparse coding based anomaly detection in stacked RNN framework. In: ICCV (2017)
15. Mahadevan, V., Li, W., Bhalodia, V., Vasconcelos, N.: Anomaly detection in crowded scenes. In: CVPR (2010)
16. Pang, J., Chen, K., Shi, J., Feng, H., Ouyang, W., Lin, D.: Libra R-CNN: towards balanced learning for object detection. In: CVPR (2019)
17. Park, H., Noh, J., Ham, B.: Learning memory-guided normality for anomaly detection. In: CVPR (2020)
18. Paszke, A., et al.: Automatic differentiation in pytorch (2017)
19. Rezende, D.J., Mohamed, S., Wierstra, D.: Stochastic backpropagation and approximate inference in deep generative models. In: ICML (2014)
20. Wang, X., Girshick, R., Gupta, A., He, K.: Non-local neural networks. In: CVPR (2018)
21. Wang, X., Kong, T., Shen, C., Jiang, Y., Li, L.: SOLO: segmenting objects by locations. In: Vedaldi, A., Bischof, H., Brox, T., Frahm, J.-M. (eds.) ECCV 2020. LNCS, vol. 12363, pp. 649–665. Springer, Cham (2020). https://doi.org/10.1007/978-3-030-58523-5_38
22. Weston, J., Chopra, S., Bordes, A.: Memory networks. arXiv:1410.3916 (2014)
23. Xiong, L., Póczos, B., Schneider, J.G.: Group anomaly detection using flexible genre models. In: NeurIPS (2011)
24. Yu, G., et al.: Cloze test helps: effective video anomaly detection via learning to complete video events. In: ACMMM (2020)
25. Zhao, H., Shi, J., Qi, X., Wang, X., Jia, J.: Pyramid scene parsing network. In: CVPR (2017)
26. Zong, B., et al.: Deep autoencoding gaussian mixture model for unsupervised anomaly detection. In: ICLR (2018)

Adaptive Anomaly Detection Network for Unseen Scene Without Fine-Tuning

Yutao Hu[1], Xin Huang[1], and Xiaoyan Luo[2](✉)

[1] School of Electronic and Information Engineering, Beihang University, Beijing, China
{huyutao,huangxx156}@buaa.edu.cn

[2] School of Astronautics, Beihang University, Beijing, China
luoxy@buaa.edu.cn

Abstract. Anomaly detection in video is a challenging task with great application value. Most existing approaches formulate anomaly detection as a reconstruction/prediction problem established on the encoder-decoder structure. However, they suffer from the poor generalization performance when the model is directly applied to an unseen scene. To solve this problem, in this paper, we propose an Adaptive Anomaly Detection Network (AADNet) to realize few-shot scene-adaptive anomaly detection. Our core idea is to learn an adaptive model, which can identify abnormal events without fine-tuning when transferred to a new scene. To this end, in AADNet, a Segments Similarity Measurement (SSM) module is utilized to calculate the cosine distance of different input video segments, based on which the normal segments will be gathered. Meanwhile, to further exploit the information of normal events, we design a novel Relational Scene Awareness (RSA) module to capture the pixel-to-pixel relationship between different segments. By combining the SSM module with RSA, the proposed AADNet becomes much more generative. Extensive experiments on four datasets demonstrate our method can adapt to a new scene effectively without fine-tuning and achieve the state-of-the-art performance.

Keywords: Anomaly detection · Scene-adaptive · Few-shot learning

1 Introduction

Anomaly detection is an important computer vision task due to its real-world applications in autonomous surveillance systems [2,6,10,17,25]. Although received considerable attention in recent years, it remains to be a challenging problem due to two factors. First, the definition of "anomaly" is ambiguous, in which any event that does not conform to normal behaviours can be considered as an anomaly. Second, it is difficult to collect abnormal samples due to the rare occurrence of anomalies. As a result, in most previous works, anomaly detection is formulated as a reconstruction/prediction problem and only normal samples are involved in the training. Consequently, in the test stage, anomaly

H. Ma et al. (Eds.): PRCV 2021, LNCS 13020, pp. 311–323, 2021.
https://doi.org/10.1007/978-3-030-88007-1_26

events will not be generated and the reconstruction error is leveraged to indicate the extent of abnormalities. However, these approaches are established on the assumption that the training and test samples are from the same scene, leading to the unsatisfactory performance when the model is transferred to a new scene.

Recently, few-shot learning, aiming to learn a model that can be quickly adapted to the new tasks with a small amount of annotated samples, has obtained great success in many computer vision tasks [4,7,9,22]. Specifically, to achieve the rapid adaption for the previously unseen scene, Lu *et al.* [12] propose few-shot scene-adaptive anomaly detection, in which the test scenes are completely different from the training ones. However, this method adopts the optimization-based framework, leading to an extra fine-tuning process when transferred to a new scene during the test stage.

In this paper, we propose an Adaptive Anomaly Detection Network (AADNet) for few-shot scene-adaptive anomaly detection, which can adapt to the previously unseen scene without extra training. Our AADNet adopts the metric-based framework, which includes a support set and a query set, respectively. During the training phase, the support set contains a group of normal samples and abnormal samples as prior information to train the model. While during the test, in order to detect the anomalies from an unseen scene, we only involve very few normal samples as the reference in the support set.

The proposed AADNet builds upon the encoder-decoder structure and adopts the prediction strategy, in which multiple video segments are fed as inputs and the prediction of the next frame is generated. The learning process within AADNet is conducted in a coarse-to-fine manner, in which the anomaly detection is performed from frame-level to pixel-level. As shown in Fig. 1, after the encoder extracting multi-level features, we design a Segments Similarity Measurement (SSM) module to calculate the cosine distance of different segments and gather the normal samples together. By doing so, SSM module learns the information of normal events in frame-level. After that, we design a novel Relational Scene Awareness (RSA) module to establish the pixel-to-pixel comparison between support and query images. To achieve this goal, in RSA module, non-local operation [23] is leveraged to learn long-range relationship across the spatial points and transmit the information of normal events from the support set to query set. Afterwards, the decoder is deployed to produce the predicted frame.

In the test, given the normal samples in the support set for the unseen scene, SSM module calculates the distance between the query samples and support ones, which is utilized as an important criterion for final judgement. Meanwhile, RSA module restrains the abnormal information and enhances the normal one in the query samples according to the support set, which contributes to generating the predicted frame without anomalies. By cooperatively performing the SSM and RSA module, AADNet shows the great generalization ability in few-shot scene-adaptive detection. Compared to the approaches toward traditional anomaly detection, AADNet utilizes the normal samples in the support set as the reference to learn normal information, which enable the model to adapt to the unseen scene. Meanwhile, we think our method is close to the real-world applications. Once the model is trained, we only need to collect very few normal samples before deploying the model to a new seen, which is easy to realize. Extensive experiments on several challenging datasets also demonstrate the proposed AADNet can detect abnormal events in an unseen scene without fine-tuning and establishes the new state-of-the-art.

In summary, this paper makes the following contributions:

- We propose an Adaptive Anomaly Detection Network (AADNet) for few-shot scene-adaptive anomaly detection, which can be applied to an unseen scene effectively without fine-tuning.
- We design a Segments Similarity Measurement (SSM) module to calculate the cosine distance of different input segments and gather the normal samples, which measures the similarity of inputs in frame-level.
- We design a Relational Scene Awareness (RSA) module to exploit the pixel-wise relationship and transmit the knowledge from the support set to query set, which learns the information of normal events in pixel-level.
- Extensive experiments show the proposed AADNet achieves the state-of-the-art performance when the model is directly applied to the previously unseen scene, outperforming the prior approaches by a large margin.

2 Related Work

2.1 Anomaly Detection

Given a video sequence, anomaly detection is the task of identifying frames that contain events deviating significantly from the norm. Due to the scarcity of abnormal samples, many approaches formulate anomaly detection as an unsupervised learning problem, where anomalies data are not available at training time [18]. Generally, these works can be divided into two groups: reconstruction-based [1,3,6,26] and prediction-based approaches [6,10,11,13,14]. Both reconstruction-based and prediction-based approaches rely on the observation that the anomalies cannot be generated accurately by a model only learned via normal data. Differently, the former one concentrates on reconstructing the input frame, while the latter one attempts to predict the next frame of input segments. Notably, our AADNet adopts the prediction-based strategy.

2.2 Few-Shot Learning

Few-shot learning aims to quickly adapt to a new task with only a few training samples, which has achieved great success in many computer vision tasks, such as few-shot classification [22], few-shot detection [4] and few-shot segmentation [7]. The researches toward few-shot learning can be divided into three categories: model-based, optimization-based and metric-based approaches. Model-based approaches train a meta-learner with memory blocks to learn novel concepts [15,21]. Optimization-based approaches learn the good initial parameters for the network, which can be adapted to a new task via fine-tuning from limited

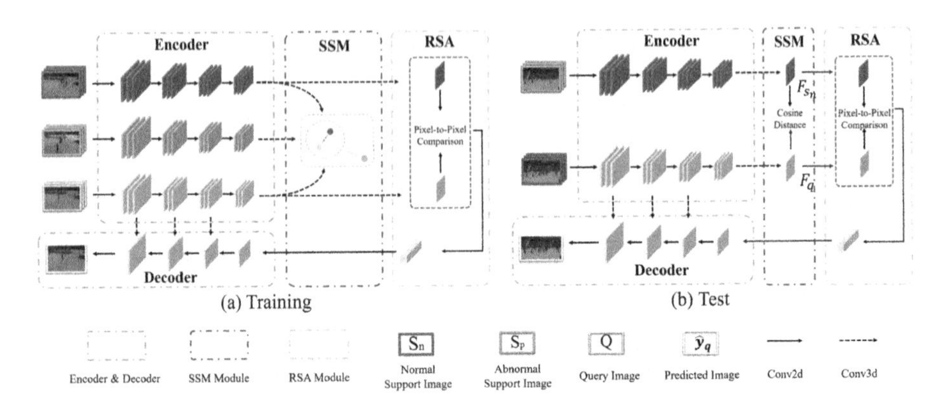

Fig. 1. The framework of Adaptive Anomaly Detection Network (AADNet). The training and test process are shown in the left and right, respectively. In the training, only a small amount of abnormal samples are involved. While in the test, only K normal samples from the unseen scene are randomly selected as the reference.

annotated samples [5,12,19]. Metric-based approaches learn a deep representation with a metric in feature space to compute the pair-wise distance among support and query images, which are the most popular methods for few-shot learning in recent years [7,16,22]. Specifically, metric-based approaches usually establish a prototypical network and calculate the similarity via various distance function. Our AADnet is a metric-based approach and can be applied to the previously unseen scene without fine-tuning.

3 Method

In our work, the training set $\mathcal{D}_{train}$ and test set $\mathcal{D}_{test}$ are constructed from two non-overlapping sets of scenes $\mathcal{C}_{seen}$ and $\mathcal{C}_{unseen}$, respectively. In the training, the AADNet is learned from $\mathcal{T}_{train} = \{S_n, S_p, Q\}$, which are episodically sampled from $\mathcal{D}_{train}$. Specifically, $S_n = \{\boldsymbol{x}^i_{s_n,t}\}$ and $S_p = \left\{\boldsymbol{x}^i_{s_p,t}\right\} (i \in \{1, ..., K\}, t \in \{1, ..., T\})$ denote normal and abnormal samples in the support set respectively, where K indicates the number of segments in S_n (or S_p) and T denotes the number of frames in each segment. Meanwhile, $\boldsymbol{x}$ represents each input frame. Moreover, in the training, the query set, $Q = \{\boldsymbol{x}_{q,t}, \boldsymbol{y}_q\}(t \in \{1, ..., T\})$, only contains normal segments without anomaly. $\boldsymbol{y}_q$ is the T+1$-th$ frame, which is utilized as the ground-truth for the predicted frame. Once the network is learned, the model can be directly deployed to an unseen scene $\mathcal{T}_{test} = \{S_n, Q\}$ without fine-tuning. In other words, in the test, we only utilize very few normal samples from the unseen scene as reference in the support set.

The whole framework of AADNet is depicted in Fig. 1. Our model is composed of four components, encoder, SSM module, RSA module and decoder. The learning process within AADNet is performed in a coarse-to-fine manner. Specifically, for each input segment, the encoder extracts multi-level feature maps $F_{1,t}$,

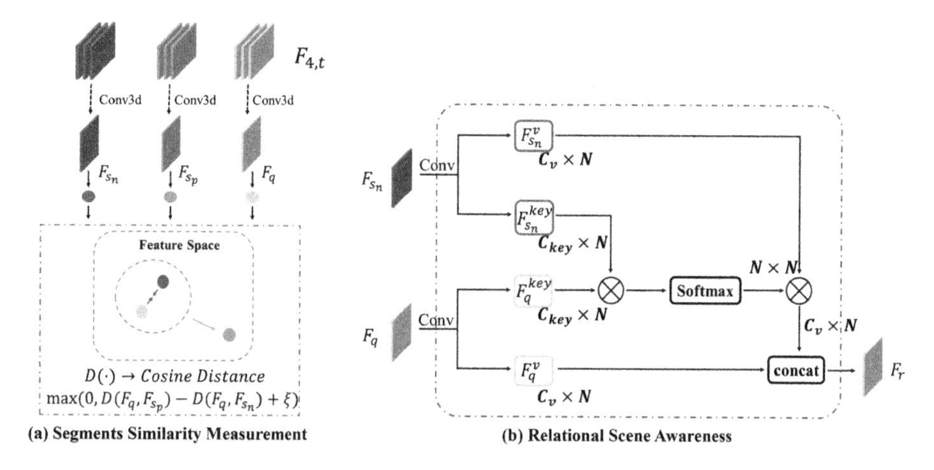

Fig. 2. (a) The structure of Segments Similarity Measurement (SSM) module. (b) The structure of Relational Scene Awareness (RSA) module.

$F_{2,t}$, $F_{3,t}$, $F_{4,t}$, in which $F_{1,t}$, $F_{2,t}$, $F_{3,t}$ are delivered to the decoder through skip connection. Meanwhile, features from the last layer, $F_{4,t}$, is sent to the SSM module, in which we obtain a temporal-aggregated feature F and calculate the cosine distance between different segments. Afterwards, the RSA module is leveraged to establish pixel-wise comparison and transmit the knowledge of normal event from the support set to query set. Finally, the decoder takes the refined feature from RSA module as input and produces the predicted frame.

3.1 Segments Similarity Measurement

After the encoder extracting multi-level features, the SSM module is utilized to measure the similarity of different segments in frame-level. The structure of SSM module is illustrated in Fig. 2(a). Specifically, given multiple features extracted from different frames, 3D-Convolution is first applied to generate temporal-aggregated feature F_{s_n}, F_{s_p} and F_q for each input segments. Then, cosine distance is employed to calculate the similarity between different features, which is denoted as:

$$D\left(F_q, F_{s_n}\right) = \frac{F_q \cdot F_{s_n}}{\sqrt{||F_q||_2^2 \cdot ||F_{s_n}||_2^2}}, \quad D\left(F_q, F_{s_p}\right) = \frac{F_q \cdot F_{s_p}}{\sqrt{||F_q||_2^2 \cdot ||F_{s_p}||_2^2}}. \tag{1}$$

Afterwards, the triplet loss is calculated as:

$$\mathcal{L}_{triplet} = \max\left(0, D\left(F_q, F_{s_p}\right) - D\left(F_q, F_{s_n}\right) + \xi\right). \tag{2}$$

In the training, $D\left(F_q, F_{s_n}\right)$ is expected to be larger than $D\left(F_q, F_{s_p}\right)$ with a least margin ξ. By doing so, SSM module gathers the normal samples together while pushes the abnormal samples away, which distinguishes the abnormal segments from several inputs.

3.2 Relational Scene Awareness

Besides the frame-level learning, to further obtain the fine-grained information of normal events, we exploit the condition of each pixel in the query features. To be more specific, since S_n only contains normal events, delicately comparing F_{s_n} with F_q contributes to exploiting the information of normal events in the current scene. Therefore, we design a novel RSA module to establish the pixel-to-pixel comparison between different inputs and extracts the long-range relationship. The detail of the RSA module is shown in Fig. 2(b).

In RSA module, several 1×1 convolutions are conducted on two input feature maps F_{s_n} and F_q to squeeze the channels and generate $F_{s_n}^{key}$, $F_{s_n}^{v}$, F_q^{key} and F_q^v. Then, they are flattened to $\hat{F}_{s_n}^{key}$, $\hat{F}_q^{key} \in \mathbb{R}^{C_{key} \times N}$ and $\hat{F}_{s_n}^{v}$, $\hat{F}_q^{v} \in \mathbb{R}^{C_v \times N}$, respectively, in which C_{key} and C_v are the number of channels, while $N = W \times H$ indicates the total spatial points. In our experiments, we set C_{key} and C_v to 128 and 512, respectively. After that, the relation matrix $V \in \mathbb{R}^{N \times N}$ is calculated as:

$$V = \left(\hat{F}_q^{key}\right)^T \times \hat{F}_{s_n}^{key} \tag{3}$$

where V denotes the matrix describing the long-range relationship between F_q and F_{s_n}, which indicates the similarity of each pixel-pair. Then, V is further normalized by the softmax function. Afterwards, matrix multiplication is performed as:

$$I_F = V \times \left(\hat{F}_{s_n}^{v}\right)^T \tag{4}$$

In this way, the RSA module enhances the information of normal events in the current scene and restrains the abnormal information. Finally, I_F is combined with F_q^v via concatenation to generate the refined feature F_r, which is regarded as the output of RSA module.

3.3 Integration

In AADNet, we deploy a weights-shared convolutional neural network as the feature extractor to obtain a sequence of deep feature maps. Specifically, the first five blocks of WideResNet-38 [24] is employed as the encoder, in which the features from $B2$, $B3$, $B4$ and $B5$ are denoted as $F_{1,t}$, $F_{2,t}$, $F_{3,t}$ and $F_{4,t}$, respectively. Among them, similar to the structure of U-Net [20], $F_{1,t}$, $F_{2,t}$ and $F_{3,t}$ are delivered to the decoder via skip-connection. Note that, before the skip connection, features extracted from different frames are fused to generate a temporal-aggregated feature through 3d-convolution. Meanwhile, $F_{4,t}$, the feature from the last layer of the encoder, is fed into the SSM and RSA module.

Afterwards, the decoder takes the refined feature F_r as the input and generates the final predicted frame $\hat{\boldsymbol{y}}$. To be more specific, the decoder has three up-sampling blocks. Each up-sampling block contains the nearest neighbor interpolation, followed by a 3×3 convolutional layer. Meanwhile, the output of each up-sampling block is concatenated to the corresponding feature map from the encoder, which brings rich spatial information [8,20]. Finally, the predicted frame

$\hat{\boldsymbol{y}}$ is produced. In the training, to generate the accurate prediction, we deploy multiple supervision on the $\hat{\boldsymbol{y}}$, which is denoted as:

$$\mathcal{L}_{rec} = 3\mathcal{L}_{MSE} + 2\mathcal{L}_{gdl} + 3\mathcal{L}_{ssim}, \tag{5}$$

where the $\mathcal{L}_{MSE}$ computes the L2 distance between the predicted frame and its ground truth:

$$\mathcal{L}_{MSE}(\hat{\boldsymbol{y}}_q, \boldsymbol{y}_q) = \frac{1}{W_q \times H_q} \left\|\hat{\boldsymbol{y}}_q - \boldsymbol{y}_q\right\|_2^2. \tag{6}$$

where W_q and H_q represent the width and height of the query image. The $\mathcal{L}_{gdl}$ denotes the gradient loss, which penalizes unclear edges and sharpens the generated frames:

$$\begin{aligned}\mathcal{L}_{gdl}(\hat{\boldsymbol{y}}_q, \boldsymbol{y}_q) = \sum_{i,j} (&\left\|\left|\hat{\boldsymbol{y}}_q^{i,j} - \hat{\boldsymbol{y}}_q^{i-1,j}\right| - \left|\boldsymbol{y}_q^{i,j} - \boldsymbol{y}_q^{i-1,j}\right|\right\| \\ &+ \left\|\left|\hat{\boldsymbol{y}}_q^{i,j} - \hat{\boldsymbol{y}}_q^{i,j-1}\right| - \left|\boldsymbol{y}_q^{i,j} - \boldsymbol{y}_q^{i,j-1}\right|\right\|),\end{aligned} \tag{7}$$

where i and j denote locations of each spatial point. Moreover, the $\mathcal{L}_{ssim}$ indicates Structural Similarity measurement, which is calculated as:

$$\mathcal{L}_{ssim}(\hat{\boldsymbol{y}}_q, \boldsymbol{y}_q) = 1 - SSIM(\hat{\boldsymbol{y}}_q, \boldsymbol{y}_q). \tag{8}$$

Then, combining triplet loss in Eq. 2, total loss for AADNet is denoted as:

$$\mathcal{L}_{AAD} = \mathcal{L}_{rec} + \mathcal{L}_{triplet}. \tag{9}$$

In the test, only very few normal samples from the new scene are involved in the support set as reference, based on which the SSM and RSA module are utilized to calculate the similarity and transmit the knowledge of normal events. Therefore, different from traditional anomaly detection method, our AADNet can extract normal information from the given samples in the support set, which enables the model to transfer to the unseen scene without extra training. Then, the decoder is expected to generate the predicted frame only containing the normal events in the current scene.

4 Experiment

4.1 Dataset

In the experiments, several challenging datasets are employed to verify the efficacy of AADNet, in which the CUHK Avenue, UR Fall and UCSD Ped2 are only utilized for the test. Meanwhile, the Shanghai Tech and UCF Crime are leveraged as the training set, respectively. As mentioned before, during the training stage, AADNet needs a small amount of abnormal samples to calculate the triplet loss, which helps to distinguish the abnormal segment. However, original training sets in Shanghai Tech and UCF Crime do not satisfy this requirement. For Shanghai Tech, the original training set only contains normal events. Therefore, for each scene, we randomly select five abnormal segments from the original test set of Shanghai Tech to enrich the training data. Additionally, similar to [12], we also leverage different scenes in Shanghai Tech as the training/test data to perform cross-scene evaluation. For the UCF Crime, the original training videos contain both normal and abnormal segments. Thus, we reserve all the normal videos and randomly select two abnormal segments for each scene when it is utilized in the training. Notably, in all the experiments, the test scenes are completely different from the training ones.

4.2 Evaluation Metric

In the test, following previous works [10,12,18], we utilize the average area under curve (AUC) to estimate the extent of abnormalities for each segment. Specifically, at first, we employ the cosine distance to indicate the extent of abnormalities in frame-level. Specifically, in SSM module, given the query feature F_q, we select its the nearest normal feature $F_{s_n;m}$ to compute the cosine distance:

$$D(F_q, F_{s_n;m}) = \frac{F_q \cdot F_{s_n;m}}{\sqrt{||F_q||_2^2 \cdot ||F_{s_n;m}||_2^2}} \tag{10}$$

where F_q and $F_{s_n;m}$ are the temporal-aggregated feature extracted from Q and S_n, respectively. Specifically, $F_{s_n;m}$ is one of the K features extracted from different support segments, which has the nearest distance from F_q. As suggested in [10], for one test video, we normalize the cosine distance of all frames to the range [0, 1] by a min-max normalization:

$$\bar{D}\left(F_q^t, F_{s_n;m}^t\right) = \frac{D\left(F_q^t, F_{s_n;m}^t\right) - \min_t D\left(F_q^t, F_{s_n;m}^t\right)}{\max_t D\left(F_q^t, F_{s_n;m}^t\right) - \min_t D\left(F_q^t, F_{s_n;m}^t\right)} \tag{11}$$

Besides, we also utilize the PSNR (Peak Signal-to-Noise Ratio) to indicate the extent of abnormalities in pixel-level. To be more specific, we denote the PSNR between the future frame $\boldsymbol{y}_q$ and the predicted frame $\hat{\boldsymbol{y}}_q$ as $P(\hat{\boldsymbol{y}}_q, \boldsymbol{y}_q)$. Then, similar to Eq. 11, $P(\hat{\boldsymbol{y}}_q, \boldsymbol{y}_q)$ is also normalized by a min-max normalization. Afterwards, combining $\bar{P}(\hat{\boldsymbol{y}}_q^t, \boldsymbol{y}_q^t)$ and $\bar{D}(F_q^t, F_{s_n;m}^t)$, the anomaly score for each video frame is denoted as:

$$S(t) = \alpha(1 - \bar{P}(\hat{\boldsymbol{y}}_q^t, \boldsymbol{y}_q^t)) + (1 - \alpha)(1 - \bar{D}(F_q^t, F_{s_n;m}^t)) \tag{12}$$

where α is regarded as a hyper-parameter. Finally, based on $S(t)$, we calculate the AUC score (%) by computing the area under the receiver operation characteristics (ROC) with varying threshold values for abnormality scores [17,18]. Notably, a higher AUC score indicates a better performance.

Table 1. Performance comparisons with Lu *et al.* [12] when the model is trained on the Shanghai Tech and UCF Crime, respectively.

Datasets	Methods	Shanghai tech			UCF Crime		
		1-shot	5-shot	10-shot	1-shot	5-shot	10-shot
UCSD Ped2	r-GAN	85.64	89.66	91.11	65.58	72.63	78.32
	meta r-GAN	91.19	91.80	92.8	83.08	86.41	90.21
	AADNet	**92.76**	**93.45**	**94.42**	**87.62**	**89.06**	**90.28**
CUHK Avenue	r-GAN	75.43	76.52	77.77	66.70	67.12	70.61
	meta r-GAN	76.58	77.10	78.29	72.62	74.68	79.02
	AADNet	**78.88**	**79.74**	**80.59**	**79.84**	**80.76**	**81.30**
UR Fall	r-GAN	73.53	74.76	76.12	71.20	72.85	74.02
	meta r-GAN	80.21	82.54	84.70	79.62	82.10	83.93
	AADNet	**83.26**	**84.85**	**85.53**	**81.31**	**83.27**	**85.88**

Table 2. Performance comparisons with Lu *et al.* [12] when the model is trained/tested on different scenes from the Shanghai Tech.

Methods	K = 1	K = 5	K = 10
r-GAN	71.61	70.47	71.59
meta r-GAN	74.51	75.28	77.36
AADNet	**75.00**	**76.86**	**77.94**

4.3 Implementation Details

In the training, all the input frames are resized 256×384 and augmented by random horizontal flipping. The initial learning rate is set to 8e-5 and drops by a factor of 0.8 every 3000 iterations. Meanwhile, ξ equals to 0.9 and K is set to 1, which means a support set only contains one normal segment and one abnormal segment in the training. In the test, after being resized to 256×384, the original image is directly fed into the network without data augmentation. Meanwhile, K is set to 1, 5, or 10, which corresponds to the k-shot setting. Besides, α in Eq. 12 is set to 0.6. Moreover, in the experiments, T is set to 3. Namely, each selected video segment contains three frames. At last, all the experiments are conducted on the NVIDIA GeForce TITAN X GPU.

4.4 Performance Comparison

In this part, we perform cross-scene experiments to evaluate the efficacy of the proposed AADNet. Since few-shot scene-adaptive anomaly detection is a newly proposed task, there are only very few prior works could be compared with. Following the recent work [12], we adopt the r-GAN and meta r-GAN as the comparisons. At first, we utilize the Shanghai Tech and UCF Crime as the

Fig. 3. The visualization of prediction result. The left three images come from the CUHK Avenue (White foreign bodies floating in the air are regarded as the anomalies), while the right ones come from the UR Fall (People falling down is regarded as the anomaly). In each group, there are three figures. **Left:** Input frame. **Middle:** Prediction error. The highlight region indicates the major differences between predicted frame and ground-truth, which is considered as the abnormal regions. **Right:** Annotated input. Pixels colored with red indicate the abnormal regions predicted by our model.

Table 3. Ablation studies of the proposed SSM and RSA modules when evaluated on the UCSD Ped2 and the CUHK Avenue, respectively.

SSM	RSA	UCSD Ped2			CUHK Avenue		
		1-shot	5-shot	10-shot	1-shot	5-shot	10-shot
×	×	87.36	88.12	90.28	75.51	76.59	77.81
✓	×	89.42	90.32	92.26	77.18	77.96	79.90
×	✓	88.83	89.57	91.11	76.34	77.08	78.73
✓	✓	**92.76**	**93.45**	**94.42**	**78.88**	**79.74**	**80.59**

training set, and test the model on the UCSD Ped2, CUHK Avenue and UR Fall, respectively. The results are listed in Table 1. Generally, AADNet produces the best results. When the model is trained on the Shanghai Tech, our model outperforms the second-best method by 1.57%, 2.30% and 3.05% on three test datasets under 1-shot setting, respectively. Meanwhile, when AADNet is trained on the UCF Crime, we also surpass the compared methods by a large margin.

Moreover, similar to [12], we also utilize different scenes in Shanghai Tech to perform the cross-scene evaluation. To be more specific, we randomly select six scenes for the training and utilize the other seven scenes in the test. The experimental results are listed in Table 2. As shown, the proposed AADNet also obtains a higher AUC score than the compared methods, which demonstrates the superiority of our method.

Additionally, to further reveal the effectiveness of AADNet, we visualize some predicted image in Fig. 3. As shown, there are obvious distortions and ambiguities around the abnormal events, demonstrating the great ability of our model in localizing the abnormal regions.

4.5 Ablation Study

To better understand the efficacy of the SSM and RSA module, we perform ablation studies by removing each component respectively. To be more specific, when removing the SSM module, we retain the 3d-convolution to obtain aggregated-temporal feature and discard the triplet loss. On the other hand, when ablating RSA module, we directly concatenate F_{s_n} and F_q without establishing pixel-to-pixel comparison. The results are listed in Table 3. Note that all the models are trained on the Shanghai Tech while evaluated on the UCSD Ped2 and CUHK Avenue. Compared to be baseline (first row), both the SSM and RSA modules improve the performance and generate a higher AUC score, demonstrating the great effectiveness of two proposed module. Meanwhile, the best performance is achieved when the SSM and RSA modules combined together, showing the two components are highly complementary.

5 Conclusion

In this paper, to solve the poor generalization ability of existing methods toward anomaly detection, we propose an Adaptive Anomaly Detection Network (AADNet), which can be applied to the previously unseen scenes without fine-tuning. To achieve this goal, in AADNet, we perform the anomaly detection in a coarse-to-fine manner. At first, Segments Similarity Measurement (SSM) module is deployed to calculate the cosine distance between different segments and compare the similarity in frame-level. Then, Relational Scene Awareness (RSA) module is designed to establish the pixel-to-pixel comparison and transmit the knowledge of normal events from the support set to query set. In this way, based on the very few given samples in the support set, AADNet can detect the anomalies in an unseen scene. Extensive experiments also show the great generalization ability of our method.

Acknowledgment. This paper was supported by the National Natural Science Foundation of China (NSFC) under grant No. U1833117.

References

1. Chalapathy, R., Menon, A.K., Chawla, S.: Robust, deep and inductive anomaly detection. In: Ceci, M., Hollmén, J., Todorovski, L., Vens, C., Džeroski, S. (eds.) ECML PKDD 2017. LNCS (LNAI), vol. 10534, pp. 36–51. Springer, Cham (2017). https://doi.org/10.1007/978-3-319-71249-9_3
2. Chandola, V., Banerjee, A., Kumar, V.: Anomaly detection: a survey. ACM Comput. Surv. (CSUR) **41**(3), 1–58 (2009)
3. Chang, Y., Tu, Z., Xie, W., Yuan, J.: Clustering driven deep autoencoder for video anomaly detection. In: Vedaldi, A., Bischof, H., Brox, T., Frahm, J.-M. (eds.) ECCV 2020. LNCS, vol. 12360, pp. 329–345. Springer, Cham (2020). https://doi.org/10.1007/978-3-030-58555-6_20
4. Fan, Q., Zhuo, W., Tang, C.K., Tai, Y.W.: Few-shot object detection with attention-RPN and multi-relation detector. In: Proceedings of the IEEE/CVF Conference on Computer Vision and Pattern Recognition, pp. 4013–4022 (2020)
5. Finn, C., Abbeel, P., Levine, S.: Model-agnostic meta-learning for fast adaptation of deep networks. In: International Conference on Machine Learning, pp. 1126–1135. PMLR (2017)

6. Gong, D., et al.: Memorizing normality to detect anomaly: memory-augmented deep autoencoder for unsupervised anomaly detection. In: Proceedings of the IEEE/CVF International Conference on Computer Vision, pp. 1705–1714 (2019)
7. Wang, H., Zhang, X., Hu, Y., Yang, Y., Cao, X., Zhen, X.: Few-shot semantic segmentation with democratic attention networks. In: Vedaldi, A., Bischof, H., Brox, T., Frahm, J.-M. (eds.) ECCV 2020. LNCS, vol. 12358, pp. 730–746. Springer, Cham (2020). https://doi.org/10.1007/978-3-030-58601-0_43
8. Hu, Y., et al.: NAS-count: counting-by-density with neural architecture search. In: Vedaldi, A., Bischof, H., Brox, T., Frahm, J.-M. (eds.) ECCV 2020. LNCS, vol. 12367, pp. 747–766. Springer, Cham (2020). https://doi.org/10.1007/978-3-030-58542-6_45
9. Karlinsky, L., et al.: Repmet: representative-based metric learning for classification and few-shot object detection. In: Proceedings of the IEEE/CVF Conference on Computer Vision and Pattern Recognition, pp. 5197–5206 (2019)
10. Liu, W., Luo, W., Lian, D., Gao, S.: Future frame prediction for anomaly detection-a new baseline. In: Proceedings of the IEEE Conference on Computer Vision and Pattern Recognition, pp. 6536–6545 (2018)
11. Lu, Y., Kumar, K.M., shahabeddin Nabavi, S., Wang, Y.: Future frame prediction using convolutional VRNN for anomaly detection. In: 2019 16th IEEE International Conference on Advanced Video and Signal Based Surveillance (AVSS), pp. 1–8. IEEE (2019)
12. Lu, Y., Yu, F., Reddy, M.K.K., Wang, Y.: Few-shot scene-adaptive anomaly detection. In: Vedaldi, A., Bischof, H., Brox, T., Frahm, J.-M. (eds.) ECCV 2020. LNCS, vol. 12350, pp. 125–141. Springer, Cham (2020). https://doi.org/10.1007/978-3-030-58558-7_8
13. Luo, W., Liu, W., Gao, S.: Remembering history with convolutional LSTM for anomaly detection. In: 2017 IEEE International Conference on Multimedia and Expo (ICME), pp. 439–444. IEEE (2017)
14. Medel, J.R., Savakis, A.: Anomaly detection in video using predictive convolutional long short-term memory networks. arXiv preprint arXiv:1612.00390 (2016)
15. Munkhdalai, T., Yu, H.: Meta networks. In: International Conference on Machine Learning, pp. 2554–2563. PMLR (2017)
16. Nichol, A., Schulman, J.: Reptile: a scalable metalearning algorithm. arXiv preprint arXiv:1803.02999, vol. 2, no. 2, p. 1 (2018)
17. Pang, G., Yan, C., Shen, C., Hengel, A.V.D., Bai, X.: Self-trained deep ordinal regression for end-to-end video anomaly detection. In: Proceedings of the IEEE/CVF Conference on Computer Vision and Pattern Recognition, pp. 12173–12182 (2020)
18. Park, H., Noh, J., Ham, B.: Learning memory-guided normality for anomaly detection. In: Proceedings of the IEEE/CVF Conference on Computer Vision and Pattern Recognition, pp. 14372–14381 (2020)
19. Ravi, S., Larochelle, H.: Optimization as a model for few-shot learning (2016)
20. Ronneberger, O., Fischer, P., Brox, T.: U-Net: convolutional networks for biomedical image segmentation. In: Navab, N., Hornegger, J., Wells, W.M., Frangi, A.F. (eds.) MICCAI 2015. LNCS, vol. 9351, pp. 234–241. Springer, Cham (2015). https://doi.org/10.1007/978-3-319-24574-4_28
21. Santoro, A., Bartunov, S., Botvinick, M., Wierstra, D., Lillicrap, T.: Meta-learning with memory-augmented neural networks. In: International Conference on Machine Learning, pp. 1842–1850. PMLR (2016)

22. Sung, F., Yang, Y., Zhang, L., Xiang, T., Torr, P.H., Hospedales, T.M.: Learning to compare: relation network for few-shot learning. In: Proceedings of the IEEE Conference on Computer Vision and Pattern Recognition, pp. 1199–1208 (2018)
23. Wang, X., Girshick, R., Gupta, A., He, K.: Non-local neural networks. In: Proceedings of the IEEE Conference on Computer Vision and Pattern Recognition, pp. 7794–7803 (2018)
24. Wu, Z., Shen, C., Van Den Hengel, A.: Wider or deeper: revisiting the resnet model for visual recognition. Pattern Recogn. **90**, 119–133 (2019)
25. Zaheer, M.Z., Mahmood, A., Astrid, M., Lee, S.-I.: CLAWS: clustering assisted weakly supervised learning with normalcy suppression for anomalous event detection. In: Vedaldi, A., Bischof, H., Brox, T., Frahm, J.-M. (eds.) ECCV 2020. LNCS, vol. 12367, pp. 358–376. Springer, Cham (2020). https://doi.org/10.1007/978-3-030-58542-6_22
26. Zong, B., et al.: Deep autoencoding gaussian mixture model for unsupervised anomaly detection. In: International Conference on Learning Representations (2018)

Facial Expression Recognition Based on Multi-scale Feature Fusion Convolutional Neural Network and Attention Mechanism

Yana Wu[1,2,3], Kebin Jia[1,2,3](✉), and Zhonghua Sun[1,2,3]

[1] Faculty of Information Technology, Beijing University of Technology, Beijing, China
kebinj@bjut.edu.cn

[2] Beijing Laboratory of Advanced Information Networks, Beijing University of Technology, Beijing, China

[3] Beijing Key Laboratory of Computational Intelligence and Intelligent System, Beijing University of Technology, Beijing, China

Abstract. Facial expression recognition plays an important role in human-computer interaction. At present, the traditional algorithm applied to facial expression recognition is not accurate enough, and deep learning methods have problems such as a large number of network model parameters, insufficient generalization ability, and high requirements on hardware devices in actual deployment. To solve these problems, a lightweight facial expression recognition model based on multi-scale feature fusion and attention mechanism is proposed in this paper. Firstly, a multi-scale deeply separable densely connected convolutional neural network is constructed, which can greatly reduce the number of parameters and computation. The convolution kernel of different scales can obtain the receptive fields of different scales, and finally integrate the features of multiple scales, which is beneficial to the learning of the features of different scales. A dense connection reduces gradient disappearance and improves model generalization performance. Secondly, group convolution is introduced to further reduce the number of parameters. Finally, the attention mechanism and multi-scale were integrated to improve the classification accuracy of the model. The network model proposed in this paper has been trained and verified on RAF-DB, FER2013, CK+ and FERPlus datasets. Experimental results show that compared with other classical convolutional neural networks, the model proposed in this paper can significantly reduce the number of parameters and calculation amount while ensuring high accuracy. In the future, it is of great significance for the practical application of facial expression recognition in medical monitoring, education assistance, traffic warning and other fields.

Keywords: Facial expression recognition · Dense connection · Group convolution · Attention mechanism · Multi-scale depth separation convolution

1 Introduction

As a research hotspot in the field of computer vision, facial expression recognition aims at analyzing and judging the expression categories of a given face. It has a wide

H. Ma et al. (Eds.): PRCV 2021, LNCS 13020, pp. 324–335, 2021.
https://doi.org/10.1007/978-3-030-88007-1_27

application prospect and great application value in the fields of educational psychology, traffic safety, medical monitoring and so on [1].

With the rapid development of deep learning technology, various studies using deep learning models for facial expression recognition have made great progress. Deep learning technology not only improves the recognition effect, but also dramatically increases the computing requirements and the hardware configuration required for model operation. However, in the production environment, the level of equipment configuration is limited by the cost, and too high configuration requirements will greatly hinder the practical application of the model. Therefore, how to reduce the number of model parameters so that the model can run normally on devices with low performance is an important research direction.

In recent years, Convolutional Neural Network (CNN) has become the most commonly used algorithm in computer vision tasks. There are two main directions for the optimization of the convolutional neural network: one is to increase the model depth by increasing the effective optimization distance of gradient descending, such as Resnet [2] for classic network structures; the other is to increase the width of the model by adding convolutional layers of different sizes for convolutional kernels when extracting feature images, such as Inception [3]. Whether the network depth or network width is increased, the improvement of accuracy will bring a large increase in the number of parameters.

When the traditional neural network is fully connected, the network has obvious redundancy. In fact, for the task of facial expression recognition, the features extracted by each layer of a neural network are limited, and the convergence of the model will not be destroyed if some network layers are randomly abandoned in the training process. In addition, the current expression recognition model often performs better for specific datasets. Considering the abandonment of the full connection mode, the generalization performance of the model can be improved to a certain extent. In view of the above problems, this paper proposes to apply dense connection to the task of expression recognition.

The main contributions of this paper are as follows:

1. The densely connected convolutional neural network is applied to the ex-pression recognition to realize the reuse of the feature map so that the network can learn more original information, improve the model accuracy and generalization performance, and reduce the number of model parameters and the amount of calculation.
2. The multi-scale deep separable convolution is introduced to integrate the features of different scales in multi-scale acquisition. The channel information can be more fully utilized to improve the model classification accuracy.
3. Attention module [4] is added to improve the accuracy of the network, and the combination method of attention module and a multi-scale module is explored, and grouping convolution is introduced to improve the calculation efficiency of the model.

2 Related Work

Deep learning algorithms do not rely on complex image preprocessing and do not need to design accurate manual features. Compared with traditional machine learning algorithms, deep learning algorithms perform better and have better robustness in the face of illumination, various poses, and occlusion, etc. Therefore, they are widely used in facial expression recognition tasks.

Hu et al. [5] used CNN network to complete expression recognition, and at the same time, multi-task learning was used to detect facial key points. In the multi-task training, the bottom layer of the CNN network is shared, and through automatic learning, the layer that needs to be shared is determined. This model is not affected by the distribution of datasets and has better robustness. Zhang et al. [6] proposed an end-to-end deep learning model based on the GAN network, which simultaneously synthesized face images and recognized facial expressions by using different poses and expressions. Different facial expressions generated by the GAN network enriched the data set and greatly improved the expression recognition rate of the model. In order to avoid complex feature extraction processes and data manipulation problems involved in traditional facial expression recognition, Li et al. [7] recognized facial expressions based on the Faster R-CNN model in the field of target detection. First, face expression images are normalized, and then the model convolutional network extracts features and reduces the dimension of feature maps. Region Proposal Networks (RPNS) generate candidate regions for detection by Faster R-CNN. Finally, softmax classifier and regression layer are used to give expression recognition classification and border coordinates respectively, which proves the generalization ability of Faster R-CNN in expression recognition. In the Emotiw competition, Liu et al. [8] used multi-modal fusion methods such as audio, video frames, video sequences and motion information of key points of faces, and used Densenet, VGG-16, LSTM, SVM and other models to process multiple modal information and finally performed multi-modal fusion. In 2020, Wang et al. proposed a simple yet efficient self-cure network (SCN) [9] that suppresses the uncertainties efficiently and prevents deep networks from over-fitting uncertain facial images.

Although the above model has made great progress in the task of facial expression recognition, effective facial expression recognition systems still face the following challenges, which are the motivation for this work.

In practical applications, firstly, the model needs to operate in a complex background environment, including complex lighting and facial occlusion. Therefore, more discriminating features should be extracted. Secondly, deep learning-based models rely on large amounts of data, which leads to complex training networks and the consumption of computing resources. So a lightweight network model should be proposed. Thirdly, most facial expression recognition methods focus on improving the ac-curacy of a particular dataset, rather than improving its generalization and robustness. To solve this problem, the model should be validated on multiple datasets.

3 Proposed Model

The network model proposed in this paper draws on Densenet's idea of dense connection, Densenet [10] is free from the traditional thinking of improving network performance by deepening network layers and widening the width. From the point of view of features, the number of network parameters is greatly reduced through feature reuse and bypass setting, and the gradient disappearance problem is alleviated to some extent. To further reduce the number of model parameters, deep separable convolution and grouping convolution are introduced. In addition, the original deep separable convolution usually uses a 3×3 single convolution kernel with limited receptive field and single extraction information, which is not conducive to model classification. In this study, the depth separable convolution with different sizes of multiple branch convolution kernels was introduced, and the channel attention module was added to assign importance weight to each channel, so as to improve the classification accuracy of the model.

3.1 The Algorithm Framework

The overall structure of the model proposed in this paper is shown in Fig. 1.

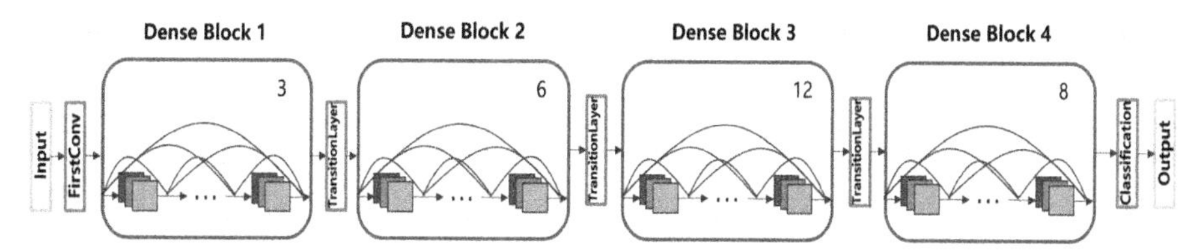

Fig. 1. Expression recognition network model based on densely connected convolutional neural network.

Consider a single image x that is passed through a convolutional network. The network comprises l layers, each of which implements a non-linear transformation $H_l(\cdot)$, where l indexes the layer. $H_l(\cdot)$ can be a composite function of operations such as batch normalization, rectified linear units, pooling, or convolution. We denote the output of the lth layer as x_l.

Traditional convolutional feed-forward networks connect the output of the lth layer as input to the $(l+1)^{th}$ layer. To further improve the information flow between layers we choose another connectivity pattern: we introduce direct connections from any layer to all subsequent layers. Figure 1 illustrates the layout of the resulting DenseNet schematically. Consequently, the lth layer receives the feature-maps of all preceding layers, $x_0,\ldots,x_{l-1}$, as input:

$$x_l = H_l([x_0, x_1, ..., x_{l-1}]) \tag{1}$$

where $[x_0, x_1, ..., x_{l-1}]$ refers to the concatenation of the feature maps produced in layers $0, ..., l-1$.

3.2 Bottleneck Layer Module

The concatenation operation used in Eq. (1) is not viable when the size of feature maps changes. However, an essential part of convolutional networks is down-sampling layers that change the size of feature maps. In order to facilitate the downward sampling in the architecture, we divide the network into four closely connected dense blocks, each of which is connected by a transition layer. The dense blocks can reduce the computing load of the network, and the downward sampling is carried out in the transition layer. Each dense block contains 3, 6, 12, and 8 custom substructure-the bottleneck layer module in turn, which can be expressed by the following equations:

$$F(x) = [P_w(M(x)), x] \tag{2}$$

$$M(x) = [M_1(x), M_2(x), ..., M_i(x)] \tag{3}$$

$$M_i(x) = D_{w_i}(G_i(x)) \tag{4}$$

where $G_i(x)$ means that the input image x is processed by 1×1 convolution layer, SE module and group convolution,i represents the index for each branch in each bottleneck layer module, $D_w(\cdot)$ is the depthwise convolution. It takes $G_i(x)$ as the input, and goes through the depthwise convolution to get $M_i(x)$.$[M_1(x), M_2(x), ..., M_i(x)]$ represents the result of depthwise convolution of each scale to carry out concatenate operation. P_w is the pointwise convolution, $[P_w(M(x)), x]$ represents the concatenation between the output obtained by the pointwise convolution and the original input x.

The specific network structure of the bottleneck layer module is shown in Fig. 2. The network consists of four branches. In the first three branches, the input image first goes through the 1×1 convolution layer for dimensionality reduction, then goes through the SE module, and then goes through the group convolution and separable convolution of different scales. For group convolution, convolution kernels are all 3×3. For separable convolution, the sizes of convolution kernels at three different scales are 1×1, 3×3 and 5×5, respectively. The last branch contains a 1×1 convolution layer, SE module, 3×3 group convolution layer and max-pooling layer. After concatenate operation connects the outputs of the four branches, pointwise convolution is carried out and finally concatenated with the original input x.

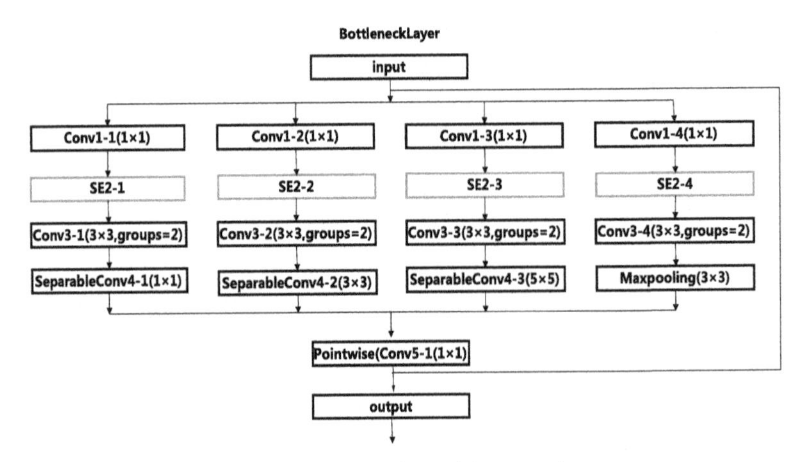

Fig. 2. The bottleneck layer module network structure.

3.3 Other Modules

FirstConv Module: After the facial expression image is input, it is first preprocessed and then passed through the FirstConv module, which contains a 3 × 3 convolution and a 2 × 2 average pooling layer to change the input image from 3 channels to the number of custom channels.

Transition Layer Module: We refer to layers between blocks as transition layers, which do convolution and pooling. The transition layers used in our experiments consist of a batch normalization layer and a 1 × 1 convolutional layer followed by a 2 × 2 average pooling layer.

Classification Module: The output classifier is designed based on the classification strategy of a full convolutional neural network. It consists of a batch normalization layer and an average pooling layer followed by a linear fully connected layer.

4 Experiments

4.1 Datasets

Our models are mainly trained on RAF-DB datasets. It contains nearly 30K images. In addition, there are two subsets of the dataset, including 7 basic expressions and 12 compound expressions (Fig. 3).

Fig. 3. The examples of RAF-DB dataset.

CK+, FER2013 and FERPLUS datasets are used in the comparative experiment. The CK+ dataset was released in 2010, consisting of 123 objects and 593 image sequences. The last frame of each image sequence is labeled with a facial motion unit. The 2013 facial expression recognition contest dataset held by FER2013 for Kaggle contains 28,709 training images, 3,589 public test images and 3,589 private test images. It is labeled into 7 categories. Due to the low accuracy of manual labeling, the FER2013 label has been proved to be inaccurate. In this case, Barsoum et al. used a crowd-sourcing approach to improve the accuracy of the labeling and added "contempt", "unknown" and "non-face" categories to the dataset known as FERPlus (Figs. 4 and 5).

Fig. 4. The examples of CK+ dataset.

Fig. 5. The examples of FER2013 dataset.

4.2 Experiments on RAF-DB Dataset

All the networks are trained using stochastic gradient descent (SGD). For all datasets, we train for 350 epochs, the learning rate is set at 0.01 at the beginning, and then drop from the 50th epoch to 90% of the last epoch every five epochs. The module is implemented using the PyTorch toolbox. All of the reported results are obtained by running the python code on an NVIDIA GeForce GTX 1080 Ti.

The research of this paper mainly considers solving the problem that the number of model parameters is too large to lead to practical application deployment and the existing model generalization performance needs to be improved. Firstly, the full connection mode of the traditional convolutional neural network is considered to be abandoned. By referring to the idea of the Densenet network, the dense connection network is applied to the facial expression recognition task to reduce the number of parameters and improve the generalization performance to ensure high accuracy. Considering that facial expression recognition requires high real-time performance in practical applications, it is necessary to further reduce the number of parameters. Model1 adds deep detachable convolution to bottleneck layer module, replacing the normal convolution in the original bottleneck layer structure. The convolution kernel is set to 1×1 and the step size is 1. The experimental results are shown in Table 1.

Table 1. Comparison of experimental results of depth separable convolution.

Model	Parameters (million)	Accuracy (%)
Densenet	0.34	81.26
Model1	0.22	81.32

As can be seen from Table 1, after the addition of the deep separable convolution, the accuracy is only slightly improved, but the number of parameters is greatly reduced, which shows the effectiveness of the deep separable convolution in the task of facial expression recognition.

However, the original depth separable convolution usually uses a 3×3 convolution. The receptive field of the single-scale convolution kernel is limited, and the extracted information is relatively single, which affects the classification performance of the model to some extent. In order to improve the richness of convolution features, the original single convolution kernel is replaced by a multi-scale convolution kernel. Model2 changes the combination of convolution kernels on the basis of Model1, and the proportions of different convolution kernels are allocated equally by default. The experimental results are shown in Table 2.

Table 2. Comparison of experimental results of multi-scale depth separable convolution.

Model	Convolution kernel combination	Parameters (million)	Accuracy (%)
Model1	3×3	0.22	81.32
Model2–1	3×3, 5×5	0.22	81.42
Model2–2	1×1, 3×3, 5×5	0.22	81.48
Model2–3	1×1, 3×3, 5×5, 7×7	0.24	81.36
Model2–4	1×1, 3×3, 5×5, max-pooling	0.22	81.52

As can be seen from Table 2, Model2–4 has the best performance. This is because 5×5 large size convolution nuclear energy is used to obtain a large receptive field, so as to facilitate the network to extract features of different scales. In addition, the error of feature extraction mainly comes from two aspects: the increase of the variance of the estimated value caused by the limited size of the neighborhood and the deviation of the estimated mean value caused by the parameter error of the convolutional layer. Max-pooling can reduce the second error and retain more texture information, which is suitable for the task of facial expression classification.

Depth of separable convolution compared to the original convolution to reduce the number of arguments, computing speed is faster, but the depthwise convolution collect only the characteristics of each channel, channel information interaction between enough will affect the model accuracy to a certain extent, in order to strengthen the interactivity between channels, at the same time avoid the sharp rise in the number of parameters, consider to join the group convolution before the depthwise convolution. On the basis of Model2–4, grouping convolution is added into the four branches in Model3. The experimental results are shown in Table 3.

Table 3. Comparison of experimental results of group convolution.

Model	Parameters (million)	Accuracy (%)
Model2–4	0.22	81.52
Model3	0.25	83.05

As can be seen from Table 3, the addition of grouping convolution enables information exchange between channels to avoid the loss of useful information, and thus greatly improves the accuracy.

Expression is an indicator of emotion projected by humans and other animals from the appearance of the body. Most of them refer to the state of facial muscles and facial features, such as smiles and anger. Recognition of facial expressions requires attention to the features of specific facial regions. The core goal of the attention mechanism in deep learning is to select the information that is more critical to the current task goal from numerous information. Therefore, considering the introduction of the attention mechanism into the model is conducive to the efficient transmission of specific information.

Model4 adds channel attention module SE on the basis of Model3, and changes the position of SE to carry out a comparative experiment. In the first three models, SE modules are added at different positions outside the branch, and in Model4–4, 1 × 1 convolution layer and SE module are added in each branch respectively. The experimental results are shown in Table 4.

Table 4. Comparison of experimental results of SE modules added in different positions of the model.

Model	SE Position	Parameters (million)	Accuracy (%)
Model3	–	0.25	83.05
Model4–1	After the average pooling layer of the transition layer	0.30	82.98
Model4–2	Before Pointwise	0.28	83.30
Model4–3	After Pointwise	0.28	83.41
Model4–4	After 1 × 1 convolutional layer in four branches	0.33	84.58

As can be seen from Table 4, Model4–4 has the best performance. This is because the function of the SE module is to assign importance weight to each channel. After dimensionality reduction of 1 × 1 convolutional layer, the input image passes through the SE module with more channels and the best effect is achieved.

Model4–4 is the final model proposed in this paper, which is named MSENet.

The model proposed in this paper was compared with the network model with high accuracy in the current facial expression recognition task and the current mainstream lightweight model. The experimental results are shown in Table 5 and Table 6, respectively.

Table 5. Comparison of experimental results of existing high precision models.

Model	Parameters (million)	Accuracy (%)
VGG16	138	88.98
Resnet18	11.17	87.78
MSENet	0.33	84.58

As can be seen from Table 5, compared with models with higher accuracy, although the accuracy of the model proposed in this paper is slightly reduced, the number of parameters is greatly reduced, which is more conducive to the practical application and deployment of the model. According to Table 6, compared with the current mainstream lightweight models, the model proposed in this paper has a higher accuracy rate.

Table 6. Comparison of experimental results of other lightweight models.

Model	Parameters (million)	Accuracy (%)
Mobilenet	3.88	80.66
Shufflenet	1.86	81.36
Densenet	0.34	81.26
MSENet	0.33	84.58

4.3 Experiments on Other Datasets

The model MSENet proposed in this paper is further verified on CK+, FER2013 and FERPlus datasets. The experimental results are shown in Table 7.

Table 7. Comparison of experimental results of different datasets.

Dataset	Model	Parameters (million)	Accuracy (%)
CK+	Mobilenet	3.88	92.21
	Densenet	0.34	87.37
	MSENet	0.33	90.10
FER2013	Resnet18	11.17	71.38
	Inception	0.39	71.60
	MSENet	0.33	71.89
FERPlus	VGG19	20.04	86.50
	Resnet18	11,17	83.35
	MSENet	0.33	82.67

The confusion matrix of FER2013 and FERPlus datasets is shown in Fig. 6.

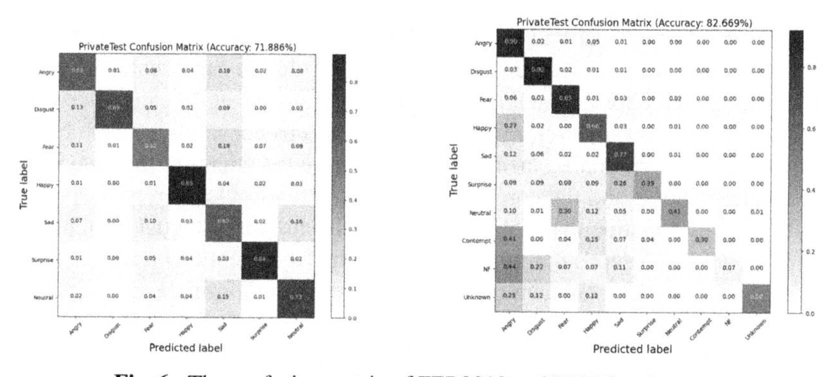

Fig. 6. The confusion matrix of FER2013 and FERPlus datasets.

As can be seen from Table 7, the lightweight model proposed in this paper, compared with other models that are currently superior in each dataset, not only ensures high accuracy but also reduces the size of the model, reduces the number of model parameters and has a better generalization performance.

5 Conclusion

In recent years, expression recognition algorithms based on deep learning have achieved good performance on public data sets. However, in the pursuit of accuracy, the increase of network depth and width often leads to a significant increase in the number of parameters and the amount of computation, which makes it difficult to deploy them in practical applications. At present, the focus pursued by the industry is gradually shifting from accuracy to speed and model size. Realizing the lightweight of the model and reducing the cost of the system in practical application are also issues worth considering in the future.

Therefore, this paper proposes an expression recognition model based on a densely connected convolutional neural network, which integrates multi-scale features and attention mechanism, and ensures a high accuracy while greatly reducing the number of parameters. The idea of dense connection comes from the Densenet network, whose main feature is to establish the connection between all the previous layers and the following layers, so as to realize the reuse of feature graph and make the network learn more original features. Deep separable convolution decomposes the standard convolution into deep convolution and separable convolution. Deep convolution is actually an extreme case of grouping convolution, where the number of groups is equal to the number of channels of the input characteristic graph. It independently performs the convolution operation on each channel of the input layer, which can greatly reduce the number of model parameters compared with the traditional convolutional neural network. On the basis of deep separable convolution, the single convolution kernel in the original convolution is replaced by a multi-scale convolution kernel. The convolution of different scales can obtain the receptive fields of different scales, and finally get the features of different scales, which is conducive to the recognition of facial expression images of different scales by the model and improve the classification accuracy. Further considering the depthwise convolution while collecting the spatial characteristics of each channel, but information interaction between channels is not enough, compared with the traditional convolution layer will lose some information, to overlay the traditional convolution layer number is too large, and grouping convolution can counteract these two points, to strengthen information interaction between channel at the same time increase in only a few parameters, therefore choose to join groups convolution in the branches. In addition, the recognition of facial expressions requires attention to the features of specific facial regions, and the core goal of the attention mechanism in deep learning is to select the information that is more critical to the current task goal from numerous information. Therefore, considering the introduction of the attention mechanism into the model is conducive to the efficient transmission of specific information.

The model proposed in this paper has achieved good results on RAF-DB, FER, FERPLUS and CK+ datasets, and there are still some areas to be improved in the follow-up work. For example, when the model is deployed in practical application, the real-time performance of the system should be considered, and the recognition speed should be further improved. At present, some more advanced attention mechanisms have been proposed, which may be applied to the facial expression recognition task to improve the model performance more greatly. In addition, for specific application scenarios, more real facial expression data sets need to be collected for further verification experiments.

Acknowledgment. This work was supported by the Beijing Natural Science Foundation under Grant No. 4212001, in part by the Basic Research Program of Qinghai Province under Grants No. 2020-ZJ-709 and No. 2021-ZJ-704.

References

1. Kanade, T., Cohn, J.F., Tian, Y.: Comprehensive database for facial expression analysis. In: Proceedings of the Fourth IEEE International Conference on Automatic Face and Gesture Recognition, Grenoble, France, pp. 46–53 (2020)
2. Li, X., Ding, L., Wang, L., Cao, F.: FPGA accelerates deep residual learning for image recognition. In: 2nd Information Technology, Networking, Electronic and Automation Control Conference (ITNEC), Chengdu, China, pp. 837–840 (2017)
3. Zhang, X., Huang, S., Zhang, X., Wang, W., Wang, Q., Yang, D.: Residual inception: a new module combining modified residual with inception to improve network performance. In: 25th IEEE International Conference on Image Processing (ICIP), Athens, Greece, pp. 3039–3043 (2018)
4. Vaswani, A., Shazeer, N., Parmar, N., et al.: Attention is all you need. In: Advances in Neural Information Processing Systems, pp. 5998–6008 (2017)
5. Guosheng, H., et al.: Deep multi-task learning to recognise subtle facial expressions of mental states. In: Ferrari, V., Hebert, M., Sminchisescu, C., Weiss, Y. (eds.) ECCV 2018. LNCS, vol. 11216, pp. 106–123. Springer, Cham (2018). https://doi.org/10.1007/978-3-030-01258-8_7
6. Zhang, F., Zhang, T., Mao, Q., et al.: Joint pose and expression modeling for facial expression recognition. In: Proceedings of the IEEE Conference on Computer Vision and Pattern Recognition (ECCV), pp. 222237 (2018). https://doi.org/10.1109/CVPR.2018.00354
7. Li, J., Zhang, D., Zhang, J., et al.: Facial expression recognition with faster R-CNN. Procedia Comput. Sci. **107**, 135–140 (2017)
8. Liu, C., Tang, T., et al. Multi-feature based emotion recognition for video clips. In: Proceedings of the 20th ACM International Conference on Multimodal Interaction, pp. 630634 (2018)
9. Wang, K., Peng, X., Yang, J., Lu, S., Qiao, Y.: Suppressing uncertainties for large-scale facial expression recognition. In: IEEE/CVF Conference on Computer Vision and Pattern Recognition (CVPR), Seattle, WA, USA, pp. 6896–6905 (2020)
10. Huang, G., Liu, Z., Laurens, V., et al.: Densely connected convolutional networks. In: Proceedings of the IEEE Conference on Computer Vision and Pattern Recognition, pp. 4700–4708 (2017)

Separable Reversible Data Hiding Based on Integer Mapping and Multi-MSB Prediction for Encrypted 3D Mesh Models

Zhaoxia Yin(✉), Na Xu, Feng Wang, Lulu Cheng, and Bin Luo

Anhui Provincial Key Laboratory of Multimodal Cognitive Computation,
School of Computer Science and Technology, Anhui University, Hefei 230601, China
yinzhaoxia@ahu.edu.cn

Abstract. Extensive research has been conducted on image-based reversible data hiding in encrypted domain (RDH-ED) methods, but these methods cannot be directly applied to other cover medium, such as text, audio, video, and 3D mesh. With the widespread use of 3D mesh on the Internet, the use of 3D mesh as cover medium for RDH has gradually become a research topic. The main challenge of studying RDH based on 3D mesh is that the data structure of 3D mesh is complex and the geometric structure is irregular. In this paper, we propose a separable RDH-ED method based on integer mapping and multiple most significant bit (Multi-MSB) prediction. Firstly, all vertices of 3D mesh are divided into "embedded" set and "reference" set, and floating-point vertex values are mapped to integers. Then, sender calculates prediction error of the "embedded" set. Data hider embeds additional data by replacing the Multi-MSB of the encrypted vertex coordinates of the "embedded" set without prediction error. According to different permissions, legal recipients can obtain the original mesh, the additional data or both of them by using the proposed separable method. Experimental results prove that the proposed method outperforms state-of-the-art methods.

Keywords: Reversible data hiding · 3D mesh · Multi-MSB prediction · Encrypted domain

1 Introduction

Reversible data hiding (RDH) methods [1–6] can recover cover medium losslessly after extracting embedded data, and are widely used in military, medical, remote sensing, law enforcement and other sensitive fields. The classic RDH methods are mainly based on lossless compression [1,2], difference expansion [3,4] and histogram shifting [5,6]. With the rapid development of cloud computing, the demand for privacy protection is growing. In order to store or share files safely, the original content is converted to unreadable ciphertext using encryption, and then additional data is embedded directly in the ciphertext. In some scenarios,

H. Ma et al. (Eds.): PRCV 2021, LNCS 13020, pp. 336–348, 2021.
https://doi.org/10.1007/978-3-030-88007-1_28

it is vital to restore the original content losslessly after decrypting and extracting the additional data. This privacy protection scenario triggers RDH-ED to manage ciphertext data. RDH-ED methods can be mainly classified into two categories: vacating room after encryption (VRAE) [7–9] and reserving room before encryption (RRBE) [10,11].

Zhang et al. [7] proposed a VRAE method for the first time. The encrypted image was divided into several non-overlapping blocks, and additional data was embedded by flipping the least significant bit (LSB) of the encrypted data. With the emergence of new application requirements in different scenarios, Zhang et al. [8] proposed a separable RDH-ED method for the first time. In this method, the least significant bit (LSB) of the encrypted image was losslessly compressed by a specific matrix multiplication, and the separability of data extraction and image restoration was realized. Compared with VRAE method, RRBE method has better performance in the accuracy of data extraction and image restoration. Ma et al. [10] first proposed RRBE method, which achieved data extraction error-free and lossless image restoration, and truly achieved reversibility. This method released a part of the original image by shifting the histogram, and embedded data by replacing the LSB of the encrypted pixel. Then, Puteaux et al. [11] predicted the MSB of the pixel, and embedded additional data through bit replacement according to the position indicator map.

Image-based RDH methods have been extensively studied, but these methods cannot be directly applied to other cover medium, such as text, audio, video and 3D mesh. Due to the wide application and huge inherent capacity of 3D mesh, 3D mesh was considered as a potential covering medium for RDH. At present, the research was still in its infancy, and many scientific and technological problems need to be solved. According to the literature, the existing RDH methods based on 3D model were mainly divided into four domains: spatial domain [12–14], transform domain [15], compressed domain [16] and encrypted domain [17–19].

The method in [17] was the first work on reversible data hiding in encrypted 3D mesh. Jiang et al. [17] flipped the LSBs of each vertex to embed one bit of data. The recipient used a smoothness estimation function for data extraction and mesh recovery. In order to improve the embedding capacity, Moshin et al. [18] proposed a two-layer embedding scheme based on homomorphic encryption. The sender of the first-tier used the histogram shifting to embed the additional data, and the second-tier cloud manager used the self-blind property of homomorphic encryption to embed the additional data into the marked encrypted mesh. However, due to the large ciphertext expansion and high computational complexity of Paillier cipher system, this method [18] was not efficient in practical application. Recently, Tsai et al. [19] used spatial coding technology with embedding threshold to embed additional data, and the embedding capacity has been improved. Due to the bit error rate in data extraction, the application of this method has certain limitations. In order to make full use of the model local correlation, we embedded additional data on the x-axis, y-axis, and z-axis of the vertices, replacing the n MSB bits of each coordinate axis with n bits of additional data. Compared with [17–19], the proposed method vacated more room for additional data embedding.

In this paper, we propose a separable RDH method for encrypted 3D mesh based on integer mapping and Multi-MSB prediction. The main contributions of this paper are as follows: (1) Multi-MSB embedding strategy is adopted to obtain higher embedding capacity. (2) By making full use of the correlation of adjacent vertices in natural mesh, the recipient can recover the Multi-MSB of the "embedded" vertex by Ring-prediction, so as to achieve lossless recovery mesh. (3) The proposed method can ensure the data extraction is error-free and separable, which is of great significance to privacy protection.

In this paper, Sect. 2 introduces the proposed method. Section 3 presents the analysis of experimental results. Section 4 concludes this paper and describes the future work.

2 Proposed Method

Figure 1 shows the framework of the proposed method. Firstly, the sender divides the original 3D mesh into "embedded" set and "reference" set. The sender analyzes the vertex information with prediction errors in the "embedded" set and records it as auxiliary information. The data hider uses the Multi-MSB bit replacement strategy to embed additional data into the encrypted 3D mesh $E(M)$, and can obtain the marked encrypted mesh $E(M)w$. According to different permissions, legal recipients can obtain the original mesh, the additional data or both of them by using the proposed separable method.

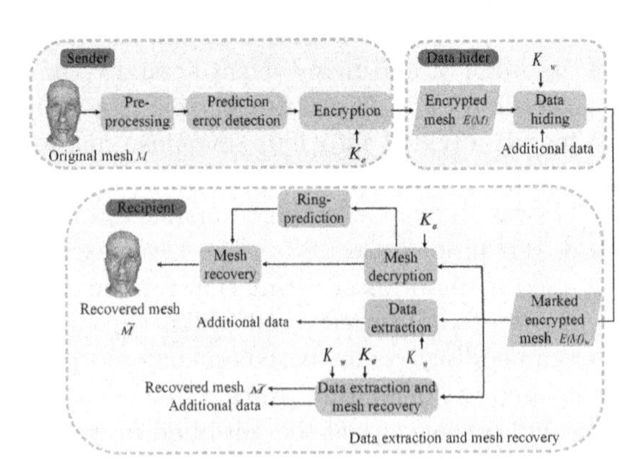

Fig. 1. Framework of the proposed method.

2.1 Pre-processing

According to the suggestion of [20], sender can perform lossy compression in different application scenarios. According to the different precision m, the corresponding integer value is between -10^m and 10^m, where $m \in [1, 33]$. Normalizing floating point coordinates $v_{i,j}$ to integer coordinates $\bar{v}_{i,j}$ as

$$\bar{v}_{i,j} = \lfloor v_{i,j} \times 10^m \rfloor, \tag{1}$$

where i is the ith vertex, $j \in \{x, y, z\}$, $v_{i,j}$ is the original set of floating point vertices and $\bar{v}_{i,j}$ is the set of integer vertices.

Recipient can convert the processed integer coordinates back to floating point coordinates by Eq. (2).

$$\hat{v}_{i,j} = \bar{v}_{i,j}/10^m, \tag{2}$$

the value of m corresponds to the bit-length l of integer coordinates as

$$l = \begin{cases} 8, & 1 \leq m \leq 2 \\ 16, & 3 \leq m \leq 4 \\ 32, & 5 \leq m \leq 9 \\ 64, & 10 \leq m \leq 33. \end{cases} \tag{3}$$

2.2 Prediction Error Detection

The "embedded" set s_e is used to embed additional data, and the "reference" set s_n is used to recover the mesh without modifying the vertices during the whole process. We traverse all the vertices contained in the face data in ascending order, and assume that $F = (f_1, f_2 \dots f_M)$ represents the face data sequence, where f_i=($v_{i,x}$, $v_{i,y}$, $v_{i,z}$), M is the number of face data. Assuming that f_n=($v_{n,x}$, $v_{n,y}$, $v_{n,z}$) is the next face sequence to be traversed, and both s_e and s_n are initially 0. If there is no vertex in f_n in s_e or s_n, we choose the first vertex in f_n to add $f_{n,x}$ to s_e, and add $f_{n,y}$ and $f_{n,z}$ to s_n. Figure 2 shows the close view of the Cow mesh and its vertex connections. The blue vertices in Fig. 2 are the "embedded" set vertices, the vertices marked in yellow are the "reference" vertices, and the red vertices are the vertices to be traversed.

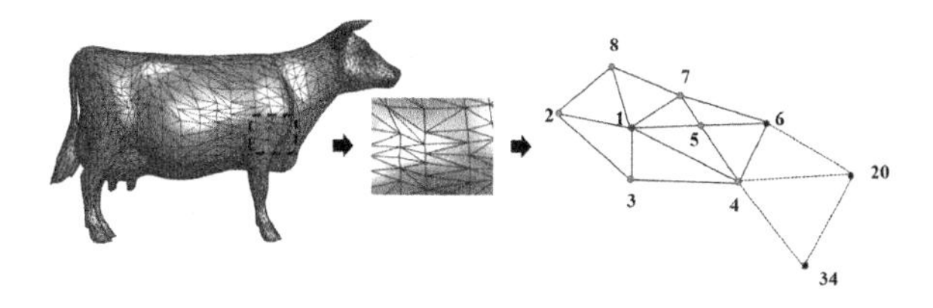

Fig. 2. The close view of Cow mesh and its vertex connection relationship. (Color figure online)

An example shows that when the m is 6, the process of selecting the maximum embedding length L without prediction error is as follows: we predict each bit of the prediction coordinates in the order of reference coordinates from MSB to LSB until a certain bit has prediction error. As shown in Fig. 3, the x coordinates of an "embedded" vertex numbered 1 has the MSB 0. The sender counts the number of 0 and 1 occurrences of the MSB of the "reference" vertex coordinates numbered 2, 3, 4, 5, 7, 8. If the number of 0s is greater than or equal to the number of 1s, the MSB of the "embedded" vertex coordinates numbered 1 is predicted to be 0. Then, the prediction of 2-MSB and 3-MSB are counted until the maximum embedding length $k1$ is found. According to Fig. 3, it can be found that the prediction error occurred in the bit prediction when embedding length is $t1$=17. Therefore, it can be concluded that the maximum embedding length $t1$ is 16 of x coordinates when the m is 6 on the Cow mesh. The maximum embedding length of vertex coordinate x is calculated as $t1$. Similarly, we calculate the maximum embedding length of vertex coordinate y, z axises as $t2$ and $t3$. At this time, the maximum embedding length of this vertex is $\min\{t1, t2, t3\}$. In the data embedding stage, the final maximum embedding length L of the embedded vertex is the minimum embedding length of all "embedded" vertex coordinates. After the prediction error detection of the x-axis, y-axis and z-axis coordinates of the vertex coordinates, if $n \geq 1$, we call vertex numbered 1 as the "embedded" vertex without prediction error in s_e. Otherwise, the vertex index information records as auxiliary information.

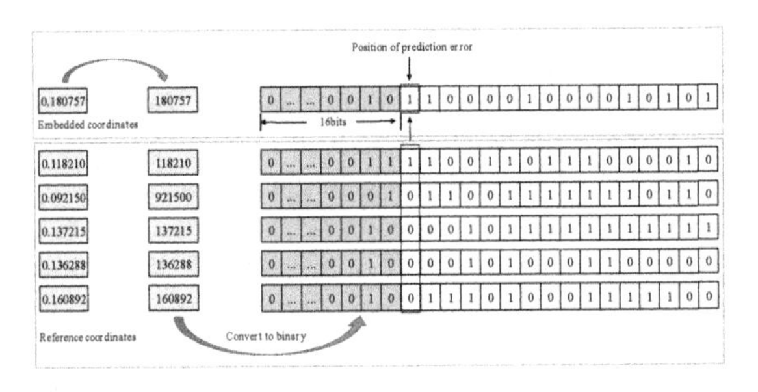

Fig. 3. An example of prediction error detection for vertex 1 of Cow mesh.

2.3 Encryption

The sender uses Eq. (4) to convert the pre-processed vertex integer coordinates into binary.

$$b_{i,j,u} = \lfloor \bar{v}_{i,j}/2^u \rfloor \quad mod \quad 2, \qquad u = 0, 1 \dots l-1, \tag{4}$$

where $\lfloor . \rfloor$ is a floor function and $l \leq i \leq N$ and $j \in \{x, y, z\}$, the l of the coordinate can be obtained by Eq. (3).

The sender encrypts the bit stream of the original 3D mesh $b_{i,j,u}$ by stream cipher function generated pesudo-random bits $c_{i,j,u}$, and obtains the encrypted coordinate binary stream $e_{i,j,u}$.

$$e_{i,j,u} = b_{i,j,u} \oplus c_{i,j,u}, \tag{5}$$

where $\oplus$ stands for exclusive OR.

The encrypted mesh can be obtained by Eq. (6)

$$E_{i,j} = \sum_{u=0}^{l-1} e_{i,j,u} \times 10^m, \tag{6}$$

where $E_{i,j}$ are the integral values of coordinates.

2.4 Data Hiding

The data hider first calculates s_e and s_n, and embeds the data into the n-MSB of the vertices in s_e without prediction error. The n-MSB of x, y, and z coordinates is replaced by n bits respectively. Finally, each vertex in s_e uses Eq. (7) to embed $3n$ bits.

$$v_{i,j}'' = s_1 \times 2^{l-1} + s_2 \times 2^{l-2} + \ldots + s_n \times 2^{l-n} + v_{i,j}' mod 2^{l-(n+1)}, \quad (7)$$

where s_k is additional data and $l \leq k \leq n$, n is embedding length and $1 \leq n \leq L$, $v_{i,j}' \in s_e$ is the vertex after pre-processing and encryption, $v_{i,j}''$ is the vertex of marked encrypted mesh.

2.5 Data Extraction and Mesh Recovery

Extraction with only Data Hiding Key. The n-MSB is extracted from the vertex coordinates of the s_e, and then the data hiding key Kw can be used to obtain the plaintext additional data.

$$s_k = v_{i,j}''/2^{l-k}, \qquad 1 \leq k \leq n \quad (8)$$

where $v_{i,j}'' \in s_e$ is vertex of the marked encrypted mesh.

Mesh Recovery with only Encryption Key. The recipient can recover the $E(M)w$ to get the M with Encryption Key Ke. After mesh decryption and Ring-prediction, the M can be obtained.

The pseudorandom bits $c_{i,j,u}$ are generated by the encryption key Ke, and is used to perform XOR function with $e''_{i,j,u}$ to decrypt the $E(M)w$.

$$b''_{i,j,u} = e''_{i,j,u} \oplus c_{i,j,u}, \quad (9)$$

where $e''_{i,j,u}$ is the binary stream of the marked encrypted mesh, $b''_{i,j,u}$ is the binary stream of the decrypted mesh with additional data and $u = 0, 1 \ldots l-1$.

After decrypting the $E(M)w$, since the s_n has not been modified in the whole process, the coordinate value after decryption is the original coordinate value. In order to recover the coordinate values in the s_e, after decryption of the $E(M)w$, the recipient can predict the n-MSB of the embedded vertices by embedding the n-MSB of adjacent vertices around the vertices, which is called Ring-prediction.

Extraction and Mesh Recovery with both Keys. In this case, the recipient can extract the additional data and recover the original 3D mesh perfectly. Note that data extraction step needs to be performed before mesh restoration.

3 Experimental Results and Discussion

In this section, we analyze the embedding capacity and reversibility of the proposed method, and compare the performance with state-of-the-art methods [17–19]. The experimental environment is MATLAB R2018b. Four standard original meshes, Beetle, Mushroom, Mannequin, and Elephant are used to demonstrate the experimental performance. Two data sets used for performance evaluation are: The Princeton Shape Retrieval and Analysis Group[1] and The Stanford 3D Scanning Repository[2]. In Sect. 3.1, we analyze the key indicator embedding capacity of the proposed method. In Sect. 3.2, Hausdorff distance and signal-to-noise ratio (SNR) are used to evaluate the geometric and visual quality of the proposed method, that is to evaluate reversibility. In Sect. 3.3, the performance comparison of the proposed method and the state-of-the-art methods is given. The additional data is a randomly generated 0/1 sequence.

3.1 Embedding Capacity

The embedding rate (ER) is defined as the ratio of the number of embedding bits to the number of vertices in the mesh model, that is, the number of bits per vertex (bpv). This section mainly discusses the impact on embedding rate under different m and embedding length n. Figure 4 show that when m takes a fixed value, with the increase of n, the embedding rate first increases and reaches the maximum embedding rate, and then shows a decreasing trend. Taking the Beetle as an example, when $m = 2$ and $n = 5$, the embedding rate is 3.70 bpv. When $m = 3$ and $n = 8$, the embedding rate is 7.74 bpv. When $m = 5$ and $n = 17$, the maximum embedding rate is 16.51 bpv. When $m = 6$ and $n = 14$, the embedding rate is 13.64 bpv, and the embedding rate shows a downward trend. Similarly, when $m = 5$ and $n = 15$, the maximum embedding rate of the Mushroom is 16.72 bpv. When $m = 5$ and $n = 15$, the maximum embedding rate of the Mannequin is 13.66 bpv. When $m = 5$ and $n = 18$, the maximum embedding rate of Elephant is 18.12 bpv.

3.2 Geometric and Visual Quality

For differences that cannot be distinguished by the naked eye, Hausdorff distance and SNR can be used to measure the geometric distortion. Hausdorff distance is defined as follows:

$$H(A,B) = max(h(A,B), h(B,A)), \tag{10}$$

$$h(A,B) = max(a \in A)min(b \in B) \parallel a - b \parallel, \tag{11}$$

$$h(B,A) = max(b \in B)min(a \in A) \parallel b - a \parallel, \tag{12}$$

[1] http://shape.cs.princeton.edu/benchmark/index.cgi.
[2] http://graphics.stanford.edu/data/3Dscanrep/.

where $\| \cdot \|$ is the distance between point a of set A and point b of set B (such as L2), a and b are the number of elements in the set.

The signal-to-noise ratio (SNR) is defined as follows: $SNR=$

$$10 \times \lg \frac{\sum_{i=1}^{N}[(v_{i,x}-\overline{v}_x)^2+(v_{i,y}-\overline{v}_y)^2+(v_{i,z}-\overline{v}_z)^2]}{\sum_{i=1}^{N}[(g_{i,x}-\overline{v}_x)^2+(g_{i,y}-\overline{v}_y)^2+(g_{i,z}-\overline{v}_z)^2]}, \tag{13}$$

where $\bar{v}_x$,$\bar{v}_y$,$\bar{v}_z$ are the averages of the mesh coordinates,$v_{i,x}$,$v_{i,y}$,$v_{i,z}$ are the original coordinates, $g_{i,x}$, $g_{i,y}$, $g_{i,z}$ are the modified mesh coordinates, N is the number of vertices.

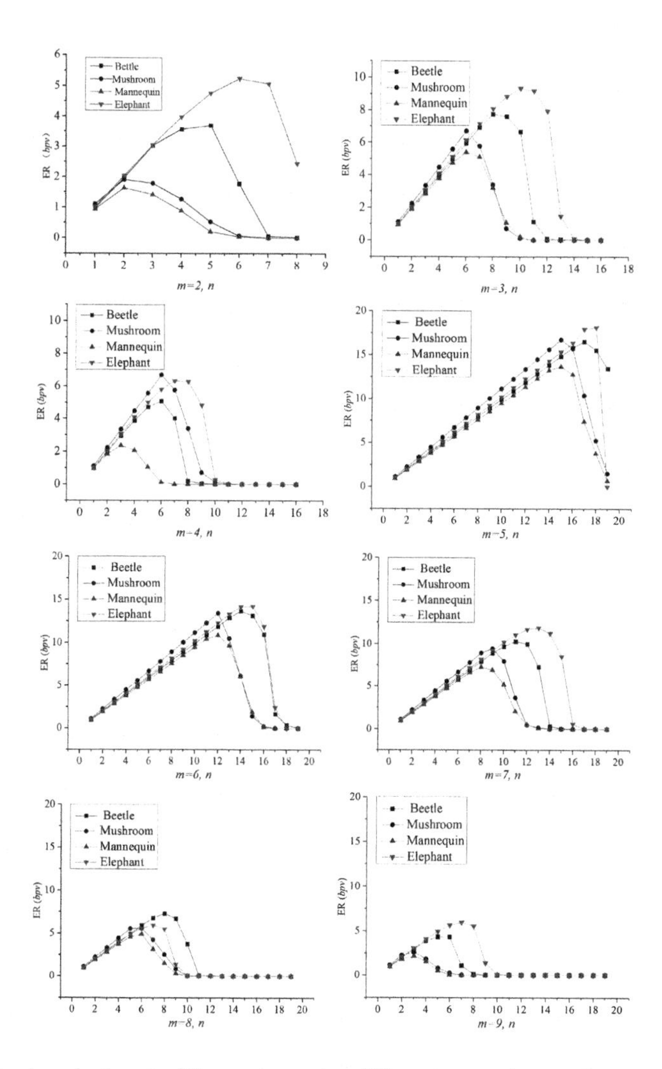

Fig. 4. The trend of embedding rate under different m and n on four test meshes.

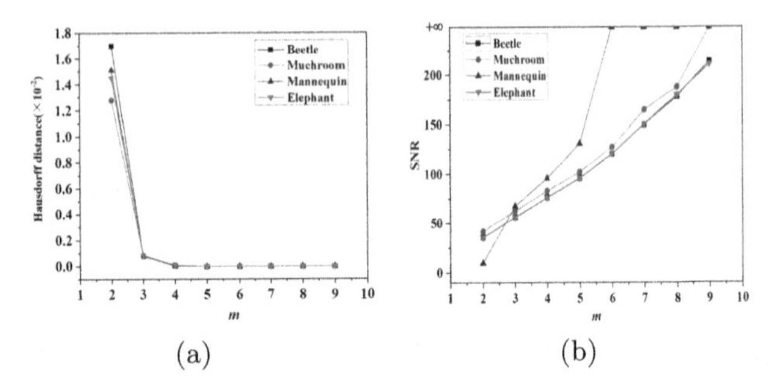

Fig. 5. The Hausdorff distance and SNR on different m under maximum embedding rate on four test meshes. (a) Hausdorff distance, (b) SNR.

Fig. 6. Illustrative examples showing the appearance of the mesh of different stages when $m = 5$. From left to right is the original mesh, encrypted mesh, marked encrypted mesh and recovered mesh.

Figure 5(a) shows that with the increase of m, the recipient gets a higher quality recovered mesh. As shown in Fig. 5(b), as m increases, SNR gradually increases and tends to ∞, which indicates that the recovered meshes becomes higher. Thus, the proposed method achieves reversibility by adjusting m. Figure 6 shows the visual effects of each stage when $m = 5$, including the original mesh, the encrypted mesh, the marked encrypted mesh, and the recovered mesh. Figure 6 shows that in terms of visual quality, there is no visually perceptible difference between the original mesh and the recovered mesh obtained by the proposed method, that is, the method proposed does not introduce perceptible distortion.

3.3 Performance Comparison

In this section, we compared the embedding rate of our method with methods [17–19]. Jiang et al. [17] embeds 1bit into each vertex by flipping the three

LSBs of the vertex in the data hiding stage. Since the embedding rate is limited by the mesh connectivity, the embedding rate is lower than 0.5bpv. Mohsin's method [18] after two layers of embedding, the embedding rate reaches 6 bpv. Tsai et al. [19] uses spatial coding technology with embedding threshold to embed additional data into the vertices, and the embedding rate is about 7.68 bpv. The proposed method replaces the nMSB bits of the vertices in the "embedded" set with the n bits of the additional data, so each vertex is embedded with $3n$ bits. The proposed method makes full use of the local correlation of the model, and the embedding rate is improved. Figure 7(a) shows that the embedding rate of the proposed method on Beetle is 16.51 bpv, while the embedding rate of Jiang et al., Mohsin and Tsai et al. are 0.35 bpv, 6.00 bpv and 7.68 bpv, respectively.

In order to verify the effectiveness of the experiment, we tested the performance of embedding rate on the Princeton Shape Retrieval and Analysis Group data set. It can be seen from Fig. 7(b) that the average embedding rate of the proposed method is 14.25 bpv, while the average embedding rate of Jiang et al., Mohsin et al. and Tsai et al. method is 0.36 bpv, 6.00 bpv and 7.68 bpv, respectively. In summary, the experimental results show that the embedding rate of the proposed method is higher than that of the state-of-the-art methods [17–19].

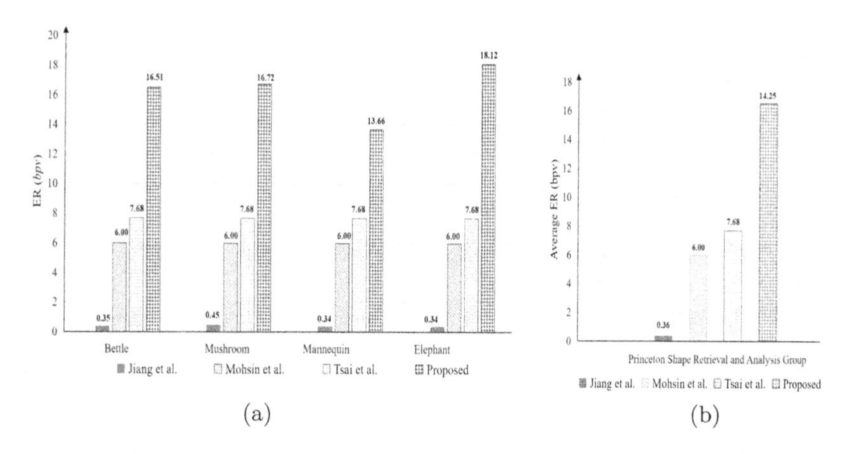

Fig. 7. Comparison of the proposed method and the state-of-the-art methods on embedding rate. (a) Maximum embedding rate on four test meshes. (b) Average embedding rate on Princeton Shape Retrieval and Analysis Group data set.

3.4 Feature Comparison

Table 1 shows the feature comparison of the proposed method and the state-of-the-art methods [17–19]. Jiang et al. [17] method and Mohsin et al. [18] method must decrypt the mesh when extracting additional data, both of which are inseparable methods. The proposed method realizes that data extraction and mesh recovery are separable. The average bit error rate of the data extracted by the method of Jiang et al. is 4.22%. In the method of Tsai et al. [19], the data extraction error is relatively large. The proposed method can extract data completely without error. In addition, by combining the value of m and Ring-prediction, the proposed method achieves error-free in mesh recovery stage. E_d represents error-free in data extraction. E_r represents error-free in mesh recovery.

Table 1. Feature comparison of the proposed method and the state-of-the-art methods.

Methods	Features		
	Separable	E_d	E_r
Jiang et al. [17]	×	×	×
Mohsin et al. [18]	×	✓	✓
Tsai et al. [19]	✓	×	×
Proposed	✓	✓	✓

3.5 Performance Analysis on Dense Meshes

The experimental results in Table 2 show that the experiments on dense meshes show the applicability and effectiveness of the method. Taking Dragon as an example, we can find that the maximum embedding rate is 17.3890 bpv, while the Hausdorff distance is 0.0086 (10^{-3}), the SNR is 101.1375. The data shows that this method also achieves a higher embedding rate on dense meshes. The recipient obtains a higher-quality recovered mesh through the Ring-prediction.

Table 2. Performance analysis on dense meshes

Test meshes	Embedding rate (bpv)	Hausdorff distance (10^{-3})	SNR
Dragon	17.3890	0.0086	101.1375
Armadillo	14.8133	0.0004	104.1417
Happyvrip	19.6532	0.0008	100.4581

4 Conclusions

In this paper, a separable RDH method for encrypted 3D mesh based on integer mapping and Multi-MSB prediction is proposed. The proposed method achieves a balance between capacity and distortion. To get large embedding capacity while ensuring data extraction and mesh recovery separable and error-free, Multi-MSB embedding strategy is used. To obtain high-quality recovered mesh, Ring-prediction is adopted. The results prove that the proposed method improves the embedding capacity compared with the state-of-the-art methods. A major limitation is that the embedding capacity is still limited by mesh connectivity. How to design a more effective algorithm to improve the embedding capacity could be the future work.

Acknowledgment. This research work is partly supported by National Natural Science Foundation of China (61872003, 61860206004).

References

1. Fridrich, J., Goljan, M., Rui, D.: Lossless data embedding for all image formats. SPIE Secur. Watermarking Multimed. Contents **4675**, 572–583 (2002)
2. Celik, M.U., Sharma, G., Tekalp, A.M., Saber, E.: Lossless generalized-LSB data embedding. IEEE Trans. Image Process. **14**(2), 253–266 (2005)
3. Tian, J.: Reversible data embedding using a difference expansion. IEEE Trans. Circ. Syst. Video Technol. **13**(8), 890–896 (2003)
4. Hu, Y., Lee, H.-K., Chen, K., Li, J.: Difference expansion based reversible data hiding using two embedding directions. IEEE Trans. Multimed. **10**(8), 1500–1512 (2008)
5. Li, X., Zhang, W., Gui, X., Yang, B.: Efficient reversible data hiding based on multiple histograms modification. IEEE Trans. Inf. Forensics Secur. **10**(9), 2016–2027 (2015)
6. Wang, J., Ni, J., Zhang, X., Shi, Y.: Rate and distortion optimization for reversible data hiding using multiple histogram shifting. IEEE Trans. Cybern. **47**(2), 315–326 (2016)
7. Zhang, X.: Reversible data hiding in encrypted image. Signal Process. Lett. IEEE **18**(4), 255–258 (2011)
8. Zhang, X.: Separable reversible data hiding in encrypted image. IEEE Trans. Inf. Forensics Secur. **7**(2), 826–832 (2011)
9. Qian, Z., Zhang, X.: Reversible data hiding in encrypted images with distributed source encoding. IEEE Trans. Circ. Syst. Video Technol. **26**(4), 636–646 (2015)
10. Ma, K., Zhang, W., Zhao, X., Nenghai, Yu., Li, F.: Reversible data hiding in encrypted images by reserving room before encryption. IEEE Trans. Inf. Forensics Secur. **8**(3), 553–562 (2013)
11. Puteaux, P., Puech, W.: An efficient MSB prediction-based method for high-capacity reversible data hiding in encrypted images. IEEE Trans. Inf. Forensics Secur. **13**(7), 1670–1681 (2018)
12. Wu, H., Cheung, Y.: A reversible data hiding approach to mesh authentication. In: IEEE/WIC/ACM International Conference on Web Intelligence, Compiegne, France (2005)
13. Fei, P., Bo, L., Min, L.: A general region nesting based semi-fragile reversible watermarking for authenticating 3D mesh models. IEEE Trans. Circ. Syst. Video Technol. **PP**(99), 1 (2021)
14. Jiang, R., Zhang, W., Hou, D., Wang, H., Nenghai, Yu.: Reversible data hiding for 3D mesh models with three-dimensional prediction-error histogram modification. Multimed. Tools Appl. **77**(5), 5263–5280 (2018)
15. Wu, H., Cheung, Y.: A reversible data hiding approach to mesh authentication. In: IEEE/WIC/ACM International Conference on Web Intelligence (2005)
16. Li, L., Li, Z., Liu, S., Li, H.: Rate control for video-based point cloud compression. IEEE Trans. Image Process. **29**, 6237–6250 (2020)
17. Jiang, R., Zhou, H., Zhang, W., Nenghai, Yu.: Reversible data hiding in encrypted three-dimensional mesh models. IEEE Trans. Multimed. **20**(1), 55–67 (2017)
18. Shah, M., Zhang, W., Honggang, H., Zhou, H., Mahmood, T.: Homomorphic encryption-based reversible data hiding for 3D mesh models. Arab. J. Sci. Eng. **43**(12), 8145–8157 (2018)

19. Tsai, Y.: Separable reversible data hiding for encrypted three-dimensional models based on spatial subdivision and space encoding. IEEE Trans. Multimed. **23**, 2286–2296 (2020)
20. Deering, M.: Geometry compression. In: Proceedings of the 22nd Annual Conference on Computer Graphics and Interactive Techniques, pp. 13–20, New York, United States, September 1995

MPN: Multi-scale Progressive Restoration Network for Unsupervised Defect Detection

Xuefei Liu, Kaitao Song, and Jianfeng Lu(✉)

Nanjing University of Science and Technology, Nanjing, China
{xuefei_liu,kt.song,lujf}@njust.edu.cn

Abstract. Defect detection is one of the most challenging tasks in the industry, as defects (*e.g.*, flaw or crack) in objects usually own arbitrary shapes and different sizes. Especially in practical applications, defect detection usually is an unsupervised issue, since it is difficult to collect enough labeled defect samples in the industrial scenario. Although a lot of works have achieved remarkable progress with the help of labeled data, how to effectively detect defects with only positive samples (*i.e.*, clean image without any defects) for training is still a troublesome problem. Therefore, in our paper, we adopt the image restoration strategy to address the unsupervised defect detection task. More specifically, we enable model to restore the original image from the defect samples first and then calculate the differences between the restored image and defect image as the final results. To deal with the unsupervised scenario, we first introduce a self-synthesis component to generate pseudo defects for training by Poisson editing, and the generated pseudo defects will be used for training. Targeting at the unsupervised defect detection, we introduce a novel Multi-scale Progressive restoration Network (**MPN**), which utilizes multi-scale information to detect defects progressively. More specifically, we choose an invariant scale convolution network as the restoration network for image reconstruction. However, it is difficult to detect defects with once restoration. Thus, we adopt the iterative restorations and our model is conditioned on the previous results for progressive defect detection. Our progressive detection is maintained by using a recurrent neural network to memory previous states. Considering that the defects can be arbitrary size, we incorporate the top-down and bottom-up structure into our model to extract multi-scale semantics better. Experimental results on multiple datasets demonstrate that our model achieves better performance than previous methods.

Keywords: Defect detection · Image restoration · Self-synthetic samples · Multi-scale · Recurrent Neural Network

1 Introduction

In recent years, benefiting from the rapid development of the deep neural network, computer vision has observed significant progress in many different challenging tasks [6,11,12] (*e.g.*, object detection, semantic segmentation, image

H. Ma et al. (Eds.): PRCV 2021, LNCS 13020, pp. 349–359, 2021.
https://doi.org/10.1007/978-3-030-88007-1_29

generation, and *etc.*) and drawn enormous attention from the industry. Specifically, defect detection [2,9,14] is a representative industrial task, which owns wide application prospects. Different from other detection tasks in computer vision, industrial defects are usually too subtle to discriminate and localize. Meanwhile, the shapes of most defects are usually arbitrary (*e.g.*, flaw, crack, deformation, irregular or abnormal shape), which brings additional challenges for model training. More severely, due to the expensive human annotations, defect detection usually suffers from data scarcity, which further raises the task difficulties. Therefore, how to address these issues effectively in defect detection is still a troublesome problem for the industry and academia.

Early works for defect detection are usually in a supervised way. For example, [9] adopted discrete fourier transform to convert the spatial signal into the frequency domain and used neural network for crack identification. [2] attempted to use gabor filter and principal component analysis to extract feature descriptors for defect detection. Recently, benefiting from the advances of deep learning, some pure neural-network-based works have been proposed to address defect detection and achieved significant performance. Most of them [13] considered semantic segmentation, such as FCN [12], SegNet [1], U-Net [11] and DeepLabV3+ [3], to address defect detection. More specifically, when providing supervised data (*i.e.*, the annotation of defect location), we enable model to identify whether the label of every pixel is defect or not. However, most existing works only conducted defect detection under the supervised scenario, while defect detection usually suffers from data scarcity which requires expensive human labor. Additionally, as the industrial defect can be arbitrary shapes and scales, it is difficult to capture all defects precisely with only single stage (*i.e.*, only detect defects once), especially under the unsupervised scenario. How to improve the model with previous detected results to obtain more precise defects is a research direction of our paper.

In this paper, we mainly investigate the unsupervised defect detection task, which means we can only utilize positive samples (*i.e.*, clean images without any defects) to detect defects. To handle the unsupervised scenario of our task, we introduce a self-synthesis module, which leverages a post-editing operation to generate pseudo defect over the positive samples as the training data. Based on our synthetic samples, we adopt an image restoration strategy to deal with the defect detection task, which means we reconstruct the original image from the defective sample and minimize the difference between the reconstructed image and the original image during the training. At the inference stage, we calculate the difference between the input image and the reconstructed image as the detected defects. To further obtain defects with better accuracy, we introduce MPN, a multi-scale progressive restoration network, a novel approach to improve defect detection from two perspectives, which are iteratively restoration and multi-scale feature fusion, respectively. More specifically, considering that only once detection cannot detect all defects very well, we introduce a progressive restoration module to refine the results based on the previous detections. Since defects can be arbitrary scales, we incorporate a top-down and bottom-up structure into our model to integrate the advances of multi-scale features for better prediction. Benefiting from two of our well-designed strategies, our model is able to detect arbitrary shape and different scale defects with better accuracy. Experimental results on multiple industrial defect detection datasets demonstrate that our method achieves better performance than previous methods.

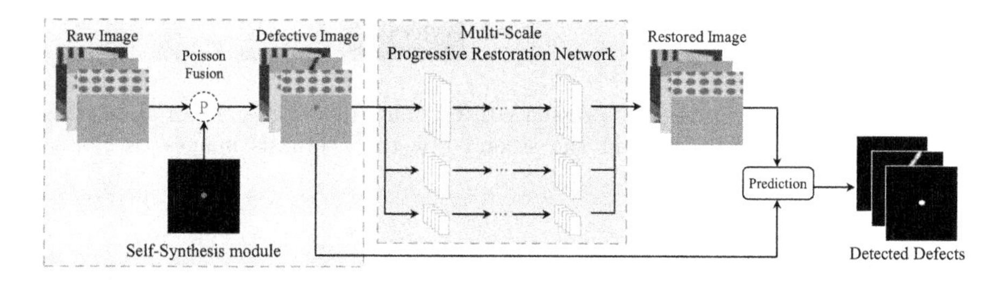

Fig. 1. The pipeline of our Multi-scale Progressive restoration Network (MPN). Our model first uses poisson editing to synthesize the defective image. Then the defective image is feed into the progressive restoration network to obtain the restored image. And finally, we use mean square error and structural similarity to minimize the similarity distance between the original image without defects and the restored image at the training stage. When switching to the inference stage, the differences between the input image and restored image are considered as the detected defects.

Overall, the contributions of our paper can be summarized as follow:

1. To deal with the data scarcity of annotated defects, our task is treated as an unsupervised learning task, and introduce a self-synthesis module with poisson editing to generate high-quality detective samples based on the positive samples to imitate supervised training.
2. In our paper, we model defect detection as the image restoration task, and propose a multi-scale progressive restoration network to detect defects with more precise accuracy from two different perspectives, progressively detecting defects based on the previous results by recurrent neural network and using a top-down and bottom-up structure to utilize multi-scale features.
3. Experimental results on multiple datasets demonstrate that our method outperforms previous models by a large margin, and validates the effectiveness of our model in defect detection.

2 Method

In this section, we will introduce the detailed design of our approach for unsupervised defect detection, including the training pipeline, self-synthesis module, progressive restoration network and multi-scale feature fusion.

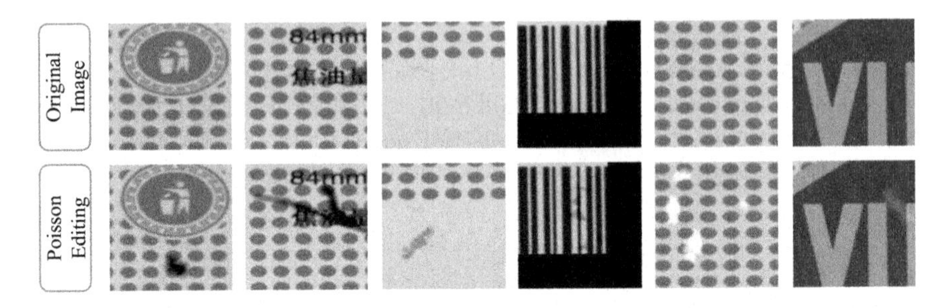

Fig. 2. Self-synthesis samples by Poisson editing. The first row is the original image (*i.e.*, image without defects), and the second raw the synthesis images by Poisson editing.

2.1 Pipeline

As aforementioned, in our paper, we mainly investigate how to detect defects under the unsupervised scenario (only positive samples can be used for training). Target at this issue, we design a self-synthesis module to generate pseudo defective samples for training. After obtaining the generated samples, we feed them into our multi-scale progressive restoration network to iteratively reconstruct the original image, which applies a recurrent neural network to memory the states of the previous stage for progressively defect detection. Since defects can be arbitrary scales, to enable our model to detect defects at any scales adaptively, we configure our model with a top-down and bottom-up structure to better extract multi-scales feature maps for prediction. Figure 1 illustrates the detailed training pipeline of our model. More details are introduced in the next subsections.

2.2 Self-synthesis Module

Defective samples usually do not have uniform shapes, while they can be arbitrary deformable shapes and scales. Specifically, the ratio of defects to the original image is usually too extremely small, which makes defects too difficult to distinguish and causes us cannot directly use generation models like GAN [5] to construct pseudo data. Previous works have validated the effectiveness of Poisson editing [10] in image fusion. Inspired by the successes of Poisson editing, we conduct the defective image synthesis based on Poisson editing by selecting different kinds of noise (*e.g.*, circle, ellipse, line segment, curve, polygon, and other geometric shapes) to guarantee the diversity of generated defects. Figure 2 shows some synthesis images by our proposed module. We define the synthetic image obtained by the self-synthesis module as $\boldsymbol{x}$ and feed it into the master network (*i.e.*, multi-scale progressive restoration network) to predict defects.

2.3 Progressive Restoration Network

Previous works [13,14] usually considered semantic segmentation as the solution of defect detection, which converts the identification of each pixel as binary

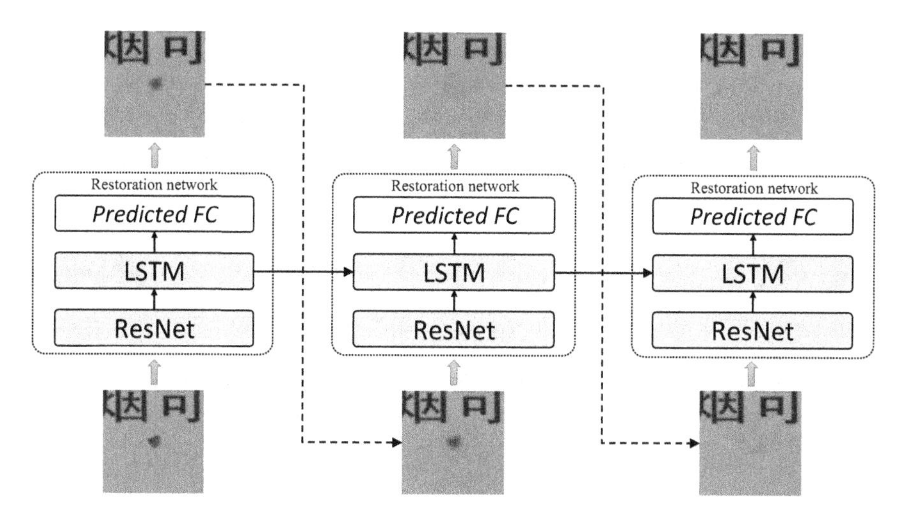

Fig. 3. Progressive Restoration Network. Each stage for image restoration will use previous results as the input, except the first stage which uses the defective image. Parameters are reused across each stage.

classification (*i.e.*, whether the pixel is defect or not). However, as aforementioned, the ratio of defect to the input image is usually too small, which leads to a serious imbalanced classification problem in semantic segmentation, and thus affects the final prediction. Therefore, in our paper, we borrow the idea of image restoration into our work to address defect detection. More specifically, we use an invariant scale convolution network as the restoration network, which adopts ResNet [6] as the basic feature extractor, to reconstruct the original image $\boldsymbol{y}$ from the pseudo sample. The objective function is to minimize the distance between the original image $\boldsymbol{y}$ and the restored image $\boldsymbol{x}$ by calculating the weighted sum of the mean square error (MSE) and structural similarity (SSIM) [15]. The formulation of the final objective function is as:

$$\mathrm{L}(\boldsymbol{o}, \boldsymbol{y}) = \alpha \mathrm{MSE}(\mathrm{o}, \mathrm{y}) + (1 - \alpha)(1 - \mathrm{SSIM}(\mathrm{o}, \mathrm{y})), \tag{1}$$

where α is a hyper-parameter to balance MSE and SSIM loss, and is set as 0.7 in our experiments tuned on the validation set. Hence our model is targeted at minimizing the value of $\mathrm{L}(\boldsymbol{o}, \boldsymbol{y})$. However, we regard that the restored image with only one detection is not enough to reconstruct the original image very well. Therefore, we raise a simple hypothesis: is it possible to apply multiple restorations to obtain defects with better accuracy? Based on our hypothesis, we extend our network to a progressive restoration network, which tries to restore the original image based on previous results, except the first stage uses the synthetic image $\boldsymbol{x}$. To enhance the connection between each stage, we employ a recurrent neural network [7] after the ResNet, to memory the current state and previous restored state. Such strategy enables our model to utilize previous results for better restoration and detect defects progressively. In our experiments, we notate the number of iterative stages as t and set t as 5. Figure 3 illustrates the principle of our progressive restoration network. At the inference stage, when given the input image, we first obtain the restored image by our progressive restoration network, and then calculate the difference between the input image and the restored image as the detected defects.

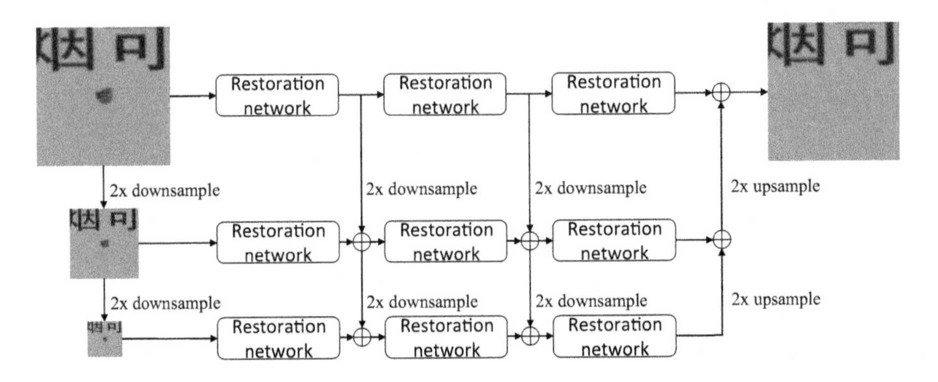

Fig. 4. The top-down and bottom-up structure in our model for using multi-scale features. "Restoration network" is our progressive restoration network as described in Subsect. 2.3. "2× downsample" means downsample 2 times in resolution and "2× upsample" means upsample 2 times in resolution. "⊕" means the element-wise plus operation in the channel dimension.

2.4 Multi-scale Feature Fusion

Multi-scale feature fusion has been confirmed as a successful technique, like feature pyramid network (FPN) [8], in addressing computer vision tasks. Inspired by the success of FPN, we apply a top-down and bottom-up module into our model to fuse multi-scale features for better prediction. Figure 4 illustrates the detailed architecture of our model in using multi-scale features. More specifically, we sample three scales of the original image, which are $1\times$, $\frac{1}{2}\times$, $\frac{1}{4}\times$ times of the original resolution. For each scale, we conduct a progressive resolution network to obtain the image descriptors for the image reconstruction. To enhance the correlations between different scales, we add an element-wise plus operation in the channel dimension from the high-resolution into the low-resolution features to integrate the advances of multi-scale semantics. And finally, we apply an upsampling operation over the lower-resolution features, and add it for the higher-resolution features for fusion. The features from different scales will be summed up for the final prediction. Benefiting from our designed top-down and bottom-up structure, our model is able to utilize high-level semantic with different scales and can detect defects at arbitrary scales.

3 Experiments

In this section, we will introduce the empirical setting, results and ablation studies of our method.

3.1 Empirical Setting

Dataset. We choose three datasets to evaluate our method, which are Kolektor [13][1], RSDDs [4] and "ZHENGTU Cup"[2]. Specifically, Kolektor and RSDDs are classical surface-defect datasets. Kolektor dataset includes a total of 399 high-quality images with a resolution of 1408×512 pixels and RSDDs dataset includes 128 images with a resolution of 1024 pixels. "ZHENGTU Cup" is a standard dataset for industrial defect detection. This dataset consists of three parts: gray image, infrared image and color image. Each part contains 100 large clean images without any defects with a high resolution of 4096×2048 for training, and some small images with a resolution of 128×128 for validation. During the training, we randomly crop 128×128 image patches from the original image as the training data. Our experiments are deployed on single NVIDIA TITAN XP GPU, and the batch size is set as 8.

Metrics. In binary classification, there are 4 different predicted categories for sample, which are true positive (**TP**), true negative (**TN**), false positive (**FP**), false negative (**FN**). To better evaluate the performance of our method, we choose four standard metrics, which are precision (**Pre**), accuracy (**Acc**), recall (**Rec**) and mean intersection-over-union (**mIOU**) respectively. Each metric is computed as follow:

Table 1. Evaluation metrics

Metric	Formulation
Pre	TP/(TP + FP)
Acc	(TP + TN)/(TP + TN + FP + FN)
Rec	TP/(TP + FN)
mIOU	TP/(FP + FN + TP)

[1] http://www.vicos.si/Downloads/KolektorSDD.
[2] https://www.marsbigdata.com/competition/details?id=5293671830016 "ZHENGTU Cup" is the national campus machine vision application competition for defect detection.

Baselines. To further highlight the advantages of our method, we choose three classical semantic segmentation approaches as our baselines for detecting defects, which are U-Net [11], SegNet [1] and DeepLabV3+ [3]. In particular, U-Net uses a contracting path and expansive path to obtain each pixel feature for segmentation, SegNet uses a convolution neural network and removes pooling operation to keep high-frequency features, DeepLabV3+ adopts an encoder-decoder structure with spatial pyramid pooling module for better prediction. All of the methods are trained on the synthetic samples generated by our self-synthesis module to make a fair comparison.

Table 2. Comparisons between different methods. The meanings of Pre, Acc, Rec and mIOU refer to Table 1.

Datasets	Method	Pre	Acc	Rec	mIOU
Kolektor	U-Net	76.36	85.87	72.81	67.74%
	SegNet	79.56	86.67	71.33	69.56%
	DeepLabV3+	83.38	87.73	75.63	73.86%
	MPN	**84.24**	**88.75**	**76.89**	**75.58%**
RSDDs	U-Net	72.58	82.47	75.43	70.82%
	SegNet	78.63	83.85	75.55	73.28%
	DeepLabV3+	84.87	88.34	82.45	77.15%
	MPN	**85.38**	**89.27**	**83.74**	**80.79%**
Zhengtu cup	U-Net	78.23	81.43	74.86	70.54%
	SegNet	81.57	84.59	78.62	74.82%
	DeepLabV3+	84.27	87.38	82.36	79.48%
	MPN	**85.78**	**89.75**	**83.64**	**81.23%**

3.2 Results

Our results are shown on Table 2. From Table 2, we observe that our model achieves the best performance in each task and outperforms other baselines, including U-Net, SegNet and DeepLabV3+, in terms of all metrics, which also reflects the effectiveness of our method in detecting defects. Please note that DeepLabV3+ is a powerful semantic segmentation while our method is still superior to it. To further reveal the advances of our approach, we also select some detected results between different methods for visualization. The results are displayed in Fig. 5. We note that our MPN can precisely detect the correct defects, while other methods misclassify some complex background as the defects. These significant improvements also indicate that the effectiveness of our model in detecting industrial defects.

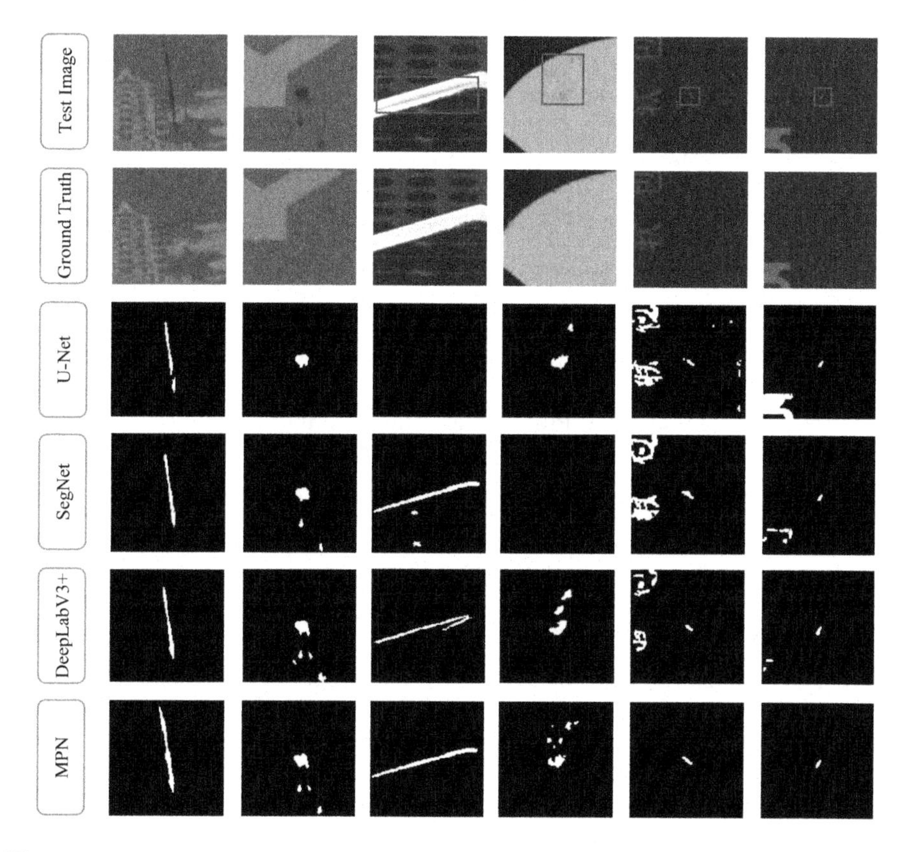

Fig. 5. The visualization of the detected results between different methods. The first row is the test image, the second row is the ground truth. The third, fourth, fifth and sixth rows are corresponding to the detected results of U-Net, SegNet, DeepLabV3+ and our MPN respectively.

3.3 Ablation Studies

To better analyze the necessity of each component of our model, we also conduct a series of ablation studies for validation. In detail, we respectively disable multi-scale fusion and recurrent neural network to evaluate the importance of these components. The results are reported in Table 3. From Table 3, we observe that disabling recurrent layer harms model accuracy a lot, especially in precision and accuracy metric. When disabling multi-scale feature fusion, it still affects model performance slightly, especially in terms of mIOU. Therefore, these ablation studies also validate the necessity of recurrent neural network and multi-scale feature fusion in our model.

Table 3. Ablation study of our MPN model on "Zhengtu Cup". "−LSTM" means removing recurrent network in our restoration network and "−Multi-scale" means removing multi-scale feature fusion.

Method	Pre	Acc	Rec	mIOU
MPN	**85.78**	**89.75**	**83.64**	**81.23%**
− LSTM	82.76	84.28	80.32	75.46%
− Multi-scale	84.34	86.28	80.47	77.18%

Table 4. Ablation study of different iteration stage t on "Zhengtu Cup". $t = 5$ is equivalent to the results of our MPN.

Method	Pre	Acc	Rec	mIOU
$t = 1$	76.53	79.56	73.37	71.64%
$t = 3$	80.65	86.42	76.68	75.48%
$t = 5$	**85.78**	**89.75**	**83.64**	**81.23%**

3.4 Effects of Iterative Stage t

In order to demonstrate the effectiveness of our progressive strategy, we further conduct experiments to study the effect of the different number of the iterative stage t. The results are reported in Table 4. From Table 4, we observe that, when using more iterative stages, our model also receives more gains in model performance in terms of all metrics. This phenomenon also indicates the effectiveness of our progressive strategy (*i.e.*, restore images iteratively) in defect detection.

4 Conclusion

In this paper, we propose a novel model for solving industrial defect detection in practical application, named as the multi-scale progressive restoration network. Our method first designs a self-synthesis module to generate defective samples. And then we incorporate a recurrent neural network into our model to reconstruct the original image with multiple stages and detect defects progressively. Furthermore, we also configure our model with a top-down and bottom-up structure to leverage high-level semantic at multiple scales. Experimental results demonstrate that our method achieves better performance in unsupervised defect detection than existed segmentation approaches. In the future, we expect to apply more efficient and advanced structures to our model for better feature extraction and efficient training.

References

1. Badrinarayanan, V., Kendall, A., Cipolla, R.: SegNet: a deep convolutional encoder-decoder architecture for image segmentation. IEEE Trans. Pattern Anal. Mach. Intell. **39**(12), 2481–2495 (2017)

2. Bissi, L., Baruffa, G., Placidi, P., Ricci, E., Scorzoni, A., Valigi, P.: Automated defect detection in uniform and structured fabrics using Gabor filters and PCA. J. Vis. Commun. Image Representation **24**(7), 838–845 (2013)
3. Chen, L.-C., Zhu, Y., Papandreou, G., Schroff, F., Adam, H.: Encoder-decoder with atrous separable convolution for semantic image segmentation. In: Ferrari, V., Hebert, M., Sminchisescu, C., Weiss, Y. (eds.) ECCV 2018. LNCS, vol. 11211, pp. 833–851. Springer, Cham (2018). https://doi.org/10.1007/978-3-030-01234-2_49
4. Gan, J., Li, Q., Wang, J., Yu, H.: A hierarchical extractor-based visual rail surface inspection system. IEEE Sens. J. **17**(23), 7935–7944 (2017)
5. Goodfellow, I.J., et al.: Generative adversarial nets. In: Proceedings of the 27th International Conference on Neural Information Processing Systems, vol. 2, pp. 2672–2680 (2014)
6. He, K., Zhang, X., Ren, S., Sun, J.: Deep residual learning for image recognition. In: 2016 IEEE Conference on Computer Vision and Pattern Recognition, CVPR 2016, Las Vegas, NV, USA, 27–30 June 2016, pp. 770–778 (2016)
7. Hochreiter, S., Schmidhuber, J.: Long short-term memory. Neural Comput. **9**, 1735–1780 (1997)
8. Lin, T.Y., Dollar, P., Girshick, R., He, K., Hariharan, B., Belongie, S.: Feature pyramid networks for object detection. In: Proceedings of the IEEE Conference on Computer Vision and Pattern Recognition (CVPR), July 2017
9. Paulraj, M.P., Shukry, A.M.M., Yaacob, S., Adom, A.H., Krishnan, R.P.: Structural steel plate damage detection using DFT spectral energy and artificial neural network. In: 2010 6th International Colloquium on Signal Processing its Applications, pp. 1–6 (2010)
10. Pérez, P., Gangnet, M., Blake, A.: Poisson image editing. In: ACM SIGGRAPH 2003 Papers, pp. 313–318 (2003)
11. Ronneberger, O., Fischer, P., Brox, T.: U-Net: convolutional networks for biomedical image segmentation. In: Navab, N., Hornegger, J., Wells, W.M., Frangi, A.F. (eds.) MICCAI 2015. LNCS, vol. 9351, pp. 234–241. Springer, Cham (2015). https://doi.org/10.1007/978-3-319-24574-4_28
12. Shelhamer, E., Long, J., Darrell, T.: Fully convolutional networks for semantic segmentation. IEEE Trans. Pattern Anal. Mach. Intell. **39**(4), 640–651 (2017)
13. Tabernik, D., Sela, S., Skvarc, J., Skocaj, D.: Segmentation-based deep-learning approach for surface-defect detection. J. Intell. Manuf. **31**(3), 759–776 (2020)
14. Zhao, Z., Li, B., Dong, R., Zhao, P.: A surface defect detection method based on positive samples. In: PRICAI 2018: Trends in Artificial Intelligence - 15th Pacific Rim International Conference on Artificial Intelligence, Nanjing, China, 28–31 August 2018, Proceedings, Part II, pp. 473–481 (2018)
15. Wang, Z., Bovik, A.C., Sheikh, H.R., Simoncelli, E.P.: Image quality assessment: from error visibility to structural similarity. IEEE Trans. Image Process. **13**(4), 600–612 (2004)

Scene-Aware Ensemble Learning for Robust Crowd Counting

Ling Xu[1], Kefeng Huang[2], Kaiyu Sun[2], Xiaokang Yang[1], and Chongyang Zhang[1](✉)

[1] School of Electronic Information and Electrical Engineering, Shanghai Jiao Tong University, Shanghai 200240, China
sunny_zhang@sjtu.edu.cn

[2] Shanghai Jianke Engineering Consulting Co., Ltd., Shanghai, China

Abstract. Crowd counting models usually suffer from sharp fluctuations when trained on one specific dataset but tested on the other one with huge scene deviation. To alleviate this problem, we propose a Scene-aware Ensemble Learning CNN model (SEL-CNN) for robust cross-scene crowd counting. Firstly, crowd scenes are divided into three levels: low, middle, and high level, and then one three-parallel-branch based ensemble learning framework is developed to regress density maps for three-level scenes, respectively. Secondly, one scene-aware branch is introduced to learn global weight parameters, which performs as a scene adaptive weighting scheme that dynamically fuses three outputs to generate a final prediction map. Moreover, a Combination-Division-Iteration-based three-stage training strategy is also applied to ensure the learning efficiency of the proposed multi-branch CNN model. Extensive intra-dataset and cross-dataset experiments demonstrate the advantages of our method in terms of scene adaptability and robustness.

Keywords: Crowd counting · Scene adaptability · Ensemble learning

1 Introduction

Crowd counting has been actively studied over the past decades with widespread applications in social security, scene monitoring, etc. [8,30]. With the development of deep learning techniques and available training datasets, the performance of crowd counting method has been greatly improved.

However, precisely estimating the number of crowds in different scenes is still a challenging task. The main problem is the huge deviation of scenes, including large variations of density distributions, head scales and backgrounds. When a crowd counting network is trained with one specific dataset, it will work well on the scenes consistent with those in the training dataset, but for a new unseen scenario, its performance may drop dramatically [28]. The problem of insufficient robustness in cross-scene situations will severely restrict the application of existing models.

H. Ma et al. (Eds.): PRCV 2021, LNCS 13020, pp. 360–372, 2021.
https://doi.org/10.1007/978-3-030-88007-1_30

There are many previous works aiming to develop more robust crowd counting models. In [28], when a model is trained with a fixed dataset, training samples which are similar to the target scene according to the perspective information are retrieved from the training dataset to fine-tune the model. In recent works [3,16], one or a small number of labeled data is needed to adapt the crowd counting model to a new test scene. Although these methods aim to solve the task of scene adaptability, prior knowledge of the target scene is still required, which means one or more labeled samples and extra offline model retraining are necessary. Obviously, this inconvenience will limit the flexible application of the mentioned models.

Thus, what we can do to deal with the cross-scene problem? Our goal is to build a scene-adaptive model without extra target domain data or offline training. In order to achieve this goal, we should consider the following two aspects. 1) The scene attribute chosen to distinguish application scenes is density level. The main factors affecting the density map estimation are the head scale and density level [28], and the head scale is often directly related to the density level: the denser the scene, the smaller the head scale. Considering this aspect, crowd scenes can be classified into a few levels according to the density degree. 2) In real-world counting applications, the crowd density varies enormously in spatial domains. Since it is difficult for a model to learn the common feature representations for various scenes, can we ensemble multiple models/branches [2,18,29] and make each model/branch have their own strengths on different crowd densities? With scene perception, the multiple outputs of different scene-specific regressors can be selected or fused dynamically, and thus one accurate counting result can be got.

Therefore, we design a Scene-aware Ensemble Learning CNN model (SEL-CNN) for robust crowd counting. Specifically, we firstly divide crowd scenes into three levels to represent different congested degrees. We propose to learn not only the scene-specific base-regressor according to the corresponding training data but an ensemble of several base-regressors to obtain more robust results. Subsequently, a Scene-aware Density Map Fusion branch is introduced to learn different importance weights, which guide the combination of three outputs to generate one final density map. Additionally, a Combination-Division-Iteration-based three-stage training strategy is applied to guarantee the learning efficiency of the whole network. To verify the effectiveness of this work, extensive intra-dataset and cross-dataset experiments are conducted.

Our main contributions are summarized as follows:

- We propose one novel robust crowd counting network, in which the cross-scene prediction ability can be significantly improved through an ensemble of multiple branches combined with scene-aware adaptive fusion.
- By Combination-Division-Iteration-based three-stage training strategy, each stage makes the training of the network more targeted. The inner-outer-update method is applied to achieve better density regression and scene adaptation under the guidance of global information.
- Extensive experiments on three challenging datasets demonstrate the superiority on cross-scene task of our proposed approach.

2 Related Work

2.1 Density-Based Crowd Counting

Deep convolutional networks are widely adopted for counting by estimating density maps from images [1,6,8,11,15]. MCNN [29] uses a multi-column CNN to encourage different columns to respond to crowds at different scales. [18] classifies the input images into three different branch according to density level, and then processes the images with corresponding regressor. Recently, multi-task networks have shown to reduce the counting error. CP-CNN [22] proposes extra classification networks to combine global and local density level to achieve performance improvements. DecideNet [5] constructs a model applying two branches for detection and regression task respectively, and the two intermediate results are fused using quality-aware attention module. RAZNet [10] adds an extra head localization task and adaptively zooms in the regions that are difficult to recognize to improve counting accuracy. In this paper, we also investigate counting by density map estimation but focus on cross-scene counting.

2.2 Scene Adaptation Task for Crowd Counting

In the research of scene adaptation, [28] proposes the cross-scene crowd counting with deep learning. To apply the model to unseen scenes, when given a new test scene, scenes with similar attributes such as perspectives would be found in the training dataset to fine-tune the model. This method requires scene retrieval, which is more dependent on datasets. In work [3], the author builds a pre-training model that can quickly adapt to the current test scene. The goal of this work is to use the single sample to adapt the crowd counting model to the specific scene. After that, [16] introduces a meta-learning method. In the test scene, a very small number of labeled samples would be selected to conduct one or few training iterations on the model. This work aims to use few labeled data and few gradient descent steps to extract the crowd density features of an unseen scene to achieve scene adaptation.

In our work, we learn three parallel scene-specific regressors according to density level and then utilize a scene-aware ensemble of these regressors to obtain more robust results. Different from other cross-scene method, our approach has no need for prior knowledge.

3 Methodology

The architecture of our proposed Scene-aware Ensemble Learning CNN model (SEL-CNN) is shown in Fig. 1. It is composed of three modules: Frontend Encoder, Three parallel Regressors and Scene-aware Density Map Fusion branch (SDF). Frontend Encoder and Three parallel Regressors together are named as Density Map Generator.

Frontend Encoder is used to extract general multi-level feature maps from crowd images and feeds them into the following modules. Three parallel Regressors

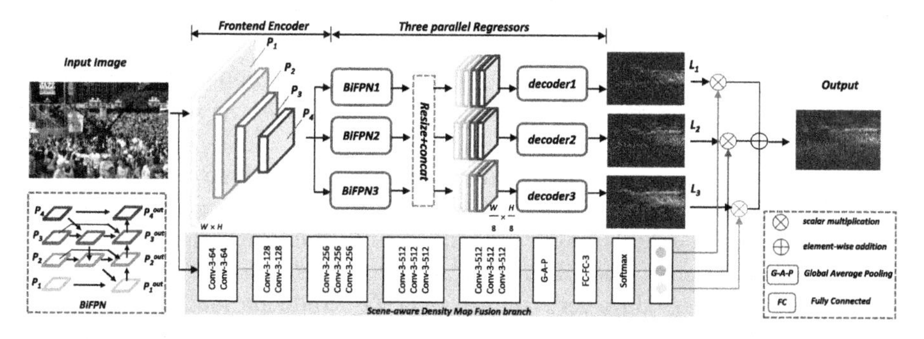

Fig. 1. Overview of our proposed Scene-aware Ensemble Learning CNN model (SEL-CNN). It consists of Frontend Encoder, Three parallel Regressors and Scene-aware Density Map Fusion branch (SDF).

learn to customize its own scene-specific regressor. SDF learns the combination weights of three intermediate density maps to obtain the final output. In addition, to assist the learning of the whole framework, we design a Combination-Division-Iteration-based three-stage training strategy.

3.1 Scene-Oriented Multi-branch Density Map Generator

Following most previous works [6,9,12], we directly adopt the first 10 layers of VGG-16 [20] with three pooling layers as the frontend encoder which generates the feature maps.

To pursue effective fusion of rich features from different layers, various feature fusion strategies including direct fusion [24], top-down fusion [17], bidirectional fusion [25] have been employed in crowd counting. Inspired by the success of [26], one straightforward solution is to utilize the bidirectional feature pyramid network (BiFPN). Our feature fusion module is adapted from the famous BiFPN originally designed for object detection. Considering that layers in crowd counting do not need to be as deep as object detection, we simplify the BiFPN to make it suitable for crowd counting by taking four-level features as inputs and aggregating them in a both top-down and bottom-up manner shown in Fig. 1. Extra learnable weights are added to ensure that each branch can take advantage of unequal contributions of input features.

These outputs of BiFPN will be resized and concatenated as an input of the decoder structure. The detailed structure of the decoder module is: {C(256, 256,3,2)-R-C(256,128,3,2)-R-C(128,64,3,2)-R-C(64,32,3,2)-R-C(32,8,3,2)-R-C(8, 1,1,1)}, where $C(N_i, N_o, k, d)$ means 2D convolutional layer with N_i input channels, N_o output channels, $k \times k$ filter size, d dilation rate, and R means ReLU layer. We employ three parallel scene-oriented regressors including the BiFPN and decoder module as an ensemble to train powerful crowd counting models for new scenes. Our multiple regressors are different models with a specific set of parameters values, which will be adapted to the assigned diverse density level through divided training described later in Sect. 3.3.

3.2 Scene-Aware Density Map Fusion Branch

In order to incorporate the distributions of three intermediate density maps on the final output across spatial dimension, we aim to model the selection process for optimal counting estimations.

Herein, we have described the details about generating three different density maps: $\boldsymbol{D}_i^1$, $\boldsymbol{D}_i^2$ and $\boldsymbol{D}_i^3$ for a given image I_i. CNN based scene-specific counting networks are usually suitable for images with different density levels. For example, the low-density based map $\boldsymbol{D}_i^1$ could get relatively precise counts in sparse density scenes. However, counting via $\boldsymbol{D}_i^1$ could be inaccurate on congested occasions and on the contrary, the high-density based map $\boldsymbol{D}_i^3$ prefers crowded scenes. Intuitively, one may think that fusing three density maps by applying average may obtain better performance. Taking into account the needs to adapt to various scenes, it is reasonable to dynamically assign weights to intermediate density maps with different density levels.

We design an Scene-aware Density Map Fusion (SDF) to capture the different importance weights of three intermediate density maps by dynamically assessing the density level. VGG16 network is adapted as the Scene-aware Density Map Fusion branch by keeping the first 10 layers with only three pooling layers. The operation of Global Average Pooling (G-A-P) on Conv4 features aggregates globally relevant and discriminative features. G-A-P is followed by two fully connected layers and then 3-class softmax weights are gained. The final density map can be obtained by the weighted sum of three intermediate density maps.

3.3 Loss Function and Network Training

We regress the total four density maps in our whole network. For density map supervision, a typical loss function is the mean squared error loss which sums up the pixel-wise Euclidean distance:

$$\mathcal{L}_k = \frac{1}{N}\sum_{i=1}^{N} \|\hat{\boldsymbol{D}}_i^k - \boldsymbol{D}_i\|_2^2 \tag{1}$$

$$\mathcal{L}_o = \frac{1}{N}\sum_{i=1}^{N} \|\hat{\boldsymbol{D}}_i^o - \boldsymbol{D}_i\|_2^2 \tag{2}$$

where $\hat{\boldsymbol{D}}_i^k$, $\boldsymbol{D}_i$ and $\hat{\boldsymbol{D}}_i^o$ are the estimated density map of regressor k, ground truth and the final output of i-th image, respectively, and N refers to the number of training samples.

In order to make full use of all training data and guarantee that each branch regressor has a unique representation of the corresponding scene, one Combination-Division-Iteration based three-stage training strategy is adopted to ensure the learning efficiency of the proposed multi-branch framework.

Stage 1: Combined Pretraining. Density Map Generator is trained with all the training data. The final output is set as the average value of three intermediate results. The loss function of this stage is $\mathcal{L}_c = \mathcal{L}_1 + \mathcal{L}_2 + \mathcal{L}_3$.

Algorithm 1: A Combination-Division-Iteration-based three-stage training strategy

Input: N training images with ground truth density maps D and density level labels l

Output: one final density map

1 **for** *Stage 1:* $t = 1$ *to* T_c *epochs* **do**
2   *update Density Map Generator parameters*;
3     $\theta \leftarrow \theta - \alpha \frac{\partial L_c}{\partial \theta}$
4 **end**
5 **for** *Stage 2:* $t = 1$ *to* T_d *epochs* **do**
6   **if** *density level* l $==$ k **then**
7   *keep frontend encoder frozen, update the regressor k parameters* ;
8         $\theta_k \leftarrow \theta_k - \alpha \frac{\partial L_k}{\partial \theta_k}$
9   **end**
10 **end**
11 *keep frontend encoder and three BiFPNs frozen*;
12 **for** *Stage 3:* $t = 1$ *to* T_i *epochs* **do**
13   **for** $m = 1$ *to inner-update epochs* T_u **do**
14   *freeze the scene-aware branch, update three decoders parameters* ;
15         $\theta_d \leftarrow \theta_d - \beta \frac{\partial L_{den}}{\partial \theta_d}$
16   **end**
17   *freeze decoders parameters, update scene-aware branch parameters* ;
18     $\theta_w \leftarrow \theta_w - \alpha \frac{\partial L_w}{\partial \theta_w}$
19 **end**

Stage 2: Divided Training. To make sure each feature fusion network can be scene-specific, all training data is divided into three groups on the basis of groundtruth density level which is defined as the ratio of people number to the area of image (namely $height \times width$). Each regressor is fine-tuned on its own subset. The parameters of the frontend encoder keep frozen. The low-level, middle-level and high-level crowds are used as the training labels of the branch Regressor 1, Regressor 2 and Regressor 3 using the corresponding loss function $\mathcal{L}_1$, $\mathcal{L}_2$, and $\mathcal{L}_3$, respectively.

Stage 3: Iterative Training. Rather than learning multiple regressors, we propose to use them as the ensemble and also learn how to combine them for best performance automatically with fine-grained hyper-parameters [14] via iterative training. The inner-loop jointly fine-tunes three decoders with the third stage loss function $\mathcal{L}_{den}$, and the outer-loop then optimizes the Scene-aware Density Map Fusion branch with $\mathcal{L}_w = \mathcal{L}_o$. The loss function $\mathcal{L}_{den}$ of this stage is weighted by a hyper-parameter:

$$\mathcal{L}_{den} = \mathcal{L}_o + \lambda(\mathcal{L}_1 + \mathcal{L}_2 + \mathcal{L}_3) \quad (3)$$

4 Experiments

4.1 Implementation and Settings

In our experiment, we initialize the VGG-16 backbone with corresponding pre-trained weights from the ImageNet Classification [7] challenge datasets. For the rest of the parameters, the initial values come from a Gaussian initialization with mean 0 and standard deviation 0.01. Stochastic gradient descent is applied with fixed learning rate α of 1e−6 during training. An inner-loop gradient descent is computed with the meta-learning rate β of 1e−7 during iterative training. The mentioned hyper-parameters T_c, T_d, T_i, T_u and λ are set as 50, 50, 100, 5 and 1e-3 during training.

Groundtruth Generation. We generate ground-truth density maps and corresponding segmentation maps derived from the point annotations. As for density maps, following [29], each labeled head p_j is substituted with a Gaussian kernel $\mathcal{N}(p_i, \sigma^2)$, where σ is the average distance of p_j and its 3 nearest neighbors.

Evaluation Metrics. We evaluate the performance via the mean absolute error (MAE) and mean squared error (MSE) are commonly used in crowd counting task. Small MAE and MSE values represent good performance.

4.2 Datasets

Extensive experiments have been conducted on three benchmark datasets to evaluate the performance of our framework: WorldExpo'10 [28], ShanghaiTech [29], UCF_CC_50 [4]. The detailed comparisons are described in the following subsections.

WorldExpo'10. It includes 3,980 labeled images from 1,132 video sequences based on 108 surveillance cameras with different locations in Shanghai 2010 WorldExpo. 103 scenes are considered for training and the remaining 5 scenes are for testing.

ShanghaiTech. It consists of PartA and PartB. PartA has 300 training images and 182 testing images with relatively high density. PartB has 400 training images and 316 testing images with relatively low density.

UCF_CC_50. It has 50 images with 63,974 head annotations in total. The head counts range between 94 and 4,543 per image. The small dataset size and large count variance make it a very challenging dataset. We perform 5-fold cross validations to report the average test performance.

4.3 Results on WorldExpo'10

To make a overall comparison with other cross-scene frameworks, we employ WorldExpo'10 commonly used in [3,16,28]. Since the Region Of Interest (ROI) are provided, the previous fashion [23] is followed to only calculate persons within the ROI area. The results using MAE of our proposed approach on each test scene and the comparisons to other cross-scene methods are listed in Table 1. Pre-trained CSRNet means the model parameters of CSRNet [29] are trained with all images in the training set, and then evaluated directly on images in the new target scene without any adaption following [16]. Fine-tuned CSRNet[29] is similar to pre-trained CSRNet but when testing one image is used for fine-tuning. Since [16] is few-shot problem, we choose results of $K = 1$ for better comparison.

Table 1. Results on WorldExpo'10 [28] with other cross-scene approaches. Red, green and blue indicate the first, the second and the third best result, respectively. Best viewed in color.

Method	S1	S2	S3	S4	S5
Fine-tuned CC [28]	9.8	14.1	14.3	22.2	3.7
pre-trained CSRNet [29]	5.55	24.07	35.54	23.95	10.70
fine-tuned CSRNet [29]	5.45 ± 0.03	22.74 ± 0.47	33.89 ± 0.26	15.69 ± 0.28	8.9 ± 0.05
FSCC [16]	3.19 ± 0.03	11.17 ± 1.01	8.07 ± 0.23	9.39 ± 0.26	3.82 ± 0.05
SEL-CNN (ours)	1.32	11.87	9.46	18.06	2.93

Table 2. Results on ShanghaiTech [29] and UCF_CC_50 [4]. SHA means ShanghaiTech Part A, SHB means ShanghaiTech Part B, UC50 means UCF_CC_50. Red, green and blue indicate the first, the second and the third best result, respectively. Best viewed in color.

Method	Intra-dataset						Cross-dataset					
	SHA		SHB		UC50		SHA→SHB		SHB→SHA		SHA→UC50	
	MAE	MSE	MAE	MSE	MAE	MSE	MAE	MSE	MAE	MSE	MAE	MSE
MCNN [29]	110.2	173.2	26.4	41.3	377.6	509.1	85.2	142.3	221.4	357.8	397.7	624.1
CMTL [21]	101.3	152.4	20.0	31.1	322.8	397.9	–	–	–	–	–	–
L2R [13]	101.3	152.4	13.7	21.4	279.6	388.9	–	–	–	–	337.6	434.3
Switch-CNN [18]	90.4	135.0	21.6	33.4	318.1	439.2	59.4	130.7	–	–	1117.5	1315.4
D-ConvNet [19]	73.5	112.3	18.7	26.0	288.4	404.7	49.1	99.2	140.4	226.1	364.0	545.8
CSRNet [9]	68.2	115.0	10.6	16.0	266.1	397.5	–	–	–	–	–	–
L2SM [27]	64.2	98.4	7.2	11.1	188.4	315.3	21.2	38.7	126.8	203.9	332.4	425.0
HACNN [24]	62.9	94.9	8.1	13.4	256.2	348.4	29.1	74.1	–	–	339.8	463.2
Ours	69.6	109.1	8.8	18.1	246.0	312.0	18.4	31.5	97.2	159.0	338.9	445.0

Compared to one/few-shot learning method, our method obtains competitive performance. Errors on S1 and S5 significantly decrease to 1.32 and 2.93, which show our great adaptability on sparse scenes. The reason that errors on other scenes are a little bit higher may lie on the fact that people in these scenes majorly gather crowdedly, which poses great challenge for our approach to adapt to these scenes without any prior knowledge. However, our SEL-CNN obtains almost three minimum MAE errors when compared to other approaches. This shows that our method have a good generalization ability and prediction robustness on different scenes.

4.4 Results on ShanghaiTech and UCF_CC_50

Table 2 both shows the evaluation results trained on one source dataset and tested on the same and another target dataset of ShanghaiTech [29] and UCF_CC_50 [4]. Our approach obtains competitive crowd counting performances and outperforms the state-of-the-art methods on cross-dataset test. Besides, it can be observed that the proposed work is top-three in most scenes.

Table 3. Ablation Study of SEL-CNN architectures and training strategy on ShanghaiTech dataset [29]. *i.t.* means Iterative Training.

Method	Part A → Part A		Part A → Part B	
	MAE	MSE	MAE	MSE
One regressor	75.8	120.8	20.9	34.1
SEL-CNN w/o SDF	72.9	118.1	19.0	31.8
SEL-CNN w/o *i.t.*	72.4	**108.9**	23.7	32.3
SEL-CNN (ours)	**69.6**	109.1	**18.4**	**31.5**

Table 4. Test performance of each regressor for different density levels on ShanghaiTech PartA [29] after divided training. d represents the ground-truth density level which is calculated as the ratio of the crowd count and the area.

Density level	D^1	D^2	D^3
Low (1.0e−4 $\leq d \leq$ 5.0e−4)	**44.87**	47.54	85.75
Middle (5.0e−4 $< d <$ 1.0e−3)	61.87	**45.15**	75.48
High (1.0e−3 $\leq d \leq$ 4.9e−3)	140.72	115.57	**109.52**

Specifically, for ShanghaiTech, our method reduces 2.8 points transferring from PartA sub-dataset to PartB and 29.6 points transferring from PartB to PartA. Our MAE and MSE on PartA and PartB are slightly higher than the best one but still have competitive performance. These results suggest that our method can be well applied in both crowded and sparse scenes. Meanwhile, our method works effectively transferring either from sparse to crowded or from crowded to sparse.

For UCF_CC_50, as shown in Table 2, we achieve the second best MAE and the best MSE on intra-dateset test and both the second best MAE and MSE on cross-dataset performance. Although this dataset images are very finite, our method can still obtain a promising performance, which shows its robustness despite very limited labeled training samples.

4.5 Ablation Study

To gain further insight into the proposed SEL-CNN, we conduct ablation experiments on the widely adopted ShanghaiTech dataset [29] to verify its effectiveness.

We first conduct experiments with alternative backbone settings. One regressor means network with only one branch feature fusion and decoder, and SEL-CNN w/o SDF means three intermediate density maps are selected as final density map. Our SEL-CNN outperforms other architecture designs. With the help of iterative training, we can gain significant improvements on the test dataset ShanghaiTech Part A in Table 3. Even when the proposed method adapts to an unseen dataset ShanghaiTech Part B, our proposed SEL-CNN with iterative training performs significantly better than SEL-CNN without iterative training. Our iterative learning reduces the count error MAE from 23.7 to 18.4 and MSE from 32.3 to 31.5 in Table 3.

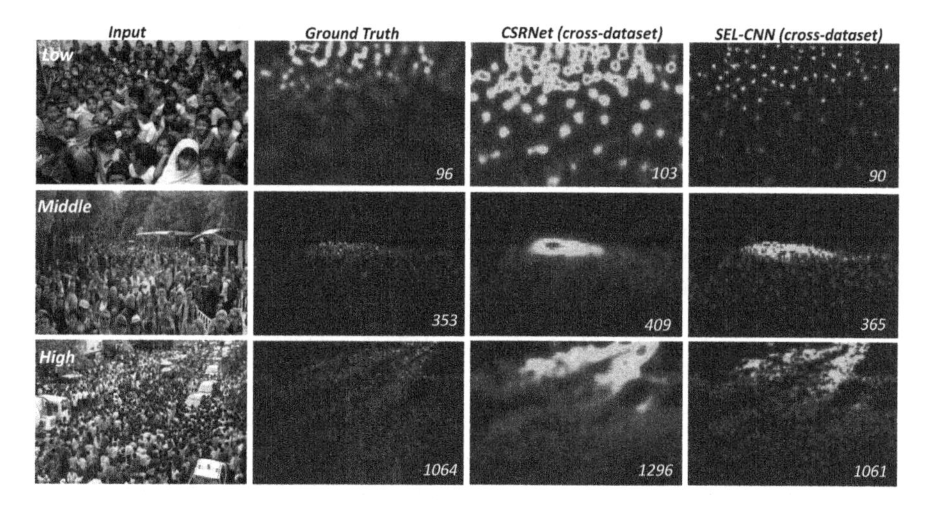

Fig. 2. Visualization of predicted density maps from CSRNet and SEL-CNN for three test images from ShanghaiTech PartA with different density levels.

Referring to Sect. 3.3, $\boldsymbol{D}^1$, $\boldsymbol{D}^2$ and $\boldsymbol{D}^3$ should focus more on low, medium and high density levels, respectively. From Table 4, we can see each regressor works best at a certain density level after divided training to complete the amplification of regressor differences. Specifically, the lowest MAEs for $\boldsymbol{D}^1$, $\boldsymbol{D}^2$ and $\boldsymbol{D}^3$ on the corresponding density levels are 44.87, 45.15 and 109.52, respectively.

4.6 Visual Results on ShanghaiTech Samples

To give an intuitive evidence on how our network works, we visualize the cross-scene prediction comparisons between reproduced CSRNet[29] and our SEL-CNN in Fig. 2. Our samples is chosen from ShanghaiTech PartA sub-dataset when the two models are trained on the PartB.

In Fig. 2, we show the original images with different density levels, ground truth density maps, predicted density maps from CSRNet [29] and final density maps from our SEL-CNN in four columns, respectively. The true and predicted crowd counts are also put in corresponding images for a direct comparison. It can be viewed that the distribution of our final density map is very close to the groundtruth and the positions of heads are clearer than the third column.

5 Conclusion

We propose a Scene-aware Ensemble Learning CNN model (SEL-CNN) to investigate the cross-scene task for crowd counting. We learn not only scene-specific regressor but an ensemble to obtain robust results with the help of scene-aware fusion module. A novel Combination-Division-Iteration based training strategy is applied for more efficient training. Extensive intra-dataset and cross-dataset experiments demonstrate the effectiveness.

References

1. Chen, X., Bin, Y., Sang, N., Gao, C.: Scale pyramid network for crowd counting. In: 2019 IEEE Winter Conference on Applications of Computer Vision (WACV), pp. 1941–1950. IEEE (2019)
2. Cheng, Z.Q., Li, J.X., Dai, Q., Wu, X., He, J.Y., Hauptmann, A.G.: Improving the learning of multi-column convolutional neural network for crowd counting. In: Proceedings of the 27th ACM International Conference on Multimedia, pp. 1897–1906 (2019)
3. Hossain, M.A., Kumar, M., Hosseinzadeh, M., Chanda, O., Wang, Y.: One-shot scene-specific crowd counting. In: BMVC, p. 217 (2019)
4. Idrees, H., Saleemi, I., Seibert, C., Shah, M.: Multi-source multi-scale counting in extremely dense crowd images. In: Proceedings of the IEEE Conference on Computer Vision and Pattern Recognition, pp. 2547–2554 (2013)
5. Jiang, L., Gao, C., Meng, D., Hauptmann, A.G.: DecideNet: counting varying density crowds through attention guided detection and density estimation. In: 2018 IEEE/CVF Conference on Computer Vision and Pattern Recognition (CVPR) (2018)
6. Jiang, X., et al.: Attention scaling for crowd counting. In: Proceedings of the IEEE/CVF Conference on Computer Vision and Pattern Recognition, pp. 4706–4715 (2020)
7. Krizhevsky, A., Sutskever, I., Hinton, G.E.: ImageNet classification with deep convolutional neural networks. In: Advances in Neural Information Processing Systems, vol. 25, pp. 1097–1105 (2012)

8. Lempitsky, V., Zisserman, A.: Learning to count objects in images. In: Advances in Neural Information Processing Systems, vol. 23, pp. 1324–1332 (2010)
9. Li, Y., Zhang, X., Chen, D.: CSRNet: dilated convolutional neural networks for understanding the highly congested scenes. In: Proceedings of the IEEE Conference on Computer Vision and Pattern Recognition, pp. 1091–1100 (2018)
10. Liu, C., Weng, X., Mu, Y.: Recurrent attentive zooming for joint crowd counting and precise localization. In: Proceedings of the IEEE/CVF Conference on Computer Vision and Pattern Recognition, pp. 1217–1226 (2019)
11. Liu, N., Long, Y., Zou, C., Niu, Q., Pan, L., Wu, H.: ADCrowdNet: an attention-injective deformable convolutional network for crowd understanding. In: Proceedings of the IEEE/CVF Conference on Computer Vision and Pattern Recognition, pp. 3225–3234 (2019)
12. Liu, W., Salzmann, M., Fua, P.: Context-aware crowd counting. In: Proceedings of the IEEE Conference on Computer Vision and Pattern Recognition, pp. 5099–5108 (2019)
13. Liu, X., Van De Weijer, J., Bagdanov, A.D.: Leveraging unlabeled data for crowd counting by learning to rank. In: Proceedings of the IEEE Conference on Computer Vision and Pattern Recognition, pp. 7661–7669 (2018)
14. Liu, Y., Sun, Q., Liu, A.A., Su, Y., Schiele, B., Chua, T.S.: LCC: learning to customize and combine neural networks for few-shot learning. arXiv preprint arXiv:1904.08479 (2019)
15. Oñoro-Rubio, D., López-Sastre, R.J.: Towards perspective-free object counting with deep learning. In: Leibe, B., Matas, J., Sebe, N., Welling, M. (eds.) ECCV 2016. LNCS, vol. 9911, pp. 615–629. Springer, Cham (2016). https://doi.org/10.1007/978-3-319-46478-7_38
16. Reddy, M.K.K., Hossain, M., Rochan, M., Wang, Y.: Few-shot scene adaptive crowd counting using meta-learning. In: The IEEE Winter Conference on Applications of Computer Vision, pp. 2814–2823 (2020)
17. Sam, D.B., Babu, R.V.: Top-down feedback for crowd counting convolutional neural network. In: Proceedings of the AAAI Conference on Artificial Intelligence, vol. 32 (2018)
18. Sam, D.B., Surya, S., Babu, R.V.: Switching convolutional neural network for crowd counting. In: 2017 IEEE Conference on Computer Vision and Pattern Recognition (CVPR), pp. 4031–4039. IEEE (2017)
19. Shi, Z., et al.: Crowd counting with deep negative correlation learning. In: Proceedings of the IEEE Conference on Computer Vision and Pattern Recognition, pp. 5382–5390 (2018)
20. Simonyan, K., Zisserman, A.: Very deep convolutional networks for large-scale image recognition. arXiv preprint arXiv:1409.1556 (2014)
21. Sindagi, V.A., Patel, V.M.: CNN-based cascaded multi-task learning of high-level prior and density estimation for crowd counting. In: 2017 14th IEEE International Conference on Advanced Video and Signal Based Surveillance (AVSS), pp. 1–6. IEEE (2017)
22. Sindagi, V.A., Patel, V.M.: Generating high-quality crowd density maps using contextual pyramid CNNs. In: Proceedings of the IEEE International Conference on Computer Vision, pp. 1861–1870 (2017)
23. Sindagi, V.A., Patel, V.M.: A survey of recent advances in CNN-based single image crowd counting and density estimation. Pattern Recogn. Lett. **107**, 3–16 (2018)
24. Sindagi, V.A., Patel, V.M.: HA-CCN: hierarchical attention-based crowd counting network. IEEE Trans. Image Process. **29**, 323–335 (2019)

25. Sindagi, V.A., Patel, V.M.: Multi-level bottom-top and top-bottom feature fusion for crowd counting. In: Proceedings of the IEEE/CVF International Conference on Computer Vision, pp. 1002–1012 (2019)
26. Tan, M., Pang, R., Le, Q.V.: EfficientDet: scalable and efficient object detection. In: Proceedings of the IEEE/CVF Conference on Computer Vision and Pattern Recognition, pp. 10781–10790 (2020)
27. Xu, C., Qiu, K., Fu, J., Bai, S., Xu, Y., Bai, X.: Learn to scale: generating multipolar normalized density maps for crowd counting. In: Proceedings of the IEEE International Conference on Computer Vision, pp. 8382–8390 (2019)
28. Zhang, C., Li, H., Wang, X., Yang, X.: Cross-scene crowd counting via deep convolutional neural networks. In: Proceedings of the IEEE Conference on Computer Vision and Pattern Recognition, pp. 833–841 (2015)
29. Zhang, Y., Zhou, D., Chen, S., Gao, S., Ma, Y.: Single-image crowd counting via multi-column convolutional neural network. In: Proceedings of the IEEE Conference on Computer Vision and Pattern Recognition, pp. 589–597 (2016)
30. Zhao, M., Zhang, J., Zhang, C., Zhang, W.: Leveraging heterogeneous auxiliary tasks to assist crowd counting. In: Proceedings of the IEEE Conference on Computer Vision and Pattern Recognition, pp. 12736–12745 (2019)

Complementary Temporal Classification Activation Maps in Temporal Action Localization

Lijuan Wang, Suguo Zhu(✉), Zhihao Li, and Zhenying Fang

Media Intelligence Laboratory, School of Computer Science and Technology, Hangzhou Dianzi University, Hangzhou 310018, China
zsg2016@hdu.edu.cn

Abstract. Weakly-supervised temporal action localization aims to correctly predict the categories and temporal intervals of actions in an untrimmed video by using only video-level labels. Previous methods aggregate category scores through a classification network to generate temporal class activation map (T-CAM), and obtain the temporal regions of the object action by using a predetermined threshold on generated T-CAM. However, class-specific T-CAM pays too much attention to those regions that are more discriminative for classification tasks, which ultimately leads to fragmentation of localization results. In this paper, we propose a complementary learning strategy for weakly-supervised temporal action localization. It obtains the erasure feature by masking the high activation value position of the original temporal class activation map, and takes it as input to train an additional classification network to produce complementary temporal class activation map. Finally, the fragmentation problem is alleviated by merging two temporal class activation map. We have conduct sufficient experiments on the THUMOS'14 and ActivityNet1.2, and the experimental results show that the localization performance of the proposed method has been greatly improved compared with the existing methods.

Keywords: Temporal action localization · Weakly supervision · Temporal class activation map · Complementary learning

1 Introduction

Temporal action localization in an untrimmed video is an important challenging task. It aims to identify the category of each action that occurs in the input video and detect the start and end time corresponding to each action. In order to achieve excellent performance, previous methods are mostly based on full supervision [5,6,15], which means that in the training phase of the model, the dataset used must contain both the category annotations and the temporal annotations

L. Wang—Student as the first author.

H. Ma et al. (Eds.): PRCV 2021, LNCS 13020, pp. 373–384, 2021.
https://doi.org/10.1007/978-3-030-88007-1_31

of the corresponding action. Although the fully-supervised method has achieved good performance, it is extremely expensive and time-consuming to accurately annotate the temporal boundaries of each action in a large-scale video dataset, which makes the practicability of the fully-supervised method limited to a certain extent. Therefore, weakly-supervised method, which needs only category annotations without temporal annotations, has attracted more and more attention from researchers.

Due to the similarity of task objects, some of previous weakly-supervised temporal action localization methods are mostly inspired by weakly-supervised object detection methods (e.g. [3,8], etc.). In the field of weakly-supervised object detection, the class activation map (CAM), which proposed by Zhou et al. [23], has attracted the attention of many researchers. This method aggregates category scores through a classification network to generate class activation map (CAM) corresponding to each category, and then uses a preset threshold on the generated CAM to obtain the object region corresponding to the object. Inspired by CAM, Nguyen et al. [9] designed a sparse temporal pooling network (STPN), which migrate the CAM to the filed of weakly-supervised temporal action localization, and proposed temporal class activation maps (T-CAM). At the same time, an additional attention module is designed to better supervise the sparsity of actions in the temporal dimension. Since then, more and more researchers have tried to use T-CAM to implement weakly-supervised temporal action localization, and have spawned many excellent methods (e.g. AutoLoc [12], w-TALC [10], etc.). However, there is an implicit flaw in methods based on T-CAM: the T-CAM obtained by aggregating the frame-level class scores focuses too much on those regions that are discriminative for the classification task, which eventually leads to fragmentation of the detection results and inability to obtain complete action localization results.

In this paper, we propose a complementary learning strategy, which includes two sub-branch classification networks (named branch A and branch B), to solve the above mentioned defects. Firstly, branch A is used to produce temporal class activation maps corresponding to all categories, and in this temporal class activation maps, the discriminative regions in the action segment for the current classification task will get a higher activation response. Then, a preset threshold is used to erase the high-response regions in the T-CAM generated by the branch A, and obtain the erased feature. Then, branch B takes the erased feature as input, and generate a complementary T-CAM relative to the branch A. Finally, by merging two temporal class activation maps generated from two sub-branch, the fragmentation problem is alleviated, and the localization performance of our model is effectively improved.

To sum up, the main contribution of our work are as follows:

We propose a complementary learning strategy for the weakly-supervised temporal action localization. By using the erased features to force the complementary network to have a higher activation response for segments with insignificant features in a complete action instance, a more complete localization is obtained, and the fragmentation problem caused by using a single temporal class activation map is greatly alleviated.

2 Related Work

Fully-Supervised Temporal Action Localization (TAL). In recent years, driven by the development of deep learning, TAL has made great progress. Among all TAL methods, Shou et al. [13] first proposed a method of extracting temporal proposals by using sliding window, but limited of fixed temporal scales, its performance is not satisfactory enough. Yang et al. [19] replaced the temporal convolution operation by designing temporal preservation convolution, so that the model can get the same size of the receptive field as the temporal convolution after the pooling operation without pooling in the temporal dimension, and effectively avoids the loss of temporal information in the process of down-sampling and then up-sampling. Gao et al. [4] used cascaded boundary regression to solve the shortcoming that the temporal proposals obtained by sliding window may only contain a partial fragment of an action. Zhao et al. [21] proposed a temporal actionness grouping (TAG) algorithm, which breaks away from the limitation of sliding window settings, and can better utilize the local semantic information in the video sequence to generate more flexible temporal proposal.

Weakly-Supervised Temporal Action Localization (W-TAL). Temporal action localization in a weakly-supervised setting can train the locator only by using all the action classes that occur in the video. It does not have to provide labeled information about each area in the video where the action occurs during training. And when tested, it can locate the type of action that occurs in the input video and the corresponding start and end times for each action, like a fully supervised temporal action locator. At present, there have been many researches on weakly-supervised temporal action localization [11,18]. And some scholars have migrated the class activation map (CAM) method in weakly-supervised object localization algorithm to weakly-supervised temporal action localization, and each action temporal region is located through the generation of temporal class activation maps (T-CAMs). One problem is that the temporal class activation map has the problem of fragmentation of localization results. Therefore, we proposed a method to generate temporal class activation map online, and based on this online generation method, we designed a complementary learning mechanism for temporal class activation map to alleviate the fragmentation problem of localization results.

3 Our Approach

In this section, the proposed weakly-supervised temporal action localization framework is described in detail, as shown in Fig. 1. The introduction of the whole framework will be divided into four parts: feature extraction layer, feature embedding layer, complementary temporal class activation layer and the optimization method.

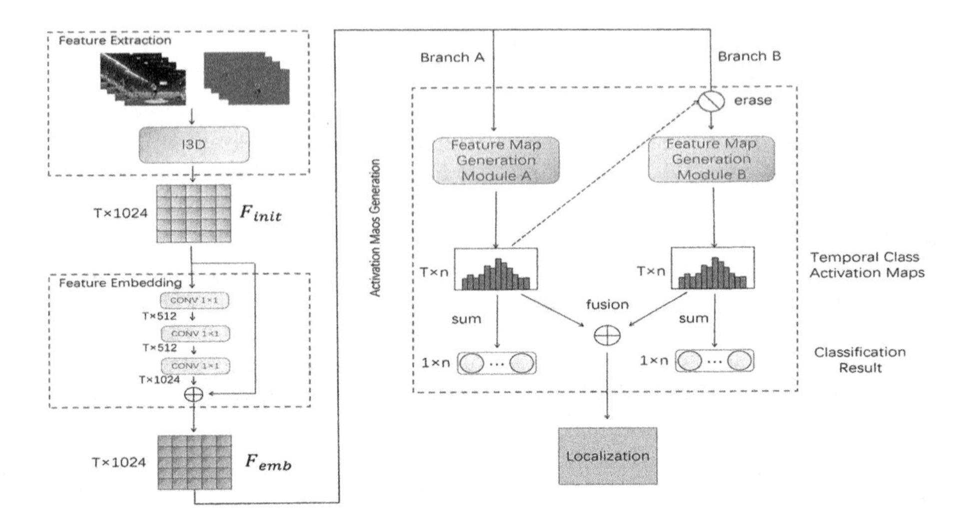

Fig. 1. The basic structure of weakly-supervised temporal action localization. A complete input video will be through the feature extraction layer, feature embedding layer and the unique temporal class activation map complementary learning layer to complete the temporal action localization. Here T represents the number of frames and m represents the number of action classes in the dataset.

3.1 Feature Extraction Layer

We used the I3D [1] network trained on the Kinetics dataset as the feature extraction network. The I3D model introduces 3D convolutional layer and 3D maximum pooling layer to learn the temporal semantic information between video image sequences. In addition, using the optical flow field modal extracted from the video sequence as a certain and priori temporal information input into the network, a large improvement in action recognition performance can still be achieved. Therefore, the I3D model introduces the 3D convolutional layer and the 3D maximum pooling layer, and also based on the dual-stream structure to separately model the RGB image and the optical flow field to extract more optimal video spatio-temporal features.

For the extraction of video data RGB image sequence, we use OpenCV library to extract images in the input video at a rate of 10 frames per second. For the video optical flow field mode extraction, we use TV-L1 [20] algorithm to obtain. After extraction, in the training and prediction process of weakly-supervised network, continuous RGB image sequence and optical flow field sequence are respectively used as the input of RGB branch and optical flow field branch in I3D network to extract the temporal features of video data. Before the video image and optical flow data are input into the feature extraction network, the corresponding data enhancement is needed to improve the data diversity and ensure the generalization performance of the weakly-supervised localization model.

3.2 Feature Embedding Layer

In the task of weakly-supervised temporal action localization, the temporal features of video obtained by the above feature extraction layer are not the most suitable. Therefore, we design a new feature embedding layer based on the video temporal features output by the feature extraction module. The feature embedding layer is based on the residual-connection structure similar to that in the ResNet network. The $l_i \times 1024$ dimensional video temporal features output by the feature extraction module are used as the input. The embedded feature dimension obtained by using the feature embedding layer will not change, which is also $l_i \times 1024$.

3.3 Complementary Temporal Class Activation Map

The weakly-supervised temporal action localization method based on temporal class activation map can better locate the occurrence area of action, but there are problems of localization result fragmentation and missing location. To solve these two problems, we propose a complementary generation model. This model based on temporal class generated online activation map, and it supervises the elimination of the temporal features corresponding to the locations with higher activation values in the output of feature extraction module, and trains additional action classification networks. The temporal class activation map output by the action classification network can generate higher activation for those action locations with lower activation value before feature elimination.

3.4 Model Optimization

The weakly-supervised temporal action localization model can achieve end-to-end training by jointly optimizing the classification loss and sparse loss of classifier A and B. The total loss function is shown in Eq. (1).

$$L = L_{sparse,A} + L_{cls,A} + L_{sparse,B} + L_{cls,B} \tag{1}$$

$L_{sparse,A}$ and $L_{sparse,B}$ correspond to the sparsity loss in classifiers A and B respectively, and $L_{cls,A}$ and $L_{cls,B}$ correspond to the classification loss in classifiers A and B respectively.

The sparse loss is obtained by calculating the L_1 regular corresponding to the attention weight output by the attention mechanism module. Since Sigmoid is used as the activation function in the output and L_1 is a regular constraint, the combined effect of the two will make the final weight value of attention close to 0 or 1. The idea of this sparse constraint mainly comes from the priori that most of the action fragments in the video will not appear continuously.

An input video may contain multiple actions that occur, so the classification model is modeled as a many to one label classification work. For each class, the binary cross-entropy loss is used to calculate the loss value of the current class, and the final classification loss value is obtained by averaging the binary cross-entropy loss of all classes. The specific mathematical expression is shown in Eq. (2).

$$L_{cls} = \frac{\sum_{c} y_c log y_c^{conv} + (1 - y_c) log(1 - y_c^{conv}}{C} \quad (2)$$

C is the total number of action classes, and y_c is the true label of class c, with a value of 0 or 1. y_c^{conv} is the prediction probability value of the classification model for class c.

4 Experiments and Analysises

4.1 Datasets

In this experiment, we use two public datasets, THUMOS'14 [7] and ActivityNet-1.2, to verify the performance of the proposed algorithm. The THUMOS'14 dataset is mainly used for video action recognition and temporal action localization. ActivityNet-1.2 dataset is a very challenging dataset. Compared with the THUMOS'14 dataset, the number of action in each video in the ActivityNet-1.2 dataset is less, with an average of 1.5 actions in each video according to statistics, which brings great challenges for weak supervision and localization.

4.2 Implementation Details

For videos from two different datasets, both of them extracted the video into RGB images at 10 frames per second, and then extracted the temporal features of the video using the I3D network. For the extracted RGB image, the following operations are performed: First, the shortest edge is scaled to 256 pixels while maintaining aspect ratio. Then, the region of 224×224 pixel is clipped in the center as the input of the feature extraction network. For optical branch, we use TV-L1 algorithm to extract the optical flow field of video. Then, after the same operation as the RGB image, 224×224 subimages are used as the network input.

During the training, the end-to-end training is achieved by combining the classification loss and sparse loss of the two classifiers, and the model parameters are updated by Adam optimizer with a learning rate of 0.0001. For post-erased-temporal features to be used by the complementary classifier B, the position with an activation value higher than 0.53 is regarded as the position with high contribution degree in the temporal class activation map corresponding to the real action class output by the classifier A. The embedded features of these high contribution positions are erased and used for the training of classifier B. During the test, since the actual action class could not be obtained, the action whose probability value was higher than 0.1 in the prediction results of all action classes by the classifier A was taken as the pseudo-label of the action class in the video, and then the erasification operation was performed.

The localization results were obtained from the temporal class activation map. The fusion of RGB branch and optical flow branch is needed to process the temporal class activation map. RGB branches are mainly used: firstly, the temporal class activation map corresponding to the action class with RGB prediction probability value higher than 0.1 is processed. Secondly, the weighted average operation is carried out for the corresponding positions in the activation map of temporal classes with the ratio of $RGB : Flow = 0.7 : 0.3$. Finally, the first part of localization results were obtained by combining the continuous regions with activation values higher than 0.51. Optical flow field branches are mainly used: firstly, it deals with the temporal class activation map corresponding to the action class whose optical flow guessing probability is higher than 0.1. Secondly, the weighted average operation is carried out for the corresponding positions in the activation map of temporal classes with the ratio of $RGB : Flow = 0.3 : 0.7$. Finally, the regions above the threshold value of 0.51 were combined to obtain the localization results of the second part. In fusion, the first and second part localization results are combined, and then non-maximum suppression with a threshold of 0.9 is performed to remove the regions with low confidence when the overlap degree is high.

4.3 Ablation Studies

We verified the effectiveness of the proposed modules on the THUMOS' 14 dataset, Fig. 2 shows the comparisons between our baselines, the full model and the performance comparison of each module's test results are listed in Table 1. The basic model is the model that only uses the online temporal class activation map method without feature embedding and complementary learning. mAP $(avg : 0.1 - 0.5)$ represents the mean value of all mean accuracy means under the threshold of 0.1 to 0.5.

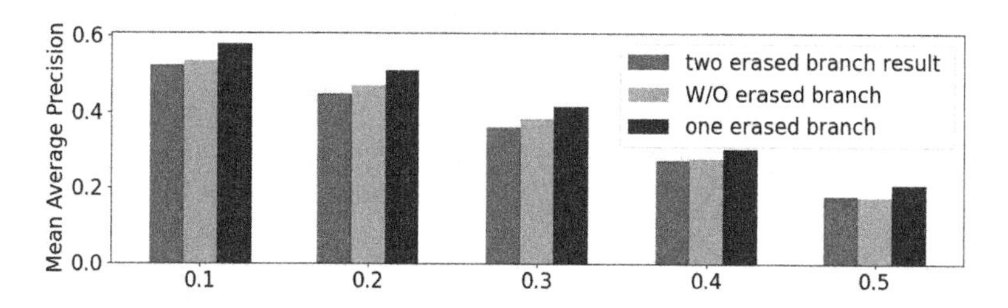

Fig. 2. Performance with respect to architectural variations. We can clearly observe that, regardless of the IoU value, the experiment with one branch erasure is better than the experiment with no branch erasure and the experiment with two branches erasure.

As can be seen from Table 1, when the basic model without any components is compared with the STPN model, the basic model we use is similar to the STPN model in localization performance, because the parameter matrix of the convolution is the same in mathematical expression form as the representation of the full connection layer classification. On the other hand, due to the use of convolution to replace the full connection layer, the temporal class activation map can be output directly by the convolution, which improves the flexibility of the generation of the temporal class activation map, and enables it to be easily

Table 1. Results of ablation experiment.

	Features embedding	Complementary learning	mAP (avg:0.1–0.5)
STPN [9]			35.0
Basic models			34.6
	✓		35.6
		✓	39.4
	✓	✓	40.2

integrated into a more complex network structure as a module. Although the performance of our basic model is similar to that of STPN, it is more flexible, simple and convenient than the generation method of temporal class activation map in STPN.

4.4 Experimental Results on the THUMOS'14 Dataset

In Table 2, $avg(0.1:0.5)$ represents the mean value of all mean accuracy means between the crossover ratio threshold of 0.1 and 0.5, and "–" represents the results under the crossover ratio threshold that were not given in the original paper.

Table 2. Performance comparison on the THUMOS'14 dataset

Supervision methods	Methods	mAP@tIoU					
		0.1	0.2	0.3	0.4	0.5	0.6
Full supervision	S-CNN [13]	47.7	43.5	36.3	28.7	19	–
	R-C3D [17]	54.5	51.5	44.8	35.6	28.9	–
	SSN [21]	60.3	56.2	50.6	40.8	29.1	–
	TAL-Net [2]	59.8	57.1	53.2	48.5	42.8	33.8
Weak supervision	Hide-and-Seek [14]	36.4	27.8	19.5	12.7	6.8	–
	UntrimmedNet [16]	44.4	37.7	28.2	21.1	13.7	–
	Zhong et al. [22]	45.8	39	31.1	22.5	15.9	–
	STPN [9]	52	44.7	35.5	25.8	16.9	9.9
	AutoLoc [12]	53.5	46.8	37.5	29.1	19.9	12.3
	W-TALC [10]	55.2	49.6	40.1	31.1	22.8	–
	Our method	57.9	50.8	41.4	30.1	20.6	12.5

As can be seen from Table 2, the method proposed in this paper achieves the best localization performance at present with the average results of all mean accuracy averages under the $0.1:0.5$ threshold value. At the same time, the performance of the proposed algorithm is very close to that of most fully supervised localization methods, although the performance of the fully supervised temporal action localization method is somewhat lower than that of the best fully supervised action localization method. This trend is particularly obvious under the threshold of low crossover ratio.

We use an additional feature embedding layer to learn the temporal feature representation that is more beneficial to the temporal action localization task under weak supervision, and we also use a complementary learning module based on online feature elimination to alleviate the fragmentation problem of weakly-supervised localization results. Due to the addition of the above two modules, the method proposed in this paper has a great improvement in performance compared with other existing weakly-supervised temporal action weakly-supervised localization algorithms in Table 2. At the same time, the model we used has high flexibility. After calculating the spatial and temporal characteristics of the input video, all modules can be flexibly integrated together to achieve end-to-end training. Compared with other methods, it is more excellent in weakly-supervised localization speed and model flexibility.

4.5 Experimental Results on ActivityNet-1.2 Dataset

We also tested on ActivityNet-1.2 dataset with more video data. The test results of the existing weakly-supervised localization model were compared with the existing weakly-supervised localization model, as shown in Table 3. "–" refers to the mean accuracy under the specified crossover ratio threshold that was not given in the original paper corresponding to the algorithm, and UNT method is the abbreviation of Untrimmednet algorithm.

Table 3. On ActivityNet-1.2 dataset.

Supervision methods		Weak supervision				
Methods		UNT	Zhong et al.	Auto-Loc	W-TALC	Our methods
mAP@tIoU	0.5	7.4	27.3	27.3	37	34.5
	0.55	6.1	–	24.9	–	31.5
	0.6	5.2	–	22.5	–	28.7
	0.65	4.5	–	19.9	–	25.7
	0.7	3.9	–	17.5	14.6	23
	0.75	3.2	14.7	15.1	–	20.4
	0.8	2.5	–	13	–	17.4
	0.85	1.8	–	10	–	13
	0.9	1.2	–	6.8	–	8.9
	0.95	0.7	2.9	3.3	–	4.8
	Avg	3.6	15.6	16	18	20.8

As can be seen from the localization results listed in Table 3, under the high crossover ratio threshold, the performance of our method is better than that of other weakly-supervised temporal action localization models. This performance improvement is due to the use of additional complementary classifiers to locate low contribution temporal locations, thus alleviating the problem of fragmentation of weakly-supervised localization results by a single classifier.

4.6 Results Analysis

The localization performance of the model proposed in this paper and the direction of future improvement are intuitively analyzed through the visualization of localization results.

As shown in Fig. 3, compared with STPN, our proposed weakly-supervised localization method obviously has a higher generalization ability and can accurately locate the motion region in the video.

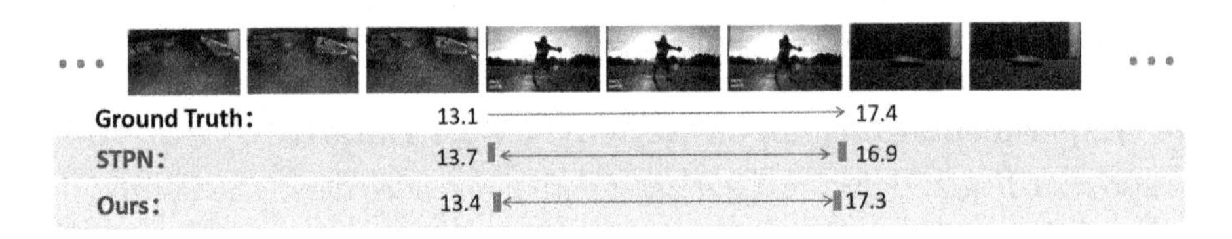

Fig. 3. Performance with respect to architectural variations. Some comparisons between the localization results of the proposed weakly-supervised temporal action localization algorithm, the real action region labeling and STPN are visualized on the THUMOS'14 dataset. Compared to STPN, our approach is significantly better.

5 Conclusion

We propose a method of weakly-supervised temporal action localization which based on complementary learning of temporal class activation maps. Using this method, the complementary temporal class activation map obtained by the complementary classifier can be complementary to the original temporal class activation map to alleviate the problem of fragmentation of weakly-supervised localization results. We tested the performance of the localization algorithm on two public datasets of THUMOS'14 and ActivityNet-1.2. Through the analysis of the weakly-supervised localization results, it is proved that the proposed weakly-supervised temporal action localization model has better localization performance and higher generalization.

Acknowledgements. This work was supported by the National Natural Science Foundation of China, with grant numbers 61902101 and 61806063 respectively.

References

1. Carreira, J., Zisserman, A.: Quo vadis, action recognition? A new model and the kinetics dataset. In: Proceedings of the IEEE Conference on Computer Vision and Pattern Recognition, pp. 6299–6308 (2017)
2. Chao, Y.W., Vijayanarasimhan, S., Seybold, B., Ross, D.A., Deng, J., Sukthankar, R.: Rethinking the faster R-CNN architecture for temporal action localization. In: Proceedings of the IEEE Conference on Computer Vision and Pattern Recognition, pp. 1130–1139 (2018)
3. Cheng, G., Yang, J., Gao, D., Guo, L., Han, J.: High-quality proposals for weakly supervised object detection. IEEE Trans. Image Process. **29**, 5794–5804 (2020)
4. Gao, J., Yang, Z., Nevatia, R.: Cascaded boundary regression for temporal action detection. arXiv preprint arXiv:1705.01180 (2017)
5. Heilbron, F.C., Barrios, W., Escorcia, V., Ghanem, B.: SCC: semantic context cascade for efficient action detection, pp. 3175–3184 (2017)
6. Heilbron, F.C., Niebles, J.C., Ghanem, B.: Fast temporal activity proposals for efficient detection of human actions in untrimmed videos. In: Proceedings of the IEEE Conference on Computer Vision and Pattern Recognition, pp. 1914–1923 (2016)
7. Idrees, H., et al.: The thumos challenge on action recognition for videos "in the wild". Comput. Vis. Image Underst. **155**, 1–23 (2017)
8. Mai, J., Yang, M., Luo, W.: Erasing integrated learning: a simple yet effective approach for weakly supervised object localization, pp. 8766–8775 (2020)
9. Nguyen, P., Liu, T., Prasad, G., Han, B.: Weakly supervised action localization by sparse temporal pooling network. In: Proceedings of the IEEE Conference on Computer Vision and Pattern Recognition, pp. 6752–6761 (2018)
10. Paul, S., Roy, S., Roy-Chowdhury, A.K.: W-TALC: weakly-supervised temporal activity localization and classification. In: Ferrari, V., Hebert, M., Sminchisescu, C., Weiss, Y. (eds.) ECCV 2018. LNCS, vol. 11208, pp. 588–607. Springer, Cham (2018). https://doi.org/10.1007/978-3-030-01225-0_35
11. Shen, Z., Wang, F., Dai, J.: Weakly supervised temporal action localization by multi-stage fusion network. IEEE Access **8**, 17287–17298 (2020)
12. Shou, Z., Gao, H., Zhang, L., Miyazawa, K., Chang, S.-F.: AutoLoc: weakly-supervised temporal action localization in untrimmed videos. In: Ferrari, V., Hebert, M., Sminchisescu, C., Weiss, Y. (eds.) ECCV 2018. LNCS, vol. 11220, pp. 162–179. Springer, Cham (2018). https://doi.org/10.1007/978-3-030-01270-0_10
13. Shou, Z., Wang, D., Chang, S.F.: Temporal action localization in untrimmed videos via multi-stage CNNs. In: Proceedings of the IEEE Conference on Computer Vision and Pattern Recognition, pp. 1049–1058 (2016)
14. Singh, K.K., Lee, Y.J.: Hide-and-seek: Forcing a network to be meticulous for weakly-supervised object and action localization. In: 2017 IEEE International Conference on Computer Vision (ICCV), pp. 3544–3553 (2017)
15. Tran, D., Bourdev, L., Fergus, R., Torresani, L., Paluri, M.: Learning spatiotemporal features with 3D convolutional networks, pp. 4489–4497 (2015)
16. Wang, L., Xiong, Y., Lin, D., Van Gool, L.: UntrimmedNets for weakly supervised action recognition and detection. In: Proceedings of the IEEE Conference on Computer Vision and Pattern Recognition, pp. 4325–4334 (2017)
17. Xu, H., Das, A., Saenko, K.: R-C3D: region convolutional 3D network for temporal activity detection. In: Proceedings of the IEEE International Conference on Computer Vision, pp. 5783–5792 (2017)

18. Xu, M., Perez-Rua, J.M., Zhu, X., Ghanem, B., Martinez, B.: Low-fidelity end-to-end video encoder pre-training for temporal action localization. arXiv preprint arXiv:2103.15233 (2021)
19. Yang, K., Qiao, P., Li, D., Lv, S., Dou, Y.: Exploring temporal preservation networks for precise temporal action localization **32**(1) (2018)
20. Zach, C., Pock, T., Bischof, H.: A duality based approach for realtime tv-l 1 optical flow. In: Joint Pattern Recognition Symposium, pp. 214–223 (2007)
21. Zhao, Y., Xiong, Y., Wang, L., Wu, Z., Tang, X., Lin, D.: Temporal action detection with structured segment networks. In: Proceedings of the IEEE International Conference on Computer Vision, pp. 2914–2923 (2017)
22. Zhong, J.X., Li, N., Kong, W., Zhang, T., Li, T.H., Li, G.: Step-by-step erasion, one-by-one collection: a weakly supervised temporal action detector. In: Proceedings of the 26th ACM International Conference on Multimedia, pp. 35–44 (2018)
23. Zhou, B., Khosla, A., Lapedriza, A., Oliva, A., Torralba, A.: Learning deep features for discriminative localization. In: Proceedings of the IEEE Conference on Computer Vision and Pattern Recognition, pp. 2921–2929 (2016)

Improve Semantic Correspondence by Filtering the Correlation Scores in both Image Space and Hough Space

Shihua Xiong and Yonggang Lu(✉)

School of Information Science and Engineering, Lanzhou University, Lanzhou 730000, China
ylu@Lzu.edu.cn

Abstract. Semantic correspondence is an important and challenging task in computer vision due to background clutter, especially for analyzing images from the same category but with large intra-class variations. The methods using regularized Hough matching only consider global offset consensus to enhance geometric consistency, where the filtering in Hough space is usually used to aggregate the scores of the similar offsets. However, the matched objects between two images are usually located in a local region of the images. Therefore, in this paper, we propose an improved method for establishing semantic correspondence by introducing regional offset consensus, which is implemented by filtering the correlation scores in both Hough space and image space, where the filtering in image space is used to enforce the regional offset consensus. Experiments show that the proposed method can produce superior results than the state-of-the-art methods, when applied on four benchmark datasets including a large-scale SPair-71k dataset.

Keywords: Semantic correspondence · Consensus analysis · Spatial filtering · Hough space

1 Introduction

Establishing dense semantic correspondence between images is a basic task in computer vision, which is different from traditional corresponding tasks. Traditional tasks like optical flow [5,12] aims to find the corresponding relationship between the images of the same target and the same scene. However, semantic correspondence can find the corresponding relationship between the images of different targets that have intra-class variation. This task can be applied in many fields such as image editing, 3D reconstruction [1], object detection, and scene understanding.

Traditional corresponding tasks mainly use hand-crafted local features, such as SIFT [19] and HOG [2,8]. However, large intra-class variations and background clutter make the semantic correspondence task more challenging, and traditional manual features are no longer applicable. Most recent methods use more

H. Ma et al. (Eds.): PRCV 2021, LNCS 13020, pp. 385–396, 2021.
https://doi.org/10.1007/978-3-030-88007-1_32

robust features extracted by convolutional neural networks, and both single-layer features [13,22,23] and multi-layer features [17,20] have shown excellent performance. The supervised training model [9,15] requires ground-truth correspondences between training image pairs, but the manual annotation is very expensive, so the training data is difficult to prepare. Rocco et al. [22,23] have adopted a weak supervision method and proposed an end-to-end trainable neural network structure that includes feature extraction and regression networks to estimate the parameters of geometric models, such as affine transformation and thin-plate spline transformation [4] between image pairs. This method does not require manual annotation, but only finds the geometric correspondence in the image through affine transformation and thin-plate spline transformation, which is not suitable for complex scenes. HPF [20] and SCOT [17] only require a small number of validation sets and use the regularized Hough matching to enforce geometric consistency in matching. Regularized Hough matching only considers global offset consensus, however, the matched objects between two images are usually located in a local region of the images.

In this work, we proposed an improved method which uses the Hough transform to enhance geometric consistency in the semantic matching process. Different from the global offset consensus used in regularized Hough matching, regional offset consensus is introduced in this work to semantic correspondence task, which is implemented by filtering the correlation scores in both Hough space and image space. Experiments show that the proposed method can better handle scenes with complex changes than other methods, and can achieve state-of-the-art performance on multiple benchmark datasets.

2 Related Work

The semantic correspondence task aims to establish dense semantic correspondence between different targets of the same class images. In HPF [20], the multi-layer features produced by deep neural networks is applied to the semantic correspondence task for the first time. Multi-layer features composed of high-level features and low-level features can be used to produce better performance than single-layer features. The first process is to select a small feature set from the features of n-layers $(f_0, f_1, f_2, ..., f_{n-1})$ which are composed of the neural network intermediate outputs. To search the optimal feature set from Multi-layer features, a breadth-first algorithm with limited memory is used [20], and a small number of verification sets are used to evaluate the effectiveness of each feature set. Finally, the k-layer features in the optimal feature set are up-sampled to the same size and are concatenated as the hyperpixel feature of the image:

$$F = [f_{n1}, \varphi(f_{n2}), \varphi(f_{n3}), ..., \varphi(f_{nk})] \tag{1}$$

where φ is used to up-sample the features to the same size as f_{n1}. The hyperpixel features $F_s \in R^{h_s \times w_s \times D}$ of the source image and $F_t \in R^{h_t \times w_t \times D}$ of the target image are used to calculate the cosine similarity between them, which gives the correlation map C:

$$C = \frac{F_s \cdot F_t^{\top}}{\| F_s \| \| F_t \|} \in R^{h_s \times w_s \times h_t \times w_t} \tag{2}$$

The correlation map $C(i, j, k, l)$ represents the similarity between the feature with the coordinate (i, j) in the source image and the feature with the coordinate (k, l) in the target image. The correlation mapping C is calculated by individual feature matching, but the mutual relation between features is not considered, so the optimization of the correlation map is very necessary. SCOT [17] regards semantic correspondence as an optimal transport problem to strengthen the correspondence between the matched objects in images. After getting the correlation map, SCOT uses CAM to compute μ_s and μ_t to assign different weights to the foreground pixels and the background pixels. The specific process is to add the following constraints to the correlation map:

$$\begin{gathered} T^* = argmax \sum_{ij} T_{ij} C_{ij}, \\ T\mathbf{1}_{nt} = \mu_s, T^{\top}\mathbf{1}_{nt} = \mu_t \end{gathered} \tag{3}$$

where T is the matching probability between matching pairs. The size of T is the same as the size of the correlation map C. $\sum_{ij} T_{ij} C_{ij}$ denotes the global correlation. The main purpose of SCOT is to obtain the global optimal match T^* by maximizing the global correlation. The value of μ_s indicates the importance of each hyperpixel in the source image. The value of μ_t indicates the importance of each hyperpixel in the target image. The row sum and column sum of T are constrained to μ_s and μ_t.

Finally, HPF and SCOT both use the regularized Hough matching algorithm [20]. The main idea is to use the voting results of the Hough space to re-weight the correlation scores to enhance geometric consistency. Let $D = (R_s, R_t)$ represent a set of hyperpixel positions in the two images, for a hyperpixel position $r_s \in R_s$ and a hyperpixel position $r_t \in R_t$, $m = (r_s, r_t)$ represents the matching between the hyperpixel at r_s and the hyperpixel at r_t. The confidence score $P(m|D)$ of matching m can be calculated by the following formula:

$$P(m|D) \propto P(m_a) \sum_{x \in \chi} P(m_g|x) \sum_{m} P(m_a) P(m_g|x) \tag{4}$$

where $P(m_a)$ is the performance matching confidence. $P(m_a) = C$ is used in HPF, and $P(m_a) = T^*$ is used in SCOT. $P(m_g|x)$ represents the geometric matching probability of a given offset x, and χ represents the set of offsets. After obtaining the performance matching confidence $P(m_a)$, the maximum correlation scores indicate the best matching which can be used to obtain the hyperpixel flow I.

Then there is the process of semantic flow formation. Given a point x_p in the source image, $\mathcal{H}(x_p)$ is the set of the hyperpixels whose receptive field contains x_p. The distance between x_p and the center of the receptive field of each hyperpixel in $\mathcal{H}(x_p)$ is $d(x_q)_{x_q \in \mathcal{H}(x_p)}$. The offset $I(x_q)$ of the hyperpixel feature x_q can be obtained by the hyperpixel flow I, so the semantic flow of x_p can be expressed as the average value of $(I(x_q) + d(x_q))_{x_q \in \mathcal{H}(x_p)}$. After obtaining the semantic correspondence between the images, the annotations in the dataset are used to evaluate the method.

3 The Proposed Method

The proposed method can be divided into three stages (see Fig. 1):

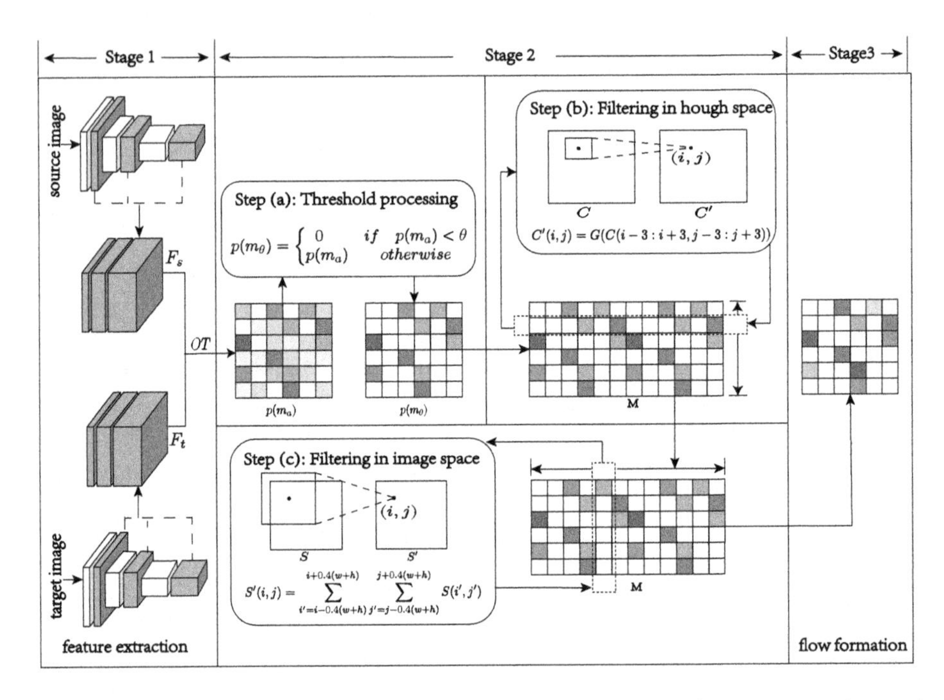

Fig. 1. Overall architecture of the proposed method, stage 1 is feature extraction and the last stage is flow formation. The intermediate stage is divided into three steps, including threshold processing, flitering in hough space and filtering in image space. For details, see text.

- Stage 1: Extracting the features to calculate the correlation mapping and using the optimal transport to optimize;
- Stage 2: Constructing the Hough space, and being used to optimize the correlation score;
- Stage 3: Calculating the corresponding flow.

$$p(m|D) = p(m_a) \sum_{x \in \chi, r_s \in R_s} p(m_g|x, r_s), p(x, r_s|D) \tag{5}$$

Stage 1 and Stage 3 are the same as the corresponding processes in SCOT or HPF. The main contribution of the work is realized in Stage 2. The key idea is to enforce geometric consistency by using regional offset consensus in Stage 2, which is implemented with three steps: Step (a) threshold processing, Step (b) filtering in Hough space, and Step (c) filtering in image space which is mainly used to achieve regional offset consensus, which are described in detail in the following.

Step (a) Threshold Processing. A threshold θ of the correlation score is given first. $p(m_\theta)$ is the correlation score matrix after filtering. It can be calculated by the following formula:

$$p(m_\theta) = \begin{cases} 0 & if \quad p(m_a) < \theta \\ p(m_a) & otherwise \end{cases} \tag{6}$$

In this work, the performance matching confidence $p(m_a)$ is set to the global optimal match T^* obtained from Stage 1. For a given threshold θ, we set the correlation scores less than the threshold to zero and keep the correlation scores greater than the threshold. Through this step, the unimportant matchings are screened out.

Step (b) Filtering in Hough Space. Hough space is constructed by constructing two-dimensional offset bins with size of $h_{offset} \times w_{offset}$. Hough voting is used to obtain the correlation score matrix M that is another form of correlation score matrix $p(m_\theta)$. M represents the correlation score of each hyperpixel in the source image under each offset. Therefore, the size of M is $(h_s \times w_s) \times (h_{offset}, w_{offset})$, where h_s and w_s are the height and width of the hyperpixel feature map extracted from the source image, and h_{offset} and w_{offset} are the height and width of the Hough space. The row of M contains the correlation score of a certain hyperpixel in the source image under each offset, while the column of M contains the correlation scores of all hyperpixel in the source image under a certain offset.

This step filters each row of the matrix M after transformation to aggregate the scores of the similar offsets. Each row of M is transformed into a two-dimensional matrix C which has the same size as the Hough space, and then a Gaussian filter G is used to re-weight the value in C. The purpose of this operation is to optimize correlation scores by encouraging matchings with similar offsets.

$$C'(i,j) = G(C(i-3:i+3, j-3:j+3)) \tag{7}$$

C' is the optimized value from C. Return C' to M for subsequent operations.

Step (c) Filtering in Image Space. This step filters each column of the matrix M after transformation to achieve regional offset consensus.

The confidence of matching $P(m|D)$ can be calculated by the following formula:

where $p(m_g|x, r_s)$ represents the geometric matching probability of a hyperpixel located at r_s in the source image under offset x. $p(x, r_s|D)$ is the geometric prior probability of hyperpixel located at r_s in the source image under offset x. Regularized Hough matching only considers the global offset consensus, all matchings with offset x have the same geometric prior probability. Our work introduces regional offset consensus by introducing location information r_s in the geometric prior probability, which is used to distinguish matchings at different positions with the same offset. $p(m_g|x, r_s)$ is obtained by comparing $r_s - r_t$ and x under a certain r_s.

The geometric prior probability $p(x, r_s|D)$ is set to the sum of the correlation scores of all hyperpixels in the neighborhood $N(r_s)$ under a given offset x.

$$p(x, r_s|D) \propto \sum_{m=(r_s, r_t), r_s^{'} \in N(r_s)} p(m_\theta) p(m_g|x, r_s^{'}) \tag{8}$$

where $N(r_s)$ is the hyperpixel set in the neighborhood of r_s. The specific process is to transform each column of M into a two-dimensional matrix S. The size of S is the same as the size of the hyperpixel feature map extracted from the target image. Then the following filtering is performed on the matrix S.

$$S'(i.j) = \sum_{i'=i-0.4(w+h)}^{i+0.4(w+h)} \sum_{j'=j-0.4(w+h)}^{j+0.4(w+h)} S(i', j') \tag{9}$$

The purpose of this operation is to achieve regional offset consensus by filtering in image space. S' is the optimized correlation score matrix from S. The computed values in S' are used to update the corresponding column in matrix M.

4 Experiments

In this section, the benchmark datasets and the implementation details are introduced, followed by the evaluation of the proposed methods on four benchmark datasets.

4.1 Benchmarks and Evaluation Metric

Evaluation Metric. Use Percentage of Correct Keypoints PCK as the evaluation metric. For the keypoint r_s, if the Euclidean distance d between the predicted point r_t and the ground truth point r_{gt} is less than the threshold value $\alpha \times max(w, h)$ where α is the matching accuracy, w and h are the width and height respectively of the image (α_{img}) or the object bounding box (α_{bbox}), the key point r_s is considered to be the correct key point. Our method is evaluate by the following four datasets:

PF-PASCAL. [8] PF-PASCAL contains 1351 pairs of images in 20 different classes, and each image pair contains 4 to 17 sets of key point annotations. PCK(α_{img}) is used to estimate methods on PF-PASCAL.

PF-WILLOW. [7] PF-WILLOW contains 4 different classes, in a total of 900 pairs of images, and each image pair contains 10 sets of keypoint annotations. PCK(α_{bbox}) is used to evaluate methods on TSS.

TSS. [26] TSS consists of three parts: FG3DCar, JODS, and PASCAL, with a total of 400 pairs of images. Each pair of images provides ground-truth flow, so PCK can be calculated through the dense pixels from matched objects. PCK(α_{img}) is used to evaluate methods on TSS.

SPair-71k. [21] SPair-71k contains 70958 pairs of images in 18 different classes, and each image pair contains 3 to 30 sets of annotations. Compared with other datasets, it has a much larger number and rich annotations. It shows greater variability in terms of viewpoint, scale, occlusion, and truncation. In the evaluation process, the variation factors of keypoints annotations can be used for classification evaluation. PCK(α_{img}) is used to estimate methods on SPair-71k.

4.2 Implementation Details

We use the pre-trained networks ResNet50 and ResNet101 [10] on the ImageNet [3] as feature extraction networks. For PF-PASCAL and PF-WILLOW, in order to compare with the HPF method and the SCOT method, we also use the pre-trained networks Res101-FCN [18] on PSACAL VOC 2012 [6].

In the feature extraction stage, different feature layer set is selected for each dataset. For PF-PASCAL and PF-WILLOW, the hyperpixel layer set is {2,7,11,12,13} with ResNet-50 and {2,22,24,25,27,28,29} with ResNet-101, and {2,4,5,18,19,20,24,32} with Res101-FCN. For Spair-71k, the hyperpixel layer set is {0,11,12,13} with ResNet-50 and {0,19,27,28,29} with ResNet-101. For TSS, the hyperpixel layer set is {2,22,24,25,27,28,29} with RexNet-101.

In Eq. (6), The threshold θ is set to be the 99.5th percentile of all the values in $p(m_a)$. In step (2), 7×7 Gaussian filtering are performed four times. In Eq. (8), The neighborhood $N(r_s)$ is a square area centered on r_s, the side length of the square is $0.4(w + h)$, where w and h are the width and height of the hyperpixel feature map.

4.3 Evaluation Results

Matching Results on PF-PASCALL and PF- WILLOW. The comparison between our method and the other thirteen methods on PF-PASCALL and PF- WILLOW is shown in Table 1. For each method, evaluation metrics are computed with α setting to three different thresholds, where the small threshold represents the strict evaluation. It is shown that the proposed method achieves the best results on PF-PASCALL and PF- WILLOW with three different α values. For HPF, SCOT, and the proposed method, using Resnet101-FCN as backbone has better performance than using Resnet101 as backbone. For both Resnet101-FCN and Resnet101, the proposed method produces better results than both HPF and SCOT. These results show that the proposed method is effective for both PF-PASCALL and PF-WILLOW.

Matching Results on TSS. The results of the comparison with nine methods on TSS are in Table 2. The proposed method produces better results than both HPF and SCOT in all parts of TSS. For part FG3DCar, the proposed method

Table 1. PCK results on PF-PASCALL and PF- WILLOW. Numbers in bold indicate the best performance. Subscripts of the method names indicate backbone networks used.

Methods	PF-PASCAL			PF-WILLOW		
	0.05	0.1	0.15	0.05	0.1	0.15
$\mathrm{PF}_{\mathrm{HOG}}$ [7]	31.4	62.5	79.5	28.4	56.8	68.2
$\mathrm{CNNGeo}_{\mathrm{res101}}$ [22]	41.0	69.5	80.4	36.9	69.2	77.8
$\mathrm{A2Net}_{\mathrm{res101}}$ [25]	42.8	70.8	83.3	36.3	68.8	84.4
$\mathrm{DCTM}_{\mathrm{CAT\text{-}FCSS}}$ [14]	34.2	69.6	80.2	38.1	61.0	72.1
$\mathrm{Weakalign}_{\mathrm{res101}}$ [23]	49.0	74.8	84.0	37.0	70.2	79.9
$\mathrm{NC\text{-}Net}_{\mathrm{res101}}$ [24]	54.3	78.9	86.0	33.8	67.0	83.7
$\mathrm{DCCNet}_{\mathrm{res101}}$ [11]	–	82.3	–	43.6	73.8	86.5
$\mathrm{RTNs}_{\mathrm{res101}}$ [13]	55.2	75.9	85.2	41.3	71.9	86.2
$\mathrm{SCNet}_{\mathrm{VGG16}}$ [9]	34.2	69.6	80.2	38.1	61.0	72.1
$\mathrm{NN\text{-}Cyc}_{\mathrm{res101}}$ [15]	55.1	85.7	94.7	40.5	72.5	86.9
$\mathrm{SFNet}_{\mathrm{res101}}$ [16]	–	78.7	–	–	74.0	–
$\mathrm{HPF}_{\mathrm{res101}}$ [20]	60.1	84.8	92.7	45.9	74.4	85.6
$\mathrm{HPF}_{\mathrm{res101\text{-}FCN}}$ [20]	63.5	88.3	95.4	48.6	76.3	88.2
$\mathrm{SCOT}_{\mathrm{res101}}$ [17]	63.1	85.4	92.7	47.8	76.0	87.1
$\mathrm{SCOT}_{\mathrm{res101\text{-}FCN}}$ [17]	67.3	88.8	95.4	50.7	78.1	89.1
$\mathrm{Ours}_{\mathrm{res101}}$	64.3	86.7	93.4	48.2	76.8	87.9
$\mathrm{Ours}_{\mathrm{res101\text{-}FCN}}$	**69.8**	**89.6**	**96.2**	**51.9**	**78.9**	**89.7**

Table 2. PCK results on TSS. Numbers in bold indicate the best performance. Subscripts of the method names indicate backbone networks used.

Methods	FG3D	JODS	PASC	Avg
$\mathrm{CNNGeo}_{\mathrm{res101}}$ [22]	90.1	76.4	56.3	74.3
$\mathrm{DCTM}_{\mathrm{CAT\text{-}FCSS}}$ [14]	89.1	72.1	61.0	74.0
$\mathrm{Weakalign}_{\mathrm{res101}}$ [23]	90.3	76.4	56.5	74.4
$\mathrm{RTNs}_{\mathrm{res101}}$ [13]	90.1	78.2	**63.3**	77.2
$\mathrm{NC\text{-}Net}_{\mathrm{res101}}$ [24]	94.5	81.4	57.1	77.7
$\mathrm{DCCNet}_{\mathrm{res101}}$ [11]	93.5	**82.6**	57.6	77.9
$\mathrm{HPF}_{\mathrm{res101}}$ [20]	93.6	79.7	57.3	76.9
$\mathrm{SCOT}_{\mathrm{res101}}$ [17]	95.3	81.3	57.7	78.1
$Ours_{res101}$	**95.7**	81.5	58.3	**78.5**

achieves the best result compared to the other eight methods. For parts JODS and PASCAL, the proposed method achieves the second-best and the third-best results, respectively. The average result of the proposed method is the best, which is show that the proposed method is effective for TSS.

Table 3. Per-class PCK results on SPair-71k. Numbers in bold indicate the best performance. All methods use ResNet101 as the backbone network. *CNNGeo*, *DCTM*, *Weakalign* and *RTNs* are further finetuned on SPair-71k [20].

class	CNNGeo [22]	A2net [25]	Weakalign [23]	NC-Net [24]	HPF [20]	SCOT [17]	Ours
aero	23.4	22.6	22.2	17.9	25.2	34.9	**36.1**
bike	16.7	18.5	17.6	12.2	18.9	20.7	**21.4**
bird	40.2	42.0	41.9	32.1	52.1	63.8	**64.1**
boat	14.3	16.4	15.1	11.7	15.7	21.1	**21.9**
bottle	36.4	37.9	38.1	29.0	38.0	43.5	**43.9**
bus	27.7	**30.8**	27.4	19.9	22.8	27.3	27.5
car	26.0	26.5	**27.2**	16.1	19.1	21.3	22.8
cat	32.7	35.6	31.8	39.2	52.9	63.1	**63.8**
chair	12.7	13.3	12.8	9.9	17.9	20.0	**20.3**
cow	27.4	29.6	26.8	23.9	33.0	**42.9**	**42.9**
dog	22.8	24.3	22.6	18.8	32.8	**42.5**	42.2
horse	13.7	16.0	14.2	15.7	20.6	**31.1**	30.5
moto	20.9	21.6	20.0	17.4	24.4	29.8	**30.1**
person	21.0	22.8	22.2	15.9	27.9	35.0	**35.9**
plant	17.5	20.5	17.9	14.8	21.1	27.7	**28.7**
sheep	10.2	13.5	10.4	9.6	15.9	**24.4**	24.3
train	30.8	31.4	32.2	24.2	31.5	**48.4**	**48.4**
tv	34.1	36.5	35.1	31.1	35.6	40.8	**43.2**
all	20.6	22.3	20.9	20.1	28.2	35.6	**36.2**

Table 4. PCK analysis by variation factors on SPair-71k, numbers in bold indicate the best performance. All methods use ResNet101 as the backbone network.

Variation factor		CNNGeo [22]	A2net [25]	Weakalign [23]	NC-Net [24]	HPF [20]	SCOT [17]	Ours
View-point	easy	28.8	30.9	29.3	26.1	35.6	42.7	**43.5**
	medi	12.0	13.3	11.9	13.5	20.3	28.0	**28.3**
	hard	6.4	7.4	7.0	10.1	15.5	**23.9**	23.8
Scale	easy	24.8	26.1	25.1	24.7	33.0	41.1	**41.6**
	medi	18.7	21.1	19.1	17.5	26.1	33.7	**34.5**
	hard	10.6	12.4	11.0	9.9	15.8	21.4	**21.5**
Truncation	none	23.7	25.0	24.0	22.2	31.0	39.0	**39.5**
	src	15.5	17.4	15.8	17.1	24.6	32.4	**33.3**
	tgt	17.9	20.5	18.4	17.5	24.0	30.0	**30.2**
	both	15.3	17.6	15.6	16.8	23.7	30.0	**31.1**
Occlusion	none	22.9	24.6	23.3	22.0	30.8	39.0	**39.5**
	src	16.4	17.2	16.4	16.3	23.5	30.3	**30.8**
	tgt	16.4	17.2	16.4	16.3	22.8	28.1	**28.9**
	both	14.4	16.4	15.7	15.2	21.8	26.0	**26.9**
ALL		20.6	22.3	20.9	20.1	28.2	35.6	**36.2**

Matching Results on SPair-71k. The comparison between the proposed method and the other six methods on SPair-71k is shown in Table 3 and Table 4. The classification analysis of the image class is shown in Table 3. Table 3 shows that for the eighteen classes in SPair-71k, the evaluation results of the proposed method in thirteen classes are better than the other six methods. The classification analysis of variation factors is shown in Table 4. Table 4 shows that our method can better handle variation factors include scale, truncation, and occlusion with various difficulty levels.

Only for the variation of view-point whose difficulty level is hard, the proposed method is weaker than SCOT. But the proposed method can effectively deal with the variation of view-point whose difficulty level is easy or medium. The comparison of the average performance of each method on the overall dataset SPair-71k indicates that our method has obtained the best result. Figure 2 shows some examples where our method finds more reliable correspondences than other methods.

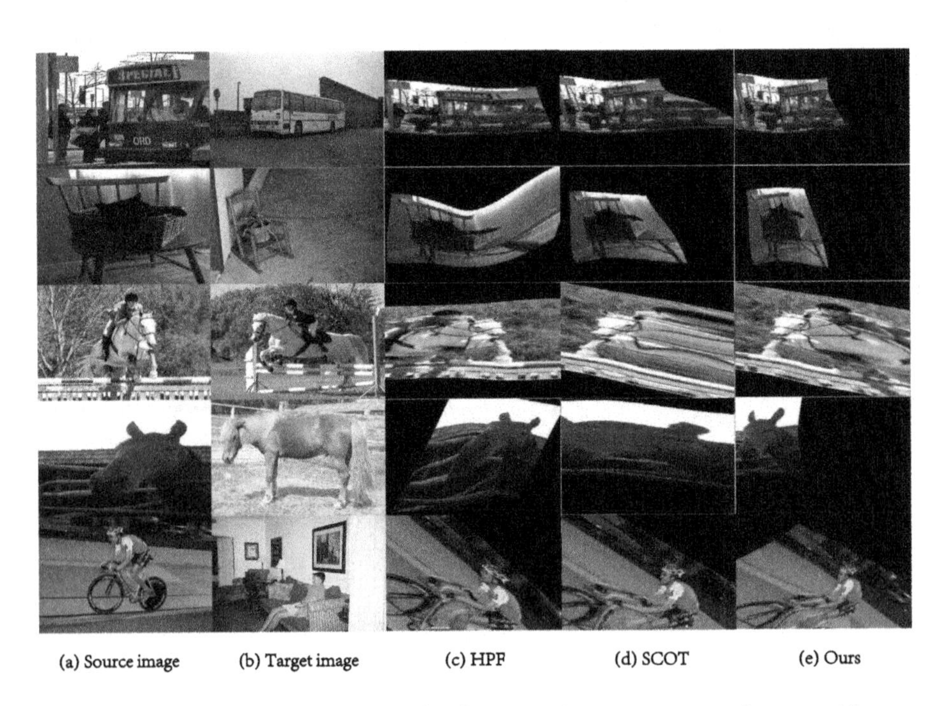

Fig. 2. Qualitative results on SPair-71k. The source images are warped to target images using its semantic correspondence.

The evaluation results of the proposed method on the above four datasets include PF-PASCALL, PF- WILLOW, TSS, and SPair-71k have achieved the best performance. These results show that it is effective to introduce regional consensus into semantic correspondence tasks. Especially the performance on the large-scale dataset SPair-71k proves the universality of the proposed method.

5 Conclusion

We propose an improved method for establishing semantic correspondence by introducing regional offset consensus, which is implemented by filtering the correlation scores in both Hough space and image space. Experiments show that the proposed method can produce superior results than the state-of-the-art methods, when applied on four benchmark datasets. It can be seen that introducing regional offset consensus into semantic correspondence tasks is beneficial. For the future work, the regional offset consensus can be further improved by the matched objects offset consensus, which considers the shape information of the matched objects.

References

1. Agarwal, S., et al.: Building Rome in a day. Commun. ACM **54**(10), 105–112 (2011)
2. Dalal, N., Triggs, B.: Histograms of oriented gradients for human detection. In: 2005 IEEE Computer Society Conference on Computer Vision and Pattern Recognition (CVPR 2005), vol. 1, pp. 886–893 (2005)
3. Deng, J.: ImageNet: A large-scale hierarchical image database. In: Proceedings CVPR 2009 (2009)
4. Donato, G., Belongie, S.: Approximate thin plate spline mappings. In: Heyden, A., Sparr, G., Nielsen, M., Johansen, P. (eds.) ECCV 2002. LNCS, vol. 2352, pp. 21–31. Springer, Heidelberg (2002). https://doi.org/10.1007/3-540-47977-5_2
5. Dosovitskiy, A., et al.: FlowNet: learning optical flow with convolutional networks. In: Proceedings of the IEEE International Conference on Computer Vision (ICCV), pp. 2758–2766, December 2015
6. Everingham, M., Eslami, S., Gool, L.V., Williams, C., Winn, J., Zisserman, A.: The pascal visual object classes challenge: a retrospective. Int. J. Comput. Vision **111**(1), 98–136 (2015)
7. Ham, B., Cho, M., Schmid, C., Ponce, J.: Proposal flow. In: Proceedings of the IEEE Conference on Computer Vision and Pattern Recognition (CVPR), pp. 3475–3484, June 2016
8. Ham, B., Cho, M., Schmid, C., Ponce, J.: Proposal flow: semantic correspondences from object proposals. IEEE Trans. Pattern Anal. Mach. Intell. **40**(7), 1711–1725 (2017)
9. Han, K., et al.: ScNet: learning semantic correspondence. In: Proceedings of the IEEE International Conference on Computer Vision (ICCV), pp. 1831–1840, October 2017
10. He, K., Zhang, X., Ren, S., Sun, J.: Deep residual learning for image recognition. IEEE (2016)
11. Huang, S., Wang, Q., Zhang, S., Yan, S., He, X.: Dynamic context correspondence network for semantic alignment. In: Proceedings of the IEEE/CVF International Conference on Computer Vision (ICCV), pp. 2010–2019, October 2019
12. Ilg, E., Mayer, N., Saikia, T., Keuper, M., Dosovitskiy, A., Brox, T.: FlowNet 2.0: evolution of optical flow estimation with deep networks. In: Proceedings of the IEEE Conference on Computer Vision and Pattern Recognition (CVPR), pp. 2462–2470, July 2017

13. Kim, S., Lin, S., Jeon, S., Min, D., Sohn, K.: Recurrent transformer networks for semantic correspondence. In: Neural Information Processing Systems (NeurIPS) (2018)
14. Kim, S., Min, D., Lin, S., Sohn, K.: DCTM: discrete-continuous transformation matching for semantic flow. In: Proceedings of the IEEE International Conference on Computer Vision (ICCV), pp. 4529–4538, October 2017
15. Laskar, Z., Tavakoli, H.R., Kannala, J.: Semantic matching by weakly supervised 2D point set registration. In: 2019 IEEE Winter Conference on Applications of Computer Vision (WACV), pp. 1061–1069 (2019)
16. Lee, J., Kim, D., Ponce, J., Ham, B.: SFNet: learning object-aware semantic correspondence. In: Proceedings of the IEEE/CVF Conference on Computer Vision and Pattern Recognition (CVPR), pp. 2278–2287, June 2019
17. Liu, Y., Zhu, L., Yamada, M., Yang, Y.: Semantic correspondence as an optimal transport problem. In: Proceedings of the IEEE/CVF Conference on Computer Vision and Pattern Recognition (CVPR), pp. 4463–4472, June 2020
18. Long, J., Shelhamer, E., Darrell, T.: Fully convolutional networks for semantic segmentation. In: Proceedings of the IEEE Conference on Computer Vision and Pattern Recognition (CVPR), June 2015
19. Low, D.G.: Distinctive image features from scale-invariant keypoints. Int. J. Comput. Vision **60**, 91–110 (2004)
20. Min, J., Lee, J., Ponce, J., Cho, M.: Hyperpixel flow: semantic correspondence with multi-layer neural features. In: Proceedings of the IEEE/CVF International Conference on Computer Vision (ICCV), pp. 3395–3404, October 2019
21. Min, J., Lee, J., Ponce, J., Cho, M.: Spair-71k: a large-scale benchmark for semantic correspondence (2019)
22. Rocco, I., Arandjelovic, R., Sivic, J.: Convolutional neural network architecture for geometric matching. In: Proceedings of the IEEE Conference on Computer Vision and Pattern Recognition (CVPR), pp. 6148–6157, July 2017
23. Rocco, I., Arandjelovic, R., Sivic, J.: End-to-end weakly-supervised semantic alignment. In: Proceedings of the IEEE Conference on Computer Vision and Pattern Recognition (CVPR), pp. 6917–6925, June 2018
24. Rocco, I., Cimpoi, M., Arandjelovic, R., Torii, A., Pajdla, T., Sivic, J.: NCNet: neighbourhood consensus networks for estimating image correspondences. IEEE Trans. Pattern Anal. Mach. Intell. **PP**, 1–14 (2020). https://doi.org/10.1109/TPAMI.2020.3016711
25. Seo, P.H., Lee, J., Jung, D., Han, B., Cho, M.: Attentive semantic alignment with offset-aware correlation Kernels. In: Ferrari, V., Hebert, M., Sminchisescu, C., Weiss, Y. (eds.) ECCV 2018. LNCS, vol. 11208, pp. 367–383. Springer, Cham (2018). https://doi.org/10.1007/978-3-030-01225-0_22
26. Taniai, T., Sinha, S.N., Sato, Y.: Joint recovery of dense correspondence and cosegmentation in two images. In: Proceedings of the IEEE Conference on Computer Vision and Pattern Recognition (CVPR), pp. 4246–4255, June 2016

A Simple Network with Progressive Structure for Salient Object Detection

Boyi Zhou[1,2], Gang Yang[1(✉)], Xin Wan[1,2], Yutao Wang[1], Chang Liu[1], and Hangxu Wang[1]

[1] Northeastern University, Shenyang 110819, China
yanggang@mail.neu.edu.cn
[2] DUT Artificial Intelligence Institute, Dalian 116024, China

Abstract. Recently, most CNNs-based Salient Object Detection (SOD) models have achieved great progress through various feature aggregation strategies. But most of them usually introduce a large number of features into a module for fusion, which results in information dilution. In this paper, we propose a Progressive Fusion Network (PFNet) to solve this problem. The PFNet alleviates the information dilution through a Progressive Fusion Architecture (PFA), which aggregates the features extracted from encoder in a progressive fusion manner. In addition, we use a simple Feature Fusion Module (FFM) that utilize high-level features to enhance the semantic information of low-level features, thereby ensuring the effective fusion of features. Finally, we leverage an Enhanced Loss to guide the optimization process of the network and obtain high-quality saliency maps. The whole network is trained end-to-end without any pre-processing and post-processing. The quantitative and qualitative experimental results on five benchmark datasets demonstrate that the superiority of the proposed approaches.

Keywords: Salient object detection · CNNs · Progressive Fusion Network

1 Introduction

Salient Object Detection (SOD) aims to detect the visually distinctive regions or objects in a scene [3]. It is widely used in many computer vision tasks [27], such as person re-identification [23], scene classification [21], object tracking [13] and video compression [6].

The traditional SOD methods directly obtain salient regions through hand-crafted features [10, 39]. However, due to the lack of high-level semantic information, these methods perform poorly in many complex scenarios. Recently, after the emergence of Convolutional Neural Networks (CNNs) with powerful feature extraction capability, many CNNs-based SOD methods have been proposed [8, 9, 35, 37]. The CNNs-based SOD methods obtain multi-level features containing different information from encoder. Among them, low-level features have affluent detailed information, while high-level features have semantic information for locating. Therefore, the key to SOD is how to fuse multi-level features while suppressing interference, which ensures the SOD model can

H. Ma et al. (Eds.): PRCV 2021, LNCS 13020, pp. 397–408, 2021.
https://doi.org/10.1007/978-3-030-88007-1_33

utilize semantic information and detailed information to obtain high-quality saliency maps. To address this issue, many existing SOD methods propose various feature aggregation strategies [9, 19, 25, 32, 33]. Hou et al. [9] and Wang et al. [25] use short connections to introduce high-level features into low-level features, as shown in Fig. 1(a). These methods use the semantic information of high-level features to guide low-level features and thus suppress background interference. However, plenty of short connections will not only increase the complexity of the model [29], but also lead to information dilution [2]. Shown in the circle in Fig. 1(a), the detailed information of the low-level features will be diluted by four high-level features, resulting in blurred edges of saliency maps. Pang el al. [19] choose to merge with adjacent features, which alleviates information dilution to a certain extent, as shown in Fig. 1(b). However, the obtained high-level features are sent to the low-level features with poor quality in the decoder. And the direct fusion of features with large span will increase the difficulty of extracting useful information. In Fig. 1(b), a high-level feature containing semantic information and a little detailed information is directly fused with a low-level feature containing a large amount of interference information, which will reduce the quality of the fusion feature and finally cause performance degradation. Most existing methods ignore the problem of information dilution and information confusion, and ultimately reduce the quality of the saliency maps.

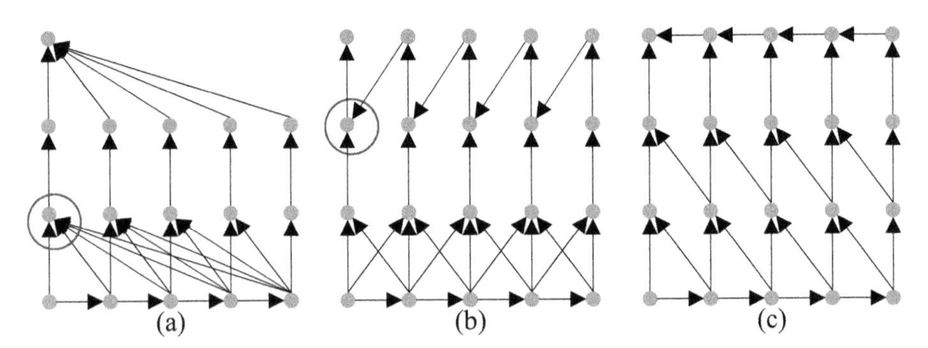

Fig. 1. Feature aggregation strategies. (a)PFPN [25]. (b)MINet [19]. (c)PFNet.

To solve the above problems and obtain high-quality saliency maps, we propose a Progressive Fusion Network (PFNet) for SOD, which aggregates features through a Progressive Fusion Architecture (PFA) to alleviate information dilution. The structure of PFNet is shown in Fig. 1(c). In addition, we design a simple Feature Fusion Module (FFM) to ensure the effective fusion of features. The flexible application of FFM in PFA and decoder ensures that the semantic information of high-level features and the detailed information of low-level features can fuse fully. Finally, we use an Enhanced Loss to guide the optimization process of the proposed network and ultimately obtain high-quality saliency maps.

Our main contributions are summarized as follows:

- We propose a Progressive Fusion Network (PFNet) for salient object detection, which aggregates the features by Progressive Fusion Architecture (PFA) to alleviate information dilution.
- We design a simple Feature Fusion Module (FFM) for PFA and decoder to ensure effective fusion of feature.
- The results of quantitative and qualitative experiments conducted on five benchmark datasets demonstrate the superiority of the proposed approaches.

2 Related Works

Recently, after the emergence of Fully Convolutional Network (FCN) [16] and U-shape Network [22], such end-to-end structures are rapidly applied to SOD. The key to SOD lies in how to integrate high-level semantic information and detailed information while suppressing background interference. To solve this problem, existing methods propose various feature integration strategies. Hou et al. [9] and Wang et al. [25] introduce high-level features into low-level features to enhance semantic information of low-level features, thereby improving the accuracy of locating. Pang et al. [19] choose to merge with adjacent features to alleviate the information dilution to a certain extent. Other methods extract useful information from features through well-designed modules or mechanisms. Zhao et al. [36] utilize channel attention for high-level features and spatial attention for low-level features to filter the information. Liu et al. [14] propose a PPM that generates multi-scale features through top-level features to improve the quality of saliency maps. Zhao et al. [37] utilize Gated Mechanism to filter the information from encoder to decoder.

Most of the existing methods ignore the information dilution, which will cause the current feature information to be replaced by other features. Other methods of designing complex modules will make the network too complex.

3 Methodology

The architecture of the PFNet is shown as Fig. 2, which contains an encoder to extract features, a PFA to improve the quality of features, and a decoder to obtain high-quality saliency maps. The features in encoder, PFA and decoder are respectively denoted as $\mathbf{E_i}$, $\mathbf{f_i}$, $\mathbf{D_i}$ (i $\in \{1, 2, 3, 4, 5, 6\}$ indexes different levels).

3.1 Overview of the Proposed Network

Encoder Network. In this paper, we use ResNet-50 [7] to extract features. To obtain saliency maps with more details, we borrow the first two convolution layers from VGG-16 [24] to obtain the original size (320×320) features, which denoted as $\mathbf{E_1}$. The features extracted from ResNet-50 are denoted as $\mathbf{E_2}$, $\mathbf{E_3}$, $\mathbf{E_4}$, $\mathbf{E_5}$, $\mathbf{E_6}$ respectively.

Progressive Fusion Architecture (PFA). For the sake of reducing the consumption of memory and computing, we only process features $\mathbf{E_2}$–$\mathbf{E_6}$. We firstly employ a convolution layer to compress the channel of features to k. In this work k is set to 64.

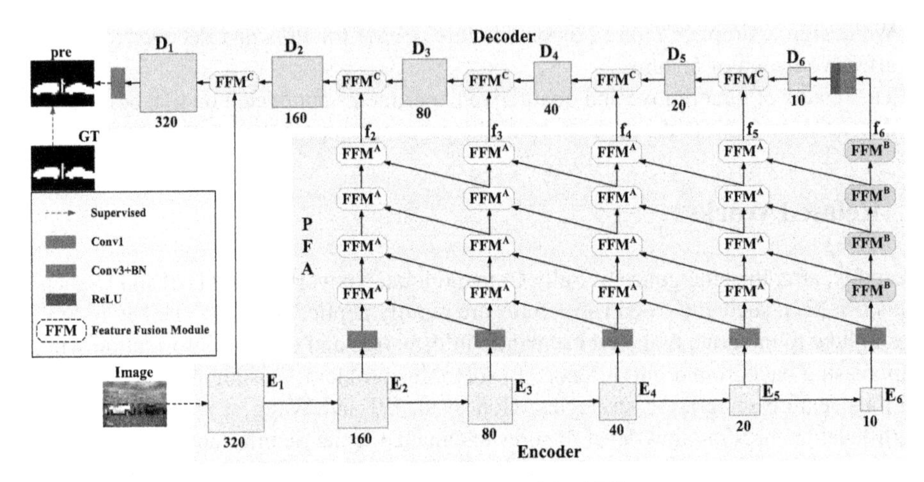

Fig. 2. The architecture of the PFNet.

Decoder Network. We adopt a layer-by-layer decoding method to obtain the saliency maps. E1 comes from the encoder, $\mathbf{f_2}$–$\mathbf{f_6}$ from PFA, and they will be sent to the decoder and then fused by **FFM-C**.

3.2 Progressive Fusion Architecture

The multi-level features extracted from the encoder are not good enough to obtain high-quality saliency maps directly through the decoder. Thus, some methods design various modules to improve the quality of the features from encoder [19, 28, 32, 37]. But most of them ignore information dilution and result in inadequate fusion of features. To obtain high-quality fusion features, we process the features from encoder with a Progressive Fusion Architecture (PFA). The PFA is shown in Fig. 2.

In order to obtain high-quality features while avoiding information dilution as much as possible, we gradually fuse two adjacent features. As shown in Fig. 2, the high-level features are gradually introduced into the low-level features, thus guiding the low-level features to suppress the background interference with semantic information. Thanks to two points, this mechanism works. 1) Fusing two features at once can avoid information dilution and thus ensure effective fusion. 2) There is no big gap between adjacent features, so it is easy to merge the feature. According to this structure, the best semantic information in the top-level features can be transfered to bottom-level features. In addition, because the top-level features are of good quality, we allow the top-level features to optimize themselves in the forward. Based on the above considerations, we believe that the output features of PFA are excellent for obtaining high-quality saliency maps.

3.3 Feature Fusion Module

The key to SOD is the fusion of semantic information and detailed information, thus many existing methods design some well-designed modules to achieve this goal. But most of them ignore the ability of convolution to extract information. In this work, benefiting from the use of PFA, we only fuse two features at a time. In other words, we can complete the fusion by simple convolution instead of complicated mechanisms. We design a simple Feature Fusion Module (FFM) to fuse features. To make FFM work better in different cases, we design three different structures of FFM, which denote as **FFM-A**, **FFM-B** and **FFM-C** respectively in Fig. 2. The structures of them are show in Fig. 3.

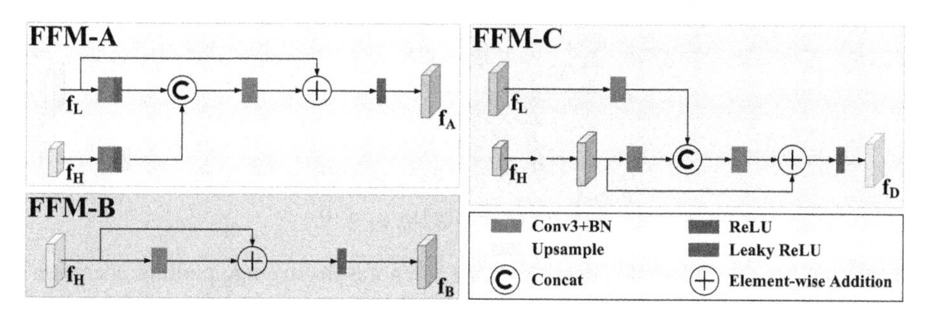

Fig. 3. Structure of FFM. $\mathbf{f_L}, \mathbf{f_H}$ indicates low-level features and high-level features respectively, $\mathbf{f_A}, \mathbf{f_B}, \mathbf{f_D}$ indicates the output features of FFM-A, FFM-B and FFM-C respectively.

The **FFM-A** and **FFM-B** are used in PFA to ensure the features sent to the decoder are good enough. In **FFM-A**, we first apply a convolutional layer and a Leaky ReLU to capture context knowledge from $\mathbf{f_L}$ and $\mathbf{f_H}$, and upsample the $\mathbf{f_H}$ to the size as $\mathbf{f_L}$ at the same time. Then, we concatenate two features, and use a convolution layer to fuse the features while compressing the channel to k. Finally, we introduce the original low-level feature $\mathbf{f_L}$ through residual connection. The **FFM-B** only contains a convolution layer, residual connection and ReLU. Above operation can be formulated as

$$f_A = ReLU\big(Conv\big(Cat\big[Leaky(Conv(f_L)), Up(Leaky(Conv(f_H)))\big]\big) + f_L\big) \quad (1)$$

$$f_B = ReLU(Conv(f_H) + f_H) \quad (2)$$

where $Conv(.)$ denotes the operation of convolution and batch normalization, $Up(.)$ denotes the operation of upsample, and $Cat(.)$ means concatenating features in the channel dimension.

The **FFM-C** used in decoder is a little different from **FFM-A**. On the one hand, it removes the activation function after the first convolution, which ensures the information of features can be utilize adequately. On the other hand, the residual connection introduces the high-level features after upsample operation. Because the layer-by-layer decoding method is a process of continuously refining the edges, so the high-level features are fundamental in decoder. The operation can be formulated as

$$f_D = ReLU\big(Conv\big(Cat\big[Conv(f_L), Conv(UP(f_H))\big]\big) + UP(f_H)\big) \quad (3)$$

3.4 Enhanced Loss

In order to obtain more excellent saliency maps, we use an Enhanced Loss to guide the optimization process of PFNet. The Enhanced Loss contains three parts, a BCE loss, an edge loss and IOU loss. The BCE loss is defined as:

$$L_{bce} = \frac{\sum_{H}\sum_{W}(g \times log(p) + (1-g) \times log(1-p))}{H \times W} \tag{4}$$

where p denotes the saliency maps, g denotes the corresponding ground-truth, H and W denote the high and width of the ground-truth respectively. The edge loss can instruct network to obtain sharp edges. We use the edge loss from the [18], which is defined as

$$L_e \frac{\sum_{H}\sum_{W}(e \times |p-g|)}{\sum_{H}\sum_{W} e}$$
$$e = \begin{cases} 0\, if\, (g - P(g))_{[h,w]} = 0, \\ 1\, if\, (g - P(g))_{[h,w]} \neq 0, \end{cases} \tag{5}$$

where e denotes the edge of ground-truth, $P(.)$ denotes the average pooling operation. The IOU loss is used to improve the accuracy for locating, which is defined as

$$L_{IOU} = 1 - \frac{\sum_{H}\sum_{W}(p \times g)}{\sum_{H}\sum_{W}(p + g - p \times g)} \tag{6}$$

The total loss is defined as

$$L_{EL} = L_{bce} + \alpha_1 \times L_e + \alpha_2 \times L_{IOU} \tag{7}$$

where α_1 and α_2 is the parameters that balance loss, we simply set to 1 in this paper.

4 Experiments

Datasets and Evaluation Metrics. We evaluate our model on five widely used benchmark datasets, including DUTS [26], DUT-O [31], ECSSD [30], PASCAL-S [12] and HKU-IS [11]. The training set of DUTS is used to train our model. To quantitatively evaluate the performance of our model, we compare the proposed method with 13 State-of-the-arts in terms of five metrics including precision- recall, F-measure curves, average F-measure (F_{avg}) [1], mean absolute error (MAE) and E-measure (E_m) [4].

Implementation Details. The proposed network is implemented based on an NVIDIA GTX 2080Ti GPU and PyTorch deep learning framework. The batch size is set to 10 and the epoch is set to 30. For the optimizer, we adopt the stochastic gradient descent (SGD). The momentum, weight decay, and learning rate are respectively set as 0.9, 0.0005 and 0.005. The parameters of the feature extraction network are initialized by

Table 1. Comparisons of our method and other 13 methods on five benchmark datasets, where **bold** and *italic* indicate the best and second-best performance, respectively

Method	DUTS-TE			DUT-O			ECSSD			PASCAL-S			HKU-IS		
	F_{avg}	MAE	E_m	F_{avg}	MAE	E_m	F_{avg}	MAE	E_m	F_{avg}	MAE	E_m	F_{avg}	MAE	E_m
BMPM [32]	.745	.049	.863	.692	.064	.839	.868	.045	.916	.770	.074	.847	.871	.039	.938
PAGR [34]	.784	.055	.883	.711	.071	.843	.894	.061	.917	.808	.093	.854	.886	.047	.939
PiCANet [15]	.749	.054	.865	.710	.068	.842	.885	.046	.926	.804	.077	.862	.870	.042	.938
AFNet [5]	.793	.046	.895	.739	.057	.860	.908	.042	.942	.829	.071	.887	.888	.036	.947
BASNet [20]	.791	.048	.884	*.756*	.056	.869	.879	.037	.921	.781	.077	.853	.895	.032	.946
CPD [29]	.805	.043	.904	.747	.056	*.873*	.917	.037	.949	.831	.072	.888	.891	.034	.950
MLMSNet [28]	.801	.045	.899	.744	.056	.865	.914	.038	.948	.840	.068	.894	.893	.034	.950
PoolNet [14]	.809	.040	.904	.747	.056	.869	.915	.039	.945	.830	.075	.878	.899	.033	.954
CAGNet [17]	*.841*	.040	.913	.754	*.054*	.861	.922	.036	.944	*.848*	.067	.896	*.912*	.030	.949
ITSD [38]	.804	.041	.898	*.756*	.061	.867	.894	*.035*	.932	.799	.065	.868	.899	.031	.953
GateNet [37]	.807	.040	.903	.746	.055	.868	.916	.040	.943	.832	.069	.884	.899	.033	.953
GCPANet [2]	.817	.038	.913	.748	.056	.869	.919	*.035*	*.952*	.841	*.062*	*.901*	.898	.031	*.956*
MINet [19]	.828	*.037*	*.917*	*.756*	.056	*.873*	*.924*	**.033**	**.953**	.842	.064	.899	.909	*.029*	**.960**
PFNet(ours)	**.848**	**.036**	**.926**	**.770**	**.052**	**.880**	**.925**	*.035*	.948	**.857**	**.061**	**.907**	**.916**	**.028**	**.960**

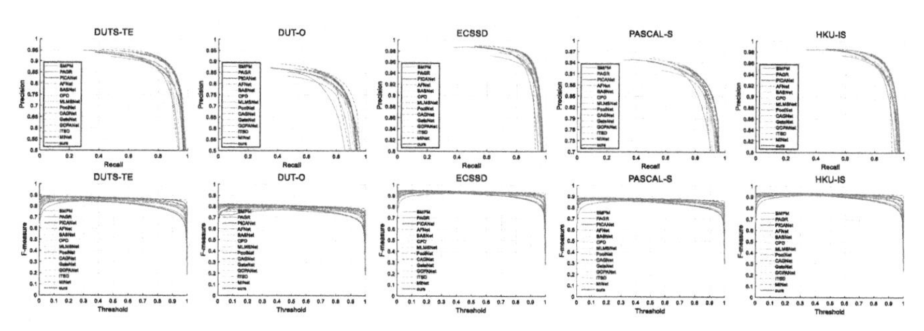

Fig. 4. The precision-recall curve and F-measure curve of 14 methods under five benchmark datasets.

the pre-trained ResNet-50 on ImageNet [7], while the other convolutional layers are randomly initialized. All training and test images are resized to 320 × 320. In order to alleviate the over-fitting problem, we utilize simple random horizontal flipping and random color jittering and normalization for RGB images.

4.1 Comparison with State-of-the-Arts

We compare our PFNet with other 13 Sate-of-the-arts SOD methods, including BMPM [32], PAGR [34], PiCANet [15], AFNet [5], BASNet [20], CPD [29], MLMSNet [28], PoolNet [14], CAGNet [17], ITSD [38], MINet [19], GateNet [37] and GCPANet [2]. For fair comparison, we directly evaluate saliency maps provided by authors.

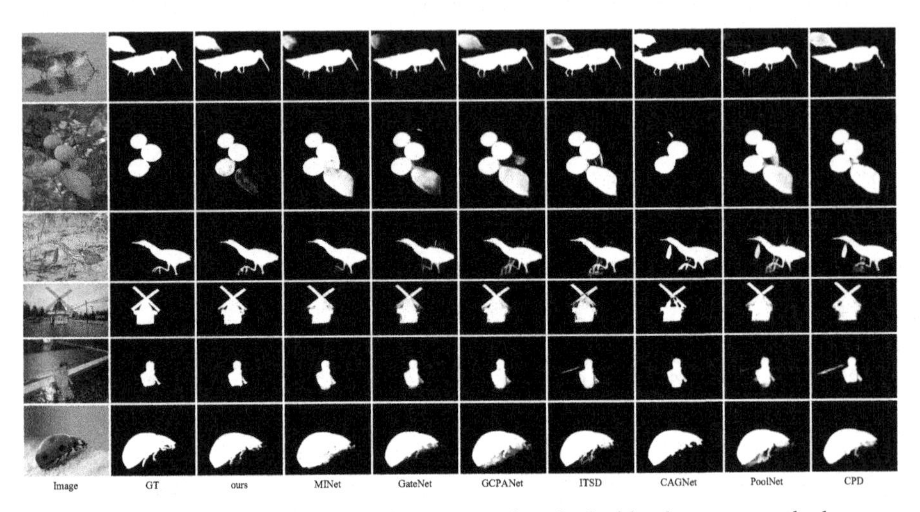

Fig. 5. Qualitative comparison of the proposed method with other seven methods.

Quantitative Evaluation. The results of different methods are shown in Table 1. Our method achieves the best performance in four datasets including DUTS-TE, DUT-O, HKU-IS and PASCAL-S. Though the metrics of our method in ECSSD dataset are not the best, the gap between our method and the best method is tiny. In addition, we can observe that our method improves average F-measure F_{avg} by a gain of 0.7%, 1.4%, 0.1%, 0.9%, 0.4% on DUTS-TE, DUT-O, ECSSD, PASCAL-S and HKU-IS datasets, respectively. The precision-recall curves and F-measure curves are shown in Fig. 4, which indicates our method achieves better performance in different threshold.

Qualitative Evaluation. In order to evaluate the proposed PFNet, we visualize some saliency maps produced by our method and other 7 SOTA methods in Fig. 5. The first column is RGB image, the second column is Ground-Truth, the third column is saliency map produced by our method and the last 7 columns are SOTA methods. The comparison results of the first two rows verify our method's ability to detect multiple objects, while other methods will lose the part of the salient region. The first three rows indicate that our method has a certain ability to suppress the background interference. The fourth and fifth rows show the locating accuracy of our method and the last row demonstrates that the advantage of our method in detecting edge detail.

4.2 Ablation Studies

In order to verify the effectiveness of different modules in PFNet, we conduct ablation studies on PASCAL-S and DUTS-TE dataset and the results are shown in Table 2. All of the tests are conducted under the same training setting. The first two lines show the importance of the Enhanced Loss, which improves the training efficiency of the network, so that better performance can be obtained under the same setting. After that, PFA helps the network achieve better performance. Finally, we employ the FFM in PFA and Decoder to further improve the performance of network. In the last row of Table 2, although the performance improvement on DUTS-TE dataset is small, the performance improvement on PASCAL-S dataset is sufficient to prove the effectiveness of FFM.

Table 2. Ablative studies of the proposed modules on two datasets. The best statistic results are highlighted in **bold**

Baseline	PFA	FFM	Enhanced loss	DUTS-TE			PASCAL-S		
				F_{avg}	MAE	E_m	F_{avg}	MAE	E_m
√				0.772	0.045	0.882	0.816	0.071	0.876
√			√	0.828	0.038	0.914	0.830	0.066	0.888
√	√		√	0.845	**0.036**	0.923	0.850	0.065	0.900
√	√	√	√	**0.848**	**0.036**	**0.926**	**0.857**	**0.061**	**0.907**

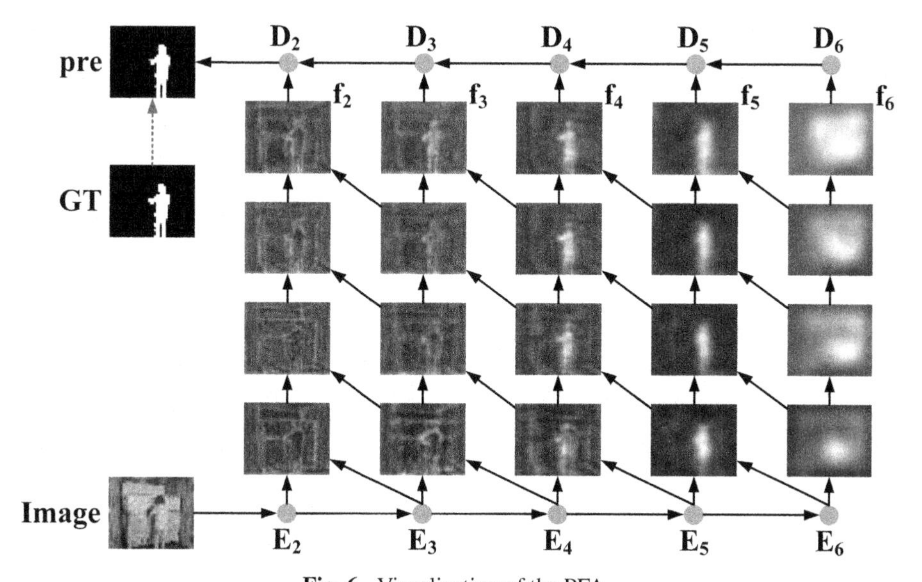

Fig. 6. Visualization of the PFA.

4.3 Visualization

To show the process of PFA improving feature quality more intuitively, in this section we visualize the output features of each layer in PFA. The results are shown in Fig. 6.

In Fig. 6, the last two columns are the high-level features containing semantic information for locating, and the first three columns are the low-level features containing detailed information. We can observe that the high-level features will maintain high-quality in the forward process. And the low-level features will suppress the background information under the guidance of the high-level features. Obviously, the features after PFA are better than those extracted by the encoder. Naturally, decoding using these features will obtain the better saliency maps.

5 Conclusion

In this paper, we propose a simple PFNet with a progressive structure for salient object detection. The PFNet integrates semantic information and detailed information through progressive fusion of adjacent features, thus effectively avoiding information dilution and finally obtaining the high-quality saliency maps. Quantitative and qualitative experimental results demonstrate the superiority of the method and the effectiveness of the Progressive Fusion Architecture (PFA).

Acknowledgements. This work is supported by National Natural Science Foundation of China under Grant No. 62076058.

References

1. Achanta, R., Hemami, S., Estrada, F., Susstrunk, S.: Frequency-tuned salient region detection. In: Proceedings of the IEEE Conference on Computer Vision and Pattern Recognition, pp. 1597–1604 (2009)
2. Chen, Z., Qianqian, X., Cong, R., Huang, Q.: Global context-aware progressive aggregation network for salient object detection. Proc. AAAI Conf. Artif. Intell. **34**(07), 10599–10606 (2020)
3. Cong, R., Lei, J., Fu, H., et al.: Review of visual saliency detection with comprehensive information. IEEE Trans. Circuits Syst. Video Technol. **29**(10), 2941–2959 (2019)
4. Fan, D., Gong, C., Cao, Y., Ren, B., Cheng, M.M., Borji, A.: Enhanced-alignment measure for binary foreground Map Evaluation (2018). https://arxiv.org/abs/1805.10421
5. Feng, M., Lu, H., Ding, E.: Attentive feedback network for boundary-aware salient object detection. In: Proceedings of the IEEE Conference on Computer Vision and Pattern Recognition, pp. 1623–1632 (2019)
6. Guo, C., Zhang, L.: A novel multiresolution spatiotemporal saliency detection model and its applications in image and video compression. IEEE Trans. Image Process. **19**(1), 185–198 (2010)
7. He, K., Zhang, X., Ren, S., Sun, J.: Deep residual learning for image recognition. In: Proceedings of the IEEE Conference on Computer Vision and Pattern Recognition, pp. 770–778 (2016)

8. He, S., Lau, R., Liu, W., Yang, Q.: Supercnn: a superpixelwise convolutional neural network for salient object detection. Int. J. Comput. Vision **115**(3), 330–344 (2015)
9. Hou, Q., Cheng, M.M., Hu, X., Borji, A., Tu, Z., Torr, P.H.: Deeply supervised salient object detection with short connections. In: Proceedings of the IEEE Conference on Computer Vision and Pattern Recognition, pp. 3203–3212 (2017)
10. Hou, X., Zhang, L.: Saliency detection: a spectral residual approach. In: Proceedings of the IEEE Conference on Computer Vision and Pattern Recognition, pp. 1–8 (2007)
11. Li, G., Yu, Y.: Visual saliency based on multiscale deep features. In: Proceedings of the IEEE Conference on Computer Vision and Pattern Recognition, pp. 5455–5463 (2015)
12. Li, Y., Hou, X., Koch, C., Rehg, J.M., Yuille, A.L.: The secrets of salient object segmentation. In: Proceedings of the IEEE Conference on Computer Vision and Pattern Recognition, pp. 280–287 (2014)
13. Liang, P., Pang, Y., Liao, C., Mei, X., Ling, H.: Adaptive objectness for object tracking. IEEE Signal Process. Lett. **23**(7), 949–953 (2016)
14. Liu, J., Hou, Q., Cheng, M.M., et al.: A simple pooling-based design for real-time salient object detection. In: Proceedings of the IEEE Conference on Computer Vision and Pattern Recognition, pp. 3912–3921 (2019)
15. Liu, N., Han, J., Yang, M.H.: PiCANet: learning pixel-wise contextual attention for saliency detection. In: Proceedings of the IEEE Conference on Computer Vision and Pattern Recognition, pp. 3089–3098 (2018)
16. Long, J., Shelhamer, E., Darrell, T.: Fully convolutional networks for semantic segmentation. In: Proceedings of the IEEE Conference on Computer Vision and Pattern Recognition, pp. 3431–3440 (2015)
17. Mohammadi, S., Noori, M., Bahri, A., Majelan, S.G., Havaei, M.: CAGNet: content-aware guidance for salient object detection. Pattern Recogn. **103**, 107303 (2020)
18. Pang, Y., Zhang, L., Zhao, X., Huchuan, L.: Hierarchical dynamic filtering network for RGB-D salient object detection. In: Vedaldi, A., Bischof, H., Brox, T., Frahm, J.-M. (eds.) ECCV 2020. LNCS, vol. 12370, pp. 235–252. Springer, Cham (2020). https://doi.org/10.1007/978-3-030-58595-2_15
19. Pang, Y., Zhao, X., Zhang, L., et al.: Multi-scale interactive network for salient object detection. In: Proceedings of the IEEE Conference on Computer Vision and Pattern Recognition, pp. 9413–9422 (2020)
20. Qin, X., Zhang, Z., Huang, C., Gao, C., Dehghan, M., Jagersand, M.: BASNet: boundary-aware salient object detection. In: Proceedings of the IEEE Conference on Computer Vision and Pattern Recognition, pp. 7479–7489 (2019)
21. Ren, Z., Gao, S., Chia, L., et al.: Region-based saliency detection and its application in object recognition. IEEE Trans. Circuits Syst. Video Technol. **24**(5), 769–779 (2014)
22. Ronneberger, O., Fischer, P., Brox, T.: U-Net: convolutional networks for biomedical image segmentation. In: Navab, N., Hornegger, J., Wells, W.M., Frangi, A.F. (eds.) MICCAI 2015. LNCS, vol. 9351, pp. 234–241. Springer, Cham (2015). https://doi.org/10.1007/978-3-319-24574-4_28
23. Rui, Z., Ouyang, W., Wang, X.: Unsupervised salience learning for person re-identification. In: Proceedings of IEEE Conference on Computer Vision and Pattern Recognition, pp. 3586–3593 (2013)
24. Simonyan, K., Zisserman, A.: Very deep convolutional networks for large-scale image recognition (2014). https://arxiv.org/abs/1409.1556
25. Wang, B., Chen, Q., Zhou, M., Zhang, Z., Jin, X., Gai, K.: Progressive feature polishing network for salient object detection. Proc. AAAI Conf. Artif. Intell. **34**(07), 12128–12135 (2020)

26. Wang, L., et al.: Learning to detect salient objects with image-level supervision. In: Proceedings of the IEEE Conference on Computer Vision and Pattern Recognition, pp. 136–145 (2017)
27. Wang, W., Lai, Q., Fu, H., et al.: Salient object detection in the deep learning era: an in-depth survey (2019). https://arxiv.org/abs/1904.09146
28. Wu, R., Feng, M., Guan, W., Wang, D., Lu, H., Ding, E.: A mutual learning method for salient object detection with intertwined multi-supervision. In: Proceedings of IEEE Conference on Computer Vision and Pattern Recognition, pp. 8150–8159 (2019)
29. Wu, Z., Su, L., Huang, Q.: Cascaded partial decoder for fast and accurate salient object detection. In: Proceedings of the IEEE Conference on Computer Vision and Pattern Recognition, pp. 3907–3916 (2019)
30. Yan, Q., Xu, L., Shi, J., Jia, J.: Hierarchical saliency detection. In: Proceedings of the IEEE Conference on Computer Vision and Pattern Recognition, pp. 1155–1162 (2013)
31. Yang, C., Zhang, L., Lu, H., Ruan, X., Yang, M.H.: Saliency detection via graph-based manifold ranking. In: Proceedings of the IEEE Conference on Computer Vision and Pattern Recognition, pp. 3166–3173 (2013)
32. Zhang, L., Dai, L., Lu, H., et al: A bi-directional message passing model for salient object detection. In: Proceedings of the IEEE Conference on Computer Vision and Pattern Recognition, pp. 1741–1750 (2018)
33. Zhang, P., Wang, D., Lu, H., Wang, H., Ruan, X.: Amulet: aggregating multi-level convolutional features for salient object detection. In: Proceedings of the IEEE International Conference on Computer Vision (2017)
34. Zhang, X., Wang, T., Qi, J., Lu, H., Wang, G.: Progressive attention guided recurrent network for salient object detection. In: Proceedings of the IEEE Conference on Computer Vision and Pattern Recognition, pp. 714–722 (2018)
35. Zhao, R., Ouyang, W., Li, H., Wang, X.: Saliency detection by multi-context deep learning. In: Proceedings of the IEEE Conference on Computer Vision and Pattern Recognition, pp. 1265–1274 (2015)
36. Zhao, T., Wu, X.: Pyramid feature attention network for saliency detection. In: Proceedings of the IEEE Conference on Computer Vision and Pattern Recognition, pp. 3085–3094 (2019)
37. Zhao, X., Pang, Y., Zhang, L., Huchuan, L., Zhang, L.: Suppress and balance: a simple gated network for salient object detection. In: Vedaldi, A., Bischof, H., Brox, T., Frahm, J.-M. (eds.) ECCV 2020. LNCS, vol. 12347, pp. 35–51. Springer, Cham (2020). https://doi.org/10.1007/978-3-030-58536-5_3
38. Zhou, H., Xie, X., Lai, J., et al.: Interactive two-stream decoder for accurate and fast saliency detection. In: Proceedings of the IEEE Conference on Computer Vision and Pattern Recognition, pp. 9141–9150 (2020)
39. Zhu, W., Liang, S., Wei, Y., Sun, J.: Saliency optimization from robust background detection. In: Proceedings of the IEEE Conference on Computer Vision and Pattern Recognition, pp. 2814–2821 (2014)

Feature Enhancement and Multi-scale Cross-Modal Attention for RGB-D Salient Object Detection

Xin Wan[1,2], Gang Yang[1](✉), Boyi Zhou[1,2], Chang Liu[1], Hangxu Wang[1], and Yutao Wang[1]

[1] Northeastern University, Shenyang 110819, China
yanggang@mail.neu.edu.cn
[2] DUT Artificial Intelligence Institute, Dalian 116024, China

Abstract. RGB-D Salient Object Detection (SOD) methods utilize the information provided by the depth map to assist detection. To make effective use of depth information, existing methods face two challenges: (1) How to obtain valuable cross-modal complementary information. (2) How to reduce the negative effects of unreliable depth maps for RGB-D SOD. To prevent unreliable depth maps from interfering with the detection results, we design a feature enhancement module (FEM). The module first uses a feature guidance unit (FGU) to guide the depth feature to restore the saliency information, and then uses a feature selection unit (FSU) to obtain the higher quality depth feature. Finally, the representation of depth features is deeply enhanced through progressive steps of guidance and selection. To obtain valuable cross-modal complementary information, we design a multi-scale cross-modal attention module (MCA). We first use multiple parallel dilated convolutional layers to capture global and local information, and then we design a cross-modal attention unit (CAU) to generate cross-modal attention maps. These maps can eliminate redundant information and suppress background interference. Quantitative and qualitative experiments on five datasets demonstrate the superiority of the proposed approaches.

Keywords: RGB-D salient object detection · Cross-modal attention · Feature enhancement

1 Introduction

Salient Object Detection (SOD) aims to allow computers to imitate the human visual attention mechanism to find the most salient objects or regions in an image. As a preprocessing technology, it is widely used in many computer vision tasks, such as image understanding [31], action recognition [19], semantic segmentation [21], target tracking [11], and pedestrian re-identification [27].

The mainstream SOD methods mainly use RGB images for research. Although these methods have achieved high accuracy on the RGB datasets, it is still difficult to

H. Ma et al. (Eds.): PRCV 2021, LNCS 13020, pp. 409–420, 2021.
https://doi.org/10.1007/978-3-030-88007-1_34

obtain accurate detection results in some difficult scenarios, such as confusing background, similar foreground-background and complex object appearance. To solve this problem, researchers use depth maps as a supplement to RGB images. The depth map contains a wealth of object contour information and spatial structure information, which can provide useful clues for SOD. Early RGB-D SOD methods use image priors and hand-crafted features for detection [4, 20]. With the development of convolutional neural network (CNN), researchers have proposed CNN-based algorithms to improve the accuracy of RGB-D SOD. Chen et al. [2] utilize a complementarity-aware fusion module to progressively fuse multi-level and cross-modal features. Zhu et al. [30] use two networks with different structures to extract RGB features and depth features separately, and simply concatenate the top-level features of two networks. Zhao et al. [28] propose a single-stream network for early fusion. These studies design different fusion strategies to integrate the multi-modal features extracted by CNN.

While the CNN-based methods have achieved good performance, most studies suffer from two problems. Firstly, they ignore the interference of low-quality depth maps. The quality of the depth map is influenced by many factors, such as sensor temperature, lighting conditions, distance to the object, and intensity of reflections. Low-quality depth maps often have fuzzy foregrounds and incomplete object areas. The unreliable depth maps will bring a lot of noise to the original RGB features. Many studies [2, 5, 23] make use of depth features without screening, ignoring the noise carried by low-quality depth maps, and often fail to obtain better detection results. Secondly, they ignore the redundant information and background noise in high-level features. The high-level features extracted by CNN contain rich position information while some studies use simple fusion methods such as concatenation [25, 30], summation [16], and multiplication [24] to roughly integrate them. These approaches will introduce a lot of redundant information and background noise, resulting in inaccurate localization and insufficient fusion. Therefore, it is necessary to design an effective fusion method to obtain valuable complementary information from high-level features.

To solve the above problems, we propose a novel RGB-D SOD method. Firstly, to alleviate interference from low-quality depth maps, a feature enhancement module (FEM) is designed to optimize depth features. The FEM includes a feature guidance unit (FGU) and a feature selection unit (FSU), the units can filter noise interference and enhance the representation of the depth features. Secondly, to capture valuable multi-modal complementary information, a multi-scale cross-modal attention module (MCA) is designed in this paper. The cross-modal attention unit (CAU) in MCA allows multi-modal high-level features to focus on each other's useful information, filter background interference from different scales, and capture meaningful complementary information.

Our major contributions are summarized as follows:

1) We propose a feature enhancement module (FEM) to filter noise interference in low-quality depth maps and enhance the representation of depth features.
2) We propose a multi-scale cross-modal attention module (MCA) to suppress background interference from multi-modal features and obtain valuable cross-modal complementary information.
3) A large number of experiments on five datasets prove that our algorithm has good performance.

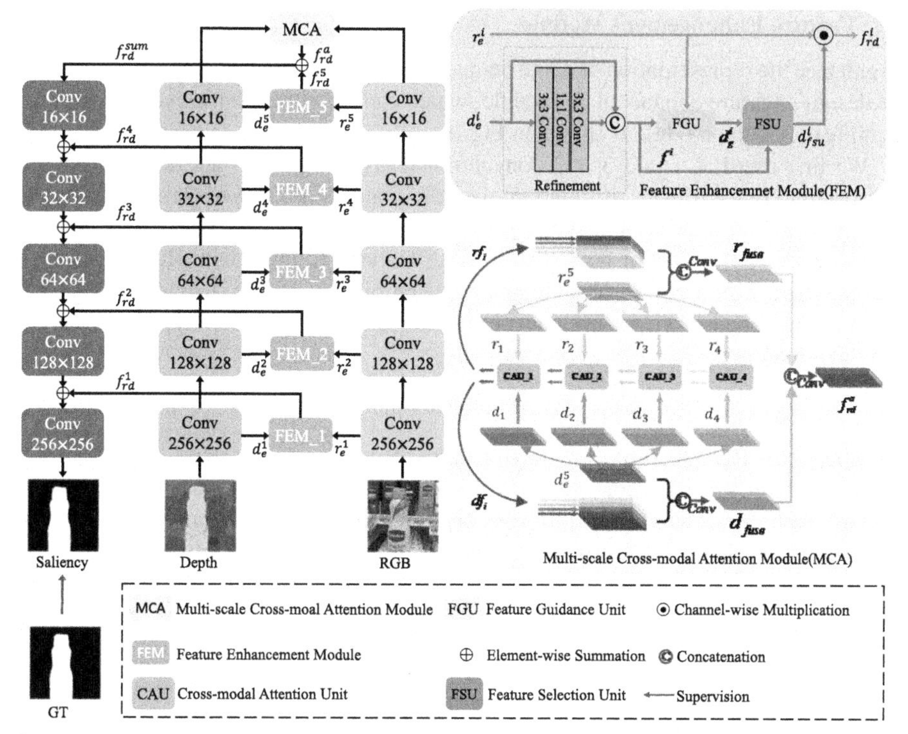

Fig. 1. The architecture of our network. First, the two-stream encoder extract features from the RGB image and the depth image. Then these features are fed into FEM and MCA for fusion. The fusion features are decoded to obtain the saliency map.

2 Our Method

In this section, we illustrate the proposed network in detail. As is shown in Fig. 1, it consists of a two-stream encoder, FEM and MCA. We describe the architecture of our network in Sect. 2.1. In Sects. 2.2 and 2.3, we show the details of FEM and MCA separately. In Sect. 2.4, we introduce the loss function.

2.1 Overview of Network Architecture

The RGB image and depth map are encoded separately through the two-stream VGG-16 [22] to generate multi-modal features. These features are then fed to FEMs to clean up depth features and enhance the RGB features, and the output features of FEMs are used for decoding. The MCA takes the top-level features as input to suppress background interference and accurately locate salient objects. The output features of the fifth FEM and MCA are summed to obtain fusion features which are decoded layer by layer to obtain a saliency map.

2.2 Feature Enhancement Module

To enhance the representation of depth features and improve the robustness of the model, we design a feature enhancement module, which can filter the interference noise of the depth features, the module is shown in Fig. 1.

We first use 1×1 and 3×3 convolutional layers to refine the input depth features to highlight detailed information, then the features are passed through FGU and FSU in turn, and their structures are shown in Fig. 2. The FGU is based on the idea of image registration [8], it guides poor depth features to align with RGB features to restore saliency information. The alignment operation contains two elements: offset and warping. Inspired by [1], we use convolution to learn the offset between deep features and RGB features:

$$\phi^i = conv_3(cat(r_e^i, f^i)) \tag{1}$$

where r_e^i is RGB feature, f^i is depth feature, ϕ^i is the offset, $conv_3(*)$ is 3×3 convolutional layer, $cat(*)$ is concatenation, $i \in \{1, 2, 3, 4, 5\}$. We then use the differential image sampling mechanism [12] to warp depth features. This mechanism maps a standard sampling grid into a deformed grid which is used to bilinearly interpolate the depth features to obtain the guided depth features. The above process can be summarized as:

$$G_o^i(x_o, y_o) = G_s^i(x_s + \phi^i(x_s), y_s + \phi^i(y_s)) \tag{2}$$

$$d_g^i = \varphi(f^i, \omega_G^i) \tag{3}$$

where G_s^i is the standard sampling grid, G_o^i is the deformed grid, (x, y) is a point on the grid, $\varphi(*)$ is the bi-linear interpolation, ω_G^i is the bilinear kernel weights which is estimated from G_o^i. d_g^i is the output depth feature of FGU.

The FSU aims to select the best quality depth feature from original depth feature f^i and the guided feature d_g^i. It concatenates the features and then generates a weight vector through a convolutional layer. The weight vector controls the ratio of the two-type features at the fusion stage. The above operations can be formulated as:

$$\alpha = conv_1(cat(f^i, d_g^i)) \tag{4}$$

$$d_{fsu}^i = conv_3(\alpha \odot d_g^i + (1 - \alpha) \odot f^i) \tag{5}$$

where $\alpha \in R^{C \times 1 \times 1}$ is the weight vector, $conv_1(*)$ is 1×1 convolutional layer, $\odot$ is concatenation, d_{fsu}^i is the output depth feature of the FSU. It can be seen from the formula that when the quality of f^i is not good, it will be assigned a smaller weight and d_g^i will has a larger proportion, and vice versa.

The above-mentioned steps based on guidance and selection progressively enhance the presentation of depth features. Finally, to make full use of the depth feature, we multiply the depth features and the RGB features to highlight the salient area of the RGB features.

$$f_{rd}^i = d_{fsu}^i \odot r_e^i \tag{6}$$

where r_e^i is the input RGB feature of FEM, f_{rd}^i is the output RGB feature of FEM.

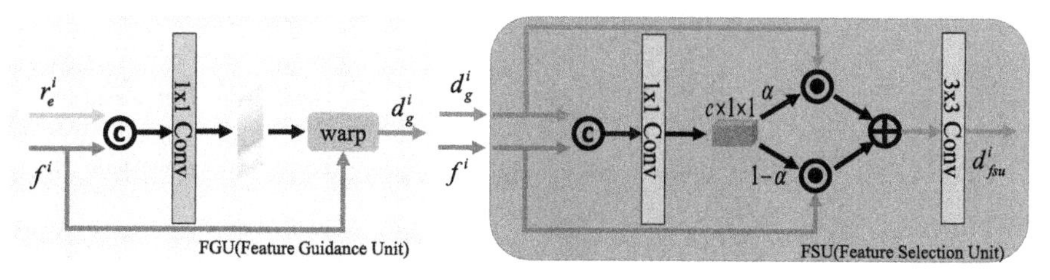

Fig. 2. The architecture of FGU and FSU.⊙ is channel-wise multiplication. ⊕ is element-wise summation. The letter c denotes concatenation.

2.3 Multi-scale Cross-Modal Attention Module

Among the features extracted by CNN, high-level features contain rich semantic information that is crucial for the localization of salient objects. Some studies have used a simple approach to fuse multi-modal high-level features, which introduces a large amount of redundant information and background interference, resulting in inaccurate localization. To address this problem, we design a multi-scale cross-modal attention module (MCA) which is shown in Fig. 1.

In this paper, we comprehensively take into account the background interference from different scales. To obtain multi-scale features, the top-level RGB feature r_e^5 and depth feature d_e^5 are passed through four dilated convolutional layers with different dilated ratios:

$$r_i = dconv(r_e^5; \theta_i, rate_i) \tag{7}$$

$$d_i = dconv(d_e^5; \theta_i, rate_i) \tag{8}$$

where $\{r_i, d_i\}_{i=1,2,3,4}$ are multi-scale features generated by r_e^5 and d_e^5. $dconv(*; \theta_i, rate_i)$ is the 3×3 dilated convolution with parameters θ_i and dilated rate $rate_{i=1,2,3,4} \in \{1, 2, 4, 8\}$.

Then, we let the RGB and depth feature pairs with the same scale pass through the cross-modal attention unit (CAU) to filter redundant information and suppress background interference. The structure of CAU is shown in Fig. 3, it allows cross-modal features to pay attention to each other's useful information with the help of cross-modal attention. Specifically, in the unit, the depth features d_i and the RGB features r_i are passed through the convolutional layer respectively to generate two attention maps $att_d \in R^{1\times H\times W}$, $att_r \in R^{1\times H\times W}$. Then we multiply the attention map att_r generated by the RGB features with the depth features d_i to help the depth features filter background interference, att_d and r_i are merged in the same way. The above process is summarized as:

$$df_i = conv_1(cat(d_i, d_i * att_r)) \tag{9}$$

$$rf_i = conv_1(cat(r_i, r_i * att_d)) \tag{10}$$

where $*$ is element-wise multiplication, df_i and rf_i are the output depth features and RGB features of CAU.

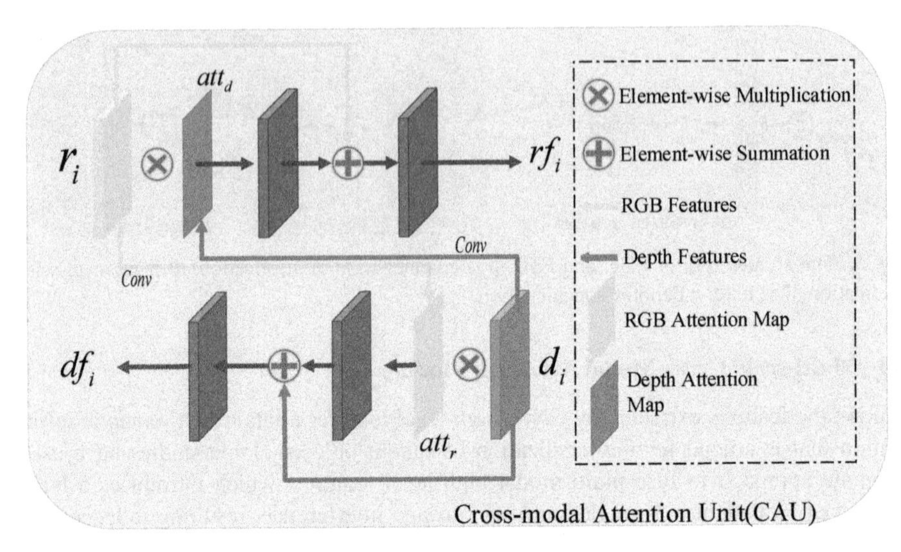

Fig. 3. The architecture of CAU

Through the CAUs, multi-modal features provide each other with valuable complementary information. Finally, we concatenate the outputs of all CAUs to obtain two multi-modal features with pure contextual information:

$$d_{fuse} = conv_1(cat(df_1, df_2, df_3, df_4, d_e^5)) \tag{11}$$

$$r_{fuse} = conv_1(cat(rf_1, rf_2, rf_3, rf_4, r_e^5)) \tag{12}$$

Then we integrate them to obtain fusion feature:

$$f_{rd}^a = conv_1(cat(r_{fuse}, d_{fuse})) \tag{13}$$

where f_{rd}^a is the output of MCA.

2.4 Loss Function

We employ a hybrid loss to train our network which is composed of cross-entropy loss [17], IOU loss [7], and edge loss [14].

The cross-entropy loss is a pixel-level loss, it calculates the loss value independently for each position on the saliency map. IOU is used to measure the similarity between two sets and IOU loss is a region-wise loss. Edge loss pays attention to edge pixels.The total loss is defined as:

$$L = \lambda_1 L_{bce}(S, G) + \lambda_2 L_{iou}(S, G) + \lambda_3 L_{edge}(S, G) \tag{14}$$

where S is saliency map, G is ground truth. L_{bce} is the binary cross-entropy loss, L_{iou} is the IOU loss, L_{edge} denotes edge loss. $\lambda_1 = \lambda_2 = \lambda_3 = 1$.

3 Experiments

Implementation Details. The proposed network is implemented based on an NVIDIA GTX 2080Ti GPU and PyTorch deep learning framework. We select 1485 images from NJUD [9] and 700 images from NLPR [15] as the training set. In both the training and test stage, the images are resized to 256 × 256. We use VGG-16 as our backbone. During training, random rotation and random horizontal flipping are used to reduce overfitting. The initial learning rate is set to 0.01, the batch size is set to 2, the learning rate adjustment strategy is the poly strategy. We use SGD as the optimizer and stop training after 30 epochs.

Datasets and Evaluation Metrics. We evaluate our method on five widely used benchmark datasets, including STEREO [13], NJUD [9], NLPR [15], SSD [29], LFSD [10]. Then we compare the proposed method with 11 state-of-the-arts in terms of three metrics including maximum F-measure (MaxF), mean absolute error (MAE), and structure measure ($\mathrm{S_m}$).

3.1 Comparisons with State-of-the-Arts

We compare the proposed method with 11 state-of-the-art deep learning-based RGB-D SOD methods, including DF [18], PCFN [2], PDNet [30], MMCI [3], CPFP [26], AFNet [23], DMRA [16], ATSA [24], D3Net [6], PGAR [5], FRDT [25].

Quantitative Evaluation. The Table 1 shows that the evaluation metrics of our method outperform the others. The PR curves in Fig. 5 also show that the proposed method achieves better performance.

Qualitative Evaluation. Figure 4 shows the visual comparisons of the proposed method with the state-of-the-art method. As can be seen, our method can accurately detect salient objects in simple scenes (rows 1 and 4), and can also highlight salient objects with clear detail in some complex scenes, such as multiple small objects (row 2), cluttered background (row 3), similar foreground and background (row 5), and low contrast between the foreground and background (row 6).

Effectiveness of Proposed Modules. We quantitatively show the contributions of FEM and MCA to our model in Table 2. It can be seen that both the FEM and MCA can improve the performance of the baseline, and the combination of the two modules can achieve better performance. Furthermore, to visually demonstrate the effectiveness of the proposed FEM, we visualize the features of FEM in Fig. 6. As can be seen from Fig. 6, the original depth feature d_e^4 is confused by a lot of background noise and does not pay attention to the salient area, while the optimized depth feature d_{fsu}^4 highlight the foreground and the color of its background becomes diluted. In addition, the salient area of the output RGB feature f_{rd}^4 is obviously more noticeable than that of the input feature r_e^4. The above observations prove that FEM is effective.

Table 1. A quantitative comparison of our method and other 11 other methods on five benchmark datasets. The best results are shown in **bold**.

Methods	Datasets														
	NJUD			NLPR			SSD			LFSD			STEREO		
	MaxF	S_m	MAE	MaxF	S_m	MAE	MaxF	S_m	MAE	MaxF	S_m	MAE	MaxF	S_m	MAE
Ours	**0.914**	**0.900**	**0.043**	**0.905**	**0.905**	**0.032**	**0.859**	**0.853**	**0.055**	**0.858**	**0.832**	**0.087**	**0.899**	0.887	**0.048**
D3Net20	0.897	0.890	0.055	0.882	0.896	0.040	0.825	0.826	0.078	0.852	0.830	0.101	0.887	0.882	0.060
FRDT20	0.892	0.883	0.059	0.890	0.894	0.039	0.827	0.817	0.085	0.821	0.805	0.115	0.889	0.881	0.059
ATSA20	0.889	0.881	0.060	0.876	0.885	0.044	0.834	0.821	0.082	0.841	0.821	0.107	0.892	0.882	0.060
PGAR20	0.886	0.885	0.059	0.894	0.902	0.037	0.834	0.820	0.081	0.856	0.828	0.109	0.898	**0.891**	0.058
AFNet19	0.893	0.853	0.105	0.891	0.846	0.082	0.831	0.793	0.124	0.838	0.806	0.155	0.889	0.858	0.105
DMRA19	0.871	0.868	0.071	0.869	0.881	0.045	0.823	0.815	0.089	0.824	0.809	0.123	0.876	0.866	0.071
$CPFP_{19}$	0.872	0.876	0.050	0.865	0.886	0.037	0.805	0.802	0.085	0.820	0.823	0.097	0.878	0.875	0.055
MMCI19	0.848	0.845	0.087	0.860	0.853	0.056	0.782	0.785	0.084	0.773	0.785	0.135	0.862	0.875	0.068
PDNet19	0.853	0.842	0.092	0.853	0.857	0.060	0.799	0.782	0.124	0.777	0.759	0.155	0.850	0.843	0.091
PCFN18	0.862	0.848	0.078	0.858	0.835	0.050	0.790	0.803	0.096	0.780	0.767	0.141	0.872	0.877	0.083
DF17	0.807	0.807	0.122	0.776	0.802	0.089	0.678	0.687	0.194	0.790	0.771	0.160	0.813	0.813	0.116

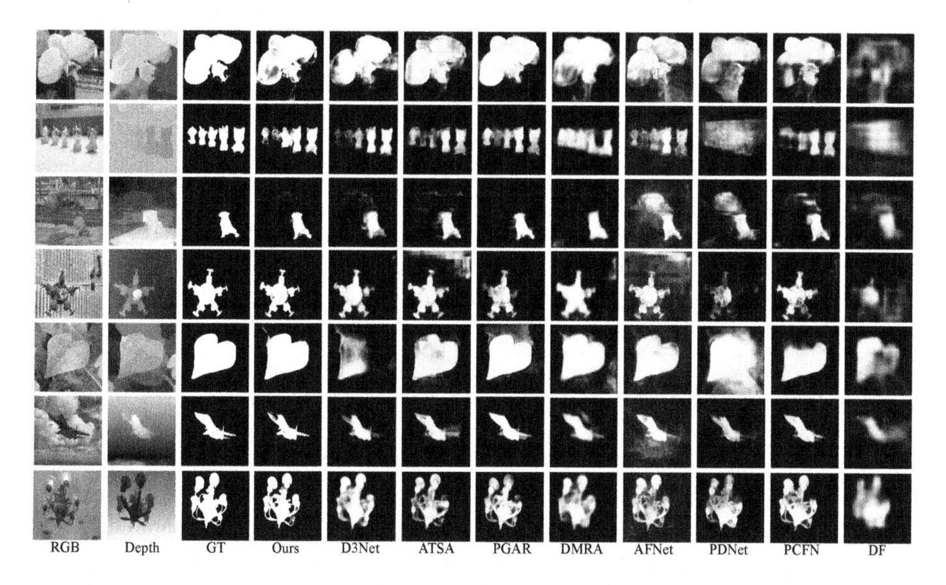

Fig. 4. Visual comparisons between our results and state-of-the-art methods

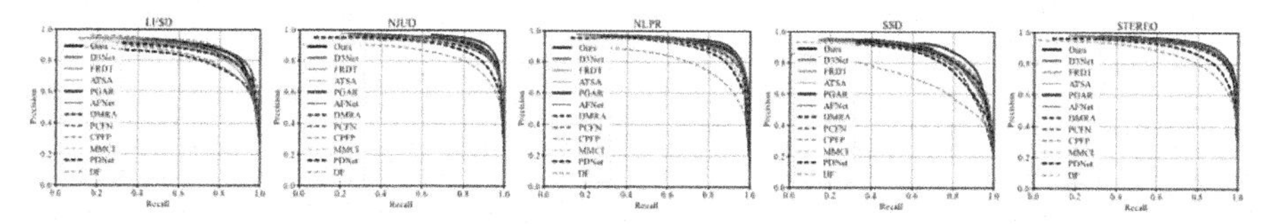

Fig. 5. PR curves on four popular RGB-D SOD datasets

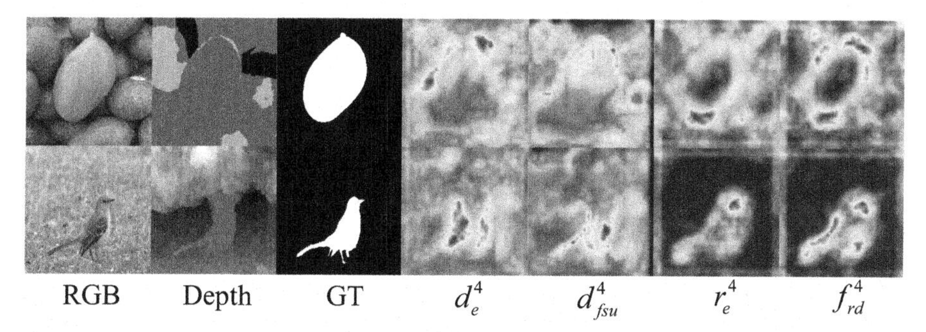

Fig. 6. Visualization of features in FEM. d_e^4 is input depth feature of the fourth FEM. d_{fsu}^4 is output feature of FSU. r_e^4 is input RGB feature of FEM and f_{rd}^4 is output RGB feature of FEM.

Table 2. Ablation on the NJUD datasets

Baseline	FEM	MCA	Hybrid loss	MaxF	MAE
✔				0.876	0.066
✔	✔			0.898	0.056
✔		✔		0.899	0.054
✔	✔	✔		0.910	0.050
✔	✔	✔	✔	0.914	0.043

Effect of Different Variants. We provide results in Table 3 to further verify the effectiveness of the proposed modules. From the table, we can see that removing any component of each module will cause performance degradation. The performance degradation of removing FIU or FSU proves that they are effective for enhancing depth features and filtering noise. The performance degradation of removing the four parallel dilated convolution branches proves that it is necessary to obtain multi-scale information. The performance degradation of removing CAU proves that it can suppress background interference and capture complementary information.

Table 3. Comparisons of different variants of FEM and MCA. w/o FIU: removing FIU, w/o FSU: removing FSU, w/o MS: removing four dilated convolutional layers, w/o CAU: removing CAU. The best results are shown in **bold**.

Variants	NJUD			NLPR		
	MaxF	S-measure	MAE	MaxF	S-measure	MAE
FEM (Ours)	**0.898**	**0.887**	**0.056**	**0.891**	**0.893**	**0.039**
w/o FIU	0.886	0.880	0.060	0.880	0.888	0.042
w/o FSU	0.887	0.883	0.059	0.883	0.890	0.040
MCA (Ours)	**0.899**	**0.891**	**0.054**	**0.874**	**0.889**	**0.041**
w/o MS	0.885	0.879	0.062	0.868	0.879	0.045
w/o CAU	0.893	0.887	0.054	0.872	0.884	0.043

4 Conclusion

In this paper, we propose a novel network to solve the problems of RGB-D SOD. It includes a feature enhancement module (FEM) and a multi-scale cross-modal attention module (MCA). The FEM enhances the presentation of depth features and filter noise by guidance and selection. The MCA can suppress background interference and obtains useful cross-modal complementary information. Experimental results on five datasets demonstrate the effectiveness of our method.

Acknowledgements. This work is supported by National Natural Science Foundation of China under Grant No. 62076058.

References

1. Alexey, D., Philipp, F., Eddy, L., Philip, H., Caner, H., Vladimir, G.: FlowNet: learning optical flow with convolutional networks. In: Proceedings of 2015 IEEE International Conference on Computer Vision, pp. 2758–2766 (2015)
2. Chen, H., Li, Y.: Progressively complementarity-aware fusion network for RGB-D salient object detection. In: Proceedings of 2018 IEEE Conference on Computer Vision and Pattern Recognition, pp. 3051–3060 (2018)
3. Chen, H., Li, Y.F., Su, D.: Multi-modal fusion network with multi-scale multi-path and cross-modal interactions for RGB-D salient object detection. Pattern Recognit. **86**, 376–385 (2019)
4. Chen, H.T.: Preattentive co-saliency detection. In: Proceedings of 2010 IEEE International Conference on Image Processing, pp. 1117–1120 (2010)
5. Chen, S., Yun, F.: Progressively guided alternate refinement network for RGB-D salient object detection. In: Vedaldi, A., Bischof, H., Brox, T., Frahm, J.-M. (eds.) ECCV 2020. LNCS, vol. 12353, pp. 520–538. Springer, Cham (2020). https://doi.org/10.1007/978-3-030-58598-3_31
6. Fan, D.P., Lin, Z., Zhang, Z., Zhu, M.L., Cheng, M.M.: Rethinking RGB-D salient object detection: models, data sets, and large-scale benchmarks. IEEE Trans. Neural Netw. Learn. Syst. **32**(5), 2075–2089 (2020)
7. Gellért, M., Luo, W.J., Raquel, U.: DeepRoadMapper: extracting road topology from aerial images. In: Proceedings of 2017 IEEE International Conference on Computer Vision, pp. 3458–3466 (2017)
8. Grant, H., Uwe, K., Yan, P.K.: Deep learning in medical image registration: a survey. Mach. Vis. Appl. **31**, 1–2 (2020)
9. Ju, R., Ge, L., Geng, W.J., Ren, T.W., Wu, G.S.: Depth saliency based on anisotropic center-surround difference. In: Proceedings of 2014 IEEE International Conference on Image Processing, pp. 1115–1119 (2014)
10. Li, N., Jinwei Ye, Y., Ji, H.L., Jingyi, Y.: Saliency detection on light field. IEEE Trans. Pattern Anal. Mach. Intell. **39**(8), 1605–1616 (2017)
11. Mahadevan, V., Vasconcelos, N.: Saliency-based discriminant tracking. In: Proceedings of 2009 IEEE Conference on Computer Vision and Pattern Recognition, pp. 1007–1013 (2009)
12. Max, J., Karen, S., Andrew, Z., Koray, K.: Spatial transformer networks (2016). https://arxiv.org/abs/1506.02025
13. Niu, Y.Z., Geng, Y.J., Li, X.Q., Liu, F.: Leveraging stereopsis for saliency analysis. In: Proceedings of 2012 IEEE Conference on Computer Vision and Pattern Recognition, pp. 454–461 (2012)
14. Pang, Y., Zhang, L., Zhao, X., Huchuan, L.: Hierarchical dynamic filtering network for RGB-D salient object detection. In: Vedaldi, A., Bischof, H., Brox, T., Frahm, J.-M. (eds.) ECCV 2020. LNCS, vol. 12370, pp. 235–252. Springer, Cham (2020). https://doi.org/10.1007/978-3-030-58595-2_15
15. Peng, H., Li, B., Xiong, W., Hu, W., Ji, R.: RGBD salient object detection: a benchmark and algorithms. In: Fleet, D., Pajdla, T., Schiele, B., Tuytelaars, T. (eds.) ECCV 2014. LNCS, vol. 8691, pp. 92–109. Springer, Cham (2014). https://doi.org/10.1007/978-3-319-10578-9_7
16. Piao, Y.R., Ji, W., Li, J.J., Zhang, M., Lu, H.C.: Depth-induced multi-scale recurrent attention network for saliency detection. In: Proceedings of 2019 IEEE International Conference on Computer Vision, pp. 7253–7262 (2019)

17. Pieter-Tjerk, D.B., Dirk, P., Shie, M., Reuven, Y.: A tutorial on the cross-entropy method. Ann. Oper. Res. **134**(1), 19–67 (2005)
18. Qu, L.Q., He, S., Zhang, J., Tian, J., Tang, Y., Yang, Q.: RGBD salient object detection via deep fusion. IEEE Trans. Image Process. **26**(5), 2274–2285 (2017)
19. Rapantzikos, K., Avrithis, Y., Kollias, S.: Dense saliency-based spatiotemporal feature points for action recognition. In: Proceedings of 2009 IEEE Conference on Computer Vision and Pattern Recognition, pp. 1454–1461 (2009)
20. Ren, J.Q., Gong, X.J., Lu, Y., Zhou, W.H., Yang, M.Y.: Exploiting global priors for RGB-D saliency detection. In: Proceedings of IEEE Conference on Computer Vision and Pattern Recognition Workshops, pp. 25–32 (2015)
21. Shimoda, W., Yanai, K.: Distinct class-specific saliency maps for weakly supervised semantic segmentation. In: Leibe, B., Matas, J., Sebe, N., Welling, M. (eds.) ECCV 2016. LNCS, vol. 9908, pp. 218–234. Springer, Cham (2016). https://doi.org/10.1007/978-3-319-46493-0_14
22. Simonyan, K., Zisserman, A.: Very deep convolutional networks for large-scale image recognition. In: Proceedings of 2015 International Conference on Learning Representations, pp. 1–14 (2015)
23. Wang, N.N., Gong, X.J.: Adaptive fusion for RGB-D salient object detection. IEEE Access **7**, 55277–55284 (2019)
24. Zhang, M., Fei, S.X., Liu, J., Shuang, X., Piao, Y., Huchuan, L.: Asymmetric two-stream architecture for accurate RGB-D saliency detection. In: Vedaldi, A., Bischof, H., Brox, T., Frahm, J.-M. (eds.) ECCV 2020. LNCS, vol. 12373, pp. 374–390. Springer, Cham (2020). https://doi.org/10.1007/978-3-030-58604-1_23
25. Zhang, M., Zhang, Y., Piao, Y.R., Hu, B.Q., Lu, H.C.: Feature reintegration over differential treatment: a top-down and adaptive fusion network for RGB-D salient object detection. In: Proceedings of the 28th ACM International Conference on Multimedia, pp. 4107–4115 (2020)
26. Zhao, J.X., Cao, Yang., Fan, D.P., Cheng, M.M., Li, X.Y., Zhang, L.: Contrast prior and fluid pyramid integration for RGBD salient object detection. In: Proceedings of 2019 IEEE Conference on Computer Vision and Pattern Recognition, pp. 3922–3913 (2019)
27. Zhao, R., Oyang, W., Wang, X.G.: Person re-identification by saliency learning. IEEE Trans. Pattern Anal. Mach. Intell. **39**(2), 356–370 (2016)
28. Zhao, X., Zhang, L., Pang, Y., Huchuan, L., Zhang, L.: A single stream network for robust and real-time RGB-D salient object detection. In: Vedaldi, A., Bischof, H., Brox, T., Frahm, J.-M. (eds.) ECCV 2020. LNCS, vol. 12367, pp. 646–662. Springer, Cham (2020). https://doi.org/10.1007/978-3-030-58542-6_39
29. Zhu, C.B., Li, G.: A three-pathway psychobiological framework of salient object detection using stereoscopic technology. In: Proceedings of 2017 IEEE International Conference on Computer Vision Workshops, pp. 3008–3014 (2017)
30. Zhu, C.B., Cai, X., Huang, K., Li, T., Li, G.: PDNet: Prior-model guided depth-enhanced network for salient object detection. In: Proceedings of 2019 IEEE International Conference on Multimedia and Expo, pp. 199–204 (2019)
31. Zhu, J.Y., Wu, J.J., Xu, Y., Chang, E., Tu, Z.W.: Unsupervised object class discovery via saliency-guided multiple class learning. In: Proceedings of 2012 IEEE Conference on Computer Vision and Pattern Recognition, pp. 3218–3225 (2012)

Improving Unsupervised Learning of Monocular Depth and Ego-Motion via Stereo Network

Mu He, Jin Xie, and Jian Yang(✉)

PCA Lab, Key Lab of Intelligent Perception and Systems for High-Dimensional Information of Ministry of Education, Nanjing University of Science and Technology, Nanjing, China
{muhe,csjxie,csjyang}@njust.edu.cn

Abstract. Unsupervised learning of monocular depth and ego-motion is a challenging task, which uses the photometric loss as the supervision to train the networks. Although existing unsupervised methods can get rid of expensive annotations, they are still limited in estimation accuracy. In this paper, we explore the use of stereo depth network for improving the performance of monocular depth estimation and ego-motion estimation. To this end, we propose a novel two-stage unsupervised learning framework. Specifically, in the first stage, we jointly train the stereo depth network and ego-motion network in an unsupervised manner, in order to get a more accurate ego-motion estimator. Then we transfer and freeze the ego-motion network to the second stage, and only train the monocular depth network in this stage. Moreover, we propose a dense feature fusion module to further enhance the expressive ability of monocular depth network without increasing the number of network parameters. Extensive experiments on the KITTI and Make3D datasets demonstrate that our proposed method achieves superior performance on both monocular depth estimation and ego-motion estimation to existing unsupervised methods.

Keywords: Unsupervised learning · Monocular depth estimation · Ego-motion estimation · Stereo depth network

1 Introduction

Inferring depth from a single image is a classic problem in computer vision, and estimating ego-motion from a continuous video sequence is likewise indispensable in robotics [1,6,7,28]. Both tasks have a wide range of applications such as autonomous vehicles, robot navigation, 3D modelling.

Recently, many deep learning based methods estimate monocular depth and ego-motion in a supervised manner. Although these supervised methods have made great progress, they are still limited by expensively-collected annotations (*e.g.*, the ground truth depth annotations from expensive lasers and depth cameras). By contrast, unsupervised monocular depth and ego-motion estimation methods do not require ground truth annotations, and have gained continuous

H. Ma et al. (Eds.): PRCV 2021, LNCS 13020, pp. 421–433, 2021.
https://doi.org/10.1007/978-3-030-88007-1_35

attention in recent years. The main idea of these unsupervised methods is to use the estimated depth and ego-motion for image reconstruction, and then employ the photometric loss as the supervision to train the monocular depth network and ego-motion network simultaneously. However, jointly learning of monocular depth and ego-motion from scratch is a very challenging task, as both networks are inaccurate at the beginning of training, which cannot guarantee the final performance of the models. Recent works have shown that the performance of stereo depth estimation is much better than that of monocular depth estimation, since stereo depth network can infer the actual and accurate depth by learning the pixel matching relationships between the calibrated stereo images. Therefore, this work aims to use the advantages of stereo depth network to improve the performance of monocular depth estimation and ego-motion estimation.

In this paper, we propose a novel two-stage unsupervised learning framework to employ the stereo depth network for improving the learning of monocular depth and ego-motion. Specifically, in the first stage, we jointly train the stereo depth network and ego-motion network by using the photometric loss as the supervisory signal. Thanks to the advantages of stereo depth estimation, we can obtain a more accurate pose estimator than previous methods. Then we transfer the ego-motion network trained in the first stage to the second stage and freeze it. In the second stage, we use the frozen ego-motion network to provide relative camera pose and train a monocular depth network to estimate the depth. We use the provided pose and estimated depth for image warping, and then employ the photometric warp loss to just optimize the monocular depth network. During inference, we use the monocular depth network trained in the second stage to estimate monocular depth, and the ego-motion network trained in the first stage to estimate relative camera pose. Since we use stereo videos for training, the estimated depth and pose are in a common and actual scale (set by the known stereo camera baseline). Moreover, we propose a dense feature fusion module to enable the monocular depth network to recover more spatial details during decoding, which can also slightly reduce the number of network parameters.

In summary, the main contributions of this paper are as follows:

1. We propose a novel two-stage unsupervised learning framework to improve the performance of monocular depth and ego-motion via stereo depth network.
2. We develop a dense feature fusion module to recover the spatial details of the monocular depth network without increasing the number of parameters.
3. Extensive experiments demonstrate that our proposed method achieves satisfactory performance on both monocular depth estimation and ego-motion estimation.

2 Related Work

2.1 Supervised Monocular Depth and Ego-Motion Estimation

With the rapid development of deep neural networks, supervised learning based methods have been proposed for learning monocular depth and ego-motion.

Eigen *et al.* [2] first propose a learning-based method for monocular depth estimation, which transform the depth estimation into a pixel-by-pixel regression problem. Laina *et al.* [13] propose a deeper fully convolutional network to learn single-view depth from RGBD datasets, which can improve the depth results obviously. Fu *et al.* [3] regard the monocular depth estimation as a classification problem and achieve excellent performance. For ego-motion estimation, Kendall *et al.* [10] first use the convolutional network to regress the 6-DoF relative camera pose from RGB images. Ummenhofer *et al.* [24] exploit the convolutional network to learn monocular depth and ego-motion simultaneously in a supervised manner. However, these supervised methods are still limited by expensive annotated data and show poor generalization ability.

2.2 Unsupervised Monocular Depth and Ego-Motion Estimation

Recently, unsupervised learning methods for monocular depth estimation and ego-motion estimation have gained more attention. Garg *et al.* [4] learn monocular depth from synchronized stereo image pairs in an unsupervised manner. Following this work, Godard *et al.* [6] introduce a left-right consistency term to improve depth results, and Poggi *et al.* [20] propose trinocular stereo assumptions to further enhance depth performance. Zhou *et al.* [28] first propose a fully unsupervised framework to jointly learn monocular depth and ego-motion from monocular videos. The principle is that image reconstruction can be completed by using the estimated depth and relative camera pose, and then employ the photometric loss as the supervision signal to jointly train the depth network and ego-motion network. Following this framework, Mahjourian *et al.* [17] introduce an additional 3D geometric constraint to learn depth and ego-motion. Zhan *et al.* [26] use stereo videos for training, as a result, the estimated monocular depth and pose are under a common and real-world scale. Bian *et al.* [1] propose a geometry consistency constraint to enforce the scale-consistency of the estimated depth and pose. Godard *et al.* [7] propose an advanced unsupervised framework called Monodepth2, which introduce a novel minimum reprojection loss and an auto-mask term to solve the problem of occlusions and moving objects. In particular, Monodepth2 [7] can be trained on three types of data, *i.e.*, monocular videos, stereo pairs, or stereo videos. In this paper, we choose Monodepth2 trained on stereo videos as our baseline.

Stereo depth estimation is also called stereo matching, and its goal is to estimate the disparity (inverse depth) map given stereo pairs. Zhou *et al.* [27] propose an unsupervised framework to learn stereo matching by iterative left-right consistency check. Kim *et al.* [11] combine stereo depth estimation and optical flow in an unsupervised manner and achieves superior results on both tasks. Godard *et al.* [6] have shown that the performance of stereo depth estimation is far better than that of monocular depth estimation. However, stereo depth estimation still requires calibrated stereo images during inference. In our work, we aim to use stereo depth network for improving the monocular depth estimation and ego-motion estimation.

3 Method

3.1 Problem Formulation

In our framework, we use the photometric loss as the supervisory signal at the training stage to convert the problem of depth and ego-motion estimation into a problem of image reconstruction. Our proposed framework uses stereo videos for training, each training instance contains a target image $\boldsymbol{I}_l$ and three source images $\boldsymbol{I}_s$ ($s \in \{l-1, l+1, r\}$), where $\boldsymbol{I}_{l-1}$ and $\boldsymbol{I}_{l+1}$ are temporal adjacent frames to $\boldsymbol{I}_l$, and $\boldsymbol{I}_r$ is a opposite stereo view to $\boldsymbol{I}_l$. With the predicted depth map (named as $\boldsymbol{D}_l$) of $\boldsymbol{I}_l$, and the relative camera pose (named as $\boldsymbol{T}_{l\to s}$) from $\boldsymbol{I}_l$ to $\boldsymbol{I}_s$, the image reconstruction process from the source view s to the target view l can be expressed as:

$$\boldsymbol{I}_{s\to l} = \boldsymbol{I}_s \langle proj(\boldsymbol{D}_l, \boldsymbol{T}_{l\to s}, \boldsymbol{K})\rangle, \tag{1}$$

where $\boldsymbol{K}$ is the known camera intrinsic matrix, $proj()$ is a coordinate projection operation from target image pixels to source image pixels [28] and $\langle\rangle$ is the differentiable bilinear sampling operator [9]. Note that, the relative pose $\boldsymbol{T}_{l\to r}$ between the stereo cameras is real and known, not estimated. The use of $\boldsymbol{T}_{l\to r}$ constrains the estimated depth ($\boldsymbol{D}_l$) and relative poses ($\boldsymbol{T}_{l\to l-1}$ and $\boldsymbol{T}_{l\to l+1}$) to be in a unified, actual scale. As a result, our estimated depth and pose can get rid of the issue of scale ambiguity. Following [7], we combine L1 and SSIM [25] to compute the photometric errors pe, which is given by:

$$pe(\boldsymbol{I}_a, \boldsymbol{I}_b) = \frac{\alpha}{2}(1 - SSIM(\boldsymbol{I}_a, \boldsymbol{I}_b)) + (1-\alpha)\left\|\boldsymbol{I}_a - \boldsymbol{I}_b\right\|_1. \tag{2}$$

Here α is empirically set to 0.85. The per-pixel minimum reprojection loss in [7] is an effective way to handle occlusion, so we adopt it as our photometric loss:

$$\mathcal{L}_{ph} = \min_{s} pe(\boldsymbol{I}_l, \boldsymbol{I}_{s\to l}). \tag{3}$$

In addition, to deal with disparity discontinuities, we follow the form of edge-aware smoothness loss in [7]:

$$\mathcal{L}_{sm} = \left|\partial_x d_l^*\right| e^{-\left|\partial_x I_l\right|} + \left|\partial_y d_l^*\right| e^{-\left|\partial_y I_l\right|}, \tag{4}$$

where ∂_x and ∂_y represent the gradients in the horizontal and vertical direction respectively, and $d_l^* = d_t/\overline{d_t}$ denotes the mean-normalized inverse depth with the goal to prevent shrinking of the predicted depth [7].

Regardless of training in the first stage or in the second stage, we combine the pixel minimum photometric loss (refer to Eq. (3)) and smoothness loss (refer to Eq. (4)) as the final loss function:

$$\mathcal{L}_{fi} = \mu\mathcal{L}_{ph} + \lambda L_{sm}, \tag{5}$$

where μ and λ are the hyper-parameters of photometric loss and smoothness loss, respectively.

3.2 Two-Stage Unsupervised Learning Framework

Our goal is to improve the performance of monocular depth estimation and ego-motion estimation via stereo depth network. To this end, we propose a novel two-stage unsupervised learning framework, as shown in Fig. 1. We will describe the details of each stage.

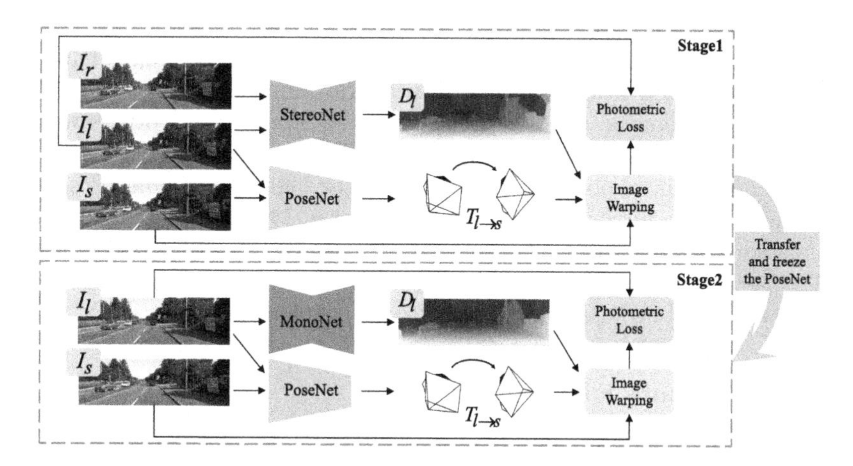

Fig. 1. Overview of our proposed two-stage unsupervised learning framework for monocular depth and ego-motion estimation, where the ego-motion network (PoseNet) is trained in the first stage and the monocular depth network (MonoNet) is trained in the second stage.

The First Stage. In order to obtain a more accurate ego-motion estimator, we jointly train the stereo depth network (named as StereoNet) and the ego-motion network (named as PoseNet) in an unsupervised manner. The StereoNet inputs a pair of stereo images $(\boldsymbol{I}_l, \boldsymbol{I}_r)$, and outputs the depth map $\boldsymbol{D}_l$ of the target image $\boldsymbol{I}_l$. The PoseNet inputs two temporal adjacent images, *i.e.*, the target image $\boldsymbol{I}_l$ and source image ($\boldsymbol{I}_{l-1}$ or $\boldsymbol{I}_{l+1}$), and then estimates the relative camera pose between the two images. Note that the relative camera pose $\boldsymbol{T}_{l\to r}$ is known rather than estimated. With the $\boldsymbol{D}_l$ and $\boldsymbol{T}_{l\to s}$ ($s \in \{l-1, l+1, r\}$), we warp $\boldsymbol{I}_s$ to synthesize $\boldsymbol{I}_l$ through Eq. (1), and then use the final loss (refer to Eq. (5)) as the supervision signal to train the StereoNet and PoseNet simultaneously. Thanks to the advantages of stereo network, we have obtained high-performance pose estimator with actual scale in the first stage, which also lays the foundation for the training of the monocular depth network in the second stage.

The Second Stage. In order to use the high-performance pose estimator of the first stage, we only train the monocular depth network (named as MonoNet) in the second stage. Specifically, we use the PoseNet trained in the first stage to provide relative camera pose and freeze the PoseNet in the second stage, then train the MonoNet to estimate depth. With the provided pose and estimated depth, image reconstruction as in Eq. (1) can be completed, then the final loss (see Eq. (5)) is used as the supervision signal to only optimize the MonoNet. The MonoNet inputs $\boldsymbol{I}_l$ and outputs $\boldsymbol{D}_l$. Since we use stereo videos for training in both stages, our estimated depth and pose are in a common and actual scale. In the inference phase, we use the MonoNet trained in the second stage and the PoseNet trained in the first stage to evaluate the performance of monocular depth estimation and ego-motion estimation.

Compared with prior methods of jointly training the monocular depth network and ego-motion network, the advantages of our framework are two-fold. In terms of performance, thanks to the advantages of stereo depth estimation, the PoseNet jointly trained with the stereo depth network in the first stage can estimate more accurate pose. Then we transfer the high-performance PoseNet to the second stage to further promote the training of MonoNet. In terms of optimization, since we freeze the PoseNet and only use the loss function to optimize the MonoNet in the second stage, which is easier than optimizing the monocular depth network and ego-motion network from scratch simultaneously.

3.3 Dense Feature Fusion

Existing monocular depth network is based on the U-Net architecture [21], which utilizes the skip-connection to fuse the low-level features from the encoder during decoding. Since the low-level features in the encoder contain rich spatial information, such as object shapes and boundaries, making full use of these features is beneficial for depth estimation. However, the skip-connection only fuses the features of the same spatial resolution between the encoder and decoder, resulting in poor feature fusion.

Inspired by [14,29], as shown in Fig. 2, we propose a dense feature fusion module to better recover the spatial information and enhance the ability of the network to handle details structures. Let $\boldsymbol{E}_i$ denote the feature after each downsampling in the encoder, $\boldsymbol{D}_i$ denote the feature after each upsampling in the decoder and $\boldsymbol{S}_i$ represent the dense feature fusion during decoding, we perform the dense feature fusion at the i-th layer of the decoder:

$$\boldsymbol{E}_i' = \mathrm{Conv}_{1\times 1}\left(\boldsymbol{E}_i\right), \tag{6}$$

$$\boldsymbol{E}_i'' = \mathrm{Conv}_{3\times 3}\left(\sum_{j=1}^{4} \mathrm{Inte}\left(\boldsymbol{E}_j'\right)\right), \tag{7}$$

$$\boldsymbol{S}_i = \boldsymbol{E}_i'' + \boldsymbol{D}_i, \tag{8}$$

where i enumerates the index of layers in descending order $\{4, 3, 2, 1\}$, $\mathrm{Conv}_{1\times 1}(\cdot)$ is an 1×1 convolution and $\mathrm{Conv}_{3\times 3}(\cdot)$ is a 3×3 convolution. $\mathrm{Inte}(\cdot)$ denotes the nearest neighbor interpolation term, which is represented by the colored lines in Fig. 2. First, for each $\boldsymbol{E}_i$, we use an 1×1 convolution to adjust the number of channels to get $\boldsymbol{E}_i'$. During the dense feature fusion at the i-th layer in the decoder, we first use nearest neighbor interpolation to scale each

$\boldsymbol{E}'_j$ ($j \in \{1, 2, 3, 4\}$) to the same spatial resolution as $\boldsymbol{D}_i$, and then sum them up through element-wise addition, and use a 3×3 convolution to get $\boldsymbol{E}''_i$, the number of channels of $\boldsymbol{E}''_i$ is the same as $\boldsymbol{D}_i$. Finally, we sum $\boldsymbol{E}''_i$ and $\boldsymbol{D}_i$ up to complete the dense feature fusion in the i-th layer of the decoder.

In particular, during the dense feature fusion, since we use the convolution operation to reduce the number of channels of the original features from the encoder and use summation instead of concatenation for feature fusion, the final total parameters are even reduced, see Sect. 4.3 for more details. As a result, our proposed dense feature fusion module improves the model performance with fewer network parameters.

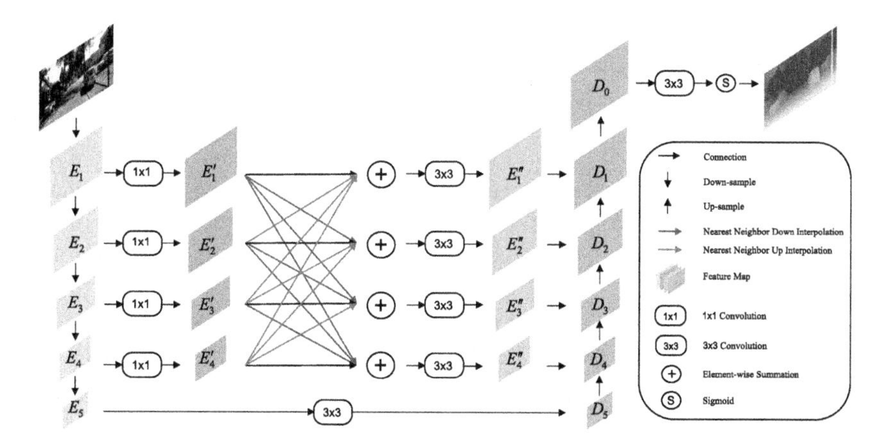

Fig. 2. Illustration of the dense feature fusion module in the monocular depth network. In this module, the skip connection in the traditional U-Net architecture [21] is redesigned with the purpose to enhance the ability of the network to handle details structures.

4 Experiments

4.1 Implementation Details

Settings. Our proposed approach is implemented using Pytorch [19] and trained on a single Titan RTX GPU. Similar to [7], we initialize our networks with weights pretrained on ImageNet [22], and use Adam optimizer [12] with $\beta1 = 0.9$ and $\beta2 = 0.999$. Regardless of training in the first stage or in the second stage, the depth estimation network and the ego-motion estimation network are trained for 20 epochs, with a batch size of 12 and an input resolution of 640×192, and the learning rate for the first 15 epochs is 10^{-4} and then dropped to 10^{-5} for the last 5 epochs. During training, we adopt $\mu = 1.0$, $\lambda = 0.001$ in Eq. (5), and set the weight of SSIM to $\alpha = 0.85$. Following [1], we only compute the loss on the final (single-scale) output of the depth network, instead of on the multiple-scale outputs of the depth network. The single-scale loss calculation greatly reduces training time without affecting model performance.

Network Architecture. In our two-stage framework, for depth estimation network, the StereoNet adopts general U-Net architecture while the MonoNet uses encoder-decoder structure with our proposed dense feature fusion module, and both StereoNet and MonoNet use ResNet50 [8] as encoder. For the ego-motion estimation network (PoseNet), we use the same architecture as Monodepth2 [7], which contains ResNet18 [8] followed by several convolutional layers.

Table 1. Quantitative monocular depth estimation results on the KITTI Eigen split [5]. We have divided four categories of comparison methods based on the training data, where D denotes depth ground truth, M denotes monocular videos, S denotes stereos, and MS denotes stereo videos. In each category, the best results are in **bold** and second best are underlined. R50: ResNet50; †: newer results from github.

Methods	Train	Error metric ↓				Accuracy metric ↑		
		AbsRel	SqRel	RMSE	RMSElog	$\delta<1.25$	$\delta<1.25^2$	$\delta<1.25^3$
Eigen *et al.* [2]	D	0.203	1.548	6.307	0.282	0.702	0.890	0.958
Liu *et al.* [15]	D	0.201	1.584	6.471	0.273	0.680	0.898	0.967
DORN [3]	D	**0.072**	**0.307**	**2.727**	**0.120**	**0.932**	**0.984**	**0.994**
Zhou *et al.* [28]†	M	0.183	1.595	6.709	0.270	0.734	0.902	0.959
EPC++ [16]	M	0.141	1.029	5.350	0.216	0.816	0.941	0.976
SC-SfMLearner R50 [1]†	M	0.115	**0.814**	4.705	0.191	0.873	0.960	0.982
Monodepth2 R50 [7]	M	**0.110**	0.831	**4.642**	**0.187**	**0.883**	**0.962**	**0.982**
Garg *et al.* [4]†	S	0.152	1.226	5.849	0.246	0.784	0.921	0.967
Monodepth R50 [6]†	S	0.133	1.142	5.533	0.230	0.830	0.936	0.970
3Net R50 [20]	S	0.129	0.996	5.281	0.223	0.831	0.939	0.974
Monodepth2 R50	S	**0.106**	**0.863**	**4.835**	**0.205**	**0.873**	**0.951**	**0.976**
Zhan FullNYU [26]	MS	0.135	1.132	5.585	0.229	0.820	0.933	0.971
EPC++ [16]	MS	0.128	0.935	5.011	0.209	0.831	0.945	0.979
Monodepth2 R50 [7]	MS	0.105	0.811	4.702	0.195	0.879	0.957	0.980
Ours	MS	**0.101**	**0.744**	**4.540**	**0.189**	**0.882**	**0.960**	**0.981**

4.2 Main Results

KITTI Eigen Split. We evaluate the performance of monocular depth estimation on the Eigen split [2] of the KITTI dataset [5]. Following Zhou *et al.* [28], we remove static frames during training, and use 39810, 4424, and 697 images for training, validation, and test respectively. Both stages are trained on the KITTI Eigen split, and then we compare the performance of our model (MonoNet) with existing state-of-the-art approaches. During evaluation, we adopt the conventional metrics in [2] and standardly cap depth to 80m [7]. Note that since we use stereo videos for training, we utilize a real scale from the known camera baseline [7] during evaluating instead of performing median scaling [28]. The results in Table 1 show that our proposed method achieves superior results, as we outperform the advanced Monodepth2 [7] significantly. The qualitative results in Fig. 3 shows that compared with other unsupervised approaches, our proposed method can obtain more refined depth maps.

Fig. 3. Qualitative comparison results of monocular depth estimation on the KITTI Eigen split [5].

KITTI Odometry. For ego-motion estimation, we report the visual odometry performance on the KITTI odometry dataset [5] using the PoseNet trained in the first stage. Following [7], we use the 00-08 sequences for training and 09-10 sequences for testing. We adopt the standard evaluation metrics in [1], as they are commonly used and have more reference value. In Table 2, we compare our method with ORB-SLAM [18] system and existing unsupervised approaches [1, 7, 26, 28] and achieve impressive results. Figure 4 shows qualitative results for all the methods. These unsupervised methods [1, 7, 26, 28] all use the monocular depth network and pose network for jointly training, while we jointly train the pose network with the stereo depth network in the first stage. The results show that the pose network can obtain significant gains from the stereo depth network.

Table 2. Visual odometry results evaluated on KITTI odometry dataset [5].

Methods	Seq.09		Seq.10	
	$t_{err}(\%)$	$r_{err}(°/100\ \text{m})$	$t_{err}(\%)$	$r_{err}(°/100\ \text{m})$
ORB-SLAM [18]	15.30	0.26	3.68	0.48
Zhou *et al.* [28]	17.84	6.78	37.91	17.78
Zhan *et al.* [26]	11.92	3.60	12.62	**3.43**
SC-SfMLearner [1]	7.03	3.00	8.07	5.01
Monodepth2 [7]	12.07	2.85	10.13	4.33
Ours	**5.48**	**1.48**	**7.07**	3.51

4.3 Ablation Study

As shown in Table 3, we perform ablation experiments on the KITTI Eigen split [5] to demonstrate the effectiveness of our proposed framework, which includes the two-stage unsupervised learning framework and dense feature fusion module.

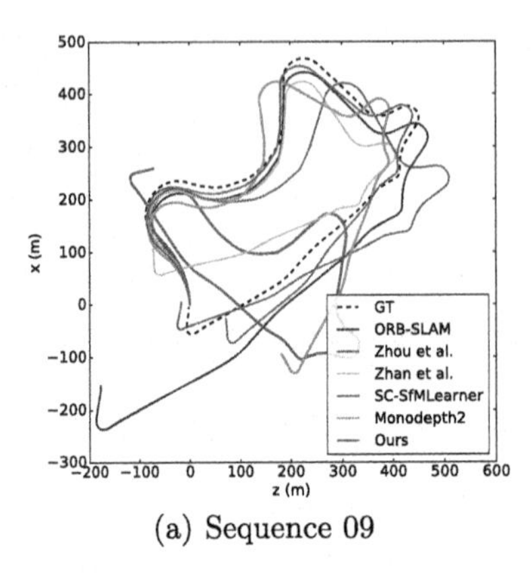

(a) Sequence 09

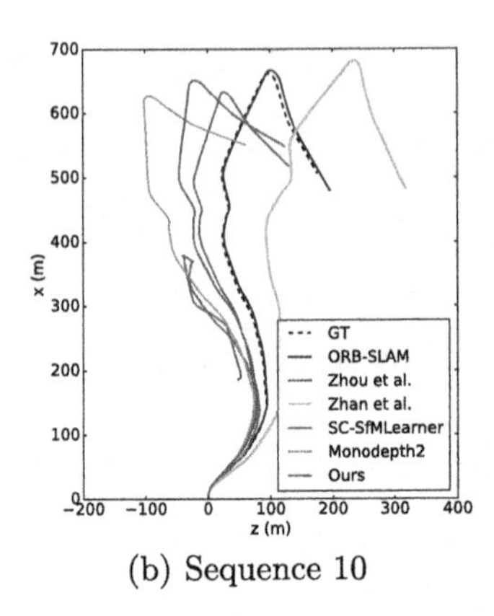

(b) Sequence 10

Fig. 4. Qualitative comparison results on KITTI odometry dataset [5].

Table 3. Ablation study results of the proposed components, the best results are in **bold** and second best are underlined.

	Two-stage framework	Dense feature fusion	#Para	Error metric ↓		Accuracy metric ↑
				AbsRel	RMSE	δ<1.25
Baseline			34.57M	0.105	4.702	0.879
with TSF	√		34.57M	<u>0.102</u>	4.628	<u>0.880</u>
with DFF		√	33.43M	0.103	<u>4.602</u>	<u>0.880</u>
Ours	√	√	33.43M	**0.101**	**4.540**	**0.882**

We choose Monodepth2 [7] trained on stereo videos as our baseline model. TSF denotes the two-stage unsupervised framework and DFF denotes the dense feature fusion module, and #Para denotes the parameters of the monocular depth network. The results show that our proposed two-stage training strategy and dense feature fusion module can bring obvious improvement individually. Finally, we combine them together and improve the performance of monocular depth estimation significantly, and reduce the number of parameters.

4.4 Make3D

In Table 4, we use our model trained on KITTI to test on the Make3D dataset [23]. We use the same evaluation criteria and crop method as Monodepth2 [7]. Note that, the methods trained on monocular videos (M) get benefit from median scaling. As our model is trained on stereo videos (MS), we directly evaluate the unmodified network prediction results instead of using median scaling. Type D represents supervised methods, and the best results in each type are in bold. We outperform the advanced Monodepth2 [7] trained on stereo videos, which shows that our model has more robust generalization performance.

Table 4. Quantitative monocular depth estimation results on Make3D dataset [23]. All M methods use median scaling, while S and MS use the original network prediction.

Methods	Type	AbsRel	SqRel	RMSE	RMSElog
Liu *et al.* [15]	D	0.475	6.562	10.05	0.165
Laina *et al.* [13]	D	**0.204**	**1.840**	**5.683**	**0.084**
Monodepth [6]	S	0.544	10.94	11.760	0.193
Zhou *et al.* [28]	M	0.383	5.321	10.470	0.478
Monodepth2 [7]	M	**0.322**	**3.589**	**7.417**	**0.163**
Monodepth2 [7]	MS	0.372	3.707	8.365	**0.207**
Ours	MS	**0.369**	**3.529**	**8.218**	**0.207**

5 Conclusion

In this paper, we exploited the advantages of the stereo depth network to improve the unsupervised learning of monocular depth and ego-motion. Specifically, we proposed a novel two-stage unsupervised learning framework to enable both monocular depth estimation and ego-motion estimation to benefit from the stereo depth network. Moreover, we proposed a dense feature fusion module for the monocular depth network, which can further boost the performance of the monocular depth network without increasing the number of parameters. Extensive experiments showed that our proposed method can yield satisfactory results in terms of monocular depth estimation and ego-motion estimation.

References

1. Bian, J.W., et al.: Unsupervised scale-consistent depth and ego-motion learning from monocular video. In: NeurIPS (2019)
2. Eigen, D., Puhrsch, C., Fergus, R.: Depth map prediction from a single image using a multi-scale deep network. arXiv preprint arXiv:1406.2283 (2014)
3. Fu, H., Gong, M., Wang, C., Batmanghelich, K., Tao, D.: Deep ordinal regression network for monocular depth estimation. In: CVPR (2018)
4. Garg, R., Vijay Kumar, B.G., Carneiro, G., Reid, I.: Unsupervised CNN for single view depth estimation: geometry to the rescue. In: Leibe, B., Matas, J., Sebe, N., Welling, M. (eds.) ECCV 2016. LNCS, vol. 9912, pp. 740–756. Springer, Cham (2016). https://doi.org/10.1007/978-3-319-46484-8_45
5. Geiger, A., Lenz, P., Stiller, C., Urtasun, R.: Vision meets robotics: the kitti dataset. Int. J. Robot. Res. **32**(11), 1231–1237 (2013)
6. Godard, C., Mac Aodha, O., Brostow, G.J.: Unsupervised monocular depth estimation with left-right consistency. In: CVPR (2017)
7. Godard, C., Mac Aodha, O., Firman, M., Brostow, G.J.: Digging into self-supervised monocular depth estimation. In: ICCV (2019)

8. He, K., Zhang, X., Ren, S., Sun, J.: Deep residual learning for image recognition. In: CVPR (2016)
9. Jaderberg, M., Simonyan, K., Zisserman, A., Kavukcuoglu, K.: Spatial transformer networks. arXiv preprint arXiv:1506.02025 (2015)
10. Kendall, A., Grimes, M., Cipolla, R.: PoseNet: a convolutional network for real-time 6-DOF camera relocalization. In: ICCV (2015)
11. Kim, T., Ryu, K., Song, K., Yoon, K.J.: Loop-Net: joint unsupervised disparity and optical flow estimation of stereo videos with spatiotemporal loop consistency. IEEE Robot. Autom. Lett. **5**(4), 5597–5604 (2020)
12. Kingma, D.P., Ba, J.: Adam: a method for stochastic optimization. arXiv preprint arXiv:1412.6980 (2014)
13. Laina, I., Rupprecht, C., Belagiannis, V., Tombari, F., Navab, N.: Deeper depth prediction with fully convolutional residual networks. In: 3DV. IEEE (2016)
14. Li, X., Chen, H., Qi, X., Dou, Q., Fu, C.W., Heng, P.A.: H-DenseuNet: hybrid densely connected UNet for liver and tumor segmentation from CT volumes. IEEE Trans. Med. imaging **37**(12), 2663–2674 (2018)
15. Liu, F., Shen, C., Lin, G., Reid, I.: Learning depth from single monocular images using deep convolutional neural fields. IEEE Trans. Pattern Anal. Mach. Intell. **38**(10), 2024–2039 (2015)
16. Luo, C., et al.: Every pixel counts++: joint learning of geometry and motion with 3D holistic understanding. IEEE Trans. Pattern Anal. Mach. Intell. **42**(10), 2624–2641 (2019)
17. Mahjourian, R., Wicke, M., Angelova, A.: Unsupervised learning of depth and ego-motion from monocular video using 3D geometric constraints. In: CVPR (2018)
18. Mur-Artal, R., Montiel, J.M.M., Tardos, J.D.: ORB-SLAM: a versatile and accurate monocular SLAM system. IEEE Trans. Rob. **31**(5), 1147–1163 (2015)
19. Paszke, A., et al.: Automatic differentiation in pytorch (2017)
20. Poggi, M., Tosi, F., Mattoccia, S.: Learning monocular depth estimation with unsupervised trinocular assumptions. In: 3DV. IEEE (2018)
21. Ronneberger, O., Fischer, P., Brox, T.: U-Net: convolutional networks for biomedical image segmentation. In: Navab, N., Hornegger, J., Wells, W.M., Frangi, A.F. (eds.) MICCAI 2015. U-Net: convolutional networks for biomedical image segmentation, vol. 9351, pp. 234–241. Springer, Cham (2015). https://doi.org/10.1007/978-3-319-24574-4_28
22. Russakovsky, O., et al.: ImageNet large scale visual recognition challenge. Int. J. Comput. Vis. **115**, 211–252 (2015)
23. Saxena, A., Sun, M., Ng, A.Y.: Make3D: learning 3D scene structure from a single still image. IEEE Trans. Pattern Anal. Mach. Intell. **31**(5), 824–840 (2008)
24. Ummenhofer, B., et al.: DeMoN: depth and motion network for learning monocular stereo. In: CVPR (2017)
25. Wang, Z., Bovik, A.C., Sheikh, H.R., Simoncelli, E.P.: Image quality assessment: from error visibility to structural similarity. IEEE Trans. Image Process. **13**(4), 600–612 (2004)
26. Zhan, H., Garg, R., Weerasekera, C.S., Li, K., Agarwal, H., Reid, I.: Unsupervised learning of monocular depth estimation and visual odometry with deep feature reconstruction. In: CVPR (2018)
27. Zhou, C., Zhang, H., Shen, X., Jia, J.: Unsupervised learning of stereo matching. In: ICCV (2017)

28. Zhou, T., Brown, M., Snavely, N., Lowe, D.G.: Unsupervised learning of depth and ego-motion from video. In: CVPR (2017)
29. Zhou, Z., Rahman Siddiquee, M.M., Tajbakhsh, N., Liang, J.: UNet++: a nested U-Net architecture for medical image segmentation. In: Stoyanov, D., et al. (eds.) DLMIA/ML-CDS -2018. LNCS, vol. 11045, pp. 3–11. Springer, Cham (2018). https://doi.org/10.1007/978-3-030-00889-5_1

A Non-autoregressive Decoding Model Based on Joint Classification for 3D Human Pose Regression

Yuhang Guo[1], Dongmei Fu[1,2](✉), and Tao Yang[1,2]

[1] Beijing Engineering Research Center of Industrial Spectrum Imaging, School of Automation and Electrical Engineering, University of Science and Technology Beijing, Beijing 100083, China
fdm_ustb@ustb.edu.cn

[2] Shunde Graduate School of University of Science and Technology Beijing, Fo Shan 528300, China

Abstract. Recently, the graph convolution networks (GCN) has been widely applied in 3D human pose regression and has showed encouraging performance. One limitation of this method is that it only models the semantic correlation between 2D joints feature, but ignores the variability of the semantic correlation between 2D joints feature. To address this limitation, we propose a non-autoregressive decoding model based on joint classification (JC-NARD), which realizes the 3D joints regression with the method of sequence analysis. The model splits joints into several joint sub-sequences according to the connection and semantic correlation, and then models the correlation between 3D-3D joints feature and 2D-3D joints feature by attention mechanism in each sub-sequence to establish 3D spatial constraint between joints. In order to verify the accuracy and generalization of the model, we combine our model with several 3D human pose regression networks, and the performance of the models are all improved by 1.2–4.5 mm.

Keywords: Correlation · Classification · Constraint · Non-autoregressive · Attention

1 Introduction

3D human pose regression [1,6,17] has a wide application in many fileds. Thus, the research of 3D human pose regression has developed rapidly in the past decades. However, it's a challenging task to further improve the accuracy due to the self-occlusion between joints and the depth ambiguity of 2D to 3D.

Recently, Graph convolutional network (GCN) [2,3,8,16,23] has been widely applied in the research of 3D human pose regression, which establishes the constraints among 2D joints feature to overcome above difficulties through modeling the semantic correlation [11,12,22,23]. However, these method is suboptimal, because the semantic correlation among 2D joints will be changed as the perspective changes in 2D space, as shown in Fig. 1.

H. Ma et al. (Eds.): PRCV 2021, LNCS 13020, pp. 434–446, 2021.
https://doi.org/10.1007/978-3-030-88007-1_36

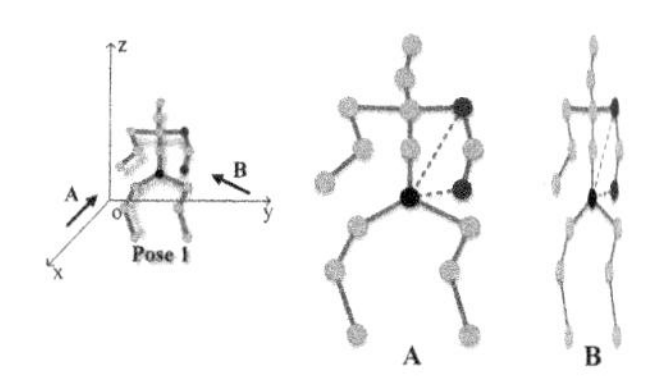

Fig. 1. The **Pose1** in 3D space obtains two 2D projections from the perspectives of A and B; The red joint is the joint to be studied, and the relative positions of the two blue joints to the red joints are very different in two perspectives. (Color figure online)

To address the above limitation, we propose a non-autoregressive [4,5] decoding model based on joint classification (JC-NARD) by utilizing the sequential analysis method [19]. Since joints coordinates is no longer affected by perspective in 3D space, the correlation among 3D joints feature would be more reliable. Specifically, we will model the semantic correlation of 3D-3D joint feature and 2D-3D joint feature rather than 2D-2D feature. Considering of the disorder of the joint sequence, we adopt a pre-trained 3D pose regression network as the reference model [5] to provide 2D joints feature and reference 3D joints feature to JC-NARD to predict corresponding 3D feature instead of directly predicting 3D feature based on the regression results of the first N joints as usual sequential analysis [19] methods. And then the model would learn their inherent correlation among 3D feature to approximate the real 3D pose.

Given reference 3D joints feature, we firstly classify the correlation among joints into multiple correlation levels according to the semantic correlation and the connection, the joints with stronger correlation will be proposed to form several sub-sequences. And then 3D spatial constraint among 3D-3D joints feature and among 2D-3D joints feature will be established in each sub-sequence. Finally, JC-NARD will output predicted 3D joints result in parallel by adaptively aggregating 2D and 3D feature. Compared with the previous model [2,23], our model establishes a more reliable constraint between the joints and overcomes the disorder of joint sequence to output 3D joints sequence in parallel. In sum, the contributions of this paper are in following:

1. By establishing constraints among 3D-3D joints feature and among 2D-3D joints feature, we provide more reliable constraints for regression joints and solve the problem that constraints among 2D joints vary with perspectives.
2. By classifying and reorganizing human joints, the 2D-3D pose regression problem is transformed into a 3D joints sequence output problem. Thus, the disorder of joint sequence is overcome and the parallel output of 3D joint sequence is realized.

2 Related Work

3D Human Pose Estimation. The development of 3D human pose regression has gone through a long time. At the beginning, it is studied as a pure image procressing problem and the researchers rebuilt 3D human pose by analyzing geometric constraints [1,6] among joints. In recent years, Martinez et al. [14] was the first to propose a deep neural network to realize 3D human pose regression, which improved the state-of-the-art performance significantly. After that, most researchers [9,13,17,18,21,23,24] start to exploit the neural network for end-to-end learning in the field of 3D human pose regression and pushed forward the performance rapidly. Zhou et al. [24] introduced a weakly-supervised transfer learning method to make full use of mixed 2D and 3D labels. Yang et al. [21] proposed an adversarial [10] network framework, which distilled the 3D human pose structures learned from the fully annotated dataset to in-the-wild images with only 2D annotations.

Up to now, 3D pose regression task has been divided into two sub-tasks: 2D human pose regression from images [15] and the 3D pose regression from 2D pose. Our research falls into the latter. Most related works apply GCNs [2,23] for 3D pose regression by analyzing the correlation among the joints according to the semantic correlation among 2D feature. Their form is similar to ours, but we further analyzed the semantic correlation among 3D-3D feature and among 2D-3D feature.

3 Network Architecture

The network architecture is shown in Fig. 2. It consists of a reference 3D regression model and non-autoregressive decoding model based on joint classification (JC-NARD), where the former is a pre-trained 3D human pose regression model and the latter is the main part of our research in this paper. In this paper, the *SemGCN* [23] is chosen as our reference model. Specifically, the decoding model is mainly composed of two residual block, the input of JC-NARD is a set of reference 3D joints feature $X_{ref} \in \mathbb{R}^{J \times 3}$ and 2D joints feature $X_{2D} \in \mathbb{R}^{J \times 128}$ provided by reference model, which will be processed by a novel method of joints sequence analysis. Finally the output is the corresponding 3D coordinates of joints. The model is of the end-to-end training.

4 Joint Classification

In the field of sequential structured output learning, it is essential to model the correlation [19] between outputs feature at N moments before and after as well as to model the correlation between outputs and input feature. For joint sequences without a causal relationship, it is not effective to model the correlation sequentially among previously predicted joints feature and the joint feature to be predicted. Thus, we propose a method to classify joints and create sub-sequences of strongly correlated joints, where we will split joints into J sub-sequences and the jth sub-sequence corresponds to the jth joint to be predicted, in particular, we assume that there is a causal relationship between the strongly correlated joints.

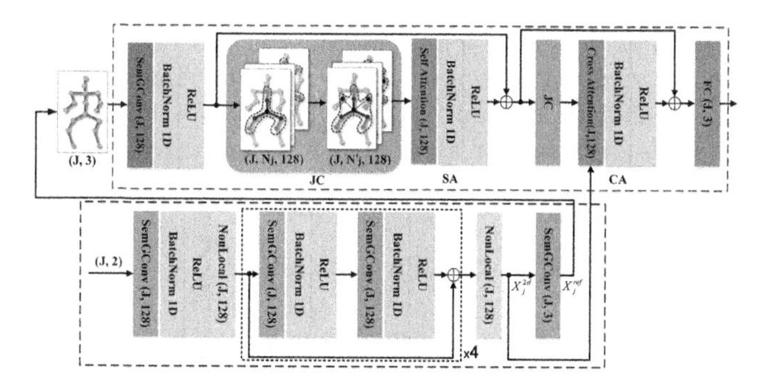

Fig. 2. An example of non-autoregressive decoding network based on joint classification for 3D human pose regression. The model in the red dotted frame is a reference 3D human pose regression model. And the model in blue frame is JC-NARD, which mainly composed of the Joint Classification module (JC), Self-Attention [12,20] module [19,20] (SA) and Cross-Attention [19] module (CA). The function of the JC is to split the joints into multiple joint sub-sequences, where N_j and $N_j^{'}$ are the number of the joints in j^{th} sub-sequence based on connection and semantic correlation respectively. And the latter two modules will model the correlation among 3D-3D joints and among 2D-3D joints feature respectively. (Color figure online)

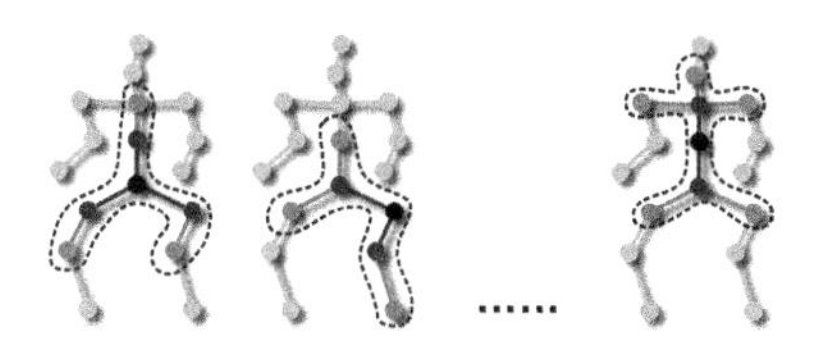

Fig. 3. The red joints is the joints to be studied. According to the connection among joints, the dark green joint is direct correlated to the red one, the light green joints is indirect to the red one, and the gray joint is weakly correlated to the red one (Color figure online)

4.1 Classification Based on Connection

The correlations among joints could be classified by the connection between joints (Fig. 3), including self-correlation, direct-correlation, indirect-correlation and weak-correlation. $A_{self} \in \mathbb{R}^{J \times J}$, $A_{direct} \in \mathbb{R}^{J \times J}$, $A_{indirect} \in \mathbb{R}^{J \times J}$, $A_{weak} \in \mathbb{R}^{J \times J}$ are the adjacency matrices representing the correlations above respectively. Then, we uniformly define the joints with the first three correlations as strong correlated joints with the joints to be predicted. The adjacency matrix representing strong correlation $A_{strong} \in \mathbb{R}^{J \times J}$ could be calculated by Eq. 1:

$$A_{strong} = A_{self} + A_{direct} + A_{indirect} \tag{1}$$

4.2 Global Correlation Based on Multi-head Attention Mechanism

In addition to the joints with above three connections, the other joints might also have strong correlation with the joints to be predicted, as Fig. 4 (Middle) shown. In order to avoid useful joints information missing, we further classify the joints based on semantic correlation among joints. And we introduce multi-head attention mechanism into our model. The joint to be predicted is treated as query $\boldsymbol{q}$ and the other are treat as keys $\boldsymbol{k}$, then a scaled dot-product [19] attention mapping can be computed as:

$$Attention\left(\mathbf{Q},\mathbf{K}\right) = softmax\left(\frac{\mathbf{Q}\mathbf{K}^{\top}}{\sqrt{d_k}}\right) \tag{2}$$

where matrices $\mathbf{Q}$ and $\mathbf{K}$ denote a set of queries and keys respectively, $d_k = 128$ is the dimension of $\mathbf{Q}$ and $\mathbf{K}$. Moreover, to extend the capacity of exploring different sub-spaces, the attention could be extended to case with multi-head:

$$MultiHead\left(\mathbf{Q},\mathbf{K}\right) = softmax\left(Concat(head_1, head_2, \dots, head_h)\right) \tag{3}$$

where $head_i$ is the attention mapping in sub-space i, computed as below:

$$head_i = \frac{\left(\mathbf{Q}\mathbf{W}_i^Q\right)\left(\mathbf{K}\mathbf{W}_i^K\right)}{\sqrt{d_k^h}} \tag{4}$$

where linear transformations W_i^Q and W_i^K are the parameter matrices of sub-space i, h is the number of sub-spaces, and $d_k^h = d_k/4 = 32$ is the dimension of each sub-space in our implementation.

Finally, we would obtain a global semantic correlation matrix $M \in \mathbb{R}^{J\times J}$ among the joints.

4.3 Dynamic Classification Based on Global Correlation

According to the semantic correlation matrix M, we could further classify joints, as Fig. 4 (Right). In particular, the joints which had stronger semantic correlation with the joint to be predicted will be also defined as the joint with direct correlation or the joint with indirect correlation, as Eq. 5:

$$\widetilde{A}_{strong} = A_{strong} + A_{direct}^{extra} + A_{indirect}^{extra} \tag{5}$$

where $\widetilde{A}_{strong} \in \mathbb{R}^{J\times J}$ and $A_{direct(indirect)}^{extra} \in \mathbb{R}^{J\times J}$ are the dynamic strong correlated adjacency matrices and the supplementary adjacency matrix with direct or indirect correlations respectively.

The supplementary adjacency matrix A_{direct}^{extra} is constructed based on the global semantic correlation matrix M. Specifically, each element $A_{direct}^{extra}(i,j)$ conforms a Bernoulli distribution, where corresponding global semantic correlation value $M(i,j)$ serves as probability:

$$\begin{cases} M_{direct} = M \odot A_{direct}; M_{weak} = M \odot A_{weak} \\ m_{direct} = min\left(M_{direct}\right) \\ A_{direct}^{extra}(i,j) \sim B(m_{direct}, M_{weak}) \end{cases} \tag{6}$$

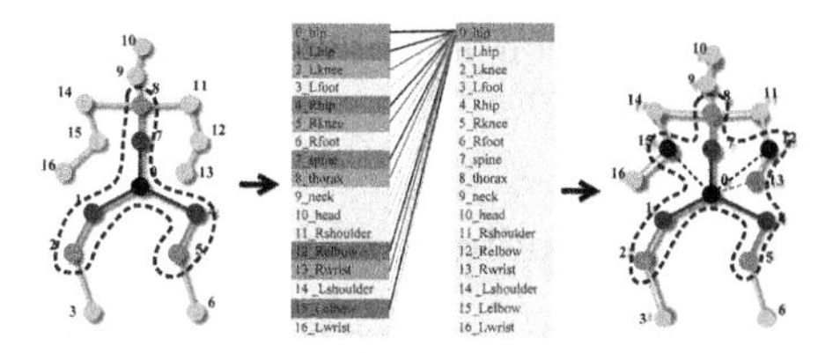

Fig. 4. Take an example. **Left**: joint 0(red) is the joint to be studied. Based on the connection among joints, the sub-sequence consisting of joints 0, 1, 2, 4, 5, 7, 8(green) could be created. **Middle**: By comparing the strength of correlations between joints, it is proved that few joints (blue box) with other connection had stronger semantic correlations with joints 0. **Right**: The joints with stronger semantic correlations (blue) are added to the sub-sequence, and the new sub-sequence consist of joint 0, 1, 2, 4, 5, 7, 8, 12, 13, 15. (Color figure online)

where $M_{direct} \in \mathbb{R}^{J \times J}$ and $M_{weak} \in \mathbb{R}^{J \times J}$ are the semantic correlations matrices of the joints with direct connection and the ones with weak correlations respectively; m_{direct} is the minimum value of M_{direct}; B is Bernoulli distribution. During regression, the A^{extra}_{direct} will change dynamically according to Bernoulli distribution, which means that the classification of joints will also change dynamically. It is the same way for the joints with indirect correlation. It is the same way to obtain $A^{extra}_{indirect}$.

5 Non-autoregressive Decoding Based on 3D Spatial Constraints

5.1 3D Spatial Constrains

In 2D space, the relative coordinate between joints will change as the perspective changes. Thus, it is not sufficient to analyze correlations between joints in 2D space alone. According to the absoluteness of 3D coordinates, we proposed a modeling method based on semantic correlation among 3D feature to make up for the shortcomings of the above methods. As shown in Fig. 2, joints sub-sequences are input into the SA and CA module in turn, and then we will obtain semantic correlation matrix $M_{self} \in \mathbb{R}^{J \times J}$ and $M_{cross} \in \mathbb{R}^{J \times J}$, which denote the semantic correlation among 3D-3D joints feature and among 2D-3D joints feature respectively. That is computed by the element-wise product of $M^{global}_{self(cross)}$ and $\widetilde{A}_{strong}$:

$$\begin{cases} M^{global}_{self(cross)} = MultiHead\left(X_{3D(2D)}, X_{3D(3D)}\right) \\ M_{self(cross)} = M^{global}_{self(cross)} \odot \widetilde{A}_{strong} \end{cases} \tag{7}$$

where $X_{3D} \in \mathbb{R}^{J \times 128}$ denotes the 3D joints feature, $M^{global}_{self(cross)} \in \mathbb{R}^{J \times J}$ denotes global semantic correlation among 3D-3D joints feature (among 2D-3D joints feature), $\odot$ denotes the element-wise product. In particular, the jth row of

$M_{self(cross)}$ denotes the correlation among the joints feature in the jth sub-sequence.

5.2 Non-autoregressive Decoding and Decouple Correlation

Non-autoregressive Decoding. Similar to the regular data sequence, the joints that are next to each other also have strong correlation in sub-sequence. Thus, we assume that there is a causal relationship between joints in each sub-sequence. Specifically, joint j would be treated as the joint to be predicted in present in j^{th} sub-sequence, and the others would be treated as previously predicted joints in first $N_j^{'}$ moments. Finally, 3D predicted joints feature will be output in parallel by adaptively aggregating [2] 2D and 3D joints features, as Eq. 7 shown:

$$X_{out} = \mathop{\|}_{i=1}^{J} \left(\widetilde{M}(i,:) \mathop{\|}_{j=1}^{J} (X_{in}(j,:)W_j) \right) \tag{8}$$

where $X_{out} \in \mathbb{R}^{J\times 128}$ and $X_{in} \in \mathbb{R}^{J\times 128}$ are the representation of joints feature before and after attention module; $W_j \in \mathbb{R}^{128\times 128}$ denote the linear transformation matrix of joint j; $\|$ denotes the channel-wise concatenation. In particular, X_{in} denotes 3D joints feature in SA module, but denotes 2D joints feature in CA module. Similarly, $\widetilde{M}$ denotes $\widetilde{M_{self}}$ in CA module, but denotes $\widetilde{M_{cross}}$ in SA module.

Decouple Correlation. After joint classification, different correlation levels were determined between the other joints and the joints to be studied in the sub-sequences. In following research, we found that joints with different correlation levels had different feature relations with the joints to be predicted in sub-sequence. This motivates us to decouple their feature transformations between the joints with different correlation types and that to be studied:

$$\begin{cases} \widetilde{M_1} = \widetilde{M} \odot A_{self} \\ \widetilde{M_2} = \widetilde{M} \odot (A_{direct} + A_{direct}^{extra}) \\ \widetilde{M_3} = \widetilde{M} \odot (A_{indirect} + A_{indirect}^{extra}) \end{cases} \tag{9}$$

where $\widetilde{M}_{1,2,3}$ denotes the correlation matrix of the joints with self-correlation, direct-correlation and indirect-correlation with the joints to be predicted respectively. Finally, the non-autoregressive decoding by correlation decoupling is shown in below:

$$X_{out} = \sum_{k=1}^{3} \left(\mathop{\|}_{i=1}^{J} \left(\widetilde{M_k}(i,:) \mathop{\|}_{j=1}^{J} (X_{in}(j,:)W_j^k) \right) \right) \tag{10}$$

where W_j^k denotes the linear transformation matrix of joint j under the kth correlation level;

5.3 Loss Function

The loss function is as follows:

$$L = \sum_{j=1}^{J} \|\hat{y}_j - y_j\|_2^2 \tag{11}$$

where $\hat{y}_j$ and y_j denote the 3D regressive result and 3D ground truth of joints j respectively.

6 Experiments

In this section, we firstly introduce the setting of parameters and the detail of model during training and validation, and then conduct an ablation study on components in our method, finally the comparison results are reported with state-of-the-art methods.

6.1 Implementation Details

During training, we employ the Adam as optimization function with initial learning rate 0.004, and use the minimum batch of size 512. The learning rate is dropped with a decay of 0.85 as the loss gets saturated on validation. During model training, we input 3D ground truth into JC-NARD as reference 3D pose for learning correct correlation among joints, and then input the output 3D pose of reference model into JC-NARD as reference 3D pose during validation.

6.2 Dataset and Evaluation Metric

Human3.6M [7] is the currently largest public dataset applied in the field of 3D human pose regression. The dataset contains 3,600,000 sets of indoor human pose data, including image data, 2D pose data and 3D pose data, where seven professional actors offered 15 different daily poses, such as walking, sitting, taking photos, talking on the phone and so on. 2D pose data and 3D pose data are used in our model training. The evaluation metric is the Mean Per Joint Position Error (MPJPE) in millimeters, which is the mean error between groundtruth and the predicted 3D joint coordinates across all cameras and joints after aligning the pre-defined root joints (the pelvis joint). In the following experiments, we applied this metric to evaluate the performance of our model.

6.3 Ablation Study

We conduct ablation study to compare the three joints classification methods and two joints sequence regression methods. To avoid influence of 2D human pose detector, we utilize 2D groundtruth as the input for all models.

Joints Classification Methods: In this experiment, we uniformly utilize the SemGCN model pretrained on the Human3.6m as the reference model. And we perform non-autoregressive decoding for 3D pose based on non-classification strategy, connective classification strategy, correlation classification strategy and mixed classification strategy respectively.

The result is shown in Table 1. We can see that the performance of all models with different joints classification strategy is better than baseline model's, and joints classification strategies could further benefit from decoupling correlations among joints. This suggests that the joints classification strategies can improve performance of 3D human pose regression model effectively, and decoupling correlations in joints classification strategies is important for 3D human pose regression. Thus, we will use decoupled mixed joint classification strategy in the remaining experiments.

Table 1. Ablation study on the impact of decoupling correlation and different classification strategy; Δ denotes the difference of performance between our models and baseline model(SemGCN). All errors are measured in millimeters (mm)

Model	Decouple?	Dimension	MPJPE	Δ
SemGCN (baseline)	No	128	43.8	–
Connectivity classification	No	128	40.82	2.98
Connectivity-classification	**Yes**	128	39.91	**3.89**
Correlation-classification	No	128	41.13	2.65
Correlation-classification	**Yes**	128	40.38	**3.42**
Mixed-classification	No	128	40.93	1.87
Mixed-classification	**Yes**	128	38.99	**4.81**

Regression Methods: In this experiment, we utilize the autoregressive decoding model (ARD) and non-autoregressive decoding model (NARD) to predict the 3D human pose respectively to verify that the non-autoregressive decoding strategy is more effective for the 3D pose regression. Table 2 shows the result. It is obvious that the performance of model with autoregression strategy is much worse than that with non-autoregression strategy, meanwhile the former is further less efficient than the latter.

Table 2. Ablation study on the impact of different regression strategy. All errors are measured in millimeters (mm)

Model	Classification strategy	FPS (single GPU)	MPJPE
ARD	None	7.3	55.32
JC-NARD	Mixed	**31.4**	**38.99**

6.4 Comparison with the State-of-the-Art

Following Zhao et al. [23], we use 2D poses provided by a pre-trained 2D stacked hourglass model detector (SH) for benchmark evaluation. We utilized **FCN** from Martinez et al. [14], **SemGCN** from Zhao et al. [23], and **LCN** from Ci et al. [2] as reference 3D human pose regression model to provide JC-NARD model with a set of 3D reference human pose and 2D joints feature. Meanwhile we use mixed joints classification strategy to classify joints in non-autoregression decoding model. Besides, a dropout with a factor of 0.2 is added after each module in JC-NARD model to avoid overfitting in addition to the setting in the Sect. 5.1.

Table 3. Quantitative comparisons on the Human 3.6M dataset. † denotes that this model is the reference model of our model in following experiments. The bold result denotes the best result in this column. All errors are measured in millimeters (mm)

Model	Direct	Discuss	Eating	Greet	Phone	Photo	Pose	Purch	Sitting	SittingD	Smoke	Wait	WalkD	Walk	WalkT	Avg
Zhou et al. [24]	87.4	109.3	87.1	103.2	116.2	143.3	106.9	99.8	124.5	199.2	107.4	118.1	114.2	79.4	97.7	113
Pavlakos et al.	67.4	71.9	66.7	69.1	72.0	77.0	65.0	68.3	83.7	96.5	71.7	65.8	74.9	59.1	63.2	71.9
Mehta et al.	52.6	64.1	55.2	62.2	71.6	79.5	52.8	68.6	91.8	118.4	65.7	63.5	49.4	76.4	53.5	68.6
Zhou et al. [24]	54.8	60.7	58.2	71.4	62.0	65.5	53.8	55.6	75.2	111.6	64.1	66	51.4	63.2	55.3	64.9
Sun et al. [18]	52.8	54.8	54.2	54.3	61.8	53.1	53.6	71.7	86.7	61.5	62.3	59.1	65.1	49.5	52.4	62.9
Yang et al. [21]	51.5	58.9	50.4	57.0	62.1	65.4	49.8	52.7	69.2	85.2	57.4	58.4	43.6	60.1	47.7	58.6
Pavlakos et al.	48.5	54.4	54.4	52.0	59.4	65.3	49.9	52.9	65.8	71.1	56.6	52.9	60.9	44.7	47.8	56.2
Liu et al.	46.3	54.4	47.3	50.7	55.5	67.1	49.2	46.0	60.4	71.1	51.5	50.1	54.5	40.3	43.7	52.7
Martinez et al. †[14] (**FCN**)	51.8	56.2	58.1	59.0	69.5	78.4	55.2	58.1	74.0	94.6	62.3	59.1	65.1	49.5	52.4	62.9
Zhao et al. †[23] (**SemGCN**)	48.2	60.8	51.8	64.0	64.6	53.6	51.1	67.4	88.7	**57.7**	73.2	65.6	48.9	64.8	51.9	60.8
Ci et al. †[2] (**LCN**)	46.8	52.3	**44.7**	**50.4**	**52.9**	68.9	49.6	**46.4**	**60.2**	78.9	**51.2**	50.0	54.8	40.4	43.3	52.8
Ours (**FCN**)	54.6	60.8	55.1	58.3	62.0	70.8	61.2	56.6	66.1	68.6	59.5	62.0	61.8	52.0	53.9	60.2
Ours (**SemGCN**)	48.7	55.2	55.6	55.8	63.5	75.7	51.7	53.9	72.2	89.7	59.9	57.1	60.0	45.8	48.0	59.6
Ours (**LCN**)	**45.1**	**47.9**	48.5	51.2	57.3	**68.0**	**45.6**	46.7	62.0	80.4	52.1	**49.3**	**53.6**	**40.4**	**41.8**	**52.6**

The result of comparison is shown in Table 3. Compared to the performance of reference model, each model's performance of ours is better than corresponding reference model's. But it can be seen that the performance of the model with **LCN** as the reference model has only a slight improvement. That's because the performance of the **LCN** I built based on [2] is only 54.3mm, far below the performance in [2]. Thus, I infer that the performance of our model could further improve if the performance of **LCN** is same as that in [2]. Even that, the performance of our last model outperform most state-of-the-art model, which prove that our model can improve the performance of 3D human pose regression effectively and it's very generalizable. Figure 5 demonstrates some qualitative results of our model on the Human3.6M dataset.

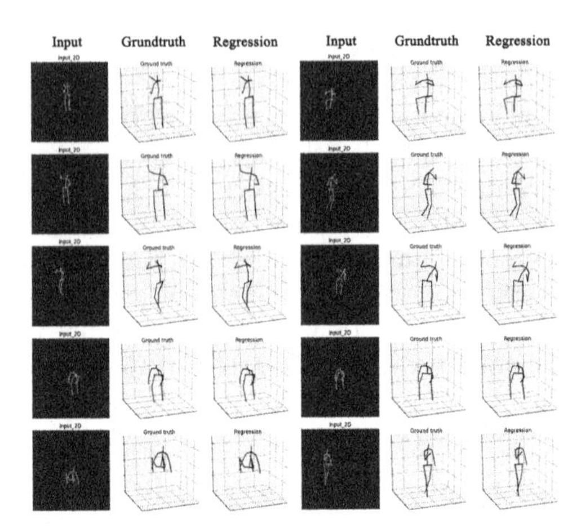

Fig. 5. Qualitative results on Human 3.6M. The column of 'Input' denotes 2D groundtruth pose, the column of 'Groundtruth' denotes the 3D groundtruth pose, and the last column denotes the predicted 3D poses by our model. Comparing the result of last two column, it's obvious that the error is all very small even of occluded joints.

7 Conclusions

In this paper, we propose the non-autoregressive decoding model based on joint classification(JC-NARD) to realize the 3D human pose regression by utilizing the method of sequential analysis. Our model establishes the 3D spatial constraints among joints by modelling the correlation among 2D-3D and among 3D-3D feature to address the limitation of insufficient constraints among 2D feature and realize the parallel regression of 3D pose. The performance of our model exceeds the results of most models on human3.6m dataset. Therefore, it is proved that the JC-NARD model can deal with the limitation of recent researches in this filed effectively and provide a new idea for the research of 3D human pose regression.

References

1. Agarwal, A., Triggs, B.: Recovering 3D human pose from monocular images. IEEE Trans. Pattern Anal. Mach. Intell. **28**(1), 44–58 (2005)
2. Ci, H., Wang, C., Ma, X., Wang, Y.: Optimizing network structure for 3D human pose estimation. In: Proceedings of the IEEE/CVF International Conference on Computer Vision, pp. 2262–2271 (2019)
3. Gori, M., Monfardini, G., Scarselli, F.: A new model for learning in graph domains. In: Proceedings 2005 IEEE International Joint Conference on Neural Networks, 2005. vol. 2, pp. 729–734. IEEE (2005)
4. Gu, J., Bradbury, J., Xiong, C., Li, V.O., Socher, R.: Non-autoregressive neural machine translation. arXiv preprint arXiv:1711.02281 (2017)

5. Huang, L., Tan, J., Liu, J., Yuan, J.: Hand-transformer: non-autoregressive structured modeling for 3D hand pose estimation. In: Vedaldi, A., Bischof, H., Brox, T., Frahm, J.-M. (eds.) ECCV 2020. LNCS, vol. 12370, pp. 17–33. Springer, Cham (2020). https://doi.org/10.1007/978-3-030-58595-2_2
6. Ionescu, C., Li, F., Sminchisescu, C.: Latent structured models for human pose estimation. In: 2011 International Conference on Computer Vision, pp. 2220–2227. IEEE (2011)
7. Ionescu, C., Papava, D., Olaru, V., Sminchisescu, C.: Human3.6M: large scale datasets and predictive methods for 3D human sensing in natural environments. IEEE Trans. Pattern Anal. Mach. Intell. **36**(7), 1325–1339 (2013)
8. Kipf, T.N., Welling, M.: Semi-supervised classification with graph convolutional networks. arXiv preprint arXiv:1609.02907 (2016)
9. Li, C., Lee, G.H.: Generating multiple hypotheses for 3D human pose estimation with mixture density network. In: Proceedings of the IEEE/CVF Conference on Computer Vision and Pattern Recognition, pp. 9887–9895 (2019)
10. Lin, C.H., Yumer, E., Wang, O., Shechtman, E., Lucey, S.: ST-GAN: spatial transformer generative adversarial networks for image compositing. In: Proceedings of the IEEE Conference on Computer Vision and Pattern Recognition, pp. 9455–9464 (2018)
11. Liu, J., Liang, Z., Li, Y., Guan, Y., Rojas, J.: A graph attention spatio-temporal convolutional networks for 3D human pose estimation in video. arXiv e-prints pp. arXiv-2003 (2020)
12. Liu, R., Shen, J., Wang, H., Chen, C., Cheung, S.c., Asari, V.: Attention mechanism exploits temporal contexts: real-time 3D human pose reconstruction. In: Proceedings of the IEEE/CVF Conference on Computer Vision and Pattern Recognition, pp. 5064–5073 (2020)
13. Luvizon, D.C., Picard, D., Tabia, H.: 2D/3D pose estimation and action recognition using multitask deep learning. In: Proceedings of the IEEE Conference on Computer Vision and Pattern Recognition, pp. 5137–5146 (2018)
14. Martinez, J., Hossain, R., Romero, J., Little, J.J.: A simple yet effective baseline for 3D human pose estimation. In: Proceedings of the IEEE International Conference on Computer Vision, pp. 2640–2649 (2017)
15. Qiu, Z., Qiu, K., Fu, J., Fu, D.: DGCN: dynamic graph convolutional network for efficient multi-person pose estimation. In: Proceedings of the AAAI Conference on Artificial Intelligence, vol. 34, pp. 11924–11931 (2020)
16. Scarselli, F., Gori, M., Tsoi, A.C., Hagenbuchner, M., Monfardini, G.: The graph neural network model. IEEE Trans. Neural Networks **20**(1), 61–80 (2008)
17. Sun, X., Shang, J., Liang, S., Wei, Y.: Compositional human pose regression. In: Proceedings of the IEEE International Conference on Computer Vision, pp. 2602–2611 (2017)
18. Sun, X., Xiao, B., Wei, F., Liang, S., Wei, Y.: Integral Human Pose Regression. In: Ferrari, V., Hebert, M., Sminchisescu, C., Weiss, Y. (eds.) ECCV 2018. LNCS, vol. 11210, pp. 536–553. Springer, Cham (2018). https://doi.org/10.1007/978-3-030-01231-1_33
19. Vaswani, A., et al.: Attention is all you need. arXiv preprint arXiv:1706.03762 (2017)
20. Veličković, P., Cucurull, G., Casanova, A., Romero, A., Lio, P., Bengio, Y.: Graph attention networks. arXiv preprint arXiv:1710.10903 (2017)
21. Yang, W., Ouyang, W., Wang, X., Ren, J., Li, H., Wang, X.: 3D human pose estimation in the wild by adversarial learning. In: Proceedings of the IEEE Conference on Computer Vision and Pattern Recognition, pp. 5255–5264 (2018)

22. Yu, C., Ma, X., Ren, J., Zhao, H., Yi, S.: Spatio-temporal graph transformer networks for pedestrian trajectory prediction. In: Vedaldi, A., Bischof, H., Brox, T., Frahm, J.-M. (eds.) ECCV 2020. LNCS, vol. 12357, pp. 507–523. Springer, Cham (2020). https://doi.org/10.1007/978-3-030-58610-2_30
23. Zhao, L., Peng, X., Tian, Y., Kapadia, M., Metaxas, D.N.: Semantic graph convolutional networks for 3D human pose regression. In: Proceedings of the IEEE/CVF Conference on Computer Vision and Pattern Recognition, pp. 3425–3435 (2019)
24. Zhou, X., Huang, Q., Sun, X., Xue, X., Wei, Y.: Towards 3D human pose estimation in the wild: a weakly-supervised approach. In: Proceedings of the IEEE International Conference on Computer Vision, pp. 398–407 (2017)

Multimedia Processing and Analysis

Multiple Semantic Embedding with Graph Convolutional Networks for Multi-Label Image Classification

Tong Zhou and Songhe Feng(✉)

Key Laboratory of Machine Intelligence and Advanced Computing, School of Computer and Information Technology, Beijing Jiaotong University, Beijing, China
{19120457,shfeng}@bjtu.edu.cn

Abstract. Multi-label image classification focuses on predicting a set of object labels presented in an image, which is a practical yet challenging task since labels are normally co-occurred in an image while mutual interactions among labels and the correspondence between a given image and the corresponding labels are rarely considered in the existing methods. To address above challenges, we propose a Multiple Semantic Embedding model with Graph Convolutional Networks (**MSEGCN**) for multi-label image classification by capturing important dependencies related labels. Specifically, the proposed MSEGCN leverages graph structure to guide label co-occurrence propagation among different categories to obtain appropriate label representations. Then, by formulating multi-label classification problem as a label ranking problem with the aid of end-to-end convolutional neural network framework, we focus on learning a transform matrix to seek the image-label relevance relations in embedding space. Furthermore, an adaptive weighting strategy is introduced to effectively improve the classification performance. Experimental studies across a wide range of benchmark datasets show that our method achieves highly competitive performances against other state-of-the-art approaches.

Keywords: Multi-label image classification · Graph convolutional networks · Visual-semantic embedding

1 Introduction

Multi-label image classification (MLIC) is arguably one of the most important problems in computer vision, which focuses on predicting all existing labels in a given image since labels are normally co-occurred. Aside from its usefulness in many downstream tasks, MLIC is an important tool for understanding the underlying semantics of images, as well as a model for categorization in cognitive science. A plethora of MLIC methods have been developed and successfully employed in various fields, including image retrieval [4], scene recognition [18] and etc. Among various MLIC methods, Convolutional Neural Networks (CNNs)

H. Ma et al. (Eds.): PRCV 2021, LNCS 13020, pp. 449–461, 2021.
https://doi.org/10.1007/978-3-030-88007-1_37

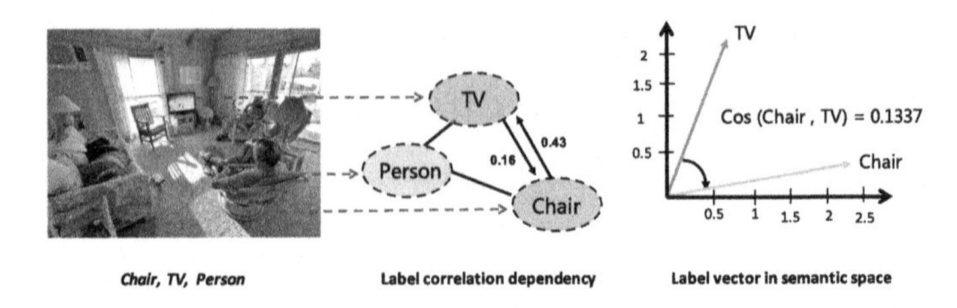

Fig. 1. Illustration of label co-occurrence in multi-label classification problem. This example indicates the co-occurrence of *Chair* and *TV* on the MS-COCO dataset.

has been widely used because of its flexibility. By formulating the state-of-the-art deep network with a n-way softmax output layer, where n is the number of classes, CNNs-based MLIC methods predicts potential confidence of each label and the high label confidence indicates the corresponding object presents on the image. However, the distinction between labels is blurred by the growth of the number of labels, and it is difficult to obtain sufficient numbers of training samples for words which could rarely appear in images [24].

To overcome the aforementioned drawbacks, visual semantic embedding methods attempt to leverage semantic knowledge from text domain to improve image classification performance. Most existing visual semantic embedding methods project the image and labels into the embedding space and their similarities [7] are measured to indicate their label relations to images. By encoding the prior external knowledge into the embedding vectors of labels, semantically similar labels are being close to each other in the new embedding space. Most existing methods only focuses on the semantic relationship between labels, while ignoring the co-occurrence relationships between labels which are not similar in semantic space. Actually, it is desirable to model co-occurrence relationships among labels since they are normally co-occurred in an image. In practice, when people recognize the label of a given image, not only the image feature, but also some commonsense knowledge is used adequately [13]. As illustrated in Fig. 1, the words *Chair* and *TV* are very likely to appear in same image while their cosine similarity is much small.

Inspired by above cognitive approach, we propose a Multiple Semantic Embedding model with Graph Convolutional Networks (MSEGCN) for multi-label image classification by capturing important dependencies related labels for better performance. Specifically, we propose to leverage graph structure to learn improved feature representations for a specific multi-label classification problem. We regard the label semantic representations as the nodes of the graph, and the edges describe the relations between nodes. Thus, global graph structured knowledge is encoded into the label embedding vectors with the aid of Graph Convolutional Networks (GCNs). In this way, our framework not only explores the relationships between images and labels, but also leverages label co-occurrence relationships for guiding the model training to promote the discriminability for multi-label classification.

Meanwhile, by extending semantic embedding methods to multi-label problems, one image can be regarded as a unique channel for building the relationship of labels. Therefore, we convert multi-label classification problem into a label ranking problem. Specifically, we leverage an image to learn a transformation matrix, which can be seen as a joint direction guide for word vector offsets. According to joint direction guide, labels are projected from the label space to the new embedding space which is able to divide the labels into two parts based on their relevances to the input image. The proposed framework has the advantage of applying visual-semantic embedding that effectively utilize the semantic acknowledge to facilitate the classification process, and it also leverages the graph structure to capture and explore the label correlation dependency. Our contributions can be summarized as follows:

- A novel end-to-end trainable multi-label image classification framework is proposed by employing GCNs to serve as the basis for the mapping operations in order to explicit model label correlations and utilizing a joint direction guide to discover the relationship between classes and images.
- An effective adaptive weighted ranking loss function is introduced that is much easier calculate and allows us to achieve a better ranking performance.
- Extensive experiments are conducted on widely-used large-scale multi-label datasets, NUS-WIDE and Microsoft COCO, and results demonstrate the superiority of our proposed method in terms of both accuracy and efficiency over other competitive methods.

2 Related Work

In this section, we briefly introduce some recent developments in multi-label image classification. A naive solution to address multi-label classification problem is to degrade it into single-label classification problem, such as single-label sub-tasks learning [6], single-label multi-instance learning [5] and etc. However, these methods do not model the explicit correlation between labels, thus they lack the capacity to capture the label relations.

In order to overcome this challenge, the most of existing research on multi-label learning tried to capture inter-label correlations. To effectively capture label dependencies, Wang et al. [19] proposed a unified CNN-RNN framework that CNNs focus on learning image representations and then RNNs are introduced to characterize the semantic label dependency and relevance. However, CNN-RNN may ignore the specific associations between semantic labels and the image content [22]. Recently, some work attempted to utilized LSTM to capture the label correlation among different regions [1,22,23]. For example, Chen et al. [1] introduced a framework through recurrently discovering attentional regions to model their global dependencies to capture label correlation. Yazici et al. [23] proposed two alternative losses which adaptively rank the labels based on the prediction of the LSTM model to prevent duplicate generation problem. Furthermore, attention modeling provided powerful tools to address the multi-label classification task. Huynh et al. [10] presented a shared multi-attention mechanism which guides the attention to focus on diverse and relevant image regions. On the other hand, some work utilized ranking loss to discovery the pairwise relations among labels (e.g., the order between a relevant label and an irrelevant label). Gong et al. [8] investigated several different ranking-based loss functions, and found that the weighted approximated-ranking loss is particularly suitable for multi-label classification problems. Yeh et al. [24] aimed to model complex image-to-label dependencies via a simple infrastructure, and related labels and irrelevant labels are separated by a transformation.

Additionally, graph structure was also widely applied to discover the label correlation multi-label image classification due to its flexible and effective local structure capacity. Lee et al. [12] attempted to describe the label correlations by incorporating a knowledge graph in the label space. Chua et al. [3] modeled label relationships by graph convolution network. Wang et al. [21] proposed a label graph superimposing framework to improve the graph structure developed for multi-label classification. Chen et al. [2] incorporated a statistical label occurrence graph, where each node is represented by a category-related feature vector. And it adopted a gated recurrent update mechanism to propagate message through the graph. In this paper, we propose a new multiple embedding framework with the aid of graph structure among labels that incorporates the label correlation dependency to propagate both visual and semantic space information to effectively improve the multi-label image classification performance.

3 The Proposed Method

In this section, we present the Multiple Semantic Embedding model with Graph Convolutional Networks, MSEGCN, which consists of direction-guided feature embedding learning, GCNs-based semantic embedding learning and adaptive weighted label ranking to accomplish multi-label image classification. The overall framework is given in Fig. 2.

3.1 Notations and Problem Formulation

Before we present our method, we first introduce some notations and basic concepts. The goal of the multi-label image classification is to assign relevant labels to an image. Accordingly, we use I to denote an input image with ground-truth labels $y = [y_1, y_2, ..., y_C]$, where y_i is a binary indicator. $y_i = 1$ indicates the presence of the label j in the image I, $y_i = 0$ otherwise. C is the number of all possible labels in the dataset. To brevity, we overload the notations y_i and $\bar{y}_j$ to respectively denote the relevant label and the irrelevant label of the training image I.

3.2 Direction-Guided Feature Embedding Learning

In our approach for multi-label image classification, we convert multi-label classification problem into a label ranking problem and we hope that the positive

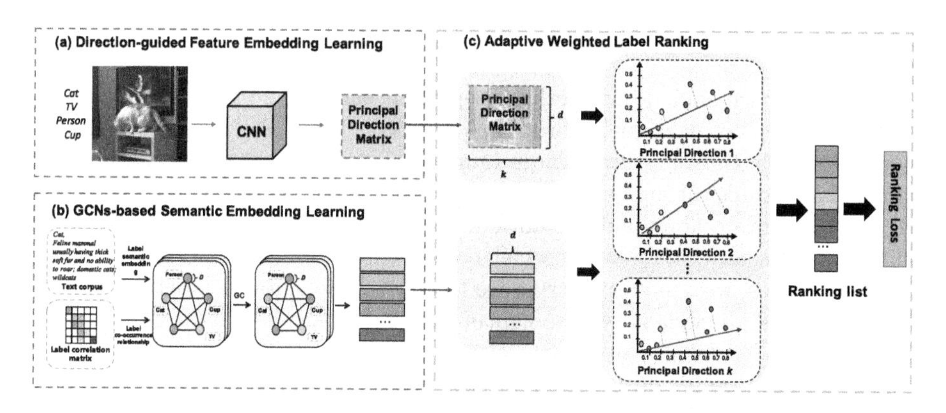

Fig. 2. The framework of the proposed MSEGCN model, which is composed of three key components: (a) Direction-guided feature embedding learning aims to get the label principal direction matrix, (b) GCNs-based semantic embedding learning focuses to learning the high quality d-dim semantic representation of labels, and (c) Adaptive weighted label ranking works to linearly map the learned label representations to the new multidimensional embedding space by exploiting the principal direction matrix.

labels will always be ranked higher than the negative labels. And inspired by [25], we propose to solve the multi-label image classification problem by estimating the principal direction, along which an image divides all labels into two parts according to image-dependent relevance. As the multi-label ranking problem predicts the labels in a sequential fashion, it is necessary to consider both ranking list of all labels and the margin between positive and negative labels when performing image label prediction. In order to make the margin as large as possible, the intuition is that it is easier to find the optimal direction guidance using a linear combination of multiple principal directions than a single principal direction.

Different from existing deep learning methods focus on learning single image feature representation, our method learns a principal direction matrix $M \in \mathbb{R}^{k \times d}$, which aims to separate the relevant and irrelevant labels of the given image. Specifically, we obtain the mapping matrix M by feeding the image into the convolutional neural network. We aim to derive the transformation matrix M by feeding an image I into the CNNs model $f_\theta(\cdot)$:

$$f_\theta(I) = M \in \mathbb{R}^{k \times d} \tag{1}$$

where I is the input image, matrix M can be regarded as a linear mapping from a d-dim feature space to a k-dim new multidimensional embedding space. This implies that the relevant label y_i rank ahead of the irrelevant label $\bar{y}_j$ in new space under the joint guidance of k principal directions, i.e.,

$$\|My_i\|_2 < \|M\bar{y}_j\|_2 \tag{2}$$

3.3 GCNs-Based Semantic Embedding Learning

General semantic embedding methods represent each label of the word space as an embedding. For example, Word2Vec [15] learns word distributed representations for labels that capture a large number of precise syntactic and semantic word relationships through a shallow neural network. The word embeddings from GloVe [16] contain abundant semantic knowledge, but it is difficult to learn the suitable representation for words which could be less common in text data. Based on above issues, we use graph structure to learn more suitable label representations, which is an efficient way to capture the topological structure in the label space.

Specifically, we introduce a simple and well-behaved layer-wise propagation rule for neural network models which operate directly on graphs [11]. By modeling the correlation dependencies between labels through a graph $\mathcal{G} = \{X, A\}$, GCNs [11] takes features $X = H^{(0)}$ and adjacent matrix A as inputs, then it conducts layer-wise propagation in hidden layers as follows:

$$H^{(l+1)} = \sigma(\tilde{D}^{-\frac{1}{2}} \tilde{A} \tilde{D}^{-\frac{1}{2}} H^{(l)} W^{(l)}) \tag{3}$$

where $\tilde{A} \in \mathbb{R}^{C \times C}$ is the adjacent matrix A of graph $\mathcal{G}$ with added self-connections (where C denotes the number of nodes). D is a diagonal matrix with $\tilde{D}_{ii} = \sum_j \tilde{A}_{ij}$ and $W^{(l)}$ is a layer-trainable weight matrix to be learned. $H^{(l+1)}$ denotes the node feature of updates in the l-th layer and $\sigma(\cdot)$ denotes an activation function, such as the $\mathrm{ReLU}(\cdot) = \max(0, \cdot)$.

We introduce two-layer GCNs network into our label representations learning branch for better label representations that absorb the information of its neighbors. Then, the layer-wise graph propagation in Eq. (3) can be improved by:

$$H^{(1)} = \sigma(\tilde{D}^{-\frac{1}{2}} \tilde{A} \tilde{D}^{-\frac{1}{2}} X W^{(l)}) \tag{4}$$

$$H^{(2)} = \sigma(\tilde{D}^{-\frac{1}{2}} \tilde{A} \tilde{D}^{-\frac{1}{2}} H^{(1)} W^{(l)}) \tag{5}$$

$$V = \mu(H^{(2)}) \tag{6}$$

where $\mu(\cdot)$ is the normalize operator, $V = [v_1, ..., v_C] \in \mathbb{R}^{d \times C}$ indicates the final label representations obtained by the two-layer graph convolutions. We utilize GloVe [16] to obtain the initial label representations, which is denoted as X. A denotes the corresponding adjacent matrix among labels, $H^{(1)} \in \mathbb{R}^{d' \times C}$ and $H^{(2)} \in \mathbb{R}^{d \times C}$ indicate the hidden representations of the labels which are obtained from the first graph convolutional layer and the second convolutional layer.

In contrast with existing methods which focus on considering the relationship between labels and images, our method exploits the correlations between labels and images while also taking the inter dependencies among labels into consideration. More formally, we adopt GloVe [16] trained on Wikipedia corpus as the initialization of label representation X, then we obtain the new label representation v_i through two-layer GCNs network and embedding the output of the last layer from GCNs into the k-dim new multidimensional embedding space. Thus, the distance metric in Eq. (2) can be written as:

$$\|Mv_i\|_2 < \|M\bar{v}_j\|_2 \tag{7}$$

In this multidimensional embedding space, a distance metric is applied on these embeddings to discover relevancy between an image and labels such that labels with high relevancy could be closer to the original one, while those with weak relevancy could be moved far from the origin one.

3.4 Adaptive Weighted Label Ranking

To form a more rational ranking list, existing methods utilize the log-sum-exp pairwise function to optimize the ranking in the result list. However, these methods share the same drawback that the process of sampling over the positive-negative pairs may lead the model confront with large computational quantity, especially when the number of categories is large. Thus, we propose a novel loss function which considers the distance between each label vector and the origin separately to avoid the issue of sampling over the positive-negative pairs. Inspired by [20], we adopt an adaptive weighted strategy which adaptively gives larger penalties to false prediction of ranking positive labels. For example, if an image has m ground truth labels, the predicted rank of each positive label should be within the top-m. The adaptive weight gives a greater penalty if the constraint is not satisfied. The label ranking loss function is formally defined as follows:

$$\mathcal{L}_{lr} = \sum_{i \in S} f\left(\left\lceil \frac{R_i}{|S|} \right\rceil\right) \|Mv_i\|_2^2 + \sum_{j \in \bar{S}} [1 - \|Mv_j\|_2^2]_+ \tag{8}$$

where $\lceil \cdot \rceil$ is the ceiling function, R_i is the rank of the positive label i over all classes given by $\|Mw^i\|$. $[\cdot]$ is the hinge function $\max(0, \cdot)$ and S indicates the related label set, where$|S|$ denotes its cardinality, $\bar{S}$ denotes the irrelated label set. $f(T) = \sum_{t=1}^{T} \frac{1}{t}$ is a re-weighting function.

This formulation encourages the model to directly consider that the relevant label should be close to the origin one, and the irrelevant label should be apart from the origin one. Equation (7) considers the relationship of labels and the image by the distance between the label vector and the origin. However, one significant limitation of this loss is that the relationships among labels are not constrained from the perspective of ranking. More specifically, we constrain the order of labels so that the distance from the positive label vector to the origin should be smaller than the distance from the negative label vector to the origin. In order to make the margin between relevant and irrelevant labels as large as possible, we restrict the maximum distance of relevant label vector and the minimum distance of irrelevant label vector. The formulation is defined as follows:

$$\mathcal{L}_{lc} = [1 + (\max \|Mv_i\|_2^2 - \min \|Mv_j\|_2^2)]_+ \tag{9}$$

By considering above label ranking loss $\mathcal{L}_{lr}$ and label constraint term $\mathcal{L}_{lc}$, the overall objective of MSEGCN is given by

$$\mathcal{L} = \mathcal{L}_{lr} + \alpha \mathcal{L}_{lc} \tag{10}$$

where α is a trade-off parameter. To optimize the objective function, we use mini-batch stochastic gradient descent as the optimizer. Dropout is an effective solution to prevent neural networks from overfitting. We propose to employ two widely used dropout methods: message dropout and node dropout in our model.

4 Experiments

In this section, we evaluate the proposed MSEGCN in terms of both accuracy and efficiency over other competitive stat-of-the-art baselines.

4.1 Experimental Setup

Datasets. Two widely-used datasets, Microsoft COCO [14] and NUS-WIDE [5], are used to evaluate the model efficiency. Microsoft COCO is a large scale benchmark dataset for several vision task. There are in total 123,287 images for training and validation with 80 object concepts annotated and each image contains approximately 2.9 labels on average. The NUS-WIDE [5] is another popular dataset for multi-label classification. It contains 209,347 images from 81 object categories. After removing the non-annotated image, the training set and the test set contain 125,449 and 83,898 images respectively.

Metrics. Following the previous works [3,24], We use the average of overall/per-class precision (OP/CP), overall/per-class recall (OR/CR), and overall/per-class F1-score (OF1/CF1) as our evaluation metrics. The metrics CP, CR, CF1, OP, OR, OF1 are defined as:

$$\begin{aligned} CP &= \frac{1}{C}\sum_i \frac{N_i^c}{N_i^p}, \quad CR = \frac{1}{C}\sum_i \frac{N_i^c}{N_i^g}, \quad CF1 = \frac{2 \times CP \times CR}{CP + CR}, \\ OP &= \frac{\sum_i N_i^c}{\sum_i N_i^p}, \quad OR = \frac{\sum_i N_i^c}{\sum_i N_i^g}, \quad OF1 = \frac{2 \times OP \times OR}{OP + OR}, \end{aligned} \tag{11}$$

Implementation Details. In our experiments, the input images are randomly cropped into 488×488 with random horizontal flips for data augmentation. We select ResNet-101 [9] as the backbone of our proposed method. The CNNs is first pre-trained on the ImageNet [6] and further fine-tuned on the multi-label classification dataset. In addition, the GCNs of our model built from two graph convolution layers with output dimensionality of 1024 and 300. We use LeakyReLU with the negative slope of 0.2 as the non-linear activation function for graph convolution layers. We adopt 300-dim GloVe [16] trained on WiKipedia dataset to initialize the label representations. We train our model using stochastic gradient descent (SGD) with momentum of 0.9. The initial learning rate is 10^{-6}, and it is divided by 10 every 10 epochs until 60, the total number of training epochs. The hyperparameter α of Eq. (10) is set to be 0.05, in all of the experiments. We implement the network based on PyTorch.

Table 1. Performance comparisons (%) on the MS-COCO dataset.

Method	MS-COCO					
	CP	CR	CF1	OP	OR	OF1
WARP [8]	55.5	57.4	54.8	59.6	61.5	–
CNN-RNN [19]	66.0	55.6	60.4	69.2	**66.4**	67.8
MDVSE [24]	63.0	61.0	62.0	62.9	65.0	63.0
MESGCN(ours)	**68.8**	**61.3**	**64.2**	**74.2**	64.7	**69.1**

Table 2. Performance comparisons (%) on the NUS-WIDE dataset.

Method	Top-3						Top-5					
	CP	CR	CF1	OP	OR	OF1	CP	CR	CF1	OP	OR	OF1
WARP [8]	31.7	35.6	33.5	48.6	60.5	53.9	22.3	52.0	31.2	36.2	75.0	48.8
CNN-RNN [19]	40.5	30.4	34.7	49.9	61.7	55.2	–	–	–	–	–	–
MIE [17]	37.7	40.2	38.9	52.2	65.0	57.9	28.3	59.8	38.4	39.0	80.9	52.6
MDVISE [24]	**46.6**	**56.6**	**51.1**	55.2	**68.6**	61.2	33.7	**72.9**	46.1	40.6	**84.2**	54.8
MSEGCN(ours)	42.3	42.5	42.4	**63.8**	67.2	**65.5**	**37.5**	60.1	**46.2**	**55.6**	80.8	**66.1**

4.2 Experiment Results

Results on MS-COCO. As shown in Table 1, MSEGCN significantly outperforms all comparing methods on MS-COCO dataset. Especially, when evaluating with respect to the class precision of first three ranked tags, we see our method yields around 5.8% and 2.8% improvement over MDVSE and CNN-RNN on MS-COCO dataset, respectively. In addition, the proposed method also outperforms WARP. The results demonstrate that the proposed method has superior capacity of classification against other methods based on semantic-visual embedding strategy. We attribute such success to that it can utilize the label co-occurrence to obtain the more suitable label representations.

Results on NUS-WIDE. Table 2 reports comparison results for the proposed algorithm and four baselines on NUS-WIDE dataset. We can see that MSEGCN also significantly outperforms most baselines on NUS-WIDE dataset. When evaluating with respect to the first three ranked tags, the overall precision of MSEGCN is 8.6% higher than MDVSE and 11.6% higher than MIE on

NUS-WIDE dataset. When compared with the CNN-RNN and WARP, it also can achieve 13.9% and 15.2% higher performance on the NUS-WIDE dataset, respectively. The experimental results prove that the proposed model is effective for label ranking. Figure 3 provides examples of top-5 annotations generated by MSEGCN and MDVSE for the NUS-WIDE dataset.

Table 3. Comparisons(%) with different depths of GCNs in our model

#Layer	MS-COCO					
	CP	CR	CF1	OP	OR	OF1
2-layer	68.8	**61.3**	**64.2**	74.2	**64.7**	**69.1**
3-layer	69.3	58.4	63.4	75.3	63.6	68.9
4-layer	**69.5**	57.1	62.7	**75.9**	62.5	68.5

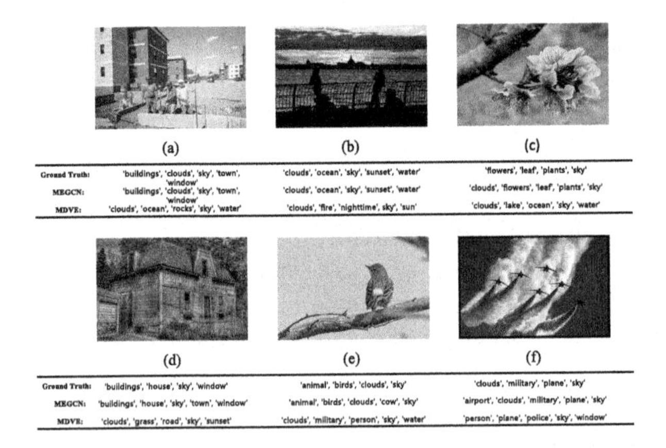

Fig. 3. Examples of test images from NUS-WIDE dataset with top-5 annotations generated by MSEGCN and MDVSE. (Correct labels are in blue, and mistaken labels are in red.) (Color figure online)

4.3 Ablation Experiment

We conduct ablation studies on MS-COCO dataset to further investigate the model behavior.

The Number of GCN Layers. To explore the effectiveness of different number of GCN layers, we conduct experiments by applying GCN with different number of layers on the MS-COCO dataset. Three types (2-layer, 3-layer and 4-layer) are used in our experiments. Specifically, for the 3-layer GCNs model, we set the output dimensionality of each layer to 1024, 512 and 300, respectively. For the 4-layer GCNs model, the output dimensions of each layer are 2048, 1024,

512 and 300 respectively. As shown in Table 3, the model with fewer layers has better ability to explore the local graph structure. More prior knowledge can be added into the framework as the number of GCN layers tends to be small, while less prior knowledge contributes to the learning process when the number of GCN layers becomes larger. We conjecture that repeatedly applying Laplacian smoothing may mix the feature of labels from different clusters and make them indistinguishable.

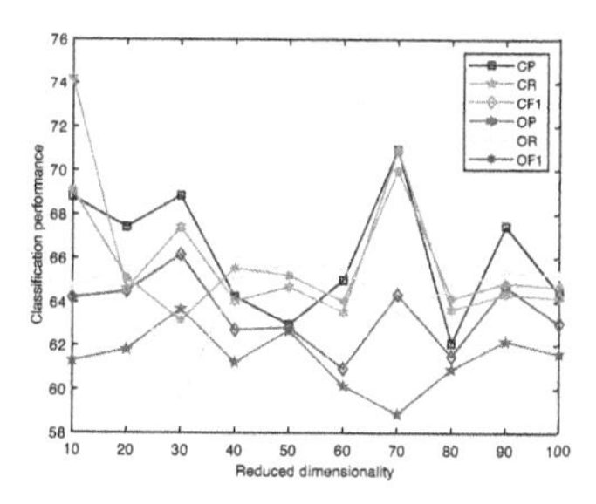

Fig. 4. Effect of transformed dimension parameter k

The Transformed Dimension Parameter k. The transformed dimension parameter controls the number of principal directions, which plays a key role on the effectiveness of prediction algorithm. Figure 4 illustrates the performance of MSEGCN on MS-COCO dataset, as the number of transformed dimension parameter k increases from 10 to 100 with the step-size 10. We can see that the proposed method is more resilient to the value of k. We further employ $k = 1$ to evaluate the effectiveness of joint direction guidance. As shown, the experimental results show that the per-class precision is dropped by 6.3% without considering multiple principal directions.

5 Conclusion

In this paper, we propose a MSEGCN model for multi-label image classification, which employs GCN to capture the label correlations and utilizes a joint direction guide to discover the relationship between labels and images. For efficient model training, we propose an adaptive weighted loss function which is much easier to calculate than existing general ranking loss functions. The extensive experimental results on MS-COCO and NUS-WIDE datasets show our model outperforms all baselines.

Acknowledgement. This work was supported in part by the National Natural Science Foundation of China (No. 61872032) and the Beijing Natural Science Foundation (No. 4202058).

References

1. Chen, T., Wang, Z., Li, G., Lin, L.: Recurrent attentional reinforcement learning for multi-label image recognition. In: AAAI, pp. 6730–6737 (2018)
2. Chen, T., Xu, M., Hui, X., Wu, H., Lin, L.: Learning semantic-specific graph representation for multi-label image recognition. In: ICCV, pp. 522–531 (2019)
3. Chen, Z., Wei, X., Wang, P., Guo, Y.: Multi-label image recognition with graph convolutional networks. In: CVPR, pp. 5177–5186 (2019)
4. Chua, T., Pung, H., Lu, G., Jong, H.: A concept-based image retrieval system. In: HICSS, pp. 590–598 (1994)
5. Chua, T., Tang, J., Hong, R., Li, H., Luo, Z., Zheng, Y.: Nus-wide: a real-world web image database from national university of Singapore. In: CIVR, pp. 1–9 (2009)
6. Deng, J., Dong, W., Socher, R., Li, L., Li, K., Li, F.: Imagenet: a large-scale hierarchical image database. In: CVPR, pp. 248–255 (2009)
7. Frome, A., et al.: Devise: a deep visual-semantic embedding model. In: NIPS, pp. 2121–2129 (2013)
8. Gong, Y., Jia, Y., Leung, T., Toshev, A., Ioffe, S.: Deep convolutional ranking for multilabel image annotation. In: ICLR (2013)
9. He, K., Zhang, X., Ren, S., Sun, J.: Deep residual learning for image recognition. In: CVPR, pp. 770–778 (2016)
10. Huynh, D., Elhamifar, E.: A shared multi-attention framework for multi-label zero-shot learning. In: CVPR, pp. 8773–8783 (2020)
11. Kipf, T., Welling, M.: Semi-supervised classification with graph convolutional networks. arXiv preprint arXiv:1609.02907 (2016)
12. Lee, C., Fang, W., Yeh, C., Wang, Y.: Multi-label zero-shot learning with structured knowledge graphs. In: CVPR, pp. 1576–1585 (2018)
13. Li, Y., Song, Y., Luo, J.: Improving pairwise ranking for multi-label image classification. In: CVPR, pp. 1837–1845 (2017)
14. Lin, T.-Y., et al.: Microsoft COCO: common objects in context. In: Fleet, D., Pajdla, T., Schiele, B., Tuytelaars, T. (eds.) ECCV 2014. LNCS, vol. 8693, pp. 740–755. Springer, Cham (2014). https://doi.org/10.1007/978-3-319-10602-1_48
15. Mikolov, T., Sutskever, I., Chen, K., Corrado, G., Dean, J.: Distributed representations of words and phrases and their compositionality. In: NIPS 26, pp. 3111–3119 (2013)
16. Pennington, J., Socher, R., Manning, C.: Glove: global vectors for word representation. In: EMNLP, pp. 1532–1543 (2014)
17. Ren, Z., Jin, H., Lin, Z., Fang, C., Yuille, A.: Multiple instance visual-semantic embedding. In: British Machine Vision Conference (2017)
18. Shao, J., Loy, C., Kang, K., Wang, X.: Slicing convolutional neural network for crowd video understanding. In: CVPR, pp. 5620–5628 (2016)
19. Wang, J., Yang, Y., Mao, J., Huang, Z., Huang, C., Xu, W.: CNN-RNN: a unified framework for multi-label image classification. In: CVPR, pp. 2285–2294 (2016)
20. Wang, M., Luo, C., Hong, R., Tang, J., Feng, J.: Beyond object proposals: random crop pooling for multi-label image recognition. IEEE Trans. Image Process. **25**(12), 5678–5688 (2016)
21. Wang, Y., et al.: Multi-label classification with label graph superimposing. In: AAAI, pp. 12265–12272 (2020)
22. Wang, Z., Chen, T., Li, G., Xu, R., Lin, L.: Multi-label image recognition by recurrently discovering attentional regions. In: ICCV, pp. 464–472 (2017)

23. Yazici, V., Garcia, A., Ramisa, A., Twardowski, B., Weijer, J.: Orderless recurrent models for multi-label classification. In: CVPR, pp. 13437–13446 (2020)
24. Yeh, M., Li, Y.: Multilabel deep visual-semantic embedding. IEEE Trans. Pattern Anal. Mach. Intell. **42**(6), 1530–1536 (2019)
25. Zhang, Y., Gong, B., Shah, M.: Fast zero-shot image tagging. In: CVPR, pp. 5985–5994 (2016)

AMEN: Adversarial Multi-space Embedding Network for Text-Based Person Re-identification

Zijie Wang[1], Jingyi Xue[1], Aichun Zhu[1], Yifeng Li[1(✉)], Mingyi Zhang[1], and Chongliang Zhong[2]

[1] School of Computer Science and Technology, Nanjing Tech University, Nanjing, China
[2] ZKTeco Co., Ltd., Dongguan, China

Abstract. Many of the existing methods manage to extract modality-invariant features from both modalities by learning a joint latent space, in which the visual/textual feature vector can be better aligned. However, though misaligned information can be removed when mapping features from two high-dimensional spaces into a common space, discriminative clues as well may be lost. To this end, merely embedding features into a joint latent space may not be sufficient to give satisfactory performance, and the utilization of both visual and textual high-level spaces deserves more in-depth exploration. In this paper, we proposed a novel Adversarial Multi-space Embedding Network (AMEN) to learn and match embeddings in multiple spaces. Following an encoder-decoder manner, the inter-modal reconstruction paradigm works in concert with the intra-modal reconstruction paradigm to properly embedding a feature into the opposite modality space while learning a strong common space. A consistency constraint is adapted to ensure that the learned visual and textual spaces are trained jointly and work consistently. To enhance both the common space learning and feature reconstruction, the adversarial mechanism is utilized. Our proposed AMEN is evaluated on CUHK-PEDES, which is currently only accessible dataset for text-base person re-identification task. Extensive experimental results demonstrate that AMEN achieves the state-of-the-art performance.

Keywords: Multi-space embedding · Text-based person re-identification · Cross-modal retrieval · Adversarial learning

1 Introduction

Person re-identification aims to search for the corresponding pedestrian image according to a given query, which can be an image, a video, a set of attributes or a text description. Considering that in most of the scenes, text descriptions of a target person are much more accessible than any other type of queries,

Z. Wang–This is a student paper.

H. Ma et al. (Eds.): PRCV 2021, LNCS 13020, pp. 462–473, 2021.
https://doi.org/10.1007/978-3-030-88007-1_38

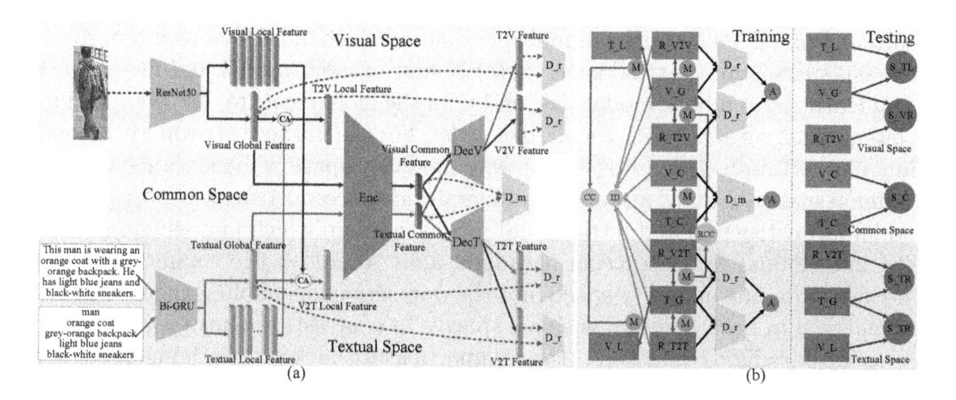

Fig. 1. The overall architecture of our proposed Adversarial Multi-space Embedding Network (AMEN) is illustrated in (a). Following an encoder-decoder manner, three spaces including a common space, a visual space and a textual space are learned to complement each other. Inter-modal (which forms $V2T$ and $T2V$ features) and intra-modal (which forms $V2V$ and $T2T$ features) reconstruction paradigms are utilized to realize the cross modal embedding while learning a stronger common space. Losses for training and similarities for testing AMEN are shown in (b). ID, M, A, CC and RCC denote the proposed ID loss, triplet ranking loss, adversarial loss, consistency constraint and reconstruction consistency constraint, respectively. S_C, S_{TL}, S_{VL}, S_{VR} and S_{TR} are the five corresponding cosine similarities employed to test AMEN.

text-based person re-identification [7,10,11,13,16,17,23] has drawn remarkable attention. The task of text-based person re-identification involves handling multimedia data, which can be regarded as a subtask of cross-modal retrieval [8,9,14,15,20,25]. Instead of containing various categories of objects in an image, however, each image cared by text-based person re-identification task only contains one pedestrian while its corresponding text description offers more clues. This particularity of the text-based person re-identification task causes that many previous methods proposed on common cross-modal retrieval benchmarks (e.g. MSCOCO [12] and Flickr30k [18]) generalize poorly on it.

The main challenge of text-based person re-identification is to effectively extract and match feature vectors from both visual and textual modalities. Many of the existing methods [13,16] manage to extract modality-invariant features from both modalities by learning a joint latent space, in which the visual/textual feature vector can be better aligned. Intuitively, bringing feature vectors into a specific space can be conducive to the following matching process, hence striving to learn a stronger joint latent space makes sense. However, though misaligned information can be removed when mapping features from two high-dimensional spaces into a common space, discriminative clues as well may be lost. To this end, merely embedding features into a joint latent space may not be sufficient to give satisfactory performance, and the utilization of both visual and textual high-level spaces deserves more in-depth exploration.

In this paper, we proposed a novel Adversarial Multi-space Embedding Network (AMEN) (shown in Fig. 1), which follows an encoder-decoder manner to learn and match embeddings in multiple spaces, which includes a common space, a visual space and a textual space. AMEN first extracts global and fine-grained local features from both modalities, and then the global features are mapped into a latent common space with a shared encoder. For the purpose of properly embedding one certain feature into the opposite modality space while learning a strong common space, AMEN contains two different ways of reconstruction, namely inter-modal reconstruction and intra-modal reconstruction, which play different roles. The inter-modal reconstruction paradigm aims to embed the common features encoded from one modality space into the opposite one, enabling the features to be matched in both high-dimensional spaces. In contrast, the intra-modal reconstruction paradigm reconstructs the common feature back to the original modality. By minimizing the differentiate between the original and reconstructed features, a stronger common space can be learned. To adequately exploit fine-grained clues, a cross-modal attention (CA) mechanism [17,23] is utilized to match a local feature matrix from one modality space with the global feature in the other. In the mean time, when performing visual-to-textual (V2T) and textual-to-visual (T2V) embeddings simultaneously, it is crucial that the two high-level spaces are learned consistently and jointly. Thus, we introduce a consistency constraint into the training process of AMEN to avoid the situation where the visual and textual spaces develop and work independently, or even oppositely. Moreover, we utilize adversarial mechanism to enhance the performance of AMEN. A modality discriminator is used to determine whether a feature in the common space is encoded from the visual or textual modality. Meanwhile, a reconstruction discriminator is proposed to distinguish reconstructed features from original ones. As long as AMEN deceives the discriminators successfully, much more discriminative feature vectors can be extracted and generated.

The main contributions can be summarized as fourfold: (1) We proposed a novel Adversarial Multi-space Embedding Network (AMEN) to learn and match embeddings in multiply spaces. Following an encoder-decoder manner, the inter-modal reconstruction paradigm works in concert with the intra-modal one to properly embedding a feature into the opposite modality space while learning a strong common space. (2) We adapt a consistency constraint to ensure that the learned visual and textual spaces are trained jointly and work consistently. (3) We utilize adversarial mechanism to enhance both the common space learning and feature reconstruction. (4) We evaluate our proposed AMEN on CUHK-PEDES [11], which is currently only accessible dataset for the text-base person re-identification task. Extensive experimental results demonstrate that AMEN outperforms previous methods and achieves the state-of-the-art performance on CUHK-PEDES.

2 Related Works

2.1 Person Re-identification

Person re-identification has drawn increasing attention in both academical and industrial fields. This technology addresses the problem of matching pedestrian images across disjoint cameras. The key challenges lie in the large intra-class and small inter-class variation caused by different views, poses, illuminations, and occlusions. Existing methods can be grouped into handed-crafted descriptors, metric learning methods and deep learning methods, and deep learning methods generally plays a major role in current state-of-the-art works. Yi et al. [26] firstly proposed deep learning methods to match people with the same identification. Hou et al. [6] proposed an Interaction-and-Aggregation (IA) Block, which consists of a Spatial Interaction-and-Aggregation (SIA) Module and a Channel Interaction-and-Aggregation (CIA) Module to strengthen the representation capability of the deep neural network. Xia et al. [24] proposed the Second-order Non-local Attention (SONA) Module to learn local/non-local information in a more end-to-end way. Sun et al. [21] proposed a visibility-aware part model to significantly improve the learned representation and the achieving accuracy by considering a few parts of the Re-ID scenes combined with the self-supervising model of some feature observations to perceive the visibility of the region.

2.2 Text-Based Person Re-identification

Text-based person re-identification searches for the corresponding pedestrian image according to a given text query. This task is first put forward by Li et al. [11] and they further take an LSTM to handle the input image and text. An efficient patch-word matching model [3] is proposed to capture the local similarity between image and text. Jing et al. [7] utilize pose information as soft attention to localize the discriminative regions. Niu et al. [17] propose a Multi-granularity Image-text Alignments (MIA) model exploit the combination of multiple granularities. Nikolaos et al. [16] propose a Text-Image Modality Adversarial Matching approach (TIMAM) to learn modality-invariant feature representation by means of adversarial and cross-modal matching objectives. Besides that, in order to better extract word embeddings, they employ the pre-trained publicly-available language model BERT. Wang et al. [23] proposed an IMG-Net model to incorporate inner-modal self-attention and cross-modal hard-region attention with the fine-grained model for extracting the multi-granular semantic information. Liu et al. [13] generate fine-grained structured representations from images and texts of pedestrians with an A-GANet model to exploit semantic scene graphs.

3 Methodology

In this section, we describe the proposed Adversarial Multi-space Embedding Network (AMEN) in detail (shown in Fig. 1). First, we explain how the local and global visual/textual features are extracted in Sect. 3.1. Then we introduce the two reconstruction paradigms which are adopted to reconstruct features back into high-level spaces in Sect. 3.2. The proposed loss functions and training strategy are detailed in Sect. 3.3.

3.1 Feature Extraction

A ResNet-50 [5] backbone pretrained on ImageNet is utilized to extract global/local visual features. Given an image I, the global feature $V_G \in \mathbb{R}^P$ is obtained by down-scaling the output before the last pooling layer of ResNet-50 to a vector $\in \mathbb{R}^{1\times1\times2048}$ with an average pooling layer and then passing it through a group normalization (GN) layer followed by a fully-connected (FC) layer. The same output is first horizontally k-partitioned by pooling it to $k \times 1 \times 2048$, and then the local strips are separately passed through a GN and two FCs with a ReLU layer between them to form k P-dim vectors, which are finally concatenated to obtained the local visual feature matrix $M^V \in \mathbb{R}^{k\times P}$. To obtain global/local textual features, a whole sentence and the n phrases extracted from it are used as text materials. We employ a bi-directional GRU (bi-GRU) to handle the text materials, whose last hidden states of the forward and backward GRUs are concatenated to form both the processed global and local P-dim feature vectors. The P-dim vector got from a whole sentence is passed through a GN followed by an FC to give the global textual feature $T_G \in \mathbb{R}^P$ With each certain input phrase, the corresponding output P-dim vector is processed consecutively by a GN and two FCs with a ReLU layer between them and then concatenated with each other to form the local textual feature matrix $M^T \in \mathbb{R}^{n\times P}$.

To adequately utilize the fine-grained local information, we adopt a cross-modal attention (CA) mechanism to covert each local feature matrix to the V2T local feature V_L or T2V local feature T_L according to the global feature in the opposite modality:

$$\alpha_i^X = \frac{exp(S(M_i^X, Y_G))}{\sum_{j=1}^{q} exp(S(M_j^X, Y_G))}, \tag{1}$$

$$X_L = \sum_{\alpha_i^X > \frac{1}{q}} \alpha_i^X \cdot M_i^X, \tag{2}$$

where (X, Y) can be (V, T) or (T, V) which denotes the corresponding pair of visual and textual features and q can be k or n denoting the number of local strips or phrases. $S(\cdot, \cdot)$ is the cosine similarity. The visual and textual global features are mapped into a latent common space with a shared encoder, which is implemented as a two-layered perceptron to encode the features from visual and textual spaces to B-dim common features V_C and T_C, respectively.

3.2 Feature Reconstruction

AMEN proposes two different reconstruction paradigms, namely inter-modal and intra-modal reconstruction, aiming to properly embed one certain feature into the opposite modality space while learning a strong common space. To do this, we employ a visual decoder and a textual decoder which decode the common features into the visual and textual spaces respectively. The inter-modal reconstruction paradigm embeds the common features encoded from one modality space into the opposite space, which gives the P-dim V2T feature R_{V2T} and T2V feature R_{T2V}, enabling the features to be matched in both high-dimensional spaces. In contrast, the intra-modal reconstruction paradigm reconstructs the common feature back into the original modality space to get V2V feature R_{V2V} and T2T feature R_{T2T}. By minimizing the differentiate between the original and reconstructed features (which will be detailed in Sect. 3.3), a stronger common space can be learned.

3.3 Loss Functions and Training Strategy

Losses for training and similarities for testing AMEN are illustrated in Fig. 1(b). The complete training process includes 3 stages.

Stage-1. We first fix the parameters of the visual backbone and train the left parts of AMEN with the identification (ID) loss

$$L_{id}(X) = -log(softmax(W_{id} \times GN(X)) \tag{3}$$

to cluster person images into groups according to their identification, where $W_{id} \in \mathbb{R}^{Q \times P}$ is a shared transformation matrix implemented as a FC layer without bias and Q is the number of different people in the training set.

As global features can provide more complete information for clustering, only V_G and T_G are utilized here:

$$L_{ID} = L_{id}(V_G) + L_{id}(T_G). \tag{4}$$

And the entire loss in Stage-1 is

$$L_{Stage1} = L_{ID}. \tag{5}$$

Stage-2. In this stage all the parameters of AMEN without the two decoders are fine-tuned together. Besides ID loss, three more loss functions are adopted here.

First, a triplet ranking loss

$$\begin{aligned} L_{rk}(X_1, X_2) = &\sum_{\widehat{X_2}} max\{\alpha - S(X_1, X_2) + S(X_1, \widehat{X_2}), 0\} \\ &+ \sum_{\widehat{X_1}} max\{\alpha - S(X_1, X_2) + S(\widehat{X_1}, X_2), 0\} \end{aligned} \tag{6}$$

is utilized to more accurately constrain the matched pairs to be closer than the mismatched pairs with a margin α, where $(X_1, \widehat{X_2})$ or $(\widehat{X_1}, X_2)$ denotes a mismatched pair. Instead of using the furthest positive and closest negative sampled pairs, we adopt the sum of all pairs within each mini-batch when computing the loss following [4]. The proposed matching loss in Stage-2 is

$$L_M = L_{rk}(V_C, T_C) + L_{rk}(V_G, T_L) + L_{rk}(V_L, T_G). \tag{7}$$

When transforming a local feature matrix into the opposite modality via the cross-modal attention mechanism, a consistency constraint loss

$$L_{CC} = MSE(S(V_G, T_L), S(V_L, T_G)) \tag{8}$$

is proposed to ensure that the two high-level spaces are learned consistently and jointly, where MSE denotes the Mean Square Error.

To learn modality-invariant features in the common space, we employ a modality discriminator D_m to predict which modality the input feature comes from. An adversarial loss

$$L_{ADM} = \mathop{\mathbb{E}}_{V_C^i \sim V_C}[log D_m(V_C^i)] + \mathop{\mathbb{E}}_{T_C^i \sim T_C}[1 - log D_m(T_C^i)] \tag{9}$$

is adopted to optimize D_m and the loss for training AMEN is

$$L_{AM} = -L_{ADM}. \tag{10}$$

After being able to successfully deceive the D_m, much more discriminative modality-invariant features can be extracted by AMEN. The complete loss in this stage is

$$L_{Stage2} = L_{Stage1} + L_M + L_{CC} + L_{AM}. \tag{11}$$

Stage-3. Then the two reconstruction paradigms are added to train AMEN. Instead of being superficially look-alike with the target feature, the reconstructed feature ought to be discriminative for a proper matching. Thus, rather than utilizing the traditional Euclidean Distance to guide the reconstruction, the reconstruction matching loss

$$L_{RM} = L_{rk}(R_{V2V}, V_G) + L_{rk}(R_{T2T}, T_G) + L_{rk}(R_{T2V}, V_G) + L_{rk}(R_{V2T}, T_G) \tag{12}$$

is adopted. Besides, a reconstruction ID loss

$$L_{RID} = L_{id}(R_{V2V}) + L_{id}(R_{T2T}) + L_{id}(R_{T2V}) + L_{id}(R_{V2T}) \tag{13}$$

is used to ensure that the reconstructed feature can be correctly related to the corresponding person. While performing V2T and T2V reconstruction simultaneously, we also utilize a consistency constraint loss

$$L_{RCC} = MSE(S(R_{T2V}, V_G), S(R_{V2T}, T_G)) \tag{14}$$

to avoid the situation where visual and textual spaces develop and work independently, or even oppositely. A reconstruction discriminator D_r is employed to distinct a reconstructed feature with the original one, for which the loss is

$$L_{DAR} = \sum_{(X,Y)\in\Omega} \mathop{\mathbb{E}}_{X^i \sim X}[log D(X^i)] + \mathop{\mathbb{E}}_{Y^i \sim Y}[1 - log D(Y^i)], \tag{15}$$

where $\Omega = \{(V_G, R_{V2V}), (T_G, R_{T2T}), (V_G, R_{T2V}), (T_G, R_{V2T})\}$. The reconstruction adversarial loss for AMEN is

$$L_{AR} = \sum_{Y \in \Upsilon} \mathop{\mathbb{E}}_{Y^i \sim Y} [logD(Y^i)] \tag{16}$$

where $\Upsilon = \{R_{V2V}, R_{T2T}, R_{T2V}, R_{V2T}\}$. The final loss for training AMEN is

$$L_{Stage3} = L_{Stage2} + L_{RID} + L_{RM} + L_{RCC} + L_{AR}. \tag{17}$$

For testing, the combined similarity is defined as

$$S_{AMEN} = S(V_C, T_C) + \frac{1}{2}(S(V_G, T_L) + S(V_L, T_G)) + \frac{1}{2}(S(V_G, R_{T2V}) + S(R_{V2T}, T_G)). \tag{18}$$

4 Experiments

4.1 Experimental Setup

Dataset and Metrics. At present CUHK-PEDES [11] is the only dataset for text-based person re-identification task. Follow the official data split approach, the training set contains 34054 images, 11003 persons and 68126 textual descriptions. The validation set contains 3078 images, 1000 persons and 6158 textual descriptions while the testing set has 3074 images, 1000 persons and 6156 descriptions. Almost every image has two descriptions, and each sentence is generally no shorter than 23 words. After dropping words that appear less than twice, the word number is 4984. The performance is evaluated by the top-k accuracy. Given a query description, all test images are ranked by their similarities with this sentence. If any image of the corresponding person is contained in the top-k images, we call this a successful search. We report the top-1, top-5, and top-10 accuracy for all experiments.

Implementation Details. The feature dimension P is set to 1024 while $B = 512$. The phrases of each sentence are obtained with the Natural Language ToolKit (NLTK) by syntactic analysis, word segmentation and part-of-speech tagging. The total number of noun phrases n obtained from each sentence is kept flexible with an upper bound 26. Number of local strips k is set to 6. An Adam optimizer is used to train AMEN with a batch size of 32. The margin α of ranking losses is set to 0.2. In training stage-1, AMEN is trained with a learning rate of 1×10^{-3} for 10 epochs with the ResNet-50 backbone fixed. In stage-2, the learning rate is initialized as 2×10^{-4} to optimize all parameters of AMEN including the visual backbone except the two decoders for 3 epochs. Then in stage-3 the two reconstruction paradigms are added and we train the complete AMEN for extra 25 epochs. The learning rate is down-scaled by $\frac{1}{10}$ every 10 epochs.

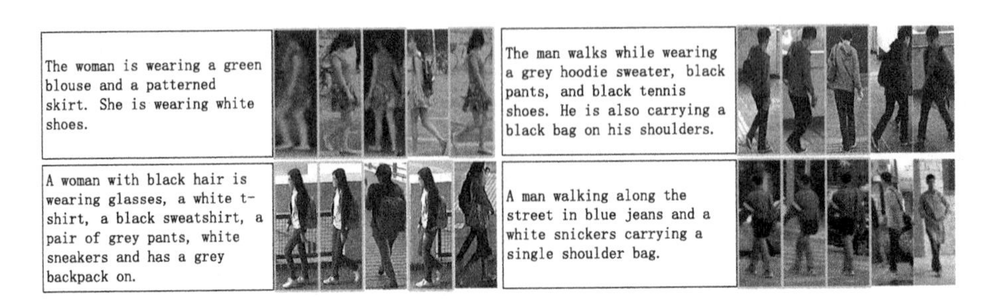

Fig. 2. Examples of top-5 text-based person re-identification results by AMEN. Images of the target pedestrian are marked by green rectangles.

Table 1. Ablation analysis of AMEN

No.	V2V	T2T	T2V	V2T	CC	RCC	D_m	D_r	S_C	S_{TL}	S_{VL}	S_{VR}	S_{TR}	Top-1	Top-5	Top-10
1	×	×	×	×	×	×	×	×	✓	✓	✓	×	×	53.08	76.54	84.86
2	✓	✓	×	×	×	×	✓	✓	✓	✓	✓	×	×	54.89	77.48	85.20
3	×	×	✓	✓	×	×	✓	✓	✓	✓	✓	✓	✓	54.93	77.83	84.93
4	✓	✓	✓	✓	×	×	✓	✓	✓	✓	✓	✓	✓	55.38	77.94	85.27
5	✓	✓	✓	✓	✓	×	✓	✓	✓	✓	✓	✓	✓	55.81	78.12	85.44
6	✓	✓	✓	✓	×	✓	✓	✓	✓	✓	✓	✓	✓	55.78	78.16	85.59
7	✓	✓	✓	✓	✓	✓	×	×	✓	✓	✓	✓	✓	54.98	77.81	85.01
8	✓	✓	✓	✓	✓	✓	✓	×	✓	✓	✓	✓	✓	55.99	78.27	85.48
9	✓	✓	✓	✓	✓	✓	×	✓	✓	✓	✓	✓	✓	56.15	78.31	85.43
10	✓	✓	✓	✓	✓	✓	✓	✓	✓	×	×	×	×	47.82	72.35	81.64
11	✓	✓	✓	✓	✓	✓	✓	✓	×	✓	×	×	×	48.99	73.65	82.16
12	✓	✓	✓	✓	✓	✓	✓	✓	×	×	✓	×	×	50.84	75.41	83.76
13	✓	✓	✓	✓	✓	✓	✓	✓	×	×	×	✓	×	50.26	74.11	83.07
14	✓	✓	✓	✓	✓	✓	✓	✓	×	×	×	×	✓	50.23	74.46	83.53
15	✓	✓	✓	✓	✓	✓	✓	✓	✓	✓	✓	×	×	55.49	78.07	85.46
16	✓	✓	✓	✓	✓	✓	✓	✓	✓	✓	✓	✓	✓	**57.16**	**78.64**	**86.22**

4.2 Ablation Analysis

Extensive ablation experiments are carried out to further investigate several key components of AMEN (shown in Table 1). ✓ and × denote whether the corresponding component or similarity is used. Comparing the results in line 2, 3 and 4, it can be concluded that the inter-modal reconstruction paradigm works in concert with the intra-modal one can properly embedding a feature into the opposite modality space while learning a strong common space. Via the consistency constraints, as shown from line 4, 5, 6 and 16, AMEN can learn the two high-level spaces consistently and jointly and improve in performance. As can be seen from line 7 and 16, the top-1 accuracy increases by 1.56% with the assistance of the tow proposed adversarial mechanisms, which indicates that being able to deceive the discriminator successfully, more discriminative representation vectors can be extracted and generated.

Table 2. Comparison with other state-of-the-art methods

Method	Top-1	Top-5	Top-10
CNN-RNN [19]	8.07	–	32.47
Neural Talk [22]	13.66	–	41.72
GNA-RNN [11]	19.05	–	53.64
IATV [10]	25.94	–	60.48
PWM-ATH [3]	27.14	49.45	61.02
Dual Path [27]	44.40	66.26	75.07
GLA [2]	43.58	66.93	76.26
MIA [17]	53.10	75.00	82.90
A-GANet [13]	53.14	74.03	81.95
GALM [7]	54.12	75.45	82.97
TIMAM [16]	54.51	77.56	84.78
IMG-Net [23]	56.48	76.89	85.01
CMAAM [1]	56.68	77.18	84.86
AMEN (ours)	**57.16**	**78.64**	**86.22**

4.3 Comparison with Other State-of-the-Art Methods

The proposed AMEN is compared against 13 previous SOTA methods, including CNN-RNN [19], Neural Talk [22], GNA-RNN [11], IATV [10], PWM-ATH [3], Dual Path [27], GLA [2], MIA [17], A-GANet [13], GALM [7], TIMAM [16], IMG-Net [23] and CMAAM [1]. As shown in Table 2, AMEN achieves the best performance under top-1, top-5 and top-10 accuracy metrics in the text-based person retrieval task on CUHK-PEDES, which approves the effectiveness of our proposed method. Compared with the best competitor MIA, the AMEN model significantly outperforms it by 3.58% under top-1 metric, indicating the effectiveness of the cross-modal attention. Pose-guided joint global and attentive local matching network (GALM) utilizes pose information to help localize the discriminative regions. With ResNet-50 as visual backbone, AMEN surpasses GALM by about 3% without suffering from the deviations of the pose estimation and the large computation consumption, which proves the effectiveness of two different reconstruction paradigms, namely inter-modal and intra-modal reconstruction.

Some examples of top-5 text-based person re-identification results by AMEN are displayed in Fig. 2. Images of the target pedestrian are marked by green rectangles.

5 Conclusion

In this paper, we proposed a novel Adversarial Multi-space Embedding Network (AMEN) to learn and match embeddings in multiply spaces. Following an encoder-decoder manner, the inter-modal reconstruction paradigm works in concert with the intra-modal one to properly embedding a feature into the opposite modality space while learning a strong common space. A consistency constraint is adapted to ensure that the learned visual and textual spaces are trained jointly and work consistently. To enhance both the common space learning and feature reconstruction, the adversarial mechanism is utilized. We evaluated our proposed AMEN on the CUHK-PEDES dataset, which is currently only accessible dataset for text-base person re-identification task. Extensive experimental results demonstrate that AMEN outperforms previous methods and achieves the state-of-the-art performance.

Acknowledgment. This work is partially supported by the National Natural Science Foundation of China (Grant No. 61972016 and 61802176), China Postdoctoral Science Foundation (Grant No.2019M661999) and Natural Science Foundation of Jiangsu Higher Education Institutions of China (19KJB520009).

References

1. Aggarwal, S., Radhakrishnan, V.B., Chakraborty, A.: Text-based person search via attribute-aided matching. In: Proceedings of the IEEE/CVF Winter Conference on Applications of Computer Vision, pp. 2617–2625 (2020)
2. Chen, D., et al.: Improving deep visual representation for person re-identification by global and local image-language association. In: Proceedings of the European Conference on Computer Vision (ECCV), pp. 54–70 (2018)
3. Chen, T., Xu, C., Luo, J.: Improving text-based person search by spatial matching and adaptive threshold. In: 2018 IEEE Winter Conference on Applications of Computer Vision (WACV), pp. 1879–1887 (2018)
4. Faghri, F., Fleet, D.J., Kiros, J.R., Fidler, S.: VSE++: improving visual-semantic embeddings with hard negatives. arXiv preprint arXiv:1707.05612 (2017)
5. He, K., Zhang, X., Ren, S., Sun, J.: Deep residual learning for image recognition. In: Proceedings of the IEEE Conference on Computer Vision and Pattern Recognition, pp. 770–778 (2016)
6. Hou, R., Ma, B., Chang, H., Gu, X., Shan, S., Chen, X.: Interaction-and-aggregation network for person re-identification. In: Proceedings of the IEEE Conference on Computer Vision and Pattern Recognition, pp. 9317–9326 (2019)
7. Jing, Y., Si, C., Wang, J., Wang, W., Wang, L., Tan, T.: Pose-guided multi-granularity attention network for text-based person search. arXiv preprint arXiv:1809.08440 (2018)
8. Karpathy, A., Fei-Fei, L.: Deep visual-semantic alignments for generating image descriptions. In: Proceedings of the IEEE Conference on Computer Vision and Pattern Recognition, pp. 3128–3137 (2015)
9. Lee, K.H., Chen, X., Hua, G., Hu, H., He, X.: Stacked cross attention for image-text matching. In: Proceedings of the European Conference on Computer Vision (ECCV), pp. 201–216 (2018)
10. Li, S., Xiao, T., Li, H., Yang, W., Wang, X.: Identity-aware textual-visual matching with latent co-attention. In: Proceedings of the IEEE International Conference on Computer Vision, pp. 1890–1899 (2017)
11. Li, S., Xiao, T., Li, H., Zhou, B., Yue, D., Wang, X.: Person search with natural language description. In: Proceedings of the IEEE Conference on Computer Vision and Pattern Recognition, pp. 1970–1979 (2017)

12. Lin, T.-Y., et al.: Microsoft COCO: common objects in context. In: Fleet, D., Pajdla, T., Schiele, B., Tuytelaars, T. (eds.) ECCV 2014. LNCS, vol. 8693, pp. 740–755. Springer, Cham (2014). https://doi.org/10.1007/978-3-319-10602-1_48
13. Liu, J., Zha, Z.J., Hong, R., Wang, M., Zhang, Y.: Deep adversarial graph attention convolution network for text-based person search. In: Proceedings of the 27th ACM International Conference on Multimedia, pp. 665–673 (2019)
14. Liu, Y., Guo, Y., Bakker, E.M., Lew, M.S.: Learning a recurrent residual fusion network for multimodal matching. In: Proceedings of the IEEE International Conference on Computer Vision, pp. 4107–4116 (2017)
15. Nam, H., Ha, J.W., Kim, J.: Dual attention networks for multimodal reasoning and matching. In: Proceedings of the IEEE Conference on Computer Vision and Pattern Recognition, pp. 299–307 (2017)
16. Sarafianos, N., Xu, X., Kakadiaris, I.A.: Adversarial representation learning for text-to-image matching. In: ICCV, pp. 5813–5823 (2019)
17. Niu, K., Huang, Y., Ouyang, W., Wang, L.: Improving description-based person re-identification by multi-granularity image-text alignments. IEEE Trans. Image Process. **29**, 5542–5556 (2020)
18. Plummer, B.A., Wang, L., Cervantes, C.M., Caicedo, J.C., Hockenmaier, J., Lazebnik, S.: Flickr30k entities: collecting region-to-phrase correspondences for richer image-to-sentence models. In: Proceedings of the IEEE International Conference on Computer Vision, pp. 2641–2649 (2015)
19. Reed, S., Akata, Z., Lee, H., Schiele, B.: Learning deep representations of fine-grained visual descriptions. In: Proceedings of the IEEE Conference on Computer Vision and Pattern Recognition, pp. 49–58 (2016)
20. Sun, C., Song, X., Feng, F., Zhao, W.X., Zhang, H., Nie, L.: Supervised hierarchical cross-modal hashing. In: Proceedings of the 42nd International ACM SIGIR Conference on Research and Development in Information Retrieval, pp. 725–734 (2019)
21. Sun, Y., et al.: Perceive where to focus: Learning visibility-aware part-level features for partial person re-identification. In: Proceedings of the IEEE Conference on Computer Vision and Pattern Recognition, pp. 393–402 (2019)
22. Vinyals, O., Toshev, A., Bengio, S., Erhan, D.: Show and tell: a neural image caption generator. In: Proceedings of the IEEE Conference on Computer Vision and Pattern Recognition, pp. 3156–3164 (2015)
23. Wang, Z., Zhu, A., Zheng, Z., Jin, J., Xue, Z., Hua, G.: Img-net: inner-cross-modal attentional multigranular network for description-based person re-identification. J. Electron. Imaging **29**(4), 043028 (2020)
24. Xia, B.N., Gong, Y., Zhang, Y., Poellabauer, C.: Second-order non-local attention networks for person re-identification. In: Proceedings of the IEEE International Conference on Computer Vision, pp. 3760–3769 (2019)
25. Yan, F., Mikolajczyk, K.: Deep correlation for matching images and text. In: Proceedings of the IEEE Conference on Computer Vision and Pattern Recognition, pp. 3441–3450 (2015)
26. Yi, D., Lei, Z., Liao, S., Li, S.Z.: Deep metric learning for person re-identification. In: 2014 22nd International Conference on Pattern Recognition, pp. 34–39. IEEE (2014)
27. Zheng, Z., Zheng, L., Garrett, M., Yang, Y., Xu, M., Shen, Y.D.: Dual-path convolutional image-text embeddings with instance loss. ACM Trans. Multimedia Comput. Commun. Appl. (TOMM) **16**(2), 1–23 (2020)

AFM-RNN: A Sequent Prediction Model for Delineating Building Rooftops from Remote Sensing Images by Integrating RNN with Attraction Field Map

Zeping Liu, Hong Tang(✉), and Wei Huang

State Key Laboratory of Remote Sensing Science, Faculty of Geographical Science, Beijing Normal University, Beijing 100875, China
{tanghong,hongtang}@bnu.edu.cn

Abstract. Accurately and automatically delineating the rooftops of buildings from remote sensing images is very essential to many fields. Recently, several Sequent Prediction Models have been proposed to deal with this task. These models use CNN encoder to recognize the boundary fragments by edge and vertex probability maps in order to guide RNN decoder find a set of sequent vertexes linking the boundaries into object regions. However, the recognition result of edge looks like large buffers around the real edges or vertexes of an object instance, which significantly influences the accuracy of predicted polygons. In this paper, we present a novel framework named AFM-RNN, in which a dual representation of lines called Attraction Field Map (AFM) is embedded by Neural Discriminative Dimensionality Reduction Layer (NDDR Layer). Consequently, the problem of over-smooth edge recognition can be effectively solved by directly modeling line segments. Experimental results over three datasets show that the proposed method outperforms other closely related methods. Code is available at https://github.com/Zeping-Liu/AFMRNN-pytorch.

Keywords: Rooftop delineation · Sequent Prediction Model · CNN · RNN

1 Introduction

The rooftops of buildings extracted from remote sensing images are essential to cartography, urban planning and quick response to natural disasters and so on. While the maps of well-established urban areas provide precise building contour, the building rooftops' information of them may be outdated or unavailable in a timely manner. In this case, it is motivated to create an automatic and accurate way to extract the building rooftops from new available remote sensing images.

To alleviate this problem, a number of models have been proposed [1, 3, 4, 14, 16], which can be technically categorized into two groups, i.e., semantic segmentation and instance segmentation. Since the current segmentation methods capture buildings well, except for occasional inaccuracies alone some of the building footprints [8], directly

H. Ma et al. (Eds.): PRCV 2021, LNCS 13020, pp. 474–485, 2021.
https://doi.org/10.1007/978-3-030-88007-1_39

predicting the building corners one by one under the view of instance segmentation seems to be an effective solution, such as Sequent Prediction Models like Polygon RNN [3] and Polygon RNN ++ [1].

The basic idea of Sequent Prediction Models is to use CNN encoder to generate the approximate edges and corners of buildings so as to guide RNN decoder to fit to the outer contour on buildings. Current research mainly uses binary edge recognition as the attention, due to the obvious category imbalance in the edge and background pixels' classification, the results of edge recognition are often accompanied with large buffers, which significantly influences the precision of the building corner predictions. Figure 1(d) shows the edge recognition result of Polygon RNN [3]. Obviously, further accuracy needs to be achieved.

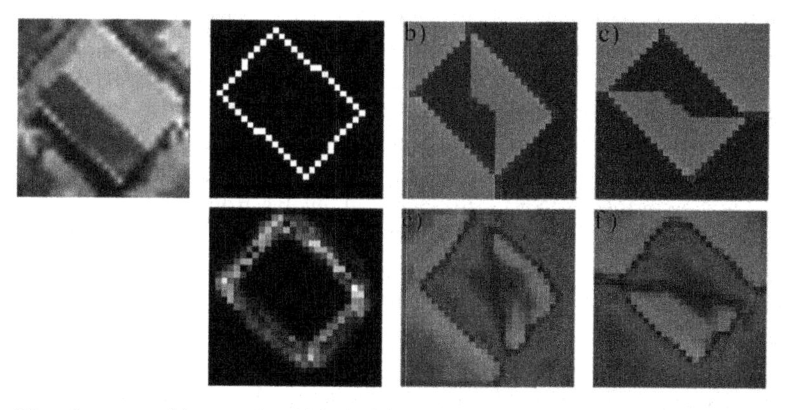

Fig. 1. The edge recognition results. (a) is the binary mask of rooftop's edge. (b) and (c) are the x and y components of AFM ground truth, respectively. (d) is the edge recognition result of Polygon RNN [3]. (e) and (f) are the predicted x and y components of AFM in AFM-RNN (ours). It can be seen that AFM makes full use of all the pixels in the image to represent the geometry of building rooftop, instead of only focusing on the edge pixels, thus getting better edge recognition result.

In this paper, a deep network named AFM-RNN is proposed to address such problem. Different from traditional Sequent Prediction Models, the proposed method uses a state-of-the-art Line Segment Detection approach called Attraction Field Map (AFM) [23] to alleviate the inaccuracy of rooftop edge recognitions. Specifically, our model uses AFM as an attention and predict it through another branch of the network. The prediction of AFM is treated as an auxiliary task to the main task of delineating the boundaries of buildings. We also carefully design a pattern for the main task and the auxiliary one to reinforce each other by NDDR Layer [7]. Finally, the skip feature fused with AFM attention is sent to the Convolutional Long Short-Term Memory (Conv-LSTM) [19] to predict the positions of corners on buildings one by one.

The rest of the paper is organized as follows. Section 2 provides a brief review about the main techniques used in our model. Section 3 will introduce the details about our approach. Section 4 includes our experiments. Finally, several conclusions are drawn in Sect. 5.

2 Related Work

The proposed method approaches the delineation of building rooftops through instance segmentation by using a novel edge representation under a multi-task framework. Related work is reviewed in this section.

2.1 Instance Segmentation

Instance segmentation methods aim to allocate a unique label for per instance. A natural approach is to use semantic segmentation results as a part of instance segmentation like Mask R-CNN [9]. However, this kind of methods still originate from pixel-wise semantic segmentation and have not apt to integrating output with shape priors directly.

Another kind of approaches of instance segmentation is to directly model the boundary of an object instance as an active contour (Contour-based Models) [8, 12, 13, 16, 17] or a polygon consisting of sequent vertexes (Sequent Prediction Models) [1, 3]. Contour-based Models firstly initialize a contour of the object and then move the contour across the energy surface until it halts by minimizing an energy function. Marcos el al. use DNN to do this work by minimizing its energy via gradient descent and proposed Deep Active Ray Network (DARNet) [16]. Gur et al. evolve from it and present a simpler framework with a fully differentiable rendering called ACDRNet [8]. Although they have been proved to be effective, the predicted contours of object instances tend to be rounded shape. This feature makes it unsuitable for the prediction of angular building contours in aerial images.

Unlike Contour-based Models, Sequent Prediction Models firstly detect boundary fragments, and then find an optimal cycle linking the boundaries into object regions using RNN [5]. Because of directly predicting vertex positions of object instances, the irregularity of predicted contours can be avoided. Hence, we follow this idea and propose AFM-RNN method.

2.2 Edge Representation Approach

It is natural to treat the edge representation as a binary classification problem, and numbers of edge detection models [15, 22] have been proposed under this perspective. HED [22] learns multi-scale and multi-level features from supervised edge binary marks by CNNs. However, edge maps estimated by CNNs are usually over-smoothed, which will lead to local ambiguities for accurate localization. In this situation, Xue et al. [23] directly model the line segments map instead of edge map, and proposed a dual representation of line segments, called Attraction Field Map (AFM). As shown in Fig. 1(b) and (c), AFM uses a regression way to represent every pixel of the image by 2D projection vector, and have been proved to be learnt well by CNNs.

Due to the imbalance in the number of pixels of edges and background, the edge recognition of Sequent Prediction Models tends to be over-smooth. Hence, our intuitive idea is to integrate the RNN with AFM.

2.3 Multi-task Learning

Multi-Task Learning (MTL) aims to improve model generalization by leveraging domain-specific information contained in the training signals of related tasks [21]. Technically, MTL approaches can be classified by encoder-focused and decoder-focused. Encoder-focused method shares the task features in the encoding stage [7], while the other shares the information in the decoding stage [24]. We follow the path of encoder-focused theory, and use it to make a sufficient use of AFM.

3 Methods

As shown in Fig. 2, given the bounding box of a building in an image, the AFM-RNN firstly extracts an image crop containing one building from the large aerial image by the 15% enlarged bounding box, and resize it to 224 × 224. Then it uses Resnet-50 [10] as backbone, and uses NDDR-CNNs to get skip features and AFM. Next, an AFM attention mechanism is applied on skip features. First corner is predicted from the skip features integrated with AFM. Finally, both the first corner and the skip features integrated with AFM are fed into recurrent decoder to delineate the rooftop of a building.

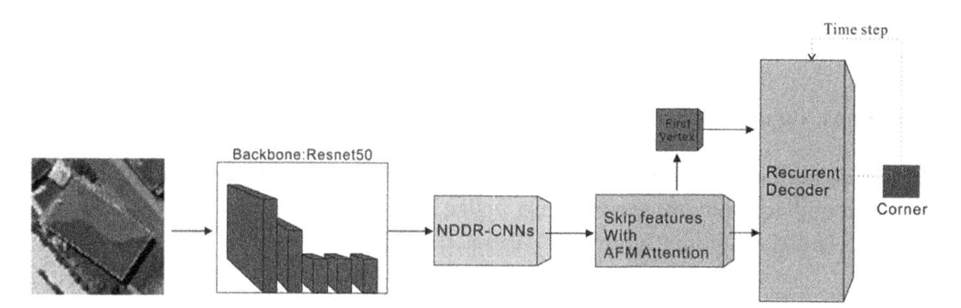

Fig. 2. The workflow of the AFM-RNN

3.1 NDDR-CNNs

In order to use the AFM of a target building as attention to guide the sequent prediction, a simple idea is to apply another branch of CNNs to predict AFM based on the extracted high-dimension feature map of Resnet-50 backbone. However, the potential link between them cannot be ignored. Therefore, we design two branches CNNs with information sharing mechanism followed by the backbone.

The way to use Resnet-50 is totally same as [1], after a series of down-sampling, up-sampling and cascading operations, a 112 × 112 × 256 feature map is obtained uniquely. Then, two branches of identical repeated convolution and batch normalization layers are utilized to produce skip features (28 × 28 × 126) and predict the AFM (28 × 28 × 2). Meanwhile, the skip features and AFM predictions are treated as two different tasks.

Because of the potential connection between these two tasks, we refer to the relevant NDDR Layer [7] and introduce it into our model to reinforce the learning process. The structure of NDDR-CNNs is shown in Fig. 3. The NDDR Layers for these two tasks can be constructed by: 1) concatenating the features from two tasks according to the channel dimension 2) using the 1×1 convolution (filter is $1 \times 1 \times 256$) and batch normalization layer to link two tasks.

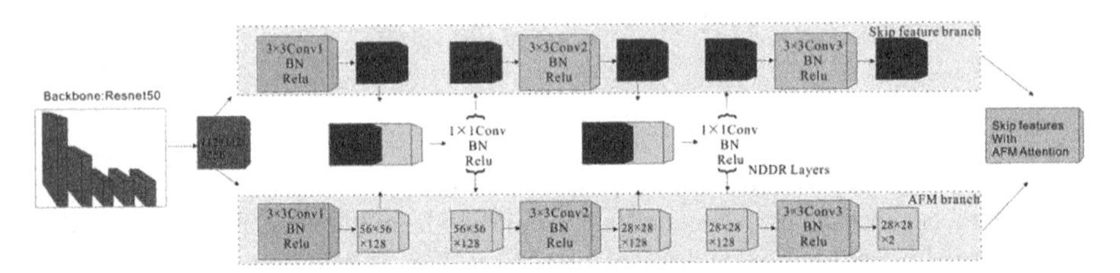

Fig. 3. The structure of NDDR-CNNs.

Intuitively, we use NDDR Layer to make the learning process no longer independent. NDDR Layers learn the common information of these two tasks and facilitate the flow of parameters adjustment in the two branches. Besides, MTL approach applies some noise in training process [21], which can significantly improve the validation accuracy.

3.2 AFM Attention Mechanisms

Attraction field map (AFM) [23] is calculated on the basis of region-partition map. Let I be a crop image containing a building. A line is denoted by $l_i = \left(x_i^s, x_i^e\right)$ with two end points x_i^s and x_i^e. Building rooftops are denoted by a set of lines $L = \{l_1, \ldots, l_n\}$.

Region-Partition Map. Computing region-partition map for L is a process of assigning every pixel in I to the nearest line segment region. Given a pixel $p \in I$ and a line $l_i \in L$, we first project p onto l_i vertically. If the projection point is not on l_i, the nearest end point of l_i is considered as the projection point. Then, the distance between them is calculated as $d(p, l_i)$:

$$d(p, l_i) = \min d(p, l_i; \alpha) = min\left\| x_i^s + \alpha \cdot \left(x_i^e - x_i^s\right) - p \right\|_2^2 \tag{1}$$

$$\alpha_p^* = arg \min d(p, l_i; \alpha) \tag{2}$$

The projection point is on line segment when $\alpha_p^* \in (0, 1)$, and is on the end point when α_p^* is 0 or 1.

The region of l_i is defined as below, and the region-partition map is denoted as $R = \{R_1, \ldots, R_n\}$.

$$R_i = \{p | p \in I; d(p, l_i) < d\left(p, l_j\right), \forall j \neq i, l_j \in L\} \tag{3}$$

Attraction Field Map. Given the region-partition map R_i corresponding to a line segment, for each pixel p in R_i, the projection point $p^{'}$ on l_i is calculated:

$$p' = x_i^s + \alpha_p^* \cdot \left(x_i^e - x_i^s\right) \tag{4}$$

For every pixel p, the 2D attraction or projection vector $a(p)$ can be calculated as below and the attraction field map is denoted as $A = \{a(p), p \in I\}$.

$$a(p) = p' - p \tag{5}$$

In AFM-RNN, we utilize a pre-processing module to generate supervised AFM for each building rooftop, and apply a branch of CNNs in Sect. 3.1 to generate the probability map of AFM. Smooth L1 loss [18] is the loss function.

Attention Mechanisms. Traditional edge maps approximate the line segment with only a few pixels in images. Different from that, AFM gives equal attention to all pixels, which contains more information about the line geometry of the building rooftop. Given AFM, F_{afm} and skip features, F_s from NDDR-CNNs, we directly concatenate them to get the skip features fused with AFM, F_{s-AFM}:

$$F_{s-AFM} = \text{Concat}\left(F_{afm}, F_s\right) \tag{6}$$

3.3 Recurrent Decoder

The model uses double-layer ConvLSTM with a 3×3 kernel with 64 and 16 channels to decode F_{s-AFM} and delineate the boundary of a rooftop. Sharp corners of building rooftops are used as supervised information during training. In a time-step, decoder needs the previous corner C_{t-1} and an implicit direction, the previous is always given by the last time step, except for the first time. So, an additional branch that contains two $D \times D$ layers is applied from F_{s-AFM} to predicts the corner probability map and generate the potential position of the first corner C_{first}. Given the one-hot encoding of the two previous corners C_{t-1}, C_{t-2} and C_{first}, F_{s-AFM} is fed into ConvLSTM at time t to get C_t (C_{t-1}, C_{t-2} are initialized as zero tensor at first). The decoding process is recurrent and the rooftop polygon is naturally drawn by the corners in order.

4 Experiments

4.1 Datasets and Evaluation Metrics

Datasets. Three datasets with different spatial resolutions are utilized to test the performance on remote sensing images, i.e., WHU Building Dataset [11], Massachusetts Building Dataset [20] and DREAM-A Building Dataset. The aerial images of WHU Building Dataset photographed from Christchurch, the size and resolution are 512×512 and 0.3 m. Massachusetts Building Dataset contains 137 aerial images of size 1500×1500 with a spatial resolution of 1 m. In DREAM-A Building Dataset, the images are from Tianditu of Beijing, Guangzhou, Hangzhou and Fuzhou in P. R. China, these images are 224×224 with a spatial resolution of 2.5 m.

Evaluation Metrics. Given the predicted polygon and ground-truth of a building, we use four metrics for evaluation, including the Intersection over Union (IoU) [6], the ratio of the predicted and the ground-truth intersection to their union; Rotation Error (RE), a value defined as the deflection angle between the orientations of predicted polygon and ground-truth; Hausdorff distance (Hd) [2], a value used to measure the maximal-minimal distance between two sets of vertex points from ground-truth and the predicted; Vertex Number Error (VNE), a value measuring the difference between the number of predicted vertices and number of ground-truth vertexes. All of them are average number.

4.2 Training Details

We use pretrained Resnet-50 as our backbone and train our model by Adam optimizer with a batch size of 8 and an initial learning rate $\lambda = 1e - 3$. The learning rate was decayed after 10 epochs by a factor of 10 and other optimizer parameters are all default values.

AFM Predictions. Considering AFM is numerically very small, which will result in the instability of training process, we apply a reversible nonlinear transformation on supervised information of AFM a:

$$a' = -sign(a) \times \log(|a| + \varepsilon) \tag{7}$$

Parameter ε is applied to avoid log (0), we set it 1e-6.

Polygon Predictions. The rooftop polygon we predicted is at resolution $D \times D$, we set D to 28. In order to not over-penalize the incorrect predictions closed to the ground truth corner, we assign non-zero probability mass to those locations that are within a distance of 2. For each time step of RNN, we apply one-hot encoding on every corner and use cross-entropy as loss. For the task of the first corner prediction, we use logistic loss to generate the probability map of each corner and pick the most possibility to be the first corner.

4.3 Results and Comparisons

We use three instance segmentation methods, i.e., Mask R-CNN, ACDRNet and Polygon RNN, to compare with our model in terms of both qualitative and quantitative performance.

Mask R-CNN [9] is an instance segmentation method for object detection, classification and semantic segmentation. We use the semantic result of the mask branch in Mask R-CNN for comparisons.

ACDRNet is a Contour-based Model. The method firstly initializes a circle with diameter of 16 pixels by taking the center of the input image crop. The number of initialized vertices is set to be 32, at which the performance of the model can reach saturation quickly as described in [8].

Polygon RNN [3] is a Sequent Prediction Model and structurally similar to AFM-RNN. The framework of it is CNN-RNN and the backbone is VGG-16. To make a fair comparison, we replace VGG-16 with Resnet-50.

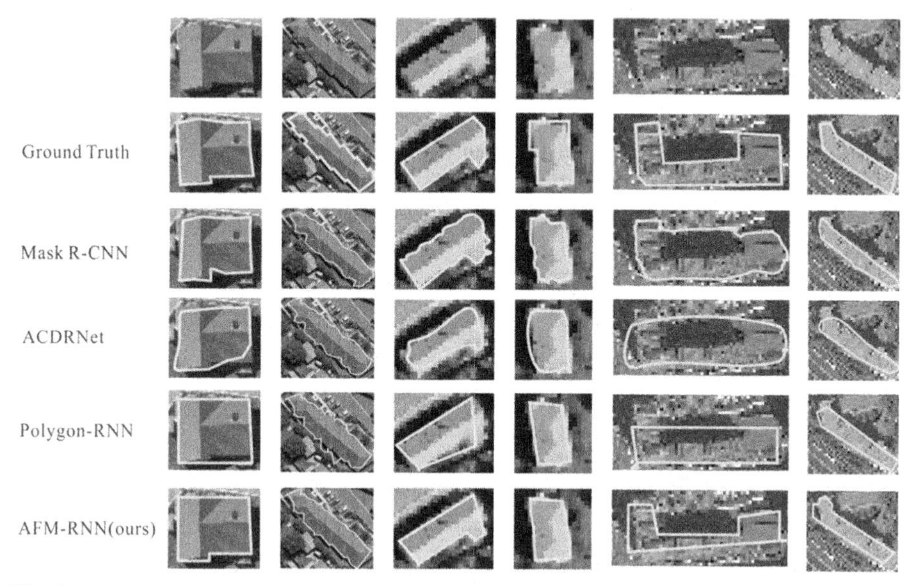

Fig. 4. Qualitative comparisons. These building instances are derived from three datasets: The first and second columns are from WHU Building Dataset, the third and fourth columns are from Massachusetts Building Dataset and the fifth and sixth columns are from DREAM-A Building Dataset. (From up to bottom) The first row is the instance images, the second row is the ground truth and from the third row to the sixth row are results of Mask R-CNN [9], ACDRNet [8], Polygon-RNN [3] and AFM-RNN (ours).

Qualitative Comparisons. We select two images from each of the three datasets for qualitative comparisons. As shown in Fig. 4, AFM-RNN can better predict the number and position of rooftops corners without severe corner deviations, our model exhibits good performance with different resolution images as input.

There are obvious differences in the delineating results of rooftops among different models in Fig. 4. Although Mask R-CNN can accurately locate the building without the provided bounding box, it can hardly identify the edge of rooftops precisely compared with our model. ACDRNet is a Contour-based Model, it firstly initializes a polygon by control points and then extend it to the shape of the building rooftop, which result in the rounded predicted corners. Polygon RNN often recognizes the number of corners inaccurately. It misses several corners in row 5, column 1 and row 5, column 4. It also recognizes more corners in row 5, column 2.

Quantitative Comparisons. The quantitative comparison results for different models are listed in Table 1. The AFM-RNN obtained the highest value in three datasets in terms of IoU. RE shows the difference in the deflection angle of polygons and Hd measures the spatial distance of two sets of corners that make up polygons. The results of AFM-RNN are all lower than others in terms of both RE and Hd. For VNE, due to ACDRNet needs to initialize a fixed number of control points, this metric is not considered for ACDRNet comparison. Since Mask R-CNN directly identifies the pixel of buildings, the number of predicted corners is significantly more than other models, which results in a higher value in VNE. Although the difference between Polygon RNN and AFM-RNN in VNE metrics is not much, AFM-RNN still shows better performance than the other.

Table 1. Quantitative comparisons of different models.

Dataset	Methods	IoU	RE	Hd	VNE
WHU Building	Mask R-CNN	0.8154	0.5055	4.45	7.67
	ACDRNet	0.8016	0.5541	5.12	—
	Polygon RNN	0.8147	0.5543	4.60	2.75
	AFM-RNN (ours)	**0.8462**	**0.4988**	**4.25**	**2.47**
Massachusetts Building	Mask R-CNN	0.7402	0.7689	3.52	6.83
	ACDRNet	0.7404	0.7085	2.74	—
	Polygon RNN	0.7473	0.7180	3.11	3.75
	AFM-RNN (ours)	**0.7874**	**0.6908**	**2.45**	**3.00**
DREAM-A Building	Mask R-CNN	0.7213	0.7621	6.87	8.75
	ACDRNet	0.6988	0.7711	6.69	—
	Polygon RNN	0.7437	0.7193	6.21	6.98
	AFM-RNN (ours)	**0.7504**	**0.6742**	**5.55**	**6.24**

4.4 Ablation Experiments

We conduct ablation studies on WHU building dataset [11]. The three combinations are evaluated, including our proposed network AFM-RNN, the method of only using NDDR Layer (i.e., + NDDR) and only AFM concatenation (i.e., + AFM). All of the experiments are trained for 15 epochs (40.5k steps).

Baseline. Our baseline is modified Polygon RNN [3]. Specifically, it follows the structure: 1) a Resnet-50 backbone used in the same way as [1] 2) only a skip feature generating branch that is identical to the "skip feature branch" in Fig. 3, with exception that the output dimension is 128. 3) the output skip feature is directly fed into a recurrent decoder which is the same as that in Sect. 3.3.

The Effectiveness of Each Component. Table 2 summarizes the evaluation results of ablation experiments. To validate the advantage of our model, the second row only applies NDDR Layer in AFM and skip features generating branch without AFM concatenation. Obviously, its result is worse than the baseline, which will be discussed later. The third row only applies AFM concatenation on skip feature without NDDR-Layer and there is some improvement in performance. The last row is the proposed method AFN-RNN, it combines NDDR Layer and AFM concatenation and yields almost 3% accuracy improvement in terms of IoU.

Table 2. The evaluation results of ablation experiments

	IoU	RE	Hd	VNE
Baseline	0.8198	0.5755	4.88	2.67
+ NDDR	0.7213	0.6116	5.10	3.18
+ AFM	0.8288	0.5441	4.75	2.55
AFM-RNN (ours)	**0.8462**	**0.4988**	**4.25**	**2.47**

Specifically, the AFM attention can improve the prediction accuracy of the first corner, as shown in Table 3.

Table 3. The percentage of the predicted first corner exactly in the ground truth corner (%)

Methods	Baseline	+ AFM	+ NDDR	AFM-RNN (ours)
Percentage	88.5%	89.6%	87.8%	**90.0%**

Besides, NDDR Layer can improve generalization ability of the model. As shown in Fig. 5, the training IoU of AFM-RNN is lower than " + AFM", but its validation IoU grows higher, which means NDDR Layer also plays the role of introducing noise.

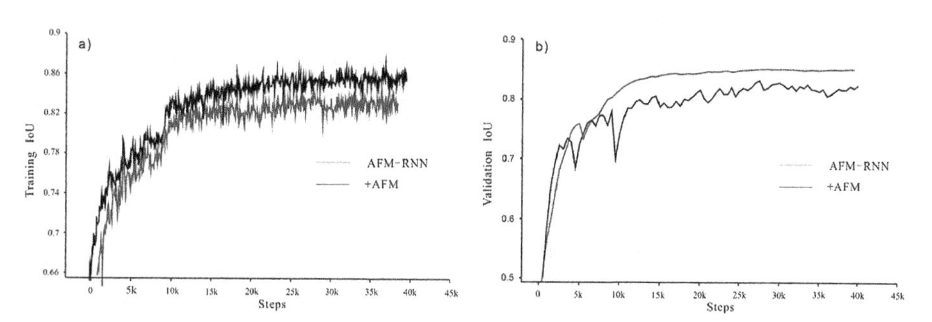

Fig. 5. Comparisons of training and validation IoU between AFM-RNN (ours) and " + AFM" in training process. Blue represents AFM-RNN (ours) and red represents " + AFM". (a) and (b) are training curve and validation curve respectively. (Color figure online)

Discussion and Summary. The network only contains NDDR Layer might be overfitting in the training process, as shown in Fig. 6. Intuitively, without the AFM attention part, the loss of RNN cannot be directly passed into AFM prediction branch, which result in the insufficient information sharing mechanism brought by NDDR Layer.

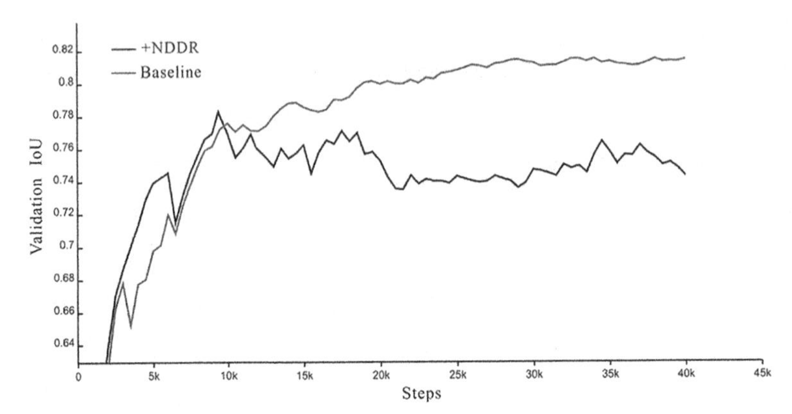

Fig. 6. The validation IoU of baseline and " +NDDR" in training process. Red is " +NDDR" and Blue is our baseline (Color figure online)

Overall, AFM attention is the attention mechanism that explicitly works on the skip feature. We apply a simple and effective fusion mechanism on skip feature to improve the prediction accuracy of the first corner. NDDR Layer is more like an attention that implicitly works on the skip feature. It can incorporate the AFM into the skip feature generating branch in training process and lead a more clearly implicit direction for RNN. Meanwhile, it also introduces noise into network, which improves the generalization ability of the model.

5 Conclusions

In this paper, we introduce AFM-RNN, a Sequent Prediction Model for accurately and automatically delineating the rooftop of buildings in remote sensing images. This model achieves much better performance than our baseline due to the reasonable solution to the over-smooth edge recognition from target instance by introducing AFM. Experiments over three remote sensing datasets show that AFM-RNN outperforms other closely related methods. In the future, further research is needed to explore the potential of AFM for edge recognitions in different tasks.

Acknowledgment. This work was supported in part by the National Natural Science Founsdation of China under Grant No. 41971280 and in part by the National Key R&D Program of China under Grant 618 No. 2017YFB0504104.

References

1. Acuna, D., et al.: Efficient interactive annotation of segmentation datasets with polygon-RNN++. In: CVPR, pp. 859–868 (2018)
2. Avbelj, J., Müller, R., Bamler, R.: A metric for polygon comparison and building extraction evaluation. IEEE Geosci. Remote Sens. Lett. **12**(1), 170–174 (2015)

3. Castrejon, L., Kundu, K., Urtasun, R., Fidler, S.C.: Annotating object instances with a polygon-RNN. In: CVPR, pp. 5230–5238 (2017)
4. Cheng, D., Liao, R., Fidler, S., Urtasun, R.: Darnet: deep active ray network for building segmentation. In: CVPR, pp. 7423–7431 (2019)
5. Duan, L., Lafarge, F.: Towards large-scale city reconstruction from satellites. In: Leibe, B., Matas, J., Sebe, N., Welling, M. (eds.) ECCV 2016. LNCS, vol. 9909, pp. 89–104. Springer, Cham (2016). https://doi.org/10.1007/978-3-319-46454-1_6
6. Everingham, M., Van Gool, L., Williams, C.K.I., Winn, J., Zisserman, A.: The pascal visual object classes (VOC) challenge. Int. J. Comput. Vis. **88**(2), 303–338 (2010)
7. Gao, Y., Ma, J., Zhao, M., Liu, W., Yuille, A.L.: NDDR-CNN: layerwise feature fusing in multi-task CNNs by neural discriminative dimensionality reduction. In: CVPR, pp. 3200–3209 (2019)
8. Gur, S., Shaharabany, T., Wolf, L.: End to End Trainable Active Contours via Differentiable Rendering (2019). https://arxiv.org/abs/1912.00367
9. He, K., Gkioxari, G., Dollar, P., Girshick, R.: Mask R-CNN. In: ICCV, pp. 2961–2969 (2017)
10. He, K., Zhang, X., Ren, S., Sun, J.: Identity mappings in deep residual networks. In: Leibe, B., Matas, J., Sebe, N., Welling, M. (eds.) ECCV 2016. LNCS, vol. 9908, pp. 630–645. Springer, Cham (2016). https://doi.org/10.1007/978-3-319-46493-0_38
11. Ji, S., Wei, S., Lu, M.: Fully convolutional networks for multisource building extraction from an open aerial and satellite imagery data set. IEEE Trans. Geosci. Remote Sens. **57**(1), 574–586 (2019)
12. Kass, M., Witkin, A., Terzopoulos, D.: Snakes: active contour models. Int. J. Comput. Vis. **1**(4), 321–331 (1988)
13. Ling, H., Gao, J., Kar, A., Chen, W., Fidler, S.: Fast interactive object annotation with curve-GCN. In: CVPR, pp. 5257–5266 (2019)
14. Li, Z., Wegner, J.D., Lucchi, A.: Topological map extraction from overhead images. In: ICCV, pp. 1715–1724 (2019)
15. Maninis, K., et al.: Convolutional oriented boundaries: from image segmentation to high-level tasks. IEEE Trans. Pattern Anal. Mach. Intell. **40**(4), 819–833 (2018)
16. Marcos, D., et al.: Learning deep structured active contours end-to-end. In: CVPR, pp. 8877–8885 (2018)
17. Peng, S., Jiang, W., Pi, H., Li, X., Bao, H., Zhou, X.: Deep snake for real-time instance segmentation. In: CVPR, pp. 8530–8539 (2020)
18. Ren, S., He, K., Girshick, R., Sun, J.: Faster R-CNN: towards real-time object detection with region proposal networks. IEEE Trans. Pattern Anal. Mach. Intell. **39**(6), 1137–1149 (2017)
19. Shi, X., Chen, Z., Wang, H., Yeung, D.Y., Wong, W.K., Woo, W.C.S.: Convolutional LSTM network: a machine learning approach for recipitation nowcasting. In: NIPS, pp. 802–810 (2015)
20. Volodymyr, M.: Machine Learning for Aerial Image Labeling. Ph.D.disseartion, University of Toronto, Canada (2013)
21. Vandenhende, S., et al.: Multi-task learning for dense prediction tasks: a survey. IEEE Transactions on Pattern Analysis and Machine Intelligence (2021). https://doi.org/10.1109/TPAMI.2021.3054719
22. Xie, S., Tu, Z.: Holistically-nested edge detection. In: CVPR, pp. 1395–1403 (2015)
23. Xue, N., Bai, S., Wang, F., Xia, G., Wu, T., Zhang, L.: Learning attraction field representation for robust line segment detection. In: CVPR, pp. 1595–1603 (2019)
24. Zhang, Z., Cui, Z., Xu, C., Jie, Z., Li, X., Yang, J.: Joint task-recursive learning for semantic segmentation and depth estimation. In: Ferrari, V., Hebert, M., Sminchisescu, C., Weiss, Y. (eds.) ECCV 2018. LNCS, vol. 11214, pp. 238–255. Springer, Cham (2018). https://doi.org/10.1007/978-3-030-01249-6_15

Attribute-Level Interest Matching Network for Personalized Recommendation

Ran Yang, Meng Jian, Ge Shi(✉), Lifang Wu, and Ye Xiang

Faculty of Information Technology, Beijing University of Technology, Beijing, China

Abstract. Personalized recommendation refers to identifying items that satisfy users' interests from large-scale item databases according to users' habits and preferences. The task is very challenging due to the complexity of user interests. Previous works use a uniform representation to model user interests, neglecting the diversity of user preferences when they adopt items. However, users consider many different attributes when choosing an item. Introducing attribute-level matching information into the model can express user interests more accurately. To achieve this goal, we propose a novel Attribute-level Interest Matching Network (AIMN) for personalized recommendation. We first adopt a knowledge representation learning method to construct spaces of different attributes, then employ a knowledge graph to extend entities as side information for representing users. Finally, we project entities and candidate items into diverse attribute spaces, match and aggregate them to realize fine-grained attribute-level information matching. Empirical results demonstrate that the proposed AIMN achieves substantial gains on several benchmarks, beating many solid baselines and achieving state-of-art performance.

Keywords: Personalized recommendation · User interest · Attribute-level matching · Knowledge graph

1 Introduction

With the development of intelligent mobile devices and multimedia technology, the information on the Internet is increasing exponentially [20]. Nowadays, information is no longer a scarce product, and users want to obtain personalized information that suits their interests [12]. Under these circumstances, personalized recommendation system comes into being an effective bridge connecting users and information [19]. As a user-oriented personalized information filtering tool, the core is user interest modeling and matching between user interests and items.

As the mainstream method of personalized recommendation, collaborative filtering (CF) leverages one fixed latent vector that transformed from personal information or historical interaction items to represent the user [13], and demonstrate their remarkable prediction performance [16]. Despite their success, we

H. Ma et al. (Eds.): PRCV 2021, LNCS 13020, pp. 486–497, 2021.
https://doi.org/10.1007/978-3-030-88007-1_40

noted that the uniform modeling methods compress user interests into a fixed-length representation, regardless of how many and what interests are, which hardly offer interpretable embedding. To be more specific, we don't know what information is encoded in particular dimensions. Present embedding functions simply employ a black-box neural network on the interaction records and output user and item representations. The match between users and items is also vague.

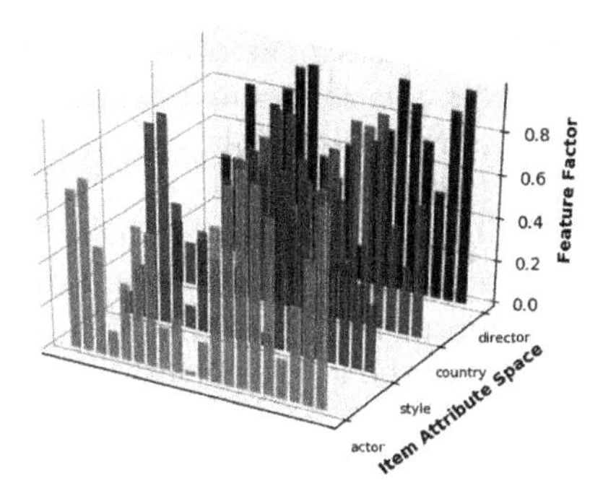

Fig. 1. User have corresponding attribute-level interest representation over different movie attributes spaces.

To represent the fine-grained and diverse user interests and match users and items more explainable, we propose attribute-level interest matching. The basic idea is that the overall personalized interest of users can be regarded as a collection of interest distribution over multiple item attributes. On every item attribute space, users will have a corresponding interest appearance. Take the movie recommendation scenario as an example in Fig. 1, the movies have attributes such as actors, directors, and genres. Naturally, in the actor space, there should be the user's interest appearance on movie actor; similarly, in the director space, there should also be the user's interest appearance on movie director. The overall user representation for movies is composed of interests with all the related attributes. When predicting the possibility of interaction between users and movies, the matching on each movie attribute should be considered.

Some researchers have also realized that a fixed latent vector is not enough to represent the complex user interests [17]. Some work that focuses on diverse user representation and dynamic matching of users and items has emerged. DIN [22] makes the user representation vary over different items with attention mechanisms to capture the diversity of user interests [10], and achieves dynamic matching of users and items. Nevertheless, the adoption of attention mechanisms also makes the computational cost very high. In recent years, generating recommendations with knowledge graph as side information has attracted considerable attention from researchers. Such an approach can not only alleviate the data sparsity problem but also provide explanations for recommended items [5]. Items and their attributes can be mapped into KG to understand the mutual relations between items. We realized that the link relations stored in KG could be considered as explicit evidence of different item attributes. Guided by the relations in KG, we could obtain corresponding user interest representation.

In this paper, we introduce an end-to-end Attribute-level Interest Matching Network (AIMN) for personalized recommendation, which obtains attribute-level user interest representations and matches user and item in each attribute space. AIMN employs KG to extend the related entities and relations, and maps the users historical interacted items to entities. The knowledge graph embedding method is Trans R [11]. AIMN devises an attribute-level interest extractor layer, this layer according to different relations to extract corresponding user interest representation, since an entity may have multiple aspects, and various relations focus on different aspects of entities. We equipped AIMN with an interest matching module. Specifically, the target item and user interests are mapping to each item attribute space by corresponding relation mapping matrices, then calculate the match between user interests and target item.

To summarize, this work has the following three main contributions:

- Compared to the previous fixed modeling methods, we emphasize the importance of obtaining attribute-level user interests, and modeling such interest distribution could bring about better user representations.
- We devise a novel framework AIMN. Under the guidance of relations in KG, AIMN obtains attribute-level user interest representations and matches user and item in each item attribute space.
- Due to the introduction of external knowledge information, the proposed AIMN has intuitive interpretability and empirically proves the efficacy of our method over several state-of-the-art baselines.

2 Related Work

The proposed AIMN method is highly relevant with collaborative filtering-based and Knowledge Graph-based recommendation algorithms. Here we introduce the related literature and focus on the differences to the proposed AIMN.

Collaborative filtering (CF) based algorithms have been proven successful in the practical application of recommender systems. The most classic Matrix factorization [9](MF) projects users and items to the same latent space and obtains the prediction score by conducting inner product between the two parts. In order to further enrich CF representation, there are also some research attempting to leverage personal history to refine CF representations, such as FISM [8], NAIS [6]and ACF [4]. On the other hand, some researchers focus on exploiting deep neural networks instead of the inner product to enhance the interaction function to capture the nonlinear interactions between users and items, for example, NCF [7], and DeepICF [18]. Despite effectiveness, we still point out that the above methods use a mixed latent vector to represent the user, which brings about blur in user representations.

The performance of CF-based algorithms has always been limited by data sparsity and cold start issues. In recent years, from the perspective of uniting the interaction-level similarity and content-level similarity, incorporating multiple types of side information like text review, social networks, item content, and external knowledge graph information has become a research focus [15]. Especially the knowledge graph (KG) can not only alleviate data sparsity but also provide interpretability for recommend results, showing promise in enhancing model performance. Such as CKE [21] model, which exploits semantic embeddings derived from Trans R [11] to enhance matrix factorization, while CFKG [1] applies Trans E [3]on the user-item interaction graph and KG, turning the recommendation task as plausibility prediction of triples. Wang et al. proposed KGAT [15], which employs an attention mechanism to achieves high-order relations (in KG) modeling under the GNN framework. Similarly, there is GC-MC [2] to apply graph network to recommendation task. Another research RippleNet [14], obtain optimized user interest representation by preference propagation along the relation path in the KG. Excellent performance of the above methods inspires us that the introduction of KG can enrich the representation and promise accurate recommendations.

Even with side information such as KG, researchers still find that a mixed latent vector is not enough to represent the complexity of user interests. Some recent research studies focus on diverse user representation, and the core lies in diverse user interests. Representative works DeepICF [18] utilize historical interaction items to model user interests and distinguish the contribution of user's historical items by introducing the attention mechanism, which realizes the dynamic representation of the user's interest along with the target item. But it still uses a single latent vector to represent the user and is insufficient to model users' diverse and attribute-level interest distribution.

3 Attribute-Level Interest Matching Network

In this section, we present the proposed attribute-level interest matching network for personalized recommendation, short for AIMN, which is illustrated in Fig. 2. AIMN takes a user u and an item v_i as input and outputs the predicted probability of user-item interaction. It consists of three components:

Primary Operations of Knowledge Representation Learning: We first use Trans R algorithm to learn entity and relation embeddings in KG, since Trans R points out that an entity may have multiple levels of information, and different relations may focus on different levels of the entity. For each triple (h, r, t), entities are projected into r-relation space as h_r and t_r with relation-specific mapping matrix M_r, and then $h_r + r = t_r$.

Attribute-Level User Interest Extraction: For the input user u, his historical interacted items are treated as seeds in KG. Naturally, the relations stored in KG can be considered as explicit evidence of different item attributes. According to different relations, aggregates entities connected with the seeds, respectively, as the representation of user interest in the corresponding item attribute space. This layer outputs the set of attribute-level user interest representations.

Interest Matching Under Each Attribute Space: After obtaining the attribute-level user interest representation, we map the item v_i to different item attribute spaces by relation-specific mapping matrix M_r, then calculate the match between user interest and item in the same attribute spaces, respectively. The final output is the concatenation of matching results in each attribute space, that is, the predicted probability of user-item interaction.

Next, we show the details of each component in our architecture.

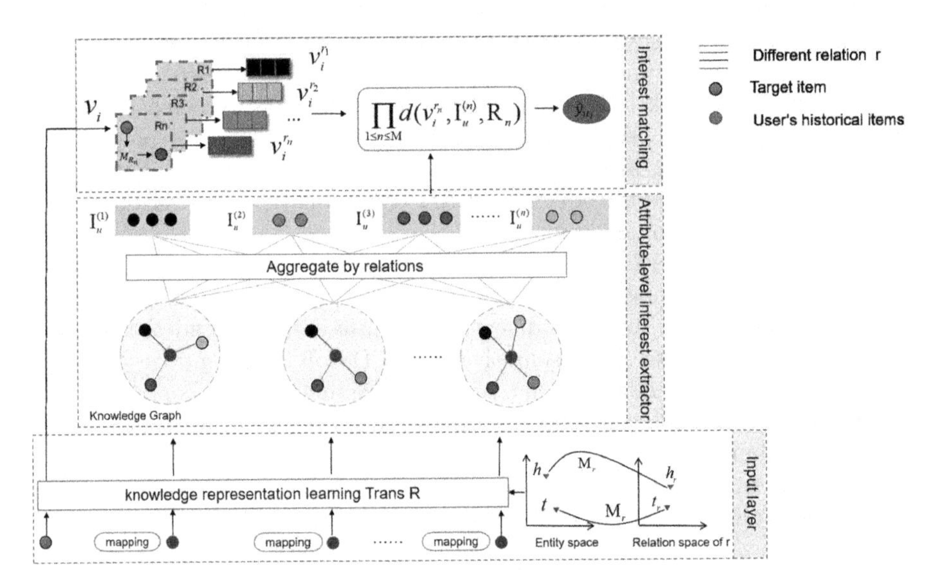

Fig. 2. Framework of the proposed Attribute-level Interest Matching Network (AIMN) for Personalized Recommendation.

3.1 Knowledge Representation Learning

The heterogeneous knowledge graph network encodes the semantic information of entities and their connected relations. The commonly used method is to embed this heterogeneous network into a continuous vector space while preserving the entity semantics and link relation in the network to capture this knowledge information. In this paper, we select the Trans R [11] algorithm as Knowledge representation learning method since Trans R points out that an entity may have multiple levels of information, and different relations may focus on different levels of the entity, two entities that are similar in relation space A may not be similar in relation space B.

In an undirected knowledge graph $G = (V, R)$, where $V = \{v_1, v_2 \dots v_N\}$ is the set of N entities in the KG, $R = \{r_1, r_2 \dots r_M\}$ is the set of M relations in the KG. A triple can be expressed as (v_h, r, v_t), where $v_h, v_t \in V$ stands for head entity and tail entity, respectively. $r \in R$ stands for the link relation. Here we point out that, the KG used in this paper is mapping items to entities, (e.g., a book item can usually be mapped to an entity describing this book), where these entities are regarded as item entities. Different from other network embedding learning methods, Trans R considering relations and entities are completely different objects and may not be represented in a common semantic space.

In this paper, for each triple (v_h, r, v_t), entities are projected into latent embedding $v_h, v_t \in R^k$, relations are embedded with $r \in R^d$, with a fully connected layer parameterized by $\{W^e, W^r\}$, where k, d denotes dimensions of entity and relation space, respectively, and $k \neq d$. For each type of relation, Trans R sets up a mapping matrix $M_r \in R^{k \cdot d}$ that transforms the entities from the entity space to the relation space: $v_h^r = v_h M_r, v_t^r = v_t M_r$, and $\| v_h^r \|_2 \leq 1 \| v_t^r \|_2 \leq 1$. The scoring function of this triple is defined as: $f_r(v_h^r, v_t^r) = \|v_h^r + r - v_t^r\|_2^2$. We argue that the link relations stored in the KG can be considered as explicit evidence of different item attributes, considering the users have corresponding interest representation for different items attributes spaces, naturally, the relations in KG can be used as a guide to obtain this attribute-level user interests.

3.2 Attribute-Level Interest Extraction

To obtain user interest representation over different item attribute spaces, we design an attribute-level interest extraction layer $f_{extractor}(\cdot)$, which takes the mapped entities collection of users historical interacted items

$$V_u = \{v_{e_j}, | j \in N_u^+, v_{e_j} \in V\} \tag{1}$$

and the whole KG structure as input, where N_u^+ denotes users historical interacted items group. The output is the set of user's attribute-level interests:

$$I_{\mathrm{u}} = f_{extractor}(V_u, G = (V, R)) \tag{2}$$

Take the entity that corresponds to the historical interacted items as the seeds, under the guidance of relation in KG, according to different relations, it aggregates entities connected with the seeds, respectively, as the representation of user interest in the corresponding item attribute space:

$$\begin{aligned}
I_u^{(1)} &= \sum_{j \in \mathrm{N}_x^+} \{v_t, v_{h_j}, | \left(v_{h_j} \mathrm{M}_{r_1}, \mathrm{r}_1, v_t \mathrm{M}_{r_1}\right) \in \mathrm{G} \text{ and } v_{h_j} \in \mathrm{V}_u\} \\
I_u^{(2)} &= \sum_{j \in \mathrm{N}_x^+} \{v_t, v_{h_j}, | \left(v_{h_j} \mathrm{M}_{r_2}, \mathrm{r}_2, v_t \mathrm{M}_{r_2}\right) \in \mathrm{G} \text{ and } v_{h_j} \in \mathrm{V}_u\} \\
&\dots \\
I_u^{(M)} &= \sum_{j \in \mathrm{N}_x^+} \{v_t, v_{h_j}, | \left(v_{h_j} \mathrm{M}_{r_M}, \mathrm{r}_M, v_t \mathrm{M}_{r_M}\right) \in \mathrm{G} \text{ and } v_{h_j} \in \mathrm{V}_u\}
\end{aligned} \tag{3}$$

At the same time we need to ensure: $r_1 \neq r_2 \neq ... \neq r_M, \quad r_1, r_2 ... r_M \in R$. Besides, it is worth highlighting that:

- In most cases, there are more than one historical interacted items for the user, taking the book Dataset as an example, the users interest representation in the author attribute space is composed of multiple entities, which makes current feature representation reflects the users comprehensive interest in author attribute of books, it may be the authors writing style, nationality, etc., not a specific author.
- When capturing the users attribute-level interests, the users historical interacted items are mapped to different item attribute spaces by corresponding relation mapping matrices, then extract the representation of user interest in each attribute space. Different relations focus on different levels of the entity, and the same entity will have different characteristics in different item attribute spaces.
- Although users prefer the items that match their interests the most, users and items are hard to match completely due to their mixed composition. In general scenarios, a user chooses an item when several specific attributes of the item attract the user, i.e., interactive behavior occurs with partially matching between the users interests and the items attributes. Observing this phenomenon, we believe that compared to using a latent, mixed vector to represent user interests, the fine-grained users' interest representations over the different item attributes spaces is a more appropriate way to model users' interests.

3.3 Attribute-Level Interest Matching Layer

After obtaining the users' interest representation in different item attribute spaces:

$$I_{\mathrm{u}} = \left\{ I_{\mathrm{u}}^{(1)}, I_{\mathrm{u}}^{(2)}, ..., I_{\mathrm{u}}^{(M)} \right\} \tag{4}$$

The proposed AIMN devises the interest matching module. To match the attribute-level user interests reasonably, we also map the target item to different item attribute spaces by corresponding relation mapping matrices, specifically:

$$v_i^{r_n} = v_i M_{r_n}, \; n = 1, 2, ..., M \tag{5}$$

The users' interest matching in each item attribute space will contribute to the final predicted score:

$$\hat{y}_{ui} = f_{score}(I_u, v_i) = \sigma\Big(\prod_{1 \leq n \leq M} (v_i^{r_n})^T I_u^n \Big) \tag{6}$$

where $\sigma(\mathrm{x}) = \frac{1}{1+\exp(-\mathrm{x})}$ is the sigmoid function to restrict the predicted interaction score in the interval $(0, 1)$. For each user, the interest matching module outputs predicted interaction scores of all items to be recommended, sort all items in descending order according to the score, and generate a recommendation list for the user.

3.4 Model Optimization

The loss function of proposed AIMN consists of two parts: the pairwise BPR loss to optimize the model parameters, specifically, it encourages the predicted score of a users interaction items to be higher than that of unobserved items:

$$\mathrm{loss_{BPR}} = \sum_{(u,i,j)\in \mathrm{Y}^{+}\cup \mathrm{Y}^{-}} -\log\left(\mathrm{y}_{ui} - \mathrm{y}_{uj}\right) + \lambda\|\theta\|^{2} \tag{7}$$

where Y^{+} and Y^{-} denote respectively positive and negative interaction pairs of users and items. Hyper-parameter λ acts to avoid over fitting with L2-regularization on the model parameter set θ, including $\left\{W^{u}, W^{i}\right\}$, $\left\{W^{e}, W^{r}\right\}$ (one-hot ID embeddings of user u and item i are projected with a fully connected layer parameterized by $\left\{W^{u}, W^{i}\right\}$).

The other is Knowledge representation learning loss:

$$\mathrm{loss_{KG}} = \sum_{(v_h, r, v_t, v_t')\in \mathrm{G}} \log\left(\sigma\left(\|v_h \mathrm{M}_r + r - v_t \mathrm{M}_r\|_2^2 - \left\|v_h \mathrm{M}_r + r - v_t' \mathrm{M}_r\right\|_2^2\right)\right) + \lambda\left\|\theta'\right\|^{2} \tag{8}$$

where (h, r, v_t) is correct triples and (h, r, v_t') is incorrect triples, $\theta^{'}$ includes $\{W^{\mathrm{e}}, W^{\mathrm{r}}\}$. During the model training, we simultaneous optimize the $\mathrm{loss_{BPR}}$ and $\mathrm{loss_{KG}}$ with Adam optimizer.

4 Experiment

4.1 Experimental Settings

To verify the validity of the proposed AIMN, we conduct experiments on two publicly available Datasets: Amazon-book(book) and Last-FM(music). We retaining users and items with at least ten interactions for both Datasets, the construction of Knowledge Graph is the same as the setting in KGAT [15]. Table 1 summarizes the detailed characteristics of Amazon-book and Last-FM Datasets. The proposed AIMN model has been implemented in the environment of TensorFlow 1.8.0 and python 3.6.5. For AIMN, the positive sample ratio of the training set and the test set is 4:1. To compare the performance fairly, the embedding size is fixed to 64 for the AIMN and other baseline methods, the batch size is set to 1024. We perform AIMN with the coefficient of L2 of 0.0001 and a learning rate of 0.001. For each user in the test set, both AIMN and baseline methods output each users prediction scores over all items, except for the positive items that appeared in the training set. The performance is quantitatively evaluated with recall at rank N (recall@N) and Normalized Discounted Cumulative Gain at rank N (ndcg@N). In this paper, we set $N = 20$ and demonstrate the average metrics performance for all users in the test set.

Table 1. Statistics of the datasets

		Amazon-book	Last-FM
User-item interactions	#Users	70,679	23,566
	#Items	24,915	48,123
	#Interactions	847,733	3,034,796
Knowledge graph	#Entities	88,572	58,266
	#Relations	39	9
	#Triplets	2,557,746	464,567

4.2 Performance Comparison

We compare the proposed AIMN for personalized recommendation with several effective methods, including ID-based (MF [9]), KG-based (RippleNet [14], CKE [21], CFKG [1], KGAT [15]), graph neural network-based (GC-MC [2]) methods. We demonstrate the personalized recommendation performance for all methods mentioned above, the performance results are presented in Table 2, it can be observed that:

- On the whole, the proposed AIMN outperforms its baseline methods on both datasets. Specifically, AIMN yields an improvement over the most substantial baseline with recall@20 by 0.67%, 15.06% and ndcg@20 by 4.08%, 18.34%, respectively, on Amazon-book and Last-FM. The performance verifies the effectiveness of the proposed AIMN in modeling the interest distribution over different attributes item space.
- The RippleNet, CKE, and CFKG are three KG-based methods, which introduce knowledge graph to enrich the representation of users and items. Both of them outperform MF by a large margin in two Datasets, which

Table 2. Performance comparison by recall@20 and ndcg@20.

	Amazon-book		Last-FM	
	recall@20	ndcg@20	recall@20	ndcg@20
MF	0.1172	0.0798	0.0688	0.1097
CFKG	0.1142	0.0770	0.0723	0.1143
CKE	0.1343	0.0885	0.0736	0.1184
GC-MC	0.1316	0.0874	0.0818	0.1253
KGAT	**0.1489**	**0.1006**	**0.0870**	**0.1352**
RippleNet	0.1336	0.0910	0.0791	0.1238
AIMN	**0.1499**	**0.1047**	**0.1001**	**0.1568**
%Improve	0.67%	4.08%	15.06%	18.34%

demonstrates the role of KG in improving recommendation performance. GC-MC is a graph neural network-based method, which employs GCN encoder on graph-structured data, and brings a certain of computational complexity compare with AIMN.

- The RippleNet obtain mixed user interest representation by preference propagation along the relation path in the KG, compared with CKE and CFKG, performance has improved. But it still underperforms AIMN due to a lack of consideration of attribute-level matching between user interest and item. It inspires us to consider the distribution of user interest in the different item attributes spaces. On the other hand, compare with AIMN, when RippleNet calculates the matching of users and items, the representation of the item is unified, which may make a bottleneck for matching users and items.
- As can be seen from the experimental results, The performance of MF is lower than the proposed AIMN. This can be regarded as an ablation experiment: Since AIMN is built on the basis of MF, the experimental results show that with the assistance of relations in KG, AIMN generates attribute-level user interest representations over different attributes item spaces is effective and worth trying.

4.3 Case Study

We now explore the interpretability of matching representations between user and item calculated by the proposed AIMN. In Amazon-book datasets, we randomly selected one user u_{243} and one item i_{340} (the item that the user will interact with), Fig. 3 shows the visualization of the multiple knowledge connections between the user's historical items and the target item $i_3 40$ (left), according

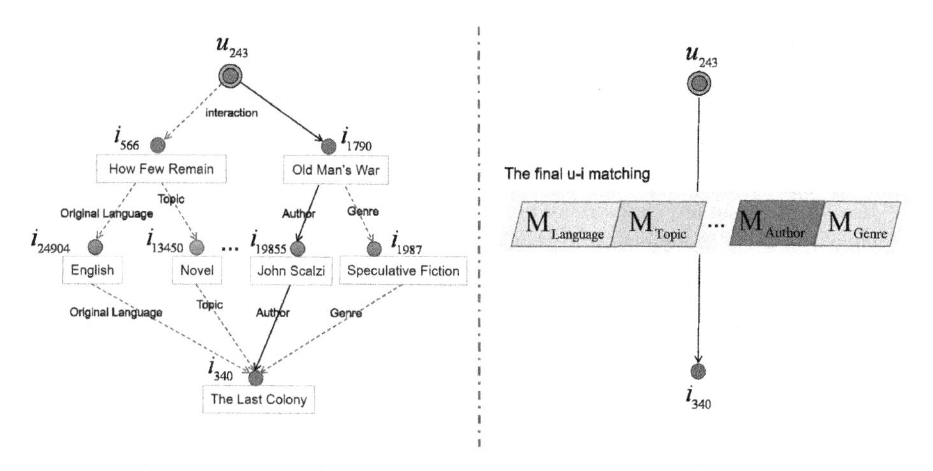

Fig. 3. A Real Example from Amazon-Book (left) and users interest matching in each item attribute space will contribute to the final predicted result (right).

to reviews, the main reason why u_{243} chooses to interact with i_{340} maybe her preference for author John Scalzi, labeled with the solid black line.

In this paper, we believe that the user's interest matching in each item attribute space will contribute to the final predicted score. In this example, the match between the user and the item under the attribute space of 'author' contributes the most to the final prediction score, as illustrated in Fig. 3 (right), labeled with the scarlet rectangle.

5 Conclusion

In this paper, we have proposed an Attribute-level Interest Matching Network (AIMN) to address personalized recommendation task, which models the interest distribution over different attributes item spaces with the assistance of relations in the knowledge graph, and matches user and item in each item attribute space, respectively. We highlight the importance of obtaining attribute-level user interest since mixed and latent factors easily result in suboptimal user representations; The matching between user interest and item in each item attribute space will contribute to the final predicted result. Compared with other personalized recommendation algorithms, experimental results confirmed that the proposed AIMN is more effective.

Acknowledgement. This work was supported by the National Natural Science Foundation of China under Grant No. 61702022, No. 61802011, No. 61976010, Beijing Municipal Education Committee Science Foundation under Grant No. KM201910005024, Inner Mongolia Autonomous Region Science and Technology Foundation under Grant NO. 2021GG0333, and Beijing Postdoctoral Research Foundation under Grant No. Q6042001202101.

References

1. Ai, Q., Azizi, V., Chen, X., Zhang, Y.: Learning heterogeneous knowledge base embeddings for explainable recommendation. Algorithms **11**(9), 137 (2018)
2. Berg, R.V.D., Kipf, T.N., Welling, M.: Graph convolutional matrix completion. In: KDD (2017)
3. Bordes, A., Usunier, N., García-Durán, A., Weston, J., Yakhnenko, O.: Translating embeddings for modeling multi-relational data. In: Advances in Neural Information Processing Systems, pp. 2787–2795 (2013)
4. Chen, J., Zhang, H., He, X., Nie, L., Liu, W., Chua, T.S.: Attentive collaborative filtering: multimedia recommendation with item- and component-level attention. In: SIGIR, pp. 335–344 (2017)
5. Guo, Q., et al.: A survey on knowledge graph-based recommender systems. CoRR (2020)
6. He, X., He, Z., Song, J., Liu, Z., Jiang, Y., Chua, T.: NAIS: neural attentive item similarity model for recommendation. IEEE Trans. Knowl. Data Eng. **30**(12), 2354–2366 (2018)
7. He, X., Liao, L., Zhang, H., Nie, L., Hu, X., Chua, T.: Neural collaborative filtering. In: The 26th World Wide Web Conference, pp. 173–182 (2017)

8. Kabbur, S., Ning, X., Karypis, G.: FISM: factored item similarity models for top-n recommender systems. In: Proceedings of the 19th ACM SIGKDD International Conference on Knowledge Discovery and Data Mining, pp. 659–667 (2013)
9. Koren, Y., Bell, R.M., Volinsky, C.: Matrix factorization techniques for recommender systems. IEEE Comput. **42**(8), 30–37 (2009)
10. Li, C., et al.: Multi-interest network with dynamic routing for recommendation at Tmall. In: Proceedings of the 28th ACM International Conference on Information and Knowledge Management, pp. 2615–2623 (2019)
11. Lin, Y., Liu, Z., Sun, M., Liu, Y., Zhu, X.: Learning entity and relation embeddings for knowledge graph completion. In: Proceedings of the Twenty-Ninth AAAI Conference on Artificial Intelligence, pp. 2181–2187 (2015)
12. Shi, G., Feng, C., Xu, W., Liao, L., Huang, H.: Penalized multiple distribution selection method for imbalanced data classification. Knowl. Based Syst. 1–9 (2020)
13. Tewari, A.S.: Generating items recommendations by fusing content and user-item based collaborative filtering. Procedia Comput. Sci. **167**, 1934–1940 (2020)
14. Wang, H., et al.: Ripplenet: propagating user preferences on the knowledge graph for recommender systems. In: Proceedings of the 27th ACM International Conference on Information and Knowledge Management, pp. 417–426 (2018)
15. Wang, X., He, X., Cao, Y., Liu, M., Chua, T.S.: KGAT: knowledge graph attention network for recommendation, pp. 950–958. ACM (2019)
16. Wang, X., He, X., Nie, L., Chua, T.: Item silk road: recommending items from information domains to social users. In: Proceedings of the 40th International ACM SIGIR Conference, 7–11 August 2017, pp. 185–194 (2017)
17. Wang, X., Jin, H., Zhang, A., He, X., Xu, T., Chua, T.: Disentangled graph collaborative filtering. In: Proceedings of the 43rd International ACM SIGIR Conference, 25–30 July 2020, pp. 1001–1010 (2020)
18. Xue, F., He, X., Wang, X., Xu, J., Liu, K., Hong, R.: Deep item-based collaborative filtering for top-n recommendation. ACM Trans. Inf. Syst. **37**(3), 33 (2019)
19. Ying, R., He, R., Chen, K., Eksombatchai, P., Hamilton, W.L., Leskovec, J.: Graph convolutional neural networks for web-scale recommender systems. In: Proceedings of the 24th ACM Conference, 19–23 August 2018, pp. 974–983 (2018)
20. Yuan, F., He, X., Karatzoglou, A., Zhang, L.: Parameter-efficient transfer from sequential behaviors for user modeling and recommendation. In: Proceedings of the 43rd International ACM SIGIR Conference, 25–30 July 2020, pp. 1469–1478 (2020)
21. Zhang, F., Yuan, N.J., Lian, D., Xie, X., Ma, W.Y.: Collaborative knowledge base embedding for recommender systems. In: The 22nd ACM SIGKDD International Conference, pp. 353–362 (2016)
22. Zhou, G., et al.: Deep interest network for click-through rate prediction. In: Proceedings of the 24th ACM SIGKDD International Conference, pp. 1059–1068 (2017)

Variational Deep Representation Learning for Cross-Modal Retrieval

Chen Yang[1], Zongyong Deng[1], Tianyu Li[1], Hao Liu[1,2], and Libo Liu[1,2(✉)]

[1] School of Information Engineering, Ningxia University, Yinchuan 750021, China
liuhao@nxu.edu.cn

[2] Collaborative Innovation Center for Ningxia Big Data and Artificial Intelligence Co-founded by Ningxia Municipality and Ministry of Education, Yinchuan 750021, China

Abstract. In this paper, we propose a variational deep representation learning (VDRL) approach for cross-modal retrieval. Numerous existing methods map the image and text to the point representations, which is challenging to model the semantic multiplicity of the sample. To address this issue, our VDRL aims to map the image and text to the semantic distributions and measure the similarity by comparing the difference between their distributions. Specifically, our VDRL network is trained under three constraints: 1) The Variational Autoencoder loss is minimized to learn the distributions of both the images in image semantic space and the texts in text semantic space. 2) The mutual information is introduced to ensure the VDRL learns the intact distribution for the sample. 3) The triplet hinge loss is incorporated to align the distributions of the images and texts at the semantic level. Consequently, the semantic multiplicity of each sample is modeled in our method. Experimental results demonstrate that our approach achieves compelling performance with state-of-the-art methods.

Keywords: Cross-modal retrieval · Representation learning · Variational autoencoder · Deep learning

1 Introduction

The research on cross-modal retrieval has become a popular topic in recent years, which aims to explore the semantic correlations between the samples from different modalities such as image, text, video, and audio [4,9,16,21]. In this paper, we focus on image-text retrieval where it aims to find the sample from the text modality that is semantically similar to the given image query, and vice versa.

Due to the heterogeneous properties (*i.e.*, inconsistent distribution and different representation structure) between image modality and text modality, the

Electronic supplementary material The online version of this chapter (https://doi.org/10.1007/978-3-030-88007-1_41) contains supplementary material, which is available to authorized users.

H. Ma et al. (Eds.): PRCV 2021, LNCS 13020, pp. 498–510, 2021.
https://doi.org/10.1007/978-3-030-88007-1_41

core problem lies in that how to effectively measure the semantic similarity between the image and text. Most methods [22,27] attempted to map images and texts to the points (feature vector) in a common semantic space. Then the similarity between one image-text pair can be measured by their point representations. However, the point representation has limitations in exploring the semantic multiplicity of the sample. Specifically, one image usually consists of multiple regions which have different semantics. Thus, the image can match with multiple texts in cross-modal retrieval as shown in Fig. 1 (a), depending on which region people concern about. At the same time, there may be multiple images corresponding to one text, as people focus on different parts of the text, just like what in Fig. 1 (b). The representation of the image or text is required to model semantic multiplicity in both image to text and text to image retrieval. Unfortunately, the point representation corresponds one specific point in semantic space expressing only the single semantic, which do not meet the above requirement. We further utilize Fig. 2 to elaborate the deficiencies of the point representation. Firstly, as shown in Fig. 2 (a), there exists an area in semantic space where any point in this area can be a candidate representation of the given image, according to semantic multiplicity of the image. Thereby a correct match occurs if the candidate point contains the information that relates to the retrieved text. Otherwise, the candidate point will be a negative representation, leading to a mismatch. Besides, as shown in Fig. 2 (b), the mismatch occurs even if the candidate point preserves the whole information of the image. It is because that a similar image-text pair often share only a portion of semantics. The extra information in obtained representation will pull down the similarity between the target pair in semantic space. In more details, the deficiencies of these methods originates from that they only pick up one candidate point in semantic space as the final representation of the sample, ignoring the correlations between the sample and other points.

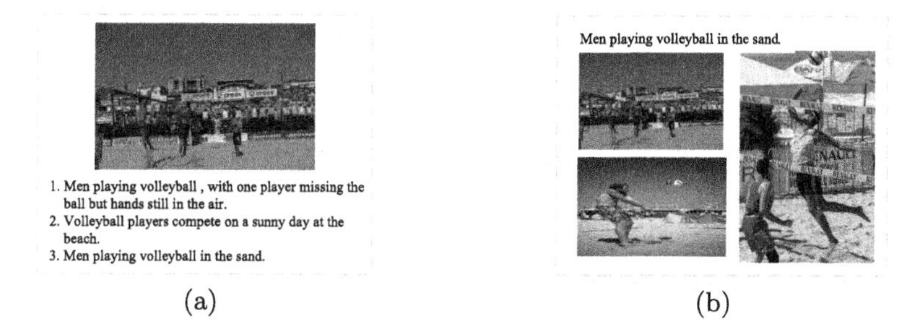

Fig. 1. The illustration of semantic multiplicity. (a) One image may correspond to multiple texts. (b) One text may correspond to multiple images. Every sample contains multiple semantics, depending on how people observe it.

To address this issue, we propose a variational deep representation learning (VDRL) approach for image-text retrieval, which utilizes the Variational Autoencoder (VAE) [11] to capture the semantic distributions of images and texts. The

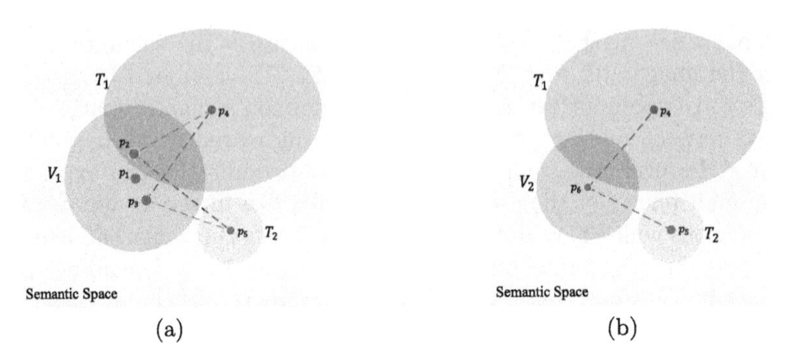

Fig. 2. The illustration of our basic idea. Suppose that each sample corresponds to an area (circle area) in semantic space, according to the semantic multiplicity. And any point in the area of the given sample can be a candidate representation. (a) For similar pair V_1-T_1 and dissimilar pair V_1-T_2, V_1 will match T_1 correctly if V_1 is mapped to point p_2 because p_4 is closer to p_2 than p_5. A mismatch will occur when V_1 is mapped to point p_3. (b) For similar pair V_2-T_1 and dissimilar pair V_2-T_2, there still exists a mismatch if all samples are mapped to the corresponding centers, which preserve the whole information of the samples. A similar image-text pair often share only a portion of semantics. Thus the irrelevant information in centers will reduce the similarity between V_2-T_1.

probability densities of the learned distribution naturally quantify the correlations between the corresponding sample and all the points in semantic space. Then the Wasserstein distance is leveraged to measure the distribution distance instead of the point distance (*e.g.*, cosine or Euclidean distance). Specifically, our VDRL network is trained under three criterions: 1) The VAE loss is minimized to guide the realized Image-Specific VAE (ISV) and Text-Specific VAE (TSV) to learn the distributions in image and text semantic space from the global representations of images and texts respectively. 2) The mutual information (MI) is maximized to enhance the dependence between the input samples and extracted global representations, which enforces the couple of VAEs to capture intact distributions from the global representations. 3) The triplet hinge loss is minimized to align the distributions from different modalities at the semantic level, which enforces the distances of similar image-text pairs to be closer than those of dissimilar pairs. Finally, we jointly optimize all parameters of our network by using the standard gradient descent method. Comprehensive experimental results demonstrate the effectiveness of distribution representation for cross-modal retrieval, including quantitative and qualitative results on Flickr30K and MS-COCO.

The main contributions of our work are summarized as follows:

(1) We propose a variational deep representation learning(VDRL) approach to achieve robust distribution representations benefiting from the Evident Lower Bound in VAE. Compared with conventional point representation, the learned feature distribution conveys the conditional probability densities of all the point representations in semantic space, so that the relationship of the given sample and all point representations is exploited.

(2) We introduce MI theory in our method to enhance the dependence between the samples and their global representations. This principally enforces the extracted global representations to preserve maximal semantic information from the samples. Experimental results show the effectiveness of the proposed method quantitatively and qualitatively.

2 Proposed Method

In this section, we propose to use VAE with MI maximization to learn the semantic distributions of images and texts. After aligning the obtained distributions from image and text spaces, we measure the similarity between one image-text pair by comparing the difference between their distributions.

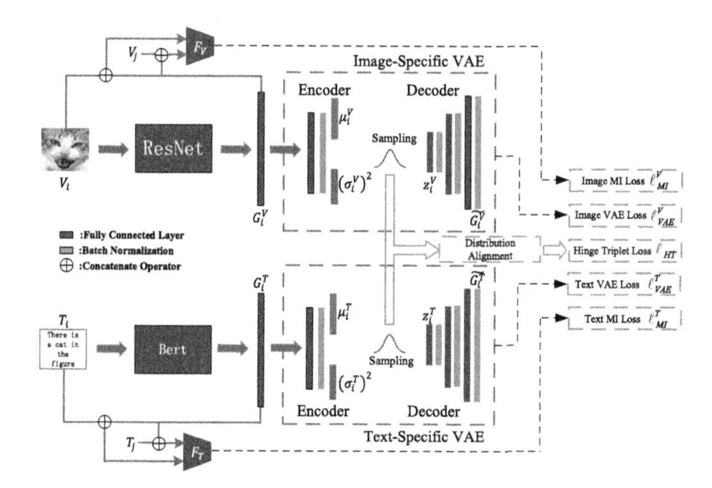

Fig. 3. The framework of the proposed VDRL. Our VDRL network mainly consists of the Image-Specific VAE, the Text-Specific VAE, and the two MI estimators. We first feed the V_i and T_i into ResNet-152 and Bert, to extract the global representations G_i^V and G_i^T. In order to make the global representations preserve all information in the samples, we introduce the MI maximization method to construct an image MI estimator F_V and a text MI estimator F_T. And the outputs of F_V and F_T for images and texts make up the image MI Loss and the text MI Loss. Secondly, we feed G_i^V into ISV to capture the semantic distribution of V_i in image semantic space and feed G_i^T into TSV to capture the semantic distribution of T_i in text semantic space. The image VAE loss and the text VAE loss are constructed to optimize the two VAEs. Finally, we utilize a bi-directional hinge triplet loss with Wasserstein distance to align the obtained image and text distributions at the semantic level.

2.1 Notations

Let $V = \{V_i\}_{i=1}^n$ denote the image set, $T = \{T_i\}_{i=1}^n$ denote the text set, and $S = \{V_i, T_i\}_{i=1}^n$ denote the set of image-text pairs where V_i and T_i are similar at the semantic level. As shown in Fig. 3, we leverage ResNet-152 [7] to obtain the global representation from V_i, which is denoted as G_i^V. Meanwhile, we extract the global representation G_i^T of T_i by using Bert [3]. Then we feed the extracted G_i^V and G_i^T into ISV and TSV respectively, in order to obtain the corresponding mean vectors and variance vectors μ_i^V, $(\sigma_i^V)^2$, and μ_i^T, $(\sigma_i^T)^2$, from the encoder of ISV and TSV. $\widetilde{G_i^V}$ and $\widetilde{G_i^T}$ are the outputs of the ISV decoder and TSV decoder. To better clarify our method, we list the meanings of the above symbols in **Section A** of the supplementary material.

2.2 Modality-Specific VAE

The goal of our framework is to estimate the semantic distribution of the given image or text. Here, we consider a real-world dataset (*e.g.*, the image or the text dataset) $X = \{x_i\}_{i=1}^n$, where each element contains some semantic information and is an i.i.d. sample from a continuous random variable x. We assume the data are generated randomly, involving a continuous random variable z in the semantic space. The point z_i can be sampled from the variable z, which represents the specific semantic information. Then the posterior density $p(z_i|x_i)$ quantitatively reflects the correlation between z_i and x_i. Hence, the semantic distribution of x_i can be denoted as the posterior distribution $z|x_i$, which reflects the correlations between x_i and all the points in semantic space. Obviously, it is intractable to directly compute $p(z|x_i)$. By following the scheme of variational inference (VI), we define a distribution function $q(z|x_i)$ to approximate the $p(z|x_i)$ by minimizing the KL divergence between them. The optimization formula is defined as:

$$\ell_{kl} = D_{KL}(q(z|x_i)||p(z|x_i)), \tag{1}$$

where $D_{KL}(\cdot||\cdot)$ denotes the KL divergence of the two distributions. However, the $p(z|x_i)$ is unknown such that we can not estimate $q(z|x_i)$ by optimizing formula 1. Then the negative Evident Lower Bound (ELBO) derived from VI can be a substitution of the right part of the equal sign in formula 1. The optimization formula is changed to:

$$\ell_{kl} = D_{KL}(q(z|x_i)||p(z)) - E_{q(z|x_i)}[logp(x_i|z)]. \tag{2}$$

Inspired by [11], the functions $q(z|x_i)$ and $p(x_i|z)$ in formula 2 can be modeled as an encoder and a decoder, constituting a VAE. The formula 2 can be further written as follows:

$$\ell_{kl} = D_{KL}(q(z|x_i)||p(z)) + ||x_i - \tilde{x}_i||_2^2, \tag{3}$$

where $\tilde{x}_i$ is generated by the decoder with the input vector z_i sampled from $z|x_i$. It represents the reconstruction of x_i. We call the two terms regularization term and reconstruction term. In general, the continuous z in VAE is assumed to obey the centered isotropic multivariate Gaussian distribution, formulated as $z \sim \mathcal{N}(0, I)$. The $z|x_i$ is assumed to obey multivariate Gaussian distribution with a diagonal covariance structure formulated as $z|x_i \sim \mathcal{N}(\mu_i, (\sigma_i)^2 I)$, where

I denotes the identity matrix, μ_i and $(\sigma_i)^2$ are the mean vector and the diagonal covariance matrix of $z|x_i$. Thus, the distribution $z|x_i$ can be represented by the output mean vector and diagonal covariance matrix of the encoder. Note that we use a variance vector to represent the diagonal covariance matrix in practice.

In our framework, we construct the Image-Specific VAE (ISV) for image modality and the Text-Specific VAE (TSV) for text modality to capture the semantic distributions of each image V_i and text T_i, by minimizing the VAE loss of them. In details, we use the global representation G_i^V as the input of the ISV and G_i^T as the input of the TSV. The semantic distribution of V_i is represented by the output mean vector μ_i^V and variance vector $(\sigma_i^V)^2$ of the ISV encoder. Similarly, the semantic distribution of T_i is represented by the output μ_i^T and $(\sigma_i^T)^2$ of the TSV encoder. Consequently, the VAE loss of image V_i is written as follows:

$$\ell_{i\text{-}VAE}^V = \beta_1 \cdot D_{KL}(q(z^V|G_i^V)||p(z^V)) + ||G_i^V - \widetilde{G_i^V}||_2^2. \quad (4)$$

And the VAE loss of text T_i can be written as:

$$\ell_{i\text{-}VAE}^T = \beta_2 \cdot D_{KL}(q(z^T|G_i^T)||p(z^T)) + ||G_i^T - \widetilde{G_i^T}||_2^2. \quad (5)$$

Then, the total VAE loss of the image-text pairs in set S is written as:

$$\ell_{VAE} = \frac{1}{n}\sum_{i=1}^{n} \ell_{i\text{-}VAE}^V + \frac{1}{n}\sum_{i=1}^{n} \ell_{i\text{-}VAE}^T, \quad (6)$$

where z^V is the random variable in the image semantic space and z^T is the random variable in the text semantic space, respectively. $\widetilde{G_i^V}$ and $\widetilde{G_i^T}$ are the outputs of the decoders in ISV and TSV. And β_1, β_2 are used to balance the influence of the terms in ISV and TSV.

2.3 MI Maximization

In our proposed framework, we enhance the dependence between V_i and G_i^V, T_i and G_i^T, by maximizing their mutual information, to make the G_i^V and G_i^T retain as much semantic information as possible from the raw inputs. Mutual information is a fundamental quantity for measuring the dependence of two random variables [1]. Given two random variables x and y, the mutual information between them is written as follows:

$$MI(x;y) = \iint p(x,y)log(\frac{p(x,y)}{p(x)p(y)})dxdy, \quad (7)$$

where $p(x,y)$ is the joint probability of x and y, $p(x)$ and $p(y)$ are the marginal probabilities. Nevertheless, it is intractable to compute the mutual information according to formula 7. As mentioned in [8], a Jensen-Shannon-based optimization objective can be the substitute of this formula. Thus we constrain the input samples and their global representations by maximizing the Jensen-Shannon-based objective instead of maximizing the former. For the input images and their global representations, the objective function is:

$$I_{JSD}(V, G^V) = E_V[-sp(-F_V(V_i, G_i^V))] - E_{V\times\widetilde{V}}[sp(F_V(V_j, G_i^V))], \quad (8)$$

where V and $\widetilde{V}$ are both the image set, V_i is the sample from V and V_j is the dissimilar sample to V_i sampled from $\widetilde{V}$. $sp(\cdot)$ is the softplus function. And $F_V(\cdot,\cdot)$ is any class of functions that outputs a MI score. Similarly, for the text inputs and their global representations, the objective function is:

$$I_{JSD}(T, G^T) = E_T[-sp(-F_T(T_i, G_i^T))] - E_{T\times\widetilde{T}}[sp(F_T(T_j, G_i^T))]. \quad (9)$$

In practice, we construct an image MI estimator and a text MI estimator by modeling the function $F_V(\cdot,\cdot)$ and function $F_T(\cdot,\cdot)$ as neural networks, respectively. During training, we define the negative $I_{JSD}(V, G^V)$ and the negative $I_{JSD}(T, G^T)$ as our image and text MI loss because we optimize the estimators by gradient descent method. Then the total MI loss is written as follows:

$$\begin{aligned} \ell_{MI} &= \ell_{MI}^V + \ell_{MI}^T \\ &= -I_{JSD}(V, G^V) - I_{JSD}(T, G^T). \end{aligned} \quad (10)$$

2.4 Semantic Alignment

Based on the above statement, we can estimate the semantic distributions of an image-text pair from ISV and TSV, respectively. In test phrase, our method measure the similarity of them by computing the Wasserstein distance [6] between their extracted distribution representations, unlike most methods that compute cosine or Euclidean distance between the extracted point representations. Following [6], since the obtained distributions in our framework are assumed to obey multivariate Gaussian distribution with a diagonal covariance structure and represented by the output mean vectors and variance vectors of the VAE encoders, the Wasserstein distance can be simplified to:

$$D_W(a, b) = (||\mu_a - \mu_b||_2^2 + ||\sigma_a - \sigma_b||_2^2)^{\frac{1}{2}}, \quad (11)$$

where a and b are samples, μ_a and μ_b denote the mean vectors of the distributions belonging to a and b, respectively. And σ_a, σ_b denote the element-wise sqrt of variance vectors.

However, the obtained image distribution and text distribution are from different semantic spaces. That is, the same value sampled from variables z^V and z^T may represent different meanings, which makes the obtained distributions of the similar image-text pair be dissimilar. Accordingly, we construct a bi-directional hinge triplet loss in training phrase to align the distributions from different modalities at the semantic level, which makes the distances of similar image-text pairs be closer than those of dissimilar pairs. Specifically, we first find the similar image-text pair and the dissimilar image-text pair for every sample in datasets V and T. For an anchor image V_i, we use the text T_i similar to V_i to construct a similar pair V_i-T_i. Inspired by the hard negative mining [5,17], we construct a dissimilar pair V_i-T_k by picking up the text which has the closest distribution distance with V_i among all the texts dissimilar to V_i. In the same way, we construct a similar pair T_i-V_i and a dissimilar pair T_i-V_l for an anchor text T_i. Then we derive the hinge triplet loss for V_i, denoted as:

$$\ell^{V}_{i\text{-}HT} = max(0, \alpha_1 + D_W(V_i, T_i) - D_W(V_i, T_k)), \tag{12}$$

and the hinge triplet loss for T_i, denoted as:

$$\ell^{T}_{i\text{-}HT} = max(0, \alpha_2 + D_W(T_i, V_i) - D_W(T_i, V_l)), \tag{13}$$

where α_1 and α_2 are the margins between the distances of similar pairs and dissimilar pairs. The total bi-directional hinge triplet loss is written as follows:

$$\ell_{HT} = \frac{1}{n}\sum_{i=1}^{n} \ell^{V}_{i\text{-}HT} + \frac{1}{n}\sum_{i=1}^{n} \ell^{T}_{i\text{-}HT}. \tag{14}$$

Finally, we optimize the proposed method by minimizing the objective denoted as:

$$\ell = \gamma_1 \ell_{VAE} + \gamma_2 \ell_{MI} + \gamma_3 \ell_{HT}, \tag{15}$$

where γ_1, γ_2 and γ_3 are scalars used to balance the influence of above loss terms. The term ℓ_{VAE} enforces our VDRL network to capture the image distributions and the text distributions from corresponding global representations. The term ℓ_{MI} denotes the dependence between the input samples and their global representations. The term ℓ_{HT} ensures that the distributions from different spaces are aligned at the semantic level.

3 Experiments

In this section, to verify the effectiveness of our method, we show the results of the quantitative and qualitative experiments on two widely used cross-modal retrieval datasets, Flickr30K [26] and MS-COCO [13].

3.1 Datasets and Evaluation Metric

MS-COCO is one of the most popular datasets on image-text retrieval task. It contains 123,287 images. Each image is described by five sentences so that five image-text pairs can be obtained. Following [10], we used 113,287 images for training (*i.e.*, 566,435 image-text training pairs), 5,000 images for validation (*i.e.*, 25,000 image-text validation pairs), and 5,000 images for testing (*i.e.*, 25,000 image-text testing pairs). Flickr30K consists of 31,000 images and every image possesses five sentences as description. Following [10], we used 1,000 images (5,000 pairs) for validation, 1,000 images (5,000 pairs) for testing, and the rest of it (145,000 pairs) for training. According to the previous works [5,22,23], we reported results by using Recall@k (R@k), which measures the fraction of queries for which the correct item is retrieved among the top k results. Additionally, We provide implementation details of our method in **Section B** of the supplementary material.

3.2 Results and Analysis

We report the quantitative results on Flickr30K and MS-COCO in the main paper, while we provide qualitative results on Flickr30K in **Section C** of the supplementary material. In quantitative experiments, we compared our method with several representative methods. For Flickr30K, we reported results on 1K test images. For MS-COCO, we reported averaged results on 5 folds of 1K test images and results on full 5K test images.

Comparisons on Flickr30K: Table 1 presents the retrieval performance of state-of-the-art methods on Flickr30K dataset in both image to text retrieval and text to image retrieval. As shown in Table 1, our method achieves 63.5%, 86.6%, 92.6% for R@1, R@5, R@10 in image to text retrieval and achieves 50.5%, 79.5%, 86.5% in text to image retrieval. From the results, our method outperforms all compared methods.

Table 1. Comparisons of the cross-modal retrieval on the Flickr30K dataset in terms of R@k.

Method	Image-to-Text			Text-to-Image		
	R@1	R@5	R@10	R@1	R@5	R@10
DSPE [22]	40.3	68.9	79.9	29.7	60.1	72.1
VSE++ [5]	52.9	80.5	87.2	39.6	70.1	79.5
CMPM [27]	49.6	76.8	86.1	37.3	65.7	75.5
JGCAR [23]	44.9	75.3	82.7	35.2	62.0	72.4
CSE [25]	44.6	74.3	83.8	36.9	69.1	79.6
TIMAM [19]	53.1	78.8	87.6	42.6	71.6	81.9
CycleMatch [15]	58.6	83.6	91.6	43.6	75.3	84.2
MIITE (MSE) [14]	56.6	83.0	88.8	42.1	71.5	80.2
DPC [28]	55.6	81.9	89.5	39.1	69.2	79.1
CyTIR-Net [2]	56.6	82.2	88.7	41.5	69.0	79.1
DMASA [12]	57.7	82.5	89.4	42.7	70.8	80.3
Ours	**63.5**	**86.6**	**92.6**	**50.5**	**79.5**	**86.5**

Comparisons on MS-COCO: For MS-COCO dataset, we evaluated our method on both 1K testing set and 5K testing set. We reported the 1K testing set results in Table 2. As shown in Table 2, our method achieves 65.7% for

Table 2. Comparisons of the cross-modal retrieval on the MS-COCO 1K testing set in terms of R@k.

Method	Image-to-Text			Text-to-Image		
	R@1	R@5	R@10	R@1	R@5	R@10
DSPE [22]	50.1	79.7	89.2	39.6	75.2	86.9
VSE++ [5]	64.6	90.0	95.7	52.0	84.3	92.0
CMPM [27]	56.1	86.3	92.9	44.6	78.8	89.0
JGCAR [23]	52.7	82.6	90.5	40.2	74.8	85.7
CSE [25]	56.3	84.4	92.2	45.7	81.2	90.6
UniVSE [24]	64.3	89.2	94.8	48.3	81.7	91.2
CycleMatch [15]	61.1	86.8	94.2	47.9	80.9	90.9
MIITE (MSE) [14]	**66.1**	91.1	**96.4**	**53.3**	84.7	92.4
DPC [28]	65.6	89.8	95.5	47.1	79.9	90.0
CyTIR-Net [2]	57.9	87.3	94.2	47.9	82.2	91.3
DMASA [12]	65.4	**91.3**	95.8	52.6	81.2	92.1
Ours	65.7	90.2	96.3	**53.3**	**86.0**	**93.6**

Table 3. Comparisons of the cross-modal retrieval on the MS-COCO 5K testing set in terms of R@k.

Method	Image-to-Text			Text-to-Image		
	R@1	R@5	R@10	R@1	R@5	R@10
VSE++ [5]	41.3	**71.1**	81.2	30.3	59.4	72.4
CMPM [27]	31.1	60.7	73.9	22.9	50.2	63.8
CSE [25]	27.9	57.1	70.4	22.2	50.2	64.4
UniVSE [24]	36.1	66.4	77.7	25.4	53.0	66.2
DPC [28]	41.2	70.5	81.1	25.3	53.4	66.4
Ours	**42.0**	**71.1**	**82.1**	**31.5**	**61.9**	**74.6**

Table 4. The effects of different backbone combinations on Flickr30K dataset.

Backbone	Image-to-Text			Text-to-Image		
	R@1	R@5	R@10	R@1	R@5	R@10
VGG, ELMO	42.4	70.9	81.6	32.5	62.9	74.7
VGG, Bert	50.6	77.5	86.6	39.1	71.3	82.3
ResNet, ELMO	52.2	80.3	88.1	39.3	69.3	79.4
ResNet, Bert	**63.5**	**86.6**	**92.6**	**50.5**	**79.5**	**86.5**

Table 5. The effects of MI loss term on our framework.

Method	Image-to-Text			Text-to-Image		
	R@1	r@5	R@10	R@1	R@5	R@10
VDRL (without MI)	62.7	86.2	91.3	48.8	78.3	**86.5**
VDRL (with MI)	**63.5**	**86.6**	**92.6**	**50.5**	**79.5**	**86.5**

Table 6. The effects of different values of β_1 and β_2.

Method	Image-to-Text			Text-to-Image		
	R@1	r@5	R@10	R@1	R@5	R@10
$\beta_1 = \beta_2 = 0.01$	63.2	**87.5**	**92.7**	49.1	78.8	86.3
$\beta_1 = \beta_2 = 0.1$	**63.5**	86.6	92.6	**50.5**	**79.5**	**86.5**
$\beta_1 = \beta_2 = 0.5$	4.7	15.7	23.7	4.9	15.4	22.7

R@1, 90.2% for R@5, 96.3% for R@10 in image to text retrieval and achieves 53.3%, 86.0%, 93.6% in text to image retrieval. Our method outperforms most methods in Table 3. We reported the 5K testing set results in Table 3, where our method achieves 42.0%, 71.1%, 82.1% for R@1, R@5, R@10 in image to text retrieval and 31.5%, 61.9%, 74.6% in text to image retrieval. We can see that our method indeed achieves pretty performance on 1K testing set. However, the performance on 5K testing set drops a lot, which indicates that our method is not ideal enough for generalization.

Analysis: We conducted the ablation study only on Flickr30K dataset. Firstly, we evaluated our method by using different backbones to extract global representations for images and texts. Specifically, we conducted the experiments by replacing ResNet and Bert with VGG19 [20] and ELMO [18], respectively. The results of different backbone combinations are shown in Table 4. We see that the combination of ResNet and Bert achieves the best performance, while others achieve pretty performances as well, which indicates that our method is effective for different backbone combinations. Meanwhile, the results also demonstrate the strong feature extraction abilities owned in ResNet and Bert. Secondly, we tested the impact of MI loss term on our framework. We only tested the MI loss term in ablation study, as the remaining loss terms are essential to our framework. Table 5 presents the results of VDRL with and without MI loss term. Obviously, the VDRL with MI achieves better performance, which indicates that enhancing the dependence between the sample and extracted global representation is helpful to capture the intact semantic distribution of the sample. Moreover, we tested the effect of different β_1 and β_2 in VAE loss term. And we set β_1 and β_2 to the same value due to the values of the image VAE loss and the text VAE loss are in the same range in experiments. Table 6 presents the impact of different β_1 and β_2. Actually, β_1 and β_2 control the impact of regularization term and reconstruction term in ISV and TSV. The overlarge β_1 and β_2 would result in a failure on capturing the semantic distributions for the two VAEs. That is why the results at $\beta_1 = \beta_2 = 0.5$ are extremely lower than others in Table 6.

4 Conclusion

In this paper, we have proposed a variational deep representation learning (VDRL) approach for robust cross-modal representation learning. Instead of computing the point distance between image-text pair, our method has exploited the similarity between the samples by comparing their distribution distance. To achieve this, we have defined two VAEs to model the semantic distributions of the samples and realized MI estimators to enhance the dependence between the input samples and their global representations. Furthermore, we have aligned the distributions from different modalities at the semantic level, which makes the similarity between the samples can be well measured, regardless of the modality. The retrieval performance on Flickr30K and MS-COCO has demonstrated the effectiveness of our idea.

Acknowledgment. This work is partially supported by the National Natural Science Foundation of China under Grant 61862050, the National Natural Science Foundation of Ningxia under Grant 2020AAC03031, and the Scientific Research Innovation Project of First-Class Western Universities under Grant ZKZD2017005.

References

1. Belghazi, M.I., Baratin, A., Rajeswar, S., et al.: MINE: mutual information neural estimation. arXiv (2018)

2. Cornia, M., Baraldi, L., Tavakoli, H.R., et al.: A unified cycle-consistent neural model for text and image retrieval. Multimedia Tools Appl. **79**(35), 25697–25721 (2020)
3. Devlin, J., Chang, M.W., Lee, K., et al.: Bert: pre-training of deep bidirectional transformers for language understanding. arXiv (2018)
4. Ding, K., Fan, B., Huo, C., Xiang, S., Pan, C.: Cross-modal hashing via rank-order preserving. IEEE TMM **19**(3), 571–585 (2016)
5. Faghri, F., Fleet, D.J., Kiros, J.R., et al.: Vse++: improving visual-semantic embeddings with hard negatives. arXiv (2017)
6. Givens, C.R., Shortt, R.M., et al.: A class of wasserstein metrics for probability distributions. Michigan Math. J. **31**(2), 231–240 (1984)
7. He, K., Zhang, X., Ren, S., et al.: Deep residual learning for image recognition. In: CVPR, pp. 770–778 (2016)
8. Hjelm, R.D., Fedorov, A., Lavoie-Marchildon, S., et al.: Learning deep representations by mutual information estimation and maximization. In: ICLR (2018)
9. Hu, D., Nie, F., Li, X.: Deep binary reconstruction for cross-modal hashing. IEEE TMM **21**(4), 973–985 (2018)
10. Karpathy, A., Fei-Fei, L.: Deep visual-semantic alignments for generating image descriptions. In: CVPR, pp. 3128–3137 (2015)
11. Kingma, D.P., Welling, M.: Auto-encoding variational bayes. arXiv (2013)
12. Li, W., Zheng, Y., Zhang, Y., et al.: Cross-modal retrieval with dual multi-angle self-attention. J. Assoc. Inf. Sci. Technol. **72**(1), 46–65 (2021)
13. Lin, T.Y., et al.: Microsoft COCO: common objects in context. In: Fleet, D., Pajdla, T., Schiele, B., Tuytelaars, T. (eds.) ECCV 2014. LNCS, vol. 8693, pp. 740–755. Springer, Cham (2014). https://doi.org/10.1007/978-3-319-10602-1_48
14. Liu, R., Zhao, Y., Wei, S., et al.: Modality-invariant image-text embedding for image-sentence matching. ACM Trans. Multimedia Comput. Commun. Appl. **15**(1), 1–19 (2019)
15. Liu, Y., Guo, Y., Liu, L., et al.: CycleMatch: a cycle-consistent embedding network for image-text matching. Pattern Recogn. **93**, 365–379 (2019)
16. Ma, X., Zhang, T., Xu, C.: Multi-level correlation adversarial hashing for cross-modal retrieval. IEEE TMM **22**(12), 3101–3114 (2020)
17. Malisiewicz, T., Gupta, A., Efros, A.A.: Ensemble of exemplar-SVMs for object detection and beyond. In: ICCV, pp. 89–96. IEEE (2011)
18. Peters, M.E., Neumann, M., Iyyer, M., et al.: Deep contextualized word representations. arXiv (2018)
19. Sarafianos, N., Xu, X., Kakadiaris, I.A.: Adversarial representation learning for text-to-image matching. In: ICCV, pp. 5814–5824 (2019)
20. Simonyan, K., Zisserman, A.: Very deep convolutional networks for large-scale image recognition. arXiv (2014)
21. Surís, D., Duarte, A., Salvador, A., Torres, J., Giró-i-Nieto, X.: Cross-modal embeddings for video and audio retrieval. In: Leal-Taixé, L., Roth, S. (eds.) ECCV 2018. LNCS, vol. 11132, pp. 711–716. Springer, Cham (2019). https://doi.org/10.1007/978-3-030-11018-5_62
22. Wang, L., Li, Y., Lazebnik, S.: Learning deep structure-preserving image-text embeddings. In: CVPR, pp. 5005–5013 (2016)
23. Wang, S., Chen, Y., Zhuo, J., et al.: Joint global and co-attentive representation learning for image-sentence retrieval. In: ACM MM, pp. 1398–1406 (2018)
24. Wu, H., Mao, J., Zhang, Y., et al.: Unified visual-semantic embeddings: bridging vision and language with structured meaning representations. In: CVPR, pp. 6609–6618 (2019)

25. You, Q., Zhang, Z., Luo, J.: End-to-end convolutional semantic embeddings. In: CVPR, pp. 5735–5744 (2018)
26. Young, P., Lai, A., Hodosh, M., et al.: From image descriptions to visual denotations: new similarity metrics for semantic inference over event descriptions. Trans. Assoc. Comput. Linguist. **2**, 67–78 (2014)
27. Zhang, Y., Lu, H.: Deep cross-modal projection learning for image-text matching. In: Ferrari, V., Hebert, M., Sminchisescu, C., Weiss, Y. (eds.) ECCV 2018. LNCS, vol. 11205, pp. 707–723. Springer, Cham (2018). https://doi.org/10.1007/978-3-030-01246-5_42
28. Zheng, Z., Zheng, L., Garrett, M., et al.: Dual-path convolutional image-text embeddings with instance loss. ACM Trans. Multimedia Comput. Commun. Appl. **16**(2), 1–23 (2020)

Vein Centerline Extraction of Visible Images Based on Tracking Method

Yufeng Zhang, Chaoying Tang(✉), Jiarui Yang, and Biao Wang

School of Automation, Nanjing University of Aeronautics and Astronautics, Nanjing 211100, China
cytang@nuaa.edu.cn

Abstract. Forearm vein is a promising biometric pattern for identification considering its security and convenience. Some methods have been proposed to uncovered vein patterns from RGB images. However, the shade and noise produced by various thicknesses of muscles, bones and tissue networks, are also captured in RGB images. Therefore, the quality of the vein images is usually low. In this paper, a novel vein centerline extraction algorithm based on clustering and Bayesian method is presented. This method makes different search strategies according to different vein configurations. Experiments show that the proposed algorithm obtains complete and continuous vein lines, and improve the recognition accuracy of the consequent vein pattern matching.

Keywords: Vein recognition · Feature extraction · K-means · Bayesian

1 Introduction

Over the past few decades, biometric technologies are increasingly becoming an important tool to enhance security and brings greater convenience. As a new physiological characteristic-based biometric, vein recognition has two distinct advantages over other biometric traits: living body identification and high security.

Most of the current available approaches for vein recognition have similar processing steps which involves: image acquisition, preprocessing, feature extraction and matching. In the whole process of vein recognition process, feature extraction is a most crucial step that selects or combines variables into features. Compared with other features, texture feature of veins can describe the shape of entire vein patterns and has a better degree of discrimination. Texture feature extraction methods can be classified into following two categories: the methods based on directly tracking and image segmentation.

The first category of approach first extracts the initial tracking points on vein region through the analysis of vein texture and topological features. The tracking rules and stop criterion are established to complete the vein extraction. Yi Yin [1] proposed a probability tracking method for retinal blood vessel extraction. Ritika [2] adopted repeated line tracking method to extract finger veins. Zhou [3] proposed a method applying Hidden Markov model to effectively trace vessel centerlines. Soodeh [4] presented a new

H. Ma et al. (Eds.): PRCV 2021, LNCS 13020, pp. 511–523, 2021.
https://doi.org/10.1007/978-3-030-88007-1_42

machine learning algorithm for joint classification and tracking of retinal blood vessels based on a hierarchical probabilistic framework.

For the second category, image segmentation is carried out to partition a digital image into multiple segments. Sohini [5] presented a novel unsupervised iterative blood vessel segmentation algorithm for fundus images. Jin [7] proposed a supervised method called Deformable U-NET, which exploits the retinal vessels' local features with a U-shape architecture. Wang [8] used a robust cascade classification framework for retinal vessel segmentation.

Compared with the small skin areas like that of hand, palm and finger, forearm obviously contains much more vein texture and topology information. In this paper we focused on the research of the forearm veins. Due to the weak penetration capability of visible light in skin compared to NIR images, traditional vein recognition is dependent on NIR cameras in a controlled environment for image acquisition. Tang [9] proposed a vein detection algorithm for RGB images based on optics and skin physiology. This method overcomes this limitation and achieved relatively good results of uncovering vein patterns from color images. In some occasions without NIR camera, this method provides feasibility for directly extracting vein patterns from RGB images. In this paper we propose a novel vein extraction method using line tracking algorithm on uncovered vein images. The proposed method is capable of completing recognition tasks in the field of the criminal investigation or the situation of non-touch biometric recognition in the context of the epidemic. The flow chart of this method is shown in Fig. 1.

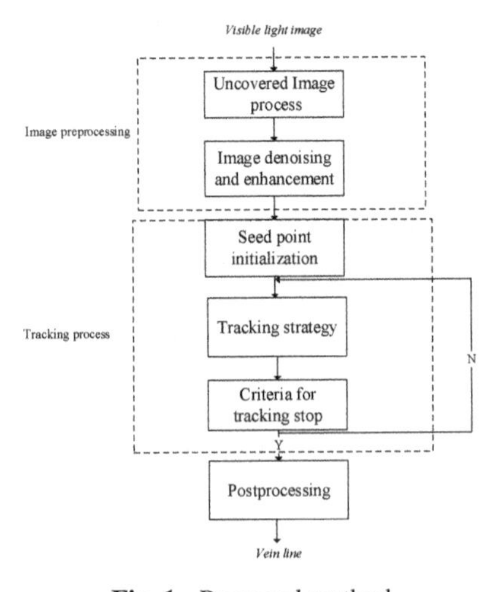

Fig. 1. Proposed method

2 Image Preprocessing

Due to the popularity of digital cameras and smart phones, RGB images can be easily acquired in our daily life. Traditionally, it is difficult to extract the vascular patterns from RGB images because they are under skin and nearly invisible to naked eyes. This paper adopts a novel method [9] to uncover vein patterns from RGB images. The pixel values of the three components (RGB) are used as the input of the neural network. And the network will output a mapped image, called the uncovered image. Some uncovered images are shown in Fig. 2.

(a)Visible light image (b) Uncovered image

Fig. 2. Uncovered results

As Fig. 2(b) shows, uncovered images can display the texture and topological structure of veins. Due to the optical blurring and skin scattering problems, the uncovered images are not always clear and contain shade and noise. Here we adopted the embedding bilateral filter in least squares model (BLF-LS) [10] with edge-preserving smoothing to remove the noise. The result of filtering is shown in Fig. 3.

(a)Original images (b)Denoised results

Fig. 3. Denoising results

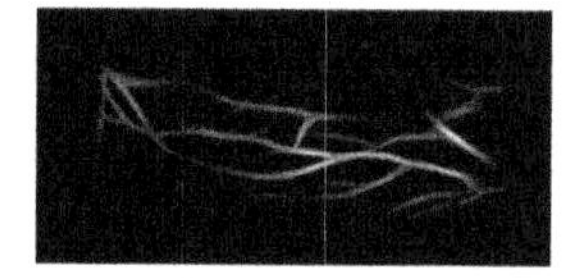

Fig. 4. Energy map

It should be mentioned that denoised uncovered images still contain some artifacts because of the uneven light intensity. For low quality images, Gabor filters [11] have been proven to be a powerful tool to capture local texture information. Our algorithm applies a filter bank made up of real parts of 16 Gabor filters with different scales and orientations to the uncovered images. And select the maximum value from the 16 filter response values as the final output of the pixel. The energy map of an uncovered image is shown in Fig. 4

3 Vein Centerline Extraction

3.1 Determination of Initial Points and Direction

The proposed method uses Gabor energy map to further extract vein centerlines. Normally, tracking method automatically determines initial seed points and tracks vein lines starting from them.

In order to enlarge the distribution of seed selection, we add horizontal and vertical grids to the energy image and the spaces are 20 and 15 pixels, respectively. As the initial candidate seed points, the local maximum gray points of Gabor energy image are searched row by row and column by column along the grid lines.

After achieving the initial seed points, it is necessary to determine their tracking direction. In this paper, an algorithm based on local gradient information [12] is used to determine the initial direction. The initial tracking directions are opposite, and the tracking process will be carried out separately until the stopping criterions are satisfied (Figs. 5 and 6).

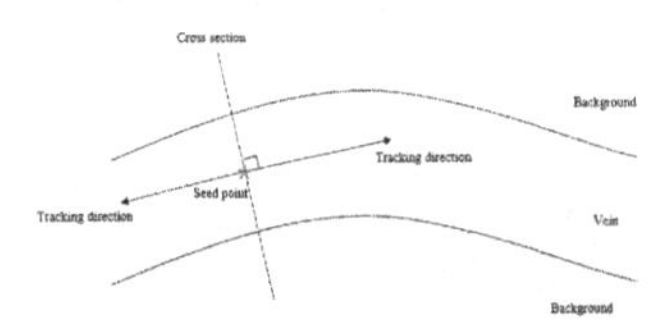

Fig. 5. Schematic diagram

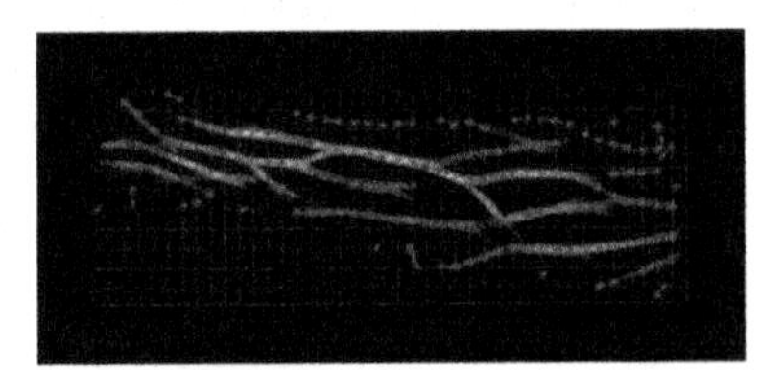

Fig. 6. Actual tracking direction

3.2 Tracking Process

Considering that vein lines have a small degree of curvature, strong continuity and more branches, we propose a vein centerline tracking strategy based on K-means algorithm and Bayesian theory. Before each tracking iteration, we need to make a preliminary determination of vein configurations. In this paper, geometric configurations of forearm veins are divided into three categories including normal, bifurcation and cross.

Direction Updating in an Iterative Process

The geometric direction is defined as the direction from the current center point to the new estimated center point. For points P^0_{k+1}, P^1_{k+1} and P_{k+1}, their geometric direction can be defined as

$$\hat{u}^i_{k+1} = \frac{P^i_{k+1} - P_k}{||P^i_{k+1} - P_k||} \tag{1}$$

The geometric direction depends entirely on the topology of the vein. Because of the continuity of the direction of vein line, we can take the geometric direction as one part of the vein direction. But for veins with larger curvature, if only geometric direction is used, the derivative at its position will be too large to result in a wrong tracking process. To solve this problem, the intensity direction needs to be considered.

Eigen analysis of the Hessian matrix is widely used to detect and analyze tubular-like structures. One of the eigenvectors corresponding to its eigenvalues is approximately along the vein direction, while the other is orthogonal to the vein. For forearm Gabor enhanced maps, vessels are brighter than background and the center point is located at the intensity valley. We define the intensity direction as the Hessian eigenvector corresponding to the smaller eigenvalue. The direction is denoted as $\hat{h}$.

$$\hat{h}_{k+1} = sign(\hat{\varphi}_k v_1)v_1 \tag{2}$$

where $\hat{\varphi}_k$ denotes the current tracking direction and v_1 denotes the eigenvector corresponding to the smaller eigenvalue. $\hat{h}_{k+1}$ denotes the intensity direction of the next point. The intensity direction naturally reveals the distribution of the pixel intensities across the vein. To consider the information of topology and intensity distribution, we combine the geometric direction and the intensity direction to get the final vein tracking direction

$$\hat{\psi}_{k+1} = \alpha\hat{u}_{k+1} + (1 - \alpha)\hat{h}_{k+1} \tag{3}$$

where α is the adjustable weighting factor ranging from 0 to 1. After many experiments, the most accurate estimation direction is obtained when the weighting factor is selected as 0.54.

Acquisition of Vein Boundary Points with Dynamic-Size Searching Windows
As mentioned above, we divided the configuration of veins into three types: normal, bifurcation and crossing. Here we design two types of search window for different vein configurations. At a given step of tracking process, we set the angle difference threshold T_{angle} of direction vectors between two boundary points. If the angle difference is greater than a threshold, it belongs to an abnormal vein configuration and a semicircular search window is suitable to track the next point. Otherwise a linear search window is adopted as Fig. 8.

In the case of a linear search window, the position of the next candidate centerline points is determined by the step size and the tracking direction. At this candidate point, we set a linear search window L_k perpendicular to the tracking direction vector. As shown in Fig. 10, O_k is the centerline point at the iteration k, $O_{k,c}$ denotes the candidate centerline point at the iteration k. $\hat{U}$ and $\hat{V}$ are the vein edge points, O is the center point, $\vec{D}$ is the vein direction and k is the index of the iteration. The points on the linear search window show the possible locations of current edge points and s is the tracking step size.

Clustering algorithm is introduced into the calculation of boundary points and current vein width. Based on the intensity profiles along the linear search window, pixels of veins and background could be separated. For a given set of n points on linear search window, they are partitioned into two clusters (V_1,V_2) with corresponding cluster centers (c_1,c_2) by minimizing the sum of the Euclidean distance of all points to their center points. After many iterations, the vein and background can be clearly distinguished. Compared with the traditional threshold segmentation method, the clustering method divides the pixels according to the gray distribution of the current position and avoids the adaptive adjustment of the gray threshold.

$$\arg \min_{(V_1,V_2),(c_1,c_2)} \sum_{i=1}^{k} \sum_{I_j \in V_i} ||I_j - c_i||^2 \tag{4}$$

$$\begin{cases} I_j \in V_2 \text{ when } I_j < c* \\ I_j \in V_1 \text{ other cases} \end{cases} \tag{5}$$

The clustered result after 50 iterations is shown in Fig. 7. The yellow dots and red dots represent pixels classified as veins and background. The blue dots represent the boundary points of normal veins.

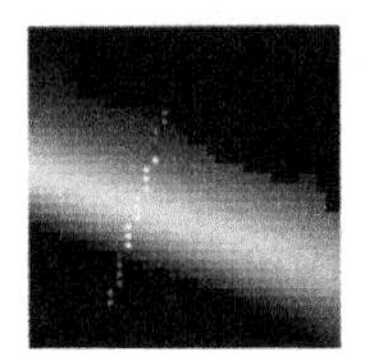

Fig. 7. Vein and non-vein areas on the linear search window (Color figure online)

For the configuration of bifurcation and crossing veins, a semi-circle search window is used to determine the position of the next point. Different from the normal configuration, the gray intensity distribution of the abnormal configuration usually does not have a clear threshold to separate the foreground from the background. Here we use a more targeted algorithm based on Bayesian theory to obtain the boundaries of vein (Fig. 9).

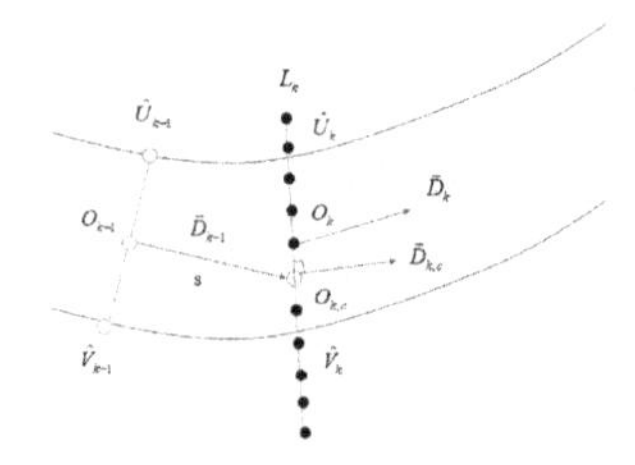

Fig. 8. Linear search window

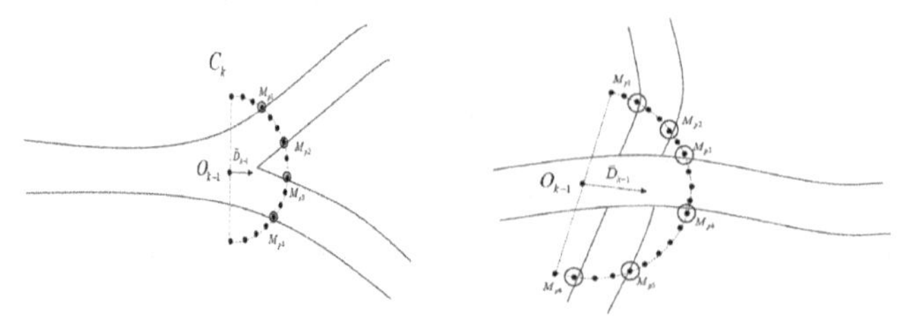

Fig. 9. Semicircle search window

On the semicircle search window, the candidate boundary points are exactly sampled in equal radians. At a given step, suppose Y represents the gray distribution of N candidate points and X represents a possible combination of candidate boundary points. A posteriori probability $P(X|Y)$ is the probability that combination X is the real blood vein in the case of current candidate point gray distribution. Our goal is to find the most accurate combination $\hat{X}$ which satisfied the following equation.

$$\hat{X} = \arg\max_X \{P(X|Y)\} = \arg\max_X \{P(Y|X)P(X)\} \tag{6}$$

The gray distribution of vein cross section can be approximated as a Gaussian curve. For a given boundary point combination X, the algorithm will construct the corresponding vein cross-section according to a proposed vein model. If the real distribution of blood vein pixels is the same as the model, then the combination of the boundary points X is the most likely to be a real boundary points combination. The gray distribution Y can be regarded as a set of N boundary candidate points where $Y = \{y_1, y_2, y_3, y_4 ... y_n\}$, the gray intensities of these N candidate points are independent of each other. The likelihood function of this normal configuration is computed as

$$P(Y|X) = \prod_{y=1}^{N} P(y_i|X) \tag{7}$$

where $P(y_i|X)$ is a conditional probability model which is defined to describe the variability of the i_{th} candidate point in the search window. For the normal configuration of veins, $P(Y|X)$ can be written as follows.

$$P(Y|X) = \prod_{i=1}^{m_1-1} P(y_i|b) \prod_{i=m_1}^{m_2} P(y_i|v) \prod_{i=m_2+1}^{N} P(y_i|b) \tag{8}$$

where b represents the background point and v represents the blood vein points.m_1 and m_2 are two edge points of combination X, $P(y_i|v)$ is the probability of y_i as a blood vein point and $P(y_i|b)$ is the probability of y_i as a background point.

For the bifurcation and crossing configurations,

$$P(Y|X) = \prod_{i=1}^{m_1-1} P(y_i|b) \prod_{i=m_1}^{m_2} P(y_i|v) \prod_{i=m_2+1}^{m_3-1} P(y_i|b) \prod_{i=m_3}^{m_4} P(y_i|v) \prod_{i=m_4}^{N} P(y_i|b) \tag{9}$$

where m_1,m_2,m_3,m_4 are the four boundary points of bifurcation configuration.

Based on the cross-section model of veins, the ideal gray value of y_i can be calculated according to Eq. (10)

$$G(y_i) = \begin{cases} (A - B)\exp(-\frac{l^2}{2\sigma^2}) + B & y_i \in vein \\ B & y_i \in background \end{cases} \tag{10}$$

where A and B are the average gray values in the local vein and at background. l is the distance between point y_i and the line passing through the central point along the vein. σ is set to be half of the vein radius.

The conditional probability model of the proposed normal configuration is

$$P(y_i|v) = \frac{1}{\sqrt{2\pi}\sigma_v} \exp(-\frac{(I(y_i) - G(y_i))^2}{2\sigma_v^2}) \tag{11}$$

$$P(y_i|b) = \frac{1}{\sqrt{2\pi}\sigma_b} \exp(-\frac{(I(y_i) - G(y_i))^2}{2\sigma_b^2}) \tag{12}$$

where I represent the practical intensity with noise.

Since the vein directions do not change dramatically in a short distance, the priori probability can be written as a Gibbs formulation (13)

$$P(X) = e^{-\lambda U(X)} \tag{13}$$

where λ is the impact factor, $U(X)$ denotes the energy function of a given configuration.

$$U(X) = d_{m1}^2 + d_{m2}^2 \tag{14}$$

where d_{m1} and d_{m2} are the distances from the candidate boundary points P_{m1}, P_{m2} to the fitting lines L_1 and L_2, as shown in Fig. 10. L_1 and L_2 are the lines fitted by the least square method at the boundary points from the previous iteration. Similarly, for the bifurcation and crossing configurations $U(X)$ can also be expressed as $d_{m1}^2 + d_{m4}^2$ and $d_{m3}^2 + d_{m4}^2$. So we can calculate the posterior probability of each combination and obtain the final boundary points corresponding to the maximum posterior probability (Fig. 11).

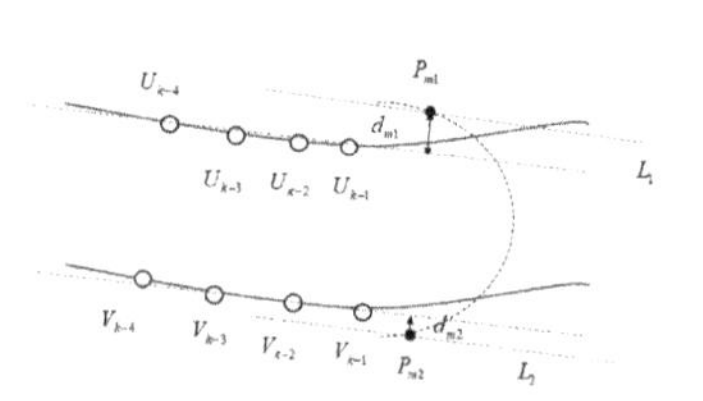

Fig. 10. Distance between candidate point and fitting line

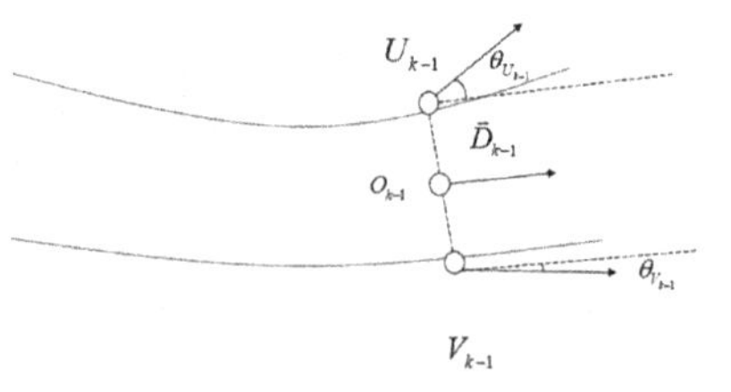

Fig. 11. Direction of boundary points

Selection of Central Points and Stopping Criterion

In normal vein configuration, a rectangular matching filter is used to select the centerline points for the linear search window. The expected pattern for vein cross-section density profile h is assumed to be a rectangle with width of $2R + 1$. The i_{th} element in h is given by

$$h[i] = \begin{cases} 1 & |i| \leq R \\ -1 & R < |i| \leq w \\ 0 & otherwise \end{cases} \tag{15}$$

where R represents the half width of current vein and $2w + 1$ represents the width of the linear search window. The vector resulting from convolution between g_l and h is denoted by r.

$$r[i] = \sum_{j=1}^{2w+1} g_l[j]h[i-j] \tag{16}$$

where $i = R+1, R+2, ..., 2w-R+1$.

The method turns the originally complicated convolution calculation into a simple addition operation. The maximum of the convolution result $r[m]$ is determined as the m_{th} pixel on linear search window corresponding to the maximum output as the candidate centerline point. If the gray intensity difference between the selected candidate point and the previous point exceeds a certain threshold, the next largest point in r is selected until the point with appropriate contrast difference is found.

In the case of bifurcation and crossing configurations, a semicircle search window is adopted since a rectangular matching filter is not suitable for selection of central points. We need to determine the location of the next tracking point according to the different vein configuration. For instance, there are two branches and two candidate tracking centerline points in the bifurcation structure. The midpoint of the boundaries from two branches is taken as two candidate tracking points. Compared with the original method [12], this method can ensure the integrity of the centerline.

The stopping criteria of the tracking algorithm includes: (1) The absolute value of the gray difference between the newly detected center point and the previous center point is larger than the threshold we set. (2) The gray value of the newly detected center point is smaller than the threshold we set. (3) The angle between the newly detected center point and the tracking direction of the previous center point is too large. (4) The distance between the center point being tracked and the center line that has been tracked is too close.

4 Experiments and Discussion

4.1 Subjective Evaluation

The performance of our method was evaluated on the dataset [9] which includes 150 RGB forearm images. The proposed method is compared with several vein extraction methods including Bayesian method [12], Line operator method [14] and U-net segmentation method [15]. Some extraction results are shown in Fig. 12. It can be found that the centerlines extracted by our proposed method is the best in terms of the integrity and robustness.

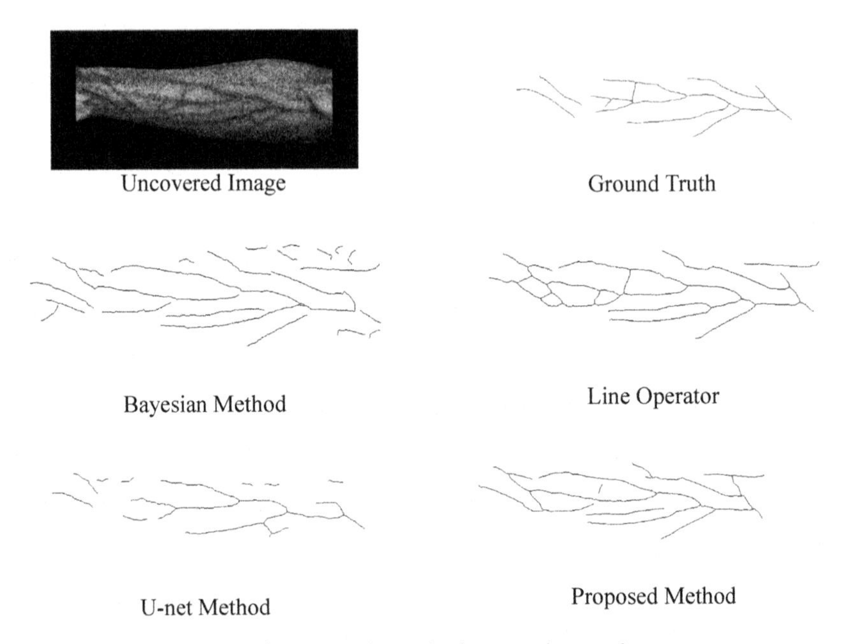

Fig. 12. Comparison of vein extraction results

4.2 Objective Evaluation

To objectively evaluate the extraction results, 150 uncovered vein images are manually labeled with line width of 10 pixels. It is regarded as the ground truth. The ROC is adopted as one of the evaluation metrics. If the position of the extracted pixel and the manually labeled pixel are coincident, this pixel is called a true positive pixel. On the contrary, it is called a false positive pixel. The ratio of the number of true positive pixel N_{TP} to the number of the actual vein pixels N_v is defined as TPR. The ratio of the number of false positive pixel N_{FP} to the number of the actual background pixel N_b is defined as FPR. After calculating the TPR and FPR values of all vein extraction results, the ROC image was drawn in Fig. 13 with FPR as the horizontal axis and TPR as the vertical axis, where each point represents an image.

As can be seen from the ROC images, the closer the points are to the top left corner, the closer the result is to the ground truth. Comparing four figures, we can find that the accuracy of our proposed method is better than others.

To have a comprehensive analysis of the methods, we further use Precision, Recall and F-measure metrics. Precision is to measure the accuracy of extracting vein lines. Recall focuses on the completeness of the extracted results. F-measure considers the accuracy and completeness at the same time. Table 1 shows the average metrics of several methods. It shows that the proposed method has the best performance in Recall and F-measure, but its precision is lower than those of U-net method and Line Operator method.

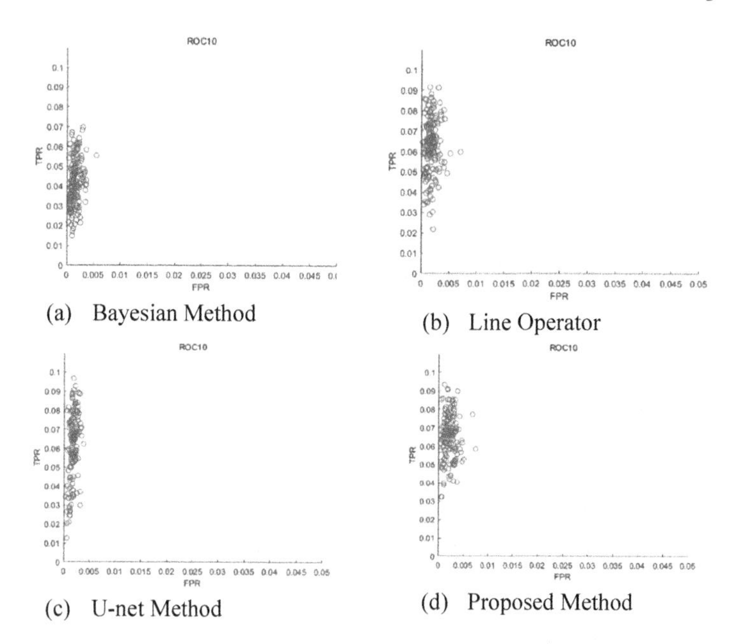

(a) Bayesian Method (b) Line Operator

(c) U-net Method (d) Proposed Method

Fig. 13. Comparison images of ROC

Table 1. Comparison of four methods' detection rate

Method	Precision	Recall	F-measure
Bayesian method	0.5388	0.0416	0.0762
Line operator	**0.6006**	0.0624	0.1120
U-net method	0.5787	0.0615	0.1107
Proposed method	0.5700	**0.0651**	**0.1155**

4.3 Evaluation by Vein Pattern Matching

Since the final purpose of vein extraction is personal identification, in this section, vein lines extracted from 150 uncovered vein images are matched for further evaluation. The manually labeled images are used as reference images in the database. The point set matching algorithm [9] is utilized. The matching results of different extraction methods are demonstrated using the cumulative matching characteristic (CMC) curve. As shown in Fig. 14, the rank-1% recognition rates of the proposed method, Line operator method, U-net method and Bayesian method are 74%, 67.33%, 48%, 46.67% respectively. It can be seen that the rank-1 matching success rate of proposed algorithm is the highest, so it has the best matching accuracy.

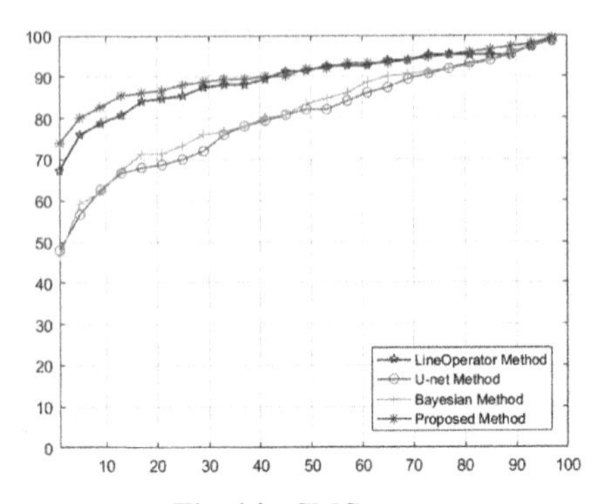

Fig. 14. CMC curve

5 Conclusion

This paper proposed a novel vein line extraction algorithm based on tracking for vein images uncovered from RGB images. Based on the ideas of K-means algorithm and Bayesian method, the tracking strategy is developed for different vein configurations. Tracking process stops until the stopping criteria according the current tracking condition and the previous condition is met. Until all the initial seed points are tracked, the algorithm ends with the output image. Experimental results show that the vein lines extracted by our method is better than the other methods.

Acknowledgment. This work was supported by the Key Research & Development Programs of Jiangsu Province (BE2018720) and the Open project of Engineering Center of Ministry of Education (NJ2020004).

References

1. Yin, Y., Adel, M.: Bourennane: retinal vessel segmentation using a probabilistic tracking method. Pattern Recogn. **45**(4), 1235–1244 (2012)
2. Chopra, R., Kaur, S.: Finger print and finger vein recognition using repeated line tracking and minutiae. Int. J. Adv. Sci. Res. **2**(2), 13–22 (2017)
3. Zhou, C., Zhang, X., Chen, H.: A new robust method for blood vessel segmentation in retinal fundus images based on weighted line detector and hidden Markov model. Comput. Methods Prog. Biomed. **187**, 105231 (2020)
4. Kalaie, S., Gooya, A.: Vascular tree tracking and bifurcation points detection in retinal images using a hierarchical probabilistic model. Comput. Methods Prog. Biomed. **151**, 139–149 (2017)
5. Roychowdhury, S., Koozekanani, D.D., Parhi, K.K.: Iterative vessel segmentation of fundus images. IEEE Trans. Biomed. Eng. **62**(7), 1738–1749 (2015)
6. Sundaram, R., Ravichandran, K.S., Jayaraman, P.: Extraction of blood vessels in fundus images of retina through hybrid segmentation approach. Mathematics **7**(2), 169 (2019)
7. Jin, Q., Meng, Z., Pham, T.D., Chen, Q., Wei, L., Su, R.: DUNet: a deformable network for retinal vessel segmentation. Knowl. Based Syst. **178**, 149–162 (2019)

8. Wang, X., Jiang, X., Ren, J.: Blood vessel segmentation from fundus image by a cascade classification framework. Pattern Recogn. **88**, 331–341 (2019)
9. Zhang, H., Tang, C., Kong, A.W.K., Craft, N.: Matching vein patterns from color images for forensic investigation. In: 2012 IEEE Fifth International Conference on Biometrics: Theory, Applications and Systems, pp. 77–84. IEEE. (2012)
10. Liu, W., Zhang, P., Chen, X., Shen, C., Huang, X., Yang, J.: Embedding bilateral filter in least squares for efficient edge-preserving image smoothing. IEEE Trans. Circuits Syst. Video Technol. **30**(1), 23–35 (2018)
11. Fogel, I., Sagi, D.: Gabor filters as texture discriminator. Biol. Cybern. **61**(2), 103–113 (1989)
12. Yin, Y., Adel, M., Bourennane, S.: Automatic segmentation and measurement of vasculature in retinal fundus images using probabilistic formulation. Comput. Math. Methods Med. **2013**, 1–16 (2013)
13. Ibañez, M.V., Simó, A.: Bayesian detection of the fovea in eye fundus angiographies. Pattern Recogn. Lett. **20**(2), 229–240 (1999)
14. Ricci, E., Perfetti, R.: Retinal blood vessel segmentation using line operators and support vector classification. IEEE Trans. Med. Imaging **26**(10), 1357–1365 (2007)
15. Ronneberger, O., Fischer, P., Brox, T.: U-net: convolutional networks for biomedical image segmentation. In: Navab, N., Hornegger, J., Wells, W.M., Frangi, A.F. (eds.) Medical Image Computing and Computer-Assisted Intervention – MICCAI 2015: 18th International Conference, Munich, Germany, October 5-9, 2015, Proceedings, Part III, pp. 234–241. Springer International Publishing, Cham (2015). https://doi.org/10.1007/978-3-319-24574-4_28

Discrete Bidirectional Matrix Factorization Hashing for Zero-Shot Cross-Media Retrieval

Donglin Zhang[1], Xiao-Jun Wu[1(✉)], and Jun Yu[2]

[1] School of Artificial Intelligence and Computer Science, Jiangnan University, Wuxi, China
dlinzzhang@gmail.com, wu_xiaojun@jiangnan.edu.cn
[2] College of Computer and Communication Engineering, Zhengzhou University of Light Industry, Zhengzhou, China
yujun@zzuli.edu.cn

Abstract. Recently, cross-modal retrieval has gained much attention due to ever-increasing multimedia data. However, existing cross-modal algorithms require that the training set and the test set share the same categories, which cannot well search data of newly emerged categories. Therefore, more practical zero-shot cross-modal retrieval (ZSCMR) has become a promising direction, which aims to search unseen classes (new classes) that never present in the training set. It is very challenging that ZSCMR needs to solve not only the inconsistent semantic between seen and unseen classes but also the semantic gap of the heterogeneous multimedia data. To mitigate these problems, a novel discrete bidirectional matrix factorization hashing method is developed for zero-shot cross-modal retrieval (DMZCR). The proposed DMZCR contains three contributions: 1) A bidirectional matrix factorization scheme is proposed in our model, more discriminative low-rank representation can be learned and the redundant information can also be removed. 2) Inspired by zero-shot learning, we build a multi-layer semantic transmission scheme to model the relationships between classes, features and attributes, then the knowledge can be transferred from seen to unseen classes. 3) The hash codes can be learned by a discrete scheme, reducing the large quantization error caused by relaxation. As far as we know, this work first employs matrix factorization scheme to solve ZSCMR task. Experiments on three popular databases illustrate the efficacy of DMZCR compared with several state-of-the-art algorithms for ZSCMR task.

Keywords: Zero-shot learning · Cross-media retrieval · Hashing

1 Introduction

With the explosive growth of multimedia data, including text, audio, image, video and others, cross-media retrieval has recently become a research hotspot.

This work was supported by NSFC [Grant 62020106012, U1836218, 61672265].

H. Ma et al. (Eds.): PRCV 2021, LNCS 13020, pp. 524–536, 2021.
https://doi.org/10.1007/978-3-030-88007-1_43

Unlike single modal retrieval approaches [17], in cross-media similarity search, one can use any modality query data to retrieve semantically relevant data of other media types. Some real-value cross-media search algorithms [14,20,31] have been presented, such as canonical correlation analysis (CCA) [14], cross-media generative adversarial (CMGA) [31] and adversarial cross-media retrieval (ACMR) [20]. However, due to the high-dimensional and large-scale properties of multimedia data, the real-value nearest neighbor search is not practical due to the large storage cost and time. To solve this issue, hashing techniques have attracted much attention, which converts the real-value multimedia data into a set of binary codes while preserving the similarity information. In that case, the search can be conducted by the XOR operation, the retrieval speed can be dramatically improved and the storage cost can also be significantly reduced. Precisely because of these merits, many hash based cross-media methods [7,24,28,29,33] have been developed.

Generally speaking, these cross-media models can be grouped into two types: unsupervised approaches and supervised ones. The former learns the binary codes without supervised semantic information. For instance, fusion similarity hashing (FSH) [9], inter-media hashing (IMH) [18], latent sparse semantic hashing (LSSH) [33], collective reconstructive hashing (CRE) [7]. The above methods can achieve good performance, however, the semantic information is not considered in the above approaches. In general, the search performance can be further improved by utilizing the label information. Typical supervised cross-media approaches include semantic preserving hashing (SePH) [8], discrete cross-modal hashing (DCH) [24], semantic correlation maximization (SCM) [30], supervised matrix factorization hashing (SMFH) [19] and label relaxation & discrete matrix factorization method (LRMF) [29]. More recently, there are also some DNN-based methods [12,20,21] are developed for cross-media tasks. Compared to the traditional models, DNN-based methods can achieve more promising results. The above supervised, unsupervised and DNN-based methods can achieve promising performance, however, there are still some limitations that we need to take into consideration.

For one thing, existing cross-media approaches require that the test set and the training set share the same categories (this scenario is called standard cross-media retrieval in this paper), which do not consider the more practical zero-shot scenario. Taking the popular Wiki [14] database as an example, the test set and the training set have the same classes from the same predefined ten classes. However, with the massive increase of new class data, it takes a lot of manpower and time to label and collect these cross-media data. Thus, the standard retrieval scenario is difficult to extend for searching the multimedia data of newly emerged classes, which limits its application in practice.

Recently, several zero-shot learning-based approaches [6,10,11] have been proposed, however, most methods focus on the unimodal scenario, which cannot deal with the multimodal data, and very few zero-shot learning algorithms [2, 23,32] are presented for cross-media retrieval. In this work, we focus on ZSCMR (zero-shot cross-media retrieval), the purpose of ZSCMR is to search across different modalities in the zero-shot scenarios, in which the unseen categories and seen categories are disjoint without overlap. Therefore, it is a very challenging task compared to standard cross-media retrieval.

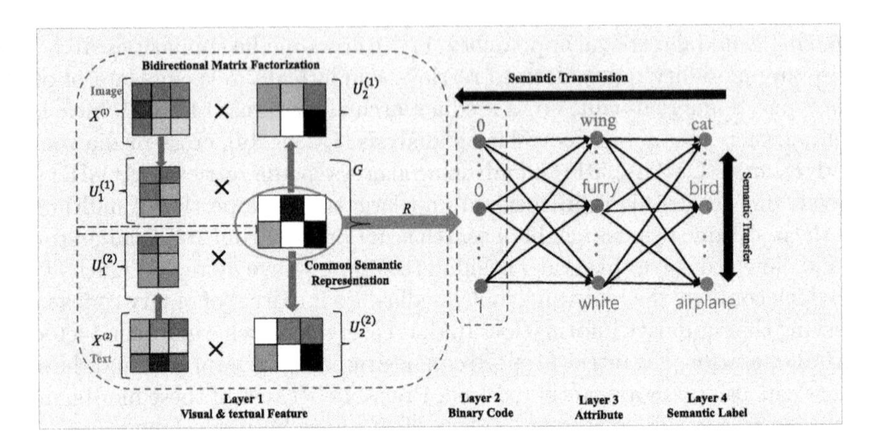

Fig. 1. The framework of DMZCR. A bidirectional matrix factorization scheme is proposed to learn the common representations, the knowledge can be transferred from seen classes to unseen classes by using the multi-layer semantic transmission network.

Moreover, some existing hashing algorithms learn the binary codes by taking rounding or relaxing scheme, which may cause large quantization errors. Besides, some matrix factorization approaches are proposed for cross-modal similarity search. For instance, CMFH [4] learns the common representations by matrix factorization scheme, LSSH [33] generates the hash codes by matrix factorization and sparse coding, and SMFH [19] learns the binary codes by constructing the pairwise Laplacian matrix. The above matrix factorization based cross-media approaches can obtain promising search results, however, the one-sided matrix factorization cannot sufficiently preserve the original features, which may cause some valuable information to be lost.

To address the above problems, we propose a novel hashing approach, named discrete bidirectional matrix factorization hashing for zero-shot cross-modal retrieval, DMZCR for short. The flowchart is shown in Fig. 1. It utilizes a bidirectional matrix factorization scheme to learn the shared semantic representations of different modalities. Specifically, we can learn the representations by the original features and the original features can also be reconstructed, which can preserve more valuable information, meanwhile, the representations are mapped to the hash codes. Moreover, DMZCR can learn the binary codes in a discrete manner, we can further reduce the large quantization error. Inspired by zero-shot learning, we propose a multi-layer semantic transmission scheme to build the relationships between hash codes, attributes and classes, then the knowledge can be transferred from the seen classes and unseen classes (newly emerged classes). Experiments conducted on three popular databases show the efficacy of DMZCR for both zero-shot retrieval and standard cross-media similarity search.

The main contributions of DMZCR can be summarized as follows:

1) We propose a bidirectional matrix factorization scheme to learn the shared representations of the heterogeneous data, the reconstructive constraint can make the representations more discriminative, which can also preserve more valuable information.

2) A multi-layer semantic transmission scheme is developed for connecting classes, attributes and hash codes, the knowledge can be transferred from seen classes to unseen classes.
3) The binary codes can be learned directly by the given discrete optimization method. We can further reduce the large quantization error. Experiments on three databases demonstrate the efficacy of DMZCR.

2 Proposed Method

2.1 Notations

In this work, we only consider two modalities (e.g., text and image). The database can be described as $D = \{D_{SD}, D_{SQ}, D_{UD}, D_{UQ}\}$, where D_{SD}, D_{SQ}, D_{UD} and D_{UQ} are seen class dataset, seen class query set, unseen class set and unseen class query set, respectively. Specifically, each set can be denoted as $D_* = \{X^{(1)}, X^{(2)}, Y_*\}$, where $X^{(1)} \in \mathbb{R}^{d_1 \times n_*}$, $X^{(2)} \in \mathbb{R}^{d_2 \times n_*}$ denote the text and image feature matrix. For ease of representation, the label $Y_* \in \mathbb{R}^{n_* \times c}$ is written as the row form. d_1 and d_2 are the dimensionality of text and image, respectively (usually $d_1 \neq d_2$), c denotes the total number of classes. It should be noted that there is no overlap classes between unseen class set $\{D_{UD}, D_{UQ}\}$ and seen class set $\{D_{SD}, D_{SQ}\}$. sgn(.) denotes the sign function. The purpose of our approach is to learn a cross-media framework by using seen categories multimodal data, which can be generalized to newly emerged categories to generate discriminative hash codes.

2.2 Objective Function

Bidirectional Matrix Factorization. Presume that there is a data matrix $X \in \mathbb{R}^{d \times n}$ used for training, we assume that $X = UG$, i.e., the matrix can be decomposed into U and G, where $U \in \mathbb{R}^{d \times r}$ denotes the basis matrix, $G \in \mathbb{R}^{r \times n}$ is the low-rank representation of X. Then the objective function can be stated as $\min_{U,G} \|X - UG\|_F^2$. By introducing $UU^T = I$, the issue can also be viewed as a special case of PCA, and by eliminating G, the above issue can be rewritten as follows:

$$\min_{U} \left\|X - U(U^T X)\right\|_F^2, \quad s.t.\ UU^T = I. \tag{1}$$

Different from the one-side matrix factorization, the inverse factorization is considered in our framework, that is, decomposing G into the original matrix X and another latent factor matrix U_2. Thus, the formulation can be defined as follows:

$$\min_{U_t,G} \|X - U_1 G\|_F^2 + \lambda \|G - U_2 X\|_F^2, \quad s.t.\ U_t U_t^T = I,\ (t = 1, 2). \tag{2}$$

In Eq. (2), we first decompose X into U_1 and G, and the learned G can also be decomposed into the original matrix X and U_2. Thus, the learned low-dimensional representation of X can be embedded into more discriminative information, which may make the retrieval performance more better. In this paper, we only consider two modalities (i.e., textual modality and visually modality), which characterize the same instance, therefore, they should have similar semantic information. According to this, it can be assumed that different modalities have a common semantic subspace, which can be obtained by the bidirectional matrix factorization. Then, we define the formulation as follows:

$$\min_{U_t^{(1)},U_t^{(2)},G} \alpha_1 \left\|X^{(1)} - U_1^{(1)}G\right\|_F^2 + \alpha_2 \left\|X^{(2)} - U_1^{(2)}G\right\|_F^2 + \lambda \left(\left\|G - U_2^{(1)}X^{(1)}\right\|_F^2 + \left\|G - U_2^{(2)}X^{(2)}\right\|_F^2\right),$$
$$s.t.\ U_t^{(1)}U_t^{(1)^T} = I,\ U_t^{(2)}U_t^{(2)^T} = I\ (t = 1, 2), \tag{3}$$

where α_1, α_2 and λ are the weight parameters.

Modeling Hashing Hierarchy. In order to learn the binary codes, it can be assumed that the hash codes can be learned from the low-dimensional representations of the original data. Then, we define the sub-optimization problem as below:

$$\min_{R,B} \|B - RG\|_F^2, \quad s.t.\ RR^T = I,\ B \in \{-1,1\}^{r\times n}, \tag{4}$$

where r denotes the length of binary code, and R represents an orthogonal matrix. The binary constraints can be kept by the orthogonal matrix R during training, which can be generated directly without relaxation. Thus, we can further reduce the quantization error. To build the relationships between classes labels, attributes and binary codes, we formulate the following loss:

$$\min_{B,V} \left\|Y - B^T VS\right\|_F^2, \tag{5}$$

where $V \in \mathbb{R}^{r\times g}$ denotes the projection matrix from hash codes to attributes. $S \in \mathbb{R}^{g\times c}$ is the attribute matrix and g denotes the number of attributes. Following the idea of [1], the attribute matrix S is obtained by extracting the word vector representation of the category name, we utilize GloVe [13] to extract the word vector in this work.

Overall Model. Combining Eq. (3), Eq. (4) and Eq. (5), the overall objective function can be formulated as

$$\begin{aligned} \min_{U_t^{(1)},U_t^{(2)},G,V,B} & \left\|Y - B^T VS\right\|_F^2 + \alpha_1 \left\|X^{(1)} - U_1^{(1)}G\right\|_F^2 \\ & + \alpha_2 \left\|X^{(2)} - U_1^{(2)}G\right\|_F^2 + \lambda \left(\left\|G - U_2^{(1)}X^{(1)}\right\|_F^2 + \left\|G - U_2^{(2)}X^{(2)}\right\|_F^2\right) \\ & + \eta \|B - RG\|_F^2 + \Omega(U_t^{(1)}, U_t^2, G, V, B), \\ & s.t.\ U_t^{(1)}U_t^{(1)^T} = I,\ U_t^{(2)}U_t^{(2)^T} = I\ (t = 1, 2),\ RR^T = I,\ B \in \{-1,1\}^{r\times n}, \end{aligned} \tag{6}$$

where λ and η are the weighting parameters, the last term is introduced for regularization, which is formulated as $\Omega(U_1^{(1)}, U_2^{(1)}, U_1^2, U_2^2, G, V, B) = \gamma \|VS\|_F^2 + \mu \left\|B^T V\right\|_F^2 + \gamma\mu \|V\|_F^2 + \beta(\|U_1^{(1)}\|_F^2 + \|U_1^{(2)}\|_F^2 + \|U_2^{(1)}\|_F^2 + \|U_2^{(2)}\|_F^2 + \|G\|_F^2)$, where β, γ and μ are the trade-off parameters.

2.3 Optimization Algorithm

The optimization problem in Eq. (6) is difficult to solve due to the nonconvexity with multiple variables. However, the problem is tractable w.r.t one variable, while keeping others fixed. Thus, an alternating iterative optimization scheme is given. In addition, to make the computation tractable, the constraints $U_t^{(1)}(U_t^{(1)})^T = I$ and $U_t^{(2)}(U_t^{(2)})^T = I$ $(t = 1, 2)$ are discarded in $U_t^{(1)}$ and $U_t^{(2)}$ subproblems. The following experimental results also demonstrate that the proposed model can also obtain promising performance when we discard those constraints in the specific subproblems. The optimization steps are shown in detail as follows:

Step-1. Update $U_1^{(1)}$. Fixing $U_2^{(1)}$, $U_1^{(2)}$, $U_2^{(2)}$, B, V, R and G, Eq. (6) can be rewritten as

$$\min_{U_1^{(1)}} \alpha_1 \left\| X^{(1)} - U_1^{(1)} G \right\|_F^2 + \beta \left\| U_1^{(1)} \right\|_F^2. \tag{7}$$

Setting the derivative w.r.t $U_1^{(1)}$ to 0, the solution can be obtained as

$$U_1^{(1)} = \alpha_1 X^{(1)} G^T (\alpha_1 G G^T + \beta I)^{-1}. \tag{8}$$

Similar to $U_1^{(1)}$, the solution of $U_1^{(2)}$ can be obtained as

$$U_1^{(2)} = \alpha_2 X^{(2)} G^T (\alpha_2 G G^T + \beta I)^{-1}. \tag{9}$$

Similarly, the solutions of $U_2^{(1)}$ and $U_2^{(1)}$ can be given by

$$U_2^{(1)} = \lambda G {X^{(1)}}^T (\lambda X^{(1)} {X^{(1)}}^T + \beta I)^{-1}. \tag{10}$$

$$U_2^{(2)} = \lambda G {X^{(2)}}^T (\lambda X^{(2)} {X^{(2)}}^T + \beta I)^{-1}. \tag{11}$$

Step-2. Similar to step-1, then the solutions of G can be obtained as,

$$\begin{aligned} G &= (\alpha_1 {U_1^{(1)}}^T U_1^{(1)} + \alpha_2 {U_1^{(2)}}^T U_1^{(2)} + \eta R^T R + (2\lambda + \beta) I)^{-1} \\ &\bullet (\alpha_1 {U_1^{(1)}}^T X^{(1)} + \alpha_2 {U_1^{(2)}}^T X^{(2)} + \lambda (U_2^{(1)} X^{(1)} + U_2^{(2)} X^{(2)}) + \eta R^T B). \end{aligned} \tag{12}$$

Step-3. Update R, when other variables are fixed, we have

$$\min_R \|B - RG\|_F^2, \quad s.t.\ RR^T = I. \tag{13}$$

Equation (13) is an orthogonal procrustes problem [16], which can be solved by using Singular Value Decomposition (SVD). Then, we first have $BG^T = P\Lambda\hat{P}^T$, thereafter, the solution can be obtained as $R = P\hat{P}^T$.

Step-4. Update B. By fixing other variables, Eq. (6) can be rewritten as

$$\begin{aligned} &\min_R \left\| Y - B^T V S \right\|_F^2 + \eta \|B - RG\|_F^2 + \mu \left\| B^T V \right\|_F^2. \\ &s.t.\ B \in \{-1, 1\}^{r \times n}. \end{aligned} \tag{14}$$

Due to the discrete constraint on the above subproblem, it is difficult to address. Many existing approaches address the problem by relaxing the binary constraint, however, the strategy may cause a large quantization error. Inspired by [17], we solve the issue in a discrete manner and the hash codes can be obtained by a bitwise strategy. Equation (14) can be further rewritten as

$$\min_{B} -2tr(Y^T B^T V S) + \left\|B^T V S\right\|_F^2 + \eta(\|B\|_F^2 - 2tr(B^T R G)) + \mu \left\|B^T V\right\|_F^2. \quad (15)$$

Equation (15) is equivalent to

$$\min_{B} -2tr(B^T Q) + \left\|B^T F\right\|_F^2 + \mu \left\|B^T V\right\|_F^2. \quad (16)$$

where $Q = VSY^T + \eta RG$ and $F = VS$. Let b denote the ith row of B^T, $\tilde{B}$ is the matrix of B^T without b. Let q_i, f_i and v_i denote the ith column of Q, F and V, respectively. $\tilde{Q}$, $\tilde{F}$ and $\tilde{V}$ are the matrices without q_i, f_i and v_i, respectively. Then, we discard the constant terms, the formulation for b can be stated as

$$\min_{b} -bq_i + b^T \tilde{B} \tilde{F} f_i^T + \beta b^T \tilde{B} \tilde{V} v_i^T, \quad (17)$$

Thereafter, the solution can be obtained as

$$b = sgn(q_i^T - \tilde{B}\tilde{F} f_i^T - \beta \tilde{B} \tilde{V} v_i^T). \quad (18)$$

2.4 Out-of-Sample Extension

In order to generate the binary codes of new coming samples. We can employ the learned matrix $U_2^{(t)}$ $(t = 1, 2)$ to generate the hash code of t-th modality $x^{(t)}$. Then, the formulation can be stated as

$$b_{query} = sgn(H^{(t)} x^{(t)}) = sgn(R(U_2^{(t)} x^{(t)})). \quad (19)$$

3 Experiments and Results Analysis

3.1 Experiment Settings

Datasets Wiki [14]. It contains 2866 text-image pairs with 10 different categories. Each textual instance is characterized by a 10-D topic vector. The visual sample is described by a 128-D SIFT feature vector.

LabelMe [15]. The database consists of 2686 text-image pairs with 8 different categories. The textual modality is described by a 366-D word frequency, and the corresponding visual modality is represented by a 512-D GIST feature.

Pascal VOC [5]. It contains 9963 text-image documents, which can be grouped into 20 classes. There are several text-image pairs are multilabled in the database. Similar to [27,32], these pairs are selected with only one label. Each text is described by a 399-D word frequency feature and the corresponding textual component is described by a 512-D vector.

Experimental Settings and Baselines. There are two different tasks on cross-media retrieval. The first task is zero-shot cross-media retrieval (ZSCMR), which follows the setting in [22]. In addition, to test the search performance on the seen categories, extra experiments are designed on the seen categories. Two typical scenarios are given in each task, i.e., Txt2Img and Img2Txt that employ a query data of one modality (text or image) to search similar samples of other media types (image or text).

For ZSCMR, we employ a new database split setting based on the original split settings. For LabelMe and Wiki, two classes are randomly chosen as the unseen classes, the remaining as seen classes. Specifically, the original training and test set are further divided into unseen class and seen class set. It should be noted that the classes in the unseen set and seen set are disjoint. For Pascal VOC, we randomly choose four classes as unseen classes. To reduce the impact of different classes combinations, the classes are randomly shuffled in the dataset. Thus, the number of test/training instances is uncertain, because it is related to the shuffle process.

It should be noted that very few works are proposed for zero-shot cross-media similarity search, and we only utilize three popular single-label datasets in this work. Some large-scale databases (e.g., NUS-WIDE [3]) are multi-label, in which the labels are mix together and the seen classes and unseen classes are difficult to separate [23], thus the standard zero-shot cross-media retrieval task is hard to perform in these datasets [26,32], the problem will be exploited in our future work. To testify the search performance of the developed method, we employ the widely-used mean average precision (mAP) as the evaluation protocol. Besides, the proposed DMZCR also compare with several competitive approaches that are originally presented for zero-shot retrieval or conventional retrieval, including CMFH [4], LSSH [33], ZSH [26], AH [25], CRE [7], CMAH [32] and CHOP [27]. The source code of CHOP is implemented by ourselves. All experiments are performed 20 times, and we take the average as the final performance.

3.2 Zero-Shot Retrieval

The mAP results of our DMZCR and other baselines are given in Fig. 2. It can be seen that our method obtain the best retrieval performance. From these results, it can be observed that the developed DMZCR can obtain the best performance in most cases, the main reason is that we employ the attribute information as the external knowledge, the inconsistent semantic information between seen and unseen classes can be alleviated. In addition, the proposed multi-layer network can guarantee the generality ability and knowledge transfer between the seen and unseen set in the learned common space, thus the performance of zero-shot retrieval can be enhanced, which shows that our method can generate more discriminative hash representations for the unseen query. These results demonstrate the efficacy of the developed model. Besides, we can see the performance of some methods reduce slightly with the increase of hash length, the reason is that the longer hash codes can contain more discriminative information and also contains noise information that affects search performance.

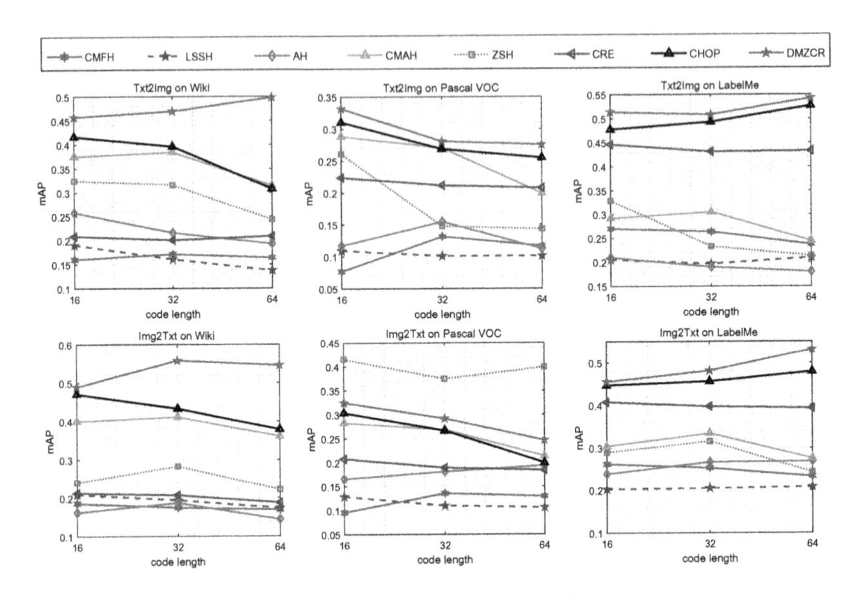

Fig. 2. mAP scores on zero-shot retrieval with varied hash code lengths on three databases.

3.3 Standard Retrieval

The mAP scores of all approaches on three databases are given in Table 1, the performance of standard retrieval is much better than that of zero-shot retrieval, which demonstrates that zero-shot retrieval is more challenging than the above standard search task. In Table 1 the proposed DMZCR outperforms all approaches on the three popular databases. From these results, we can conclude that the proposed model is effectiveness. There are three reasons: 1) our method employ a novel bidirectional matrix factorization scheme, which can preserve more valuable information of the original multimedia data. 2) the multi-layer semantic transformation scheme can not only enhance the knowledge transfer from seen classes to unseen classes but also help to narrow the semantic gap between different modalities. 3) the hash codes are learned by a discrete manner, which can reduce the quantization error. Also, we can observe that the mAP values of most methods increase when the hash length become longer, which verifies that longer hash code can embed in more valuable information. From above analysis, we can conclude that our method is competitive and effective in standard cross-media retrieval.

3.4 Ablation Study

Some ablation experiments are also further performed on Wiki database to investigate the effect of each part in DMZCR. Thus, two variants of DMZCR are designed as baselines: 1) DMZCR-1, which constructs by discarding the reconstruction part ($G \rightarrow X$, factoring G to reconstruct the original data) 2)

Table 1. MAP results on standard retrieval with different code lengths on three databases.

Task	Method	Wiki			Pascal VOC			LabelMe		
		16 bits	32 bits	64 bits	16 bits	32 bits	64 bits	16 bits	32 bits	64 bits
Img2Txt	CMFH	0.2567	0.2773	0.2804	0.2234	0.2073	0.1965	0.4236	0.3965	0.3822
	LSSH	0.2516	0.2711	0.2682	0.2931	0.3096	0.3274	0.6712	0.6881	0.7254
	ZSH	0.1807	0.1677	0.1579	0.1057	0.1069	0.1044	0.2796	0.2570	0.2779
	AH	0.1940	0.1641	0.1766	0.1106	0.1165	0.1170	0.1793	0.1859	0.1987
	CRE	0.2667	0.278	0.2664	0.2757	0.2898	0.322	0.7727	0.7813	0.8147
	CMAH	0.2324	0.2379	0.2597	0.2024	0.1917	0.2340	0.7817	0.7846	0.8039
	CHOP	0.2577	0.2642	0.2690	0.2833	0.3274	0.3313	0.6754	0.7239	0.7791
	DMZCR	**0.2743**	**0.2793**	**0.2722**	**0.3067**	**0.3420**	**0.3507**	**0.7929**	**0.7876**	**0.8286**
Txt2Img	CMFH	0.5245	0.5553	0.5726	0.5875	0.5559	0.5313	0.5310	0.4880	0.4871
	LSSH	0.5381	0.5649	0.5979	0.5886	0.6353	0.6377	0.6529	0.6716	0.7107
	ZSH	0.1426	0.1531	0.1466	0.1124	0.1157	0.1146	0.2634	0.2607	0.2716
	AH	0.2256	0.2260	0.2072	0.1291	0.1246	0.1327	0.2346	0.2595	0.2681
	CRE	0.4892	0.5188	0.5376	0.4251	0.4665	0.5338	0.8742	0.9005	0.9110
	CMAH	0.6235	0.6418	0.6286	0.8417	0.8564	0.8713	0.8726	0.8752	0.9022
	CHOP	0.6165	0.6202	0.6277	0.7721	0.8644	0.8693	0.8344	0.8823	0.9097
	DMZCR	**0.6392**	**0.6534**	**0.6865**	**0.8775**	**0.8809**	**0.8800**	**0.8887**	**0.9104**	**0.9242**

Table 2. MAP results of DMZCR and its variants.

Method/Task	Img2Txt			Txt2Img		
	16 bits	32 bits	64 bits	16 bits	32 bits	64 bits
DMZCR-1	0.2390	0.2116	0.2067	0.6127	0.6416	0.6567
DMZCR-2	0.2643	0.2699	0.2528	0.6245	0.6438	0.6659
DMZCR	**0.2743**	**0.2793**	**0.2722**	**0.6392**	**0.6534**	**0.6865**

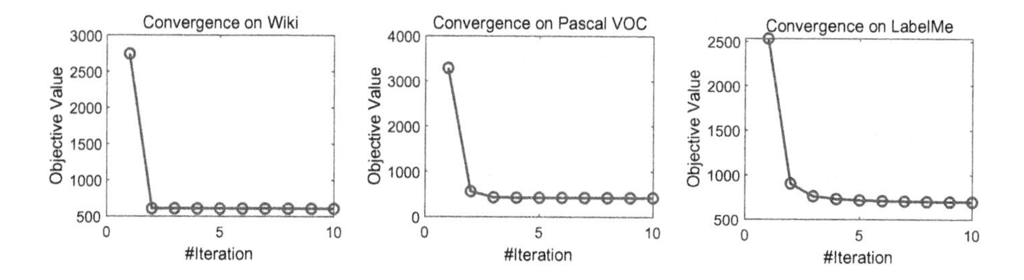

Fig. 3. Convergence analysis

DMZCR-2 is built by discarding the $X \rightarrow G$ term (decomposing X into G). Table 2 reports the experimental results. It can be observed that the search performance of DMZCR is superior to DMZCR-1 and DMZCR-2. These results further illustrates the efficacy of the developed method.

3.5 Convergence

In this part, we further perform several experiments on three databases to verify the convergence of our approach. The hash length is set to 16 bits, Fig. 3 reports the experimental results. It can be observed that DMZCR converges very fast, i.e., within 10 iterations. Thus, the proposed model is practical and efficient.

4 Conclusion

In this paper, a novel hash model (DMZCR) is developed for zero-shot cross-media similarity search. DMZCR employs a novel bidirectional matrix factorization scheme, a reconstruction constraint is exerted to the process of decomposing common semantic representations matrix from original data, then the original multimedia data can be reconstructed by the learned semantic representations. Thus the representations can be embedded in more valuable information from the original data. Besides, a multi-layer semantic transmission scheme is proposed, which can help to transfer the knowledge from seen classes to unseen classes, and the inconsistent semantics between source set and the target can also be reduced. A discrete optimization scheme is also presented in our model, which can reduce the large quantization error. Experiments on both zero-shot and standard retrieval tasks clearly demonstrate the efficacy of our model.

References

1. Changpinyo, S., Chao, W.L., Gong, B., Sha, F.: Synthesized classifiers for zero-shot learning. In: The IEEE Conference on Computer Vision and Pattern Recognition (CVPR), June 2016
2. Chi, J., Peng, Y.: Dual adversarial networks for zero-shot cross-media retrieval. In: IJCAI, pp. 663–669 (2018)
3. Chua, T.S., Tang, J., Hong, R., Li, H., Luo, Z., Zheng, Y.: NUS-WIDE: a real-world web image database from National University of Singapore. In: Proceedings of the ACM International Conference on Image and Video Retrieval, pp. 1–9 (2009)
4. Ding, G., Guo, Y., Zhou, J.: Collective matrix factorization hashing for multimodal data. In: Proceedings of the IEEE Conference on Computer Vision and Pattern Recognition, pp. 2075–2082 (2014)
5. Everingham, M., Gool, L., Williams, C.K., Winn, J., Zisserman, A.: The pascal visual object classes (VOC) challenge. Int. J. Comput. Vis. **88**(2), 303–338 (2010)
6. Guo, Y., Ding, G., Han, J., Gao, Y.: SitNet: discrete similarity transfer network for zero-shot hashing. In: IJCAI, pp. 1767–1773 (2017)
7. Hu, M., Yang, Y., Shen, F., Xie, N., Hong, R., Shen, H.T.: Collective reconstructive embeddings for cross-modal hashing. IEEE Trans. Image Process. **28**(6), 2770–2784 (2019)
8. Lin, Z., Ding, G., Hu, M., Wang, J.: Semantics-preserving hashing for cross-view retrieval. In: Proceedings of the IEEE Conference on Computer Vision and Pattern Recognition, pp. 3864–3872 (2015)
9. Liu, H., Ji, R., Wu, Y., Huang, F., Zhang, B.: Cross-modality binary code learning via fusion similarity hashing. In: Proceedings of the IEEE Conference on Computer Vision and Pattern Recognition, pp. 7380–7388 (2017)

10. Long, Y., Liu, L., Shao, L.: Towards fine-grained open zero-shot learning: inferring unseen visual features from attributes. In: 2017 IEEE Winter Conference on Applications of Computer Vision (WACV), pp. 944–952. IEEE (2017)
11. Pachori, S., Deshpande, A., Raman, S.: Hashing in the zero shot framework with domain adaptation. Neurocomputing **275**, 2137–2149 (2018)
12. Peng, Y., Qi, J., Huang, X., Yuan, Y.: CCL: cross-modal correlation learning with multigrained fusion by hierarchical network. IEEE Trans. Multimedia **20**(2), 405–420 (2018)
13. Pennington, J., Socher, R., Manning, C.: GloVe: global vectors for word representation. In: Proceedings of the 2014 Conference on Empirical Methods in Natural Language Processing (EMNLP), pp. 1532–1543 (2014)
14. Rasiwasia, N., et al.: A new approach to cross-modal multimedia retrieval. In: Proceedings of the 18th ACM International Conference on Multimedia, pp. 251–260 (2010)
15. Russell, B.C., Torralba, A., Murphy, K.P., Freeman, W.T.: LabelMe: a database and web-based tool for image annotation. Int. J. Comput. Vis. **77**(1), 157–173 (2008)
16. Schönemann, P.H.: A generalized solution of the orthogonal procrustes problem. Psychometrika **31**(1), 1–10 (1966)
17. Shen, F., Shen, C., Liu, W., Shen, H.T.: Supervised discrete hashing. In: 2015 IEEE Conference on Computer Vision and Pattern Recognition (CVPR), pp. 37–45 (2015)
18. Song, J., Yang, Y., Yang, Y., Huang, Z., Shen, H.T.: Inter-media hashing for large-scale retrieval from heterogeneous data sources. In: Proceedings of the 2013 ACM SIGMOD International Conference on Management of Data, pp. 785–796 (2013)
19. Tang, J., Wang, K., Shao, L.: Supervised matrix factorization hashing for cross-modal retrieval. IEEE Trans. Image Process. **25**(7), 3157–3166 (2016)
20. Wang, B., Yang, Y., Xu, X., Hanjalic, A., Shen, H.T.: Adversarial cross-modal retrieval. In: Proceedings of the 25th ACM International Conference on Multimedia, pp. 154–162 (2017)
21. Wang, Y., He, S., Xu, X., Yang, Y., Li, J., Shen, H.T.: Self-supervised adversarial learning for cross-modal retrieval. In: Proceedings of the 2nd ACM International Conference on Multimedia in Asia, pp. 1–7 (2021)
22. Xian, Y., Schiele, B., Akata, Z.: Zero-shot learning-the good, the bad and the ugly. In: Proceedings of the IEEE Conference on Computer Vision and Pattern Recognition, pp. 4582–4591 (2017)
23. Xu, X., Lu, H., Song, J., Yang, Y., Shen, H.T., Li, X.: Ternary adversarial networks with self-supervision for zero-shot cross-modal retrieval. IEEE Trans. Cybern. **50**(6), 2400–2413 (2019)
24. Xu, X., Shen, F., Yang, Y., Shen, H.T., Li, X.: Learning discriminative binary codes for large-scale cross-modal retrieval. IEEE Trans. Image Process. **26**(5), 2494–2507 (2017)
25. Xu, Y., Yang, Y., Shen, F., Xu, X., Zhou, Y., Shen, H.T.: Attribute hashing for zero-shot image retrieval. In: 2017 IEEE International Conference on Multimedia and Expo (ICME), pp. 133–138. IEEE (2017)
26. Yang, Y., Luo, Y., Chen, W., Shen, F., Shao, J., Shen, H.T.: Zero-shot hashing via transferring supervised knowledge. In: Proceedings of the 24th ACM International Conference on Multimedia, pp. 1286–1295 (2016)
27. Yuan, X., Wang, G., Chen, Z., Zhong, F.: CHOP: an orthogonal hashing method for zero-shot cross-modal retrieval. Pattern Recogn. Lett. **145**, 247–253 (2021)

28. Zhang, D., Wu, X.J.: Scalable discrete matrix factorization and semantic autoencoder for cross-media retrieval. IEEE Trans. Cybern. (2020)
29. Zhang, D., Wu, X.J., Liu, Z., Yu, J., Kitter, J.: Fast discrete cross-modal hashing based on label relaxation and matrix factorization. In: 2020 25th International Conference on Pattern Recognition (ICPR), pp. 4845–4850. IEEE (2021)
30. Zhang, D., Li, W.J.: Large-scale supervised multimodal hashing with semantic correlation maximization. In: Twenty-Eighth AAAI Conference on Artificial Intelligence (2014)
31. Zhang, J., Peng, Y., Yuan, M.: Unsupervised generative adversarial cross-modal hashing. In: Thirty-Second AAAI Conference on Artificial Intelligence (2018)
32. Zhong, F., Chen, Z., Min, G.: An exploration of cross-modal retrieval for unseen concepts. In: Li, G., Yang, J., Gama, J., Natwichai, J., Tong, Y. (eds.) DASFAA 2019. LNCS, vol. 11447, pp. 20–35. Springer, Cham (2019). https://doi.org/10.1007/978-3-030-18579-4_2
33. Zhou, J., Ding, G., Guo, Y.: Latent semantic sparse hashing for cross-modal similarity search. In: Proceedings of the 37th International ACM SIGIR Conference on Research & Development in Information Retrieval, pp. 415–424 (2014)

Dual Stream Fusion Network for Multi-spectral High Resolution Remote Sensing Image Segmentation

Yong Cao[1,2](✉), Yiwen Shi[3], Yiwei Liu[4], Chunlei Huo[1,2], Shiming Xiang[1,2], and Chunhong Pan[2]

[1] School of Artificial Intelligence, University of Chinese Academy of Sciences, Beijing 100049, China
yong.cao@nlpr.ia.ac.cn

[2] National Laboratory of Pattern Recognition, Institute of Automation, Chinese Academy of Sciences, Beijing 100190, China
{clhuo,smxing,chpan}@nlpr.ia.ac.cn

[3] Beijing City University Intelligent Electronic Manufacturing Research Center, Beijing 100191, China

[4] Beijing University of Civil Engineering and Architecture, Beijing 102627, China
yiwei.liu@tom.com

Abstract. Semantic segmentation is in-demand in High Resolution Remote Sensing (HRRS) image processing. Unlike natural images, HRRS images usually provide channels such as Near Infrared (NIR) in addition to RGB channels. However, in order to make use of the pre-trained model, the current semantic segmentation methods in remote sensing field usually only use the RGB channel and discard the information of other channels. In this paper, to make full use of the HRRS image information, a dual-stream fusion network is proposed to fuse the information of different channel combinations through a Feature Pyramid Network (FPN), then a Stage Pyramid Pooling (SPP) module is used to integrate the features of different scales and produce the final segmentation results. Experiments on the RSCUP competition dataset show that the proposed approach can effectively improve the segmentation performance.

Keywords: Semantic segmentation · Remote sensing · Stream fusion

1 Introduction

With the development of satellite photography, a large number of High Resolution Remote Sensing (HRRS) images have been obtained. Semantic segmentation is to assign a category label to each pixel in the image, which is a key technology to realize automatic interpretation of remote sensing images. It has important

The first author is a student.

This research was supported by the National Key Research and Development Program of China under Grant No. 2018AAA0100400, and the National Natural Science Foundation of China under Grants 62071466, 62076242, and 61976208.

H. Ma et al. (Eds.): PRCV 2021, LNCS 13020, pp. 537–547, 2021.
https://doi.org/10.1007/978-3-030-88007-1_44

application value in agriculture [1], urban planning [2], transportation [3] and other aspects.

In recent years, deep learning methods based on Fully Convolutional Network (FCN) [4] have made continuous breakthroughs in semantic segmentation and have become the mainstream method in this field. Some well-known methods in natural images, such as Pyramid Scene Parsing Network (PSPNet) [5], DeepLab series [6,7], Dual Attention Network (DANet) [8] etc. have also been widely used in semantic segmentation of HRRS images. This is because unlike hyperspectral remote sensing images, HRRS images are very close to natural images visually. The difference is that in addition to RGB channels, remote sensing images usually provide Near Infrared (NIR) or Infrared (IR) channels, such as ISPRS [9] and GID [10] datasets.

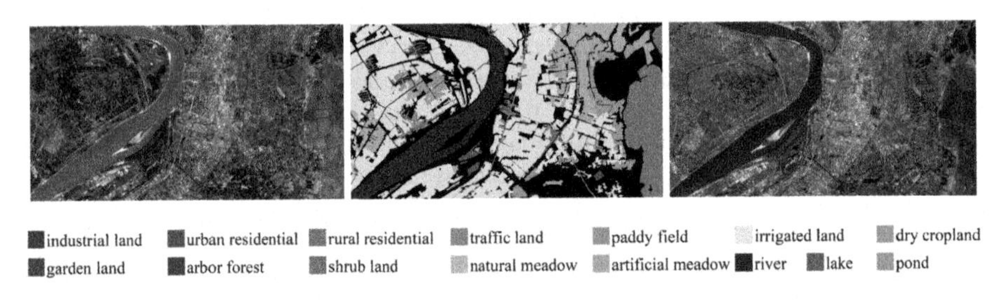

Fig. 1. Illustration of the complexity of HRRS images, the left is an RGB image, the middle is ground truth and the right is the NIR-GB image. HRRS images contain a variety of complex objects. Different objects have different separability under different spectra. Some ground objects that are not easy to distinguish in the RGB band will become easier to distinguish in the NIR band.

However, in practical applications, since the number of remote sensing images is usually relatively small, overfitting is easy to occur if the original three-channel or four-channel data are directly sent to the network. To prevent overfitting, people usually use parameters pre-trained on natural images (i.e. ImageNet) to initialize model parameters. Since natural images only contain RGB channels, the NIR channel is often discarded directly, which inevitably leads to the loss of information.

In fact, NIR channel is very important for the classification of ground objects, especially for the classification of plants, water bodies and other ground objects as shown in Fig. 1. For example, the calculation of vegetation index and water body index widely used in remote sensing is inseparable from the NIR channel. In some studies, NIR channels are used instead of R channels to obtain state of the art results on ISPRS datasets [11].

To make full use of the spectral information of remote sensing images and take advantage of the pre-trained parameters of natural images, we design a dual-stream fusion network to fuse the information of the two streams through a Feature Pyramid Network (FPN), the segmentation map is finally obtained by a Stage Pyramid Pooling module to integrate the features of different scales.

The proposed network significantly improves the segmentation accuracy and wins the second place in the RSCUP 2019 Remote Sensing Image Interpretation Competition. We also design several ablation experiments to prove the effectiveness of our method.

2 Related Work

Semantic segmentation is to classify images at pixel level. Before deep learning, traditional methods usually divide the image into many regions according to color, texture, boundary and other low-level information, and then the regions are classified into different objects. The emergence of deep neural networks, especially the FCN structure, makes end-to-end semantic segmentation possible. The current semantic segmentation framework based on deep learning can be mainly divided into three categories: encoder-decoder structure, which adopts symmetric network structure. The image is firstly downsampled to extract features, and then the feature map is restored to the size of the input image by step-by-step up-sampling. The representative works are UNet [12], SegNet [13], RefineNet [14] and so on; Another is dilated convolution and pyramid structure, such as Deeplab series [6,7] PSPNet [5], etc., and the third method is based on attention mechanism, such as CCNet [15], DANet [8], etc. Current methods often mix all three of these approaches.

Different from multi-spectral and hyperspectral remote sensing images, HRRS images have low spectral resolution and high spatial resolution, so they are more similar to natural images visually. Therefore, the semantic segmentation method in the natural image can be easily applied to HRRS images [16,17]. In [18] a multi-modal and multi-scale deep network is proposed and it is one of the earliest works to apply deep convolutional neural networks to remote sensing fields. Liu et al. [11] propose a self-cascaded convolutional neural network to aggregate multiscale information and win the ISPRS 2D image labeling competition. Li et al. [19] propose a boundary loss to improve the segmentation accuracy at the boundary of objects. However, most works [11,19] uses only three channels of data, even though the dataset provides NIR or IR channels.

Feature pyramids are a basic component in computer vision systems for extract features at different scales. A deep ConvNet computes a feature hierarchy layer by layer, and with sub-sampling layers, the feature hierarchy has an inherent multi-scale pyramidal shape. The Feature Pyramid Network (FPN) proposed by Lin et al. [20] is widely used in target detection and instance segmentation as an infrastructure. Kirillov et al. [21] extend FPN to panoramic segmentation by adding a semantic branch. Seferbekov et al. [22] apply it to the land segmentation task of remote sensing images.

3 Proposed Method

As mentioned above, the current semantic segmentation methods usually adopt three-channel input, which cannot make full use of the spectral information of remote sensing images. Therefore, we propose a dual-stream network to extract and fuse the information of different channel combinations, after that a stage pyramid pooling module is used to integrate the features of different scales and obtain the final segmentation results. The overall structure of the proposed network is shown in Fig. 2, and we will detail our method below.

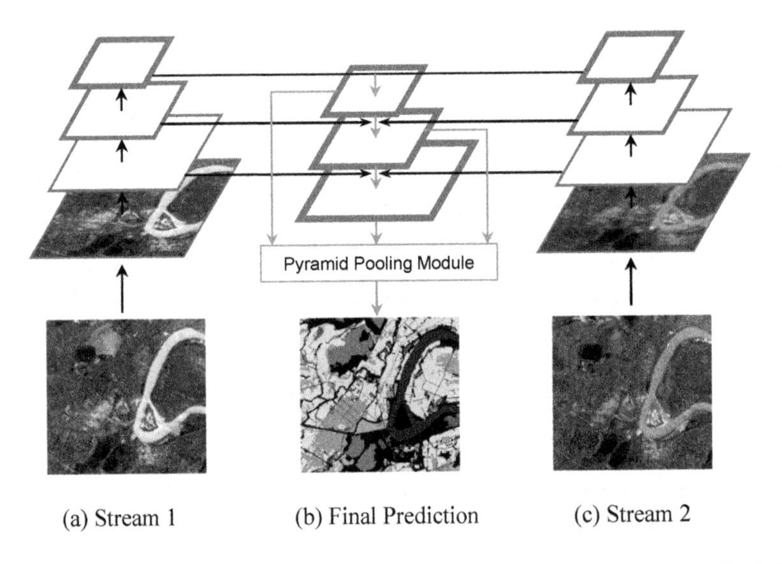

Fig. 2. The overall structure of the proposed dual-stream network. On the left and right of the figure are the NIR-GB branch and RGB branch respectively. The feature pyramid fusion module is in the middle, and the stage pyramid pooling module is in the bottom.

3.1 Dual-Stream Fusion

Dual Bottom-Up Pathway: Firstly, the four-channel (NIR-RGB) remote sensing image is divided into two three-channel (NIR-GB and RGB) combinations, and then sent to two bottom-up networks to extract features. The bottom-up pathway consists of two forward convolutional neural networks, and any existing neural network (e.g., ResNet, ResNest) can be adopted. Through the backbone network, we can get a hierarchical feature map containing four scales (referred as stages). The feature map naturally has a pyramid structure, and The feature map near the bottom usually contains more details (such as texture, edge, etc.), while the feature map near the top contains more semantic information.

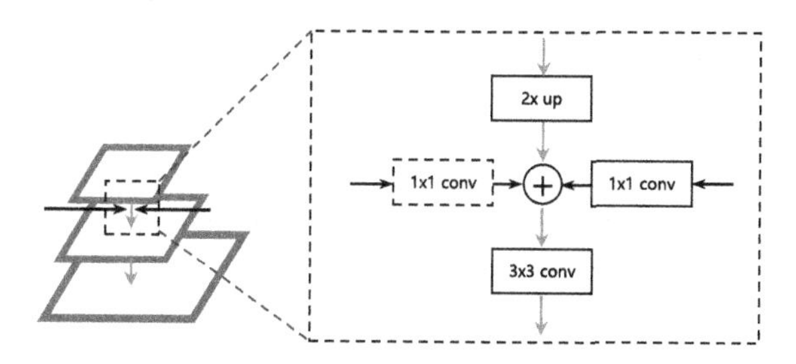

Fig. 3. Detail of the top-down fusion pathway. The 3×3 conv is used to eliminate the aliasing effect and the 2x up is a bilinear interpolation.

Top-Down Fusion Pathway: To fuse the information obtained from the two streams, a top-down feature pyramid network is designed to concat the information obtained from each stage of the backbone network through two horizontal connections. Meanwhile, the information obtained from each stage of the backbone network is upsampled to the same size as the feature map of the next stage through a bilinear interpolation, and then the three feature maps are concatenated. As shown in Fig. 3, to eliminate the aliasing effect, 3×3 convolution is added to the feature map obtained by the top stage. The final result is a $1/4$ size feature map of the original image.

3.2 Stage Pyramid Pooling

To obtain the final segmentation result and make full use of the information of all stages of the fusion module, we design a Stage Pyramid Pooling (SPP) module. Different from the spatial pyramid pooling module in PSPNet, in which different pooling kernels are used on the same feature map. Our SPP module pools the features of different stages of the fusion module respectively. As shown in Fig. 4, through stage pyramid pooling, feature maps of two scales are generated at each stage and then all of them are up-sampled to the same size and finally concat together to generate the final segmentation map.

Specifically, stage 1 to stage 3 are the feature maps generated by the dual-stream fusion module, and their scales are $1/16$, $1/8$ and $1/4$ of the original image, respectively. The proposed SPP module adopts 2×2 and 4×4 pooling kernels to pool the feature maps of the three stages and produces a series of feature maps of scales from $1/64$ to $1/8$ of the original image. Then all the feature maps are upsampled to the size of $1/8$ of the original image by bilinear interpolation. In order to further utilize the bottom information of the image, we concat the feature map generated by the SPP module with the feature map of Stage 3 as the representation. Finally, the representation is fed into a convolution layer to get the final per-pixel prediction.

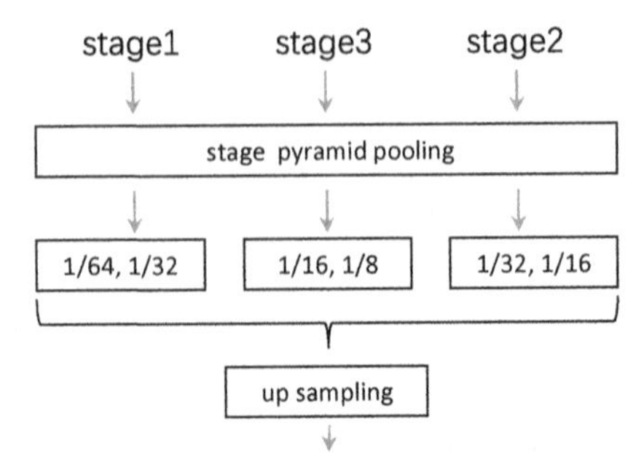

Fig. 4. Detail of the stage pyramid pooling module. Every stage generate 2 feature map at different scale, and all of them are upsampled to the same size and concated together.

4 Experiments and Analysis

4.1 Data Description

We adopt the remote sensing data provided by the semantic segmentation course of the RSCUP 2019 remote sensing image interpretation competition, including 12 finely annotated remote sensing images of China collected by Gaofen2 satellite. Each image is 7200×6800 in size and contains four bands (NIR-RGB). Among them, ground features are divided into 16 categories, including industrial land, urban residential, rural residential, traffic land, paddy field, irrigated land, dry cropland, garden land, arbor forest, shrubland, natural meadow, artificial meadow, river, lake and pond. The spatial resolution of the images is 2 m to 5 m.

4.2 Experimental Settings

Data Augmentation. Because the size of the original images is very large, which make it difficult to send to the GPUs directly. We first cut them into 520×520 patches with overlap. At the same time, because the category distribution in the original data is very unbalanced, we adopt a resampling strategy to alleviate it. That is, for the category whose proportion is less than 3%, when the number of pixels in the region to be clipped is greater than a certain threshold value (such as 200), the distribution of the upper, lower, left and right of the region will be shifted by 100 pixels to be clipped again. Finally, we get about 5000 images, of which 80% is used for training and the remaining 20% for testing. In training, 480×480 patches are randomly cut and flipped horizontally or vertically with a probability of 0.5, as well as rotated by $[0, 90, 180, 270]$ degrees.

Training Parameters. In the experiments, we use models pretrained on the VOC2012 semantic segmentation dataset as initialization parameters for the backbone network, which has better performance than the parameters of ImageNet. During the training, all models are trained on 4 Tesla-V100 GPUs with 32G memory. The training data is fed to the model with a batch size of 16 for each GPU and SyncBN is adopted. The initial learning rate of the SGD optimizer is set to be 0.01 and decreased by every iteration according to the POLY strategy with power of 2. The weight decay and momentum are set to be 1e-4 and 0.9 respectively. The learning rate of the backbone is 0.1 times of the global.

4.3 Results and Analysis

We adopt the Mean Intersection over Union (MIoU) as evaluation criterion, The formula is defined as $mIoU(P_m, P_{gt}) = \frac{1}{K}\frac{|P_m \bigcap P_{gt}|}{|P_m \bigcup P_{gt}|}$, where P_m is the prediction, P_{gt} is ground truth and K is the number of classes including background. We also report the pixel accuracy of all the models.

Table 1. Results on RSCUP 2019 dataset

Method	Mean IoU (%)	Pixel Acc. (%)
FCN+ResNet50	32.3	49.6
PSPNet+ResNet50	39.5	62.8
DeepLabV3+ResNet50	39.2	61.9
DANet+ResNet50	40.2	64.6
FCN+ResNet101	34.8	53.7
PSPNet+ResNet101	41.1	65.9
DeepLabV3+ResNet101	40.7	64.8
DANet+ResNet101	41.5	66.3
DS-FusionNet+ResNet50 (ours)	**41.7**	**66.9**
DS-FusionNet+ResNet101 (ours)	**43.6**	**70.3**

Results Comparison. We report the related results in Table 1. It can be observed that the proposed approach outperforms other widely used models in HRRS image segmentation. Among all these models, FCN has the worst performance because it is a very primitive work, and many techniques are not used. The performance of PSPNet is better than that of Deeplabv3. We believe the reason is that the context information in remote sensing images is very important, and many categories must rely on global information to determine the category, so the model with a pyramid structure has a better performance. DANet is the latest attention mechanism based network. It performs well, but it is inefficient because it has to compute the attention characteristic map twice. Our model outperforms the above three networks, more importantly, our model based on ResNet50 outperforms other models based on ResNet101. We show some visualization results in Fig. 5.

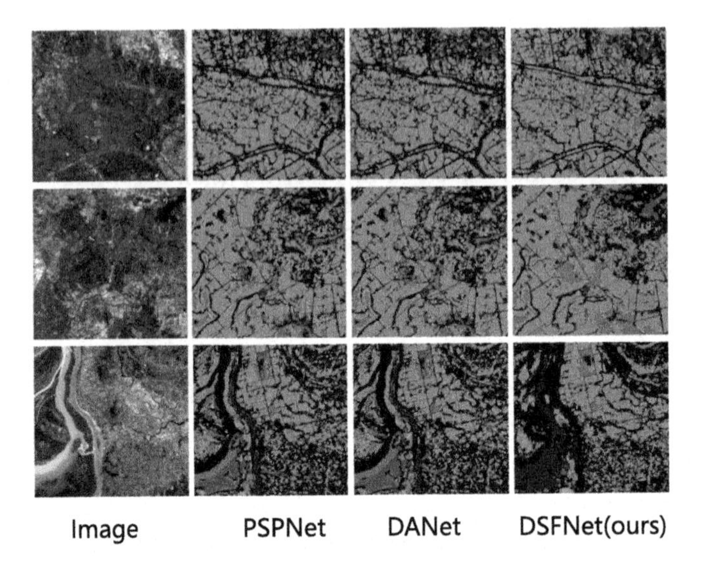

Fig. 5. Result visualization. From left to right are the results of the input image, PSPNet, DANet and our proposed Dual Stream Fusion Network.

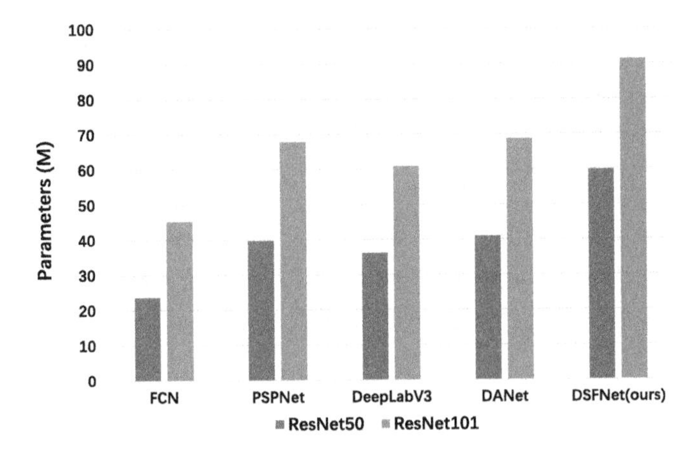

Fig. 6. Total parameters of different models, it can be found that our model based on resnet50 has higher accuracy and less computation compared with other models based on resnet101.

Efficiency Comparison. To compare model efficiency, we counted the total number of parameters with gradient in the model, which can represent the computational complexity and storage occupancy of different models. The results are shown in Fig. 6, by comparison, it can be deduced that our method with ResNet50 is less computation-intensive than other models based on Resnet101, which proves that the performance improvement of our method does not depend on the increase of the number of model parameters.

Ablation Study. To further verify the effectiveness of our method, we design an ablation experiment in which only RGB or NIR-GB are used as the input of two streams, and the final MIoU results decreased by 2.1% and 1.6% respectively. On the one hand, it shows that our model can effectively fuse the information of different channel combinations, and at the same time, it also shows that the NIR channel is very important in HRRS image segmentation. We present the visualization results of a region as shown in Fig. 7, where the arbor forest is accurately identified by introducing NIR channels.

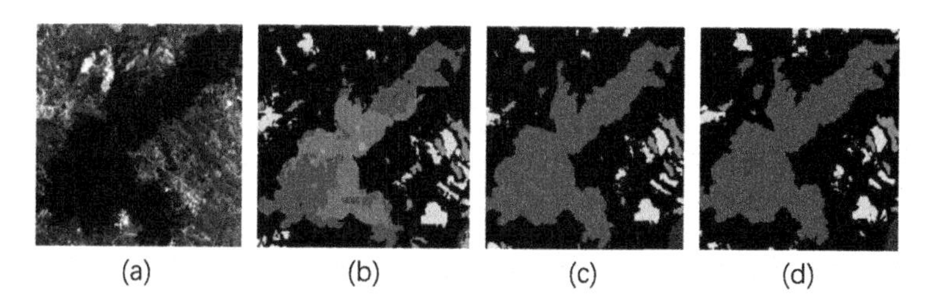

Fig. 7. Visualization of ablation study. Where (a) is the input image, (b) is the result of RGB channels, (c) is the result of NIR-GB channels and (d) is use both two channel combinations.

5 Conclusion

To make full use of the spectral information in HRRS images, a dual-stream fusion network is proposed, in which pyramid structure is widely used to fuse information of different channels and scales. Experiments demonstrate the effectiveness of our method. In future work, we will further explore the application of pyramid structure in remote sensing image interpretation.

References

1. Yin, H., Prishchepov, A.V., Kuemmerle, T., Bleyhl, B., Buchner, J., Radeloff, V.C.: Mapping agricultural land abandonment from spatial and temporal segmentation of landsat time series. Remote Sens. Environ. **210**, 12–24 (2018)
2. Zhang, Q., Seto, K.C.: Mapping urbanization dynamics at regional and global scales using multi-temporal DMSP/OLS nighttime light data. Remote Sens. Environ. **115**(9), 2320–2329 (2011)
3. Maboudi, M., Amini, J., Malihi, S., Hahn, M.: Integrating fuzzy object based image analysis and ant colony optimization for road extraction from remotely sensed images. ISPRS J. Photogramm. Remote. Sens. **138**, 151–163 (2018)
4. Long, J., Shelhamer, E., Darrell, T.: Fully convolutional networks for semantic segmentation. In: Proceedings of the IEEE Conference on Computer Vision and Pattern Recognition, pp. 3431–3440 (2015)

5. Zhao, H., Shi, J., Qi, X., Wang, X., Jia, J.: Pyramid scene parsing network. In: Proceedings of the IEEE Conference on Computer Vision and Pattern Recognition, pp. 2881–2890 (2017)
6. Chen, L., Papandreou, G., Kokkinos, I., Murphy, K., Yuille, A.L.: DeepLab: semantic image segmentation with deep convolutional nets, atrous convolution, and fully connected CRFs. IEEE Trans. Pattern Anal. Mach. Intell. **40**(4), 834–848 (2017)
7. Chen, L., Papandreou, G., Schroff, F., Adam, H.: Rethinking atrous convolution for semantic image segmentation. arXiv preprint arXiv:1706.05587 (2017)
8. Fu, J., et al.: Dual attention network for scene segmentation. In: Proceedings of the IEEE Conference on Computer Vision and Pattern Recognition, pp. 3146–3154 (2019)
9. ISPRS. https://www2.isprs.org/commissions/comm2/wg4/benchmark/semantic-labeling/. Accessed 25 Mar 2021
10. Tong, X.Y., et al.: Land-cover classification with high-resolution remote sensing images using transferable deep models. Remote Sens. Environ. (2020). https://doi.org/10.1016/j.rse.2019.111322
11. Liu, Y., Fan, B., Wang, L., Bai, J., Xiang, S., Pan, C.: Semantic labeling in very high resolution images via a self-cascaded convolutional neural network. ISPRS J. Photogramm. Remote. Sens. **145**, 78–95 (2018)
12. Ronneberger, O., Fischer, P., Brox, T.: U-Net: convolutional networks for biomedical image segmentation. In: Navab, N., Hornegger, J., Wells, W.M., Frangi, A.F. (eds.) MICCAI 2015. LNCS, vol. 9351, pp. 234–241. Springer, Cham (2015). https://doi.org/10.1007/978-3-319-24574-4_28
13. Badrinarayanan, V., Kendall, A., Cipolla, R.: Segnet: a deep convolutional encoder-decoder architecture for image segmentation. IEEE Trans. Pattern Anal. Mach. Intell. **39**(12), 2481–2495 (2017)
14. Lin, G., Milan, A., Shen, C., Reid, I.: Refinenet: multi-path refinement networks for high-resolution semantic segmentation. In: Proceedings of the IEEE Conference on Computer Vision and Pattern Recognition, pp. 1925–1934 (2017)
15. Huang, Z., Wang, X., Huang, L., Huang, C., Wei, Y., Liu, W.: CCNet: criss-cross attention for semantic segmentation. In: Proceedings of the IEEE/CVF International Conference on Computer Vision, pp. 603–612 (2019)
16. Kampffmeyer, M., Salberg, A.B., Jenssen, R.: Semantic segmentation of small objects and modeling of uncertainty in urban remote sensing images using deep convolutional neural networks. In: Proceedings of the IEEE Conference on Computer Vision and Pattern Recognition Workshops, pp. 1–9 (2016)
17. Zhao, W., Shihong, D.: Learning multiscale and deep representations for classifying remotely sensed imagery. ISPRS J. Photogramm. Remote. Sens. **113**, 155–165 (2016)
18. Audebert, N., Le Saux, B., Lefèvre, S.: Semantic segmentation of earth observation data using multimodal and multi-scale deep networks. In: Lai, S.-H., Lepetit, V., Nishino, K., Sato, Y. (eds.) ACCV 2016. LNCS, vol. 10111, pp. 180–196. Springer, Cham (2017). https://doi.org/10.1007/978-3-319-54181-5_12
19. Li, A., Jiao, L., Zhu, H., Li, L., Liu, F.: Multitask semantic boundary awareness network for remote sensing image segmentation. IEEE Trans. Geosci. Remote Sens. (2021)
20. Lin, T.-Y., Dollár, P., Girshick, R., He, K., Hariharan, B., Belongie, S.: Feature pyramid networks for object detection. In: Proceedings of the IEEE Conference on Computer Vision and Pattern Recognition, pp. 2117–2125 (2017)

21. Kirillov, A., Girshick, R., He, K., Dollár, P.: Panoptic feature pyramid networks. In: Proceedings of the IEEE/CVF Conference on Computer Vision and Pattern Recognition, pp. 6399–6408 (2019)
22. Seferbekov, S., Iglovikov, V., Buslaev, A., Shvets, A.: Feature pyramid network for multi-class land segmentation. In: Proceedings of the IEEE Conference on Computer Vision and Pattern Recognition Workshops, pp. 272–275 (2018)

Multi-scale Extracting and Second-Order Statistics for Lightweight Steganalysis

Junfu Chen[1], Zhangjie Fu[1,2(✉)], Xingming Sun[1], and Enlu Li[1]

[1] Engineering Research Center of Digital Forensics, Ministry of Education, Nanjing University of Information Science and Technology, Nanjing 210044, China
fzj@nuist.edu.cn

[2] School of Computer and Software, Nanjing University of Information Science and Technology, Nanjing 210044, China

Abstract. Steganography is a technology that modifies complex regions of digital images to embed secret messages for the purpose of covert communication, while steganalysis is to detect whether secret messages are hidden in a digital image or not. In recent years, it has become necessary to deploy computer vision based algorithms on devices that are mobile or have limited computational memories. However, the emergence of steganalysis prove the point that the more parameters available to the model, the better the presentation will be. In order to enable the model to achieve the extraction of steganographic noise with a tiny number of parameters, this paper proposes a lightweight steganalysis algorithm based on multi-scale feature extraction and the fusion of multi-order statistical properties. Compared with Yedroudj-Net, Zhu-Net and SRNet with S-UNIWARD embedding rate of 0.4 bpp, the numbers of parameters was decreased by 87.9%, 98.2%, 98.9% correspondingly and the accuracy of steganalysis was improved by 7.12%, 2.65%, 3.42% respectively. Our experimental results show that the model not only has a reduced number of parameters for existing steganalysis, but also can effectively boost the accuracy of steganalysis.

Keywords: Steganalysis · Multi-information fusion · Mobile applications

1 Introduction

Trillions of images are flooding the Internet today in all sorts of social media. People who benefit from the privilege of using Plog to express themselves are also subject to the pitfalls that arise from digital images. The misuse of steganography and the emergence of steganographic software pose a huge challenge to social stability and personal security in a real life situation. Hence, domestic and international researchers have combined deep learning with steganalysis models

Supported by Foundation item: National Key Research and Development Program of China (2018YFB1003205); National Natural Science Foundation of China (U1836110, U1836208); by the Jiangsu Basic Research Programs-Natural Science Foundation under grant numbers BK20200039.

H. Ma et al. (Eds.): PRCV 2021, LNCS 13020, pp. 548–559, 2021.
https://doi.org/10.1007/978-3-030-88007-1_45

to counter such threats, but existing steganalysis models are plagued with problems such as huge number of parameters, slow training speed and insufficient detection precision. This is why it has become a hot topic for research on how to achieve lightweight, fast and highly accurate steganalysis models for mobile applications.

Most of the existing steganalysis models are only for a fixed steganographic algorithm(such as S-UNIWARD [9], HILL [12], HOGO [4]) or for a certain scope. In 2000, Westfeld et al. [18] were the first to propose a statistical analysis for LSB steganography, and after that, researchers have proposed RS analysis, DIH analysis, and WS analysis to improve the accuracy of embedding ratio estimation. In 2005, Andrew formulated the feature histogram formula, the first dedicated steganalysis of grey-scale image for LSB [11]. In 2008, Liu et al. [13] used the correlation of the lowest two image planes as a feature to detect LSBM, which considered the effect of LSBM steganography on the lower image planes. Bohme [1] proposed a Jsteg-specific steganalysis model by transferring the LSB-specific steganalysis method to frequency domain images. In 2011, Gul et al. and Luo et al. have proposed a specific steganalysis model for HUGO [7,14]. In 2014, Xia et al. [19] designed a dedicated steganalysis analyzer for LSBM by analyzing the correlation between adjacent pixels. Tang et al. [17] have proposed a steganalysis strategy for adaptive steganography such as WOW, and this strategy can be applied to the spatial and frequency domains according to different steganography algorithms.

With the emergence of adaptive steganographic algorithms, the steganalysis resistance of various steganographic algorithms has gradually increased, which requires more and more detection capability for steganalysis. The development of steganography has evolved rapidly in recent years, and steganographic algorithms combined with deep learning have made it difficult to design tailored steganalysis. This is why a generic steganalysis based on deep learning is essential. This makes it possible to extract spatially distributed features and steganographic noise information from digital images without the need for a specialist researcher to design the feature extraction method by hand, while taking advantage of the end-to-end learning process of deep learning, which allows feature extraction and discriminators to be trained simultaneously.

In this paper, we propose a residual structure-based network called Light-Net to improve the accuracy of steganography in both the spatial and frequency domains. Our model combines several visual modules: residual structures, attention mechanisms, and second-order statistical information to achieve a lightweight steganalysis. Our proposed model performs better in terms of detection accuracy and mobility, with a relative reduction of over 90% in the number of parameters compared to other steganalysis models. We have tested the model in both the spatial and frequency domains for different steganographic methods respectively, both yielding better results, and our contribution points are summarised below:

(1) The residual structure is applied by group convolution, where the feature maps of all groups are connected and the obtained residual feature maps are fused together with 1×1 convolution blocks.
(2) The attention mechanism was used to assign weights to the combined feature blocks to avoid the problem of the distribution drifting and the model failing to converge, adding the attention mechanism intentionally can be found to speed up the convergence of the model in the experiments.
(3) The integration of the final features using second-order statistical properties and first-order statistical properties results in a significant improvement in the final detection accuracy and transferability.

2 Related Works

We classify distinct steganalysis into two categories, semi-learning models and full-learning models, based on whether their pre-processing layers undergo back-propagation or not. The semi-learning refers to the use of a fixed filter kernel as a separate pre-processing layer in the steganalysis, and the internal weight parameters are not involved in back propagation, while the other network layers rely on deep learning methods for optimisation. In 2015, Qian et al. [15] proposed a new network, called GNCNN, which has limitations in the accuracy of steganalysis due to the simplicity of the network model compared to other deep learning-based steganalysis. In 2016, Xu et al. proposed the Xu-Net [65] network. Xu-Net still follows the network architecture features of GNCNN in the network framework, for example, still using global pooling operation to reduce the loss of residual image information. A fixed high-pass filtering layer is also incorporated in the front of the network. The Xu-Net network converges the range of the feature map from a nonsensical positive and negative interval to a positive interval by adding an absolute layer to the first convolutional layer, based on the residual high-frequency noise signal that is symmetric about 0 and sign-independent after the pre-processing layer. In 2018, Yedroudj et al. [21] proposed the Yedroudj-Net. This network follows all the high-pass filtering kernels in the traditional SRM [5].

Full learning models are networks that implement end-to-end training or involve parameters from the pre-processing layer in back propagation. In 2014, Tan et al. [16] combined steganalysis with deep learning for the first time, inspiring a new wave of steganalysis based on deep learning and giving this network structure the acronym TanNet. In 2017, Ye et al. [20] proposed the Ye-Net network, which combined the filter kernels in the traditional feature extraction in SRM with a deep learning network, using 30 high-pass filter kernels of SRM to work together, and then obtained a residual superimposed image with a channel number of 30. The layer is trained in a steganalysis network, which allows the network to learn the residual information efficiently with more feature information. In 2018, Boroumand et al. [2] proposed an end-to-end steganalysis, SRNet, which uses a residual network to simulate the process of traditional SRM in filtering features. SRNet can be applied not only to the null domain, but also to the JPEG domain with good results. The success of SRNet also demonstrates that deep learning networks do not require excessive a priori knowledge. In 2019, Zhu et al. [22] proposed the Zhu-Net network. The Zhu-Net is a major improvement over previous steganalysis networks, with a modified 3×3 filter kernel in the preprocessing layer for the first time. This reduces the number of parameters in the preprocessing layer and makes it easier to adapt the model.

The full-learning model has higher detection accuracy than the traditional steganalysis and half-learning models, but requires longer training time and is more prone to overfitting. In the detection process of the full learning model, we found that the trained network is dataset specific, and if the test set and the training set are not the same dataset, then the test effect will be much weaker.

3 Proposed Method

The framework of the steganalysis proposed in this paper is shown in Fig. 1 below. The input to the network is a 256×256 PNG or JPG image, and after the calculation of the parameters a binary result is output (steog and cover, for better illustration and understanding, stego and cover represent the original image and its corresponding steganographic image respectively). The proposed model consists of a preprocessing layer, two normal convolutional layers, three grouped residual convolutional layers, three downsampled convolutional layers, first and second order pooling layers and finally a fully connected layer that passes softmax.

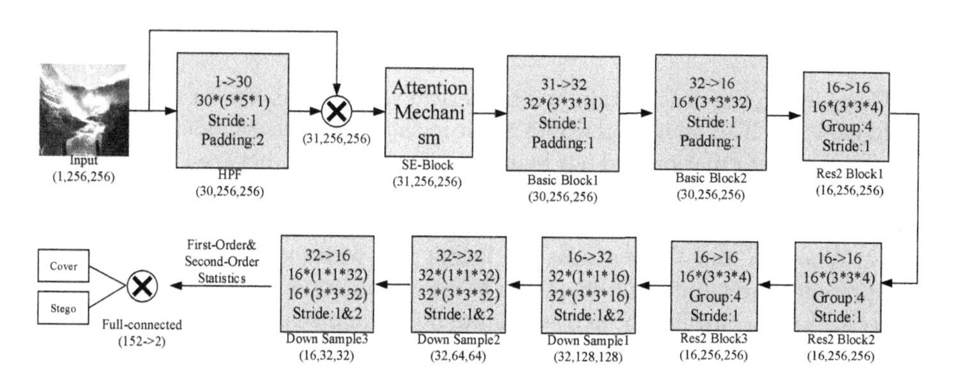

Fig. 1. The architecture of proposed LightNet. For each block, we have placed some of the necessary information within the corresponding convolution blocks to facilitate comparative experiments.

3.1 Preprocessing Layer

The steganography procedure can be considered as making subtle changes to the distribution in the image and adding a flow of noise similar to that of the adversarial examples. The proposed model is a simple two-stream network of steganographic noise streams passing through the SRM and grey-scale streams. It is due to the idea obtained by summarizing the existing steganalysis that usually a full learning model will have better learning ability and better detection accuracy, so we integrate the dual streams by channel stacking. The 30 filter kernels used in the noise stream can help the network focus on the noise itself rather than the content in the image (steganalysis does not care about the image content) while effectively extracting the steganographic noise, and the greyscale stream can be thought of as aiding the network in getting information that cannot be captured in the noise stream as a supplement.

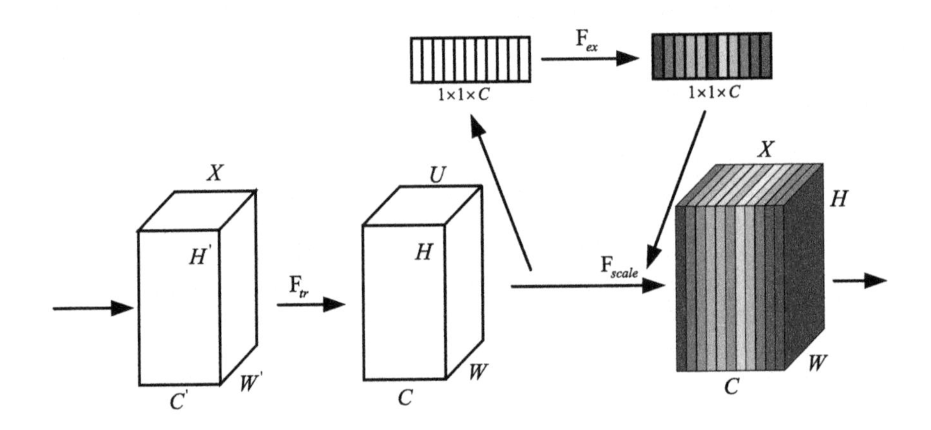

Fig. 2. The details of sequence-and-extraction moudule.

Considering the distribution and numerical variability between the two types of channels, such an operation would result in the steganalysis model getting a lot of redundant information and not being able to distinguish between the priorities in the information, using the attention mechanism [10] to assign weights to the resulting 31-layer noise map. The detailed attention mechanism as shown in Fig. 2. Where H and W represent the height and width of the feature map and C represents the number of channels, F means function of different operation, like traditional convolution, extracting operation and scale size.

3.2 Feature Extraction Layer

Therefore the introduction of a grouping residual mechanism to steganalysis for noise extraction, such a structure allows the network to perform multi-scale information extraction on finer-grained noise, while achieving a multi-scale receptive field orientation encompassing the output data, combining deep and shallow information. Not only that, the parameters of the network can be effectively decreased by this group convolution, further reducing the parameters for the purpose of parameter compression. Res2Net [6] uses the Scale hyperparameter as a control parameter for scaling, which has the advantage that the larger the

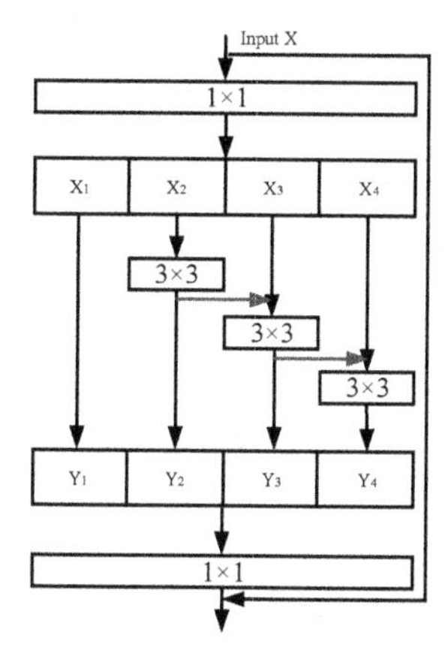

Fig. 3. Illustration of Res2Block.

Scale, the wider the field of inception, and the cascade (conact) introduces a negligible computational/memory overhead. The deatiled res2net basic block as shown in Fig. 3. However, this structure breaks parallelism to some extent and can slow down GPU operations.

3.3 Downsample Module

To avoid the problem of parameter catastrophe in the later fully-connected layers when the feature maps are oversized, we replace the original pooling operation with a convolution operation to implement the downsampling operation. As the information obtained after the high-pass filter is mostly high-frequency residual information, maximum pooling tends to lose the high-frequency residual image information, which makes the network difficult to fit. Although average pooling can be used to reduce the loss of residual information, it can destroy the signal itself and thus lose some features. Either average or maximum pooling results in feature loss. In LightNet the features are operated on separately using 1×1 and 3×3 convolution kernels with stride 2, and then the information from the two branches is combined by a add operation (as shown in Fig. 4).

3.4 Mutli-order Statistics

Most steganographic analyses are fed into the global average pooling layer after the feature extraction layer, and while this greatly reduces the number of features output, it also results in some essential information loss. Higher dimensional information can often contain more effective elements than first-order information. Inspired by the work of Deng [3] et al. we cascade first-order statistical information (mean) and second-order statistical information (covariance, as shown in Fig. 5) so that subsequent fully-connected layers can be provided

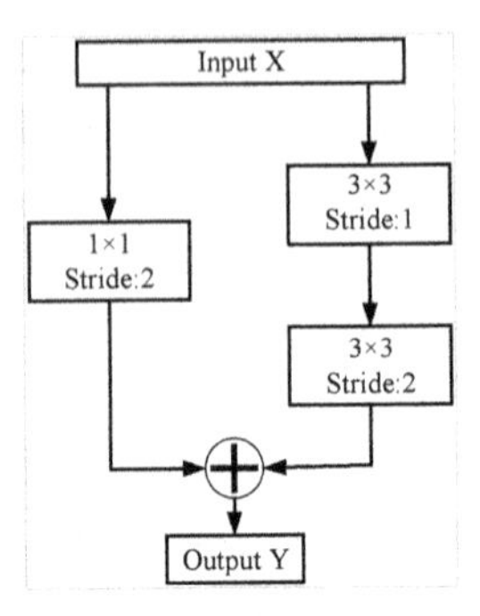

Fig. 4. The details of downsample module used in the LightNet, We use a 1×1 convolutional kernel and a 3×3 convolutional kernel to operate on the input feature maps respectively.

with sufficient discriminatory information to avoid misclassification. The implementation shows that the fusion of multi-order statistical information leads to shorter training time and better convergence of the steganalysis.

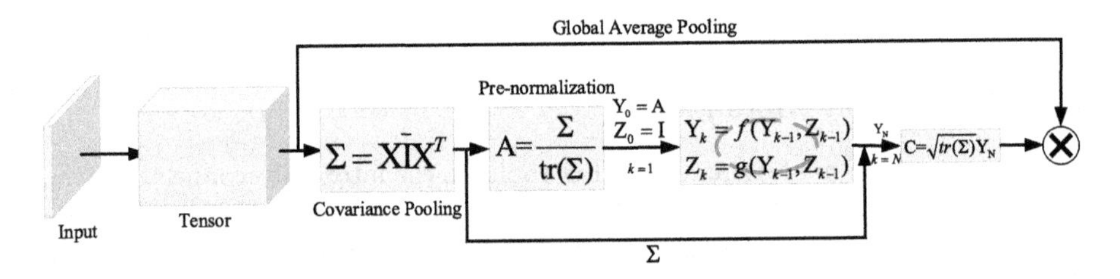

Fig. 5. Details of the global covariance pooling used in the LightNet, X denotes the input vector; C denotes the output vector; Σ denotes the covariance matrix, tr means the trace of matrix. The orange circle represents Newton-Schulz Iteration.

4 Experiments

4.1 The Environments

We acquired Cover images from the 10,000 images in the BOSSbase1.01 Natural Image Library. We set the ratio of the training dataset to the validation dataset to dataset set to 4:1:5, with no overlapping data within the three data sets. Due to our GPU computing platform and time limitation, the data used in LightNet are all grayscale images of 256×256 size. The images in each dataset are cropped using the imresize() function in MATLAB R2018b. All stegos use different steganographic methods by Matlab implementations with random embedding key. LightNet's training data is primarily data that is S-UNIWARD steganography on BOSSbase. (For a more convenient representation of such datasets, we have named them BOSS_SUNI.) All the experiments were on a Nvidia GTX 2080Ti card with 11G.

We apply a mini-batch stochastic gradient descent to train the LightNet. The momentum and weight decay of networks are set to 0.9 and 0.0005 respectively, the batch size in the training is set to 32 (16 cover/stego pair) and the learning rate is set to 0.005. All layers are initialized using Kaiming method [8].

4.2 Cross-Domain Steganalysis Experiments

Fig. 6. The four plots, from left to right, are the original Cover and its steganography (only the J-UNIWARD stego is shown here), the S-UNIWARD pixel residual map with bpp = 0.4 and the J-UNIWARD pixel residual map with 0.4 bpnzAC.

We compare the stego that obtained after frequency domain steganography, with the pixel residuals and experiment with pixel point visualisation. It can be found that although the J-UNIWARD [9] algorithm hides the secret information in the Y-channel, it still captures the steganographic noise in a pixel-wise manner (as shown the right residual map in Fig. 6). LightNet also performs the task of steganalysis of three-channel steganographic images in LSB. To solve the existing mismatch problem in the pre-processing layer, we propose two solutions: 1) the channels are operated separately, each of the three RGB channels is involved in the network as an independent grey-scale map, and the discriminant probabilities of the three channels are finally output separately. 2) The convolutional kernels with channel 1 in the 30 layers of the SRM are stacked into convolutional kernels with channel number 3 for operation. Experiments were conducted using the second experimental approach, which also achieved convincing results.

The secret information previously hidden in the Y-channel is also expressed as pixel fluctuations in the spatial domain due to the Inverse Discrete Cosine Transform, but for the network input it is all input as a pixel matrix. Therefore, we believe that even different steganographic algorithms in the steganalysis model can be used to extract steganographic noise from images by using a high-pass filtering kernel as a pre-processing layer. To demonstrate the algorithmic robustness possessed by this model, we replaced the training data with the J-UNIWARD algorithm with an embedding rate of 0.4 bpnzAC (all datasets in this

Table 1. The detection accuracy of proposed model in cross-domain.

Proposed model	S-UNIWARD	J-UNIWARD	RGB_LSB
LighNet	88.25%	89.55%	99.85%

paper default to the bossbase dataset if not otherwise specified). To ensure that the distribution of stego shows randomness, the secret messages in all steganographic methods are a random stream of bits. The details of experiment are illustrating in Table 1.

4.3 Steganalysis Experiments Result

Table 2. The numbers of parameters and detection accuracy comparison using Xu-Net, Ye-Net, Yedroudj-Net, SRNet, Zhu-Net and proposed model.

Steganalysis	Xu-Net	Ye-Net	Yedroudj-Net	SRNet	Zhu-Net	LightNet
Parameters ($\times 10^5$)	**1.4**	10.6	44.5	477.6	287.1	5.7
Accuracy	71.65%	80.05%	81.08%	87.25%	85.38%	**88.25%**

As Table 2 shows, the proposed network has significantly better performance than the other network. LightNet can achieve similar or even better results than SRNet and Zhu-Net with only 1–2% of their counts, because LighNet is more sensitive to clues such as steganographic noise. For various types of steganalysis, Xu-Net has only 5 layers of convolution and uses a single filter kernel in the SRM for residual extraction in the pre-processing layer, while our proposed model has only a fraction of the number of parameters of Xu-Net, but is about 17% more effective. The Ye-Net, Yedroudj-Net and Zhu-Net models all use all the filter kernels in the SRM in the pre-processing layer, allowing the subsequent feature discrimination network to capture enough steganographic noise while also greatly increasing the computational effort. Experiments have shown that LightNet simplifies some unnecessary operations in the model and enhances the extraction power of the network.

With the continuous development of steganalysis, researchers have found that the problem of model dismatch occurs due to the difference between the training set and the test set, but in the existing research on CNN-based steganalysis, more models are slightly lacking in migration and generality due to the difference between the distortion cost functions designed by different steganographic algorithms, different steganographic algorithms change the Cover differently, and the distribution of the added steganographic noise is also different. Among them, HUGO and MiPOd are more different compared to S-UNIWARD, but LightNet still has an advantage in terms of migrability. We believe that this is due to the structure of LightNet, assuming that different steganographic algorithms have

Table 3. Steganalysis detection accuracy comparison on different datasets.

	BOSS_SUNI	BOSS_WOW	BOSS_HUGO	BOSS_MiPOD
SRNet	87.25%	84.58%	**68.06%**	75.58%
Zhu-Net	85.38%	83.05%	68.32%	74.20%
LightNet	**88.25%**	**86.69%**	68.59%	**76.19%**

their own different distributions of space, less training rounds and fewer network layers can effectively avoid this lack of transferability due to data differences. Therefore, we will use the early stopping operation to advance the model if it converges when training the model to prevent overfitting (Table 3 and Fig. 7).

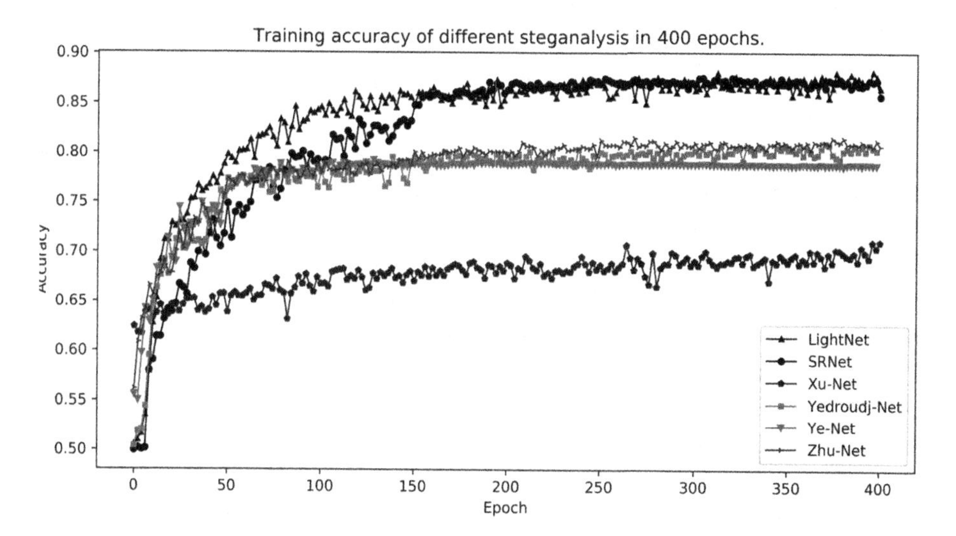

Fig. 7. The training accuracy of different steganalysis in 200 epochs comparison with LightNet, SRNet, Xu-Net, Yedroudj-Net, Ye-Net and Zhu-Net.

The semi-learning steganalysis has some advantages in terms of early convergence speed and extraction of steganographic noise, which can give decent results early in the training. LightNet converges significantly faster than the other models up to 150 rounds, and around 125 rounds LightNet achieves a better result. Semi-learning models such as Xu-Net and Yedroudj-Net have similar characteristics to LightNet but are still inadequate for handling the redundant information obtained. The use of a pre-processing layer better prevents the model from focusing on the image itself and allows the feature maps obtained from each convolution to be reused by group fusion convolution, and reduces the network parameters under the same conditions, speeding up convergence and training time.

5 Conclusion

In this paper, we propose a lightweight steganalysis based on multi-scale feature extraction and fusion of multi-order statistical features, with parameters that occupy only 1–2% of a steganalysis of the same detection capability. It has been experimentally demonstrated that LightNet can perform steganalysis tasks across domains, illustrating the commonality of steganalysis tasks. Our steganalysis can be thought of as a combined training model of full learning and half learning, balancing the convergence speed of half learning with the detection accuracy of full learning. The experimental results show that the proposed model not only outperforms existing models in terms of final results, but also converges much faster than these models. In terms of parameters, the present work can further achieve a reduction in the number of parameters by referring to the operation of the pre-processing layer in Zhu-Net.

We hope that the design concept of LightNet will inspire more researchers to take an interest in the field of information hiding, or to adapt LightNet as a noise extraction module. Steganalysis work can also be deployed in a variety of areas such as computer vision, for example, adversarial sample detection, image tampering detection, camera fingerprint forensics and many others.

In the future, our work will lead to mobile deployments and cross-domain steganalysis designs, with the aim of finding more accurate and lighter steganalysis or associated software prototypes.

References

1. Böhme, R.: Weighted stego-image steganalysis for JPEG covers. In: Solanki, K., Sullivan, K., Madhow, U. (eds.) IH 2008. LNCS, vol. 5284, pp. 178–194. Springer, Heidelberg (2008). https://doi.org/10.1007/978-3-540-88961-8_13
2. Boroumand, M., Chen, M., Fridrich, J.: Deep residual network for steganalysis of digital images. IEEE Trans. Inf. Forensics Secur. **14**(5), 1181–1193 (2018)
3. Deng, X., Chen, B., Luo, W., Luo, D.: Fast and effective global covariance pooling network for image steganalysis. In: Proceedings of the ACM Workshop on Information Hiding and Multimedia Security, pp. 230–234 (2019)
4. Filler, T., Fridrich, J.: Gibbs construction in steganography. IEEE Trans. Inf. Forensics Secur. **5**(4), 705–720 (2010)
5. Fridrich, J., Kodovsky, J.: Rich models for steganalysis of digital images. IEEE Trans. Inf. Forensics Secur. **7**(3), 868–882 (2012)
6. Gao, S., Cheng, M.M., Zhao, K., Zhang, X.Y., Yang, M.H., Torr, P.H.: Res2net: a new multi-scale backbone architecture. IEEE Trans. Pattern Anal. Mach. Intell. (2019)
7. Gul, G., Kurugollu, F.: A new methodology in steganalysis: breaking highly undetectable steganograpy (HUGO). In: Filler, T., Pevný, T., Craver, S., Ker, A. (eds.) IH 2011. LNCS, vol. 6958, pp. 71–84. Springer, Heidelberg (2011). https://doi.org/10.1007/978-3-642-24178-9_6
8. He, K., Zhang, X., Ren, S., Sun, J.: Delving deep into rectifiers: surpassing human-level performance on imagenet classification. In: Proceedings of the IEEE International Conference on Computer Vision, pp. 1026–1034 (2015)

9. Holub, V., Fridrich, J.: Digital image steganography using universal distortion. In: Proceedings of the First ACM Workshop on Information Hiding and Multimedia Security, pp. 59–68 (2013)
10. Hu, J., Shen, L., Sun, G.: Squeeze-and-excitation networks. In: Proceedings of the IEEE Conference on Computer Vision and Pattern Recognition, pp. 7132–7141 (2018)
11. Ker, A.D.: Steganalysis of LSB matching in grayscale images. IEEE Signal Process. Lett. **12**(6), 441–444 (2005)
12. Li, B., Tan, S., Wang, M., Huang, J.: Investigation on cost assignment in spatial image steganography. IEEE Trans. Inf. Forensics Secur. **9**(8), 1264–1277 (2014)
13. Liu, Q., Sung, A.H., Chen, Z., Xu, J.: Feature mining and pattern classification for steganalysis of LSB matching steganography in grayscale images. Pattern Recogn. **41**(1), 56–66 (2008)
14. Luo, X., et al.: Steganalysis of hugo steganography based on parameter recognition of syndrome-trellis-codes. Multimedia Tools Appl. **75**(21), 13557–13583 (2016)
15. Qian, Y., Dong, J., Wang, W., Tan, T.: Deep learning for steganalysis via convolutional neural networks. In: Media Watermarking, Security, and Forensics 2015, vol. 9409, p. 94090J. International Society for Optics and Photonics (2015)
16. Tan, S., Li, B.: Stacked convolutional auto-encoders for steganalysis of digital images. In: Signal and Information Processing Association Annual Summit and Conference (APSIPA), 2014 Asia-Pacific, pp. 1–4. IEEE (2014)
17. Tang, W., Li, H., Luo, W., Huang, J.: Adaptive steganalysis against WOW embedding algorithm. In: Proceedings of the 2nd ACM Workshop on Information Hiding and Multimedia Security, pp. 91–96 (2014)
18. Westfeld, A., Pfitzmann, A.: Attacks on steganographic systems. In: Pfitzmann, A. (ed.) IH 1999. LNCS, vol. 1768, pp. 61–76. Springer, Heidelberg (2000). https://doi.org/10.1007/10719724_5
19. Xia, Z., Wang, X., Sun, X., Liu, Q., Xiong, N.: Steganalysis of LSB matching using differences between nonadjacent pixels. Multimedia Tools Appl. **75**(4), 1947–1962 (2016)
20. Ye, J., Ni, J., Yi, Y.: Deep learning hierarchical representations for image steganalysis. IEEE Trans. Inf. Forensics Secur. **12**(11), 2545–2557 (2017)
21. Yedroudj, M., Comby, F., Chaumont, M.: Yedroudj-net: an efficient CNN for spatial steganalysis. In: 2018 IEEE International Conference on Acoustics, Speech and Signal Processing (ICASSP), pp. 2092–2096. IEEE (2018)
22. Zhang, R., Zhu, F., Liu, J., Liu, G.: Depth-wise separable convolutions and multi-level pooling for an efficient spatial CNN-based steganalysis. IEEE Trans. Inf. Forensics Secur. **15**, 1138–1150 (2019)

HTCN: Harmonious Text Colorization Network for Visual-Textual Presentation Design

Xuyong Yang[1], Xiaobin Xu[2], Yaohong Huang[3], and Nenghai Yu[1](✉)

[1] University of Science and Technology of China, Hefei, China
ynh@ustc.edu.cn
[2] Visnect Technology Co., Ltd., Shenzhen, China
[3] WeCar Technology Co., Ltd., Shenzhen, China
yaohong.huang@weicheche.cn

Abstract. The selection of text color is a time-consuming and important aspect in the designing of visual-textual presentation layout. In this paper, we propose a novel deep neural network architecture for predicting text color in the designing of visual-textual presentation layout. The proposed architecture consists of a text colorization network, a color harmony scoring network, and a text readability scoring network. The color harmony scoring network is learned by training with color theme data with aesthetic scores. The text readability scoring network is learned by training with design works. Finally, the text colorization network is designed to predict text colors by maximizing both color harmony and text readability, as well as learning from designer's choice of color. In addition, this paper conducts a comparison with other methods based on random generation, color theory rules or similar features search. Both quantitative and qualitative evaluation results demonstrate that the proposed method has better performance.

Keywords: Text colorization · Color harmonization · Text readability · Visual-textual presentation design

1 Introduction

In modern multimedia design, the proper combination of text and color can make the communication of information more diverse and efficient. Naturally, the need for text colorization is widespread in daily life, such as in the designing of advertisements, magazines, posters, signboards and webpages. However, it is often not easy to choose a suitable and aesthetically pleasing text color during the design process due to multiple reasons. Firstly, the number of colors is so large that it is difficult to enumerate all of them for the evaluation of coloring effectiveness. Secondly, to make the combination of selected text colors and scene colors achieve a harmonious and beautiful effect, the designer needs not only considerable professional experience, but also a certain number of trail-and-error coloring attempts. Thirdly, text colorization is different from image colorization, and it

H. Ma et al. (Eds.): PRCV 2021, LNCS 13020, pp. 560–571, 2021.
https://doi.org/10.1007/978-3-030-88007-1_46

is essential that the text is clear and easy to read in the position where it is placed. Therefore, the contrast between the text and its background should be considered. Finally, different colors can exert different effects on human psychology and emotions. Thus, in order to convey the right message, it is sometimes necessary to choose text colors that represent the right emotions and themes, such as energetic, sad, noble, etc.

The goal of this paper is to construct a model that automatically predicts harmonious, easy-to-read, and aesthetically pleasing text colors for overlapping visual-textual presentation design. Such a model makes text colorization easy, thus allowing professional designers to make a quick choice and the average person to obtain a result that satisfies basic aesthetics. Most traditional text coloring methods base on color-related design guidelines and theoretical models (e.g., [7,18]). However, their results are not optimal because theories and guidelines are often only general design guidance. Meanwhile, actual design work requires designers to make specific and detailed adjustments according to the situation.

In this paper, we propose a deep learning framework for text colorization. We model the color harmony between text and background images, the readability of text, as well as learn corresponding aesthetic knowledge from large scale design data. By considering the global color harmony of text, local text readability, and designer's choice of color, our proposed framework generates better text colorization results.

The main contributions of this paper are presented as follows:

1. We propose a global color harmony scoring network and a local text readability scoring network to measure the aesthetics of text color harmony and readability, respectively.
2. We put forward a deep learning framework HTCN (Harmonious Text Colorization Network) that integrates aesthetic color harmony and text readability, which can make good use of the inherent knowledge in big data for text colorization in visual-textual presentation design.
3. We construct a dataset called VTDSet (Visual-Textual Design Set) with 77,038 text colorization design samples and their corresponding background images, which can be capable of supporting more learning tasks.
4. Experiments demonstrate that the proposed network has better performance and can be easily applied to practical design tasks.

2 Related Works

Color Harmony. Early studies of color harmony concentrated on hue templates, describing harmonies as a theory of fixed rotational categories of the color wheel. Hue templates were employed in much of the art and design work, and were a key early theory to support color design. Itten proposed in [6] that equidistant sets of 2, 3, 4, and 6 hues on the color wheel are harmonious, and the color wheel has rotational invariance. A shortcoming of the hue templates is that they are defined independently of the color space. The source of the harmonious aesthetics dataset used in this paper, the Kuler website [1], uses the BYR color wheel, while other websites such as COLOURlovers [2] adopt the RGB color wheel, which suggest different colors despite using the same template rules. The color themes provided by Matsuda summarize a set of classic color patterns with 8 hue distributions and 10 tone distributions [11], and this color harmony model has been used in several computer vision and graphics projects [4,16,18].

Additionally, there are also many studies on color harmony which are not limited to hue templates [12,13]. Although different approaches have been empirically conducted in different color spaces, similar laws have been summarized. Such theories are fundamentally based on the fact that people visually expect smoother effects. For example, the Munsell system suggests that color combinations in HSV color space with fixed H and V but changes in saturation S are harmonious [12]. It has also been shown that colors are harmonious if they can be connected in a straight line in space, such as in the Ostwald system where it is considered harmonious when the colors have the same amount of white or black [11], because such colors can form a straight line in the color space.

There are also some studies evaluating the harmony of color themes through a data-driven approach. O'Donovan et al. [14] proposed a LASSO regression model to fit the scores of color themes (combinations of five colors). In the work of [17], Yang et al. constructed a framework that first performs maximum likelihood estimation for a color pair composed of two colors and then predicts the aesthetic score of the color theme by adopting a BPNN network.

Text Readability. There are quite a few studies concerning text readability. In the study [10], it was shown that the two forms of contrast, luminance and chromaticity, are processed independently within the visual system. Since visual acuity responds much better to changes in luminance than to changes in hue and saturation, the contrast between the luminance of the color and the background dominates when we make attempts to resolve fine details. Albers et al. [3] suggested that differences in perception of readability are difficult to measure mathematically. Zuffi et al. [21] presented an experiment on the readability of colored text on colored backgrounds, which was conducted through crowdsourcing on the web. Their goal was to contribute to the understanding of how easy it is to read text on a display under general viewing conditions. Their experiments showed that the design of the text display should be set with a 30-unit difference in brightness between the CIE text and the background. In addition, the study [20] suggested using a light and soft background for color selection, and light-colored text on a dark background is more challenging to read.

3 Proposed Method

3.1 Overall Framework

We propose a deep learning approach, Harmonious Text Colorization Network (HTCN), defining the loss function through a global harmony scoring network and a local readability scoring network, which can well introduce a large amount

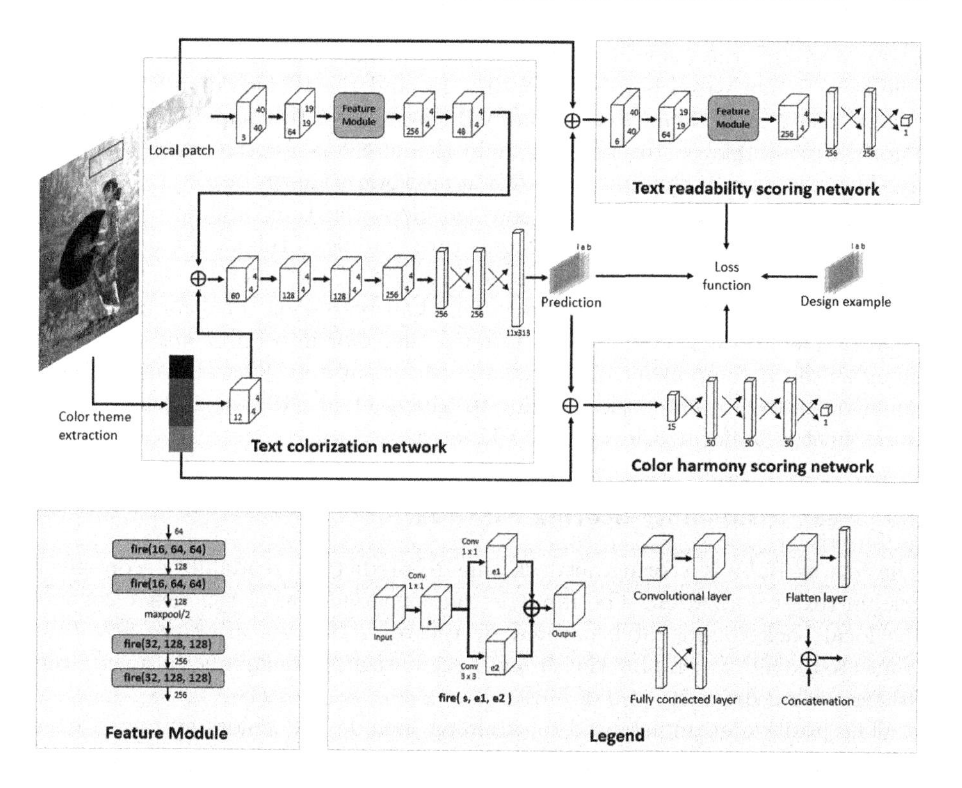

Fig. 1. HTCN network architecture.

of design experience and handle the multimodal problem of color prediction. The design of our network architecture is depicted in Fig. 1. The overall framework diagram shows that the proposed network framework provides an effective combination of global harmony and local readability as well as historical data experience. Besides, the whole network is end-to-end trainable.

In the proposed architecture, theme colors instead of the whole background image are employed as input for two reasons. One is that evaluating harmonious properties between text color and the background image is easier by abstracting the background image into theme colors and utilizing public theme color harmony datasets. The other reason is to make the HTCN architecture more generally applicable. For example, regarding signboard design or interior design, there is no well defined background image. However, theme colors can be extracted from the environment and our proposed architecture can be applied without compromise.

3.2 Text Colorization Network

The input of the text colorization network is the local background patch covered by the text and the global theme colors to be matched. The local background image is first passed through a local image feature encoding part to obtain the local features, which is concatenated with the theme color features and then passed through the color prediction part to acquire the predicted text color. The local image patch is resized to 40×40, and the theme colors can be calculated using color theme extraction methods [8,9]. In this paper, the K-Means algorithm is employed to cluster 4 main theme colors for the sake of simplicity. The structure of the text colorization network is shown in Fig. 1. To reduce parameters while preserving accuracy, we adopt the Fire module proposed in the SqueezeNet architecture [5] in the feature module of the local image feature encoding part. Based on the multimodal nature of the color prediction problem, we choose to use classification rather than regression for colorization. We separate hue and lightness by transforming the color into Lab color space, and quantize the L component and ab component separately. The ab component is quantized into 313 categories similar to Zhang et al. [19]. In addition, the L component is uniformly devided into 11 categories.

3.3 Text Readability Scoring Network

The text readability scoring network aims to predict the readability score given certain text colorization. The input is the concatenation of the text color and the local background patch covered by the text. Besides, after the first fully connected layer, we employ the dropout regularization technique to avoid overfitting, with a dropout ratio of 50%.

The positive examples used for training include text colors extracted from design works and the corresponding local background images. We synthesize negative examples based on studies in association with readability. Specifically, we use the K-Means algorithm to cluster the ab values of the local background image and randomly select a cluster center as the ab value of the text color of negative examples. According to the results of text readability experiments by Zuffi et al. [21], we randomly selected values that differed from the median of the luminance of the local background image by 30 or less as the L values of the negative examples.

3.4 Color Harmony Scoring Network

The proposed color harmony scoring network is a fully connected network as shown in Fig. 1. The second and third fully connected layers are followed by a dropout layer with a ratio of 20% to prevent overfitting. The network takes the text color and the background theme colors as input and predicts harmony rating for the color combination.

3.5 Loss Function

The optimal text color C^* should satisfy the following constraints: (1) color harmony with the color theme of the whole image, which ensures a relatively high combined aesthetic score; (2) optimal contrast between the text color and the local image area, which ensures high readability of the text. This can be modeled as maximizing an scoring function that includes two components, respectively, the global color harmony $\mathcal{H}(C,T)$, and local text readability $\mathcal{D}(C, I_{local})$.

$$C^* = \arg\max_{C}(\mathcal{D}(C, I_{local}) + \beta\,\mathcal{H}(C,T)) \tag{1}$$

where C is the text color, I_{local} is the local background image of the text block, and T is the global color theme.

Since there is no exact mathematical expression for color harmony and text readability, we design a color harmony scoring network and a text readability scoring network to model them in the energy function. Before training the text colorization network, we first trained the color harmony scoring network and the text readability scoring network. Then, we fix the parameters of these two networks before the overall training of the text colorization network. It can be noticed that when the local background is complex, it is difficult to find text colors with high harmony as well as high readability ratings. In this situation, the color C^{ref} used by the designer in the actual design work is an proper choice with comprehensive aesthetic considerations. As a result, we design the network to learn from this choice. Combining the above considerations, we design the scoring function as follows:

$$\mathcal{R}(C, C^{ref}, I_{local}, T) = \mathcal{D}(C, I_{local}) + \beta\,\mathcal{H}(C,T) + \alpha\ \mathrm{G}_{\sigma}\left(\|C - C^{ref}\|_2\right) \tag{2}$$

where α and β are weights for controlling the balance among different factors and G_{σ} is a Gaussian smoothing function with standard deviation σ, $\mathrm{G}_{\sigma}(x) = \exp(-\frac{x^2}{2\sigma^2})$.

To solve the problem of gradient transfer and use the calculation results of the loss function in order to effectively update the parameters of the text colorization network, we propose to adopt the entire predicted color classification probability distribution for loss calculation. Let the color distribution best reflecting the combined ranking of Eq. (2) on that text block be noted as p. We want the predicted distribution $\hat{p}$ to approximate p, and thus we minimize their cross entropy. The total loss is defined as follows:

$$\mathcal{L} = -\sum_{i} \mathrm{Softmax}(\mathcal{R}(C_i, C^{ref}, I_{local}, T))\,\log(\hat{p}_i) \tag{3}$$

where the softmax function maps the overall score $\mathcal{R}(C_i, C^{ref}, I_{local}, T)$ to probability of text color, $\mathrm{Softmax}(x_i) = \frac{e^{x_i}}{\sum_i e^{x_i}}$.

4 Experiments

4.1 Datasets

Color Combination Aesthetics Score Dataset. We obtained the MTurk public dataset from [14], which consists of 10,743 carefully selected color themes created by users on Adobe Kuler [1], covering a wide range of highly and poorly rated color themes, each of which rated by at least 3 random users with ratings between 1 and 5. The MTurk dataset uses Amazon Mechanical Turk[1] to collect more user ratings for the selected topics, making each topic rated by 40 users. Finally, the average score for each topic was taken as the final score.

Visual-Textual Design Works Dataset. We constructed a visual-textual design dataset called VTDSet (Visual-Textual Design Set) where 10 designers selected text colors in 5 to 7 areas on each of the total 1226 images, resulting in 77,038 designed text colors and their corresponding information. We randomly selected 10,000 design results associated with 1000 background images from the dataset as the training dataset, and 2260 design results associated with the remaining 226 background images as the testing dataset.

4.2 Implementation Details

Regarding the color harmony scoring network, we transformed the colors in the color theme into the normalized Lab color space and randomly selected 70% of the samples in the aesthetic score dataset for training with the remaining 30% as the test set. We used the Adam algorithm for optimization with a learning rate of 0.001, and $\beta_1 = 0.9$, $\beta_2 = 0.999$, and L2 regularization coefficient of 0.0005. An early termination strategy was used to avoid overfitting, and a total of 300 epochs were trained with batch size of 64. For the text readability scoring network, we employed the SGD algorithm for parameter updating with momentum of 0.9 and L2 regularization coefficient of 0.0005. We adopted a multi-step learning rate with an initial learning rate of 0.001. At both the 600th and 800th epochs, the learning rate decays by 0.1. In addition, we used an early termination strategy to avoid overfitting, and trained for a total of 1000 cycles with batch size of 8.

When training the text colorization network, we fix the parameters of the color harmony scoring network and the text readability scoring network. We set $\alpha = 0.5$, $\beta = 1$ and $\sigma = 4$. The theme color of the background image and the text color and corresponding local patches of all 5 to 7 text regions in the same design image are extracted from the design results and combined into a batch for training. We adopted the Adam algorithm for parameter updating with $\beta_1 = 0.9$, $\beta_2 = 0.999$, and the L2 regularization coefficient of 0.0005. We used a learning rate of 0.001 and an early termination strategy to avoid overfitting, and trained for a total of 30 epochs.

4.3 Effectiveness of Network Design

In order to verify the effectiveness of the proposed HTCN framework, we disassembled the proposed network structure and used only a part of it for colorization. The results obtained on the test set are shown in Fig. 2. The accuracy in the figure is obtained by calculating the Euclidean distance between the predicted color and the color chosen by the designer. To facilitate the calculation,

[1] https://www.mturk.com.

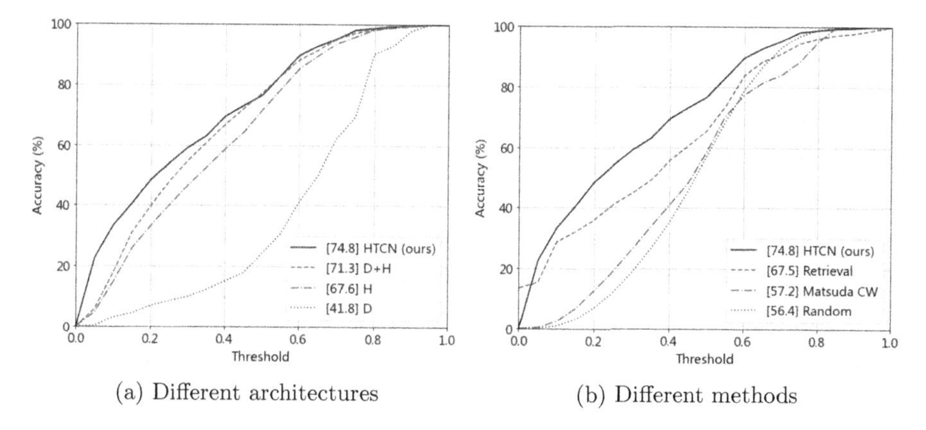

(a) Different architectures (b) Different methods

Fig. 2. Comparison of accuracy of text colorization (a) D denotes using only the local readability evaluation network, H denotes using only the global harmony evaluation network, and D+H denotes using both the local readability and global harmony evaluation networks. (b) Comparison with other methods.

we used the RGB color space and normalized the distance. As presented in the legend, we report the average accuracy of thresholds ranging from 0.05 to 1, with an interval of 0.05. According to the figure, the organic combination of local readability and global harmony can effectively improve the performance of colorization, which is in line with the designer's guidelines for selecting colors when designing, i.e., making adjustments to ensure that text is clear and easy to read while ensuring that the text color is in harmony with the background image. In addition, the complete HTCN network obtains the best colorization results with an average accuracy of 74.8% because it also learns the aesthetics implicit in the designer's design choices in difficult scenarios under the condition of ensuring local readability and global harmony.

4.4 Comparison with Other Methods

We compare the text colorization network HTCN proposed in this paper with the following three approaches:

Random Text Colorization ("Random"). A random value is selected in the RGB color space, and this baseline is used to check whether the color design of the text in the generation of the visual-textual presentation layout is arbitrary.

Text Colorization Based on Matsuda Color Wheel Theory ("Matsuda CW"). This text colorization method bases on the color wheel theory, which is also adopted in the work of Yang et al. [18]. We reproduce the method by first performing principal component analysis on the image to obtain the color theme, taking the color with the largest proportion as the base color C_d of the image, and then calculating the minimum harmonic color wheel distance between the base color C_d and the aesthetic template color set according to the constraint defined by Matsuda to obtain the optimal hue value of the text color C_r. Finally, the color mean $\mu_{h,s,v}$ of the image covered by the text area is calculated, and the optimal text color is obtained by reasonably maximizing the distance between $\mu_{h,s,v}$ and C_r in the (s, v) saturation and luminance space.

Text Colorization Based on Image Feature Retrieval ("Retrieval"). Retrieval-based strategy is frequently used in design, i.e., seeking reference among solutions of similar problems. For the text colorization problem, the original designer's color can become the recommended color when the background image and the text area are similar. As a result, we concatenate the global features of the image and the local image features of the text-covered region to obtain the K nearest neighbor recommendations for the current text coloring by the cosine distance. We used the VGG-16 network [15] pretrained on the ImageNet dataset, and selected the output of the fc6 layer as the image features. The combined feature of the text region image I_{text} on the global image I is $f = <VGG_I, VGG_{I_{text}}>$. The text color corresponding to the feature with greatest similarity in the design library is selected for colorization.

We implemented these three methods along with HTCN and compared the results on the test set of the dataset VTDSet. Figure 2 shows the accuracy curves of the HTCN network and the three methods which are compared. They are calculated in the same way as in Sect. 4.3. It can be seen that the HTCN text colorization network has obvious superiority over other methods. Typical results of each baseline method and the algorithm in this paper are shown in Fig. 3. We can intuitively observe the difference between various methods. The results of the random method are the most unsecured, while the HTCN text colorization network proposed in this paper can obtain some results that are different from the designer's work but still have good performance. In the color theory-based method, it can be found that all the text colors in the same design are close to a similar hue value, which has a very large limitation of color selection. The retrieval based approach gives historical design results. However, it is not sufficient for creativity and novelty, especially its quality depends on the quality of design examples in the database.

4.5 User Study

We performed user study for qualitative evaluation of the proposed method. Totally 20 persons aged 21 to 35 were recruited, among which eleven were females, and twelve had design-related working experience. Text colorization results of 20 design works randomly selected from the test dataset were used in the evaluation. In the readability evaluation, we let the user read the letters on the picture and kept the user's eyes 60 cm away from the screen. The images displayed on the screen were resized to the height of 15 cm.

(1) Global color harmony evaluation: participants assess the harmony of the text color with overall image in each text box, scoring from 1 to 5, respectively, as ugly, discordant, average, harmonious, and very aesthetically pleasing.

Fig. 3. Comparison of the actual effect of text colorization under various algorithms: (a) random generation of text colors, (b) method based on the Matsuda color wheel theory, (c) retrieval-based method that directly obtains corresponding color recommendations from historically similar design examples, (d) the HTCN network proposed in this paper, and (e) is the designer's original work.

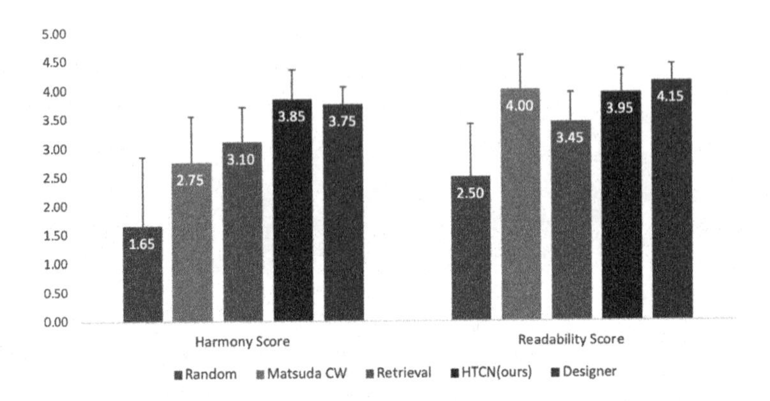

Fig. 4. Harmony and readability scores of text color with different methods

(2) Text readability evaluation: participants assess the readability of the text for each text box, scoring from 1 to 5, which are completely unreadable, partially readable, readable requiring concentration, readable with relative ease, and readable at a glance.

Figure 4 shows that the proposed method can obtain similar aesthetic harmony as the designer. In some cases, the harmony score is better than the designer, because the text color design problem is multimodal and there may exist not only one optimal color combination. By learning from a large number of design works and modeling the aesthetic metric of color themes, our method can acquire design knowledge that are not available in traditional methods, and obtain more harmonious and design-oriented text colorization results.

Obviously, the randomly generated results are most likely to produce crossover with the background because there are no basic aesthetic constraints, making it the worst readability choice and fully illustrating the necessity of other methods. The interesting finding is that the readability of the theoretically derived text color is little better than our method. The reason is that in guaranteeing readability, by explicitly stretching the saturation and the brightness difference between image and background color, a better readability can be obtained through the rule constraint while the overall color harmony is relatively lacking.

5 Conclusion

To conclude, this study introduces a deep learning framework to learn inherent aesthetic design knowledge beyond design rules from color annotation data. We propose two scoring networks that can learn well the harmony metric of color and readability metric of text in aesthetic design. By exploring the text colorization problem, we put forward a deep neural network architecture HTCN for automatic text colorization in designing of visual-textual presentation layout. Moreover, the experiments reveal that our algorithm can perform better than other methods.

References

1. Adobe Kuler. https://color.adobe.com. Accessed 19 Mar 2021
2. Colourlovers. http://www.colourlovers.com. Accessed 21 Mar 2021
3. Albers, J.: Interaction of Color. Yale University Press, London (2013)
4. Cohen-Or, D., Sorkine, O., Gal, R., Leyvand, T., Xu, Y.Q.: Color harmonization. In: ACM SIGGRAPH 2006 Papers, pp. 624–630 (2006)
5. Iandola, F.N., Han, S., Moskewicz, M.W., Ashraf, K., Dally, W.J., Keutzer, K.: SqueezeNet: AlexNet-level accuracy with 50x fewer parameters and <0.5 MB model size. arXiv preprint arXiv:1602.07360 (2016)
6. Itten, J.: The Art of Color: The Subjective Experience and Objective Rationale of Color. Translated by Ernst Van Haagen. Van Nostrand Reinhold (1973)
7. Jahanian, A., et al.: Recommendation system for automatic design of magazine covers. In: Proceedings of the 2013 International Conference on Intelligent User Interfaces, pp. 95–106 (2013)
8. Jahanian, A., Vishwanathan, S., Allebach, J.P.: Autonomous color theme extraction from images using saliency. In: Imaging and Multimedia Analytics in a Web and Mobile World 2015, vol. 9408, p. 940807 (2015)
9. Lin, S., Hanrahan, P.: Modeling how people extract color themes from images. In: Proceedings of the SIGCHI Conference on Human Factors in Computing Systems, pp. 3101–3110 (2013)
10. MacIntyre, B.: A constraint-based approach to dynamic colour management for windowing interfaces. Master's thesis, University of Waterloo (1991)
11. Matsuda, Y.: Color design. Asakura Shoten **2**(4), 10 (1995)
12. Munsell, A.H., Cleland, T.M.: A grammar of color: arrangements of Strathmore papers in a variety of printed color combinations according to the Munsell color system. Strathmore Paper Company (1921)
13. Nemcsics, A.: Coloroid colour system. Hungarian Electronic Journal of Sciences, HEJ Manuscript no.: ARC-030520-A (2003)
14. O'Donovan, P., Agarwala, A., Hertzmann, A.: Color compatibility from large datasets. In: ACM SIGGRAPH 2011 papers, pp. 1–12 (2011)
15. Simonyan, K., Zisserman, A.: Very deep convolutional networks for large-scale image recognition. arXiv preprint arXiv:1409.1556 (2014)
16. Tokumaru, M., Muranaka, N., Imanishi, S.: Color design support system considering color harmony. In: 2002 IEEE World Congress on Computational Intelligence. 2002 IEEE International Conference on Fuzzy Systems. FUZZ-IEEE 2002. Proceedings (Cat. No. 02CH37291), vol. 1, pp. 378–383. IEEE (2002)
17. Yang, B., et al.: A color-pair based approach for accurate color harmony estimation. In: Computer Graphics Forum, vol. 38, pp. 481–490. Wiley Online Library (2019)
18. Yang, X., Mei, T., Xu, Y.Q., Rui, Y., Li, S.: Automatic generation of visual-textual presentation layout. ACM Trans. Multimed. Comput. Commun. Appl. (TOMM) **12**(2), 1–22 (2016)
19. Zhang, R., Isola, P., Efros, A.A.: Colorful image colorization. In: Leibe, B., Matas, J., Sebe, N., Welling, M. (eds.) ECCV 2016. LNCS, vol. 9907, pp. 649–666. Springer, Cham (2016). https://doi.org/10.1007/978-3-319-46487-9_40
20. Zuffi, S., Brambilla, C., Beretta, G., Scala, P.: Human computer interaction: legibility and contrast. In: 14th International Conference on Image Analysis and Processing (ICIAP 2007), pp. 241–246. IEEE (2007)
21. Zuffi, S., Brambilla, C., Beretta, G.B., Scala, P.: Understanding the readability of colored text by crowd-sourcing on the web. HP Laboratories (2009)

A Fast Method for Extracting Parameters of Circular Objects

Zezhong Xu(✉), Qingxiang You, and Cheng Qian

Department of Computer and Information Engineering, Changzhou Institute of Technology, Changzhou 213032, China
xuzz@czu.cn

Abstract. A region-based method is proposed for extracting parameters of circular objects in images. The Hough space is defined by a 2-dimensional accumulator with less columns. By analysing the voting values distribution in each column of the accumulator, a quadratic function, which denotes the relationship between the voting value and the voting distance, is deduced. Regrading all columns, a linear function formula is deduced by analysing the relationship between mean voting distances and the voting angles. After all region pixels have voted for 2D accumulator, a quadratic function is fitted in each column. The circle radius is calculated based on the fitted coefficients. Then a linear function is fitted to mean voting distances corresponding to every voting angles. The circle center coordinates are computed based on the fitted coefficients. Synthetic images and real-world images are used to test the proposed region-based method. Experimental results show the proposed method is fast and accurate even in the presence of contour defects and outliers.

Keywords: Circular objects extraction · Region pixels · Hough transform

1 Introduction

Extracting parameters of circular objects in images is a fundamental task in some computer vision applications [2,3,17,18]. Existing methods for circle extraction can be classified into two categories [21]: voting based methods [22,24] and fitting based methods [19]. The voting based technique is more robust and the fitting based method is more accurate.

By using the least-squares fitting [1,7] or the estimation algorithm [5,15], fitting-based methods directly compute the circle parameters using contour points. However, these approaches are very sensitive to outliers.

The Circular Hough Transform (CHT) is a typical voting-based method for circles detection [23]. The standard CHT use a 3-dimensional (3D) accumulator

Supported by the National Natural Science Foundation of China (61602063), Jiangsu Collaborative Innovation Center for Cultural Creativity (XYN1705).

H. Ma et al. (Eds.): PRCV 2021, LNCS 13020, pp. 572–583, 2021.
https://doi.org/10.1007/978-3-030-88007-1_47

array as the Hough space, which specifies the center coordinates and the radius of a circle. A contour point in the image space votes for a cone in the Hough space. The cell with the maximum votes is searched and considered as a circle. The CHT is robust to image noise, but the computation and storage requirements are heavy.

In order to improve the computation and storage performance, many variants of standard CHT have been proposed [13,16,20]. One kind of method employs efficient sampling techniques [10,14]. Three contour points or four contour points [6] in the image space are randomly sampled and vote for one cell in the Hough space. Thus, the computation time is reduced. However, the storage requirement is still a 3D accumulator. Another kind of method exploits a two-step strategy [11,12]. The first step uses a 2D array to extract the circle center [4,9] and the second step uses a 1D histogram to compute the circle radius. The circle center is identified by taking advantage of the mid-perpendiculars of circle chords [8] or the gradient orientations of contour points [25]. The circle radius is obtained by searching the maximum of the distance histogram. Although the computational and storage performance is improved, the extraction accuracy decreases.

All these methods for circles detection are based on contour points. Contour extraction is implemented firstly. When these methods are used to extract circular objects, the results are affected by contour defects.

This paper aims at extracting circular objects fast and accurately. Region pixels are used to vote for a 2D accumulator array. A set of quadratic functions and a linear function are fitted. The circle parameters are computed using the fitted coefficients.

The rest of the paper is organized as follows. Section 2 presents Hough voting and deduces two functional relationships. Section 3 describes the extraction of circle parameters. Section 4 tests and compares three methods with synthetic and real-world images. Section 5 concludes.

2 Hough Voting

Similarly with Hough voting for lines detection, the Hough space is defined by a 2D accumulator array, which corresponds to the voting distance ρ and the voting angle θ.

2.1 Hough Space

Instead of contour points, region pixels vote for cells in the 2D accumulator array. A region pixel votes for a *sine-curve* in the 2D accumulator array. The voting formula is:

$$\rho = x \cdot \cos\theta + y \cdot \sin\theta \qquad \theta \in [0 \quad \pi) \tag{1}$$

Where θ is the voting angle and ρ is the voting distance. Only a few voting angles are used. Thus, the Hough space, which is denoted as $A(\rho, \theta)$, is a 2D accumulator

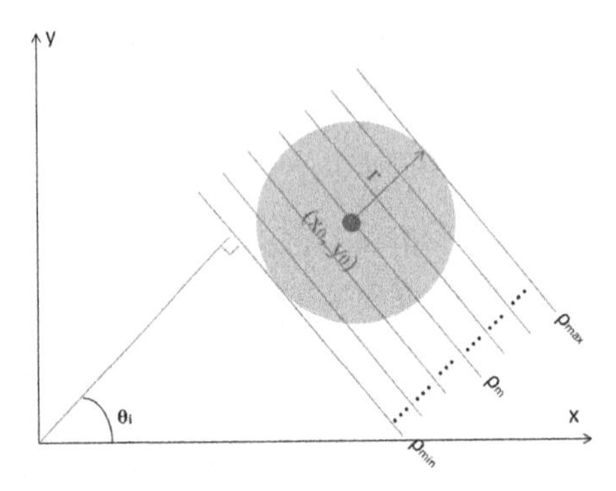

(a) Voting angles θ_i and corresponding voting distances ρ in the image space.

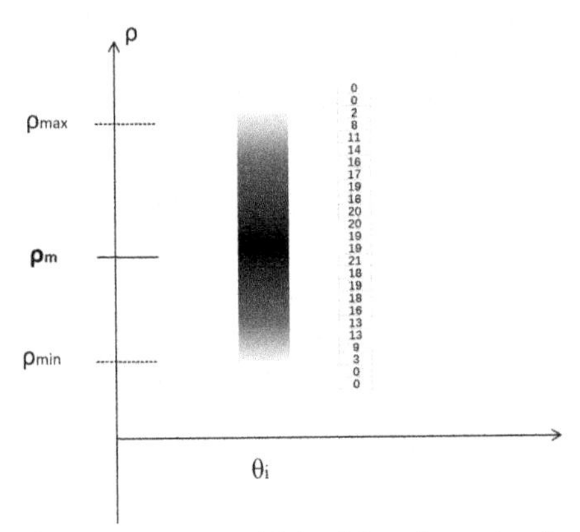

(b) Voting values $A(\rho)$ of θ_i column in the Hough space.

Fig. 1. Voting analysis in the image space and the Hough space.

array with a small number of columns. For a voting angle θ_i, there is a set of parallel lines that intersect the circular object. The computed voting distance ρ is in a limited interval. ρ_m is the voting distance which the circle center votes for. It is shown in Fig. 1 *Top*. The corresponding voting value $A(\rho)$ increases from 0 to the maximum, and then decreases to 0. ρ_m is also the voting distance that has maximum votes. It is shown in Fig. 1 *Bottom*. ρ_m is different given different voting angle θ. Let $\rho_m(\theta)$ be the mean voting distance corresponding to the voting angle θ.

The voting values $A(\rho)$ and mean voting distance $\rho_m(\theta)$ are analysed. Two functional relationships are deduced.

2.2 Functional Equation of Voting Value $A(\rho)$

In each column, the voting value $A(\rho)$ with respect to the voting distance ρ is shown in Fig. 2.

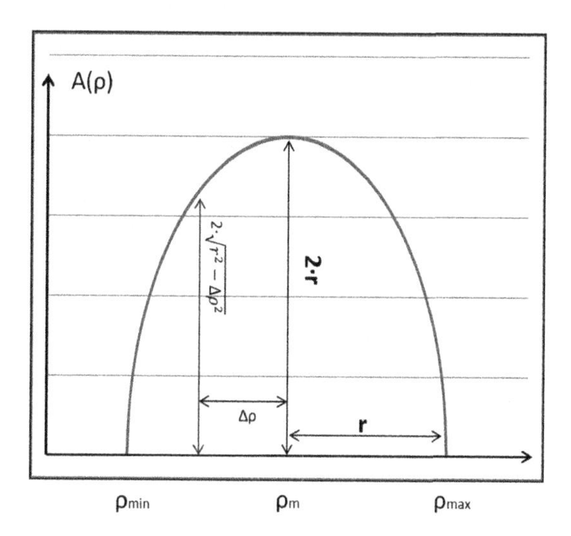

Fig. 2. The functional relationship of voting value $A(\rho)$ with respect to voting distance ρ for a voting angle θ_i.

Given a voting angle θ_i, the functional relationship of $A(\rho)$ with respect to ρ is:

$$A(\rho) = 2 \cdot \sqrt{r^2 - (\rho - \rho_m)^2} \tag{2}$$

Therefore,

$$\begin{aligned} A^2(\rho) &= 4 \cdot (r^2 - (\rho - \rho_m)^2) \\ &= -4 \cdot \rho^2 + 8\rho_m \cdot \rho + 4(r^2 - \rho_m^2) \end{aligned} \tag{3}$$

The equation denotes the functional relationship between square voting value $A^2(\rho)$ and voting distance ρ given a voting angle θ_i. It is a quadratic function, and the function coefficients rely on the circle radius r and mean voting distance ρ_m in θ_i column.

If this functional formula is known, the circle radius r can be computed based on the function coefficients.

2.3 Functional Equation of Mean Voting Distance $\rho_m(\theta)$

Following Fig. 1, because of the symmetry of the circular object, the center (x_0, y_0) of the circular object votes for the distance ρ_m for any voting angle θ. Therefore, there is the following equation:

$$\rho_m(\theta) = x_0 \cdot \cos\theta + y_0 \cdot \sin\theta \tag{4}$$

The equation denotes the functional relationship of the mean voting distance $\rho_m(\theta)$ with respect to voting angle θ. It is a *sine*-function, and the function coefficients depend on the circle center (x_0, y_0).

Theoretically, the circle center (x_0, y_0) can be calculated by solving two equations, which formed by two voting angles θ_1 and θ_2, and corresponding two $\rho_m(\theta_1)$ and $\rho_m(\theta_2)$. By considering various uncertainties, more mean voting distance $\rho_m(\theta)$ are used to fit a linear function. In order to fit it conveniently, the *sine*-function is linearized as:

$$\rho_m(\theta)/\sin\theta = x_0 \cdot \cot\theta + y_0 \tag{5}$$

Thus, the relationship between $\rho_m(\theta)/\sin\theta$ and $\cot\theta$ is a linear function, whose coefficients happen to be the circle center coordinates.

3 Parameters Extracting

By considering the computation efficiency, Only a few voting angles θ are used when region pixels vote for the 2D accumulator array. After all region pixels have voted, the circle parameters are extracted by fitting a set of quadratic functions and a linear function.

3.1 The Radius Extraction

In column θ_i of 2D accumulator array, non-zero voting cells are searched. The corresponding voting value $A(\rho)$ and voting distance ρ are recorded. According to Eq. (3), a quadratic function f_i is fitted to pairs $(A^2(\rho), \rho)$ as:

$$\begin{aligned} f_i: \quad A^2(\rho) &= f_i(\rho) \\ &\triangleq a_2 \cdot \rho^2 + a_1 \cdot \rho + a_0 \end{aligned} \tag{6}$$

An example of the fitted quadratic function in a column of 2D accumulator is illustrated in Fig. 3.

Based on Eq. (3) and Eq. (6), there are the following equations:

$$\begin{aligned} 8 \cdot \rho_m &= a_1 \\ 4 \cdot (r^2 - \rho_m^2) &= a_0 \end{aligned} \tag{7}$$

By solving the above equations, the circle radius is computed with the coefficients of the fitted quadratic function:

$$r_i = \frac{\sqrt{16 \cdot a_0 + a_1^2}}{8} \tag{8}$$

A value r_i of the radius is obtained when a quadratic function has been fitted in column θ_i. After all columns of 2D accumulator array have been processed, there are a set of values of the radius. The radius of the circular object is comprehensively computed by averaging all those r_i, which are computed in every columns.

$$r = \frac{\sum r_i}{n} \tag{9}$$

Where n is the number of fitted quadratic functions.

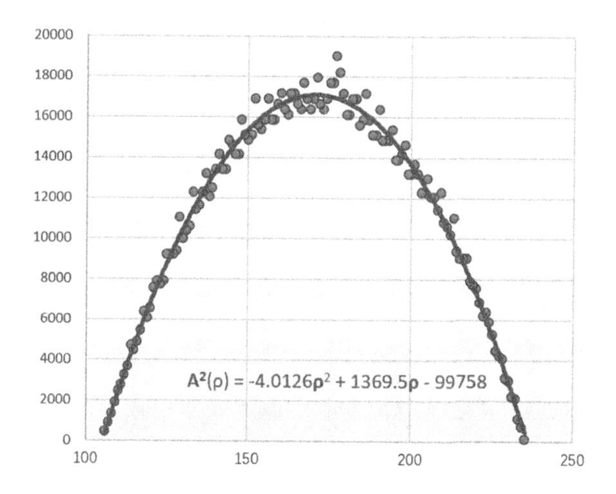

Fig. 3. The quadratic function fitting.

3.2 The Center Extraction

By solving Eq. 7, the mean voting distance $\rho_m(\theta_i)$ is computed in column θ_i as the following:

$$\rho_m(\theta_i) = \frac{a_1}{8} \tag{10}$$

After all columns of the 2D accumulator array have been processed, there are a set of mean voting distance $\rho_m(\theta)$ corresponding to every voting angle θ.

According to Eq. 5, the relationship between $\rho_m(\theta)/\sin\theta$ and $\cot\theta$ is a linear function. A linear function g is fitted to pairs $(\rho_m(\theta_i)/\sin\theta_i, \cot\theta_i)$:

$$\begin{aligned} g : \rho_m(\theta)/\sin\theta &= g(\cot\theta) \\ &\triangleq k \cdot \cot\theta + b \end{aligned} \tag{11}$$

An example of the fitted linear function is illustrated in Fig. 4.

By comparing Eq. (5) and Eq. (11), the center coordinates of the circular object are:

$$\begin{aligned} x_0 &= k \\ y_0 &= b \end{aligned} \tag{12}$$

The center coordinates of the circular object happen to be the coefficients of the fitted linear function g.

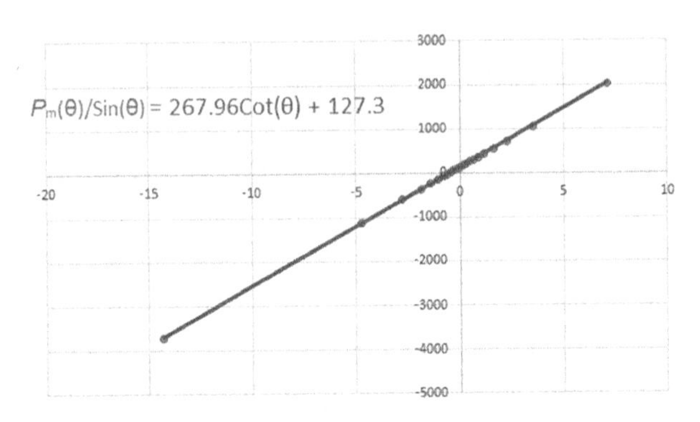

Fig. 4. The linear function fitting.

4 Experimental Results

The proposed method for extracting parameters of circular objects is tested on synthetic and real-world images.

4.1 Test with Synthetic Images

A synthetic 300×300 image consisting of a circular object is generated randomly. Three non-collinear points are chosen randomly to define a circle. The circle parameters are computed and recorded as the ground truth. In order to compare the performance, the circular object is extracted with different methods. They are fitting method, Random CHT and the proposed method.

Fitting method and Random CHT method are based on contour points, which are extracted using adaptive Canny edge detector. Regarding fitting method, a circle is directly fitted to the contour points. Regarding Random CHT method, 200 triplets of points are randomly selected from the contour points and vote for one cell in the Hough space. After voting, the cell with maximum votes defines the circle parameters. The proposed method is based on region pixels. Only 22 voting angles, which range from $8°$ to $176°$ with step $8°$, are used. 22 quadratic functions and one linear function are fitted.

The results of three methods are compared in term of computation time and extraction accuracy.

Test of Computation Time. 200 synthetic images are processed using fitting technique, Random CHT and Region-based method respectively. The mean computation times are listed in Table 1.

Random CHT method takes more computation time because a 3D accumulator array is maintained during voting and searching although contour points are randomly sampled. For fitting method, the majority of computation time is spent on extracting contour points. The proposed method only uses a 2D accumulator array with 22 columns. Contour extraction is avoided. The mean computation time is the least of three methods.

Table 1. Comparison of computation time (in ms).

	Fitting	Random CHT	Ours
Computation time	4.31	129.40	1.32

Test of Extraction Accuracy. 200 synthetic images are done using three methods. The extraction accuracy is measured by comparing the extracted circle and the ground truth.

$$Accuracy = \frac{Extracted\ Circle \bigcap Ground\ truth}{Extracted\ Circle \bigcup Ground\ truth} \tag{13}$$

The mean detection accuracies are listed in Table 2.

Table 2. Comparison of extraction accuracies.

	Fitting	Random CHT	Ours
Extraction accuracies	0.9905	0.9454	0.9964

In the absence of contour defects or outliers, the fitting method is very accurate. The random CHT is not very accurate due to quantization errors and peak spreading. Region-based method calculate circle parameters based votes from all regional pixels, and the results are very accurate.

4.2 Test on Real World Images

The proposed method for circular objects extraction is tested on real-world images. The extracted results are also compared with other two methods.

Figure 5 shows two images and extracted results. Images are binarized by using the OTSU algorithm firstly, and then the edge image is generated by using a adaptive Canny edge detector. The image is processed using three methods. Region based method is based on the binarized image, while both the fitting method and the random CHT method are based on the edge image. The raw image, the binarized image and the edge image are listed in the first row, and three extraction results are shown in the second row.

In the first image, the pedicel is looked as the outlier. The fitting method is affected and the extracted result is inaccurate because of the outlier. The random CHT is based on voting and searching the maximum, it is robust to the outlier and the extracted result is accurate. Regarding the proposed method, the votes distribution is affected by the pedicel pixels, but the influence is slight because redundant data are used to fit parameters of the melon.

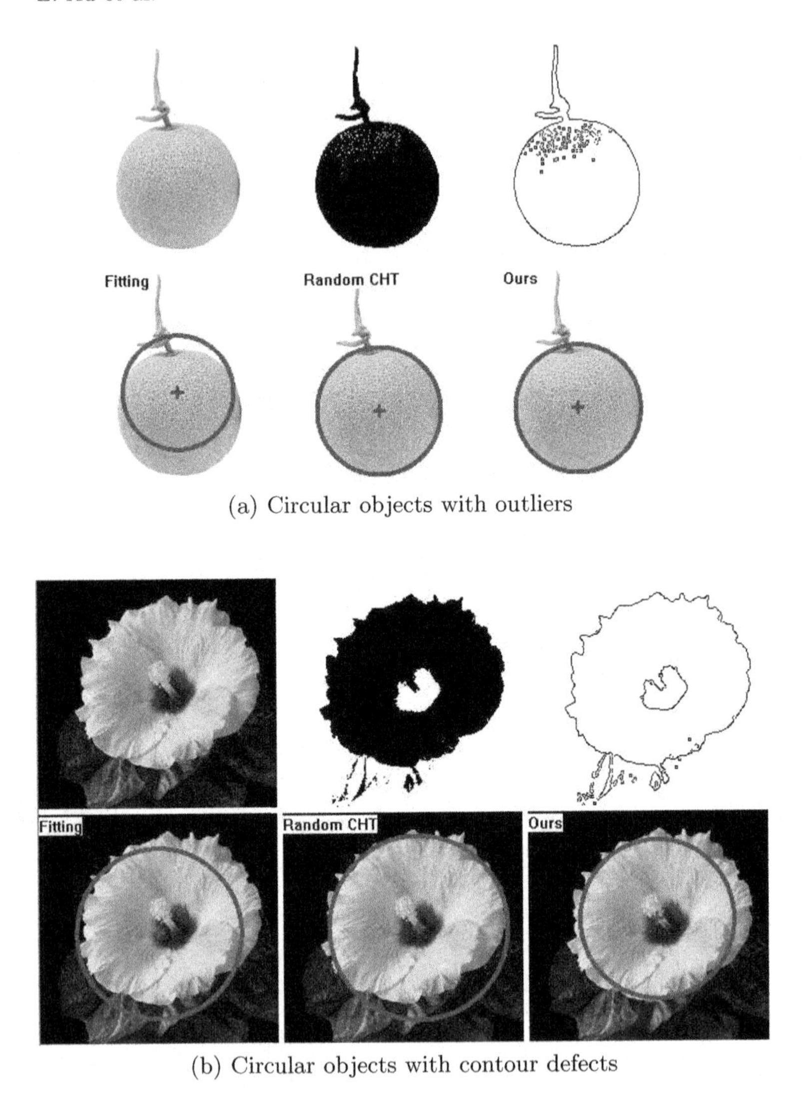

(a) Circular objects with outliers

(b) Circular objects with contour defects

Fig. 5. Extraction results of circular objects with outliers and contour defects.

In the second image, the contour of the petal is not completely a circular shape, which is considered as the contour defect. The fitting method extracts a circle that approximates globally contour points, while the random CHT method extracts a circle that approximate locally co-circular contour points. The region-based method extracts a circle that approximates the whole region. The extraction results of the region-based method is relatively accurate.

4.3 Applications in Cyclone Centers Location

The proposed method is applied in tropical cyclone centers location. Figure 6 shows two satellite images, which respectively consist of an eye tropical cyclone and a non-eye cyclone. By considering the complex background, the cloud pixels are segmented using the seed-expanding technique. The segmented cyclones are shown in the middle column. The results of center location are listed in the third column.

Although the segmented cyclone is not completely a circular shape, the center of the tropical cyclone is located accurately. The located centers are compared with the Best Track (BT). The averaged errors of center location is 16.62 km for Typhoon Saomai (No. 200608).

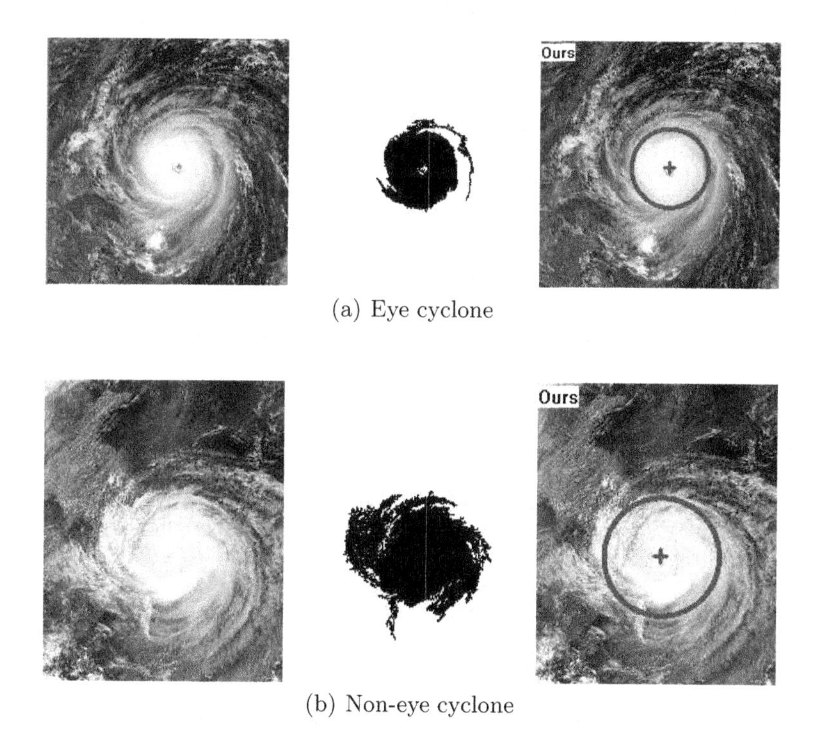

Fig. 6. Center location for tropical cyclones.

5 Conclusions

This article presents a novel method for circular objects extraction by using region pixels rather than contour points. Two functional relationships are deduced. The quadratic functional equation of voting values $A^2(\rho)$ with respect to the voting distance relies on the circle radius, and the linear functional equation of mean voting distance $\rho_m(\theta)$ with respect to the voting angle depends on the circle center. The parameters of circular objects are calculated by fitting a set of quadratic functions and a linear function. The computation efficiency is high since only a 2D accumulator with few columns is used.

References

1. Ahn, S., Rauh, W., WarnEcKe, H.: Least-squares orthogonal distances fitting of circle, sphere, ellipse, hyperbola, and parabola. Pattern Recogn. **34**(12), 2283–2303 (2001)
2. Bargoti, S., Underwood, J.P.: Image segmentation for fruit detection and yield estimation in apple orchards. J. Field Robot. **34**(6), 1039–1060 (2017)
3. Butters, L., Xu, Z., Trung, K., Klette, R.: Measuring apple size distribution from a near topcdown image. In: Pacific-Rim Symposium on Image and Video Technology, pp. 255–268 (2019)
4. Cauchie, J., Fiolet, V., Villers, D.: Optimization of an hough transform algorithm for the search of a center. Pattern Recogn. **41**(2), 567–574 (2008)
5. Chan, Y., Lee, B., Thomas, S.: Approximate maximum likelihood estimation of circle parameters. J. Optim. Theory Appl. **125**(3), 723–734 (2005)
6. Chen, T., Chung, K.: An efficient randomized algorithm for detecting circles. Comput. Vis. Image Underst. **83**(2), 172–191 (2001)
7. Chernov, N., Lesort, C.: Least squares fitting of circles. J. Math. Imaging Vis. **23**(3), 239–252 (2005)
8. Chiu, S., Liaw, J.: An effective voting method for circle detection. Pattern Recogn. Lett. **26**(2), 121–133 (2005)
9. Chung, K., Huang, Y., Wang, J., Chang, T., Liao, H.: Fast randomized algorithm for center-detection. Pattern Recogn. **43**(8), 2659–2665 (2010)
10. Jiang, L., Wang, Z., Ye, Y., Jiang, J.: Fast circle detection algorithm based on sampling from difference area. Optik **158**, 424–433 (2018)
11. Kim, H., Kim, J.: A two-step circle detection algorithm from the intersecting chords. Pattern Recogn. Lett. **22**(6), 787–798 (2001)
12. Kotyza, J., Machacek, Z., Koziorek, J.: Detection of directions in an image as a method for circle detection. IFAC-PapersOnLine **51**(6), 496–501 (2018)
13. Li, Q., Wu, M.: An improved hough transform for circle detection using circular inscribed direct triangle. In: 13th International Congress on Image and Signal Processing, BioMedical Engineering and Informatics, pp. 203–207 (2020)
14. Liang, Q., et al.: Angle aided circle detection based on randomized hough transform and its application in welding spots detection. Mathe. Biosci. Eng. MBE **16**(3), 1244–1257 (2019)
15. Ma, Z., Ho, K.C., Le, Y.: Solutions and comparison of maximum likelihood and full-least-squares estimations for circle fitting. In: Proceedings of the IEEE International Conference on Acoustics, Speech, and Signal Processing, pp. 3257–3260 (2009)
16. Manzanera, A., Nguyen, T., Xu, X.: Line and circle detection using dense one-to-one hough transforms on greyscale images. EURASIP J. Image Video Process. **2016**(1), 46 (2016)
17. Naqvi, S.S., Fatima, N., Khan, T.M., Rehman, Z.U., Khan, M.A.: Automatic optic disk detection and segmentation by variational active contour estimation in retinal fundus images. SIViP **13**(6), 1191–1198 (2019). https://doi.org/10.1007/s11760-019-01463-y

18. Okokpujie, K., Noma-Osaghae, E., John, S., Ajulibe, A.: An improved iris segmentation technique using circular hough transform. In: International Conference on Information Theoretic Security, pp. 203–211 (2017)
19. Rangarajan, P., Kanatani, K.: Improved algebraic methods for circle fitting. Electron. J. Stat. **3**(1), 1–7 (2009)
20. Singla, B., Sharma, M., Gupta, A., Mohindru, V., Chawla, S.: An algorithm to recognize and classify circular objects from image on basis of their radius. In: The International Conference on Recent Innovations in Computing, pp. 407–417 (2020)
21. Smith, E., Lamiroy, B.: Circle detection performance evaluation revisited. In: International Workshop on Graphics Recognition, pp. 3–18 (2015)
22. Su, Y., Zhang, X., Cuan, B., Liu, Y., Wang, Z.: A sparse structure for fast circle detection. Pattern Recognit. **97**, 107022 (2020)
23. Yakaiah, P., Manjunathachari, K., Reddy, K.: A novel object detection approach using circular hough transform. J. Adv. Res. Dyn. Control Syst. **9**(8), 264–273 (2017)
24. Yang, H., Luo, J., Shen, Z., Wu, W.: A local voting and refinement method for circle detection. Optik Int. J. Light Electron Optics **125**(3), 1234–1239 (2014)
25. Ye, H., Shang, G., Wang, L., Min, Z.: A new method based on hough transform for quick line and circle detection. In: 8th International Conference on Biomedical Engineering and Informatics, pp. 52–56 (2015)

GGRNet: Global Graph Reasoning Network for Salient Object Detection in Optical Remote Sensing Images

Xuan Liu[1], Yumo Zhang[2,3], Runmin Cong[2,3](✉), Chen Zhang[2,3], Ning Yang[2,3], Chunjie Zhang[2,3], and Yao Zhao[2,3]

[1] School of Software, Beijing Jiaotong University, Beijing, China
[2] Institute of Information Science, Beijing Jiaotong University, Beijing, China
{geofreey,yumozhang,rmcong,chen.zhang,ningyang,cjzhang,yzhao}@bjtu.edu.cn
[3] Beijing Key Laboratory of Advanced Information Science and Network Technology, Beijing, China

Abstract. The task of salient object detection (SOD) in optical remote sensing images (RSIs) is more challenging than the SOD in natural sensing images (NSIs) because of the unique characteristics of remote sensing images such as various object scales and background context redundancy. However, the existing methods ignore the global relationship modeling between different salient objects or different parts in one salient object. To this end, we design a Global Graph Reasoning Module (GGRM) in a lightweight and effective form, and propose a novel Global Graph Reasoning Network (GGRNet) for SOD in optical RSIs. During the graph reasoning, the GGRM considers the role of the global information. Specifically, we explore two ways to utilize the global information, including the global features and global nodes, which are ingeniously added to the interaction of graph nodes and fully integrated through iteration. Besides, we stabilize the projection channel between coordinate space and interactive space through an attention mechanism. The GGRNet outperforms the existing state-of-the-art SOD algorithms on two publicly available datasets, and the number of parameters is only 25.01 Mb.

Keywords: Salient object detection · Optical remote sensing images · Graph reasoning · Global information

1 Introduction

Salient object detection (SOD) simulates human visual characteristics and extracts salient regions (regions of human interest) in the image. Nowadays, SOD algorithms for natural scene images (NSIs) have achieved remarkable results, while only a few studies for optical remote sensing images (RSIs). Understanding the optical RSIs is very important for ground monitoring. However, the large scope of the optical RSIs may include many unimportant redundancy,

X. Liu and Y. Zhang—Equal contribution.

H. Ma et al. (Eds.): PRCV 2021, LNCS 13020, pp. 584–596, 2021.
https://doi.org/10.1007/978-3-030-88007-1_48

and thus screening the image data content in advance becomes an urgent practice to extract useful objects or regions, and realize more effective and more accurate data processing. Therefore, SOD is an effective solution to realize this pre-extraction of useful content.

Typically, optical RSIs contain diverse objects and covers a wide range with complicated background, for which SOD aims to completely segment the significant objects or regions (such as airplanes, ships, rivers and stadiums). Optical remote sensing detection is often interfered by background. For example, salient objects and background may have similar colors, and detection may be affected by shadows. To solve this problem, some studies use attention mechanism to learn more salient features and suppress background interference. In addition, saliency detection will encounter the problem of missing detection salient or incomplete structure, which they work to design more complex networks or realizing the fusion of context information. Although they have achieved satisfactory results through these methods, they ignore the reasoning relationship between background and salient objects as well as between multi-salient objects.

To this end, we propose a novel Global Graph Reasoning Network (GGRNet) for salient object detection in optical remote sensing images. The encoder-decoder structure [1] is used as our baseline network. In order to make the boundary of the detected object clearer and obtain more accurate object extraction results, the edge information and the feature map are cross-fused in the early stage. We consider using graph reasoning to infer different salient regions in feature maps so as to achieve mutual enhancement between multiple objects and to highlight the salient area and suppress the background area. Based on the graph reasoning unit (GRM) in [2], we design a novel global graph reasoning module named GGRM, in which global information is added to the process of graph reasoning with different strategies. Besides, we add the attention mechanism in the process of projection to optimize the generation of graph nodes and to stabilize the projection channel. At last, in order to fully integrate the global information, we set iterative graph reasoning.

In summary, the main contributions of this paper are as follows:

- We propose a novel Global Graph Reasoning Network (GGRNet) for salient object detection in optical remote sensing images, which considers the relationship between different regions with the guidance of global information.
- We design an effective and lightweight global graph reasoning module (GGRM) with global information, and introduce the attention mechanism to optimize the generation of the graph structure. Moreover, the graph reasoning is iteratively conducted in the whole model to improve the accuracy.
- Our model is end-to-end trained and has good engineering practicability with only 25.01 Mb parameters. The experiments show that the proposed GGRNet achieves competitive performance against state-of-the-art methods on two publicly available datasets.

2 Related Works

2.1 Salient Object Detection in NSIs

In recent years, deep learning based SOD methods have gradually emerged and developed because of its powerful feature representation capabilities, and achieved gratifying performance. Li et al. [3] combined feature maps from three different scales to get the final prediction map. After that, with the rise of Fully Convolutional Networks (FCN), more and more saliency detection methods are proposed [4–10]. Zhao et al. [11] proposed a pyramid feature attention network saliency detection method which combines high-level context features and low-level spatial structure features. Liu et al. [1] introduced two simple pool-based modules in feature pyramid network to realize real-time saliency detection. In addition, some methods use edge information to predict more accurate saliency map [12,13]. Wu et al. [12] aimed to simply refine multi-level features of salient object detection and edge detection by stacking cross refinement unit.

2.2 Salient Object Detection in Optical RSIs

For the SOD in optical RSIs, there are relatively few studies on it, many of which are based on traditional methods [14,15]. Zhang et al. [14] proposed an efficient airport extraction framework which combines vision-oriented saliency and knowledge-oriented saliency for the airport position estimation. Zhang et al. [15] proposed an adaptive multi-feature fusion model based on low-rank matrix restoration by integrating color, intensity, texture, and global contrast cues for saliency detection in RSIs. In some of the latest deep learning based methods, Li et al. [16] integrated a two-stream pyramid module and an encoder–decoder module with nested connections to achieve SOD in optical RSIs.

2.3 Graph-Based Reasoning

Graph convolutional network (GCN) [17] is a popular neural network structure in recent years, which is a good way to interact between remote regions. Based on its advantages, some scholars try to embed graph convolution into CNN network. Chen et al. [2] introduced the global reasoning unit which realized the relational reasoning by graph convolution on a small graph of interaction space. Wu et al. [18] developed a graph reasoning module (GRM) based on graph-convolutional network (GCN) to aggregate semantic information.

3 Method

3.1 Overview

In the section, we detailedly present the architecture and theory explanation of our method. The architecture of our proposed network follows an encoder-decoder architecture. In the encoder, the edge information and saliency features

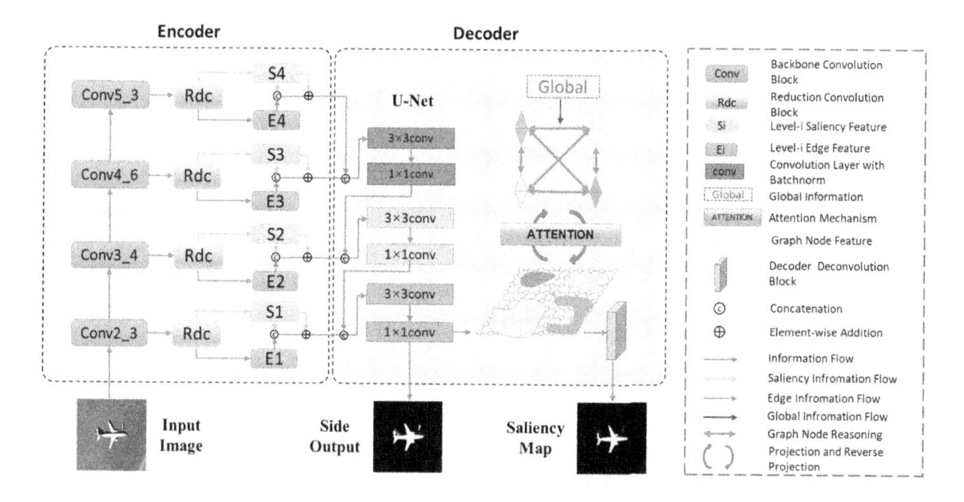

Fig. 1. The architecture of the proposed network.

of each layer in the backbone are cross fused to increase the accuracy of boundary detection. In the decoder, we integrate the multi-scale features obtained from the encoder after cross fusion. With the decoder features of the last layer, GGRM is used to model the relationship between different regions and further improve the detection accuracy and background suppression. In GGRM, global information is added to the graph reasoning process for better long-range feature interaction, and we utilize iterative graph reasoning to fully integrate global information. Moreover, the attention mechanism is introduced to optimize the generation of graph structure and stabilize the projection channel.

3.2 Encoder-Decoder Network

First, we use the ResNet50 [19] as our backbone network, and obtain the multi-level features. Then, we use the reduction convolution block (RDC) containing three consecutive convolution layers to obtain the corresponding edge features and encoder features. It is worth noting that we apply one-way cross fusion in the saliency branch, rather than the two-way cross fusion. The encoder features are updated four times after fusion in a row, while the edge features are not updated. In this way, we can save the redundant two-way cross structure and reduce the complexity of the network. The fusion of edge information and encoder features will be beneficial to the edge detail perception of optical remote sensing images. The specific operation of cross fusion is given by the following expression:

$$Si = Si + Conv(Concat(Si, Ei)), \tag{1}$$

where $i \in \{1, 2, 3, 4\}$ indexes of feature levels, S_i and E_i represent the encoder feature and edge feature of layer i respectively, *Concat* indicates the concatenation operation, and *Conv* denotes the convolution block which contains four convolution layers.

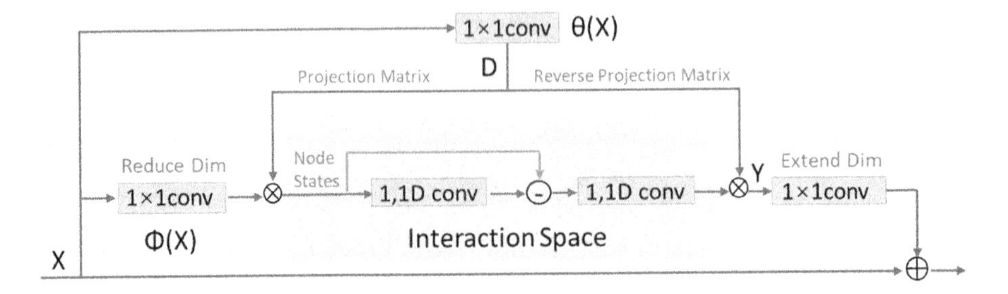

Fig. 2. Architecture of the global reasoning unit.

In the decoder stage, we first use the traditional U-Net [20] structure to fuse encoder features of different layers, thereby generating the multi-scale features with better saliency information. After that, the generated features are fed into the proposed global graph reasoning module (GGRM). In order to bring advantage of graph reasoning into full play, we put GGRM on feature decoder which contains more global information instead of feature encoder. And experimental results also show that the graph reasoning module has not played a significant role in the features of the early stage. Finally, instead of appending a 1×1 convolution to the U-Net structure to reduce the channel to 1, we obtain the final output through two deconvolution blocks. The architecture of our network is shown in Fig. 1.

3.3 Graph Reasoning with Global Information

Graph reasoning is divided into three parts. Such a three-step process is conceptually depicted in the right side of Fig. 1. In order to show the operation of the graph reasoning module more clearly, the flowchart is detailedly introduced in Fig. 2. The first step is to map the original feature to the interaction space by using the projection function. Given a set of input features $X \in \mathbb{R}^{L\times C}$, with C being the feature dimension and $L = H \times W$ locations. We hope to get new features $V = f(X) \in \mathbb{R}^{N\times C}$ in the interaction space through the projection function, where N denotes the number of the nodes. We formulate the projection function as a linear combination of original features because the new features can aggregate information from multiple regions [2]. Each node can be represented as:

$$v_i = d_i X = \sum_{\forall j} d_{ij} x_j, \tag{2}$$

where $D = [d_1, ..., d_N] \in R^{N\times L}$ is the learnable projection weights, $x_j \in R^{1\times C}$, $v_j \in R^{1\times C}$. In fact, we implement the projection function as $f(\phi(X; W_\phi))$ and $D = \Theta(X; W_\Theta)$. We model $\phi(\cdot)$ and $\Theta(\cdot)$ by two convolution layers. W_ϕ and W_Θ are the learnable convolutional kernel of each layer, and d_i is formed by the output of a convolution layer.

After that, we construct a graph with multiple feature nodes. Graph convolution is an effective convolution method to deal with graph structural features.

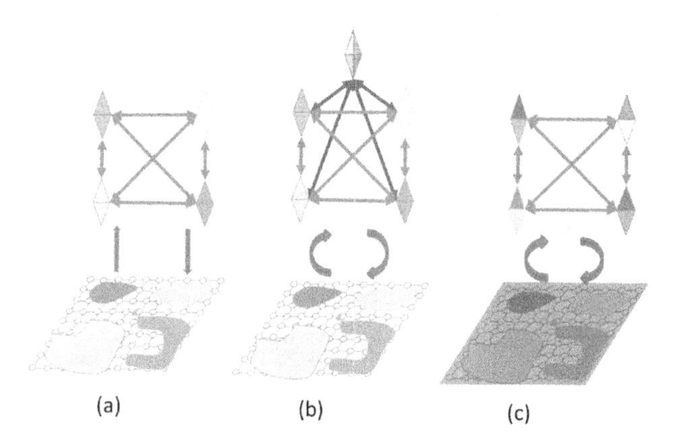

Fig. 3. Different reasoning methods. (a) Graph reasoning (b) Graph reasoning with global node information. (c) Graph reasoning with global feature information.

Through the graph convolution, each node will receive all necessary information and its state is updated. Since our graph node feature is essentially a set of one-dimensional features (channels), multiple nodes form multiple one-dimensional features. By performing convolution in the channel dimension and the node dimension respectively, we can achieve the mutual reasoning of the graph nodes. The graph convolution is formulated as:

$$GCN(X) = Conv_{1D}(Conv_{1D}(X)^T)^T, \tag{3}$$

where $Conv_{1D}$ denotes a 1-dimension convolution. The last step is to re-project the graph node features back to the coordinate space, so we need a reverse projection function. Given a graph feature matrix $G \in \mathbb{R}^{N \times C}$, $Y \in \mathbb{R}^{L \times C}$ is denoted as the output feature. Similar to the projection function, we adopt linear projection:

$$y_i = p_i G = \sum_{\forall j} p_{ij} g_j, \tag{4}$$

where the feature g_j of node j is assigned to y_i weighted by scalar p_{ij}. In practice, we set $P = D^T$. The graph reasoning has been proved to be effective, but the reasoning between different nodes lacks the guidance of global information. Introducing the global information can be conducive to the enhancement of salient region and the suppression of background region. The process of graph reasoning is shown in Fig. 3(a).

Thus, we propose to add global information to the original graph reasoning process. The nodes with global information will be more suitable for the detection requirements through reasoning, and the objects that are not easy to detect in the multi-object case will be enhanced and the background area will be prevented from interfering with other features as graph node features. There are two kinds of graph reasoning guided by global information. The first is to generate a global node from the top as shown in Fig. 3(b), which is an aggregation node generated

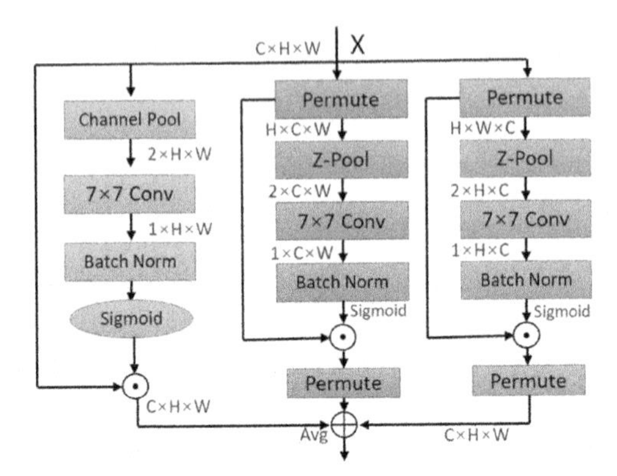

Fig. 4. Architecture of triplet attention.

by the other nodes, containing the information from each underlying region. In the space, it acts as a single node and reasons from other nodes. V^n denotes graph feature with n nodes. This process can be formulated as:

$$V^{(n+1)} = GCN(Concat(V^n, \sum_{k=1}^{n} v_k)). \tag{5}$$

The second is to generate a global region as shown in Fig. 3(c), which is contained in the projected graph nodes. The red area indicates an entire feature map. X^n represents n regions in input features X. This process is as follows:

$$V^n = GCN(f(Concat(X^n, X^1))). \tag{6}$$

In order to make the edge weights of the graph structure better optimize the relationship between each node and fully integrate the global information, we set up a total of three iterations of graph inference. The first is reasoning without global information, the second is reasoning after adding global information, and the third is reasoning without global information.

3.4 Attention Mechanism in the Graph Reasoning

In the process of projection and reverse projection of graph structure, we try to add attention to optimize the generation of graph structure. We have tried a variety of attention mechanisms, such as SKNet [21], EMANet [22], Triplet Attention [23]. And we implement the triplet attention in our model due to its more effective results. As shown in Fig. 4, the triplet network structure has three branches. Given an input tensor $F \in \mathbb{R}^{C \times H \times W}$, the first branch is the channel attention computation branch, in which the interaction between H dimension and C dimension is established. In the second branch, the interaction between C and W dimensions is established; In the third branch, the interaction between H and W dimensions is established to construct spatial attention. Finally, the output features of the three branches are added to calculate the average value.

4 Experiments and Results

4.1 Dataset and Evaluation Metrics

The ORSSD dataset [16] including 800 optical RSIs and EORSSD dataset [24] containing 2000 optical RSIs are used as our datasets. These datasets are very representative and challenging, which contain variable resolution images (such as 1264 × 987, 800 × 600, and 256 × 256) and various salient objects (such as aircraft, ships, cars, stadiums, rivers). Moreover, the number and scale of salient objects are also variable, from no salient object to multiple salient objects, from small-sized vehicles to large-sized buildings. In addition, the background generally includes shadows, trees and buildings, causing some interference to the SOD task. For evaluations metrics, we apply the Precision Recall (P-R) curve, F-measure (F_β) [25], MAE score, and S-measure (S_m) [26]. The larger F_β value and S-measure indicate the better performance, while the smaller MAE score corresponds to the better performance.

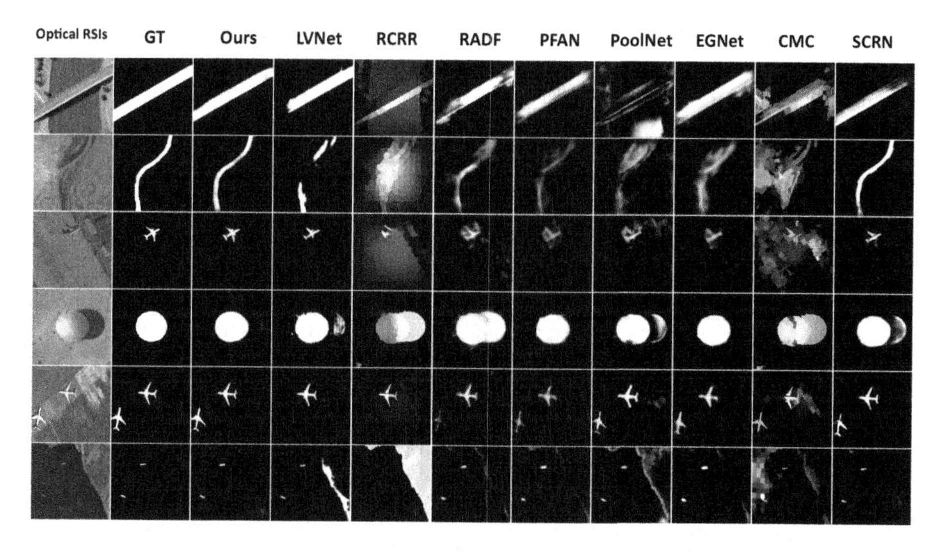

Fig. 5. Visual comparisons of different methods on two datasets.

4.2 Training Strategies and Implementation Details

For the ORSSD dataset, 600 images is used for training and 200 images for testing. For the EORSSD dataset, 1400 images is used for training and the rest 600 images is selected as the testing samples. Data augmentation techniques involving combinations of flipping and rotation are employed to improve the diversity of training samples, which produces seven variants for every single sample. Each sample is resized to the size of 352 × 352. The proposed GGRNet was implemented in Pytorch and trained by an NVIDIA GeForce RTX 2080Ti GPU. We use the stochastic gradient descent to train the network with momentum of 0.9 and weight decay of 0.0005. The batch size is set as 16. It takes 100 epochs for the whole training procedure. The learning rate is set as 0.004 and decreased by 10% at 30 epochs. The optimal evaluation weight will be saved and tested in the last. Before the final training, we make the batch size as 8, the learning rate is 0.002, a total of 30 epochs of training to compare different strategies. The number of parameters of our network is only 25.01 Mb.

4.3 Comparison with State-of-the-Art Methods

We compare our method with 15 state-of-the-art SOD methods, including HDCT [27], R3Net [28], RRWR [29], SMD [30], RCRR [31], DSS [32], RADF [33], PFAN [11], PoolNet [1], EGNet [13], CMC [34], SMFF [15], VOS [14], SCRN [12], and LVNet [16]. Figure 5 shows the visual results of the proposed method with some of state-of-the-art methods on two datasets. Compared with other methods, our method can effectively suppress the background interference and accurately locate the salient region. For a single object as shown in the second image, some methods (such as the LVNet [16]) cannot detect the complete structure, but we can make the incomplete part of the object be detected by global reasoning through other parts of the object. For multi-object detection as shown in the

Table 1. Quantitative comparisons with different methods on the testing subset of the ORSSD and EORSSD datasets.

	ORSSD Dataset			EORSSD Dataset		
	F_β	MAE	S_m	F_β	MAE	S_m
RCRR	0.5944	0.1277	0.6849	0.4495	0.1644	0.6013
HDCT	0.5775	0.1309	0.6197	0.5992	0.1087	0.5976
SMD	0.7075	0.0715	0.7640	0.6468	0.0770	0.7112
RRWR	0.5950	0.1324	0.6835	0.4495	0.1677	0.5997
DSS	0.7838	0.0363	0.8262	0.7158	0.0186	0.7874
R3Net	0.7698	0.0409	0.8092	0.7989	0.0170	0.8305
RADF	0.7865	0.0386	0.8252	0.7966	0.0162	0.8332
PoolNet	0.7911	0.0358	0.8403	0.8012	0.0209	0.8301
PFAN	0.8344	0.0543	0.8613	0.7931	0.0156	0.8446
EGNet	0.8585	0.0215	0.8780	0.8310	**0.0109**	0.8692
SCRN	0.8629	0.0224	0.8693	0.8214	0.0159	0.8564
SMFF	0.4764	0.1897	0.5329	0.5693	0.1471	0.5431
VOS	0.4168	0.2151	0.5366	0.3338	0.2096	0.5083
CMC	0.4214	0.1267	0.6033	0.3555	0.1066	0.5826
LVNet	0.8414	0.0207	0.8815	0.8213	0.0146	0.8642
Ours	**0.8847**	**0.0192**	**0.8876**	**0.8631**	0.0116	**0.8800**

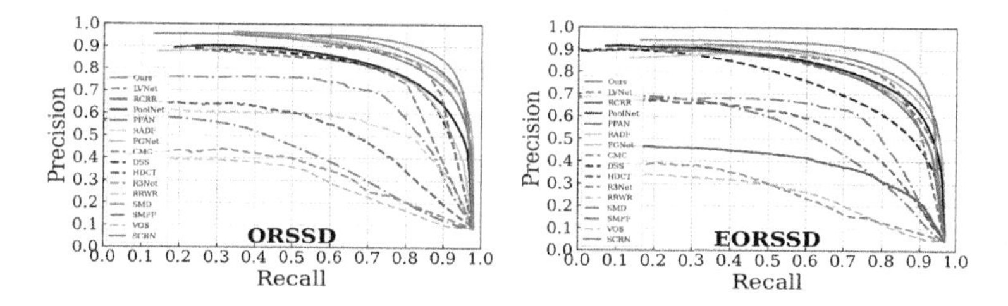

Fig. 6. P-R curves and F-M curves of different methods on the testing subset of the ORSSD datasets.

fifth image, some methods (such as LVNet [16], RADF [33]) cannot detect all the salient objects. We use the global information to detect the object accurately and suppress the non-object area through the mutual reasoning between the objects. As shown Fig. 6, the P-R curves show that the proposed method performed better than other methods. Similarly, the quantitative results reported in Table 1 demonstrate that our proposed method almost achieves the best performance in terms of the S-measure, F-measure, and MAE compared with other approaches on the two datasets.

Table 2. Ablation analysis on the EORSSD dataset.

Baseline	Cross fusion	Graph reasoning	Global node	Global feature	Triplet attention	F_β	MAE	S_m
✓						0.8467	0.0224	0.8661
✓	✓					0.8470	0.0232	0.8690
✓	✓	✓				0.8566	0.0238	0.8692
✓	✓	✓	✓			0.8623	0.0221	0.8765
✓	✓	✓		✓		0.8655	0.0214	0.8763
✓	✓	✓		✓	✓	**0.8847**	**0.0192**	**0.8876**

Table 3. Effectiveness of different GGRM positions.

Position	Res1	Res2	Res3	Res4	U-Net	U-Net × 2
F_β	0.8600	0.8696	0.8645	0.8562	**0.8847**	0.8683

4.4 Ablation Study

As shown in Table 2, it can be seen that the cross fusion of edge information and encoder features is conducive to the improvement of network performance, compared with the baseline model. After adding the graph reasoning, the F-measure is improved by about 1%. After introducing the global information, the

F-measure of the two strategies is improved by nearly 1%, which indicates that the global information is very helpful for SOD. Finally, we added triplet attention to improve GGRM, F-measure also increased by about 2%. Table 3 compares the effects of GGRM at different locations in the network. Res $\{1, 2, 3, 4\}$ means four layers of ResNet50, and U-Net indicates multi-scale fusion feature. "×2" represents the number of GGRM is two. It can be seen that the reasoning relationship is more effective based on the multi-scale fusion feature but is not applicable in the early stage of the network.

5 Conclusion

In this paper, according to the particularity of the optical remote sensing image, we propose a new graph reasoning structure, which is suitable for SOD in optical RSIs. In order to make the edge of detection effect clearer, we also carry out the cross fusion of edge information and feature map, and finally multi-scale fusion through U-Net. Compared with existing methods based on graph reasoning, we introduce global information for the first time, which has obvious effect on the improvement of detection. At the same time, we introduce attention mechanism and iterative strategy, which are helpful to the improvement of the algorithm. Experiments on two datasets show that our algorithm is superior to the latest algorithm, with smaller number of parameters.

Acknowledgement. This work was supported by the Beijing Nova Program under Grant Z2011000068 20016, in part by the National Key Research and Development of China under Grant 2018AAA0102100, in part by the National Natural Science Foundation of China under Grant 62002014, Grant 61532005, Grant 62072026, Grant U1936212, Grant 61971016, Grant U1803264, Grant 61922046, Grant 61772344, Grant 61672443, in part by Elite Scientist Sponsorship Program by the China Association for Science and Technology under Grant 2020QNRC001, in part by Beijing Natural Science Foundation under Grant JQ20022, in part by the Fundamental Research Funds for the Central Universities under Grant 2019RC039, in part by Young Elite Scientist Sponsorship Program by the Beijing Association for Science and Technology, and in part by China Postdoctoral Science Foundation under Grant 2020T130050, Grant 2019M660438.

References

1. Liu, J.-J., Hou, Q., Cheng, M.-M., Feng, J., Jiang, J.: A simple pooling based design for real-time salient object detection. In: Proceedings of the CVPR, pp. 3917–3926 (2019)
2. Chen, Y., Rohrbach, M., Yan, Z., Yan, S., Feng, J., Kalantidis, Y.: Graph-based global reasoning networks. In: Proceedings of the CVPR, pp. 433–442 (2019)
3. Li, G., Yu, Y.: Visual saliency based on multi-scale deep features. In: Proceedings of the CVPR, pp. 5455–5463 (2015)
4. Wang, L., Wang, L., Lu, H., Zhang, P., Ruan, X.: Saliency detection with recurrent fully convolutional networks. In: Leibe, B., Matas, J., Sebe, N., Welling, M. (eds.) ECCV 2016. LNCS, vol. 9908, pp. 825–841. Springer, Cham (2016). https://doi.org/10.1007/978-3-319-46493-0_50

5. Cong, R., Lei, J., Fu, H., Cheng, M.-M., Lin, W., Huang, Q.: Review of visual saliency detection with comprehensive information. IEEE Trans. Circ. Syst. Video Technol. **29**(10), 2941–2959 (2019)
6. Zhang, L., Dai, J., Lu, H., He, Y., Wang, G.: A bi-directional message passing model for salient object detection. In: Proceedings of the ICCV, pp. 1741–1750 (2018)
7. Zhang, P., Wang, D., Lu, H., Wang, H., Ruan, X.: Amulet: aggregating multi-level convolutional features for salient object detection. In: Proceedings of the ICCV, pp. 202–211 (2017)
8. Cong, R., Lei, J., Fu, H., Hou, J., Huang, Q., Kwong, S.: Going from RGB to RGBD saliency: a depth-guided transformation models. IEEE Trans. Cybern. **50**(8), 3627–3639 (2020)
9. Cong, R., et al.: An iterative co-saliency framework for RGBD images. IEEE Trans. Cybern. **49**(1), 233–246 (2019)
10. Cong, R., Lei, J., Fu, H., Huang, Q., Cao, X., Hou, C.: Co-saliency detection for RGBD images based on multi-constraint feature matching and cross label propagation. IEEE Trans. Image Process. **27**(2), 568–579 (2018)
11. Zhao, T., Wu, X.: Pyramid feature attention network for saliency detection. In: Proceedings of the CVPR, pp. 3085–3094 (2019)
12. Wu, Z., Su, L., Huang, Q.: Stacked cross refinement network for edge-aware salient object detection. In: Proceedings of the ICCV, pp. 7263–7272 (2019)
13. Zhao, J., Liu, J., Fan, D., Cao, Y., Yang, J., Cheng, M.-M.: EGNet: edge guidance network for salient object detection. In: Proceedings of the ICCV, pp. 8778–8787 (2019)
14. Zhang, Q., Zhang, L., Shi, W., Liu, Y.: Airport extraction via complementary saliency analysis and saliency-oriented active contour model. IEEE Geosci. Remote Sens. Lett. **15**(7), 1085–1089 (2018)
15. Zhang, L., Liu, Y., Zhang, J.: Saliency detection based on self-adaptive multiple feature fusion for remote sensing images. Int. J. Remote Sens. **40**(22), 8270–8297 (2019)
16. Li, C., Cong, R., Hou, J., Zhang, S., Qian, Y., Kwong, S.: Nested network with two-stream pyramid for salient object detection in optical remote sensing images. IEEE Trans. Geosci. Remote Sens. **57**(11), 9156–9166 (2019)
17. Kipf, T.N., Welling, M.: Semi-supervised classification with graph convolutional networks. In: Proceedings of the ICLR (2017)
18. Wu, Y., Jiang, A., Tang, Y., Kwan, H.K.: GRNet: deep convolutional neural networks based on graph reasoning for semantic segmentation. In: Proceedings of the VCIP, pp. 116–119 (2020)
19. He, K., Zhang, X., Ren, S., Sun, J.: Deep residual learning for image recognition. In: Proceedings of the CVPR, pp. 770–778 (2016)
20. Ronneberger, O., Fischer, P., Brox, T.: U-Net: convolutional networks for biomedical image segmentation. In: Proceedings of the MICCAI, pp. 234–241 (2015)
21. Li, X., Wang, W., Hu, X., Yang, J.: Selective kernel networks. In: Proceedings of the CVPR, pp. 510–519 (2019)
22. Li, X., Zhong, Z., Wu, J., Yang, Y., Lin, Z., Liu, H.: Expectation-maximization attention networks for semantic segmentation. In: Proceedings of the ICCV, pp. 9166–9175 (2019)
23. Diganta, M., Nalamada, T., Arasanipalai, A.U., Hou, Q.-B.: Rotate to attend: convolutional triplet attention module. In: Proceedings of the WACV (2021)
24. Zhang, Q., et al.: Dense attention fluid network for salient object detection in optical remote sensing images. IEEE Trans. Image Process. **30**, 1305–1317 (2021)

25. Achanta, R., Hemami, S., Estrada, F., Ssstrunk, S.: Frequency-tuned salient region detection. In: Proceedings of the CVPR, pp. 1597–1604 (2009)
26. Fan, D.-P., Cheng, M.-M., Liu, Y., Li, T., Borji, A.: Structure-measure: a new way to evaluate foreground maps. In: Proceedings of the ICCV, pp. 4548–4557 (2017)
27. Kim, J., Han, D., Tai, Y.-W., Kim, J.: Salient region detection via high-dimensional color transform and local spatial support. IEEE Trans. Image Process. **25**(1), 9–23 (2016)
28. Deng, Z., et al.: R^3net: recurrent residual refinement network for saliency detection. In: Proceedings of the AAAI, pp. 684–690 (2018)
29. Li, C., Yuan, Y., Cai, W., Xia, Y., Dagan Feng, D.: Robust saliency detection via regularized random walks ranking. In: Proceedings of the CVPR, pp. 2710–2717 (2015)
30. Peng, H., Li, B., Ling, H., Hu, W., Xiong, W., Maybank, S.J.: Salient object detection via structured matrix decomposition. IEEE Trans. Pattern Anal. Mach. Intell. **39**(4), 818–832 (2017)
31. Yuan, Y., Li, C., Kim, J., Cai, W., Feng, D.D.: Reversion correction and regularized random walk ranking for saliency detection. IEEE Trans. Image Process. **27**(3), 1311–1322 (2018)
32. Hou, Q., Cheng, M.-M., Hu, X., Borji, A., Tu, Z., Torr, P.H.S.: Deeply supervised salient object detection with short connections. IEEE Trans. Pattern Anal. Mach. Intell. **41**(4), 815–828 (2019)
33. Hu, X., Zhu, L., Qin, J., Fu, C.W., Heng, P.A.: Recurrently aggregating deep features for salient object detection. In: Proceedings of the AAAI, pp. 6943–6950 (2018)
34. Liu, Z., Zhao, D., Shi, Z., Jiang, Z.: Unsupervised saliency model with color Markov chain for oil tank detection. Remote Sens. **11**(9), 1–18 (2019)

A Combination Classifier of Polarimetric SAR Image Based on D-S Evidence Theory

Jiaqi Chen[1], Shuyin Zhang[1](✉), Meng Tian[1], Zhiguo Xie[2], Huan Chen[1], and Erlei Zhang[1]

[1] College of Information Engineering, Northwest A&F University, Yangling 712100, Shaanxi, China
[2] Shaanxi Academy of Forestry, Xi'an, China

Abstract. Polarimetric Synthetic Aperture radar (PolSAR) image classification plays an important role in the application of PolSAR data. In the past few years, many methods for classification of PolSAR images have been proposed. However, the pixels at the edge are easily misclassified in the image classification task. This problem causes that the classification accuracy cannot be effectively improved. Based on this, we propose a new combination classifier based on D-S evidence theory to complement each base classifier for better results. The proposed method constructs the basic probability distribution of the base classifier through confusion matrices, which are used to quantify the recognition capability of different base classifiers for different classes. Experiment conducted on real PolSAR images of Flevoland in the Netherlands shows that compared with the basic classifier, in most cases, the proposed method can achieve significant improvements, which is more obvious for edges.

Keywords: PolSAR image · Classification · D-S evidence theory

1 Introduction

In recent years, the available PolSAR data and the resolution has been continuously improved with the continuous development of the PolSAR systems. Compared with single-polarized SAR, PolSAR performs full-polarization measurements, which can be obtained with richer target information. Thus, the application of PolSAR has become more and more extensive. It is an important development direction of PolSAR technology in practical application and also an important research content of SAR image interpretation. So far, many related SAR image classification algorithms have been proposed. The classification algorithms can be divided into three categories: (1) methods based on polarization decomposition like Freeman decomposition [1], Krogager decomposition [2], Cloude decomposition [3], and Huynen decomposition [4] and so on; (2) methods based on statistical characteristics such as The proposal of Wishart distance and its application to PolSAR image classification [5, 6]; (3) methods based on machine learning related algorithms such as Support Vector Machine (SVM) [7], Stacked Auto Encoder

H. Ma et al. (Eds.): PRCV 2021, LNCS 13020, pp. 597–609, 2021.
https://doi.org/10.1007/978-3-030-88007-1_49

(SAE) [8, 9], Neural Networks (NN) [10], Convolutional Neural Networks (CNN) [11] and Random Forest Classifier(RFC) etc. [12].

Due to the complexity of the study area, many factors will affect the accuracy of image classification. However, different classifiers have their own advantages and disadvantages. Each kind of classifier has a certain range of classification accuracy and used fields. A conclusion that "no single classifier is universally applicable to various classification conditions" was proposed by Giacinto and Roli [13] through a comparative study on the classification results of several common classifiers using different image data and different regions. In addition to developing advanced classifiers or improving the existing classifier algorithms, the integration of multiple classifiers can make the information of the mode to be classified complementary to improve the accuracy of image classification. Therefore, combination or integration of multiple classifiers is a more effective method which can obtain better classification accuracy than single base classifier participating in the classification.

The combination of multiple classifiers is to combine single base classifiers. Through the classification performance of the base classifier and the independence between the base classifiers, the classification accuracy of unknown samples is improved. The weights of the classifiers are determined by constructing basic probability assignments for different base classifiers, and finally the classification results are determined. Based on this, some basic probability assignments based on confusion matrices have been proposed. (1) Xu et al. [14] proposed a method to construct the basic probability distribution based on the recognition rate, replacement rate and rejection rate of the confusion matrix. However, in the decision-making process of this method, the difference in the recognition ability of different classifiers for different classes is not considered. (2) Parikh et al. [15] proposed a method of constructing the basic probability distribution of the classifier based on the prior knowledge provided by the confusion matrix. Specifically, this method uses the precision of each actual class according to the confusion matrix to reflect the classifier's ability to recognize different classes. However, the prior knowledge contained in the confusion matrix also has the recall rate which is another important aspect that reflects the classifier's ability to recognize each class. (3) On this basis, Deng et al. [16] used the fusion of recall rate and precision rate of the confusion matrices to determine the recognition ability of each class of the base classifiers.

In this paper, we propose a combined classifier for PolSAR images based on evidence theory. The classifier can be divided into five basic steps: (1) Extract the training set and test set in the picture, and use different base classifiers to train the training set; (2) Construct a confusion matrix, calculate the recall matrix and the precision matrix and merge them which represent the recognition ability of different classifiers for different classes; (3) Use different base classifiers to test the same unknown test point, and combine the test results with the recognition ability in the second step to construct the basic probability assignment of the classifier to the test point; (4) Fuse all the basic probabilities of the unknown test point Assign, obtain the final classification result through the pignistic transformation; (5) Predict all the unknown samples in the test set according to the third and fourth steps. Compared with the classic classifier, the proposed classifier has the advantages of improving image classification final accuracy, recall rate and precision rate. It is proved by specific simulation results.

The paper is organized as follows. The preliminaries briefly introduce the notion of D-S evidence theory, SVM and SAE in Sect. 2. Section 3 develops a combination classifier of PolSAR based on D-S Evidence Theory. The applicability of the proposed method is demonstrated through several examples in Sect. 4. Lastly, this paper is concluded in Sect. 5.

2 Preliminaries

2.1 The Base Classifiers

SAE and SVM are the classic methods for the task of PolSAR image classification. Both of them are supervised classifiers which play an important role in the field of PolSAR data interpretation.

2.1.1 Support Vector Machine (SVM)

SVM method uses training error as optimization. The constraints of the problem are constructed with the minimization of the confidence range as the optimization objective, which greatly improves the generalization ability of learning. In addition, the solution of the SVM method is finally transformed into a quadratic programming problem or the solution of a linear equation system, ensuring the uniqueness of understanding and global optimality, and will not cause the problems of over-learning and local minima in traditional methods. It is precisely because of the excellent learning performance of the SVM method that its research has received more and more attention.

2.1.2 Stacked Auto Encoder (SAE)

SAE consists of two or more independent AEs. In fact, it increases the number of hidden layers and learns various expressions of the original data layer by layer, so as to better learn abstract feature vectors with different dimensions and levels from complex high-dimensional input data. Therefore, SAE can reduce the dimensionality of input data and has a strong ability to extract input feature values through learning.

The entire training process of SAE includes two steps: pre-training and fine-tuning. First, a single AE is a self-supervised algorithm that uses the output of each layer as the input of the next layer until all hidden layers are trained. The last layer uses Softmax regression as the classifier, uses the Error back propagation algorithm to fine-tune the entire network with labeled data, and finally outputs the classification results.

2.2 Dempster-Shafer Evidence Theory

The Dempster-Shafer evidence theory is widely used to handle uncertain information [17, 18]. In this theory, basic probability assignment (BPA) is used to represent the uncertain information, and Dempster's rule of combination is used to combine multiple BPAs. In Dempster-Shafer theory, a problem domain denoted by a finite nonempty set

Ω of nutually exclusive and exhaustive hypotheses is called the frame of discernment. Let 2^{Ω} denote the power set of Ω. A BPA is a mapping m: $2^{\Omega} \rightarrow [0, 1]$, satisfying

$$m(\varphi) = 0 \text{ and } \sum_{A \in 2^{\Omega}} m(A) = 1 \quad (1)$$

The rule of combination is defined as follows:

$$m(A) \begin{cases} \frac{1}{1-K} \sum_{B \cap C = A} m_1(B) m_2(C) & A \neq \varphi \\ 0, & A = \varphi \end{cases} \quad (2)$$

$$K = \sum_{B \cap C = \varphi} m_1(B) m_2(C) \quad (3)$$

where K is the normalization constant, called conflict coefficient of two BPAs. The Dempster's rule satisfies commutative and associative properties.

2.3 Confusion Matrix

Confusion matrix contains information about the actual and predicted classifications done by the classification system [19]. The confusion matrix has two dimensions: one id indexed by the actual category of object, the other if indexed by the category predicted by the classifier. Table 1 presents the form of confusion matrix with classes C_1, C_2, ..., C_l. In the confusion matrix, N_{ij} represents the number of samples actually belonging to class C_i but classified as class C_j. The table of confusion matrix is shown in Table 1.

Table 1. Confusion matrix.

		Predicted		
		C_1	$\cdots C_j \cdots$	C_l
Actual	C_1	N_{11}	$\ldots N_{1j} \cdots$	N_{1l}
	$\vdots$	$\vdots$	$\vdots$	$\vdots$
	C_i	N_{i1}	$\cdots N_{ij} \cdots$	N_{il}
	$\vdots$	$\vdots$	$\vdots$	$\vdots$
	C_l	N_{l1}	$\cdots N_{lj} \cdots$	N_{ll}

Based on the confusion matrix, many indicators of classification performance can be defined. Some common measures are given below.

Accuracy: the proportion of the total number of predictions that were correct.

$$accuracy = \sum_{i=1}^{n} N_{ii} \Big/ \sum_{i=1}^{n} \sum_{j=1}^{n} N_{ij} \quad (4)$$

Precision: a measure of the accuracy provided that a specific class has been predicted.

$$precision_i = N_{ii} \Big/ \sum_{k=1}^{n} N_{ki} \tag{5}$$

Recall: a measure of the ability of a prediction model to select instances of a certain class from a data set.

$$\mathrm{recall}_i = N_{ii} \Big/ \sum_{k=1}^{n} N_{ik} \tag{6}$$

3 The Proposed Approach

The proposed method completes PolSAR image classification by fusing several different classifiers. The core is to construct BPA by the ability of the classifier for each label according to the accuracy rate and recall rate of the confusion matrix. The final classification result is determined by fusing BPAs.

The specific steps of this method are as follows and the flow chart is shown in Fig. 1.

Step1: Extract the training set and test set in the picture, and use different base classifiers to train the training set;

The extracted training set is denoted as (D_{train}, L_{train}), where D_{train} is the data feature set extracted from the training set, L_{train} is the tags corresponding to the training set. The test set is represented as (D_{test}, L_{test}) where D_{test} is the data feature set extracted from the test set, L_{test} is the tags corresponding to the test set.

Step2: Construct a confusion matrix, calculate the recall matrix and the precision matrix and merge them which represent the recognition ability of different classifiers for different classes

(2a) For each sample x in $D_{training}$, whose actual class is R_s^{actual} in $L_{training}$, every base classifier among $\{\phi_1, \phi_2, \ldots, \phi_l\}$,indicated as ϕ_k can give a predicted class R_s^{pred}. We can repeatedly update the confusion matrix C_{ϕ_k} until every sample in $D_{training}$ has been predicted. As a result, $C_{\phi_1}, C_{\phi_2}, \ldots, C_{\phi_l}$ are obtained.

(2b) Base on the obtained confusion matrix $C_{\phi_k}, k = 1, 2, \ldots, l$, the precision matrix and the recall matrix are calculated by Eqs. (10) and (11) and are denoted as

$$C_{\phi_k}^{p} = \begin{bmatrix} r_{11}^{p} & r_{12}^{p} & \cdots & r_{1N}^{p} \\ r_{21}^{p} & r_{22}^{p} & \cdots & r_{2N}^{p} \\ \vdots & \vdots & \ddots & \vdots \\ r_{N1}^{p} & r_{N2}^{p} & \cdots & r_{NN}^{p} \end{bmatrix} \tag{7}$$

where $r_{ij}^{p} = \frac{n_{ij}}{\sum_{i=1}^{N} n_{ij}}$, r_{ij}^{p} is a measure of accuracy in the case of a particular category that has been predicted.

$$C_{\phi_k}^{r} = \begin{bmatrix} r_{11}^{r} & r_{12}^{r} & \cdots & r_{1N}^{r} \\ r_{21}^{r} & r_{22}^{r} & \cdots & r_{2N}^{r} \\ \vdots & \vdots & \ddots & \vdots \\ r_{N1}^{r} & r_{N2}^{r} & \cdots & r_{NN}^{r} \end{bmatrix} \tag{8}$$

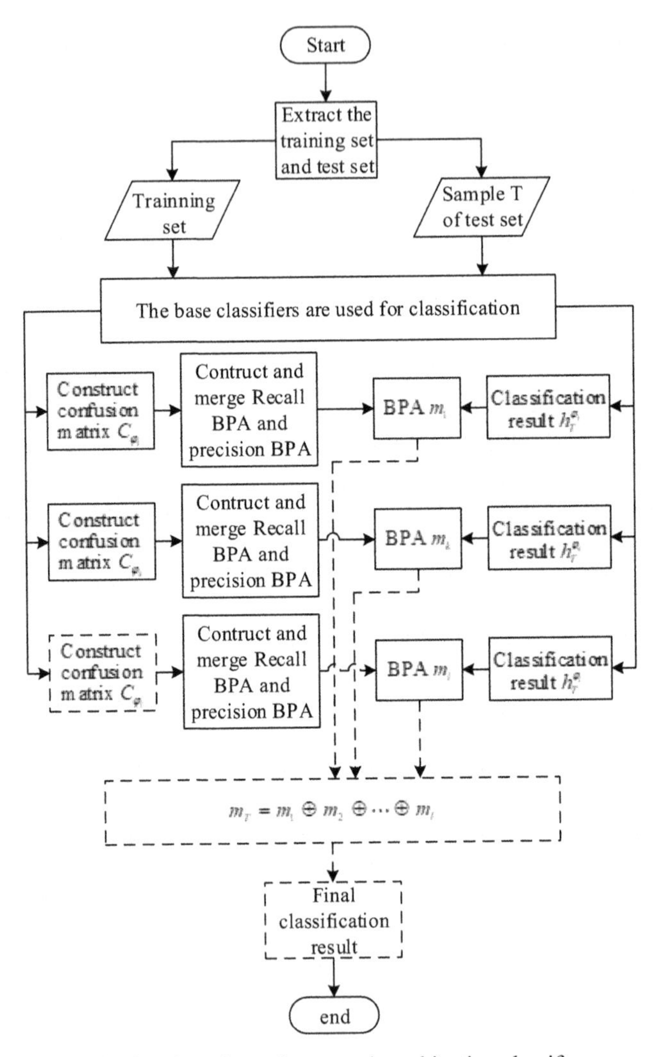

Fig. 1. Flow chart of proposed combination classifier

where $r_{ij}^r = \frac{n_{ij}}{\sum_{j=1}^{N} n_{ij}}$,$r_{ij}^r$ is a measure of a classifier's ability to select a particular instance of a class from a data set.

(2c) For each class i,$i = 1, 2, \ldots, N$, the following three formulas are used to construct the precision basic probability assign (Precision BPA).

$$m_i^p(\{c_i\}) = r_{ii} / \sum\nolimits_{j=1}^{N} r_{ji}^p,\ c_i \in \Omega \tag{9}$$

$$m_i^p(\Omega) = 1 - m_i^p(\{c_i\}),\ c_i \in \Omega \tag{10}$$

$$m_i^p(A) = 0,\ \forall A \in 2^{\Omega} \backslash \{\{c_i\}, \Omega\} \tag{11}$$

(2d) For each class i,$i = 1, 2, \ldots, N$, the following three formulas are used to construct the recall basic probability assign (Recall BPA).

$$m_i^r(\{c_i\}) = r_{ii} / \sum_{j=1}^{N} r_{ji}^r,\ c_i \in \Omega \tag{12}$$

$$m_i^r(\Omega) = 1 - m_i^r(\{c_i\}),\ c_i \in \Omega \tag{13}$$

$$m_i^r(A) = 0,\ \forall A \in 2^{\Omega} \backslash \{\{c_i\}, \Omega\} \tag{14}$$

(2e) Through the fusion rule of Dempster-Shafer evidence theory, m_i^r and m_i^p are fused to get $m_i^{\phi_k}$. That is

$$m_i^{\phi_k} = m_i^r \oplus m_i^p(A) = \frac{1}{k} \sum_{B \cap C = A} m_i^r(B) \bullet m_i^p(C) \tag{15}$$

$$K = \sum_{B \cap C \neq \Phi} m_i^r(B) \bullet m_i^p(C) = 1 - \sum_{B \cap C = \Phi} m_i^r(B) \bullet m_i^p(C) \tag{16}$$

where $B, C \in 2^{\Omega}$, K is a normalized constant. $m_i^{\phi_k}$ represents the ability of classifier ϕ_k to recognize class i.

Step3: Use different base classifiers to test the same unknown test point, and combine the test results with the recognition ability in the second step to construct the basic probability assignment of the classifier to the test point;

Each base classifier ϕ_k classifies the unclassified test sample T in D_{test}, and the predicted result is denoted as $h_T^{\phi_k}$.

Step4: Fuse all the basic probabilities of the unknown test point Assign, obtain the final classification result through the pignistic transformation;

(4a) The classification results of multiple classifiers are fused with the following formula to obtain the final BPA.

$$m_T = m_{h_T^{\phi_1}}^{\phi_1} \oplus m_{h_T^{\phi_2}}^{\phi_2} \oplus \cdots \oplus m_{h_T^{\phi_l}}^{\phi_l} \tag{17}$$

where $\oplus$ represents the BPA fusion method of Dempster-Shafer evidence theory.

(4b) The final prediction class of sample T can be determined by the formula of the pignistics transformation. That is

$$h_T^{pred} = \arg\max_x \left\{ \sum \frac{1}{|A|} \frac{m_T(A)}{1 - m_T(\varphi)} \right\},\ x \in \Omega \tag{18}$$

Step5: Predict all unknown samples in the test set according to the third and fourth steps.

In order to verify the effectiveness and feasibility of the combined classifier proposed in this paper, a simulation experiment will be performed on a real PolSAR image data. In the experiment, the classification result of the combined classifier is compared with the base classifiers, which include SVM and SAE. In the experiment, 5.76% of the total samples were selected as the training set, and the rest as the test set. The simulation of this experiment was run on Windows 10, SPI, CPU Intel(R) Core(TM) i5-9th-Gen, basic frequency 3.20 GHz, software platform Matlab R2014a, and 270 × 300 natural image Flevoland was selected for simulation.

4 Experiments with Flevoland Dataset

The PolSAR data is Flevoand, which is acquired by NASA/JPL laboratory airborne synthetic aperture radar system AIRSAR on August 16, 1989. The data is four-view L-band. The size of the original image is 750 × 1024 and the size of the captured image is 270 × 300. This image mainly contains six types of crops: Bare soil, Potatoes, Beet, Peas, Wheat, and Barley. The Pauli pseudo-color image and ground truth image are shown in the Fig. 2. The following will be explained step by step according to the method described in this paper.

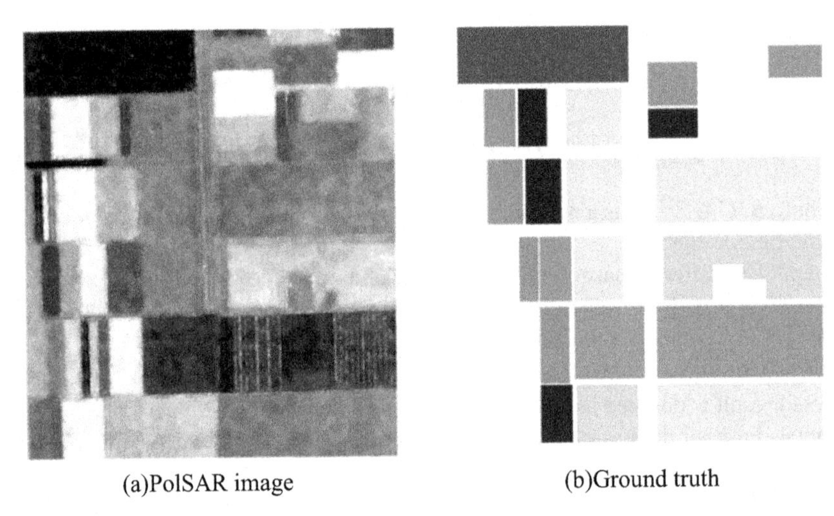

(a)PolSAR image (b)Ground truth

Fig. 2. PolSAR image and the ground truth Map.

The steps of the proposed method are as follows:

Step1: 4665 samples were selected among the 81,000 samples to form the training set. The testing set is formed by remaining samples. The 4665 training samples are trained by SVM and SAE respectively.

Step2: According to the training results, the fusion matrices of SVM and SAE are obtained respectively, which are expressed in Tables 2 and 3.

Table 2. Confusion matrix of the training set through SVM

SVM		Predicted					
		C_1	C_2	C_3	C_4	C_5	C_6
Actual	C_1	478	1	0	7	14	0
	C_2	1	552	17	6	9	8
	C_3	3	22	338	3	3	6
	C_4	0	2	8	1547	13	5
	C_5	5	9	2	22	939	6
	C_6	0	4	7	3	5	620

Table 3. Confusion matrix of the training set through SAE

SAE		Predicted					
		C_1	C_2	C_3	C_4	C_5	C_6
Actual	C_1	480	2	2	3	12	1
	C_2	2	536	24	5	11	15
	C_3	1	23	339	2	4	6
	C_4	1	7	2	1535	17	13
	C_5	6	8	6	15	945	3
	C_6	0	6	12	3	4	614

The recall matrices and precision matrices of SVM and SAE are obtained by using Eq. (7) and Eq. (8) respectively, which are denoted as:

$$C^{p}_{SVM} = \begin{bmatrix} 0.9815 & 0.0017 & 0 & 0.0044 & 0.0142 & 0 \\ 0.0021 & 0.9356 & 0.0457 & 0.0038 & 0.0092 & 0.0124 \\ 0.0062 & 0.0373 & 0.9084 & 0.0019 & 0.0031 & 0.0093 \\ 0 & 0.0034 & 0.0215 & 0.9742 & 0.0132 & 0.0078 \\ 0.0103 & 0.1525 & 0.0054 & 0.0139 & 0.9552 & 0.0093 \\ 0 & 0.0068 & 0.0019 & 0.0019 & 0.0051 & 0.9612 \end{bmatrix}$$

$$C_{SVM}^{r} = \begin{bmatrix} 0.9560 & 0.0020 & 0 & 0.0140 & 0.0280 & 0 \\ 0.0017 & 0.9309 & 0.0287 & 0.0101 & 0.0152 & 0.0135 \\ 0.0080 & 0.0587 & 0.9013 & 0.0080 & 0.0080 & 0.0160 \\ 0 & 0.0013 & 0.0051 & 0.9822 & 0.0083 & 0.0032 \\ 0.0051 & 0.0092 & 0.0020 & 0.0224 & 0.9552 & 0.0061 \\ 0 & 0.0063 & 0.0110 & 0.0047 & 0.0078 & 0.9703 \end{bmatrix}$$

$$C_{SAE}^{p} = \begin{bmatrix} 0.9796 & 0.0034 & 0.0052 & 0.0019 & 0.0121 & 0.0015 \\ 0.0041 & 0.9210 & 0.0623 & 0.0032 & 0.0111 & 0.0230 \\ 0.0020 & 0.0395 & 0.8805 & 0.0013 & 0.0040 & 0.0092 \\ 0.0020 & 0.0120 & 0.0052 & 0.9821 & 0.0171 & 0.0199 \\ 0.0122 & 0.0137 & 0.0156 & 0.0096 & 0.9517 & 0.0046 \\ 0 & 0.0103 & 0.0312 & 0.0019 & 0.0040 & 0.9417 \end{bmatrix}$$

$$C_{SAE}^{r} = \begin{bmatrix} 0.9600 & 0.0040 & 0.0040 & 0.0060 & 0.0240 & 0.0020 \\ 0.0034 & 0.9039 & 0.0404 & 0.0084 & 0.0185 & 0.0253 \\ 0.0027 & 0.0613 & 0.0613 & 0.0053 & 0.0107 & 0.0160 \\ 0.0006 & 0.0044 & 0.0044 & 0.9746 & 0.0108 & 0.0083 \\ 0.0061 & 0.0081 & 0.0081 & 0.0153 & 0.9613 & 0.0031 \\ 0 & 0.0094 & 0.0094 & 0.0047 & 0.0063 & 0.9609 \end{bmatrix}$$

Then, according to Eqs. (9) to (14), the recall BPAs and precision BPAs are obtained by the recall matrices and precision matrices. Then, the recall BPA and precision BPA are merged to get m_i^{SVM} and m_i^{SAE}, which represent the recognition ability of the classifier SVM and SAE to class i respectively.

Step3: For the sample T in the test set, SVM and SAE are used for classification. A two-tuple is formed by the prediction result and the recognition ability in step 3. The two-tuple is denoted as (h_T^{SVM}, m^{SVM}) and (h_T^{SAE}, m^{SAE}),where h_T^{SVM} and h_T^{SAE} represent the predict of classifier SVM and SAE, m^{SVM} and m^{SAE} represent the recognition ability of the classifier SVM and SAE.

Step4: Fuse SVM and SAE to obtain the final BPA denoted as $m_T = m_{h_T^{SVM}}^{SVM} \oplus m_{h_T^{SAE}}^{SAE}$, and then obtain the final prediction result through the pignistics transform.

Step5: Repeat steps 3 and 4 to predict all classification results. And the classification result of the base classifier and the classification result of the proposed combined classifier are shown in Fig. 3 and Table 4.

Three comparative experiments are used to compare with the proposed method. SVM and SAE are two base classifiers, and SSAEL is an advanced method for SAE. It can be seen that, the result of the proposed method is better in vision. The accuracy of the methods are recorded in Table 4. The highest classification accuracy of the proposed method is in black. The experimental results prove the effectiveness and feasibility of the proposed combined classifier.

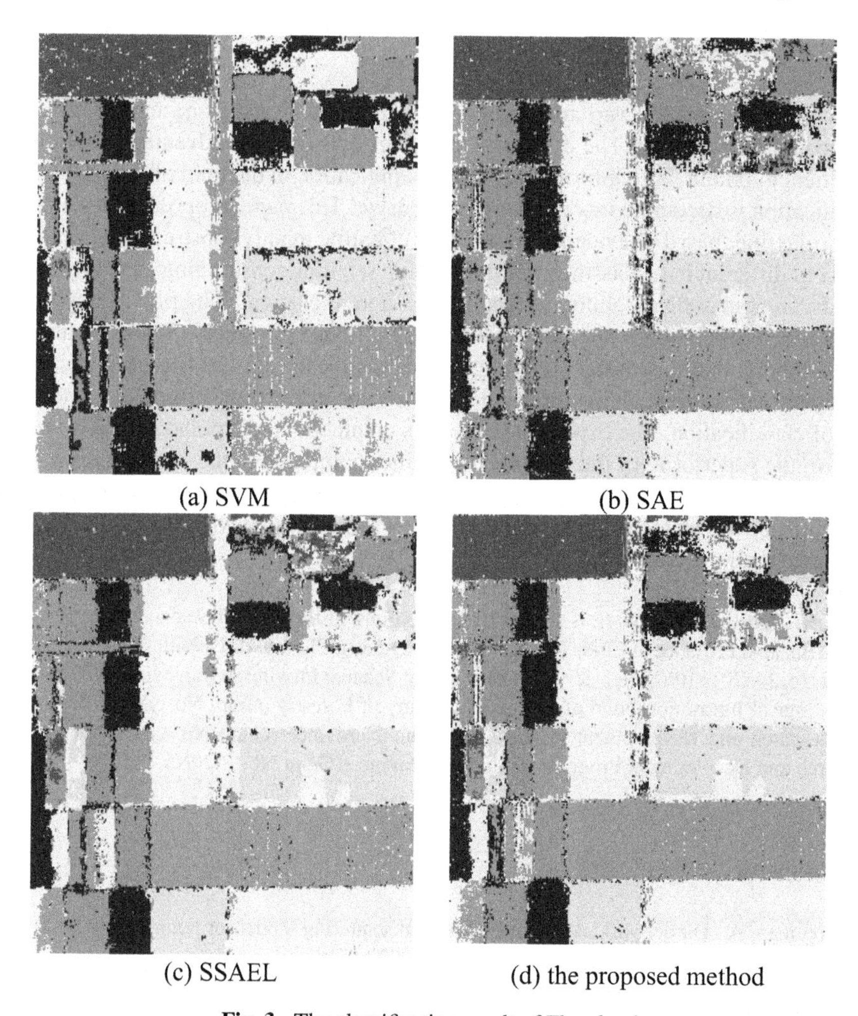

Fig. 3. The classification result of Flevoland

Table 4. The accuracy of the methods

Method	1	2	3	4	5	6	OA
SVM	0.9778	0.9160	0.8990	0.9746	0.9582	0.9731	0.9571
SAE	0.9697	0.9095	0.8663	0.9840	0.9509	0.9488	0.9515
SSAEL	0.9828	0.9304	0.9015	0.9853	0.9657	0.9407	0.9610
Proposed method	0.9818	**0.9539**	**0.9333**	0.9845	**0.9735**	**0.9723**	**0.9723**

5 Conclusion

The application range of various classifiers has limitations, and its accuracy is affected by specific applications. The combined classifier can fuse the advantages of various classifiers to obtain more information, and its application in the field of PolSAR image classification is becoming more and more extensive. This paper proposes a new combination classifier based on evidence theory. This classifier mainly constructs the confusion matrix of different base classifiers for the training set, thus constructing BPA to describe the "different classifiers' ability to recognize different classes". The final classification result is determined by fusing the final BPA of all base classifiers. The innovation of this method is that it considers the influence of the recall rate and accuracy rate in the confusion matrix on the recognition ability at the same time, thereby improving the accuracy of classification. The experimental results obtained on real polarized SAR images confirm the superiority of the combined classifier compared to the base classifier. One of the limitations is that because it is a combined classifier, each base classifier needs to make predictions on all training samples and test samples, so it will consume more time compared with the base classifier.

Acknowledgments. The work is supported by the Fund of the Natural Science Basic Research Plan in Shaanxi Province of China (Grant No. 2019JQ-539), Chinese Universities Scientific Found (Grant No. 2452018106), High-level Talents Fund of Shaanxi Province (Grant No. F2020221001), Technological Innovation Fund of Shaanxi Academy of Forestry (Grant No. SXLK2021-0215), Key Research and Development Program of Shaanxi province (Grant No. 2021NY-179), Key Research and Development Program of Shaanxi province (Grant No. 2020NY-205).

References

1. Freeman, A., Durden, S.L.: A three-component scattering model for polarimetric SAR data. IEEE Trans. Geosci. Remote Sens. **36**(3), 963–973 (1998)
2. Krogager, E.: New decomposition of the radar target scattering matrix. Electron. Lett. **26**(18), 1525–1527 (2002)
3. Cloude, S.R.: An entropy based classification scheme for polarimetric SAR data. In: Geoscience and Remote Sensing Symposium, 1995. IGARSS 1995. 'Quantitative Remote Sensing for Science and Applications'. International IEEE Xplore (1995)
4. Huynen, J.R.: Physical reality of radar targets. In: Proceedings of SPIE - The International Society for Optical Engineering (1993)
5. Jong-Sen Lee, K.W., Hoppel, S.A., Mango, A.R.M.: Intensity and phase statistics of multilook polarimetric and interferometric SAR imagery. IEEE Trans. Geosci. Remote Sens. **32**(5), 1017–1028 (1994)
6. Lee, J.S, Grunes, M.R.: Classification of multi-look polarimetric SAR imagery based on complex wishart distribution. Int. J. Remote Sens. (1992)
7. Vapnik, V.N.: The Nature of Statistical Learning Theory. Springer, New York (1995). https://doi.org/10.1007/978-1-4757-2440-0
8. Shang, R., Liu, Y., Wang, J., Jiao, L., Stolkin, R.: Stacked auto-encoder for classification of polarimetric SAR images based on scattering energy. Int. J. Remote Sens. **40**(13–14), 1–27 (2019)
9. Lecun, Y., Bengio, Y., Hinton, G.: Deep learning. Nature **2015**(521), 634–644 (2015)

10. Tzeng, Y.C., Chen, K.S.: A fuzzy neural network to SAR image classification. IEEE Trans. Geosci. Remote Sens. **36**(1), 301–307 (1998)
11. Zhao, F., Ma, G., Xie, W., Liu, H.: Semi-supervised recurrent complex-valued convolution neural network for polsar image classification. In: IGARSS 2019 - 2019 IEEE International Geoscience and Remote Sensing Symposium. IEEE (2019)
12. Peijun, D., Samat, A., Waske, B., Liu, S., Li, Z.: Random forest and rotation forest for fully polarized SAR image classification using polarimetric and spatial features. ISPRS J. Photogr. Remote Sens. **105**, 38–53 (2015)
13. Giacinto, G., Roli, F., Fumera, G.: Selection of image classifiers. Electron. Lett. **36**(5), 420 (2000)
14. Xu, L., Krzyzak, A.: Methods of combining multiple classifiers and their applications to handwriting recognition. IEEE Trans. Cybernet. **22**(3), 418–435 (1992)
15. Parikh, C.R., Pont, M.J., Jones, N.B.: Application of dempster-shafer theory in condition monitoring applications: a case study. Pattern Recogn. Lett. **22**(6–7), 777–785 (2001)
16. Deng, X., Liu, Q., Deng, Y., Mahadevan, S.: An improved method to construct basic probability assignment based on the confusion matrix for classification problem. Inf. Sci. **340–341**, 250–261 (2016)
17. Dempster, A.P.: Upper and lower probabilities induced by a multivalued mapping. Ann. Math. Stat. **38**(2), 325–339 (1967)
18. Shafer, G.: A mathematical theory of evidence. Technometrics **20** (1976)
19. Sammut, C., Webb, G.I. (eds.): Encyclopedia of Machine Learning. Springer, Boston (2010). https://doi.org/10.1007/978-0-387-30164-8

Image Tampering Localization Using Unified Two-Stream Features Enhanced with Channel and Spatial Attention

Haodong Li[1,2], Xiaoming Chen[1,2], Peiyu Zhuang[1,2], and Bin Li[1,2](✉)

[1] Guangdong Key Laboratory of Intelligent Information Processing and Shenzhen Key Laboratory of Media Security, Shenzhen University, Shenzhen 518060, China
[2] Shenzhen Institute of Artificial Intelligence and Robotics for Society, Shenzhen 518129, China
{lihaodong,libin}@szu.edu.cn

Abstract. Image tampering localization has attracted much attention in recent years. To differentiate between tampered and pristine image regions, many methods have increasingly leveraged the powerful feature learning ability of deep neural networks. Most of these methods operate on either spatial image domain directly or residual image domain constructed with high-pass filtering, while some take inputs from both domains and fuse the features just before making decisions. Though they have achieved promising performance, the gain of integrating feature representations of different domains is overlooked. In this paper, we show that learning a unified feature set is beneficial for tampering localization. In the proposed method, low-level features are firstly extracted from two input streams: one is a spatial image, and the other is a high-pass filtered residual image. The features are then separately enhanced with channel attention and spatial attention, and are subsequently subjected to an early-fusion to form a unified feature representation. The unified features play an important role under an adapted Mask R-CNN framework, achieving more accurate pixel-level tampering localization. Experimental results on five tampered image datasets have shown the effectiveness of the proposed method. The implementation is available at https://github.com/media-sec-lab/AEUF-Net.

Keywords: Image forgery localization · Feature fusion · Attention mechanism · Mask R-CNN

1 Introduction

The rapid developments of image editing technology and software have facilitated the manipulation of digital images. As a result, there is an increasing emergence of tampered images, which contain visually imperceptible artifacts. This leads to a series of security concerns and issues, including fake news, Internet rumors, falsified academic results, and fabricated court evidences [32]. Consequently, it is of great necessity to assess the image authenticity, and thus image forensics has received more and more attention.

H. Ma et al. (Eds.): PRCV 2021, LNCS 13020, pp. 610–622, 2021.
https://doi.org/10.1007/978-3-030-88007-1_50

Typically, there are three types of operations for image manipulation, namely, splicing, copy-move, and removal [16]. In order to identify the tampered regions, many tampering localization methods have been developed [27]. While most of the conventional methods are based on hand-crafted features, the recently proposed ones have resorted to deep learning (DL) techniques which have achieved outstanding success in many fields. The DL-based frameworks used in image forensics are usually adapted from the convolutional neural networks (CNN) designed for computer vision tasks, especially for object detection and semantic segmentation. For example, Zhu *et al.* [36] use a fully convolutional network (FCN) on spatial image to reveal image regions tampered with patch-based inpainting, and Salloum *et al.* [26] design a multi-task FCN to capture the traces of the surface and edge of tampered regions. Zhuang *et al.* [37] propose to incorporate dense shortcut connections into FCN to improve the tampering localization performance. Nevertheless, image forensics has its own peculiarity compared to compute vision, as the tampering traces are weak and usually imperceptible [3]. To enhance tampering traces, a typical processing in image forensics (*i.e.*, applying high-pass filtering to extract image residuals [8,24]) has been adopted in some DL-based methods [4,33]. As an alternative to simply using fixed high-pass filtering kernels for residuals extraction, a constrained convolutional layer [2] is proposed, which can adaptively learn the kernel weights. This approach has also been employed in [29,30] and achieved good results. To better capture the tampering traces from tampered regions with different sizes, Hu *et al.* [14] and Yin *et al.* [31] exploit multi-scale information. In addition, some approaches try to expose the tampered regions by analyzing inconsistencies related to image source [5,15] or editing history [20,21]. The existing methods have shown that information in both spatial and residual domains is useful for tampering localization, while most of them only extract features from either domain, meaning that the obtained features are homogeneous and have limited representational ability.

To improve the feature diversity, some works take inputs from multiple domains and learned heterogeneous features. Bappy *et al.* [1] propose a hybrid LSTM (Long and Short-Term Memory) and encoder-decoder framework for tampering localization, where spatial feature and resampling feature are involved together. Wu *et al.* [29] use convolutions with SRM kernels [8], constrained kernels [2], and conventional kernels in the first layer of the network and perform tampering localization based on features produced by the three types of convolutions. Zhou *et al.* [35] develop a two-stream Faster R-CNN network [25], called RGB-N, where both RGB information in spatial domain and image noise in residual domain are considered. Although features from different pipelines are learned in these works, they are simply chosen or fused just before making decisions. Take RGB-N for an example. Even though the features from two streams are jointly used for classification, only the features from RGB stream are used for region proposal and bounding-box prediction which achieve tampering localization. The features may not be fully exploited for locating forged regions in such a way.

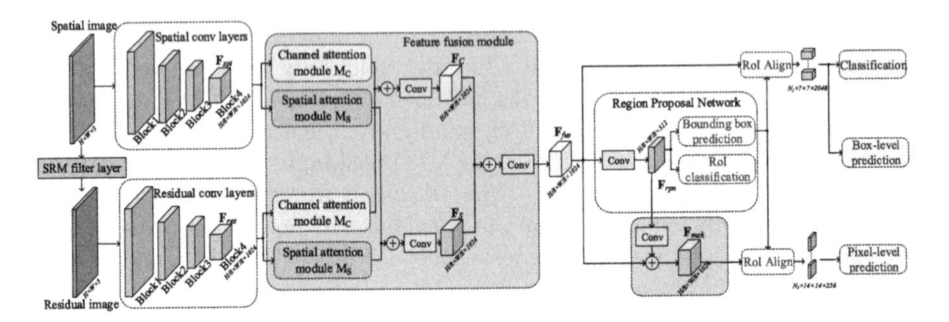

Fig. 1. The overview of the proposed model.

In this paper, we propose a tampering localization framework with unified feature representation. The proposed method adopts two typical types of input, that is, the spatial image and its filtered residual image. To obtain a unified representation, features are first separately extracted from each input stream with CNNs, and then enhanced by a dual-attention mechanism, including channel attention and spatial attention. The attention-enhanced features are finally fused together to form a unified feature representation, which is fed to an adapted Mask R-CNN [12] structure, where the features are processed by a Region Proposal Network (RPN) for tampering localization. Different from RGB-N, the proposed method learns a unified feature representation rather than using features from different pipelines separately, and it employs attention mechanism to make the learned features focus on important information. Additionally, as a pixel-level prediction branch is adopted in the adapted Mask R-CNN structure, the proposed method achieves finer localization result than RGB-N which is based on Faster R-CNN. Experimental results on five tampered image datasets have shown the superiority of the proposed method.

In the rest of this paper, we will elaborate the proposed method in Sect. 2 and present the experiments in Sect. 3. The conclusions will be drawn in Sect. 4.

2 Proposed Method

2.1 Overall Framework

The general framework of the method is shown in Fig. 1. As performed in RGB-N [35], the spatial stream extracts the spatial features $\mathbf{F}_{spt}$ from an RGB image to detect unnatural transitions at the tampered boundaries, while the residual stream extracts the residual features $\mathbf{F}_{res}$ from the filtered image to detect the statistics corrupted by image tampering operations. The feature extraction modules for both streams are constructed with the layers from `Conv1` to `Conv4_x` in ResNet-101 [13]. Different from RGB-N, we obtain a unified representation $\mathbf{F}_{fus}$ by designing an attention-based feature fusion module to combine the two-stream features together. The use of attention mechanism can motivate the fea-

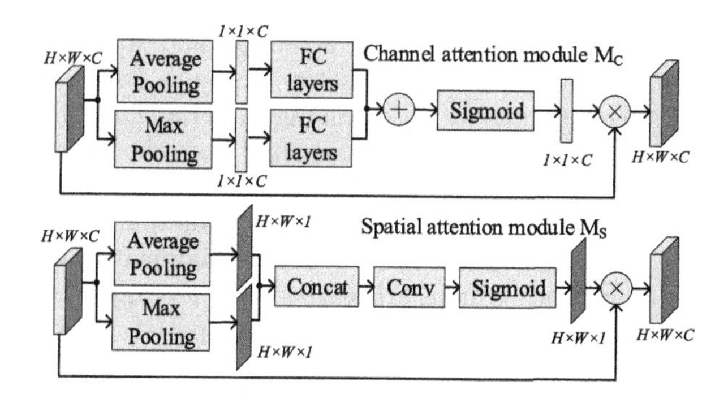

Fig. 2. Diagram of channel attention and spatial attention.

ture extraction network to focus more on the tampered regions so that more discernible features can be learned. The unified features are then sent to an adapted Mask R-CNN framework, which has three output branches at the tail, performing classification, box-level prediction, and pixel-level prediction tasks, respectively. To accomplish these tasks, region of interests (RoIs) are first extracted by an RPN, and then the features in each RoI are cropped and resized to the same size via RoI Align [12]. Similar to the constrained R-CNN [30], the numbers of extracted RoIs are $N_1 = 128$ and $N_2 = 4$ for the box-level and pixel-level prediction branches, respectively. Please note that a pixel-level prediction branch (mask branch) previously used for object segmentation is hereby applied for pixel-level tampering prediction. This is another major improvement of our method over RGB-N. Particularly, in order to facilitate the feature learning in the pixel-level prediction branch, the unified features $\mathbf{F}_{fus}$ are combined with the features $\mathbf{F}_{rpn}$ learned by RPN to form an enhanced representation $\mathbf{F}_{msk}$. For efficient network learning, we first pre-train the model with a synthetic dataset [35] by enabling only the classification and box-level prediction branches. Then, to obtain the final pixel-level localization results, we apply the pre-trained weights to the model and train all the three output branches end-to-end with realistic tampered datasets.

2.2 Attention-Based Feature Fusion Module

Using multiple-stream inputs in a single model can enhance the feature diversity but may cause distress in feature selection for different sub-tasks at the same time. Hence, we propose a feature fusion module to yield a unified representation of tampered features. As the tampering traces are usually weak and locate at only some regions within an image, it is necessary to guide the network to learn discriminative feature representations from target of interest. To this end, the attention mechanisms, which can improve the network's ability on being more aware of the target of interest by learning to emphasize or suppress information, are adopted for extracting features adaptively.

Two types of attention (*i.e.*, channel attention and spatial attention) are involved in the proposed attention-based feature fusion module. The detail structures of the two kinds of modules are shown in Fig. 2. In the channel attention module, a channel-wise weighting vector is constructed from the input feature maps, and then the feature maps are multiplied with the weighting vector. In this way, the channel attention module can make the network focus on which channels are more important. In the spatial attention module, a weighting matrix along the width and height of the input feature maps are learned, and the input feature maps are multiplied with the weighting matrix. As a result, the spatial attention module can make the network focus on which locations are more important.

In the proposed feature fusion module, the spatial features $\mathbf{F}_{spt}$ and residual features $\mathbf{F}_{res}$ are first enhanced with both attention modules, respectively. Then, the same kind of attention-enhanced features are combined via an element-wise addition followed by a convolution. Finally, the unified features $\mathbf{F}_{fus}$ is obtained by performing combination operation on two different kinds of attention-enhanced features. Such a feature fusion process can be expressed by the following formulas:

$$\mathbf{F}_c = \mathrm{Conv}\left(\mathrm{M}_c(\mathbf{F}_{spt}) + \mathrm{M}_c(\mathbf{F}_{res})\right), \tag{1}$$

$$\mathbf{F}_s = \mathrm{Conv}\left(\mathrm{M}_s(\mathbf{F}_{spt}) + \mathrm{M}_s(\mathbf{F}_{res})\right), \tag{2}$$

$$\mathbf{F}_{fus} = \mathrm{Conv}\left(\mathbf{F}_c + \mathbf{F}_s\right), \tag{3}$$

where Conv represents a 1×1 convolution, and M_c and M_s represent the operations in the channel attention and spatial attention, respectively. In this way, the features learned from both the spatial stream and residual stream can be well fused into a unified representation, namely, $\mathbf{F}_{fus}$, which is beneficial for different subsequent sub-tasks such as tampering classification and localization.

2.3 Feature Integration for Pixel-Level Prediction

In order to obtain finer localization results, a pixel-level prediction branch is added to achieve pixel-level output, which has not been considered in the previous work RGB-N [35]. Intuitively, pixel-level prediction requires more versatile features compared to box-level prediction. To improve the performance of the pixel-level prediction branch, we further enhance the unified features $\mathbf{F}_{fus}$ through integrating the features learned by the RPN (*i.e.*, $\mathbf{F}_{rpn}$) into $\mathbf{F}_{fus}$. The underlying consideration is that the role of RPN is to select suspected tampered regions as RoIs, and it ought to learn the differences between the tampered and pristine regions; thus, the knowledge learned in RPN network can be used to guide the feature learning in the pixel-level prediction branch.

There are two typical ways to perform feature integration (*i.e.*, concatenation and addition). Our experimental results (Sect. 3.3) have shown that combining $\mathbf{F}_{fus}$ and $\mathbf{F}_{rpn}$ in an additive manner is more suitable. Specifically, the number of channels of $\mathbf{F}_{rpn}$ is transited to the same of $\mathbf{F}_{fus}$ through 1×1 convolution, and then the resulting features are element-wise added to $\mathbf{F}_{fus}$, producing the enhanced features $\mathbf{F}_{msk}$. This process can be represented as:

$$\mathbf{F}_{msk} = \mathbf{F}_{fus} + \mathrm{Conv}(\mathbf{F}_{rpn}). \tag{4}$$

Such a feature integration process can increase the information interaction between the RPN network and the subsequent pixel-level prediction network, thereby improving the tampering localization accuracy.

2.4 Loss Functions

The total loss of the proposed model is a summation of the loss of the RPN network and the loss of the three branches at the tail, that is,

$$L_{out} = L_{rpn} + L_{ce} + L_{box} + L_{pix}, \tag{5}$$

where L_{ce} denotes the classification cross-entropy loss for determining whether a region has been tampered with or not, $L_{box} = \Sigma_i smooth_{L_1}(t_i, t_i^*)$ denotes the smoothed L_1 loss [9] for bounding box regression (t_i and t_i^* are respectively the predicted coordinates of each bounding box and their truth values), and L_{pix} denotes the cross-entropy loss obtained by comparing the predicted mask to the pixel-level ground-truth labels. The loss of the RPN network is defined as:

$$L_{rpn} = \frac{1}{N}\sum_{i=1}^{N} L_{cls}\left(g_i, g_i^*\right) + \frac{1}{N}\sum_{i=1}^{N} g_i^* L_{reg}\left(r_i, r_i^*\right), \tag{6}$$

where $N = 128$, i denotes the index of an anchor, L_{cls} denotes the cross-entropy loss function used for RoI classification, g_i denotes the probability for an anchor containing a tampered region, g_i^* denotes the true label of an anchor, L_{reg} denotes the smoothed L_1 loss [9] for bounding box regression, and r_i and r_i^* denote the four parameterized coordinate vectors of each anchor bounding box [10] predicted by the RPN network and their truth values, respectively.

3 Experiments

3.1 Experimental Settings

Training and Testing Protocol: As there is insufficient realistic tampered data publicly available for training a deep neural network model, we followed RGB-N [35] to pre-train the proposed model on a synthetic dataset, which was generated from the COCO dataset [18] and contained 10,761 training images. In the training phase, the weights of the spatial feature extraction module were initialized with those in ResNet-101 pre-trained on ImageNet, while the rest weights were randomly initialized. The batch size was set as 128, and the maximum number of iterations was 110,000. The learning rate was initialized as 10^{-3} and reduced to 10^{-4} and 10^{-5} after the 40,000 and 90,000 iterations, respectively. After pre-training on the COCO synthetic dataset, the model was fine-tuned and then tested on five realistic tampered image datasets, including NIST Nimble 2016 (NIST16) [11], Coverage [28], CASIA [6], Columbia [22], and IMD2020 [23]. All experiments were carried out on a NVIDIA 1080Ti GPU. The implementation of our method is available at https://github.com/media-sec-lab/AEUF-Net.

Table 1. Average precision of different network structures.

	RGB-N [35]	Block3_Fuse	Block3_PA	Block4_Fuse	Block4_PA
AP	0.765	0.782	0.794	**0.852**	0.843

Evaluation Metrics: We use average precision (AP) and F1-score as the evaluation metrics for box-level prediction and pixel-level prediction, respectively.

The AP is defined as

$$\text{AP} = \sum_{k=1}^{N} p(k)\Delta r(k), \tag{7}$$

where N denotes the number of images, $p(k)$ denotes the accuracy rate when k tampered images are recognized, and $\Delta r(k)$ denotes the amount of change in recall as the number of tampered images identified changes from $k-1$ to k.

The definition of F1-score is given by

$$\text{F}_1 = \frac{2\text{TP}}{2\text{TP} + \text{FN} + \text{FP}}, \tag{8}$$

where TP (true positive), FN (false negative), and FP (false positive) denote the numbers of correctly classified tampered pixels, misclassified tampered pixels, and misclassified pristine pixels, respectively.

3.2 Performance of the Pre-trained Model

In this experiment, we evaluate the performance of the pre-trained model with 1072 testing images in the COCO synthetic dataset. The APs obtained by different network architectures are shown in Table 1. In this table, "RGB-N" denotes the results obtained by the method [35], and "Conv3_*" and "Conv4_*" mean that the feature fusion is performed after the last layer of `Conv3_x` and `Conv4_x`, respectively. Two types of feature fusion are considered: one is the proposed dual attention based module, denoted with a suffix "Fuse"; the other one is to multiply $\mathbf{F}_{spt}$ by a weighting matrix generated from $\mathbf{F}_{res}$ with pixel-attention [34], denoted with a suffix "PA". From Table 1, it can be observed that the proposed fusion scheme with feature fusion after the fourth residual block achieves the best performance.

3.3 Localization Performance for Realistic Datasets

In this experiment, we evaluate the localization performance after fine-tuning the pre-trained model with some realistic tampered image datasets. As done in [35], for the NIST16 and Coverage datasets, the fine-tuning and testing were carried out separately on each dataset; for the CASIA v1 and Columbia datasets, the testing were performed with a model fine-tuned with the CASIA v2 dataset. The hyper-parameters used for different datasets in the fine-tuning stage are summarized in Table 2.

Table 2. Hyper-parameters used in the fine-tuning stage. "LR" means learning rate, and the format of LR decay is "LR after decay"@"the decay step".

Dataset	Max Iteration	Initial LR	LR decay #1	LR decay #2
NIST16	60k	10^{-3}	10^{-4}@30k	10^{-5}@50k
Coverage	6k	10^{-3}	10^{-4}@4k	n/a
CASIA v2	110k	10^{-3}	10^{-4}@40k	10^{-5}@90k
IMD2020	60k	10^{-3}	10^{-4}@30k	10^{-5}@50k

Table 3. F1-scores for different tampered datasets. The bold results means the bests ones, and the underlined results means they were obtained without fine-tuning.

	NIST16	Coverage	CASIA v1	Columbia	Average
ELA [17]	<u>0.236</u>	<u>0.222</u>	<u>0.214</u>	<u>0.470</u>	0.286
NOI1 [19]	<u>0.285</u>	<u>0.269</u>	<u>0.263</u>	<u>0.574</u>	0.348
CFA1 [7]	<u>0.174</u>	<u>0.190</u>	<u>0.207</u>	<u>0.467</u>	0.260
MFCN [26]	<u>0.571</u>	-	<u>0.541</u>	<u>0.612</u>	0.575
RGB-N [35]	0.722	0.437	<u>0.408</u>	<u>0.697</u>	0.566
LSTM-EnDec [1]	0.536	0.397	<u>0.302</u>	<u>0.496</u>	0.433
SPAN [14]	0.582	0.558	<u>0.382</u>	<u>0.815</u>	0.584
HLNet [31]	0.756	0.435	0.548	**0.902**	0.641
Dense-FCN [37]	0.740	0.736	<u>0.556</u>	<u>0.638</u>	0.668
Mask_fus	0.906	0.773	<u>0.584</u>	<u>0.761</u>	0.756
Mask_concat	0.905	0.789	<u>0.584</u>	<u>0.764</u>	0.761
Proposed	**0.915**	**0.794**	<u>**0.585**</u>	<u>0.787</u>	**0.770**

We conducted extensive comparisons by including 3 conventional tampering localization methods (*i.e.*, ELA [17], NOI1 [19], and CFA1 [7]) and 6 DL-based methods (*i.e.*, MFCN [26], RGB-N [35], LSTM-EnDec [1], SPAN [14], HLNet [31], and Dense-FCN [37]). The comparative results are shown in Table 3. In this table, the F1-scores of MFCN [26], RGB-N [35], SPAN [14] and HLNet [31] are those reported in the literatures [14] and [31], where "-" means that the corresponding results were not provided. For the rest methods, we used their publicly available implementations to obtain the results. As the training and testing protocols of the involved methods are not always the same, we underline the results if the model used for testing is not fine-tuned with any image in the corresponding dataset[1]. It is observed that the proposed model yields better results on the NIST16, Coverage and CASIA v1 datasets. For the Columbia dataset, the proposed method underperforms HLNet since the latter one has performed fine-tuning on this dataset, while we directly used the model fine-tuned with CASIA v2 to test the Columbia

[1] The results of conventional methods are all underlined, since these methods are unsupervised ones.

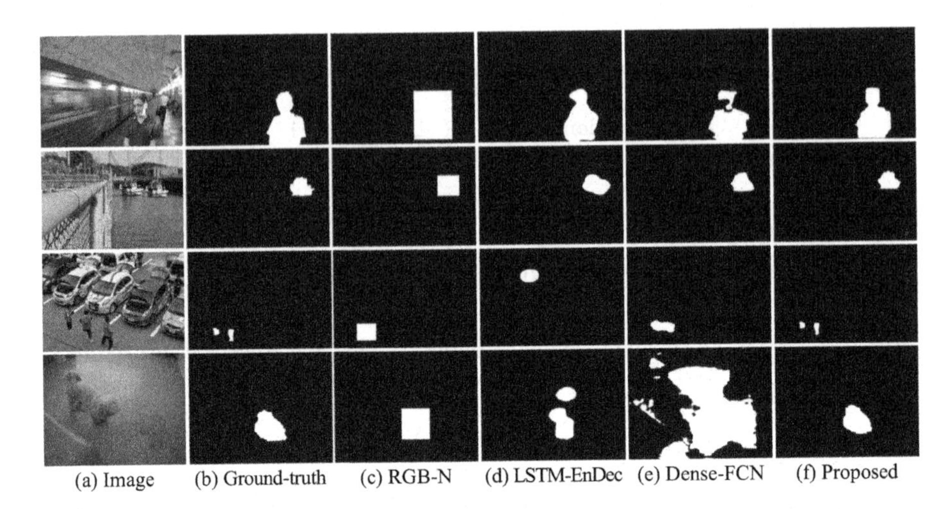

Fig. 3. Qualitative visualization of the tampering localization results.

Table 4. F1-scores for IMD2020 dataset.

	RGB-N [35]	LSTM-EnDec [1]	Dense-FCN [37]	Proposed
F1-score	0.405	0.325	0.536	**0.589**

dataset. On average, the proposed method outperforms the competitors with a margin larger than 0.1 in terms of F1-score. To visualize the tampering localization results, we show some examples in Fig. 3. Compared with RGB-N, LSTM-EnDec, and Dense-FCN, it is observed the proposed method can provided much finer and more accurate localization results.

In addition, we have compared with two variants of the features used in the pixel-level prediction branch. In Table 3, "Mask_fus" means that only the fused features $\mathbf{F}_{fus}$ are directly used for RoI Align, while "Mask_Concat" means that the concatenated features of $\mathbf{F}_{rpn}$ and $\mathbf{F}_{fus}$ are used. Through this study, we show that integrating the RPN features $\mathbf{F}_{rpn}$ and the fused features $\mathbf{F}_{fus}$ in an additive manner as shown in Fig. 1 yields better features for the pixel-level prediction task.

Since the experimental settings of different methods in Table 3 are not exactly the same, we further conducted a fairer comparison by using the IMD2020 dataset. To this end, we randomly selected 1507 tampered images from the IMD2020 dataset as the training set and 503 tampered images as the testing set, and then fine-tuned and tested the models of different methods with the same image splits. The used hyper-parameters are shown in the last row of Table 2. We have compared the proposed method with RGB-N, LSTM-EnDec, and Dense-FCN (the best one among the competitors in Table 3). As shown in Table 4, the proposed method also achieves the best F1-score in this case.

3.4 Robustness Analysis

In this experiment, we evaluate the robustness of the proposed method. We consider two common types of post-processing, *i.e.*, JPEG compression and resizing, where two JPEG quality factors (QFs) and resizing factors are included, respectively. By testing the post-processed images in NIST16 with the models fine-tuned in the previous experiment, we obtained the results as shown in Table 5 and Table 6. The experimental results show that our method outperforms the others, indicating that it is more robust to these post-processing.

Table 5. F1-scores for NIST16 dataset with post JPEG compression.

JPEG QF	100	70	50
RGB-N [35]	0.722	0.677	0.677
LSTM-EnDec [1]	0.536	0.536	0.523
Dense-FCN [37]	0.740	0.739	0.730
Proposed	**0.915**	**0.898**	**0.895**

Table 6. F1-scores for NIST16 dataset with post resizing.

Resizing factor	1.0	0.7	0.5
RGB-N [35]	0.722	0.689	0.681
LSTM-EnDec [1]	0.536	0.536	0.523
Dense-FCN [37]	0.740	0.739	0.730
Proposed	**0.915**	**0.897**	**0.882**

4 Conclusions

Feature representation is essential in image tampering localization. In this paper, we propose a method to generate a unified feature representation from two-stream input after enhancing the features with dual attention mechanism. The fused features are then used for tampering localization with the Mask R-CNN framework, achieving significant performance improvements on five realistic tampered image datasets. In the future, we will focus on designing more effective image tampering detection/localization frameworks and enhancing the feature representations among different tasks, so as to further improve the efficiency and accuracy of image tampering localization.

Acknowledgement. This work was supported in part by NSFC (Grant 61802262 and Grant 61872244), Guangdong Basic and Applied Basic Research Foundation (Grant 2019B151502001), Shenzhen R&D Program (Grant JCYJ20200109105008228), and in part by the Alibaba Group through Alibaba Innovative Research (AIR) Program.

References

1. Bappy, J.H., Simons, C., Nataraj, L., Manjunath, B., Roy-Chowdhury, A.K.: Hybrid LSTM and encoder-decoder architecture for detection of image forgeries. IEEE Trans. Image Process. **28**(7), 3286–3300 (2019)
2. Bayar, B., Stamm, M.: Constrained convolutional neural networks: a new approach towards general purpose image manipulation detection. IEEE Trans. Inf. Forensics Secur. **13**(11), 2691–2706 (2018)
3. Camacho, I.C., Wang, K.: Data-dependent scaling of CNN's first layer for improved image manipulation detection. In: 19th International Workshop on Digital-forensics and Watermarking (2020)
4. Cozzolino, D., Verdoliva, L.: Single-image splicing localization through autoencoder-based anomaly detection. In: IEEE International Workshop on Information Forensics and Security, pp. 1–6 (2016)
5. Cozzolino, D., Verdoliva, L.: Noiseprint: a CNN-based camera model fingerprint. IEEE Trans. Inf. Forensics Secur. **15**, 144–159 (2019)
6. Dong, J., Wang, W., Tan, T.: CASIA image tampering detection evaluation database. In: IEEE China Summit and International Conference on Signal and Information Processing, pp. 422–426 (2013)
7. Ferrara, P., Bianchi, T., Rosa, A., Piva, A.: Image forgery localization via fine-grained analysis of CFA artifacts. IEEE Trans. Inf. Forensics Secur. **7**(5), 1566–1577 (2012)
8. Fridrich, J., Kodovsky, J.: Rich models for steganalysis of digital images. IEEE Trans. Inf. Forensics Secur. **7**(3), 868–882 (2012)
9. Girshick, R.: Fast R-CNN. In: IEEE International Conference on Computer Vision (ICCV), pp. 1440–1448 (2015)
10. Girshick, R., Donahue, J., Darrell, T., Malik, J.: Rich feature hierarchies for accurate object detection and semantic segmentation. In: IEEE Conference on Computer Vision and Pattern Recognition (CVPR), pp. 580–587 (2014)
11. Guan, H., et al.: MFC datasets: large-scale benchmark datasets for media forensic challenge evaluation. In: IEEE Winter Conference on Applications of Computer Vision Workshops (WACVW), pp. 63–72 (2019)
12. He, K., Gkioxari, G., Dollár, P., Girshick, R.: Mask R-CNN. In: IEEE International Conference on Computer Vision (ICCV), pp. 2961–2969 (2017)
13. He, K., Zhang, X., Ren, S., Sun, J.: Deep residual learning for image recognition. In: IEEE Conference on Computer Vision and Pattern Recognition (CVPR), pp. 770–778 (2016)
14. Hu, X., Zhang, Z., Jiang, Z., Chaudhuri, S., Yang, Z., Nevatia, R.: SPAN: spatial pyramid attention network for image manipulation localization. In: Vedaldi, A., Bischof, H., Brox, T., Frahm, J.-M. (eds.) ECCV 2020. LNCS, vol. 12366, pp. 312–328. Springer, Cham (2020). https://doi.org/10.1007/978-3-030-58589-1_19
15. Huh, M., Liu, A., Owens, A., Efros, A.A.: Fighting fake news: image splice detection via learned self-consistency. In: Ferrari, V., Hebert, M., Sminchisescu, C., Weiss, Y. (eds.) ECCV 2018. LNCS, vol. 11215, pp. 106–124. Springer, Cham (2018). https://doi.org/10.1007/978-3-030-01252-6_7
16. Joseph, R., Chithra, A.: Literature survey on image manipulation detection. International Research Journal of Engineering and Technology **2**(04) (2015). 2395–0056
17. Krawetz, N., Solutions, H.F.: A picture's worth. Hacker Fact. Solutions **6**(2), 2 (2007)

18. Lin, T.-Y., et al.: Microsoft COCO: common objects in context. In: Fleet, D., Pajdla, T., Schiele, B., Tuytelaars, T. (eds.) ECCV 2014. LNCS, vol. 8693, pp. 740–755. Springer, Cham (2014). https://doi.org/10.1007/978-3-319-10602-1_48
19. Mahdian, B., Saic, S.: Using noise inconsistencies for blind image forensics. Image Vis. Comput. **27**(10), 1497–1503 (2009)
20. Mayer, O., Stamm, M.C.: Forensic similarity for digital images. IEEE Trans. Inf. Forensics Secur. **15**, 1331–1346 (2019)
21. Mayer, O., Stamm, M.C.: Exposing fake images with forensic similarity graphs. IEEE J. Sel. Top. Sig. Process. **14**(5), 1049–1064 (2020)
22. Ng, T.T., Hsu, J., Chang, S.F.: Columbia image splicing detection evaluation dataset (2009). http://www.ee.columbia.edu/ln/dvmm/downloads/authspliceddataset/authspliceddataset.htm
23. Novozamsky, A., Mahdian, B., Saic, S.: IMD2020: a large-scale annotated dataset tailored for detecting manipulated images. In: IEEE Winter Conference on Applications of Computer Vision Workshops (WACVW), pp. 71–80 (2020)
24. Qiu, X., Li, H., Luo, W., Huang, J.: A universal image forensic strategy based on steganalytic model. In: 2nd ACM Workshop on Information Hiding and Multimedia Security, pp. 165–170 (2014)
25. Ren, S., He, K., Girshick, R., Sun, J.: Faster R-CNN: towards real-time object detection with region proposal networks. IEEE Trans. Pattern Anal. Mach. Intell. **39**(6), 1137–1149 (2016)
26. Salloum, R., Ren, Y., Kuo, C.C.J.: Image splicing localization using a multi-task fully convolutional network (MFCN). J. Vis. Commun. Image Represent. **51**, 201–209 (2018)
27. Verdoliva, L.: Media forensics and DeepFakes: an overview. IEEE J. Sel. Top. Sig. Process. **14**(5), 910–932 (2020)
28. Wen, B., Zhu, Y., Subramanian, R., Ng, T.T., Shen, X., Winkler, S.: Coverage-a novel database for copy-move forgery detection. In: IEEE International Conference on Image Processing (ICIP), pp. 161–165 (2016)
29. Wu, Y., AbdAlmageed, W., Natarajan, P.: Mantra-net: manipulation tracing network for detection and localization of image forgeries with anomalous features. In: IEEE Conference on Computer Vision and Pattern Recognition (CVPR), pp. 9543–9552 (2019)
30. Yang, C., Li, H., Lin, F., Jiang, B., Zhao, H.: Constrained R-CNN: a general image manipulation detection model. In: IEEE International Conference on Multimedia and Expo (ICME), pp. 1–6 (2020)
31. Yin, Q., Wang, J., Luo, X.: A hybrid loss network for localization of image manipulation. In: 19th International Workshop on Digital-forensics and Watermarking, pp. 237–247 (2020)
32. Zampoglou, M., Papadopoulos, S., Kompatsiaris, Y.: Large-scale evaluation of splicing localization algorithms for web images. Multimedia Tools Appl. **76**(4), 4801–4834 (2017)
33. Zhang, R., Ni, J.: A dense U-Net with cross-layer intersection for detection and localization of image forgery. In: IEEE International Conference on Acoustics, Speech and Signal Processing (ICASSP), pp. 2982–2986 (2020)
34. Zhao, H., Kong, X., He, J., Qiao, Yu., Dong, C.: Efficient image super-resolution using pixel attention. In: Bartoli, A., Fusiello, A. (eds.) ECCV 2020. LNCS, vol. 12537, pp. 56–72. Springer, Cham (2020). https://doi.org/10.1007/978-3-030-67070-2_3

35. Zhou, P., Han, X., Morariu, V., Davis, L.: Learning rich features for image manipulation detection. In: IEEE Conference on Computer Vision and Pattern Recognition (CVPR), pp. 1053–1061 (2018)
36. Zhu, X., Qian, Y., Zhao, X., Sun, B., Sun, Y.: A deep learning approach to patch-based image inpainting forensics. Sig. Process. Image Commun. **67**, 90–99 (2018)
37. Zhuang, P., Li, H., Tan, S., Li, B., Huang, J.: Image tampering localization using a dense fully convolutional network. IEEE Trans. Inf. Forensics Secur. **16**, 2986–2999 (2021)

An End-to-End Mutual Enhancement Network Toward Image Compression and Semantic Segmentation

Junru Chen[1,2], Chao Yao[3], Meiqin Liu[1,2], and Yao Zhao[1,2](✉)

[1] Institute of Information Science, Beijing Jiaotong University, Beijing 100044, China
[2] Beijing Key Laboratory of Advanced Information Science and Network Technology, Beijing Jiaotong University, Beijing 100044, China
{19120289,mqliu,yzhao}@bjtu.edu.cn
[3] School of Computer and Communication Engineering, University of Science and Technology Beijing, 100083 Beijing, China
yaochao@ustb.edu.cn

Abstract. Image compression is to compress image data without compromising human vision feeling. However, the information loss through the image compression process may influence the following machine vision tasks, such as object detection and semantic segmentation. How to jointly consider the human vision and the machine vision to compress images for human and machine vision tasks is still an open problem. In this paper, we provide a multi-task framework for image compression and semantic segmentation. More specifically, an end-to-end mutual enhancement network is designed to efficiently compress the given image, and simultaneously segment the semantic information. Firstly, a uniform feature learning strategy is adopted to jointly learn the features for image compression and semantic segmentation in the encoder. Moreover, a multi-scale aggregation module in the encoder is employed to enhance the semantic features. Then, by transmitting the quantified features, both the decompressed image features and the learned semantic features can be reconstructed. Finally, we decode this information for the image compression task and the semantic segmentation task. On one hand, we can utilize the decompressed semantic features to implement semantic segmentation in the decoder. On the other hand, the quality of the decompressed image can be further improved depending on the obtained semantic segmentation map. Experimental results prove that our framework is effective to simultaneously support image compression and semantic segmentation, both in the subjective and objective evaluation.

Keywords: Learning-based compression · Video Coding for Machine · Semantic segmentation

This work was supported by National Natural Science Foundation of China (61972028, 61902022) and the Fundamental Research Funds for the Central Universities (2019JBM018, FRF-TP-19-015A1).

H. Ma et al. (Eds.): PRCV 2021, LNCS 13020, pp. 623–635, 2021.
https://doi.org/10.1007/978-3-030-88007-1_51

1 Introduction

Nowadays, a large number of image/video contents are produced and transmitted to the Internet every day. Reported from Cisco in 2018, Machine-to-Machine applications will occupy the greatest usage of Internet video traffic over the next following years. Moreover, machine learning algorithms tend to handle more contents directly instead of only by human perception. It is critical to establish the information that can be processed both by machine intelligence applications and human perception. Therefore, how to support the hybrid human-machine intelligence applications within the limited bandwidth is eager to be solved.

Recently, with the rapid development of deep learning, some learning-based compression methods [1–4] have been proposed. However, these methods are justly driven by the Rate-Distortion cost serving for the human perception, not compatible with the high-level machine vision tasks. Besides, when facing big data and high-level analysis, these methods are still questionable. Therefore, to interact the data compression with the machine intelligent analytics, a new video codec called VCM (Video Coding for Machine) [5] is organized which provides compression for machine vision as well as human-machine hybrid vision.

In this paper, we propose an end-to-end mutual enhancement network toward image compression and semantic segmentation, which not only makes the compression framework to be compatible with the semantic segmentation but also achieves the mutual enhancement to each other. The encoder consists of a base network and a multi-scale aggregation module. In particular, the multi-scale aggregation module is able to enhance the semantic features by suppressing the effect of the quantization. The decoder decompresses the latent representation obtained from the compression branch and the semantic branch, and obtains the decompressed image and the semantic segmentation map respectively. Then the enhancement module is utilized to enhance the quality of the decompressed image via the obtained semantic segmentation map. Our method is able to achieve mutual enhancement for both image compression and semantic segmentation tasks. Experimental results show that the proposed method can obtain improved decompressed image and semantic segmentation map.

In summary, the contributions of this paper are as follows:

(1) We propose a unified framework that integrates image compression with semantic segmentation to achieve mutual enhancement.
(2) We design a multi-scale aggregation module to suppress the impact of quantization in the encoder, which aims to enhance semantic features.
(3) We construct a post-enhancement module to improve the quality of the decompressed images by using the decompressed semantic segmentation map in the decoder.

2 Related Works

In this section, we briefly review some related works about learning-based image/video compression, especially several works in response to VCM.

2.1 Learning-Based Compression

Recently, lots of learning-based image/video compression methods are proposed [6]. In general, these methods can be classified into two categories based on the coding architecture. The first is to design the deep embedded modules in the traditional hybrid coding framework, and the second is the end-to-end deep compression framework.

Deep embedded modules aim to design an optimal network to replace the key parts in the traditional coding framework, such as in-loop filter [7], intra-prediction [8], inter-prediction [9], entropy coding [10], transform [11] and quantization [12]. For example, [7] proposed a post-processing learning-based method to enhance the decompressed image, instead of the in-loop filter. An intra-prediction convolutional neural network (IPCNN) was proposed in [8]. [9] utilized spatial adjacent pixels and temporal display order as additional inputs of the constructed CNN model to implement the dual prediction of video streaming. In addition, [12] proposed a fast quantization strategy of HEVC based on CNN.

The end-to-end compression architecture research starts from [1], which consists of nonlinear analysis transform, uniform quantizer and nonlinear synthesis transform. Then, many end-to-end compression methods are proposed to further improve compression performance. An end-to-end trainable image compression model based on variational autoencoder [2] was designed, where a hyper-prior potential representation was incorporated for efficiently capturing spatial dependencies. A context-adaptive entropy model that can be used for the RD optimization in the end-to-end compression architecture [3]. Furthermore, [4] introduced the discrete Gaussian mixture likelihood to parameterize the distribution of latent code and reduced the number of coding bits required. Some latest works have achieved higher compression efficiency than that of the VVC (Versatile Video Coding) [13] or HEVC (High Efficiency Video Coding) [14].

2.2 Video Coding for Machines

Traditional video coding frameworks are optimized for HVS (Human Visual System). However, with the development of AI technology, a great amount of image/video is being analyzed by machines. Hence, the target of image/video coding is not only optimized for human vision but also machine vision. Toward collaborative compression and intelligent analytics, a new codec called VCM is proposed as the next-generation video codec, which attempts to bridge the gap between feature coding for machine vision and video coding for human vision.

In response to VCM, some researchers try to integrate machine vision tasks with image compression as a uniform framework. In [15], a hybrid resolution coding framework based on a reference-based DCNN was proposed to jointly solve the problem of the interference between the resolution loss and the compression artifacts. Similarly, an end-to-end restoration-reconstruction deep neural network (RR-DNCNN) [16] based on degradation sensing technology was proposed to answer the degradation problem caused by compression and sub-sampling due to various artifacts brought by compression to the super-resolution task. Besides, some interesting works which try to combine image compression with high-level machine vision tasks have attracted various of attention. A framework called DSSLIC was proposed in [17], which combines the semantic map, coarse representation of the input image, and residuals of the input image in hierarchical coding, which can obtain a good compression reconstruction image and simultaneously facilitate other compression related computer vision tasks. A semantically structured image coding (SSIC) [18] framework was designed to generate a semantically structured bitstream (SSB), where each part of the bitstream represents a specific object, which can be directly used for various visual tasks. [19] proposed a encoder-decoder architecture that makes an image compression framework to support semantic segmentation. So far, the study on the relation between suitable compressed representations and the effectiveness of machine vision algorithms has been an active and fast-growing research area, how to standardize a bitstream format to enable both image compression and machine vision tasks will be worth noticing.

3 Proposed Method

The proposed method aims to achieve mutual enhancement for both the image compression task and the semantic segmentation task. Figure 1 shows the framework of our method which basically is an encoder-decoder structure. In the following, we will give detailed introductions.

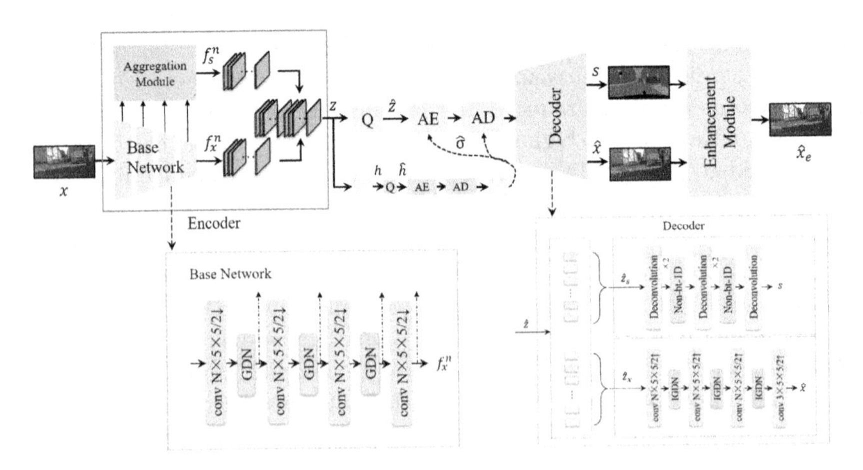

Fig. 1. The overall framework of our proposed method. "Q" denotes quantization. "AE" and "AD" mean the arithmetic encoder and decoder respectively.

3.1 Encoder

The encoder is consisted of two parts which correspond to compression branch and semantic segmentation branch respectively. One part is called base network. As shown in Fig. 1, several cascaded convolution layers are adopted to characterize the correlation between neighboring pixels, which is consistent with the hierarchical statistical properties of natural images. Here, to optimize the features for image compression, the generalized divisive normalization (GDN) transform [1] is utilized to transfer the pixel-domain feature into a divisive normalization space.

An aggregation module is designed to learn and enhance the semantic features, which is shown in Fig. 2. It is worth noticing that all of the learned features should be quantified in our unified framework, even for the semantic features. Therefore, one key issue is to suppress the impact of the quantization. We try to explore some abundant features to enhance the semantic representations. More precisely, the hierarchical features from different layers of the base network are applied to learning the high-level semantic feature. For instance, f_x^i which is from the interlayer of the base network is added into the structure feature f_x^n by a hierarchical feature fusion block (HFFB). The operation can be represented as follows:

$$f_y^{j+1} = W_{j+1} \times f_y^j + f_x^i, i = n, n-1, ..., 1, j = 1, 2, ..., n, \tag{1}$$

where, f_x^i denotes the learned features from the i-layer of base network, f_y^i is the enhanced feature from the previous layer, while $f_y^1 = f_x^n$. W_i is the learnable parameter in the current layer.

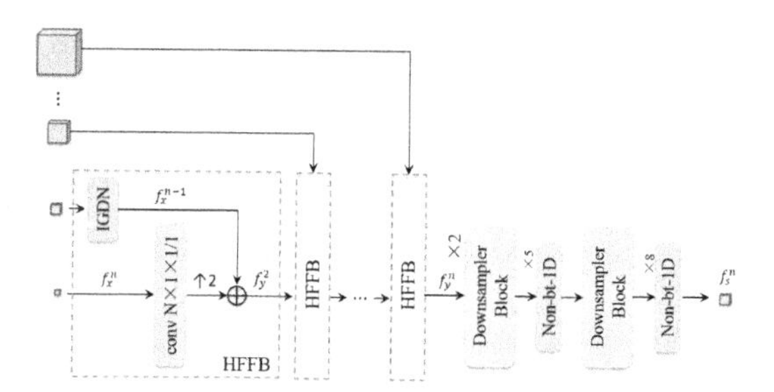

Fig. 2. The aggregation module of our proposed method, "IGDN" represents inverse GDN[20], "HFFB" represents hierarchical feature fusion block.

In the HFFB block, the feature f_x^i is first transformed to the pixel-domain by an IGDN layer associated with the GDN layer in the base network, and then the transferred feature is added to the previous fused feature f_y^i. To be noted, each HFFB block corresponds to the hierarchical features from different layers in the base network. This operation aims to suppress the additive noise by increasing the weight of the feature. To further improve the representation of semantic information, the special convolution layer non-bottleneck-1D (non-bt-1D) [21] is integrated into the HFFB blocks. Then, the features can be stretched and transformed into one-dimensional representation, which is more conducive to the subsequent pixel-level semantic classification and thus enhances the performance of the semantic segmentation task. Finally, the semantic feature f_s^n can be obtained. For the learned feature f_x^n and f_y^n, a quantization method depending on the additive noise and entropy encoding method [2] are applied to convert the learned feature into a piecewise bitstream. The bitstream is reverted to feature by entropy decoding and sent to the decoder. It is worth mentioning that quantization operation in the traditional methods is to transform continuous data into discrete data to reduce the amount of data. So quantization operation is undesirable. However, learning-based methods depend on end-to-end optimization with gradient-based techniques. Many methods have made some contribution to solve this problem. Here, we follow [1] using additive noise. Specifically, we add uniform noise to approximate quantization operation in the training stage and we round it directly in the inference stage.

3.2 Decoder

As shown in Fig. 1, the received features are firstly divided into two parts in the decoder, including the semantic feature $\hat{z}_s$ and the compression feature $\hat{z}_x$. Correspondingly, the divided features $\hat{z}_s$ and $\hat{z}_x$ are fed into different decode branches respectively. To obtain the semantic image, several deconvolution layers and non-bottleneck-1D (non-bt-1D) layers are utilized as a semantic decoder to decompress $\hat{z}_s$. The non-bt-1D layers can gather more context from the received features, and deconvolution layers can up-sample the features to match the resolution of the input image. For the image decompression, we apply inverse operations on $\hat{z}_x$ to reconstruct image $\hat{x}$, which are corresponding to the base network in the encoder. Hence, the image decoder consists of several deconvolution layers and IGDN layers.

Considering all the factors in our framework, the loss function of the whole framework can be written as follows:

$$L = \lambda D + R + CE, \tag{2}$$

where λ is one hyper parameter, D represents the distortion between the input image and the reconstruction image, R denotes the bitrate which is approximated by using the entropy of the corresponding latent representations $\hat{z}$, CE represents the cross entropy of semantic segmentation map s and the ground truth. In general, $CE = \frac{1}{N}\sum_i - \sum_{c=1}^{M} s_{ic} log(p_{ic})$. M is the number of categories, s_{ic} has value 0 or 1. If the predicted category of the sample i is the same as the ground truth (equal to c), s_{ic} should be 1, otherwise 0. p_{ic} indicates the predicted probability that the sample i belongs to the category c.

3.3 Enhancement Module

Motivated by that semantic segmentation is able to recognize the category of each pixel, we take advantage of the semantic information to enhance the decompressed images. The semantic map where each pixel is labeled by category information can provide clearer spatial structure information for human to understand or intelligent analytics.

As shown in Fig. 3, we propose a post-enhancement module to improve the details of the decompressed image $\hat{x}$. The obtained semantic segmentation map s is fed into the post-enhancement module to learn the structure information.

First, the max pooling and the average pooling operations are separately conducted along the channel dimension, whose formulation is as follows:

$$s_s = [Max(s), Avg(s)], \tag{3}$$

here, $[\cdot, \cdot]$ represents the concatenation operation. Then, the weights of spatial structure features are obtained by a convolution layer and a sigmoid activation function. Finally, the weights are utilized to multiply by the semantic features which are learned on the semantic segmentation map s, and the output is the learned spatial structure features. The process can be represented as follows,

$$s_e = W_0 W_1 W_2 W_3 \sigma(s_s), \tag{4}$$

here, W_0, W_1, W_2, W_3 represent convolution operations and σ denotes sigmoid activation function. To embed the learned spatial structure information into the decompressed image $\hat{x}$. $\hat{x}$ is mapped to the feature space by a shadow convolutional layer. Then, some residual blocks are grouped as a frequency filter to learn high-frequency information $\hat{x}_r$. Finally, we concatenate s_e and $\hat{x}_r$ to embed the spatial structure information and obtain the final reconstruction image $\hat{x}_e$.

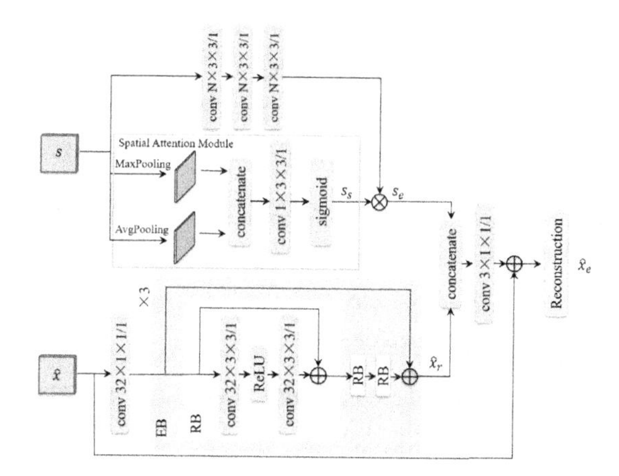

Fig. 3. The enhancement module in our proposed method. "RB" represents the residual block, "EB" refers to the enhancement block.

4 Experiments

In this section, we conduct a series of experiments to evaluate the performance of our proposed method. In our experiments, the widely used *Cityscapes* dataset is adopted. The *Cityscapes* dataset has 19 semantic labels, all 2,974 RGB images are resized to 512 × 1024. And the test dataset for the compression evaluation is constructed with 24 images from the *Kodak* image dataset [22]. For the semantic segmentation evaluation, we adopt the validation set and test set from *Cityscapes* dataset at the resolution of 1024 × 2048. The proposed framework is trained in the end-to-end way and uses different λ values (256, 512, 1024, 2048, 4096, 6144, 8192) to control the quantization step. Adam optimizer [23] with the learning rate of 0.0001 is used, which is fixed in the first 2,000,000 iterations but decreased to 0.00001 in the next 100,000 iterations. All the experiments are conducted on the NVIDIA RTX 3090 with 24 GB memory.

To objectively evaluate the compression performance of our proposed method, we conduct comparable experiments with the following previous works [17,19]. Moreover, we use Multiscale Structural Similarity (MS-SSIM) and the Peak Signal to Noise Ratio (PSNR) between the original and the decompressed images as evaluation indicators. A larger MS-SSIM or PSNR means higher fidelity. Note that MS-SSIM is applied on RGB channels and averaged over the entire test set.

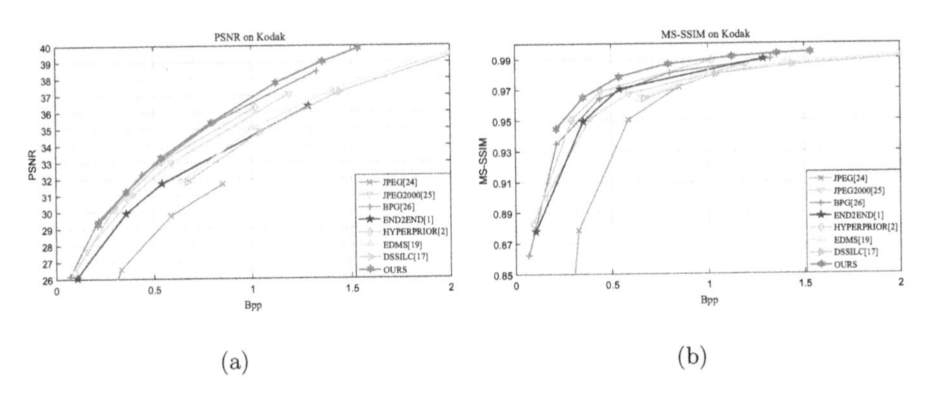

Fig. 4. Rate-distortion curves of different image compression methods using the PSNR metric and MS-SSIM metric on *Kodak* [22].

4.1 Results on Image Compression

We compare several widely used image compression algorithms [1,2,24–26] and two hybrid compression methods [17,19] with our proposed method. The performances are shown in the Fig. 4(a) and 4(b). The curves report the PSNR and MS-SSIM at different bitrates respectively, where bpp means bits-per-pixel, referring to the averaged bitrate for each pixel.

As shown in Fig. 4(a), peak signal-to-noise ratio (PSNR) is adopted as the quality metric. It is obvious that the proposed method is better than the traditional methods JPEG [24], JPEG2000 [25] and the classical end-to-end learning-based method END2END [1], HYPERPRIOR [2]. Moverover, BPG [26] has achieved the state-of-the-art performance in traditional image compression methods, our method achieves comparable performance at low bitrates and achieves better performance apparently at high bitrates over BPG. Besides that, we also compare with the semantic-based image compression methods EDMS [19] and DSSILC [17], which are proposed recently and have excellent performance. As shown in Fig. 4(a), our method is apparently superior to both of them on PSNR quality metric.

As shown in Fig. 4(b), in order to clearly show the advantages of our method over other methods, we also carry out the experiments under multiscale structural similarity (MS-SSIM) quality metric. By comparing the curves in Fig. 4(b), it can be found that our method has the best performance of all comparable methods. Especially, the results of the proposed method have huge advantages over BPG in terms of the MS-SSIM quality metric while just can be comparable under the PSNR quality metric. Then, analyzing and comparing Fig. 4(a) and 4(b) together can be easily find out that the learning-based methods perform better than traditional methods under the MS-SSIM metric.

Moreover, the compression branch of our method has a similar structure with HYPERPRIOR [2]. When it is integrated into our method, the performance of our method is better than HYPERPRIOR. It shows that the semantic embedding method is reliable and can be used to improve the reconstruction effectively. Our enhancement module can improve decompressed image by using the semantic information extracting from semantic segmentation map.

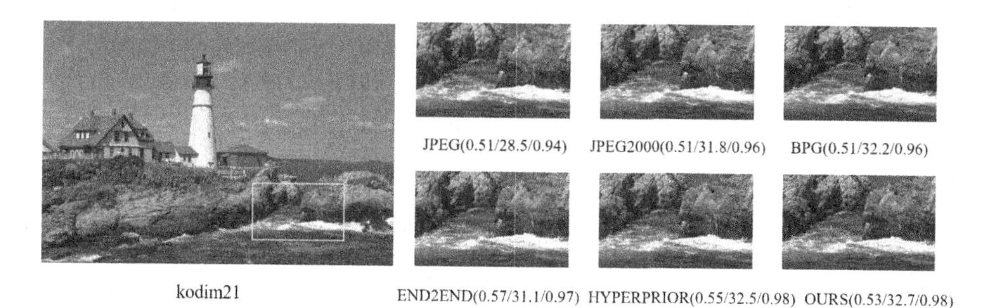

Fig. 5. Visualization of decompressed images "kodim21.png" from *Kodak* [22] and its ground truth. The numbers on the bottom of the images mean the value of (Bpp/PSNR/MS-SSIM).

To display the performance of our method intuitively, We exhibit the decompressed image of the proposed method and some competitive methods with similar bitrate in Fig. 5. A visual example of kodim21 from the *Kodak* dataset is provided. Our method obtains the best image quality at a similar bitrate. When looking carefully at the selected area, we can see that the wave of the sea in the images through JPEG and JPEG2000 methods are blurred. While the rocks in the selected area of the two methods have lots of noise and block artifacts. The more excellent traditional compression method at present BPG and the classical learning-based compression methods END2END, HYPERPRIOR are slightly better than JPEG and JPEG2000, but there are still some visible flaws. The image decompressed by our method is relatively clear in texture and color.

4.2 Results on Semantic Segmentation

In our method, the semantic segmentation branch can be compatible with many outstanding semantic segmentation networks. In this paper, ERFNet [21] is

Table 1. Results of our four semantic segmentation architectures on the *Cityscapes* Evaluation set. Per-class IoU(%) and mean classes IoU(%).

Methods	Roa	Sid	Bui	Wal	Fen	Pol	TLi	TSi	Veg	Ter	Sky	Ped	Rid	Car	Tru	Bus	Tra	Mot	Bic	Cla-IoU
Baseline	97.5	80.2	90.4	46.5	51.9	61.1	65.5	72.5	91.2	59.9	93.8	75.9	55.1	93.1	72.5	79.7	67.9	46.7	70.6	72.1
B+A	97.5	81.4	91.1	49.1	51.9	62.7	64.8	74.4	91.5	61.9	94.1	77.4	57.5	92.9	72.8	75.2	70.7	41.7	71.9	**72.8**
B+Q	97.2	80.6	90.1	51.4	52.6	59.4	63.2	72.1	91.1	61.2	92.2	75.5	55.3	92.2	66.5	75.3	63.5	44.4	70.3	71.2
B+Q+A	97.6	81.2	90.8	46.2	51.2	62.8	63.9	73.8	91.1	59.9	93.5	76.9	56.3	93.2	71.2	79.2	74.7	43.2	70.9	**72.5**

integrated into our semantic segmentation branch. Table 1 shows the segmentation results on 19 classes of the *Cityscapes* evaluation set under four conditions based on ERFNet, which are no quantization operation, only quantization operation, quantization operation plus aggregation module and only aggregation module. To conduct the segmentation experiments in four conditions, we correspondingly construct four models. We defined the original architecture of ERFNet in the semantic segmentation branch, without quantization operation and our aggregation module as baseline. Then over the architecture of baseline, only quantization is operated on the semantic segmentation branch (we called it B+Q) and only aggregation module is applied on the semantic segmentation branch (we called it B+A). The last model is our proposed model with quantization operation and aggregation module (we called it B+Q+A). As shown in Table 1, comparing baseline with B+Q, it is found that nearly 1% mean classes IoU (Cla-IoU) is declined because of quantization operation. Compared to the model B+Q, by using our aggregation module (B+Q+A), the accuracy is improved and better than baseline in the case of quantization operation. To verify the effectiveness of this aggregation module, we also compare this B+A model with the original unquantized baseline architecture. It turns out that the accuracy of our method is improved than before. Therefore, our multi-scale aggregation module is effective and the multi-scale feature information from the base network could suppress the impact of the quantization operation.

Table 2. Comparable results on *Cityscapes* Test sets.

Methods	Cla-IoU(%)	Cat-IoU(%)	Methods	Cla-IoU(%)	Cat-IoU(%)
RefineNet [27]	73.6	87.9	Dilation [28]	67.1	86.5
Adelaide-cntxt [29]	71.6	87.3	DPN [30]	66.8	86.0
LRR-4x [31]	69.7	88.2	B+A	**70.8**	**88.1**
Deeplabv2-CRF [32]	70.4	86.4	B+Q+A	**70.5**	**88.0**

Table 2 shows the semantic segmentation results of several comparable approaches. These results are obtained from the Cityscapes Test server. Baseline with the aggregation module (B+A) achieves a 70.8% mean Classes IoU (Cla-IoU) and an 88.1% mean Category IoU (Cat-IoU). The B+Q+A model achieves a 70.5% Cla-IoU and an 88.0% Cat-IoU. Cla-IoU is improved compared to LRR-4x [31], Deeplabv2-CRF [32], Dilation10 [28] and DPN [30], and Cat-IoU is improved compared to RefineNet [27], Adelaide-cntxt [29], Deeplabv2-CRF [32], Dilation10 [28] and DPN [30]. It is proved that the aggregation module extracts hierarchical features from different layers in base network could not only reduce the impact of quantization operation but also improve the quality of semantic segmentation map. In general, benefiting from the aggregation module, the semantic segmentation branch in our proposed method is much more competitive.

5 Conclusion

To achieve mutual enhancement for image compression and semantic segmentation tasks, we propose a novel end-to-end mutual enhancement network. The whole framework of our method which is based on an encoder-decoder structure contains several creative designs. A multi-scale aggregation module in the encoder is designed to improve the accuracy of the semantic segmentation and an enhancement module after the decoder is designed to enhance the reconstruction of the compression. The experimental results show that our method is effective and achieves mutual enhancement for both image compression and semantic segmentation. In the future, we would expand this framework to support more machine intelligent tasks than semantic segmentation.

References

1. Ballé, J., Laparra, V., Simoncelli, E.P.: End-to-end optimized image compression. In: 5th International Conference on Learning Representations, ICLR 2017 (2017)
2. Ballé, J., Minnen, D., Singh, S., Hwang, S.J., Johnston, N.: Variational image compression with a scale hyperprior. In: 6th International Conference on Learning Representations, ICLR 2018 (2018)
3. Lee, J., Cho, S., Beack, S.K.: Context-adaptive entropy model for end-to-end optimized image compression. In: 6th International Conference on Learning Representations, ICLR 2018 (2018)
4. Cheng, Z., Sun, H., Takeuchi, M., Katto, J.: Learned image compression with discretized gaussian mixture likelihoods and attention modules. In: Proceedings of the IEEE/CVF Conference on Computer Vision and Pattern Recognition, pp. 7939–7948 (2020)
5. Duan, L., Liu, J., Yang, W., Huang, T., Gao, W.: Video coding for machines: a paradigm of collaborative compression and intelligent analytics. IEEE Trans. Image Process. **29**, 8680–8695 (2020)
6. Liu, D., Li, Y., Lin, J., Li, H., Wu, F.: Deep learning-based video coding: a review and a case study. ACM Comput. Surv. (CSUR) **53**(1), 1–35 (2020)
7. Lin, W., et al.: Partition-aware adaptive switching neural networks for post-processing in HEVC. IEEE Trans. Multimed. **22**(11), 2749–2763 (2019)
8. Cui, W., et al.: Convolutional neural networks based intra prediction for HEVC. In: 2017 Data Compression Conference (DCC), pp. 436–436. IEEE Computer Society (2017)

9. Mao, J., Yu, L.: Convolutional neural network based bi-prediction utilizing spatial and temporal information in video coding. IEEE Trans. Circ. Syst. Video Technol. **30**(7), 1856–1870 (2019)
10. Song, R., Liu, D., Li, H., Wu, F.: Neural network-based arithmetic coding of intra prediction modes in HEVC. In: Visual Communications and Image Processing (VCIP), pp. 1–4. IEEE (2017)
11. Liu, D., Ma, H., Xiong, Z., Wu, F.: CNN-based DCT-like transform for image compression. In: Schoeffmann, K., et al. (eds.) MMM 2018. LNCS, vol. 10705, pp. 61–72. Springer, Cham (2018). https://doi.org/10.1007/978-3-319-73600-6_6
12. Alam, M.M., Nguyen, T.D., Hagan, M.T., Chandler, D.M.: A perceptual quantization strategy for HEVC based on a convolutional neural network trained on natural images. In: Applications of Digital Image Processing, vol. 9599, p. 959918. International Society for Optics and Photonics (2015)
13. Bross, B., Chen, J., Ohm, J.R., Sullivan, G.J., Wang, Y.K.: Developments in international video coding standardization after AVC, with an overview of versatile video coding (VVC). In: Proceedings of the IEEE (2021)
14. Sullivan, G.J., Ohm, J.R., Han, W.J., Wiegand, T.: Overview of the high efficiency video coding (HEVC) standard. IEEE Trans. Circ. Syst. Video Technol. **22**(12), 1649–1668 (2012)
15. Hou, D., Zhao, Y., Ye, Y., Yang, J., Zhang, J., Wang, R.: Super-resolving compressed video in coding chain. arXiv preprint arXiv:2103.14247 (2021)
16. Ho, M.M., Zhou, J., He, G.: RR-DnCNN v2.0: enhanced restoration-reconstruction deep neural network for down-sampling-based video coding. IEEE Trans. Image Process. **30**, 1702–1715 (2021)
17. Akbari, M., Liang, J., Han, J.: DSSLIC: deep semantic segmentation-based layered image compression. In: IEEE International Conference on Acoustics, Speech and Signal Processing, pp. 2042–2046. IEEE (2019)
18. Sun, S., He, T., Chen, Z.: Semantic structured image coding framework for multiple intelligent applications. IEEE Trans. Circ. Syst. Video Technol. **31**(9), 3631–3642 (2020)
19. Hoang, T.M., Zhou, J., Fan, Y.: Image compression with encoder-decoder matched semantic segmentation. In: Proceedings of the IEEE/CVF Conference on Computer Vision and Pattern Recognition Workshops, pp. 160–161 (2020)
20. Ballé, J., Laparra, V., Simoncelli, E.P.: Density modeling of images using a generalized normalization transformation. In: 4th International Conference on Learning Representations, ICLR 2016 (2016)
21. Romera, E., Alvarez, J.M., Bergasa, L.M., Arroyo, R.: ERFNet: efficient residual factorized convnet for real-time semantic segmentation. IEEE Trans. Intell. Transp. Syst. **19**(1), 263–272 (2017)
22. Kodak, E.: Kodak lossless true color image suite (PhotoCD PCD0992), vol. 6. http://r0k.us/graphics/kodak (1993)
23. Kingma, D.P., Ba, J.: Adam: A method for stochastic optimization. arXiv preprint arXiv:1412.6980 (2014)
24. Wallace, G.K.: The JPEG still picture compression standard. IEEE Trans. Consum. Electron. **38**(1), 18–34 (1992)
25. Skodras, A., Christopoulos, C., Ebrahimi, T.: The JPEG 2000 still image compression standard. IEEE Signal Process. Mag. **18**(5), 36–58 (2001)
26. Bellard, F.: Better portable graphics. https://www.bellard.org/bpg (2014)
27. Lin, G., Milan, A., Shen, C., Reid, I.: RefineNet: multi-path refinement networks with identity mappings for high-resolution semantic segmentation. arXiv preprint arXiv:1611.06612

28. Yu, F., Koltun, V.: Multi-scale context aggregation by dilated convolutions. arXiv preprint arXiv:1511.07122 (2015)
29. Lin, G., Shen, C., Van Den Hengel, A., Reid, I.: Efficient piecewise training of deep structured models for semantic segmentation. In: Proceedings of the IEEE Conference on Computer Vision and Pattern Recognition, pp. 3194–3203 (2016)
30. Krešo, I., Čaušević, D., Krapac, J., Šegvić, S.: Convolutional scale invariance for semantic segmentation. In: Rosenhahn, B., Andres, B. (eds.) GCPR 2016. LNCS, vol. 9796, pp. 64–75. Springer, Cham (2016). https://doi.org/10.1007/978-3-319-45886-1_6
31. Ghiasi, G., Fowlkes, C.C.: Laplacian reconstruction and refinement for semantic segmentation. arXiv preprint arXiv:1605.02264, vol. 4(4) (2016)
32. Chen, L.C., Papandreou, G., Kokkinos, I., Murphy, K., Yuille, A.L.: Deeplab: semantic image segmentation with deep convolutional nets, atrous convolution, and fully connected CRFs. IEEE Trans. Pattern Anal. Mach. Intell. **40**(4), 834–848 (2017)

Deep Double Center Hashing for Face Image Retrieval

Xin Fu, Wenzhong Wang(✉), and Jin Tang

School of Computer Science and Technology, Anhui University, Hefei, China
{wenzhong,tangjin}@ahu.edu.cn

Abstract. Hashing is an effective and widely used technology for fast approximate nearest neighbor search in large-scale images. In recent years, it has been combined with a powerful feature learning model, convolutional neural network(CNN), to boost the efficiency of large-scale image retrieval. In this paper, we introduce a new Deep Double Center Hashing (DDCH) network to learn hash codes with higher discrimination between different people and compact hash codes between the same person for large-scale face image retrieval. Our method uses a deep neural network to learn image features as well as hash codes. We use a deep CNN to extract image features and a multi-layer neural network as the hash function. The whole model is trained end-to-end. In order to learn compact and discriminative hash codes, we impose a compact constraint on the codes to force lower intra-class variations of the codes. Our constraint is formulated as a center-loss over the learned codes, which encourages hash codes to be near the hash center of the same class. In addition, new discrete hashing modules and multi-scale fusion are designed to capture discriminative and multi-scale information. We conduct experiments on the most popular datasets, YouTubeFaces and FaceScrub, and demonstrates the efficient performance of DDCH over the state-of-the-art face image hashing methods.

Keywords: Image retrieval · Deep hashing · Deep learning

1 Introduction

The goal of face image retrieval is to search in a gallery for images that are most similar to the query face image, and ideally these retrieved images have the same identity as the query image [1,2]. The common approach for this task is to encode each image with a fixed-length feature vector, and then perform k-nearest neighbor search in the feature space. In recent years, deep convolutional neural network (DCNN) has outperformed almost all previous hand-designed feature extraction methods, and has become the popular model for face image feature extraction. The most widely used CNN model transforms a raw image into a high-dimensional feature descriptor, typically represented as a vector of thousands of floating point numbers [3]. When each image is mapped into such a high-dimensional feature space, a linear search in a large-scale face image dataset becomes infeasible, since comparing high-dimensional vectors in a large-scale dataset requires very large memory load and computational cost.

H. Ma et al. (Eds.): PRCV 2021, LNCS 13020, pp. 636–648, 2021.
https://doi.org/10.1007/978-3-030-88007-1_52

In order to solve these difficulties, people use a hash function to construct a binary hash code for each image, and use the hash code for fast approximate nearest neighbor (ANN) search [4–6]. The length of the hash code is usually very small (only tens of bits), so the storage overhead is relatively small, meanwhile, the Hamming distance between the hash codes can be calculated very efficiently. There have been many hashing techniques for fast ANN search, such as Locality Sensitive Hashing (LSH) [7]. In recent years, data-driven method has gained more and more attention in computer vision. The core idea of data-driven method is to fit a parameterized model (usually a deep neural network) to a large dataset, and then use this model for prediction. Inspired by the success of this method in face recognition, image classification, and many other computer vision tasks, most recent state-of-the-art hashing methods use neural network as the parametric model for the hash function, and learn the parameters using large dataset. Furthermore, in the latest research, the feature extraction and feature hashing are combined in a single Deep CNN model which is trained end-to-end. These techniques (coined Deep Hashing) achieves best performance in most widely used image retrieval benchmark datasets.

However, face retrieval is more complicated than general image retrieval. First, there are large intra-identity variations in face images due to different angles, postures, lighting, expressions, makeup [8]. Such intra-class variations make same person look much different. Second, the overall shape structure and appearance of different face images are similar, such similarities between different identities could lead to small inter-identity variation. Thus, it is essential to learn features and hash codes which have small intra-identity variations and separable inter-class differences.

There are several deep hashing methods designed for face image retrieval. To be specific, Discriminative Deep Hashing (DDH) [9] proposes a divide-and-encode module to improve the discriminability of hash codes. Deep Clustering and Block Hashing (DCBH) [10] optimizes center-clustering loss on feature level and block hashing module to improve the accuracy of the same identity retrieval. Discrete Attention Guided Hashing (DAGH) [8] introduces a multi-attention module and an identity loss on hash level.

In all these methods, a Deep CNN model is used to learn face feature, and on top of it, another neural network is used to model the hash function, which maps the real-valued feature vectors to discrete hash codes. These methods treat the model training as a multi-task learning problem, and design appropriate loss functions for feature extraction and hash functions.

However, the losses for feature extraction and hash functions are independent, they cannot express the structural relation between the feature space and the hash code space. In this paper, we propose a new deep hashing method for face image retrieval. We assume that the feature space and the code space are local linear, where the features and codes for the same identity lie on compact linear subspaces, furthermore, the relation between the feature subspace and the code subspace of the same id is linear. Based on this assumption, we propose the Deep Double-Center Hashing (DDCH) method, in which two center-losses are used, one for the local feature space, and the other for the code space. The local feature subspace and code subspace is linked via the constraint imposed on the two center-losses, i.e., we require that the code center of an identity is the hash code of the feature center of the same identity. Combined with the identification losses and quantification loss, our DDCH can learn face features and hash codes that are both locally compact (i.e. small intra-class variation) and discriminative (i.e. large inter-class variation).

Our major contributions can be summarized as follows:

- We propose a new model, DDCH, to learn discriminative and compact hash codes for large-scale face image retrieval. In our model, we hash the feature center to the code space to constrain the hash function. This constraint makes the learned codes more compact and separable.
- We design a multi-layer perceptron as the hash function, which is more powerful in mapping feature space to code space than other shallower models.
- After conducting a comparison experiment on two face image retrieval data sets. Experimental results reveal that our method is more efficient than the current state-of-the-art deep hash face image retrieval methods.

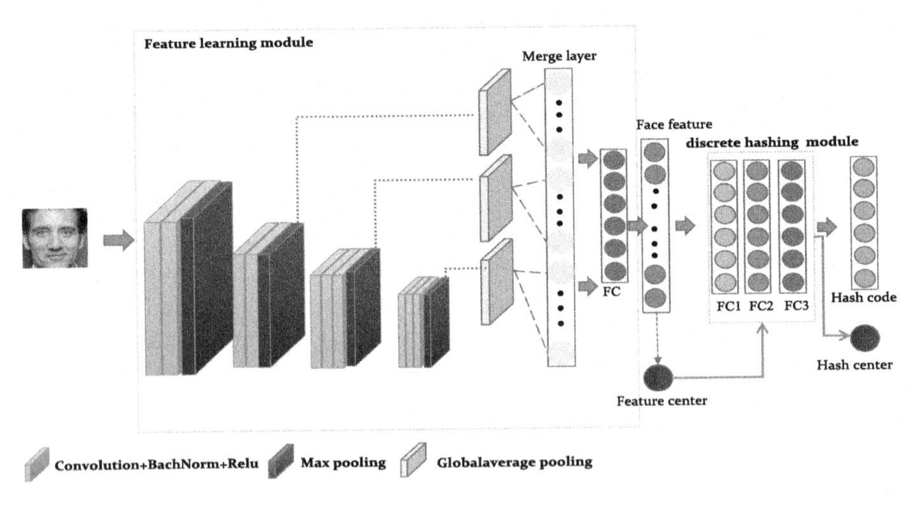

Fig. 1. The Deep Double Center Hashing (DDCH) model.

2 Related Work

In this chapter, We will present in detail the related deep hashing methods for general image retrieval and face image retrieval.

2.1 Deep Hashing in General Images Retrieval

Due to the high efficiency of deep neural networks, deep hashing methods achieve promising performances in image retrieval tasks. Compared with the traditional hash methods such as LSH [7], SH [11] and ITQ [12], the supervised hashing methods represent the hash function using a parameterized model, and learn such model using pairwise or triplet losses. The learned hash functions are then used to generate high-quality hash codes. For example, Convolutional Neural Network Hashing (CNNH) [6] learns hash function that preserves the pairwise similarities among the training images, and learn hash function by training the CNN feature function and the hash function. Deep Supervised Hashing (DSH) [13] learns compact codes from input pairs by minimizing the Hamming distances between images from the same class. Deep Neural Network Hashing (DNNH) [14] relaxed the binary hash code to real-valued code, and used Euclidean distance to form the triplet ranking loss [15] so that the model can be learned using gradient descent methods.

2.2 Deep Hashing in Face Images Retrieval

Deep hashing on face image retrieval tasks get most recently researchers attention. Discriminative Deep Hashing (DDH) [9] introduces a divide-and-encode [14] module to improve the discriminability of hash codes. Deep Hashing based on Classification and Quantization errors (DHCQ) [16] learn hash function by optimizing the two loss functions of classification loss and quantization loss, makes the error is minimized. The Discriminative Deep Quantization Hashing (DDQH) [17] integrates the hash codes learning, introduce batch normalization quantization (BNQ) module to optimize the network. Deep Clustering and Block Hashing (DCBH) [10] optimizes center-clustering loss [18] on feature level to improve the accuracy of the same identity retrieval. Discrete Attention Guided Hashing (DAGH) introduces multi-attention module to emphasize facial information and reduce background noise. However, these methods ignore the structural relationship between the hash code space and the feature space.

3 Deep Double Center Hashing

3.1 Motivations

Our goal is to learn a holistic model for both feature extraction and hashing. Let x be a face image, $z \in R^m$ is a m-dimensional feature vector of x, and $b \in (-1, 1)^k$ is a k-bit relaxed real-valued hash code generated from z. The binary code is calculated as $\tilde{b} = sign(b)$, where $sign(b_i) = 1 \ for \ b_i > 0 \ and \ -1 \ for \ b_i \leq 0, \ \forall i \in \{1, ..., k\}$. Our model is designed as:

$$z, b = F(x; \theta) \tag{1}$$

We decompose the model F into two modules as follows:

$$z = f(x; \theta_f) \tag{2}$$

$$b = h(z; \theta_h) \tag{3}$$

where f is the feature extraction module, h is the hashing function, and $\theta = \{\theta_f, \theta_h\}$ is the parameters of F.

Many recent deep hashing methods require the learned features z and hash code b to be discriminative and compact, so that the face identity can be approximately inferred from both z and b. Due to the large intra-class variation of face images, some approaches also impose a cluster-centering constraint on the learned features. Under this constraint, the learned feature z of the same face identity should be distributed closely around the feature center, that is, the intra-class variance should be small.

However, previous approaches have not attempted to learn hash codes with low intra-class distance. This paper, we introduce that an good hash code is one that minimizes the intra-class variation in the code space. Specifically, **we want to learn *locally linear* feature subspace and code subspace for each face identity, both the face features and the hash codes are locally distributed in compact regions.** This is illustrated in Fig. 2. We formalize this idea as the following criteria for z and b.

1. Compactness (Low intra-class variance) in local feature space. The variance of face feature vectors of the same face id is minimized:

$$minimize \quad V_z = \frac{1}{|C|} \sum_{x \in C} \|z(x) - c\|^2 \tag{4}$$

where C is the set of images of the same id, $z(x) = f(x; \theta_f)$ is the feature vector of face image x, $c = \frac{1}{|C|} \sum_{x \in C} z(x)$ is the mean feature of C.

2. Compactness (Low intra-class variance) in local code space. The variance of the relaxed hash codes of the same face id is minimized.

$$minimize \quad V_b = \frac{1}{|C|} \sum_{x \in C} \|b(x) - \mu\|^2 \tag{5}$$

where $b(x) = h(f(x; \theta_f); \theta_b)$ is the code vector of face image x, $\mu = \frac{1}{|C|} \sum_{x \in C} b(x)$ is the mean code of C.

3. Local linearity in feature space and code space. The features and hash codes of the same face id form a local linear sub-space in the feature space and code space, respectively. And the mapping h from local feature sub-space to the local code sub-space is linear (i.e. h is piece-wise linear). This implies the following constraint on the hashing function h:

$$\frac{1}{|C|} \sum_{x \in C} h\left(z(x)\right) = h\left(\frac{1}{|C|} \sum_{x \in C} z(x)\right) \tag{6}$$

or,

$$\mu = h(c) \tag{7}$$

3.2 The Proposed DDCH Model

The mapping F from x to z, b is modeled as a deep neural network, illustrated in Fig. 1. The model composes of two different functional modules, the feature extraction module and hash code learning module. θ is the parameters of this model.

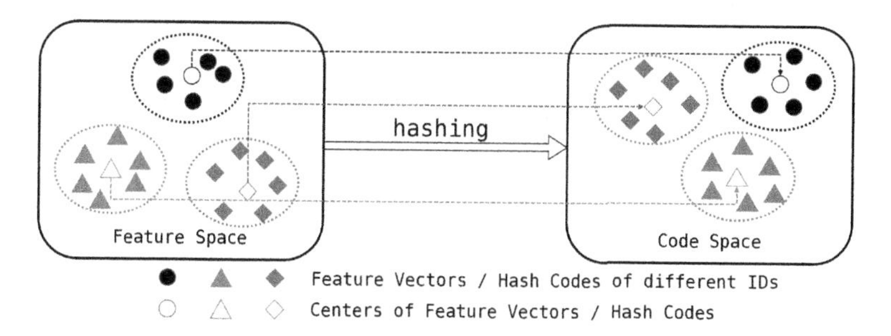

Fig. 2. Motivation: both feature descriptors and their Hash codes should distribute in compact local linear regions (dotted line circles) in the feature space and code space.

Multi-scale Facial Feature Extraction Module. Our CNN model for face feature extractor $z = f(x; \theta_f)$ is modified from the VGG net [19]. Following [10], we use the first four blocks of VGG-16. Each block uses multiple 3×3 convolution kernel to extract features, improve the stability of network training. The batch normalization layer is used before activation function layer, the activation function applies Relu function. Following the Relu layer is a 2×2 max pooling layers with stride of 2.

In order to extract the features of face images at fine scale and high level, we flatten the output feature vector of the three max-pooling layers into a vector through the global average pooling layer, and concatenate them into a large vector. This vector is then transformed into a m-dimensional ($m = 1024$ in this paper) processed by a fully connected layer.

Discrete Hashing Module. We use a three-layer fully connected neural network to approximate the hash function $h(z; \theta_h)$. The output of each dimension in each layer is 512, 256 and k (The bits of hash code). We use Relu as the non-linear activation for the first two layers, and the last layer hash codes are relaxed by $Tanh$ function. In order to accelerate the neural network training and make the training more stable, we add Batch Normalization [20] before activation function. The result of this module is a k-dimensional real-valued vector $b \in (-1, 1)^k$. The discrete binary hash code is obtained via $\tilde{b} = sign(b) \in \{-1, 1\}^k$.

3.3 Loss Functions for End-to-End Training

Losses for Feature Extractor *f*. When learning the feature extractor, we hope the feature should be discriminative enough to discern different face identity. So we classify the features into different identity classes using n-way Softmax regression:

$$\rho_j^{(i)} = \frac{e^{W_j^T z^{(i)}}}{\sum_{l=1}^{n} e^{W_l^T z^{(i)}}}, j = 1..n \tag{8}$$

where $z^{(i)}$ is the feature vector of the i-th training image: $z^{(i)} = f(x^{(i)})$, $\rho_j^{(i)}$ is the predicted probability of $z^{(i)}$ belonging to the j-th id class, and n is the total count of identities in the training set.

Let $id(x) \in \{1, ..., n\}$ denotes the id of image x. We use one-hot vector $y^{(i)}$ to represent the true class of $x^{(i)}$: $y_j^{(i)} = 1 \ if \ id(x^{(i)}) = j, \ and \ y_j^{(i)} = 0 \ otherwise. \ \forall j \in 1, ..., n$.

We use the Cross-Entropy loss to learn the parameters θ_f and $W = \{W_j\}_{j=1}^{n}$ in Eq. 8:

$$l_{id}(\theta_f, W) = -\sum_{i=1}^{N}\sum_{j=1}^{n} y_j^{(i)} log \rho_j^{(i)} \tag{9}$$

In addition to the above identification loss, we also use a center-loss [21] to minimize the intra-class variations of the learned features:

$$l_c(\theta_f) = \frac{1}{2}\sum_{i=1}^{N}\sum_{j=1}^{n} y_j^{(i)} \left\| z^{(i)} - c^{(j)} \right\|_2^2 \tag{10}$$

Losses for Hash Function *h*. We want the binary hash code to be discriminative. So we employ an identification loss on the hash codes. Let $b \in (-1, 1)^k$ be a hash code of x, we hope the identity of $x : id(x)$ can be well predicted from b. So we add a n-way softmax regression to the output of hashing module:

$$\nu_j^{(i)} = \frac{e^{V_j^T b^{(i)}}}{\sum_{l=1}^{n} e^{V_l^T b^{(i)}}}, j = 1, ..., n \tag{11}$$

where $\nu_j^{(i)}$ is the predicted probability $P(id(x^{(i)}) = j | b^{(i)})$. $V = \{V_j\}_{j=1}^{n}$ is the parameter of the classifier.

Based on the prediction, we use the cross-entropy loss for the parameter θ_h and V:

$$l_{id}(\theta_f, \theta_h, V) = -\sum_{i=1}^{N}\sum_{j=1}^{n} y_j^{(i)} log \nu_j^{(i)} \tag{12}$$

In order to learn compact local linear code subspace, we apply the center-loss on the relaxed hash codes b. We formulate the criteria that encourage hash codes to be near the hash center of the same class. The hash center-loss is defined as:

$$l_c(\theta_f, \theta_h) = \frac{1}{2}\sum_{i=1}^{N}\sum_{j=1}^{n} y_j^{(i)} \left\| b^{(i)} - h(c^{(j)}) \right\|_2^2 \tag{13}$$

We also require the relaxed hash code to closely approximate the discrete binary codes. We need to add a quantization loss to make the absolute value of the hash code as close to 1 as possible:

$$l_q(\theta_f, \theta_h) = \gamma \sum_{i=1}^{N} \left\| |b^{(i)}| - 1 \right\|_1 \tag{14}$$

where 1 is a k-dimensional vector of all ones, and γ is used to train the quantization loss optimization. γ is set as a non-negative hyperparameter. This loss function makes the training result closer to the discrete values.

The overall loss function for the whole model is defined as:

$$l_{all} = l_{id}(\theta_f, W) + l_c(\theta_f) + l_{id}(\theta_f, \theta_h, V) + l_c(\theta_f, \theta_h) + l_q(\theta_f, \theta_h) \tag{15}$$

The two modules are jointly trained by minimizing the above loss using mini-batch gradient descent algorithm.

4 Experiments

To elaborate the validity of the proposed deep hashing method, we conducted experiments on two widely used face image datasets: YouTube Faces [22] and FaceScrub [23]. In order to demonstrate the excellent performance of the DDCH method, several advanced hashing methods are also compared.

4.1 Datasets and Evaluation Metric

YouTube: There are 1595 different people in the YouTube face dataset, with a total of 3425 videos. The test set is composed of 5 faces randomly selected by each person, and the training set is composed of 40 random faces. There are 63,800 images in the training set and 7,975 images in the test set. Reset the size of all face images to 32×32.

FaceScrub: There are 106,863 face images downloaded from the Internet, the FaceScrub dataset contains 530 celebrities. The test data consists of 5 faces randomly selected by each person, and the training set comprises the remaining face images. The test data have 2,650 images, and the training set uses the remaining images. Reset the size of all face images to 32×32.

Metrics: Three metrics are used for model evaluation: the Mean Average Precision (**mAP**) calculated using the test retrieved images, Precision-Recall curves (**PR curves**), and Precision Top-K return images (**P@Top-K**).

4.2 Comparison with Baselines

To prove the effectiveness of DDCH, We conducted experiments to test the MAP on the YouTube Faces and FaceScrub dataset. We compare with some recent face retrieval methods, including DHCQ, DDH, DDQH, DCBH, DAGH as baselines, and compare the our results with these baselines. The models are trained with four different code lengths $k = 12, 24, 36, 48$. The results are presented in Table 1. Our method outperforms the other methods with all four code-lengths on the FaceScrub dataset with large margins. On the Youtube-Face dataset, our method achieved the highest mAP with $k = 24, 36$. When $k = 48$, our method

Table 1. MAP of different methods with different code-length.

	YouTube faces				FaceScrub			
	12 bits	24 bits	36 bits	48 bits	12 bits	24 bits	36 bits	48 bits
DHCQ	0.2416	0.4653	0.6927	0.7786	0.1982	0.2217	0.2534	0.2728
DDH	0.4029	0.8223	0.8457	0.9068	0.0650	0.1103	0.1437	0.1889
DDQH	0.6322	0.9720	0.9780	0.9852	0.1185	0.2682	0.3410	0.4523
DCBH	**0.9753**	0.9899	0.9914	0.9922	0.7182	0.7317	0.7696	0.7862
DAGH	0.9744	0.9926	0.9938	**0.9946**	0.7284	0.7919	0.8172	0.8204
DDCH	0.9727	**0.9931**	**0.9939**	0.9944	**0.7642**	**0.8221**	**0.8315**	**0.8460**

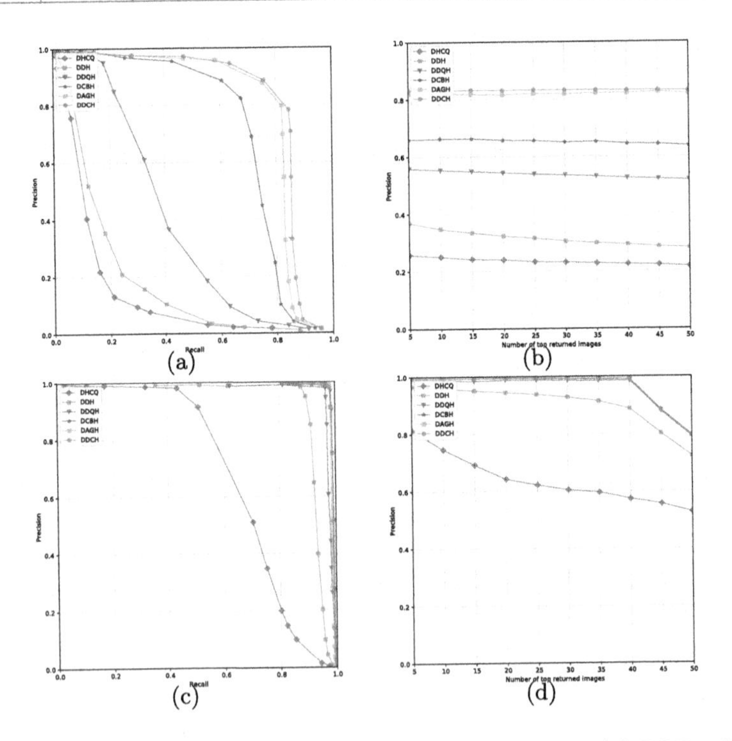

Fig. 3. The results on the two face dataset. (a) PR curves and (b) P@Top-K curves on FaceScrub; (c) PR curves and (d) P@Top-K curves on YouTube Faces

is comparable to DAGH. As shown in the table, all methods perform best at $k = 48$. In all experiments below, we evaluate all methods with $k = 48$bits. The PR-Curves and P@Top-K curves of all methods on the two datasets are shown in Fig. 3. These curves are consistent with the results in Table 1. It further validates the effectiveness of our method.

4.3 Ablation Study

We also investigate the effectiveness of different components in our model: multi-scale feature fusion, three-layer MLP for hashing, center-loss for feature extraction module and the hashing module. We reconstructed our network with different configurations as stated below, and evaluate their mAPs on the FaceScrub dataset with different code lengths.

DDCH-0 removes double center loss, replace three-layer MLP with one-layer fully connected network for the hashing function, removes the multi-scale fusion structure from the feature extraction module.

DDCH-1 adds center-loss in feature extraction module to *DDCH-0.*

DDCH-2 adds center-loss in hashing module to *DDCH-1.*

DDCH-3 using three-layer MLP as the hashing function in *DDCH-2.*

DDCH-4 adds three-level multi-scale fusion structure to *DDCH-3.*

The results are shown in Table 2. It is clear that all the components bring some improvements in the performances. This result justifies the effectiveness of our double center loss, multi-scale feature fusion, and multi-layer hashing functions.

In order to visually inspect the results of different configurations, we apply DDCH-0 ~DDCH-4 to the MNIST dataset [24]. The MNIST dataset contains hand-written digit images of 10 classes. Ideally, we want to learn a code space that consists of 10 clearly separable local sub-regions. After training on the MNIST training set, we obtained the hash codes of the MNIST test set, and visualize the codes using T-SNE in Fig. 4. As shown in the figure, a plain model without any cluster-center constraint (DDCH-0) results in cluttered code space where codes of different classes tend to overlap each other. As we add more and more components to the model, the codes of different classes are progressively more and more compact and separable from each other.

Table 2. The comparative results of different configurations.

Configuration	12 bits	24 bits	36 bits	48 bits
DDCH-0	0.6633	0.6923	0.7124	0.7546
DDCH-1	0.7118	0.7335	0.7769	0.8213
DDCH-2	0.7233	0.7582.	0.7804	0.8291
DDCH-3	0.7517	0.8138	0.8206	0.8367
DDCH-4	**0.7642**	**0.8221**	**0.8315**	**0.8460**

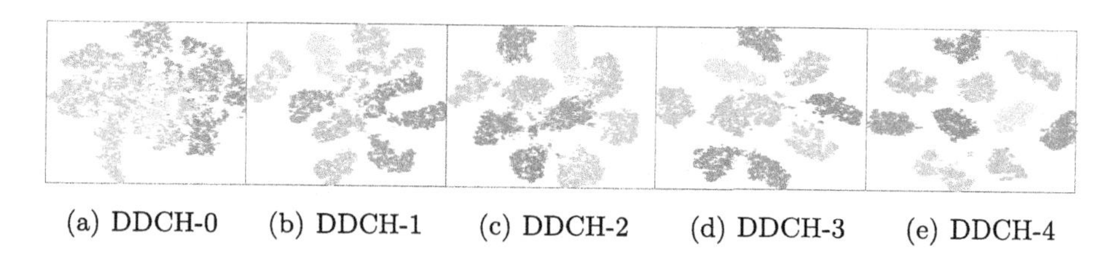

(a) DDCH-0 (b) DDCH-1 (c) DDCH-2 (d) DDCH-3 (e) DDCH-4

Fig. 4. T-SNE visualizations [25] of hash codes generated by DDCH-0 ~DDCH-4

4.4 Visualization

Using center-loss would make the learned features and codes to be compactly clustered in local regions. In order to verify the cluster phenomena, we apply DDCH to the MNIST dataset. After training DDCH on the MNIST training set, we expect to see 10 clusters in both the feature space and the code space. Since our method is closely related to the DCBH method, so we want to compare the learned features and codes from DCBH to our method. Figure 5 shows the visualizations. The DCBH uses center-loss for feature learning only, while our DDCH uses center-loss for both the feature and code learning. The learned features of DCBH and DDCH both are compact and separable, however, the DCBH-codes of different classes spread out, and the in-between boundaries are blurry. The DDCH-codes, on the other hand, are compact and separable, with clear boundaries between different classes as expected.

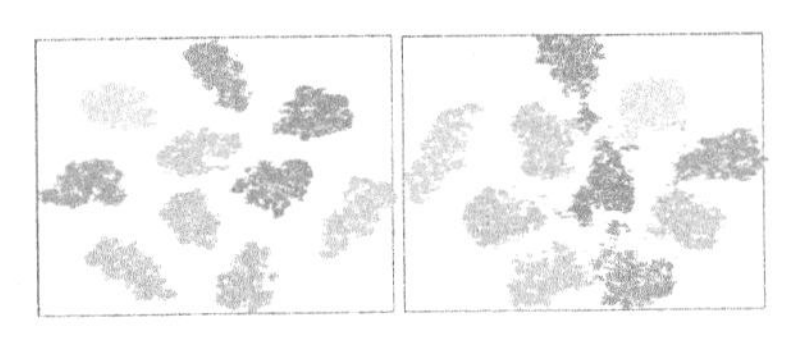

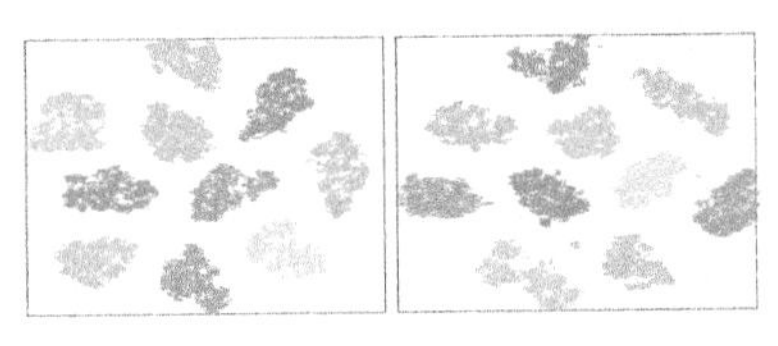

(a) DCBH-feature (b) DCBH-hash (c) DDCH-feature (d) DDCH-hash

Fig. 5. T-SNE visualizations of features and hash codes learned by DCBH (left) and DDCH (right)

Figure 6 shows three examples of top 5 images retrieved by DDCH and DCBH from the FaceScrub dataset. The results obtained by DCBH contains more errors,

Fig. 6. Top 5 images retrieved by our DDCH (top row) and DCBH (lower row).

and our DDCH performs better than DCBH. As can be seen from Fig. 6, the DCBH codes cannot distinguish subtle differences in the images of different identities. The reason is that DCBH codes are not well separable (see Fig. 5).

5 Conclusion

We propose a new deep hashing method, called Deep Double Center Hashing, for face image retrieval. In order to learn separable and more compact hash codes, we introduce the center-loss to constrain the hash function. This loss makes the learned codes for each identity to be compact and separable, enabling high-precision retrieval. We also propose a multi-scale feature fusion architecture for image feature extraction, which helps to express fine-grained face features, and a multi-layer MLP for feature hashing, which is more powerful to train optimizes the relative mapping from feature space to code space. We conduct comparative experiments with other methods on the datasets YouTube Face and Face Scrub. Experimental results prove that our design is more effective than the state-of-the-art deep hashing methods for face image retrieval.

References

1. Yandex, A.B., Lempitsky, V.: Aggregating local deep features for image retrieval. In: 2015 IEEE International Conference on Computer Vision (ICCV) (2015)
2. Guo, Y., Ding, G., Han, J.: Robust quantization for general similarity search. IEEE Trans. Image Process. **27**(2), 949–963 (2017)
3. Wang, J., Liu, W., Kumar, S., Chang, S.F.: Learning to hash for indexing big data–a survey. Proc. IEEE **104**(1), 34–57 (2015)
4. Wang, J., Kumar, S., Chang, S.F.: Semi-supervised hashing for large-scale search. IEEE Trans. Pattern Anal. Mach. Intell. **34**(12), 2393–2406 (2012)
5. Krizhevsky, A., Sutskever, I., Hinton, G.E.: ImageNet classification with deep convolutional neural networks. Adv. Neural Inf. Process. Syst. **25**, 1097–1105 (2012)
6. Xia, R., Pan, Y., Lai, H., Liu, C., Yan, S.: Supervised hashing for image retrieval via image representation learning. In: Twenty-eighth AAAI Conference on Artificial Intelligence (2014)
7. Andoni, A., Indyk, P.: Near-optimal hashing algorithms for approximate nearest neighbor in high dimensions. In: 2006 47th Annual IEEE Symposium on Foundations of Computer Science (FOCS 2006), pp. 459–468. IEEE (2006)
8. Xiong, Z., Wu, D., Gu, W., Zhang, H., Li, B., Wang, W.: Deep discrete attention guided hashing for face image retrieval. In: Proceedings of the 2020 International Conference on Multimedia Retrieval, pp. 136–144 (2020)
9. Lin, J., Li, Z., Tang, J.: Discriminative deep hashing for scalable face image retrieval. In: IJCAI, pp. 2266–2272 (2017)
10. Jang, Y.K., Jeong, D., Lee, S.H., Cho, N.I.: Deep clustering and block hashing network for face image retrieval. In: Jawahar, C.V., Li, H., Mori, G., Schindler, K. (eds.) ACCV 2018. LNCS, vol. 11366, pp. 325–339. Springer, Cham (2019). https://doi.org/10.1007/978-3-030-20876-9_21
11. Weiss, Y., Torralba, A., Fergus, R., et al.: Spectral hashing. In: NIPS, vol. 1, p. 4. Citeseer (2008)

12. Gong, Y., Lazebnik, S., Gordo, A., Perronnin, F.: Iterative quantization: a procrustean approach to learning binary codes for large-scale image retrieval. IEEE Trans. Pattern Anal. Mach. Intell. **35**(12), 2916–2929 (2012)
13. Liu, H., Wang, R., Shan, S., Chen, X.: Deep supervised hashing for fast image retrieval. In: Proceedings of the IEEE Conference on Computer Vision and Pattern Recognition, pp. 2064–2072 (2016)
14. Lai, H., Pan, Y., Liu, Y., Yan, S.: Simultaneous feature learning and hash coding with deep neural networks. In: Proceedings of the IEEE Conference on Computer Vision and Pattern Recognition, pp. 3270–3278 (2015)
15. Wu, G., Han, J., Lin, Z., Ding, G., Zhang, B., Ni, Q.: Joint image-text hashing for fast large-scale cross-media retrieval using self-supervised deep learning. IEEE Trans. Ind. Electron. **66**(12), 9868–9877 (2018)
16. Tang, J., Li, Z., Zhu, X.: Supervised deep hashing for scalable face image retrieval. Pattern Recogn. **75**, 25–32 (2018)
17. Tang, J., Lin, J., Li, Z., Yang, J.: Discriminative deep quantization hashing for face image retrieval. IEEE Trans. Neural Netw. Learn. Syst. **29**(12), 6154–6162 (2018)
18. Wen, Y., Zhang, K., Li, Z., Qiao, Yu.: A discriminative feature learning approach for deep face recognition. In: Leibe, B., Matas, J., Sebe, N., Welling, M. (eds.) ECCV 2016. LNCS, vol. 9911, pp. 499–515. Springer, Cham (2016). https://doi.org/10.1007/978-3-319-46478-7_31
19. Simonyan, K., Zisserman, A.: Very deep convolutional networks for large-scale image recognition. arXiv (2014)
20. Ioffe, S., Szegedy, C.: Batch normalization: accelerating deep network training by reducing internal covariate shift. In: International Conference on Machine Learning, pp. 448–456. PMLR (2015)
21. Glorot, X., Bengio, Y.: Understanding the difficulty of training deep feedforward neural networks. In: Proceedings of the Thirteenth International Conference on Artificial Intelligence and Statistics, pp. 249–256. JMLR Workshop and Conference Proceedings (2010)
22. Wolf, L., Hassner, T., Maoz, I.: Face recognition in unconstrained videos with matched background similarity. In: CVPR 2011, pp. 529–534. IEEE (2011)
23. Ng, H.W., Winkler, S.: A data-driven approach to cleaning large face datasets. In: 2014 IEEE International Conference on Image Processing (ICIP), pp. 343–347. IEEE (2014)
24. LeCun, Y., Bottou, L., Bengio, Y., Haffner, P.: Gradient-based learning applied to document recognition. Proc. IEEE **86**(11), 2278–2324 (1998)
25. Van der Maaten, L., Hinton, G.: Visualizing data using t-SNE. J. Mach. Learn. Res. **9**(11), 2579–2605 (2008)

A Novel Method of Cropped Images Forensics in Social Networks

Rongrong Gao[1,2], Xiaolong Li[1,2], and Yao Zhao[1,2](✉)

[1] Institute of Information Science, Beijing Jiaotong University, Beijing 100044, China
{gaorongrong,lixl,yzhao}@bjtu.edu.cn
[2] Beijing Key Laboratory of Advanced Information Science and Network Technology, Beijing Jiaotong University, Beijing 100044, China

Abstract. Operation chain forensics has achieved many advanced results, however, the spread of images in Social Networks has brought greater challenges to forensic research. The processing of images by Social Networks reduces the effectiveness of current forensics methods. An example is that if images are exchanged through Social Networks, the features they rely on will no longer be valid for detecting cropped images. To solve this problem, this paper proposes a novel method to complete the task of cropped forensics on Social Networks. By analyzing the image processing characteristics of Social Networks, we designed a special feature named "block artifacts grayscale" (BAGS) for crop detection based on block artifacts generated in JPEG compression. The proposed method is proved more accurate than some previous algorithms in cropped forensics in Social Networks, and it is also robust to serious re-compression. The proposed method modeled on Social Networks can promote other forensics issues. Moreover, the method provides a good direction for determining the process of enhanced filtering in Social Networks.

Keywords: Forensics of crop · Block artifact gray scale · Social networks · Operation chain

1 Introduction

With the rapid development of the Internet, a large number of images are spread on the Internet, many of which have been beautified or manipulated. Malicious manipulation brought serious issues in the areas of political and judicial, people's privacy and therefore threatened. Therefore, recent studies mostly concentrate on image forensics, including manipulation detection, operation chain detection, and DeepFakes, etc., and researches have achieved significant scientific results.

The operation chain is defined as a series of continuous operations, such as JPEG-crop-JPEG and copy-move [1–6], JPEG-resample-JPEG [7–10], detection of double JPEG [11–14], tampering localization [15–17], image synthesis detection [18–20] etc. In existing research, when JPEG compression is used as the last step in the operation chain, the research focus is the specific tampering detection

This work was supported by the National Natural Science Foundation of China (No. U1936212).

H. Ma et al. (Eds.): PRCV 2021, LNCS 13020, pp. 649–661, 2021.
https://doi.org/10.1007/978-3-030-88007-1_53

in the intermediate process is robust to JPEG. Mainstream methods are to use JPEG compression characteristics of the image. For JPEG images, the blocking processing introduces inter-block artifacts. This feature was utilized to analyze compression history [21]. Meanwhile, block grids will change during the process of image tampering [22]. Many researchers have altered these for their specific purposes, to complete the tampering forensics of specific operation chain.

Ideally, when the image is re-sampled, the block artifact grids (BAG) will change with the sampling factor, which is used to realize re-sampling operation chain detection [7]. This algorithm can maintain good robustness to JPEG compression, and meanwhile, theoretically estimate the re-sampling factor. The transformed block artifacts (TBAG) [10] and random feature selection [23] were proposed to conduct JPEG-resample-JPEG detection. Higher-dimensional and more complex features [24,25] were used for forensics when anti-forensics [26] appeared. There are also some methods to accomplish forensics tasks through deep learning [27,28]. The operation chain forensics can help us better distinguish the authenticity and determine the operation history of the image [29]. It is noted that crop is the most commonly used operation when performing image modification, and crop is also the basis for some complex operations such as copy-move and image synthesis. Therefore, we mainly focus on the detection of the cropped operation chain. The blocking artifact characteristics matrix (BACM) is defined in [1], BACM shows symmetrical characteristics in JPEG images, but the symmetry is destroyed after crop. There are also researches found that during image synthesis, the block grids of images have shifted because the synthetic parts come from different JPEG images. Detection of image synthesis and copy-move are realized in [30,31] based on this characteristic.

In this paper, we noticed that a large number of images spread through Social Networks in the actual scene, but few studies have considered the impact of Social Networks on forensics. The research of forensics in Social Networks has important practical significance, which is much more complicated than JPEG compression. It is found that the performance of existing methods decreases in the presence of Social Networks. To solve this problem, we proposed a new feature "block artifacts grayscale" (BAGS) to detect cropped images in Social Networks. Our proposed method is proved to be effective in the actual image propagation process, moreover, it is proven to surpass state of art techniques. The main contributions of this work are summarized as follows.

1) A novel feature BAGS is proposed to detect cropped images in Social Networks and achieved significant results.
2) The proposed method is robust against JPEG re-compression and is efficient in both aligned and non-aligned cropped forensics.
3) The proposed cropped forensics model in Social Networks can be used for other forensics problems, and further confirms the processing of Social Networks.

The rest of this paper is organized as follows. Section 2 details the background and related methods. Section 3 presents the definition of proposed BAGS in cropped forensics. Section 4 compares the performance with the advanced techniques. Finally, conclusions are given in Sect. 5.

2 Related Works

BACM was first defined in [1] for crop detection. Since the blocking processing during JPEG compression, there will be block artifacts at the boundary of each 8×8 block, it can be used to determine whether the image after JPEG compression and to distinguish different blocks. Correlations of pixel value intra-block and inter-block are calculated as follows

$$Z_{(x,y)} = |A + B - C - D| \tag{1}$$

$$Z'_{(x,y)} = |E + F - G - H| \tag{2}$$

where (x, y) is the coordinates of A in the blocks as shown in Fig. 1(a), and the coordinates of other points change followed A, $A \sim H$ is the pixel value of the corresponding position as shown in Fig. 1(b). Then, the histograms H_I, H_{II} of Z and Z' are computed, then K of the difference between H_I and H_{II} with the value $n \in [0, 255 \times 2]$ is computed as

$$K_{(x,y)}(n) = |H_I(n) - H_{II}(n)| \tag{3}$$

Finally, the matrix M called BACM are computed as

$$M(x, y) = \frac{\sum K_{(x,y)}(n)}{255 \times 2 + 1} \tag{4}$$

BACM of JPEG images and cropped images are shown in Fig. 3(b), BACM exhibits irregularity for uncompressed images, while JPEG images exhibit symmetry and symmetry is destroyed for cropped images. Thus, feature vector (14 features) is defined from the symmetry of BACM.

BAG of cropped images will have a position offset relative to original as shown in Fig. 1(c)(d), and grids crossover after re-compressed in Fig. 1(e). BAG is used to determine whether the image is cropped or not [21]. For each block, BAG is defined as

$$b(x, y) = \begin{cases} 1 & x \neq 8 \text{ } and \text{ } y \neq 8 \\ 0 & otherwise \end{cases} \quad 1 \leq x, y \leq 8 \tag{5}$$

Thus, BAG of an image $(H \times W)$ will be denoted as

$$B(x, y) = \left(1 - \sum_{h=1}^{H/T} \delta(x - hT)\right) \cdot \left(1 - \sum_{w=1}^{W/T} \delta(y - wT)\right) \quad 1 \leq x \leq H, 1 \leq y \leq W \tag{6}$$

where T indicates the periodicity of the BAG, and $T = 8$ for JPEG images, $\delta(\cdot)$is the Dirac-Delta function, which is one at zero and zero at any other value.

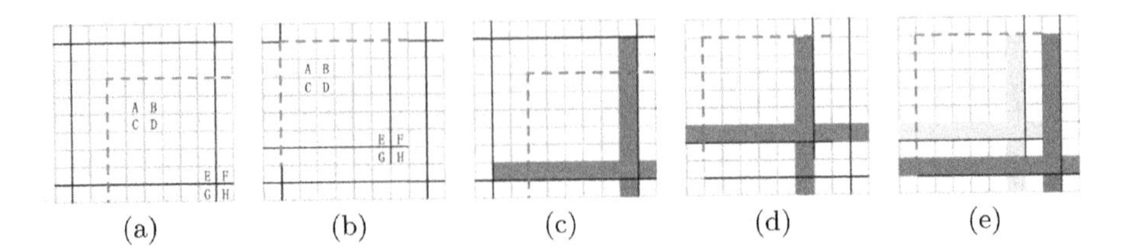

Fig. 1. The change of BACM and BAG during cropping within 8×8 block.

As mentioned above, the robustness of BACM and BAG against lossy JPEG is still poor. Meanwhile, images are spread in Social Networks, which is more complex than JPEG compression. It is imperative for us to propose a new method to accomplish the task of forensics in Social Networks. Accordingly, our work focuses on crop forensics in the presence of Social Networks, which has crucial practical significance.

3 Proposed Method

Our target is to improve the performance of cropped detection in Social Networks as shown in Fig. 2, overcome the week robustness against JPEG re-compression, including aligned and non-aligned crop. We first introduce the block artifacts for aligned and non-aligned crop and define BAGS. Then we analyze BAGS in the absence of re-compression. Finally, we describe BAGS change for Social Networks exchanged images.

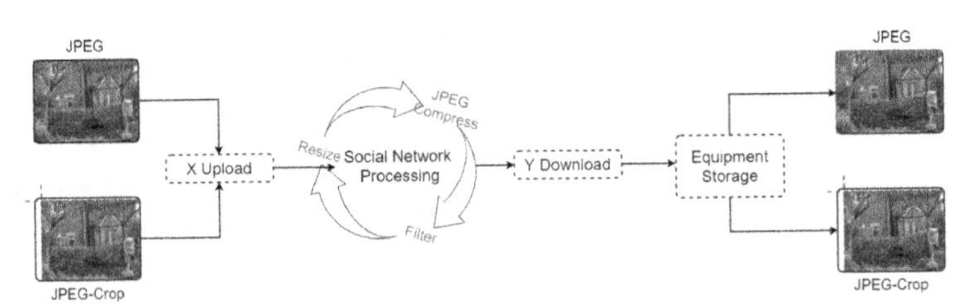

Fig. 2. The overall framework of our proposed method, in which Social Networks mainly include three types of operations, resize, compression, and enhanced filtering, and the dashed boxes represent the specific steps of WeChat.

3.1 Aligned and Non-aligned Crop

Aligned crop refers to crop an image along the block boundary, in which $B(x, y) = 0$ defined in (6), after aligned cropped and re-compression, latticed grids overlap the original grids. Although in general, images are randomly cropped and difficult to ensure crop alignment, forensic work is still needed to accurately determine in aligned cropped.

In the case of non-aligned crop, the contour matrix is shown in Fig. 3(b). The matrix of the uncompressed image shows irregularity, while JPEG images show symmetry, and cropped images show random, which is consistent with the definition. With these two distinct patterns, we can determine whether the image has been cropped or not.

In the case of aligned crop, the contour matrix of the cropped image also shows strong symmetry, due to the consistency of block grids, which is similar to JPEG images as shown in Fig. 3(a). This makes it difficult to detect cropped images. Therefore, it is suggested to use BAGS for crop detection. BAGS still distinguishable as shown in Fig. 3(c) for aligned crop. BAGS is defined as the density of BACM, in which M indicates BACM according to (4), $\theta_1 = min(M), \theta_2 = max(M)$, BAGS is defined as

$$G(x, y) = \frac{M(x, y) - \theta_1}{\theta_2 - \theta_1} \tag{7}$$

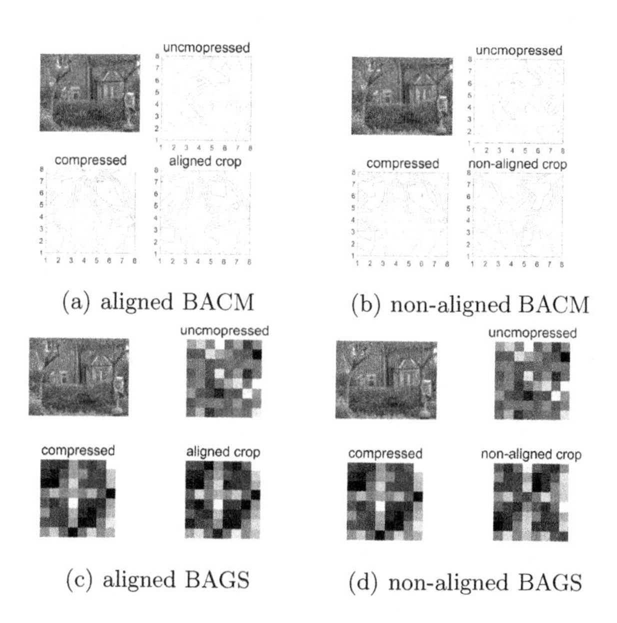

(a) aligned BACM (b) non-aligned BACM

(c) aligned BAGS (d) non-aligned BAGS

Fig. 3. BACM and BAGS of aligned and non-aligned crop image.

3.2 BAGS in Re-compression

As introduced above, BAG overlapped in aligned crop and crossover in the non-aligned crop when the image was re-compressed, BACM of cropped image in Fig. 4(a) is distinguished when JPEG re-compressed quality factor QF_2 is larger. However, it is noted that whether it is aligned or non-aligned crop, the image only retains the BAG of re-compression, covering the traces of the first compression, and also covers cropping traces when re-compression is severe. Due to the loss of cropped traces, BACM of JPEG and cropped images in Fig. 4(b) show similar symmetry, BACM may no longer determine cropped images, which also proves that BACM is less robust against JPEG. Our proposed BAGS in Fig. 4(d) is still possible to distinguish re-compressed crop images.

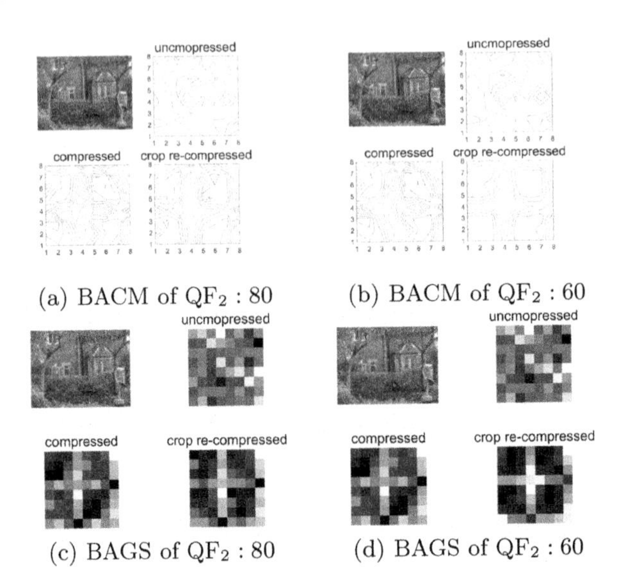

Fig. 4. BACM and BAGS in re-compressed crop images with different QF_2.

3.3 BAGS in Social Networks

Attacks on images are more complex than JPEG compression, like images spread on Social Networks, which brings great challenges to our forensic work. Thus, we mainly consider Social Networks as an attack method. In [32], Sun *et al.* find that the processing of Social Networks includes resizing, JPEG, and enhanced filtering. We extracted BACM after images are exchanged on Social Networks.

For aligned crop, the exchanged JPEG and cropped images have similar symmetry of BACM as shown in Fig. 5(a). The biggest impact on the image, in this case, is the JPEG compression of Social Networks. Due to the offset of the pixel values of the cropped image and JPEG image, BAGS can discriminate JPEG and cropped images in the presence of Social Networks as shown in Fig. 5(c)(d).

For non-aligned crop, BACM of exchanged JPEG images and cropped images in Fig. 5(b) still shown symmetry and asymmetry, it can not determine particular processing when the influence of Social Networks is weak. And BAGS of JPEG image in Fig. 5(d) still retains block artifacts, while grids crossover in the cropped image. The pixel values changed since enhanced filtering in Social Networks and brought about inconsistency for BAGS. After determining quality factor of exchanged images, we can obtain the pixel change according to BAGS, then estimate the influence of enhancement filter.

4 Experimental Results

In this section, we assess the performance of the proposed crop detection method in the absence of re-compression and various realistic scenarios of Social Networks. Our proposed algorithm BAGS is compared with advanced cropped

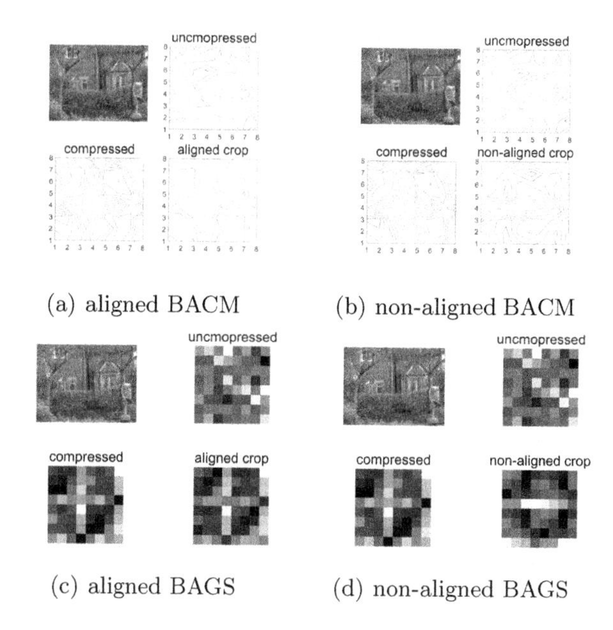

Fig. 5. BACM and BAGS of aligned and non-aligned crop images exchanged on Social Networks.

forensics methods BACM(64), feature vector from BACM(14) which is called feature14 in the following, and PCA of BACM(32), etc. Our experimental data include JPEG images, cropped images, re-compressed images, and Social Networks exchanged images. JPEG and cropped images were collected by compress and crop the uncompressed images in UCID [33], and then images were exchanged through Social Networks to get exchanged images. After collecting the experimental dataset, the data were divided into training and test set according to the ratio of 7 : 3, then the four features mentioned above are extracted separately and fed to the SVM. Accuracy $((TP + TN)/(TP + FP + TN + FN))$ and precision $(TP/(TP + FP))$ were used to evaluate the performance of the algorithm, where TP is true positive, TN is true negative, FP is false positive and FN is false negative respectively in binary classification, cropped images were taken as positive samples while JPEG images were negative samples. All results of experiments are described in the following scenarios to verify the effectiveness of our proposed method.

4.1 Experiments on Re-compression

The JPEG robustness experiment does not distinguish whether the crop is aligned or not. Our first compression quality factor QF_1 of the UCID image includes four intervals of $[50, 89]$. And the re-compression quality factor QF_2 of the cropped image ranges from 50 to 90. The accuracy curves of each feature in the four intervals are shown in Fig. 6. On the whole, the accuracy of our proposed

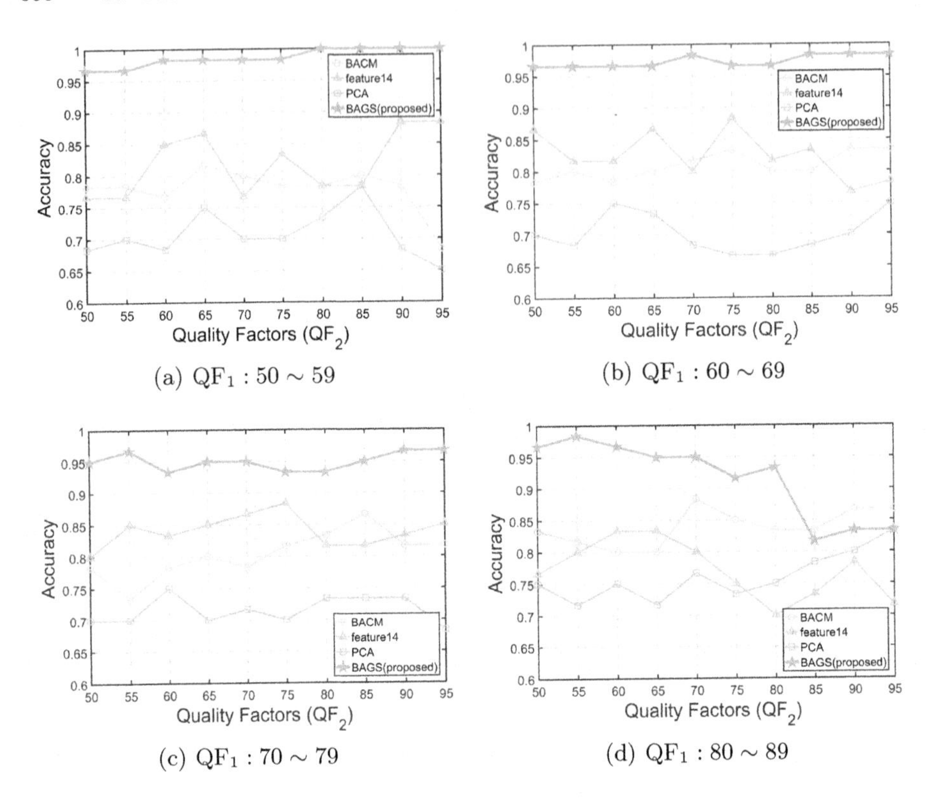

(a) $\mathrm{QF}_1 : 50 \sim 59$ (b) $\mathrm{QF}_1 : 60 \sim 69$

(c) $\mathrm{QF}_1 : 70 \sim 79$ (d) $\mathrm{QF}_1 : 80 \sim 89$

Fig. 6. Accuracy of all compared methods in four intervals of QF_1.

method is better than other methods. Meanwhile, it is founded that when QF_1 is low, especially in the range of [50, 69], the accuracy of our proposed BAGS is much higher than other features. Even when QF_2 is strong at around 50, the accuracy rate of 0.95 is maintained. This verified that our proposed method has strong robustness against JPEG re-compression.

Especially, when QF_1 is in the range of $80 \sim 89$, the accuracy rate decreases with the increase of the re-compression quality factor in Fig. 6(d). That is because when QF_2 is strong and close to QF_1, Characteristics of the first compression are retained in BAGS, while cropped features were lost, which resulting in a decrease inaccuracy.

It is also noted that when the re-compression quality factor is weak, BAGS not only retained the cropped feature but also showed the difference between the two compression. This is consistent with the description in Sect. 3.2.

4.2 Experiments on Social Networks (Facebook)

During the experiments on Facebook, uncompressed UCID images were compressed into JPEG images with four intervals of QF_1 in $50 \sim 90$. The aligned

Table 1. Accuracy of Social Networks exchanged aligned crop images

QF_1	BACM(64)	feature14(14)	PCA(32)	**BAGS(64)**
$50 \sim 59$	0.7667	0.8333	0.7667	**0.9333**
$60 \sim 69$	0.7333	0.8	0.6833	**0.9167**
$70 \sim 79$	0.6833	0.8	0.65	**0.8833**
$80 \sim 89$	0.6833	0.7667	0.5677	**0.9333**

Table 2. Precision of Social Networks exchanged aligned crop images

QF_1	BACM(64)	feature14(14)	PCA(32)	**BAGS(64)**
$50 \sim 59$	1	0.8333	1	**1**
$60 \sim 69$	**1**	0.7813	0.8235	0.963
$70 \sim 79$	0.9231	0.95	**1**	0.96
$80 \sim 89$	0.9231	0.75	0.625	**1**

Table 3. Accuracy of Social Networks exchanged non-aligned crop images

QF_1	BACM(64)	feature14(14)	PCA(32)	**BAGS(64)**
$50 \sim 59$	0.8333	0.85	0.8	**1**
$60 \sim 69$	0.8	0.85	0.7333	**0.9833**
$70 \sim 79$	0.8	0.7833	0.7667	**0.95**
$80 \sim 89$	0.8167	0.7833	0.7333	**0.95**

Table 4. Precision of Social Networks exchanged non-aligned crop images

QF_1	BACM(64)	feature14(14)	PCA(32)	**BAGS(64)**
$50 \sim 59$	1	0.8889	0.95	**1**
$60 \sim 69$	**1**	0.8	0.85	0.9677
$70 \sim 79$	1	0.9048	0.8636	**1**
$80 \sim 89$	1	0.7576	0.8889	**1**

and non-aligned cropped images were performed separately. All of the images were exchanged on Facebook. For the exchanged JPEG images and cropped images, the BAGS are extracted respectively then fed to SVM and compared with advanced methods, the accuracy and precision of all methods are shown in Table 1.

It can be seen from the Tables 1, 2, 3 and 4 that, in the four cases of QF_1, the accuracy of BAGS is relatively higher. The BACM symmetry of JPEG images is more obvious when QF_1 is weak, so it is easier to be recognized in $50 \sim 59$. In the case of the non-aligned crop, the accuracy of our proposed method is lower when QF_1 is $70 \sim 89$ than $50 \sim 69$. As stated in Sect. 3.3, this is a change brought about by the compressed process of Social Networks, it is inferred that JPEG compression quality factor of Social Networks is close to the range of 70 $\sim$ 89. In the case of aligned crop, BACM of cropped images still maintain symmetry, and the accuracy of advanced methods decreases significantly, while BAGS decreases slightly, but our proposed BAGS still maintains a higher accuracy. Meanwhile, the overall precision is better than the previous methods.

4.3 Experiments on Wechat Moments

Our proposed method was also applied in Wechat Moments. In the experiment of Wechat, it was considered that the difference between uploaded and downloaded devices as shown in Fig. 2. Through experiments, it was found that when the image was downloaded through the Android devices, the JPEG image cannot maintain symmetry of BACM in Fig. 7(a), while Apple download images hold on in Fig. 7(b). Therefore, in the following experiments, we mainly conducted experiments on the way of uploading from Apple devices and downloading from Android devices. When downloading an image using an Android device, BACM cannot determine whether the image was cropped, but BAGS retains copped traces in Fig. 7(c). Experiments were set that one user U_1 to publish Moments images through Apple devices, including JPEG images and cropped images, and another user U_2 downloaded two classes of images through the Android device, then extracted features fed to SVM. The accuracy of each feature is given in Table 5.

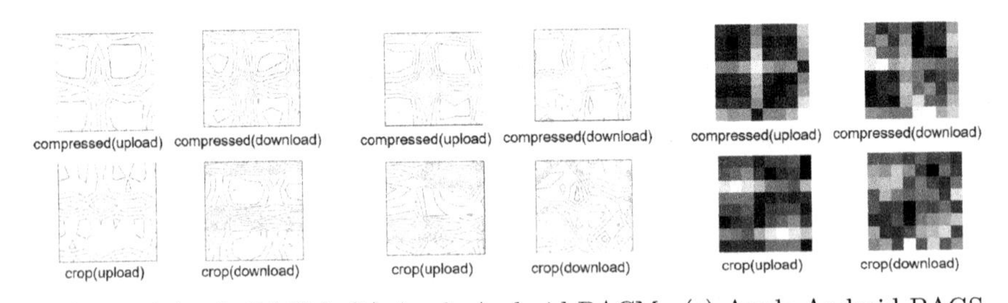

(a) Android-Apple BACM (b) Apple-Android BACM (c) Apple-Android BAGS

Fig. 7. BACM and BAGS of the exchanged image from different devices (Apple and Android devices).

It can be seen from Table 5 that the accuracy of compared methods on WeChat has decreased significantly, while our proposed method still achieves a high accuracy rate.

All in all, our proposed method has great advantages in both aligned and non-aligned crop and also has strong robustness against strong JPEG re-compression, and the cropped forensics in Social Networks has the best performance compared with existing algorithms.

Table 5. Accuracy and Precision in WeChat exchanged images

Feature	Accuracy	Precision
BACM(64)	0.6563	1
feature14(14)	0.6563	1
PCA(32)	0.625	1
BAGS(64)	**0.9375**	1

5 Conclusion

In this paper, we proposed a novel feature called BAGS to detect whether the Social Network exchanged image is cropped. Firstly, BAGS is defined based on the block artifacts grids, and then it is verified that the proposed feature is better than advanced techniques against JPEG re-compression, it is finally described the changes of cropped images in Social Networks. The BAGS we proposed is a good way to complete the cropped forensics of exchanged images in Social Networks when other features are not applicable, and the establishment of Social Networks models also promoted.

References

1. Luo, W., Qu, Z., Huang, J., Qiu, G.: A novel method for detecting cropped and recompressed image block. In: Proceedings of the IEEE International Conference on Acoustics, Speech, and Signal Processing, pp. 217–220 (2007)
2. Li, W., Yuan, Y., Yu, N.: Passive detection of doctored JPEG image via block artifact grid extraction. Signal Process. **89**(9), 1821–1829 (2009)
3. Christlein, V., Riess, C., Jordan, J., Riess, C., Angelopoulou, E.: An evaluation of popular copy-move forgery detection approaches. IEEE Trans. Inf. Forensics Secur. **7**(6), 1841–1854 (2012)
4. Fridrich, A.J., Soukal, B.D., Lukáš, A.J.: Detection of copy-move forgery in digital images. In: Proceedings of Digital Forensic Research Workshop (2003)
5. Huynh, K.T., Ly, T.N., Nguyen, P.T.: Improving the accuracy in copy-move image detection: a model of sharpness and blurriness. SN Comput. Sci. **2**(4), 1–11 (2021)
6. Barni, M., Phan, Q.T., Tondi, B.: Copy move source-target disambiguation through multi-branch CNNs. IEEE Trans. Inf. Forensics Secur. **16**, 1825–1840 (2020)
7. Bianchi, T., Piva, A.: Reverse engineering of double JPEG compression in the presence of image resizing. In: IEEE International Workshop on Information Forensics and Security, pp. 127–132 (2012)
8. Nguyen, H.C., Katzenbeisser, S.: Detecting resized double JPEG compressed images-using support vector machine. In: IFIP International Conference on Communications and Multimedia Security, pp. 113–122 (2013)
9. Peng, A., Wu, Y., Kang, X.: Revealing traces of image resampling and resampling antiforensics. Adv. Multimed. **2017**, 7130491 (2017)
10. Chen, Z., Zhao, Y., Ni, R.: Detection of operation chain: JPEG-resampling-JPEG. Signal Process. Image Commun. **57**, 8–20 (2017)

11. Bianchi, T., Piva, A.: Detection of nonaligned double JPEG compression based on integer periodicity maps. IEEE Trans. Inf. Forensics Secur. **7**(2), 842–848 (2011)
12. Pevny, T., Fridrich, J.: Detection of double-compression in JPEG images for applications in steganography. IEEE Trans. Inf. Forensics Secur. **3**(2), 247–258 (2008)
13. Huang, F., Huang, J., Shi, Y.Q.: Detecting double JPEG compression with the same quantization matrix. IEEE Trans. Inf. Forensics Secur. **5**(4), 848–856 (2010)
14. Park, J., Cho, D., Ahn, W., Lee, H.K.: Double JPEG detection in mixed JPEG quality factors using deep convolutional neural network. In: Proceedings of the European Conference on Computer Vision, pp. 636–652 (2018)
15. Bianchi, T., Piva, A.: Image forgery localization via block-grained analysis of JPEG artifacts. IEEE Trans. Inf. Forensics Secur. **7**(3), 1003–1017 (2012)
16. Chen, Y.L., Hsu, C.T.: Image tampering detection by blocking periodicity analysis in JPEG compressed images. In: IEEE Workshop on Multimedia Signal Processing, pp. 803–808 (2008)
17. Ye, S., Sun, Q., Chang, E.C.: Detecting digital image forgeries by measuring inconsistencies of blocking artifact. In: IEEE International Conference on Multimedia and Expo, pp. 12–15 (2007)
18. Zhan, F., Lu, S., Xue, C.: Verisimilar image synthesis for accurate detection and recognition of texts in scenes. In: Proceedings of the European Conference on Computer Vision, pp. 249–266 (2018)
19. Dundar, A., Sapra, K., Liu, G., Tao, A., Catanzaro, B.: Panoptic-based image synthesis. In: Proceedings of the IEEE/CVF Conference on Computer Vision and Pattern Recognition, pp. 8070–8079 (2020)
20. Zhan, F., Zhu, H., Lu, S.: Spatial fusion GAN for image synthesis. In: Proceedings of the IEEE/CVF Conference on Computer Vision and Pattern Recognition, pp. 3653–3662 (2019)
21. Fan, Z., De Queiroz, R.L.: Identification of bitmap compression history: JPEG detection and quantizer estimation. IEEE Trans. Image Process. **12**(2), 230–235 (2003)
22. Chen, Y.L., Hsu, C.T.: Detecting recompression of JPEG images via periodicity analysis of compression artifacts for tampering detection. IEEE Trans. Inf. Forensics Secur. **6**(2), 396–406 (2011)
23. Chen, Z., Tondi, B., Li, X., Ni, R., Zhao, Y., Barni, M.: Secure detection of image manipulation by means of random feature selection. IEEE Trans. Inf. Forensics Secur. **14**(9), 2454–2469 (2019)
24. Holub, V., Fridrich, J.: Low-complexity features for JPEG steganalysis using undecimated DCT. IEEE Trans. Inf. Forensics Secur. **10**(2), 219–228 (2014)
25. Chen, Z., Zhao, Y., Ni, R.: Detection for operation chain: Histogram equalization and dither-like operation. KSII Trans. Internet Inf. Syst. **9**(9), 3751–3770 (2015)
26. Kirchner, M., Bohme, R.: Hiding traces of resampling in digital images. IEEE Trans. Inf. Forensics Secur. **3**(4), 582–592 (2008)
27. Barni, M., Bondi, L., Bonettini, N., Bestagini, P., Costanzo, A., Maggini, M., Tondi, B., Tubaro, S.: Aligned and non-aligned double JPEG detection using convolutional neural networks. J. Vis. Commun. Image Represent. **49**, 153–163 (2017)
28. Liao, X., Li, K., Zhu, X., Liu, K.R.: Robust detection of image operator chain with two-stream convolutional neural network. IEEE J. Sel. Topics Signal Process. **14**(5), 955–968 (2020)
29. Conotter, V., Comesana, P., Pérez-González, F.: Forensic detection of processing operator chains: recovering the history of filtered JPEG images. IEEE Trans. Inf. Forensics Secur. **10**(11), 2257–2269 (2015)

30. Barni, M., Costanzo, A., Sabatini, L.: Identification of cut & paste tampering by means of double-JPEG detection and image segmentation. In: Proceedings of 2010 IEEE International Symposium on Circuits and Systems, pp. 1687–1690 (2010)
31. Li, Y., Zhou, J.: Fast and effective image copy-move forgery detection via hierarchical feature point matching. IEEE Trans. Inf. Forensics Secur. **14**(5), 1307–1322 (2018)
32. Sun, W., Zhou, J., Lyu, R., Zhu, S.: Processing-aware privacy-preserving photo sharing over online social networks. In: Proceedings of the 24th ACM International Conference on Multimedia, pp. 581–585 (2016)
33. Schaefer, G., Stich, M.: UCID: an uncompressed color image database. Storage Retrieval Meth. Appl. Multimed. **2004**(5307), 472–480 (2003)

MGD-GAN: Text-to-Pedestrian Generation Through Multi-grained Discrimination

Shengyu Zhang[1], Donghui Wang[1], Zhou Zhao[1], Siliang Tang[1], Kun Kuang[1], Di Xie[2], and Fei Wu[1](✉)

[1] Zhejiang University, Hangzhou, China
[2] Hikvision Research Institute, Hangzhou, China
{sy_zhang,dhwang,zhaozhou,siliang,wufei}@zju.edu.cn, xiedi@hikvision.com

Abstract. In this paper, we investigate the problem of text-to-pedestrian synthesis, which has many potential applications in art, design, and video surveillance. Existing methods for text-to-bird/flower synthesis are still far from solving this fine-grained image generation problem, due to the complex structure and heterogeneous appearance that the pedestrians naturally take on. To this end, we propose the Multi-Grained Discrimination enhanced Generative Adversarial Network, that capitalizes a human-part-based Discriminator (HPD) and a self-cross-attended (SCA) global Discriminator in order to capture the coherence of the complex body structure. A fined-grained word-level attention mechanism is employed in the HPD module to enforce diversified appearance and vivid details. In addition, two pedestrian generation metrics, named Pose Score and Pose Variance, are devised to evaluate the generation quality and diversity. We conduct extensive experiments and ablation studies on the caption-annotated pedestrian dataset, CUHK Person Description Dataset. The substantial improvement over the various metrics demonstrates the efficacy of MGD-GAN on the text-to-pedestrian synthesis scenario.

1 Introduction

Synthesizing visually authentic images from textual descriptions is a representative topic of creative AI systems. It requires a high-level understanding of natural language descriptions, usually vague and incomplete, and the imaginative ability in order to draw visual scenes. Recently, methods based on deep generative models, e.g., Generative Adversarial Networks [3], have accomplished promising outcomes in this field. With the generation quality increasing, more challenging objectives are expected, i.e., fine-grained image generation.

Fine-grained image generation, as the term implies, focuses on fine-grained categories, such as faces of a certain individual or objects within a subcategory. Existing methods for fined-grained text-to-image generation only focuses on

First author is currently a student.
Wei Wu is the corresponding author.

H. Ma et al. (Eds.): PRCV 2021, LNCS 13020, pp. 662–673, 2021.
https://doi.org/10.1007/978-3-030-88007-1_54

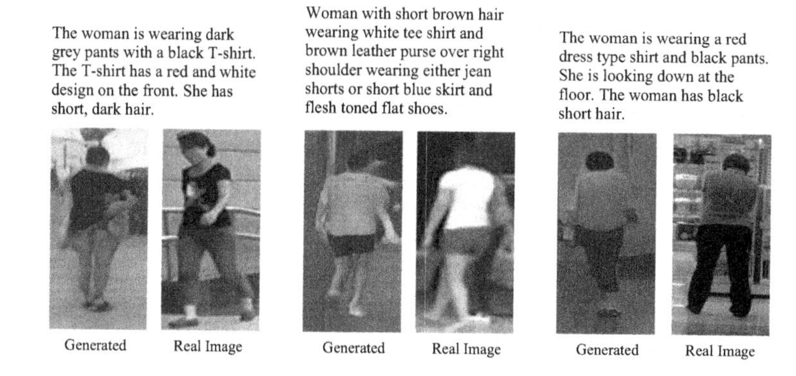

Fig. 1. Examples generated by our proposed MGD-GAN.

birds and flowers [12,15,17,24]. However, text-to-pedestrian generation (Fig. 1) is a new research direction and has many potential applications in movie making, art creation, and video surveillance – for instance, the appropriate generation of one suspect's portrait according to a witness or victim descriptions. We have carried out an experiment to demonstrate that pedestrian generation can significantly improve the text-to-person search results. Also, text-to-pedestrian generation is a rather difficult problem due to the nature of larger intra-class variances in person generation. The intra-class variances can be summarized as the inter-person variances, such as gender, age, height, and the intra-person variances, such as appearance variances, pose variances.

Not surprisingly, existing methods that are designed for a general text-to-image synthesis problem are still far from solving the pedestrian generation problem. On the one hand, existing methods fail to leverage the task-specific prior knowledge in the pedestrian generation process. To this end, we investigate the person re-id techniques and propose a human-part-based Discriminator, named HPD, which independently penalizes imperfect structure at each human part. On the other hand, as stated above, we view text-to-pedestrian generation as a fined-grained generation problem. Although there are works [15,24] that exploits a fine-grained word-level attention mechanism in the generation process, the absence of fine-grained discrimination may induce an unbalanced competition between the Generator and Discriminator since they have different levels of abilities. In essence, the Generator focuses on more fine-grained local details while the Discriminator can only capture the global coherence of the image.

To handle the problems above, we enhance the human-part-based Discriminator with a Visual-Semantic Attention module, named VISA. The difference of a regular conditional discriminator and a VISA enhanced discriminator is that rather than signifies the concatenated input of a fake image and a sentence "real" or "fake", whereas the VISA enhanced discriminator scores the importance of each word for each image region and signifies each region "real" or "fake" concerning different words. This setting can help discriminator better penalize the inconsistency between textual descriptions and generated images in a words-regions level, which makes VISA-HPD a local discriminator. As a counterpart, we construct another self-cross-attended global discriminator functioning at the sentence-image level, penalizing irregular body structure as a whole.

To the best of our knowledge, our work is the initiative to do text-to-pedestrian synthesis. The most commonly used evaluation metric for the generic image generation, the Inception Score, relies on a classification model. However, there are no specific sub-categories for pedestrian, and thus an Inception Score model trained on a generic image recognition dataset [23] may fail to capture the quality and diversity of pedestrian generation results. To this end, we propose two evaluation metrics, named Pose Score and Pose Variance, grading the generation quality and generation variety simultaneously. Based on a Pose Estimation model pre-trained on a larger pedestrian dataset, the evaluation is easy to compute in practice. To summarize, this paper makes the following key contributions:

- We extend previous fine-grained text-to-image synthesis research by advocating a more challenged pedestrian generation problem.
- We proposed a novel MGD-GAN model that exploits a visual-semantic attention enhanced human-part-based Discriminator as well as a self-cross-attended global Discriminator, alleviating the unbalance competition bottleneck between Discriminator and Generator.
- We present two novel pedestrian generation metrics, named pose score and pose variance, as a non-trivial complement to existing ones.
- The proposed MGD-GAN model achieves the best results on the challenging CUHK-PEDES dataset. The extensive ablation studies verify the effectiveness of each component within the MGD-GAN model in different aspects, i.e., generation quality, generation variety, and the consistency with the textual description.

2 Related Works

Previous **person image generation** methods mainly focus on the pose-transfer task, which aims to generate person images with the expected poses [4,10,11,25]. Recently, Zhou *et al.* [23] presents a two-stage model to manipulate the visual appearance (pose and attribute) of a person image according to natural language descriptions. Despite this progress, our work synthesizes a person image based solely on the textual descriptions, which is within the research domain of multi-modal processing [20,21].

The challenging and open-ended nature of **text-to-image generation** lends itself to a variety of diverse models. There are mainly two lines of works. The first is generating complex scenes with multiple objects [5,8]. The other is the generation on fine-grained categories. In this line of works, GAN-INT-CLS [12] firstly learns a joint embedding of text and images to capture the fine-grained relationship between them. Zhang *et al.* [17] stacks multiple generators to remedy the fine-grained details from previous generators. vmCAN [18] leverages meaningful visual cues extracted from a visual memories to enhance text-to-image

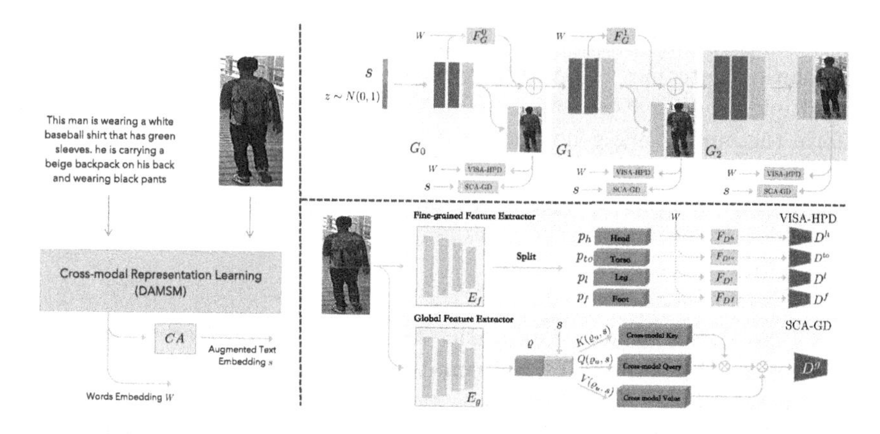

Fig. 2. The overall scheme of our proposed MGD-GAN model.

generation. HD-GAN [22] uses hierarchically-nested discriminators to generate photographic images. AttnGAN [15] can synthesize fine-grained details of subtle regions by performing an attention equipped generative network. LSIC [19] completes a broken image with natural language and thus permits grounded and controllable image completion. The main difference between our method and previous arts is two-fold. First, we advocate solving a more challenging text-to-pedestrian generation problem besides the existing birds/flowers generation. Second, we present the global-local attentional discriminators to ease the unbalanced adversarial training problem within previous works.

3 Multi-Grained Discrimination Enhanced Generative Adversarial Network

In this section, we will elaborate on each building block comprising the MGD-GAN model. As shown in Fig. 2, MGD-GAN mainly embodies three modules: the multi-stage attentional generator, the visual-semantic attention enhanced human-part-based discriminator, and the self-cross-attended global discriminator. Technically, given the sentence feature s and words feature W, we want to synthesize the desired human picture I. The sentence feature $s \in \mathbb{R}^{N_s}$ and the words feature $W = \{w_1, w_2, ..., w_T\}, w_t \in \mathbb{R}^{N_w}$ are pre-trained using a cross-modal representation learning network named DAMSM [15]. T is the specified max sequence length of sentences, N_s is the dimension of sentence features and N_w is the dimension of words features.

3.1 Multi-stage Generation Strategy

We follow a standard multi-stage baseline [17,24] to develop our generator. Rather than creating a high-resolution picture directly from the textual description s, we simplify the generation process by first creating a low-resolution picture I_0, which relies on only rough shapes and colors. To remedy fine-grained errors, we incorporate a refinement generator G_{i+1} which is conditioned on the low-resolution outcomes I_i and the words features W. The first generator G_0 and multiple refinement generators $\{G_1, ..., G_{m-1}\}$ can be stacked sequentially and form the multi-stage generation process.

In order to enforce the model to produce fine-grained details within the words feature [15] and carry out compositional modeling [2], we construct several attention modules F_G^i for adjacent generators (G_i, G_{i+1}). The structure of F_G^i is the same as VISA.

3.2 VISA-HPD: Visual-Semantic Attention Enhanced Human-Part-based Discriminator

To leverage some domain knowledge for the fine-grained text-to-pedestrian problem, we investigate some popular methods used in person re-identification research field and propose a human-part-based approach. The core idea in the behind is simple but effective. In order to tackle the complex structure of the human body, we split the encoded feature ρ from the generated image I vertically and equally into several parts, i.e., the head part p_h, the torso part p_{to}, the leg part p_l and the foot part p_f. The correctness, the coherence, and the faithfulness to the corresponding words and phrases of each human part are individually scored by a related discriminator, i.e., $D^\kappa, \kappa \in \{h, to, l, f\}$. We note that human part discriminators share a common fine-grained feature encoder E_f but do not share parameters.

Previous GAN-based models for fine-grained text-to-image generation typically equip the Generator with attention mechanisms. In order to grade the relevance of regions and words, we develop a Visual-Semantic Attention, named VISA, for each human part discriminator. The VISA module F_{D^κ} takes the words features $W \in \mathbb{R}^{N_w \times T}$ and the human part regions features $p_\kappa \in \mathbb{R}^{N_r \times N/4}$ as input. N is the number of regions in ρ and N_r is the dimension of regions.

For each image region feature ρ_u, we compute the attention weights with words features W using a Fully-Connected layer f and a softmax layer. We obtain the final words-attended region feature as the addition of original region feature and the weighted sum of words features.

$$\alpha_{u,t} = \frac{\exp(f(\rho_u, w_t))}{\sum_t \exp(f(\rho_u, w_t))}, \ \rho_u' = \sum_t \alpha_{u,t} w_t, \ r_u = \rho_u + \rho_u' \tag{1}$$

Hence, the discrimination process for generated image I can be expressed as:

$$\rho = E_f(I), \quad p_h, p_{to}, p_l, p_f = Split(\rho), \quad y_\kappa = D^\kappa(F_{D^\kappa}(W, p_\kappa)) \tag{2}$$

where $y_\kappa \in (0, 1)$ is the output of each human part discriminator, which grades the generation quality and faithfulness to attended words of each human part.

3.3 Self-cross-attended Global Discriminator

As a non-trivial counterpart of the local discriminator stated above, a self-cross-attended global discriminator is proposed to capture the coherence of the body structure as a whole. Extending beyond the Self-Attention GAN [16], we not only build a self-attention map of spatial context but also harness cross-modal context at the sentence-image level. Specifically, we build the self-cross-attention scores of cross-modal context by:

$$\beta_{v,u} = \frac{\exp(c_{uv})}{\sum_{u=1}^{N}\exp(c_{uv})}, \text{ where } c_{uv} = K(\varrho_u, s)^T Q(\varrho_v, s) \tag{3}$$

where $\varrho_u, \varrho_v \in \mathbb{R}^{N_r}$ denote region features within $\varrho = E_g(I)$ extracted by a global feature extractor E_g. K and Q are joint image-sentence feature spaces, formulated as.

$$K(\varrho_u, s) = W_k([\varrho_u, s]),\ o_v = W_z(\sum_{u=1}^{N} \beta_{v,u} V(\varrho_u, s)) \tag{4}$$

Therefore, $\beta_{v,u}$ indicates to what magnitude the uth image region is attended to synthesize the vth image region concerning the semantic context at the sentence level. The final self-cross-attention map $o = (o_1, o_2, ..., o_v, ..., o_N) \in \mathbb{R}^{N_r \times N}$ can be accordingly obtained. V is another 1x1 convolution besides K and Q.

3.4 Objective Functions

Adversarial Loss. A common practice is to employ two adversarial losses: the unconditional adversarial loss and the conditional visual-semantic adversarial loss. We further expand the conditional visual-semantic adversarial loss hierarchically, i.e., a fine-grained words-regions adversarial loss and a global sentence-image adversarial loss.

The generator G and discriminator D are alternatively trained at each learning stage of MGD-GAN. In particular, the adversarial loss of the generator G_i at the ith stage can be defined as:

$$\mathcal{L}_{G_i}^{adv} = -1/3[\underbrace{\mathbb{E}_{I_g \sim p_{G_i}} \log D_i^g(I_g)}_{\text{unconditional loss}} + \underbrace{\mathbb{E}_{I_g \sim p_{G_i}} \log D_i^g(I_g, s)}_{\text{global conditional loss}} + \underbrace{1/4 \sum_{\kappa} \mathbb{E}_{I_g \sim p_{G_i}} \log D_i^{\kappa}(I_g, W)}_{\text{local conditional loss}}] \tag{5}$$

where the D_i^g stands for the self-cross-attended global discriminator. The unconditional loss is intended to distinguish the generated image from the real image. The global conditional loss and local conditional loss are designed to determine whether the generated image is faithful to the input description at the sentence-image level and the words-regions level, respectively.

We define the loss function of two discriminators as:

$$\begin{aligned} \mathcal{L}_{D_i^g} = -\frac{1}{2}[&\mathbb{E}_{I_d \sim p_{data}} \log D_i^g(I_d) + \mathbb{E}_{I_g \sim p_{G_i}} \log(1 - D_i^g(I_g)) \\ &+ \mathbb{E}_{I_d \sim p_{data}} \log D_i^g(I_d, s) + \mathbb{E}_{I_g \sim p_{G_i}} \log(1 - D_i^g(I_g, s))] \end{aligned} \tag{6}$$

$$\mathcal{L}_{D_i^{\kappa}} = -\mathbb{E}_{I_d \sim p_{data}} \log D_i^{\kappa}(I_d, W) - \mathbb{E}_{I_g \sim p_{G_i}} \log(1 - D_i^{\kappa}(I_g, W)) \tag{7}$$

Table 1. Quantitative evaluation results of caption generation on different natural language generation metrics. The last row presents the performance of real images.

Method	Caption Generation						
	BLEU-1	BLEU-2	BLEU-3	BLEU-4	METEOR	ROUGE_L	CIDEr
Reed *et al.* [12]	0.496	0.321	0.218	0.151	0.211	0.454	0.949
StackGAN [17]	0.512	0.337	0.233	0.163	0.219	0.468	1.125
AttnGAN [15]	0.561	0.396	0.293	0.222	0.253	0.525	1.519
Ours	**0.565**	**0.401**	**0.299**	**0.229**	**0.257**	**0.530**	**1.553**
Real (upper bound)	0.599	0.443	0.340	0.268	0.282	0.565	1.855

It is worth noting that the loss function of the human-part discriminator $\mathcal{L}_{D_i^\kappa}$ does not contain an unconditional adversarial loss. We assume that a typical conditional adversarial loss does measure the visual quality of images as well as the visual-semantic correspondence and that the body structure is better captured as a whole.

Conditioning Augmentation Loss. To mitigate the discontinuity problem caused by limited training data, we corporate a regularization term, named Conditioning Augmentation [17], formulated as follows:

$$\mathcal{L}_{cond} = D_{KL}(\mathcal{N}(\mu(s), \Sigma(s))||\mathcal{N}(0, \mathrm{I})) \tag{8}$$

where D_{KL} represents the Kullback-Leibler divergence. The estimation of mean $\mu(s)$ and diagonal covariance $\Sigma(s)$ are modeled by fully-connected layers.

DAMSM Loss. As a common practice [2,15,24], we employ a cross-modal representation learning based module named DAMSM. This module provides the initial sentence and words embeddings, as well as a matching loss $\mathcal{L}_{DAMSM}$ of the generated image and conditioned text.

Therefore, the final loss function of the generator can be written as $\mathcal{L}_{G_i} = \sum_i \mathcal{L}_{G_i}^{adv} + \lambda_1 \mathcal{L}_{cond} + \lambda_2 \mathcal{L}_{DAMSM}$.

4 Pose Score and Pose Variance

In current literature, the numerical assessment approaches for GANs are not explicitly intended for pedestrian generation, such as Inception Score. Under the observation that pedestrian has no salient sub-categories like birds and flowers, the classification model for Inception Score trained on a generic image recognition dataset may fail to evaluate the generation quality and diversity competently. To this end, we propose two pedestrian-specific generation metrics named Pose Score and Pose Variance as a non-trivial complement to the Inception Score.

Specifically, we first pre-train a pose estimation model [1] on a larger COCO 2016 keypoints dataset [9]. For each generated pedestrian $I_\epsilon, \epsilon \in \{1, ..., \Xi\}$, we detect the 2D positions of body parts, represented as $B_\epsilon = \{b_{\epsilon,1}, b_{\epsilon,2}, ..., b_{\epsilon,\tau}, ..., b_{\epsilon,J}\}$ where $b_{\epsilon,\tau} \in \mathbb{R}^2$, one per part. Some parts may not

Fig. 3. Qualitative comparison of four methods.

be detected because the part is occluded or blurred. Since there is little occlusion in the CUHK-PEDES dataset, we use the average ratio of the number of detected parts and the upper-bound, i.e., $PS = \frac{1}{\Xi}\sum_{\epsilon}\frac{\#B_{\epsilon}}{18}$, as the Pose Score. As for Pose Variance, we individually compute the variance of each body part for all generated pedestrians and average them. The intuition is that the larger the variance of human poses is, the more diversified pedestrians the model can generate. Mathematically, the proposed Pose Variance can be computed as:

$$PV = \exp(\frac{1}{J*2}\sum_{\tau=1}^{J}\sum_{\iota=1}^{2}Var(\{\frac{b_{\epsilon,\tau,\iota}}{b_{max}}\}_{\epsilon=1}^{\Xi})) \tag{9}$$

where b_{max} is a normalization factor and is set to 256, which is the width or height of generated image I.

5 Experiments

Dataset. As far as we know, the CUHK-PEDES dataset [7] is the only caption-annotated pedestrian dataset. It contains 40,206 images over 13,003 persons.

Evaluation Metric. We choose to evaluate the generation result using various metrics, i.e., Inception Score [13], our newly proposed Pose Score & Variance, and Caption Generation. Caption generation is devised to reflect whether the generated image is well-conditioned on input text, we adopt a caption generation based approach [6]. The intuition behind this approach is that if the generated image is faithful to input text, a well-trained caption model on the same dataset can reconstruct the input text accordingly. We use this caption architecture [14] trained on the CUHK-PEDES dataset. We measure the similarity of generated captions and input text using four standard Natural Language Generation metrics, BLEU, METEOR, ROUGE_L and CIDEr.

Implementation Details. As stated in the previous section, we obtain the sentence embeddings and words embeddings from DAMSM and fix them during training. We have $m = 3$ training stages which generate images of resolutions 64×64, 128×128 and 256×256 individually. We set $N_w = 256$ and $N_r = 512$ as the dimension of words features and intermediate region features. The hyperparameter of generator loss is set as $\lambda_1 = 1$ and $\lambda_2 = 5$. The dimension of augmented sentence embedding N_s is set to 100.

5.1 Quantitative Evaluation

As far as we know, we are the initiative to do text-to-pedestrian generation. We compare our method with several works for text-to-bird/flower generation. [12,15,17]. Overall, the results across multiple evaluation metrics on the CUHK-PEDES test dataset consistently indicate that our proposed MGD-GAN achieves better results against the other three methods, concerning both the visually authenticity and text-image consistency.

Table 2. Quantitative evaluation results on generation visual quality. IS, PS and PV stands for the Inception Score and the proposed Pose Score/Variance metrics. The PS and PV of real images are 0.774 and 2.388.

Models	IS	PS	PV
Reed *et al.*	4.32 ± 0.194	0.275	1.935
StackGAN	4.79 ± 0.184	0.362	1.945
AttnGAN	5.07 ± 0.396	0.465	2.039
MGD-GAN	$\mathbf{5.74 \pm 0.526}$	**0.489**	**2.053**

Specifically, the Inception Score, Pose Score and Pose Variance of MGD-GAN and other methods are in Table 2. MGD-GAN achieves the best results across these three metrics. Compared with the state-of-art method on fine-grained text-to-image generation, the AttnGAN [15], MGD-GAN improved the Inception Score from 5.07 to 5.74, the Pose Score from 0.465 to 0.489 and the Pose Variance from 2.039 to 2.053. These results indicate that the proposed modules in MGD-GAN are promising directions for the fine-grained text-to-image generation. As for the evaluation of whether the generated images are well-conditioned on the input text, the natural language generation evaluation results on the generated texts are listed in Table 1. Our MGD-GAN achieves better results in terms of all the seven evaluation metrics.

5.2 Qualitative Evaluation

Figure 3 demonstrates the synthesized images produced by our MGD-GAN and three state-of-the-art models in the context of quality assessment. All samples are conditioned on text descriptions on CUHK-PEDES test set. Our MGD-GAN method produces pedestrians with a coherent structure and vivid details in most cases, comparing to the AttnGAN, the StackGAN and the Reed *et al.*. For convenience, we use C_{ij} to refer the i_{th} row and the j_{th} column example.

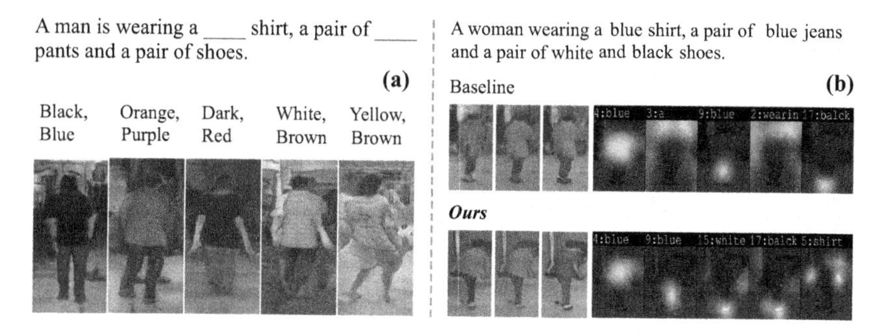

Fig. 4. (a) Controllable image generation by deliberately changing color attributes. (b) Comparison of the multi-stage generation results and the top 5 relevant words selected by attention module between the proposed method and the baseline method.

Table 3. Ablation test of different modules in our MGD-GAN.

Models	IS	PS	PV
Baseline	5.12 ± 0.470	0.275	1.979
+HPD	5.32 ± 0.357	0.486	1.981
+HPD+VISA	5.59 ± 0.462	**0.498**	2.018
+HPD+VISA+SCA	**5.74 ± 0.526**	0.489	**2.053**

Subjective Analysis. Due to the lack of attention mechanism designed for fine-grained image generation, StackGAN and Reed *et al.*generate images with vague appearance (C_{12}, C_{13}) and inconsistent body structure (C_{11}, C_{22}). The AttnGAN method, which employs an attention mechanism in the generator, achieves better results. For example, the C_{13} generated by AttnGAN is realistic and has most details described by the text, including "white t-shirt" and "black pants". By comparing MGD-GAN with AttnGAN, we can see that MGD-GAN further improves many fine-grained details, such as the facial contour in C_{21} and hairstyle in C_{22}, which depicts the femaleness. The result indicates the merit of leveraging fine-grained attention mechanism in both generator and discriminator collaboratively for generating semantically consistent images.

Controllable Image Generation. By altering the colors of wearings, Fig. 4 (b) shows the controllable image generation results. Our model can generate pedestrians with fine-grained details authentic to the input text, and diversified poses and appearances. For example, there are pedestrians wearing long pants (the first columns) and short pants (the 5_{th} column), with hands in the pocket (the first column), hands putting in front (the second column) and hands stretching on both sides (the third column).

Multi-stage Refinement and Attention Visualization. To better understand the effectiveness of our proposed modules, we visualize the multi-stage generation results and attention maps of the baseline method (without VISA-HPB and SCA-GD) and our method (Fig. 4 (b)). It can be seen that 1) The MGD-GAN can generate more details than the baseline method, such as the hands and the "white and black shoes". The body structure is also better. 2) With a more balanced training of generator and discriminator, the attended words of the generator in our model are more meaningful and more closely related to human appearance and structure. 3) Our model can progressively refine the body structure, correct irregular artifacts, and remedy missing attributes in images from the previous stages.

5.3 Ablation Study

In order to verify the efficacy of different components and gain a better understanding of the network's behavior, we conduct the ablation study by progressively removing control components. The control components are set up to HPD, VISA, and SCA, which stands for the human-part-based discriminator, the visual-semantic attention and the self-cross-attention respectively. The baseline method, denoted as BL for simplicity, is defined by removing the above modules from the MGD-GAN. The results are shown in Table 3.

Specifically, the BL achieves a slight performance boost over the AttnGAN by leveraging the VISA in generator instead of the original attention mechanism in AttnGAN. By equipping the human-part-based discriminator, the BL+PD model makes the relative improvement over the BL from 5.12 to 5.32 and 0.470 to 0.486 on Inception Score and Pose Score, respectively. It is worth noting that the VISA enhanced model (BL+PD+VISA) achieves the best performance on Pose Score metric, which verifies the effectiveness of VISA-HPD module in generating fine-grained details of different parts. Moreover, by additionally incorporating a self-cross-attended global discriminator, the MGD-GAN model (BL+PD+VISA+SCA) leads to the best performance on the Inception Score (12% improvement over BL) and Pose Variance metrics.

6 Conclusion

In this paper, we propose the MGD-GAN method for fine-grained text-to-pedestrian synthesis. We design a human-part-based discriminator equipped with the visual-semantic attention, which can be seen as a local discriminator functioning at the words-regions level, as well as a self-cross-attended global discriminator at the sentence-image level. We evaluate our model on the caption annotated person dataset CUHK-PEDES. The quantitative results across different metrics, including our proposed Pose Score & Variance, and substantial qualitative results demonstrate the efficacy of our proposed MGD-GAN method.

Acknowledgement. The work is supported by the National Key R&D Program of China (No. 2020YFC0832500), NSFC (61625107, 61836002, 62072397, 62006207), Zhejiang Natural Science Foundation (LR19F020006), and Fundamental Research Funds for the Central Universities (2020QNA5024), and the Fundamental Research Funds for the Central Universities and Zhejiang Province Natural Science Foundation (No. LQ21F020020).

References

1. Cao, Z., Simon, T., Wei, S.E., Sheikh, Y.: Realtime multi-person 2d pose estimation using part affinity fields. In: CVPR (2017)
2. Cheng, Y., Gan, Z., Li, Y., Liu, J., Gao, J.: Sequential attention gan for interactive image editing via dialogue. In: CVPR (2019)
3. Goodfellow, I., et al.: Generative adversarial nets. In: NIPS (2014)
4. Grigorev, A., Sevastopolsky, A., Vakhitov, A., Lempitsky, V.: Coordinate-based texture inpainting for pose-guided human image generation. In: CVPR (2019)
5. Hinz, T., Heinrich, S., Wermter, S.: Generating multiple objects at spatially distinct locations. In: ICLR (2019)
6. Hong, S., Yang, D., Choi, J., Lee, H.: Inferring semantic layout for hierarchical text-to-image synthesis. In: CVPR (2018)
7. Li, S., Xiao, T., Li, H., Zhou, B., Yue, D., Wang, X.: Person search with natural language description. In: CVPR (2017)
8. Li, W., et al.: Object-driven text-to-image synthesis via adversarial training. In: CVPR, pp. 12174–12182 (2019)
9. Lin, T.Y., et al.: Microsoft COCO: common objects in context. In: Fleet, D., Pajdla, T., Schiele, B., Tuytelaars, T. (eds.) ECCV 2014. LNCS, vol. 8693, pp. 740–755. Springer, Cham (2014). https://doi.org/10.1007/978-3-319-10602-1_48
10. Ma, L., Jia, X., Sun, Q., Schiele, B., Tuytelaars, T., Van Gool, L.: Pose guided person image generation. In: NIPS (2017)
11. Ma, L., Sun, Q., Georgoulis, S., Van Gool, L., Schiele, B., Fritz, M.: Disentangled person image generation. In: CVPR (2018)
12. Reed, S., Akata, Z., Yan, X., Logeswaran, L., Schiele, B., Lee, H.: Generative adversarial text to image synthesis. In: ICML (2016)
13. Salimans, T., Goodfellow, I., Zaremba, W., Cheung, V., Radford, A., Chen, X.: Improved techniques for training gans. In: NIPS (2016)
14. Xu, K., et al.: Show, attend and tell: neural image caption generation with visual attention. In: ICML (2015)
15. Xu, T., et al.: Attngan: fine-grained text to image generation with attentional generative adversarial networks. In: CVPR (2018)
16. Zhang, H., Goodfellow, I., Metaxas, D., Odena, A.: Self-attention generative adversarial networks. In: ICML (2018)
17. Zhang, H., et al.: Stackgan: text to photo-realistic image synthesis with stacked generative adversarial networks. In: ICCV (2017)
18. Zhang, S., et al.: Text-to-image synthesis via visual-memory creative adversarial network. In: PCM (2018)
19. Zhang, S., et al.: Grounded and controllable image completion by incorporating lexical semantics. CoRR (2020)
20. Zhang, S., et al.: Devlbert: learning deconfounded visio-linguistic representations. In: ACM MM (2020)
21. Zhang, S., et al.: Comprehensive information integration modeling framework for video titling. In: KDD (2020)
22. Zhang, Z., Xie, Y., Yang, L.: Photographic text-to-image synthesis with a hierarchically-nested adversarial network. In: CVPR (2018)
23. Zhou, X., Huang, S., Li, B., Li, Y., Li, J., Zhang, Z.: Text guided person image synthesis. In: CVPR (2019)
24. Zhu, M., Pan, P., Chen, W., Yang, Y.: Dm-gan: dynamic memory generative adversarial networks for text-to-image synthesis. In: CVPR (2019)
25. Zhu, Z., Huang, T., Shi, B., Yu, M., Wang, B., Bai, X.: Progressive pose attention transfer for person image generation. In: CVPR (2019)

Correction to: Pattern Recognition and Computer Vision

Huimin Ma, Liang Wang, Changshui Zhang, Fei Wu, Tieniu Tan, Yaonan Wang, Jianhuang Lai, and Yao Zhao

Correction to:
H. Ma et al. (Eds.): *Pattern Recognition and Computer Vision*, LNCS 13020, https://doi.org/10.1007/978-3-030-88007-1

For Chapter 3:
An acknowledgement for a grant from the National Natural Science Foundation of China for the paper has been added.

For Chapter 11:
A co-author (Shiping Wang) was not included in the original publication due to an oversight. This co-author has been added in this correction, reflecting the original submission of the article.

The updated version of these chapters can be found at
https://doi.org/10.1007/978-3-030-88007-1_3
https://doi.org/10.1007/978-3-030-88007-1_11

H. Ma et al. (Eds.): PRCV 2021, LNCS 13020, p. C1, 2021.
https://doi.org/10.1007/978-3-030-88007-1_55

Author Index

Printed in the United States
by Baker & Taylor Publisher Services